Angular Correlations in Nuclear Disintegration

ISBN 978-94-011-7733-7 ISBN 978-94-011-7731-3 (eBook)
DOI 10.1007/978-94-011-7731-3

Softcover reprint of the hardcover 1st edition 1971

Angular Correlations in Nuclear Disintegration

Proceedings of the International Conference on
Angular Correlations in Nuclear Disintegration
Delft, The Netherlands August 17-22, 1970

Edited by

Hans van Krugten

and

Bob van Nooijen

Both of Department of Technical Physics
Delft University of Technology
Delft, The Netherlands

1971

Springer-Science+Business Media, B.V.

THE ORGANIZING COMMITTEE

B. van Nooijen, Delft University of Technology, Chairman

H. van Krugten, Delft University of Technology

H. Verheul, Free University of Amsterdam

H. de Waard, University of Groningen

A.H. Wapstra, Institute for Nuclear Research, Amsterdam and
Delft University of Technology

THE FOREIGN ADVISORY COMMITTEE

K. Alder, Basel, Switzerland

H. Frauenfelder, Urbana, U.S.A.

T.R. Gerholm, Stockholm, Sweden

J.H. Hamilton, Nashville, U.S.A.

R.M. Steffen, Lafayette, U.S.A.

Z. Sujkowski, Warsaw, Poland

SPONSORS

Delft University Fund
Delft University of Technology
Dutch Physical Society
International Union of Pure and Applied Physics
Ministry of Education and Sciences
N.V. Philips' Gloeilampenfabrieken
I.B.M. Nederland N.V.

PREFACE

In 1969 we felt that the subject of angular correlations in nuclear disintegration had received little special attention at international conferences. It is true that perturbed angular correlations had been discussed at a small, highly - specialised meeting at Uppsala in 1963 and that in 1967 certain aspects of perturbed angular correlations had been considered at a conference on hyperfine structure at Asilomar. However, both conferences were very limited in scope from the point of view of a low - energy nuclear physicist. Besides, since these conferences were being held, the field of perturbed angular correlations had received new impetus from the still expanding application of the implantation of radioactive isotopes in suitable environments, from the construction of large superconductive magnets, etc. Meanwhile, the technique of measuring correlations between beta particles and circularly polarized gamma rays had been developed to such a degree that they lent themselves to the investigation of beta decay, nuclear structure and charge dependence of nuclear forces. The systematic study of heavy elements had increased the importance of alpha-gamma angular correlations which present their own specific problems. Theoretical internal conversion data had become available to such an extent that electron - gamma directional correlations became an important tool in the investigation of nuclear structure: in a large number of cases it is more advantageous to study electron - gamma correlations instead of the commonly measured gamma - gamma directional correlations. In addition, the development of nuclear instrumentation leading to large Ge (Li) - detectors and other solid state detectors, to refined coincidence systems, computers for on - line and off - line use, automatic systems etc. provided the possibility of considerably increasing the accuracy of experimental angular correlation results. As a consequence a more meaningful comparison between experimentally and theoretically determined quantities (for instance M1/E2 mixing ratios) became feasible.

Our feeling that time was pressing to devote a conference to the theoretical and experimental aspects of angular correlation work was shared by many physicists engaged in the field as became apparent from informal talks with colleagues. We felt then justified to send out a questionnaire to a representative number of people who were believed to be actively engaged in the field. The enthusiastic response prompted us to organize a conference on angular correlations, although we were aware of the difficulty to arrange a meeting based on an experimental method which is so extensively applied in many areas of nuclear physics. However, thanks to the active support from the members of the foreign advisory committee and of the other members of the organizing committee it was possible to come up with a well - balanced program.

From the many reactions received from participants during and after the conference, we have the impression that the conference was timely and successful.

We thank all the conference participants for their contribution to the conference. A special word of thanks is expressed to the members of the foreign advisory committee – dr. K. Alder, dr. H. Frauenfelder, dr. T.R. Gerholm, dr. J.H. Hamilton, dr. R.M. Steffen and dr. Z. Sujkowski – for their many efforts. We also want to thank the other members of the organizing committee – dr. H. Verheul, dr. H. de Waard and dr. A.H. Wapstra – for their help. We are very grateful to dr. R. van Lieshout for his excellent talk at the conference banquet. The efforts by the members of our nuclear spectroscopy group is gratefully acknowledged. Many thanks go to dr. A.V. Ramayya for his continuous help and encouragement. We thank the conference secretary Mrs. M. Mangunkusumo-Van der Maas for her invaluable assistance before, during and after the conference. It is a pleasure to thank the following agencies for their financial or moral support: the Ministry of Education and Sciences, the Delft University of Technology (particularly the Department of Technical Physics), the Delft University Fund, the International Union of Pure and Applied Physics, the Dutch Physical Society, N.V. Philips' Gloeilampenfabrieken and I.B.M. Nederland N.V. We thank Messrs. Van der Grinten Ltd. for putting an Océ copying machine at the disposal of the participants; our thanks are due to the Ministry of Public Works (Department Rijkswaterstaat) for their help in organizing the conference excursion. The cooperation of Messrs. Wolters-Noordhoff in the publication of this book is most appreciated.

June, 1971

Delft, The Netherlands.

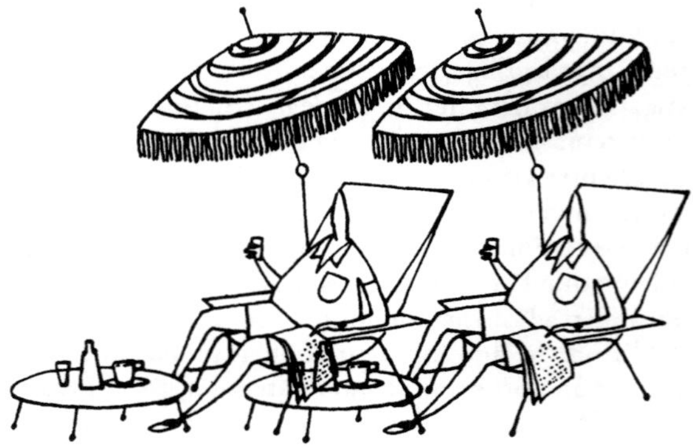

H(1)

LIST OF PARTICIPANTS

1. Y.K. Agarwal, University of Darmstadt, W. Germany
2. M.R. Ahmed, University of Uppsala, Sweden
3. J.C. Alonso, Yale University, New Haven, Conn., U.S.A.
4. C.T. Alonso, Yale University, New Haven, Conn., U.S.A.
5. K. Alder, University of Basel, Switzerland
6. G. Ambrosino, Lab. Maurice - de - Broglie, Brétigny - sur - Orge, France
7. R. Babinet, Atomic Energy Commission, Paris, France
8. A. Backer, Delft University of Technology, The Netherlands
9. Chr. Bargholtz, University of Stockholm, Sweden
10. U. Bäverstam, University of Stockholm, Sweden
11. F.A. Beck, Nuclear Research Institute, 67 Strasbourg 3, France
12. A.J. Becker, California Institute of Technology, Pasadena, U.S.A.
13. J. Becker, University of Stockholm, Sweden
14. H. Behrens, Kernforschungszentrum Karlsruhe, W. Germany
15. R. Béraud, University of Lyon, 69 Villeurbanne, France
16. H.L. Berger, Rensselaer Polytechnic Institute, Troy, N.Y., U.S.A.
17. I. Berkes, University of Lyon, 69 Villeurbanne, France
18. Chr. Bernhardt, University of Marburg (Lahn), W. Germany
19. J. Berthier, University of Grenoble, France
20. R. Bilwes, Laboratoire des basses Energies, Strasbourg, France
21. J. Bleck, Hahn - Meitner Institut, Berlin - West, W. Germany
22. S.D. Bloom, University of California, Livermore, U.S.A.
23. E. Bodenstedt, University of Bonn, W. Germany
24. F. Boehm, California Institute of Technology, Pasadena, U.S.A.
25. H. Bokhove, Delft University of Technology, The Netherlands
26. V. Bom, Delft University of Technology, The Netherlands
27. K. Bos, IKO Amsterdam, The Netherlands
28. H.E. Bosch, Universidad de Buenos Aires, Argentine
29. J.D. Bowman, University of Bonn, W. Germany
30. F. Bragança Gil, University of Lisbon, Portugal
31. W.H. Brantley, Furman University, Greenville, S.C., U.S.A.
32. J. Britz, Nuclear Research Institute, 67 Strasbourg 3, France

33. R. Broda, Joint Institute for Nuclear Research, Moscow, U.S.S.R.
34. C. Broude, The Weissman Institute, Rehovot, Israel
35. M. de Bruin, IRI Delft, The Netherlands
36. T. Byrski, Nuclear Research Institute, 67 Strasbourg 3, France
37. M. Cailliau, C.N.R.S., 91 Orsay, France
38. M.J. Canty, University of Bonn, W. Germany
39. J. Christiansen, University of Erlangen, W. Germany
40. G. Costa, Lab. des basses Energies, Strasbourg - Cronenbourg, France
41. R. Coussement, University of Leuven, Belgium
42. Y. Dar, Nuclear Physics Laboratory, Strasbourg - Cronenbourg, France
43. Mme Delaunay, Atomic Energy Commission, Paris, France
44. I. Demeter, Hungarian Academy of Sciences, Budapest 114, Hungaria
45. B. Deutch, University of Aarhus, Denmark
46. J. Dirker, Delft University of Technology, The Netherlands
47. H. Dirren, Swiss Federal Institute of Technology, Zürich, Switzerland
48. J.H.J. Distelbrink, Delft University of Technology, The Netherlands
49. H.A. Doubt, CERN, Geneva, Switzerland
50. H. Drost, University of Berlin, W. Germany
51. B.C. Dutta, Birkbeck College, London, England
52. C.W.E. van Eijk, Delft University of Technology, The Netherlands
53. M. Finger, CERN, Geneva, Switzerland
54. J. Fouan, Atomic Energy Commission, Paris, France
55. R. Foucher, C.N.R.S., 91 Orsay, France
56. R.A. Fox, University of Oxford, England
57. H. Frauenfelder, University of Illinois, Urbana, U.S.A.
58. R.M. Freeman, Nuclear Research Centre, 67 Strasbourg 3, France
59. H. Gabriël, University of Berlin, W. Germany
60. J. Gerber, Nuclear Physics Laboratory, Strasbourg - Cronenbourg, France
61. T.R. Gerholm, University of Stockholm, Sweden
62. P.W.M. Glaudemans, Rijksuniversiteit Utrecht, The Netherlands
63. J. Gomes Ferreira, University of Lisbon, Portugal
64. P. Goudsmit, IKO Amsterdam, The Netherlands
65. N.B. Gove, Oak Ridge National Laboratory, Tennessee, U.S.A.
66. Z. Grabowski, Purdue University, Lafayette, Indiana, U.S.A.
67. Chr. Günther, Niels Bohr Institute, Risø, Denmark

68. J.H. Hamilton, Vanderbilt University, Nashville, Tennessee, U.S.A.
69. W.D. Hamilton, University of Sussex, Brighton, England
70. A. Hammesfahr, University of Bonn, W. Germany
71. L. Hasselgren, University of Uppsala, Sweden
72. G. Heestand, University of Aarhus, Denmark
73. B. Herskind, University of Copenhagen, Denmark
74. P. Herzog, University of Bonn, W. Germany
75. W.H.A. Hesselink, Delft University of Technology, The Netherlands
76. S. Hofmann, University of Darmstadt, W. Germany
77. R.W. Hollander, Delft University of Technology, The Netherlands
78. L. Holmberg, University of Stockholm, Sweden
79. M.J. Holmes, University of Sussex, Brighton, England
80. D.J. Horen, Oak Ridge National Laboratory, Tennessee, U.S.A.
81. B. Hrastnik, Institute Ruder Boskovič, Zagreb, Yugoslavia
82. A. Hrynkiewicz, Institute of Nuclear Physics, Cracow, Poland
83. S. Hüfner, University of Berlin, W. Germany
84. W.J. Huiskamp, Kamerlingh Onnes Laboratory, Leiden, The Netherlands
85. S. Idzenga, Vrije Universiteit Amsterdam, The Netherlands
86. P. Inia, Rijksuniversiteit Groningen, The Netherlands
87. J.L. Irigaray, Centre d'Etudes Nucléaires de Bordeaux, Gradignan, France
88. M. van Kampen, Delft University of Technology, The Netherlands
89. B. Kardon, Hungarian Academy of Sciences, Budapest 114, Hungaria
90. E. Karlsson, University of Uppsala, Sweden
91. E.N. Kaufmann, Bell Telephone Lab., Murray Hill, N.J., U.S.A.
92. J. Konijn, IKO Amsterdam, The Netherlands
93. E.W. Koopmann, Delft University of Technology, The Netherlands
94. H. van Krugten, Delft University of Technology, The Netherlands
95. J. Krumlinde, University of Lund, Sweden
96. J. Lachkar, Lab. Maurice - de - Broglie, Brétigny - sur - Orge, France
97. J. Lange, University of Bonn, W. Germany
98. Leang Chin Fan, C.N.R.S., 91 Orsay, France
99. H.J. Leisi, Swiss Federal Institute of Technology, Switzerland
100. M. Levanoni, Technion Haifa, Israel
101. R. van Lieshout, ZWO, The Hague, The Netherlands
102. H. Lindeman, Technion Haifa, Israel

103. E.W. Lingeman, IKO Amsterdam, The Netherlands
104. A. Lopéz García, University of Bonn, W. Germany
105. W. Lourens, Delft University of Technology, The Netherlands
106. H.-E. Mahnke, University of Berlin, W. Germany
107. H.L. Mang, University of München, W. Germany
108. J.C. Manthuruthil, Dayton, Ohio, U.S.A.
109. B. Martin, University of Heidelberg, W. Germany
110. B.H.J. McKellar, University of Sydney, Australia
111. H.A. Medicus, Rensselaer Polytechnic Institute, Troy, N.Y., U.S.A.
112. P. Mennrath, Laboratoire de Physique Nucléaire, Gradignan, France
113. G. van Middelkoop, Rijksuniversiteit Utrecht, The Netherlands
114. D.E. Murnick, Bell Telephone Lab., Murray Hill, N.J., U.S.A.
115. B. van Nooijen, Delft University of Technology, The Netherlands
116. L.-O. Norlin, University of Uppsala, Sweden
117. S. Ogaza, Instityt Fisyki Jadrowej w Krakowic, Cracow, Poland
118. B. Orre, University of Uppsala, Sweden
119. B. Povh, University of Heidelberg, W. Germany
120. A.V. Ramayya, Vanderbilt University, Nashville, Tennessee, U.S.A.
121. R.L. Rasera, University of Pennsylvania, Philadelphia, U.S.A.
122. E. Recknagel, Hahn - Meitner Institut, Berlin - West, W. Germany
123. A.C. Rester, Delft University of Technology, The Netherlands
124. J.D. Rogers, Universidade de Porto Alegre, Brasil
125. S.S. Rosenblum, University of Berlin, W. Germany
126. B. Rosner, Technion Haifa, Israel
127. M. Rots, University of Leuven, Belgium
128. R. Rougny University of Lyon, 69 Villeurbanne, France
129. S.S. RupRai, Kingston Polytechnic, Kingston - upon - Thames, Surrey, England
130. M. Sakai, University of Tokyo, Japan
131. R. Samama, Atomic Energy Commission, Paris, France
132. L. Samuelsson, University of Linköping, Sweden
133. K.S.R. Sastry, University of Massachussetts, Amherst, U.S.A.
134. H. Schmidt, Institut für angewandte Kernphysik, Karlsruhe, W. Germany
135. K.P. Schmidt, University of Leuven, Belgium
136. F.J. Schroeder, University of Bonn, W. Germany
137. H. Seyfarth, Kernforschungsanlage, 517 Jülich, W. Germany

138. G. da Silva, Instituto Engenharia Nuclear, Rio de Janeiro, Brasil
139. R. Silverans, University of Leuven, Belgium
140. P.C. Simms, Purdue University, Lafayette, Indiana, U.S.A.
141. B. Skaali, University of Copenhagen, Denmark
142. Norma de Sousa, University of Stockholm, Sweden
143. R.M. Steffen, Purdue University, Lafayette, Indiana, U.S.A.
144. N.J. Stone, University of Oxford, England
145. Z. Sujkowski, University of Warsaw, Poland
146. P. Taras, University of Montreal, Quebec, Canada
147. H.W. Taylor, University of Toronto, Canada
148. P. Thieberger, Brookhaven National Laboratory, Upton, N.Y., U.S.A.
149. W.J. Thompson, University of North Carolina, Chapel Hill, U.S.A.
150. M.J. Throop, University of Liverpool, England
151. J.E. Thun, University of Uppsala, Sweden
152. H. Toschinski, University of Bonn, W. Germany
153. A. Troncy, University of Lyon, 69 Villeurbanne, France
154. P.J. Twin, University of Liverpool, England
155. L. Valentin, C.N.R.S., 91 Orsay, France
156. A. Vasquez, University of Rio Grande do Sul, Brasil
157. H. Verheul, Vrije Universiteit Amsterdam, The Netherlands
158. M.J.C. Visscher, Delft University of Technology, The Netherlands
159. J.P. Vivien, Nuclear Physics Laboratory, Strasbourg - Cronenbourg, France
160. H. van de Voorde, University of Leuven, Belgium
161. H. de Waard, Rijksuniversiteit Groningen, The Netherlands
162. A.H. Wapstra, Delft University of Technology, The Netherlands
163. L. Westerberg, University of Uppsala, Sweden
164. G. Weyer, University of Berlin, W. Germany
165. R. Wischhusen, Institut für experimentelle Kernphysik, Karlsruhe, W. Germany
166. J.L. Yntema, Argonne National Laboratory, Illinois, U.S.A.
167. F.C. Zawislak, University of Rio Grade do Sul, Brasil
168. M. Zupancic, Boris Kidrič Institute, Belgrade, Yugoslavia

CONTENTS

*Invited Paper

ANGULAR DISTRIBUTIONS AND CORRELATIONS OF NUCLEAR RADIATIONS IN NUCLEAR SPECTROSCOPY *

R.M. Steffen

Purdue University, Lafayette, Indiana, U.S.A.

1. The Role of Angular Distributions and Correlations in Nuclear Spectroscopy

It was exactly 30 years ago, in 1940, that Dunworth[1]), in a paper discussing coincidence techniques, mentioned the possibility of 'spatial correlations of successively emitted γ-rays'. In this paper the following communication by M.H.L. Pryce was quoted: 'Where successive γ-rays are emitted by a nucleus there will, in general, be a certain amount of correlation between the two directions of emission. The distribution in relative angle is determined by the value of the total angular momentum in the three nuclear states involved in the process of emitting of the two γ-rays. In the worst case – that when the angular momenta are 0, 1, 0, respectively – the distribution is proportional to $\cos^2 \vartheta$; the number in opposite directions is then three times what would be expected on the assumption of isotropy. In most cases, however, this ratio will be much nearer to unity'. Hence, M.H.L. Pryce was the first who predicted the existence of directional correlations of nuclear radiations and furthermore he recognized already the important rôle of nuclear angular momenta in angular correlation phenomena. In the very same year, D.R. Hamilton[2]) developed a theory for γ-γ directional correlations using the language of perturbation theofy. Although Hamilton's classic work contained many of the important features of angular correlation theory, it was limited in its applicability and the full generality of the underlying general principles was not recognized until about ten years later group theoretical methods and the full formalism of the quantum theory of angular momentum was applied to the description of angular correlation phenomena (Gardner[3]), Racah[4]), Fano[5]), Lloyd[6]), Alder[7]), Coester and Jauch[8]), Biedenharn and Rose[9])). Since then the theory of angular correlations and distributions[10]) of nuclear radiations including extranuclear perturbations (Alder[7]), Abragam and Pound[11]), Alder et al.[12]), Steffen and Frauenfelder (review) [13]), Gabriel[14]), Matthias et al.[15])) has been fully developed and it is well understood today. It is probably the best and most comprehensive theory of nuclear phenomena, because it is based on very general symmetry principles, i.e., on three-dimensional rotation and reflection symmetries, which lead to conservation of angular momentum and parity, respectively. It is the isotropy of three-dimensional space with its implication of conservation of angular momentum that is the fundamental reason for the existence of anisotropic angular correlations and distributions.

It is the complete mastery of angular correlation theory that makes the angular correlation methods so important and indispensable in experimental nuclear spectroscopy, since the information on nuclear properties that are extracted from angular correlation and angular distribution observations is model independent, if the nuclear states involved in the measurements have sharp angular momentum, a condition that is always fulfilled in radioactive decay experiments.

After the pioneer theoretical work of Hamilton in 1940 several attempts to observe angular correlations were made but it was not until the inefficient and slow Geiger - Müller counters were replaced by the more efficient and faster scintillation detectors that angular correlation measurements became a general and important method of nuclear spectroscopy (Brady and Deutsch[16]), 1947). The development of the NaI (Tl) scintillation counter with its energy selection capabilities and the development of multichannel electronics resulted in a flood of experimental gamma-gamma directional correlation papers in the 1950ies and later in that decade the first reliable β-γ directional correlation measurements were reported[17]). The arrival of the high-resolution solid state detectors (Ge(Li), Si(Li)) on the scene in 1963 stimulated a renaissance of angular correlation experiments which lasts until today. These high resolution detectors make it possible, for the first time, to achieve very accurate and reliable angular correlation measurements on nuclei with very complicated and complex decay schemes.

The observation of the directional correlation of two successively emitted nuclear radiations gives direct information of the angular momenta of the nuclear states involved and of the multipole character (angular momentum) of the emitted radiation field. The fact that angular distributions depend on interference terms between different multipole amplitudes of the radiation field, makes the angular correlation method particularly useful for the determination of the relative amount of different angular - momentum components in mixed multipole transitions. Conversion - electron - gamma directional correlations and gamma - gamma polarization - directional correlations also give information about the relative parities of nuclear states.

The great majority of accurate values of E2/M1 and M2/E1 amplitude mixing ratios of gamma transitions have been determined by directional correlation methods. Similarly, from combinations of β - γ directional correlations, β - γ circular polarization correlations and β - shape measurements the amplitudes (nuclear β - matrix elements) of the various multipole components contributing to first - forbidden beta transitions have been extracted[17]).

Measurements of angular correlations involving internal conversion electrons in conjunction with internal conversion coefficient studies have revealed the existence of effects caused by the penetration of the atomic electrons into the nuclear volume[18]).

Since angular correlation phenomena are based on symmetry considerations, they can also be used to test the fundamental assumptions of the symmetries in nature (e.g., parity - impurity and time - reversal experiments[19])).

Angular correlations are influenced by extranuclear fields that interact with the nuclear moments of the intermediate state involved in an angular correlation measurement and hence, cause a reorientation of this state. If this state has a sufficiently long lifetime and/or if the extranuclear fields are sufficiently strong for the extranuclear perturbation to manifest itself in an observable effect on the angular correlation, the static electromagnetic moments (magnetic dipole moment μ, electric quadrupole moment Q, etc.) of the intermediate nuclear state can be determined, if the magnitudes of the external perturbing fields are known[13,20]). The study of perturbed angular correlations (PAC) has been particularly fruitful, especially since the megagauss hyperfine fields acting on impurity nuclei in ferromagnetic hosts are being employed for g - factor measurements[20]). Recently, experiments using radiofrequency fields to alter angular correlations have been successful[21]). The field of detecting nuclear magnetic resonance by angular correlation techniques is still in its infancy, but its outlook is very promising, especially since the complete theory of these effects has been developed recently[15]).

During the past 30 years the field of angular correlations of nuclear radiations has made major and unique contributions to nuclear spectroscopy and has provided a wealth of information for the elucidation of the structure of nuclei. An illustration of the vigor of angular correlation research is the fact that well over a thousand papers relating to angular correlations and distributions of nuclear radiations have been published during this period.

In the following we present a short introduction into the general theory of angular correlation phenomena stressing symmetry considerations, and pointing out in very general terms, the type of information that can be obtained from the observation of angular correlation phenomena.

2. General Theory of Angular Distributions and Correlations

2.1 Angular Distribution of Radiation Emitted from an Oriented State

The problem of angular distributions and correlations can be formulated in a very general way. Density matrices and statistical tensors will be used to describe the nuclear and radiation states which consist of ensembles of nuclei and radiation quanta, respectively. The angular distribution of the radiation is then obtained by summing over the unobserved nuclear magnetic states, i.e., the angular distributions can be expressed in terms of the trace of a product of matrices in nuclear spin space. The general formalism of the quantum theory of angular momentum will be employed to further reduce the general expressions to practical formulae.

First, the angular distribution of a nuclear radiation X emitted from an oriented nuclear state of angular momentum I_1 leading to a final (unobserved) nuclear state I_2 is considered.

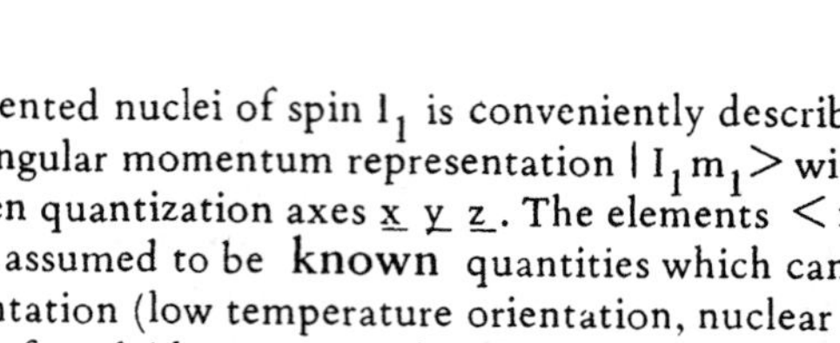

An ensemble of oriented nuclei of spin I_1 is conveniently described by a density matrix $\rho(I_1)$ in the angular momentum representation $|I_1 m_1\rangle$ with respect to some appropriately chosen quantization axes $\underline{x}$ $\underline{y}$ $\underline{z}$. The elements $\langle m_1|\rho(I_1)|m_1'\rangle$ of the density matrix are assumed to be known quantities which can be computed from the method of orientation (low temperature orientation, nuclear reactions, etc.). If the oriented ensemble of nuclei has an axis of cylindrical symmetry $\underline{a}$ (axially symmetric oriented ensemble), and if the quantization axis $\underline{z}$ coincides with $\underline{a}$ the density matrix $\rho(I_1)$ is diagonal,

$$\langle m_1 | \rho(I_1) | m_1' \rangle = P(m_1)\, \delta_{m_1 m_1'} \tag{1}$$

The diagonal elements are the relative populations $P(m_1)$ of the $| I_1 m_1 \rangle$ states.

It is convenient to construct the statistical tensors ρ^λ_q (I) of rank λ from the elements of a density matrix $\rho(I)$:

$$\rho^\lambda_q (I) = \sum_{m,m'} (-1)^{I+m} \langle I - m I m' | \lambda q \rangle \langle m | \rho (I) | m' \rangle \tag{2}$$

where $\langle I - m I m' | \lambda q \rangle$ is a Clebsch - Gordan coefficient. Equation (2) has the significance of a vector sum $\underline{\lambda} = \underline{I}' - \underline{I}$ ($|\underline{I}'| = |\underline{I}|$). The maximum rank of a statistical tensor is $\lambda_{max} = 2I$. The statistical tensors have more convenient transformation properties under rotations of the quantization coordinate system than the density matrix; they are irreducible tensors and as such transform under a three - dimensional rotation R of the representation axes according to

$$\rho^\lambda_q(I)_{Rot} = \sum_{q'} \rho^\lambda_{q'}(I)\, D^{(\lambda)}_{q'q} (\alpha \beta \gamma) \tag{3}$$

where $D^{(\lambda)}_{q'q} (\alpha \beta \gamma)$ is the rotation matrix[22]), i.e., the statistical tensors ρ^λ_q (I) transform like the spherical harmonics $Y_{\lambda q} (\vartheta,\varphi)$. The rotation R is given by the Euler angles α, β, γ.

If the state described by ρ^λ_q (I) is axially symmetric with respect to the quantization axis $\underline{z}$, it is invariant under a rotation α about this symmetry axis:

$$\rho^\lambda_q(I) = \rho^\lambda_q(I)_{Rot} = \sum_{q'} \rho^\lambda_{q'}(I)\, D^{(\lambda)}_{q'q} (\alpha, 0, 0) = \sum_{q'} \rho^\lambda_{q'} (I) e^{-iq\alpha}\, \delta_{q'q} \tag{4}$$

and $q = q' = 0$ for $\alpha \neq 0$. Hence, for states with axial symmetry only the ρ^λ_0 (I) components of the statistical tensors are non-vanishing.

The inverse of eq. (2) is:

$$\langle m | \rho | m' \rangle = \sum_{\lambda} (-1)^{I+m} \langle I - m I m' | \lambda q \rangle \rho^\lambda_q(I) \tag{5}$$

This equation corresponds to an expansion of the density matrix in terms of the unit spherical tensors U^λ_q with matrix elements

$$\langle I m' | U^\lambda_q | I m \rangle = (-1)^{I+m} \langle I - m\, I m' | \lambda q \rangle \tag{6a}$$

or, using the Wigner 3 - j symbol

$$<Im'|U_q^{\lambda}|\,Im> = (-1)^{I+m'}(2\lambda+1)^{1/2}\begin{pmatrix} I & I & \lambda \\ -m' & m & q \end{pmatrix} \tag{6b}$$

These matrices are hermitian in the sense that $U_q^{\lambda\dagger} = (-1)^q\, U_{-q}^{\lambda}$ and they are mutually orthonormal under the trace operation

$$\mathrm{Tr}\; U_q^{\lambda} U_{q'}^{\lambda'\dagger} = \sum_{m,m'} <I-m\,Im'|\lambda q><I-m\,Im'|\lambda' q'> = \delta_{\lambda\lambda'}\,\delta_{qq'} \tag{7}$$

The tensors U_q^{λ} form a complete basis set

$$\sum_{\lambda q} <Im|\,U_q^{\lambda}|\,Im'>\;<m''|\,U_q^{\lambda\dagger}\,|\,m'''> = \delta_{mm'''}\,\delta_{m'm''} \tag{8a}$$

or

$$\sum_{\lambda q} U_q^{\lambda\dagger}\Big|\,U_q^{\lambda} = E \qquad (E = \text{Unit operator}) \tag{8b}$$

With this notation eq. (2) can be written in the compact form

$$\rho_q^{\lambda}(I) = \mathrm{Tr}\,(\rho(I)\,U_q^{\lambda}) \tag{9}$$

The orientation of the state I_1 is now described by the statistical tensors $\rho_q^{\lambda}(I_1)$, which will be called the **orientation tensors** of the state I_1. They can be easily constructed from eq. (2) or (9) if the orientation mechanisms (low temperature methods, nuclear reactions, etc.) are known.

The emission of radiation X is described by the transition matrix H(X), whose elements are the amplitudes $< m_2\,|\,H(X)\,|\,m_1 >$ with $H(X)_{op}$ being the operator for the emission of the radiation field X. The density matrix of the final state I_2 is then

$$\rho(I_2) = H(X)\;\rho(I_1)\;H^{\dagger}(X) \tag{10}$$

The angular distribution of X is obtained by summing over the $|\,I_2m_2>$ states

$$W(\varphi,\vartheta,\psi) = \sum_{m_2,m_1,m_1'} <m_2|\,H(X)m_1><m_1|\,\rho(I_1)m_1'><m_1'|\,H^{\dagger}(X)|\,m_2> \tag{11a}$$

or

$$W(\varphi,\vartheta,\psi) = \mathrm{Tr}(H(X)\,\rho\,(I_1)\,H^{\dagger}(X)) \tag{11b}$$

The angles φ, ϑ, ψ are the Euler angles that describe the polarization $\underline{P}$ and propagation direction $\underline{k}$ of the radiation with respect to the quantization coordinate system $\underline{x}\ \underline{y}\ \underline{z}$ (laboratory system) in which $\rho(I_1)$ is represented (and also $H(X)$).

In order to evaluate the trace one uses eqs. (8) and writes

$$\begin{aligned} W(\varphi,\vartheta,\psi) &= \mathrm{Tr}\Big(\sum_{\lambda q} H(X)\,\rho\,(I_1)\,U_q^{\lambda\dagger}\Big|\,U_q^{\lambda}\,H^{\dagger}(X)\Big) \\ &= \sum_{\lambda q} \rho_q^{\lambda\dagger}(I_1)\ \mathrm{Tr}(H(X)\,U_q^{\lambda}\,H^{\dagger}(X)) \\ &= \sum_{\lambda q} \rho_q^{\lambda\dagger}(I_1)\,A_q^{\lambda}(X) \end{aligned} \tag{12}$$

where the statistical tensor $A_q^{\lambda}(X)$ of the radiation has been introduced by

$$A_q^{\lambda}(X) = \mathrm{Tr}\,(H\,(X)\,U_q^{\lambda}\,H^{\dagger}\,(X)) \tag{13}$$

These tensors are still referred to the 'laboratory' quantization system $\underline{x}\ \underline{y}\ \underline{z}$. Naturally, these tensors have a much simpler structure if the polarization $\underline{P}$ and propagation directions $\underline{k}$ of the radiation $\underline{k}\ \underline{P}$ - system are used for their representation. Matrices and tensors that are represented with respect to the $\underline{k}\ \underline{P}$ - system are indicated by an index 0, e.g., $A_q^{\lambda}(X)_0$. The $A_q^{\lambda}(X)$ in the 'laboratory' system are easily obtained from the $A_q^{\lambda}(X)_0$ in the $\underline{k}\ \underline{P}$ - system on the basis of eq. (3):

$$A_q^{\lambda}(X) = \sum_{q'} A_{q'}^{\lambda}(X)_o\, D_{qq'}^{(\lambda)^*}(\underline{x}\ \underline{y}\ \underline{z} \longrightarrow \underline{k}\ \underline{P}) \tag{14}$$

where the unitarity of the D - matrix has been used $D^{(\lambda)-1} = D^{(\lambda)\dagger}$. The $A_q^{\lambda}(X)_0$ are the **angular distribution coefficients** of the radiation X. General expressions for the angular distribution coefficients $A_q^{\lambda}(X)_0$ will be derived in Section 2.4.

Hence, one obtains for the angular distribution of the radiation X:

$$W(\varphi,\vartheta,\psi) = \sum_{\lambda,q,q'} \rho_q^{\lambda\dagger}(I_1)\ A_{q'}^{\lambda}(X)_o\cdot D_{qq'}^{(\lambda)^*}(\underline{x}\ \underline{y}\ \underline{z} \rightarrow \underline{k}\ \underline{P}) \tag{15}$$

If only the propagation direction $\underline{k}$ and no linear or transverse polarization $\underline{P}$ of the radiation is observed:

$$A_{q'}^{\lambda}(X)_o = A_o^{\lambda}(X)_o\,\delta_{oq'} = A_{\lambda}(X)\,\delta_{oq'} \tag{16}$$

because of the axial symmetry about the propagation direction $\underline{k}$. The $A_{\lambda}(X) \equiv A_0^{\lambda}(X)$ are the **directional distribution coefficients** of the radiation X.

For **axially symmetric orientation** of the initial nuclear state I_1, only the $\rho^{\lambda\dagger}_0(I_1)$ are different from zero. This follows directly from eqs. (1) and (2), or more generally from eq. (4).

Often the notation

$$B_\lambda(I_1) \equiv (2I_1 + 1)^{1/2} \rho^{\lambda\dagger}_0(I_1) \tag{17}$$

is used. The $B_\lambda(I_1)$ are the **orientation parameters** which have been introduced by Blin - Stoyle and Grace[23]. The factor $(2I_1 + 1)^{1/2}$ results in the normalization $B_0(I_1) = 1$.

Hence, the **directional distribution** of radiation X emitted from an axially symmetric oriented state is

$$W(\vartheta) = \sum_\lambda B_\lambda(I_1)\, A_\lambda(X)\, P_\lambda(\cos\vartheta) \tag{18}$$

where $D^{(\lambda)}_{00}(\varphi, \vartheta, \psi) = P_\lambda(\cos\vartheta)$ has been used. The angle ϑ describes the angle between the propagation direction $\underline{k}$ of the radiation and the orientation axis $\underline{a} = \underline{z}$ of the ensemble of oriented nuclei.

2.2 *Angular Correlations of Successively Emitted Radiations*

The successive emission of two radiations X_1 and X_2 between nuclear states of angular momentum I_0, I_1 and I_2, respectively, will be treated in this section.

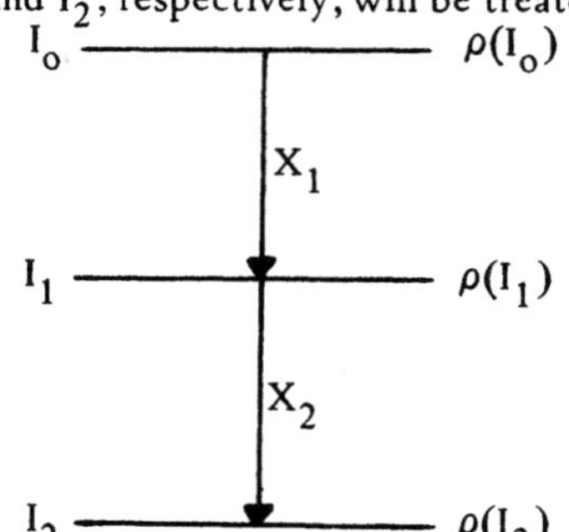

The density matrix of the intermediate state I_1 is

$$\rho(I_1) = H(X_1)\, \rho(I_0)\, H^\dagger(X_1) \tag{19}$$

Assuming that the state $\rho(I_1)$ does not change during its (short) lifetime τ, i.e., neglecting extranuclear perturbations, the density matrix $\rho(I_2)$ of the final state I_2 is

$$\rho(I_2) = H(X_2)H(X_1)\rho(I_0)H^\dagger(X_1)H^\dagger(X_2) \tag{20}$$

The angular correlation of the two radiations is again obtained by the trace operation in the $|I_2 m_2\rangle$ spin space. This operation is executed in the same manner as in eq. (12):

$$W(1,2) = \mathrm{Tr}\,(\,H(X_2)\,H(X_1)\,\rho(I_o)\,H^\dagger(X_1)\,H^\dagger(X_2))$$

$$= \sum_{\lambda,\,q} B_q^{\lambda\dagger}(X_1)\,\mathrm{Tr}\,(\,H(X_2)\,U_q^\lambda\,H^\dagger(X_2))$$

or

$$W(1,2) = \sum_{\lambda,\,q} B_q^{\lambda\dagger}(X_1)\,A_q^\lambda\,(X_2) \tag{21}$$

The statistical tensor $B_q^\lambda\,(X_1)$ of the first radiation is defined by:

$$B_q^\lambda\,(X_1) = \mathrm{Tr}\,(\,H(X_1)\,\rho(I_o)\,H^\dagger(X_1)\,U_q^\lambda) \tag{22}$$

The statistical tensor $A_q^\lambda\,(X_2)$ of the second radiation is again given by eq. (13) and eq. (14) can be used to express the angular correlation in terms of the angular distribution coefficients $A_q^\lambda\,(X_2)_0$ which are represented in the $\underline{k}_2\underline{P}_2$ - system of radiation X_2.

$$W(1,2) = \sum_{\lambda,q,q''} B_q^{\lambda\dagger}(X_1)\,A_{q''}^\lambda(X_2)_o\,D_{q\,q''}^{(\lambda)*}\,(\underline{x}\,\underline{y}\,\underline{z} \rightarrow \underline{k}_1\underline{P}_1) \tag{23}$$

Equation (23) is the same as eq. (12) for the angular distribution of radiation from oriented nuclei. The statistical tensor $B^{\lambda\dagger}_q\,(X_1)$ describes the orientation of the state I_1 that is created by the emission of radiation X_1 in the same way as $\rho_q^\lambda\,(I_1)$ described the orientation of the state I_1 that is produced by low temperature methods or nuclear reactions.

It is assumed now, that the initial state I_0 is randomly oriented (ordinary angular correlation measurement), hence $\rho(I_o)$ is simply proportional to the unit matrix.

$$\rho(I_o) = (2\,I_o + 1)^{-1}.\,1 \tag{24}$$

Neglecting irrelevant common factors, the statistical tensor for the first radiation is then

$$B_q^\lambda\,(X_1) = \mathrm{Tr}\,(\,H(X_1)\,H^\dagger(X_1)\,U_q^\lambda) \tag{25}$$

The structures of $A_q^\lambda\,(X_2)$ and $B^{\lambda\dagger}_q\,(X_1)$ in eq. (23) are similar, but not identical. This is due to the fact that X_1 is emitted from a random state I_0, but X_2 is emitted from an oriented state leading to a final unobserved (random) state I_2. As far as the general structure of $A_q^\lambda\,(X_2)$ and $B^{\lambda\dagger}_q\,(X_1)$ is concerned, one is essentially the time - reversed of the other. This relationship implies that the transition matrix elements of H(X) enter with somewhat different phases into $A_q^\lambda\,(X_2)$ and $B^{\lambda\dagger}_q\,(X_1)$,respectively, if emission matrix elements are used consistently.

Again the transformation property (3) of the statistical tensors can be used, to express $B_q^{\lambda\dagger}(X_1)$ in terms of the simpler tensors $B_q^{\lambda\dagger}(X_1)_0$, that are represented in the $\underline{k}_1\underline{P}_1$ - system:

$$B_q^{\lambda\dagger}(X_1) = \sum_{q''} B_{q'}^{\lambda\dagger}(X)_0\, D_{q'q}^{\lambda *}(\underline{k}_1\underline{P}_1 \rightarrow \underline{x}\,\underline{y}\,\underline{z}) \qquad (26)$$

The $B_{q'}^{\lambda\dagger}(X)_0$ are the **angular orientation coefficients** for which explicit expressions will be derived in Section 2.4.

The angular correlation is now easily obtained from eqs. (23) and (26):

$$W(1,2) = \sum_{\lambda,q,q',q''} B_{q'}^{\lambda\dagger}(X_1)_0\, A_{q''}^{\lambda}(X_2)_0\, D_{q'q}^{(\lambda)*}(\underline{k}_1\underline{P}_1 \rightarrow \underline{x}\,\underline{y}\,\underline{z})\, D_{qq''}^{(\lambda)*}(\underline{x}\,\underline{y}\,\underline{z} \rightarrow \underline{k}_2\underline{P}_2) \qquad (27)$$

or, using the group - property of the D - matrices

$$W(1,2) = \sum_{\lambda q'q''} B_{q'}^{\lambda\dagger}(X_1)_0\, A_{q''}^{\lambda}(X_2)_0\, D_{q'q''}^{(\lambda)*}(\underline{k}_1\underline{P}_1 \rightarrow \underline{k}_2\underline{P}_2) \qquad (28)$$

The angular correlation (including polarizations) of any two nuclear radiations emitted successively from a randomly oriented state is given by a series of D - functions which depend on the Euler angles that describe the rotation from the $\underline{k}_1\underline{P}_1$ - system to the $\underline{k}_2\underline{P}_2$ - system. The coefficients of the D - functions are the products of the angular orientation coefficients $B_{q''}^{\lambda\dagger}(X_1)_0$ of the first radiation X_1, and of the angular distribution coefficients $A_q^{\lambda}(X_2)_0$ of the second radiation X_2.

If the angular correlation is observed with polarization insensitive detectors, i.e., if only the directional correlation of the two radiations is measured, the axial symmetry about the directions $\underline{k}_1$ and $\underline{k}_2$ lead to $q' = 0$ and $q'' = 0$, respectively. Hence, the directional correlation function is

$$W(\vartheta) = \sum_{\lambda} B_{\lambda}(X_1)\, A_{\lambda}(X_2)\, P_{\lambda}(\cos\vartheta) \qquad (29)$$

where ϑ is the angle between the propagation directions $\underline{k}_1$ and $\underline{k}_2$. The **directional orientation coefficient** is defined by

$$B_{\lambda}(X_1) \equiv B_0^{\lambda\dagger}(X_1)_0 = \mathrm{Tr}\,(H(X_1)_0\, H^{\dagger}(X_1)_0\, U_0^{\lambda\dagger}) \qquad (30)$$

and the **directional distribution coefficient** $A_{\lambda}(X_2)$ has been defined in eq.(16).

2.3 Extranuclear Perturbations of Angular Correlations

The orientation of the intermediate state I_1 involved in an angular correlation measurement will, in general, be altered as soon as it is subjected to interactions of either the magnetic dipole moment μ with an extranuclear magnetic field B or of the electric quadrupole moment Q with electric field gradients $V_{x_i x_j}$. The Hamiltonian for a static magnetic interaction involving a magnetic field B that can be described classically is:

$$K(B) = -\underline{\mu} \cdot \underline{B} = -\mu_z B \tag{31}$$

This Hamiltonian K is diagonal in the $| I_1 m_1 >$ representation, if $\underline{B}$ is chosen parallel to the z - axis of the 'laboratory' system $\underline{x}\ \underline{y}\ \underline{z}$

$$< I_1 m_1 | K(B) | I_1 m_1' > = E_{m_1} \delta_{m_1 m_1'} = \omega_L \hbar \cdot m_1 \cdot \delta_{m_1 m_1'} \tag{32}$$

where ω_L is the Larmor frequency

$$\omega_L = -\frac{B \cdot \mu}{\hbar I} = -g \cdot \mu_0 \frac{B}{\hbar} \tag{33}$$

Similarly for a static quadrupole interaction K(Q) involving an axially symmetric electric field gradient V_{zz} with its symmetry axis parallel to the $\underline{z}$ - axis of the 'laboratory' system,the Hamilton matrix is:

$$< I_1 m_1 | K(Q) | I_1 m_1' > = E_{m_1} \delta_{m_1 m_1'} = \frac{3m_1^2 - I_1(I_1+1)}{4(2I_1 - 1)} e Q V_{zz} \delta_{m_1 m_1'} \tag{34}$$

The effect of an extranuclear interaction on the intermediate state I_1 is described by the evolution of the density matrix $\rho(I_1)$ with time t. The state I_1 is formed by the emission of radiation X_1 at the time $t = 0$ and is described by

$$\rho(I_1; 0) = H(X_1) H^{\dagger}(X_1) \tag{35}$$

for a random initial state I_0. The time evolution of the density matrix is given by

$$\rho(I_1; t) = \Lambda(t)\ \rho(I_1; 0)\ \Lambda^{\dagger}(t) \tag{36}$$

where $\Lambda(t)$ is the time evolution operator:

$$\Lambda(t) = e^{-i K t/\hbar} \tag{37a}$$

for K not explicitly dependent on the time t (static interactions). For a time - dependent interaction K(t) the time evolution operator is

$$\Lambda(t) = e^{-i \int_0^t K(t) dt} \tag{37b}$$

where the integral must be evaluated according to Feynman's rules for ordered operators. The matrix elements of $\Lambda(t)$ in the $| I_1 m_1 >$ representation are (for static interactions):

$$< I_1 m_1 | \Lambda(t) | I_1 m_1' > = e^{-i < I_1 m_1 | K | I_1 m_1' > t/\hbar} \tag{38}$$

If the static interaction is axially symmetric and the symmetry axis is chosen as the quantization axis, the evolution matrix $\Lambda(t)$ is diagonal:

$$\Lambda(t) = e^{-iE\,t/\hbar} \tag{39}$$

where E is the energy eigenvalue matrix. This general statement is easily verified by considering the invariance of the tensor $A^{\lambda}_{q}(t) = \mathrm{Tr}\,(\Lambda(t)\,U^{\lambda}_{q})$ under a rotation α about the quantization axis (see eq. (4)).

After a time t the second radiation X_2 is emitted from the state $\rho(I_1;t)$ and the density matrix of the final state is:

$$\rho(I_2;t) = H(X_2)\,\rho(I_1;t)\,H^{\dagger}(X_2) = H(X_2)\,\Lambda(t)\,H(X_1)\,H^{\dagger}(X_1)\,\Lambda^{\dagger}(t)\,H^{\dagger}(X_2) \tag{40}$$

The angular correlation of X_1 and X_2 is again obtained by the trace of $\rho(I_2;t)$ which is evaluated in the same manner as displayed in eq. (12):

$$\begin{aligned} W(1,2;t) &= \mathrm{Tr}(H(X_2)\,\Lambda(t)\,H(X_1)\,H^{\dagger}(X_1)\,\Lambda^{\dagger}(t)\,H^{\dagger}(X_2)) \\ &= \sum_{\lambda_1 q_1} B^{\lambda_1 \dagger}_{q_1}(X_1)\,\mathrm{Tr}(H(X_2)\,\Lambda(t)\,U^{\lambda_1}_{q_1}\Lambda^{\dagger}(t)\,H^{\dagger}(X_2)) \\ &= \sum_{\lambda_1 q_1 \lambda_2 q_2} B^{\lambda_1 \dagger}_{q_1}(X_1)\,\mathrm{Tr}(H(X_2)\,\Lambda(t)U^{\lambda_1}_{q_1}\Lambda^{\dagger}(t)U^{\lambda_2 \dagger}_{q_2}\big|U^{\lambda_2}_{q_2}H^{\dagger}(X_2)) \\ &= \sum_{\lambda_1 q_1 \lambda_2 q_2} B^{\lambda_1 \dagger}_{q_1}(X_1)\; G^{\lambda_1\lambda_2}_{q_1 q_2}(t)\;\mathrm{Tr}(H(X_2)\;U^{\lambda_2}_{q_2}\,H^{\dagger}(X_2)) \end{aligned} \tag{41}$$

or finally using eq. (13)

$$W(1,2;t) = \sum_{\lambda_1 q_1 \lambda_2 q_2} B^{\lambda_1 \dagger}_{q_1}(X_1)\;G^{q_1 q_2}_{\lambda_1\lambda_2}(t)\;A^{\lambda_2}_{q_2}(X_2) \tag{42}$$

Here, the **perturbation coefficients**

$$G^{q_1 q_2}_{\lambda_1\lambda_2}(t) = \mathrm{Tr}(\,\Lambda(t)\;U^{\lambda_1}_{q_1}\;\Lambda^{\dagger}(t)\;U^{\lambda_2 \dagger}_{q_2}\,) \tag{43}$$

have been introduced. Writing out eq. (43) explicitly shows that the definition of $G^{q_1 q_2}_{\lambda_1\lambda_2}(t)$ is identical with that of Steffen and Frauenfelder[13]), except for an (in most cases) irrelevant factor $(-1)^{\lambda_1+\lambda_2}$. The perturbation coefficients $G^{q_1 q_2}_{\lambda_1\lambda_2}(t)$ completely describe the influence of the extranuclear perturbation on angular correlations. They are the coefficients in the expansion of the statistical tensor $B^{\lambda_2 \dagger}_{q_2}(X_1;t)$, the angular orientation coefficients of the intermediate state I_1 at the time t, in terms of the angular orientation coefficients $B^{\lambda_1 \dagger}_{q_1}(X_1;0)$ at the time t = 0:

$$B^{\lambda_2 \dagger}_{q_2}(X_2;t) = \sum_{\lambda_1 q_1} G^{q_1 q_2}_{\lambda_1\lambda_2}(t)\;B^{\lambda_1 \dagger}_{q_1}(X_1;0) \tag{44}$$

Obviously, the relationship (44) is also valid for the orientation tensors $\rho^{\lambda\dagger}_{q}(I_1)$ of a state oriented by low - temperature methods or in nuclear reactions. Hence, if $B^{\lambda_1\dagger}_{q_1}(X_1)$ is replaced by the orientation tensors $\rho^{\lambda_1\dagger}_{q_1}(I_1)$ eq. (42) also represents the effects of extranuclear perturbations on angular distributions of radiations emitted from oriented nuclei.

The perturbed angular correlation function of two radiations X_1 and X_2 is explicitly:

$$W(1,2;t) = \sum_{\substack{\lambda_1 q_1 q_1' \\ \lambda_2 q_2 q_2'}} B^{\lambda_1\dagger}_{q_1'}(X_1)_0 G^{q_1q_2}_{\lambda_1\lambda_2}(t) A^{\lambda_2}_{q_2'}(X_2)_0 \times D^{(\lambda_1)}_{q_1q_1'}(\varphi_1\vartheta_1\psi_1) D^{(\lambda_2)*}_{q_2q_2'}(\varphi_2\vartheta_2\psi_2) \tag{45}$$

where the Euler angles $\varphi_i\vartheta_i\psi_i$ represent the propagation and polarization directions of the radiations X_i with respect to the 'laboratory' system $\underline{x}\ \underline{y}\ \underline{z}$ in which the perturbation coefficients $G^{q_1q_2}_{\lambda_1\lambda_2}(t)$ are represented.

The **perturbed directional correlation** is given by:

$$W(k_1,k_2;t) = 4\pi \sum_{\lambda_1 q_1 \lambda_2 q_2} B^{\dagger}_{\lambda_1}(X_1)_0 A_{\lambda_2}(X_2)_0 G^{q_1q_2}_{\lambda_1\lambda_2}(t) \times \frac{Y^*_{\lambda_1 q_1}(\vartheta_1\varphi_1) Y_{\lambda_2 q_2}(\vartheta_2\varphi_2)}{[(2\lambda_1+1)\quad(2\lambda_2+1)]^{1/2}} \tag{46}$$

The relation $D^{(\lambda)}_{qo}(\varphi\,\vartheta\,\psi) = (4\pi)^{1/2}\ (2\lambda+1)^{-1/2}\ Y^*_{\lambda q}(\vartheta,\varphi)$ has been used to express the perturbed directional correlation in terms of the spherical harmonics $Y_{\lambda q}(\vartheta,\varphi)$.

The perturbation coefficients $G^{q_1q_2}_{\lambda_1\lambda_2}(t)$ are not irreducible tensors. They can, however, be expanded in terms of irreducible tensors $G^{\Lambda}_{Q}(t)$ by (compare with eqs. (2) and (5)!)

$$G^{q_1q_2}_{\lambda_1\lambda_2}(t) = \sum_{\Lambda} (-1)^{\lambda_1+q_1} \langle \lambda_1 -q_1 \lambda_2 q_2 | \Lambda Q \rangle G^{\Lambda}_{Q}(t) \tag{47}$$

where the perturbation tensor $G^{\Lambda}_{O}(t)$ is given by

$$G^{\Lambda}_{Q}(t) = \sum_{q_1q_2} (-1)^{\lambda_1+q_1} \langle \lambda_1 - q_1 \lambda_2 q_2 | \Lambda Q \rangle\ G^{q_1q_2}_{\lambda_1\lambda_2}(t) \tag{48}$$

The tensor $G^{\Lambda}_{Q}(t)$ transforms under a three - dimensional rotation of the 'laboratory' system with $D^{(\Lambda)}_{QQ'}(\alpha\,\beta\,\gamma)$ (see eq. (3)).

If the perturbing interaction on the ensemble of nuclei as a whole is invariant under a three - dimensional rotation (isotropic interaction), i.e., if the overall effect on the ensemble is such that it does not introduce a privileged direction in space (liquid and powder sources in the absence of an externally applied field) the corresponding perturbation tensor $G_Q^\Lambda(t)$ must be invariant under rotations:

$$\bar{G}_Q^\Lambda(t)_{Rot} = \sum_{Q'} \bar{G}_{Q'}^\Lambda(t)\, D_{Q'Q}^{(\Lambda)}(\alpha\beta\gamma) = \bar{G}_Q^\Lambda(t) \tag{49}$$

for all angles $\alpha\,\beta\,\gamma$. Equation (49) can only be satisfied if $\Lambda = Q = Q' = 0$, which implies $\lambda_1 = \lambda_2$ and $q_1 = q_2$ in eqs. (48), (45) and (46). In this case the summation over $q_1 = q_2$ can be performed and the group property of the D - matrices can be used:

$$W(1,2;t) = \sum_{\lambda_1 q_1' q_2'} B_{q_1'}^{\lambda\,\dagger}(X_1)_0\, G_{\lambda\lambda}(t)\, A_{q_2'}^{\lambda}(X_2)_0\, D_{q_1' q_2'}^{(\lambda)\,*}(\underline{k}_1\underline{P}_1 \to \underline{k}_2\underline{P}_2) \tag{50}$$

or, for the **perturbed directional correlation** $(q_1' = q_2' = 0)$:

$$W(\vartheta;t) = \sum_\lambda B_\lambda(X_1)\, A_\lambda(X_2)\, G_{\lambda\lambda}(t)\, P_\lambda(\cos\vartheta) \tag{51}$$

where the **attenuation coefficient** for isotropic perturbations

$$G_{\lambda\lambda}(t) = G_{\lambda\lambda}^{qq}(t) \qquad \text{for any } q \tag{52}$$

has been introduced. This is a very general result. The effect of an isotropic interaction on **any** angular distribution is completely described by the attenuation coefficients $G_{\lambda\lambda}(t)$ i.e., the angular distribution anisotropies are only diminished, but the angular distribution is not basically altered as is possible in the case of a non - isotropic perturbation.

If the perturbing interaction has an axis $\underline{a}$ of cylindrical symmetry (external magnetic field, axially symmetric electrostatic gradient) and the z - axis of the 'laboratory' system is chosen parallel to this symmetry axis $\underline{a}$, the perturbation tensor is invariant under a rotation α of the 'laboratory' coordinate about the z - axis. Hence, only the $Q = 0$ components of the perturbation tensors $G_Q^\Lambda(t)$ are non - vanishing (compare with eq. (4)), and from eq. (47) it follows that $q_1 = q_2$. Hence, for **axially symmetric interactions**

$$G_{\lambda_1\lambda_2}^{q_1q_2}(t) = G_{\lambda_1\lambda_1}^{q_1q_1}(t)\, \delta_{q_1q_2} \tag{53}$$

If the axis of symmetry $\underline{a} = \underline{z}$ is parallel to one of the propagations directions $\underline{k}_i$ in a directional correlation experiment, $\vartheta_i = 0$ and $Y_{\lambda_i,q_i}(0,\varphi) = \delta_{q_i0}\,[\,(2\lambda_i + 1)/4\pi\,]^{1/2}$. This leads with $q_1 = q_2 = 0$ to

$$G_{\lambda_1\lambda_2}^{00}(t) = \mathrm{Tr}\,(\,\Lambda(t)\, U_0^{\lambda_1} \Lambda^\dagger(t)\, U_0^{\lambda_2\dagger})$$

$$G_{\lambda_1\lambda_2}^{00}(t) = \mathrm{Tr}\,(\,\Lambda(t)\,\Lambda^\dagger(t)\, U_0^{\lambda_1} U_0^{\lambda_2\dagger}\,) = \mathrm{Tr}\,(\,U_0^{\lambda_1} U_0^{\lambda_2\dagger})$$

$$G^{00}_{\lambda_1\lambda_2}(t) = \delta_{\lambda_1\lambda_2} \tag{54}$$

because the matrix $\Lambda(t)$ is diagonal for axially symmetric interactions with the symmetry axis $\underline{a}$ as quantization axis $\underline{z}$. (See eq. (40)). Hence, static interactions that are axially symmetric with respect to one of the propagation directions have **no effect** on **directional** distributions or correlations.

Of particular interest are perturbations by an externally applied magnetic field $\underline{B}$. The evolution matrix $\Lambda(t)$ is diagonal if the direction of $\underline{B}$ is chosen as quantization axis. From eqs. (39) and (43) one obtains

$$\begin{aligned} G^{q_1q_2}_{\lambda_1\lambda_2}(t) &= e^{-i\omega q_1 t}\, \mathrm{Tr}\,(U^{\lambda_1}_{q_1} U^{\lambda_2\dagger}_{q_2}) \\ &= e^{-\omega q_1 t}\, \delta_{q_1q_2}\, \delta_{\lambda_1\lambda_2} = D^{(\lambda_1)}_{q_1q_1}(\omega t,0,0)\, \delta_{q_1q_2}\, \delta_{\lambda_1\lambda_2} \end{aligned} \tag{55}$$

where the D - matrix $D^{(\lambda)}_{qq}$ $(\omega t, 0, 0)$ represents a rotation about the $\underline{B} = B\,\underline{z}$ axis through an angle ωt. In this case the perturbed angular correlation (45) can be written in the form

$$\begin{aligned} W(1,2;t,\underline{B}) = \sum_{\lambda_1 q_1 q_1' q_2} & B^{\lambda_1\dagger}_{q_1'}(X_1)_0\, A^{\lambda_1}_{q_2'}(X_2)_0\, D^{(\lambda_1)*}_{q_1'q_1}(\underline{k}_1\underline{P}_1 \to \underline{x}\,\underline{y}\,\underline{z}) \\ & \times D^{(\lambda_1)*}_{q_1q_1}(0,0,-\omega t)\, D^{(\lambda_1)*}_{q_1q_2'}(\underline{x}\,\underline{y}\,\underline{z} \to \underline{k}_2\underline{P}_2) \end{aligned} \tag{56}$$

or, using the group property of the D - matrices:

$$W(1,2,t,\underline{B}) = \sum_{\lambda_1 q_1' q_2'} B^{\lambda\dagger}_{q_1'}(X_1)_0\, A^{\lambda}_{q_2'}(X_2)_0\, D^{(\lambda)*}_{q_1'q_2'}(\underline{k}_1\underline{P}_1 \to -\omega t \to \underline{k}_2\underline{P}_2) \tag{57}$$

where the rotation $\underline{k}_1\underline{P}_1 \to -\omega t \to \underline{k}_2\underline{P}_2$ represents the rotation of (a) the $\underline{k}_1\underline{P}_1$ - coordinate system into the 'laboratory' system with the direction of $\underline{B}$ as z - axis, then (b) a rotation of $-\omega t$ about $\underline{B}$ and finally (c) a rotation from the rotated laboratory system into the $\underline{k}_2\underline{P}_2$ system. For a transverse magnetic field $\underline{B}$, i.e., if $\underline{B}$ is perpendicular to $\underline{k}_1$ and $\underline{k}_2$ the directional correlation function is of a particularly simple form:

$$W(\vartheta; t, \underline{B}_\perp) = \sum_{\lambda} B_\lambda(X_1)\, A_\lambda(X_2)\, P_\lambda(\cos(\vartheta - \omega t)) \tag{58}$$

where ϑ is again the angle from $\underline{k}_1$ to $\underline{k}_2$. (The sign of ϑ is important in this case; ϑ has the same sign as $\underline{z} \cdot (\underline{k}_1 \times \underline{k}_2)$). The directional correlation (58) is the basis of g - factor measurements on excited nuclear states.

The perturbation of angular distributions and correlations by radiofrequency fields (NMR/PAC) seems to offer promising possibilities of very accurate g - factor measurements. The interaction of the radiofrequency field $\underline{B}_1$ (t) in the presence of a strong static magnetic field $\underline{B}_0$ can, at resonance, either destroy the anisotropic angular distribution if $\underline{B}_0$ is parallel to one of the radiation directions k_i (or to the symmetry axis $\underline{a}$ in case of an oriented state) or restore the anisotropic angular distribution if $\underline{B}$ is perpendicular to $\underline{k}_i$ (or $\underline{a}$). The perturbation coefficients $G^{q_1q_2}_{\lambda_1\lambda_2}(t)$ for this NMR interaction have been computed by Matthias, Olsen, Shirley, Steffen and Templeton[15]. Using as quantization axes a coordinate system with $\underline{z}$ parallel to $\underline{B}_0$ and $\underline{x}$ parallel to $\underline{B}_1$ (0), the direction of the linearly polarized radiofrequency field $\underline{B}_1(t) = 2\,\underline{B}_1 \cdot \cos(\omega_1 t + \Delta)$, the perturbation coefficient is

$$G^{q_1q_2}_{\lambda_1\lambda_2}(t) = \delta_{\lambda_1\lambda_2}\, e^{-q_2\omega_1 t}\, e^{i(q_2-q_1)\Delta} \sum_{p=-\lambda_1}^{+\lambda_1} e^{-ip\omega_{eff}t}\, d^{(\lambda_1)}_{q_1p}(\beta)\, d^{(\lambda_1)}_{q_2p}(\beta) \tag{59}$$

where $d^{\lambda}_{qp}(\beta)$ are the 'small' D - matrices[22] and β and ω_{eff} are given by

$$\tan\beta = \frac{B_1}{\left(1-\frac{\omega_1}{\omega_0}\right)B_0} \tag{60}$$

$$\omega_{eff} = -\frac{g\cdot\mu_N}{\hbar}\left[\left(1-\frac{\omega_1}{\omega_0}\right)B_0{}^2 + B_1{}^2\right]^{1/2} \tag{61}$$

$$\omega_0 = -\frac{g\cdot\mu_N}{\hbar}\,B_0 \tag{62}$$

Relaxation interactions have been neglected. Equation (59) for $G^{q_1q_2}_{\lambda_1\lambda_2}(t)$ corresponds to a periodic function of t in which a fast oscillation $e^{-q_2\omega_1 t}$ is amplitude modulated by slowly varying components $e^{-ip\omega_{eff}t}$.

NMR/PAC experiments are usually done in the time - integrated mode with random phases Δ. This situation is described by the time - integrated perturbation coefficient

$$G^{q_1q_2}_{\lambda_1\lambda_2}(t) = \delta_{q_1q_2}\,\delta_{\lambda_1\lambda_2} \sum_{p=-\lambda_1}^{\lambda_1} \frac{1-i(p\omega_{eff}+q_1\omega)\tau}{1+(p\omega_{eff}+q_1\omega)^2\tau^2}\, d^{(\lambda_1)}_{q_1p}{}^2(\beta) \tag{63}$$

The expression (59) for the NMR/PAC perturbation coefficient can be written entirely in terms of products of D - matrices $D^{(\lambda_1)}$:

$$G^{q_1q_2}_{\lambda\lambda}(t) = \sum_{q_1{}',p_1p',q_2{}'} D^{(\lambda)*}_{q_1q_1{}'}(-\Delta,0,0)\, D^{(\lambda)*}_{q_1{}'p}(0,\beta,0)$$

$$\times\; D^{(\lambda)*}_{pp'}(-\omega_{eff}t,0,0)\, D^{(\lambda)*}_{p'q_2{}'}(0,-\beta,0)\, D^{(\lambda)*}_{q_2{}'q_2}(-\omega_1 t+\Delta,0,0)$$

or

$$G^{q_1q_2}_{\lambda\lambda}(t) = D^{(\lambda)*}_{q_1q_2}(R(t)) \tag{64}$$

where R (t) is the result of the following rotations of the 'laboratory system' $\underline{x}_0\underline{y}_0\underline{z}_0$ at t = 0

1) $-\Delta$ about $\underline{z}$ $\rightarrow$ $\underline{x}_1\,\underline{y}_1\,\underline{z}_1$
2) β about $\underline{y}_1$ $\rightarrow$ $\underline{x}_2\,\underline{y}_2\,\underline{z}_2$
3) $-\omega_{\text{eff}}t$ about $\underline{z}_2$ $\rightarrow$ $\underline{x}_3\,\underline{y}_3\,\underline{z}_3$
4) $-\beta$ about $\underline{y}_3$ $\rightarrow$ $\underline{x}_4\,\underline{y}_4\,\underline{z}_4$
5) $-w_1 t+\Delta$ about $\underline{z}_4$ $\rightarrow$ $\underline{x}_t\,\underline{y}_t\,\underline{z}_t$

The axes of the 'transformed' laboratory system at t are given by $\underline{x}_t\,\underline{y}_t\,\underline{z}_t$. The NMR-perturbed angular correlation of two radiations X_1 and X_2 is now given by (see eq. (42)),

$$W(1,2;\,t,\text{NMR}) = \sum_{\lambda_1 q_1, q_1' q_2, q_2'} B^{\lambda\dagger}_{q_1'}(X_1)_0\, A^{\lambda}_{q_2'}(X_2)_0$$
$$\times\; D^{(\lambda)*}_{q_1' q_1}(\underline{k}_1\underline{P}_1 \rightarrow \underline{x}_0\underline{y}_0\underline{z}_0)\, D^{(\lambda)*}_{q_1 q_2}(\underline{x}_0\underline{y}_0\underline{z}_0 \rightarrow \underline{x}_t\underline{y}_t\underline{z}_t)$$
$$\times\; D^{(\lambda)*}_{q_2 q_2}(\underline{x}_t\underline{y}_t\underline{z}_t \rightarrow \underline{k}_2\underline{P}_2)$$

or

$$W(1,2;\,t,\text{NMR}) = \sum_{\lambda q_1' q_2'} B^{\lambda\dagger}_{q_1'}(X_1)_0\, A^{\lambda}_{q_2'}(X_2)_0\, D^{(\lambda)*}_{q_1' q_2'}(\underline{k}_1\underline{P}_1 \overset{\sigma}{\rightarrow} \underline{k}_2\underline{P}_2) \tag{65}$$

where the transformation $\underline{k}_1\underline{P}_1 \overset{\sigma}{\rightarrow} \underline{k}_2\underline{P}_2$ signifies the rotation of the $\underline{k}_1\underline{P}_1$ - system into $\underline{k}_2\underline{P}_2$ via the rotation $\underline{x}_0\,\underline{y}_0\,\underline{z}_0 \rightarrow \underline{x}_t\,\underline{y}_t\,\underline{z}_t$ of the laboratory system.

The NMR - perturbed directional correlation is simply

$$W(\underline{k}_1,\underline{k}_2;\text{NMR},t) = \sum_{\lambda} B_\lambda(X_1)\,A_\lambda(X_2)\,P_\lambda(\cos(\underline{k}_1 \overset{\sigma}{\rightarrow} \underline{k}_2)) \tag{66}$$

where the symbol $\overset{\sigma}{\rightarrow}$ has the same meaning as above.

The influence of time - dependent fluctuating (relaxation) interactions on angular correlations is described by **attenuation coefficients** since the interaction as a whole is isotropic. The attenuation coefficients are of the form[11])

$$G_{\lambda\lambda}(t) = e^{-\Lambda_\lambda t} \tag{67}$$

where the relaxation constant Λ_λ depends on the strength of the fluctuating fields and on the characteristic correlation time describing the rapidity of the fluctuation. Under certain (rare) circumstances the attenuation factors must be represented by a series of exponentials[24,25]):

$$G_{\lambda\lambda}(t) = \sum_{\mathcal{H}} e^{-\Lambda_\lambda^{(\mathcal{H})} t} \tag{68}$$

A general theory of the influence of electronic relaxation processes on angular correlations has been developed by Gabriel[14]).

2.4 *The Angular Distribution and Orientation Coefficients*

2.4.1 *General Nuclear Radiation*

The angular distribution and orientation coefficients are obtained by evaluating the statistical tensors (13) and (25), respectively, referring to the $\underline{k}\underline{P}$ - system of the radiation. The emission operator $H(X)_{op}$ is expanded in terms of multipole operators $H^{(\pi)}_{LM}$ that describe the emission of radiation of angular momentum L, projection M and parity π. The radiation field is not observed in this angular momentum state, but in its plane wave state (linear momentum eigenstate). Hence, the operators that must be used are the operators $H(\underline{k}, p)$ that describe the emission of the radiation X with a (plane) wave vector $\underline{k}$ and a projection p of the spin of the radiation quanta with respect to $\underline{k}$. Any polarization state can then be constructed by coherent superpositions of the plane wave states $|\underline{k}\ p\rangle$ with different p values and amplitudes a_p. For electromagnetic radiation $p = \pm 1$, for fermions $p = \pm 1/2$, for α - particles $p = 0$.

Hence, the multipole expansion of the transition operator H(X) is of the form[25]):

$$H(X)_0 = \sum_p a_p H(\underline{k}, p)_0 = \sum_p a_p \sum_{LM\pi} \langle \underline{k}\ p | LM\pi\rangle_0^* H^{(\pi)\dagger}_{LM} \tag{69}$$

The index 0 is a reminder that the elements $\langle \underline{k}\ p | LM\pi\rangle_0$ of the unitary transformation T, i.e., $|\underline{k}\ p\rangle \longrightarrow |LM\pi\rangle$ must be evaluated using $\underline{k}$, $\underline{P}$ as quantization axes ($\underline{k}\ \underline{P}$ - system).

The matrix elements $\langle m_2 | H^{(\pi)}_{LM} | m_1\rangle$ are further reduced by using the Wigner - Eckart theorem in the form:

$$\langle m_2 | H^{(\pi)}_{L-M} | m_1\rangle = (-1)^{I_2 - m_2} \begin{pmatrix} I_2 & L & I_1 \\ -m_2 & -M & m_1 \end{pmatrix} \langle m_2 \| H^{(\pi)}_L \| m_1\rangle \tag{70}$$

The **reduced emission matrix element** $\langle m_2 \| H^{(\pi)}_L \| m_1\rangle$ contains the **model - dependent information** about the nuclear states $|I_1 m_1\rangle$ and $|I_2 m_2\rangle$ involved in the transition.

The transition amplitudes can now be written using eqs. (69) and (70)

$$\langle I_2 m_2 | H(X) | I_1 m_1\rangle_0 = \sum_{p,LM\pi} a_p \langle \underline{k}\ p | LM\pi\rangle_0 (-1)^{I_2 - m_2} \begin{pmatrix} I_2 & L & I_1 \\ -m_2 & -M & m_1 \end{pmatrix} \times (-1)^{L+M+n} \langle I_2 \| H^{(\pi)}_L \| I_1\rangle \tag{71}$$

and the angular distribution and orientation coefficients are easily constructed from eqs. (13) and (25), respectively:

$$A^\lambda_q(X)_0 = \mathrm{Tr}\,(H(X)_0\, U^\lambda_q\, H^\dagger(X)_0) \tag{72a}$$

$$B^{\lambda\dagger}_q(X)_0 = \mathrm{Tr}\,(H(X)_0\, H^\dagger(X)_0\, U^{\lambda\dagger}_q) \tag{72b}$$

or explicitly, e.g.

$$A_q^\lambda(X)_0 = \sum_{m_2,m_1,m_1',p,L,M,p',L',M'} a_p a_{p'}^* <\underline{k}\ p|\ LM\pi>_0^* <L'M'\pi'|\ \underline{k}\ p'>_0$$

$$\times (-1)^{I_1+m_1+I_2-m_2+I_2-m_2+L+L'+M+M'}(2\lambda+1)^{1/2}\begin{pmatrix} I_1 & I_1 & \lambda \\ -m_1 & m_1' & q \end{pmatrix}\begin{pmatrix} I_2 & L & I_1 \\ -m_2 & -M & m_1 \end{pmatrix}$$

$$\times\begin{pmatrix} I_2 & L' & I_1 \\ -m_2 & -M' & m_1' \end{pmatrix} <I_2||H_L^{(\pi)}||I_1> <I_2||H_{L'}^{(\pi')}||I_1>^* \tag{73}$$

and similarly for the angular orientation coefficients (72b). The summation over m_1, m_1' and m_2 can be performed using the methods of Racah algebra (see eq. (6.2.8) in Edmonds[26]).

$$A_q^\lambda(X)_0 = \sum_{LM\pi,L'M'\pi'} <LM\pi|\ \underline{k}p>_0\ a_p\ a_{p'}^* <\underline{k}p'|\ L'M'\pi'>_0^*$$

$$\times\ (-1)^{L+M'+I_1+I_2+L'}(2\lambda+1)^{1/2}\begin{pmatrix} L' & L & \lambda \\ -M' & +M & -q \end{pmatrix}\begin{Bmatrix} L & L' & \lambda \\ I_1 & I_1 & I_2 \end{Bmatrix}$$

$$\times\ <I_2||H_L^{(\pi)}||I_1> <I_2||H_{L'}^{(\pi')}||I_1>^* \tag{74}$$

$$A_q^\lambda(X)_0 = \sum_{LL'\pi\pi'}\begin{Bmatrix} L & L' & \lambda \\ I_1 & I_1 & I_2 \end{Bmatrix}(-1)^{I_1+I_2+L'+\lambda}<I_2||H_L^{(\pi)}||I_1>$$

$$\times\ <I_2||H_{L'}^{(\pi')}||I_1>^*\ \epsilon_{+q}^\lambda(L,L')_0 \tag{75}$$

where the **radiation efficiency tensor** $\epsilon_q^\lambda(L'L)$ for the detection of radiation of multipole character L′, L has been introduced[27].

$$\epsilon_q^\lambda(L\pi, L'\pi')_0 = \mathrm{Tr}\,(\epsilon(L,L')_0\ U_q^\lambda(L',L)) \tag{76}$$

The **radiation efficiency matrix** $\epsilon(L'\pi', L\pi)$ in eq. (76) is defined by

$$\epsilon(L\pi, L'\pi')_0 = T(L\pi)_0\ \epsilon(\underline{P})\ T^\dagger(L'\pi')_0 \tag{77}$$

where $T(L\pi)_0$ is the matrix of the unitary transformation from the angular momentum eigenstates $|LM\pi>$ of the radiation to the plane wave state $|\ \underline{k}\ p>$ in which the radiation is observed. The **detector efficiency matrix** $\epsilon(\underline{P})$ with elements

$$<p|\epsilon(\underline{P})|p'> = a_p a_{p'}^* \tag{78}$$

describes the response of the detector to plane waves in the polarization state $<p|p'>$. For polarization insensitive detectors $<p|\epsilon(0)p'> = 1/2\,\delta_{pp'}$, and the detector efficiency matrix is diagonal. For detectors responding to right or left circularly polarized gamma rays $\epsilon(+) = [\begin{smallmatrix}1 & 0\\ 0 & 0\end{smallmatrix}]$, or $\epsilon(-) = [\begin{smallmatrix}0 & 0\\ 0 & 1\end{smallmatrix}]$, respectively. For a detector responding to gamma rays linearly polarized in a direction $\underline{P}$, the detector efficiency matrix is

$$e(\underline{P}) = \frac{1}{2}\begin{pmatrix} 1 & -e^{-2ia} \\ -e^{+2ia} & 1 \end{pmatrix} \tag{79}$$

where a is the angle between $\underline{P}$ ($\underline{P}\cdot\underline{k} = 0$) and the polarization axis Q (direction of maximum response) of the detector.

Equation (75) clearly shows the general structure of the angular distribution coefficient $A_q^\lambda(X)_0$. It is given by the weighted sum of the efficiency tensors $\epsilon_{+q}^\lambda(L\pi, L'\pi')$ which describe the response of the detector to the radiation field of angular momentum and parity $L\pi$ and $L'\pi'$. The weighing factors contain the relative intensities of the multipole components, characterized by the reduced matrix elements $<I_2\,||\,H_L^{(\pi)}\,||\,I_1>$ $<I_2\,||\,H_{L'}^{(\pi')}\,||\,I_1>^*$. These are the, in general unknown, quantities that are deduced most frequently from angular distribution and correlation measurements. The fact that interference terms between L and L′ are present makes angular distribution measurements more sensitive to multipole mixtures than any type of non - directional intensity measurement (e.g., internal conversion coefficients, shapes of beta spectra). The other part of the weighing factor is the 6 - j symbol $\{\begin{smallmatrix} L & L' & \lambda \\ I_1 & I_1 & I_2\end{smallmatrix}\}$ which describes the recoupling of the angular momenta from $\underline{\lambda} = \underline{I}_1 - \underline{I}_1' = (\underline{L}_1 + \underline{I}_2) - (\underline{L}_1' + \underline{I}_2')$ to $\underline{\lambda} = \underline{L}_1 - \underline{L}_1'$.

2.4.2 Electromagnetic Radiation

The emission operator $H_{LM}^{(\pi)}$ for electromagnetic multipoles (π = e for electric multipoles, π = m for magnetic multipoles) is $\underline{j}_N \underline{A}_{LM}^{(\pi)*}$ where $\underline{j}_N$ is the nuclear current operator. It is important that the electromagnetic multipole fields $\underline{A}_{LM}^{(\pi)}$ be explicitly defined in order to facilitate the comparison of experimental and theoretical results of relative multipole amplitudes (mixing ratios): Throughout this discussion the definition

$$\underline{A}_{LM}^{(e)}(k, \underline{r}) = \frac{i^L}{k}\,\frac{\underline{\nabla}\times\underline{L}}{[L(L+1)]^{1/2}}\, j_L(kr)\, Y_{LM}(\hat{r}) \tag{80a}$$

$$\underline{A}_{LM}^{(m)}(k, \underline{r}) = i^L\,\frac{\underline{L}}{[L(L+1)]^{1/2}}\, j_L(kr)\, Y_{LM}(\hat{r}) \tag{80b}$$

will be used. The vector fields (80) satisfy the equation

$$\underline{A}_{LM}^{(\pi)*} = (-1)^{L-M+1}\,\underline{A}_{L-M}^{(\pi)} \tag{81}$$

A plane electromagnetic wave of helicity p (projection of the photon spin onto the propagation direction $\underline{k}$) travelling along the $+\underline{z}$ axis is represented by:

$$\underline{A}_p(\underline{k})_0 = \underline{e}_p\, e^{ikz} = -\sqrt{2\pi}\sum_{LM}(2L+1)^{1/2}\,(\underline{A}_{LM}^{(e)} + p\underline{A}_{LM}^{(m)})_0\,\delta_{Mp} \tag{82}$$

where $\underline{e}_\pm = \mp 2^{-1/2}(\underline{x} \pm i\underline{y})$ ($|\underline{x}| = |\underline{y}| = 1$). It is convenient to introduce the quantity Λ_π

$$\Lambda_e = 0 \qquad \Lambda_m = 1$$

The Hamiltonian operator for emission of electromagnetic radiation into the plane wave state $|\underline{k}\, p>_0$ can then be written in the form:

$$H(\underline{k}, p)_0 = -\underline{j}_N\, \underline{A}_p(\underline{k})_0^* = \sqrt{2\pi} \sum_{LM\pi} (2L+1)^{1/2}\, p^{\Lambda\pi}\, \delta_{Mp}\, \underline{j}_N\, \underline{A}_{LM}^{(\pi)*}$$

$$H(\underline{k}, p)_0 = \sqrt{2\pi} \sum_{LM\pi} (2L+1)^{1/2}\, p^{\Lambda\pi}\, \delta_{Mp}\, H_{LM}^{(\pi)\dagger}(\gamma, \text{emission}) \tag{83}$$

where the operator for emission of $LM\pi$ gamma multipole radiation is:

$$H^{\pi}_{LM}(\gamma, \text{emission}) = (\underline{j}_N \underline{A}_{LM}^{(\pi)})^{\dagger} = \underline{j}_N \underline{A}_{LM}^{(\pi)*} = (-1)^{L-M+1}\, \underline{j}_N \underline{A}_{L-M}^{(\pi)} \tag{84}$$

where eq. (81) has been used. For local nuclear forces the nuclear current operator $\underline{j}_N$ is proportional to the nuclear momentum and spin operators $\underline{p}$ and $\underline{\sigma}$, respectively, which transform under the time - reversal operation T as $T\underline{p}\,T^{-1} = -\underline{p}$ and $T\underline{\sigma}\,T^{-1} = -\underline{\sigma}$. Hence the interaction operators $\underline{j}_N \underline{A}_{LM}^{(\pi)}$ transform under time - reversal according to $T\,(\underline{j}_N \underline{A}_{LM}^{(\pi)})\,T^{-1} = (-1)^{L-M}\,(\underline{j}_N \underline{A}_{L-M}^{(\pi)})$. If the nuclear eigenstates $|Im>$ are chosen with their phases such that $T\,|Im> = (-1)^{I-m}\,|I-m>$, which is always possible, the matrix elements $<I_f m_f|\,\underline{j}_N \underline{A}_{LM}^{(\pi)}\,|I_i m_i>$ are real.

The reduced gamma emission matrix elements are defined by the Wigner - Eckart theorem of eq. (70):

$$<I_2 m_2|\,\underline{j}_N \underline{A}_{LM}^{(\pi)*}\,|I_1 m_1> = (-1)^{L-M+1}\,(-1)^{I_2-m_2} \begin{pmatrix} I_2 & L & I_1 \\ -m_2 & -M & m_1 \end{pmatrix} \tag{85}$$

$$\times <I_2||\,\underline{j}_N \underline{A}_L^{(\pi)}||\,I_1>$$

The (real) reduced matrix elements are related to the reduced multipole matrix elements of Bohr and Mottelson[28]) and Alder et al.[29]) by

$$<I_f||\,\underline{j}_N \underline{A}_L^{(E)}||\,I_i> = \frac{k^L}{(2L+1)!!} \left(\frac{L+1}{L}\right)^{1/2} <I_f||\,i^L \mathcal{M}(EL)||\,I_i>_{BM} \tag{86a}$$

$$<I_f||\,\underline{j}_N \underline{A}_L^{(M)}||\,I_i> = -\frac{k^L}{(2L+1)!!} \left(\frac{L+1}{L}\right)^{1/2} <I_f||\,i^{L-1} \mathcal{M}(ML)||\,I_i>_{BM} \tag{86b}$$

A comparison of eqs. (83) and (69) shows that the elements of the transformation matrix T in (77) are given by:

$$<LM\pi|\,\underline{k}\; p>_0 = \sqrt{2\pi}\; p^{\Lambda\pi}\,(2L+1)^{1/2}\,\delta_{+Mp} \tag{87}$$

The radiation efficiency tensors $\epsilon^{\lambda}_{-q}(L,L')$ can be constructed from eqs. (76) – (78) and eq. (87):

$$\epsilon^{\lambda}_{+q}(L\pi, L'\pi') = 2\pi \sum_{p,p'=\pm1} p^{\Lambda\pi}\, p'^{\Lambda'\pi'}\,(-1)^{L+p} \begin{pmatrix} L & L' & \lambda \\ +p & -p' & +q \end{pmatrix}$$

$$\times\, [(2\lambda+1)(2L+1)(2L'+1)]^{1/2}\, a_p\, a_{p'}^* \tag{88}$$

For polarization insensitive detectors (directional observation only) $\epsilon = 1/2 \begin{bmatrix} 1 & 0 \\ 0 & 1 \end{bmatrix}$ or $a_p a_{p'}^* = 1/2 \delta_{pp'}$, hence $q = 0$ (axial symmetry!). The efficiency tensor for directional measurements is therefore:

$$\epsilon_0^\lambda (L,L') = \frac{1}{2} (-1)^{L+1} [(2\lambda+1)(2L+1)(2L'+1)]^{1/2}$$

$$\times \left\{1 + (-1)^{\Lambda_\pi + \Lambda_{\pi'} + L + L' + \lambda}\right\} \begin{pmatrix} L & L' & \lambda \\ 1 & -1 & 0 \end{pmatrix} \tag{89}$$

where the symmetry property $\begin{pmatrix} j_1 & j_2 & j_3 \\ m_1 & m_2 & m_3 \end{pmatrix} = (-1)^{j_1+j_2+j_3} \begin{pmatrix} j_1 & j_2 & j_3 \\ -m_1 & -m_2 & -m_3 \end{pmatrix}$ of the 3 - j symbols has been used.

If the gamma multipole radiation L, L′ connects nuclear states whose **parities** are **perfect** quantum numbers

$$\Lambda_\pi + \Lambda_{\pi'} + L + L' = \text{even} \tag{90}$$

and $\epsilon_0^\lambda (L,L')$ is different from zero **only for λ = even**. The directional distribution coefficients are now easily obtained from eqs. (75), (85) and (89) (omitting irrelevant factors):

$$A_\lambda (\gamma) = \sum_{LL'\pi\pi'} F_\lambda (LL'I_2I_1) \langle I_2 \| j_N A_L^{(\pi)} \| I_1 \rangle \langle I_2 \| j_N A_{L'}^{(\pi')} \| I_1 \rangle \tag{91}$$

The F - coefficient F_λ $(LL'I_2I_1)$ is defined by:

$$F_\lambda(LL'I_2I_1) = (-1)^{I_1+I_2-1} [(2\lambda+1)(2L+1)(2L'+1)(2I_1+1)]^{1/2}$$

$$\times \begin{pmatrix} L & L' & \lambda \\ 1 & -1 & 0 \end{pmatrix} \times \begin{Bmatrix} L & L' & \lambda \\ I_1 & I_1 & I_2 \end{Bmatrix} \tag{92}$$

The F - coefficients have the properties

$$F_0(LL'I_2I_1) = \delta_{LL'} \qquad F_\lambda(LL'I_2I_1) = F_\lambda(L'LI_2I_1)$$

but

$$F_\lambda(LL'I_2I_1) \neq F_\lambda(LL'I_1I_2)$$

In most gamma transitions only one multipole L = L′ contributes, (pure multipole transition) and the directional distribution coefficient is simply:

$$A_\lambda(\gamma) = F_\lambda(LLI_2I) \tag{93}$$

or two multipoles with L′ = L + 1 are involved (mixed multipole transition e.g. E2 + M1) and the directional distribution coefficient is:

$$A_\lambda (\gamma) = \frac{F_\lambda(LLI_2I_1) + 2\delta(\gamma) F_\lambda(L\,L{+}1\,I_2I_1) + \delta^2(\gamma) F_\lambda(L{+}1\,L{+}1\,I_2I_1)}{1 + \delta^2(\gamma)} \tag{94}$$

The directional distribution coefficients (93) and (94) have been normalized to $A_0(\gamma) = 1$. The mixing amplitude ratio $\delta(\gamma)$ in eq. (94) is defined by

$$\delta(\gamma) = \frac{\langle I_2 \| j_N A_{L+1}^{(\pi')} \| I_1 \rangle}{\langle I_2 \| j_N A_L^{(\pi)} \| I_1 \rangle} \tag{95}$$

In the same way the directional orientation coefficients are obtained from eq. (25), e.g. for a mixed $L_1\pi_1$ and $L_1'\pi_1'$ transition

$$B_\lambda(\gamma) = \frac{F_\lambda(L_1 L_1 I_0 I_1) + 2(-1)^{L_1+L_1'}\delta(\gamma_1) F_\lambda(L_1 L_1' I_0 I_1) + \delta^2(\gamma_1) F_\lambda(L_1' L_1' I_0 I_1)}{1 + \delta^2(\gamma_1)} \tag{96}$$

where

$$\delta(\gamma_1) = \frac{\langle I_1 \| \underline{j}_N \underline{A}_{L'}^{(\pi')} \| I_0 \rangle}{\langle I_1 \| \underline{j}_N \underline{A}_{L}^{(\pi)} \| I_0 \rangle} \tag{97}$$

If the angular distribution is observed with a detector that responds only to the $p = 1$ (-1) helicity component of the radiation ($p = +1$ (-1) corresponds to right (left) circular polarization) the angular distribution coefficient is obtained from eqs. (75) - (78) with $\epsilon(+) = \left[\begin{smallmatrix}1 & 0\\ 0 & 0\end{smallmatrix}\right]$ and $\epsilon(-) = \left[\begin{smallmatrix}0 & 0\\ 0 & 1\end{smallmatrix}\right]$, respectively:

$$A_0^{\lambda=\mathrm{odd}}(\gamma,\pm) = \pm \sum_{L\pi L'\pi'} F_\lambda(LL'I_2I_1) \langle I_2 \| \underline{j}_N \underline{A}^{(\pi)} \| I_1 \rangle \langle I_2 \| \underline{j}_N \underline{A}_{L'}^{(\pi')} \| I_1 \rangle \tag{98}$$

for good parity states. Again only the $q = 0$ components are non - vanishing, because of the axial symmetry of circularly polarized radiation with respect to the propagation direction.

The angular distribution coefficients for linear polarization $\underline{P}$ are obtained from eqs. (75) - (79). After some calculation one finds

$$A_0^{(\lambda)}(\gamma, \underline{P}) = A_\lambda(\gamma) \qquad \text{with } \lambda = \text{even}$$

$$A_{\pm 1}^{(\lambda)}(\gamma, \underline{P}) = 0 \qquad \text{for all } \lambda \tag{99}$$

$$A_{\pm 2}^{(\lambda)}(\gamma, \underline{P}) = \frac{1}{2} \sum_{L\pi L'\pi'} (-1)^{\Lambda \pi'} A_\lambda^{\perp}(\gamma) \cdot e^{\pm 2ia} \qquad \lambda = \text{even}$$

The linear polarization coefficient $A_\lambda^{\perp}(\gamma)$ is defined by

$$A_\lambda^{\perp}(\gamma) = \sum_{L\pi L'\pi'} F_\lambda(LL'I_2I_1) f_\lambda(LL') \langle I_2 \| \underline{j}_N \underline{A}_L^{(\pi)} \| I_1 \rangle \langle I_2 \| \underline{j}_N \underline{A}_L^{(\pi)} \| I_1 \rangle \tag{100}$$

where

$$f_\lambda(LL') = - \frac{\begin{pmatrix} L & L' & \lambda \\ 1 & 1 & -2 \end{pmatrix}}{\begin{pmatrix} L & L' & \lambda \\ 1 & -1 & 0 \end{pmatrix}} \left[\frac{(\lambda-2)!}{(\lambda+2)!}\right]^{1/2} \tag{101}$$

If the linear polarization efficiency Q of the detector is defined by

$$Q = \frac{R(a=0) - R(a=\frac{\pi}{2})}{R(a=0) + R(a=\frac{\pi}{2})} \tag{102}$$

where $R(a)$ is the response of the detector with its polarization axis $\underline{Q}$ at an angle a with respect to the polarization vector $\underline{P}$ (direction of electric vector $\underline{E}$) the linear polarization distribution of radiation emitted from an axially oriented state (e.g., produced by observation of the direction of a preceding radiation) can be expressed in the form:

$$W(\vartheta,\varphi) = W(\vartheta) + Q(-1)^{\Lambda_{\pi'}} W_{\perp}(\vartheta,\varphi) \qquad (103)$$

where

$$W(\vartheta) = \sum_{\lambda} B_{\lambda}(I_1)\ A_{\lambda}(\gamma)\ P_{\lambda}(\cos\vartheta) \qquad (104)$$

with $A_0(\gamma) = 1$

and

$$W_{\perp}(\vartheta,\varphi) = \sum_{\lambda=2} B_{\lambda}(I_1)\ A_{\lambda}^{\perp}(\gamma)\ P_{\lambda}^{(2)}(\cos\vartheta)\cos 2\varphi \qquad (105)$$

The angle φ describes the position of the polarization axis $\underline{Q}$ of the detector with respect to the ($\underline{z}$ - $\underline{k}_2$) or ($\underline{k}_1$ - $\underline{k}_2$) plane. As seen from eq. (103) the polarization distribution depends on the parity π' of the $L'\pi'$ component of the radiation field.

If the parity of the nuclear states involved in a gamma transition is not perfect, i.e., if the state contains small admixtures of the 'wrong' parity, the relation of eq. (90) is not correct (e.g. M1 + E1 mixtures may exist) and the efficiency tensors with λ = odd do not vanish (see eq. (89)). Hence, the directional distribution of such transitions (from polarized states, i.e. $B_{\lambda=\mathrm{odd}}(I_1) \neq 0$) contains $P_{\lambda=\mathrm{odd}}(\vartheta)$ terms.

2.4.3 Alpha Particles

Since the alpha particles have no spin $\epsilon(0) = 1$ and the radiation efficiency tensors are very simply

$$\epsilon_q^{\lambda}(LL')_0 = \mathrm{Tr}\,(T^{\dagger}(L')_0\ T(L)_0\ U_q^{\lambda})$$

or

$$\epsilon_q^{\lambda}(LL') = e^{i(\Delta(L')-\Delta(L))}(2L'+1)(2L+1)^{1/2}(-1)^L(2\lambda+1)^{1/2}\begin{pmatrix} L & L' & \lambda \\ 0 & 0 & 0\end{pmatrix}\delta_{q0} \qquad (106)$$

The alpha particles moving in the screened Coulomb field of the (daughter) nucleus are not free and the plane wave is distorted. This results in different phase shifts $\Delta(L)$ and $\Delta(L')$ for the different partial waves of angular momentum L and L', respectively. Since the alpha particles are spinless L and L' are the orbital angular momenta carried away by the alpha particle. Because of parity conservation L + L' = even, and since $\begin{pmatrix} L & L' & \lambda \\ 0 & 0 & 0\end{pmatrix}$ vanishes, unless L + L' + λ = even, λ also must be even. From eqs. (75) and (106) the directional distribution coefficients are easily constructed

$$A_{\lambda}(\alpha) = \sum_{LL'} (-1)^{I_1+I_2}[(2L+1)(2L'+1)(2\lambda+1)]^{1/2}\begin{Bmatrix} L & L' & \lambda \\ I_1 & I_1 & I_2\end{Bmatrix}\begin{pmatrix} L & L' & \lambda \\ 0 & 0 & 0\end{pmatrix} \qquad (107)$$

$$\times\ e^{i(\Delta(L')-\Delta(L))}\langle I_2\|H_L(\alpha)\|I_1\rangle\ \langle I_2\|H_{L'}(\alpha)\|I_1\rangle^*$$

After some manipulation eq. (107) can be written in the form:

$$A_{\lambda}(\alpha) = \sum_{L,L'} F_{\lambda}(LL'I_2I_1)\ b_{\lambda}(LL';\alpha)\langle I_2\|H_L(\alpha)\|I_1\rangle\langle I_2\|H_{L'}(\alpha)\|I_1\rangle^* \qquad (108)$$

where the **particle parameters for alpha - particles** b_λ (LL'; α) have been introduced.

$$b_\lambda(LL';\alpha) = \frac{2[L(L+1)L'(L'+1)]^{1/2}}{L(L+1)+L'(L'+1)-\lambda(\lambda+1)} e^{i(\Delta(L')-\Delta(L))} \qquad (109)$$

Obviously the Coulomb phase shifts $\Delta(L)$ enter only in the interference term as $\cos(\Delta(L')-\Delta(L))$.

2.4.4 Conversion Electrons

The angular distribution and orientation coefficients for internal conversion electrons are much more complex than the corresponding expressions for gamma radiation. This is due to the fact that the coupling of the radiation field amplitudes $|LM\pi>$ to the electron states $|j\mu>$ must be taken into account. Also, the initial bound state $|j_o\mu_o>$ and the final states $|j_e\mu_e>$ of the electron must be treated relativistically. The angular distribution and orientation coefficients depend on the various angular momentum recoupling coefficients and on radial matrix elements of the relativistic electron wavefunctions. A complete discussion of angular distributions of conversion electrons has been presented in a previous review paper[18]).

2.4.5 Beta Particles

The angular distribution and orientation coefficients for beta particles are even more complex than those for conversion electrons. This complexity is caused by several factors: (1) the lepton field emitted in a β - transition consists of two particles, an electron (positron) β and an antineutrino (neutrino) ν, of which only the β - particle is observed, the total angular momentum of the radiation field (multipolarity) is $\underline{L}=\underline{j}_\beta+\underline{j}_\nu$. (2) The β - Hamiltonian is more complicated, since two basic couplings V and A are involved corresponding to different couplings of the spins of the emitted electron and antineutrino. (3) Parity is not conserved in beta decay and hence interference terms between pseudoscalar H_β^- and scalar H_β^+ matrix elements are present which lead to non - vanishing directional orientation and distribution coefficients $B_\lambda(\beta)$ and $A_\lambda(\beta)$, respectively with λ = odd.

A general derivation of expressions of the angular orientation coefficients for allowed and first - forbidden beta transitions in the Konopinski - Uhlenbeck approximation has been presented in reference 13. More accurate expressions for $B_q^\lambda(\beta)$ including higher order nuclear matrix elements, finite nuclear size effects and screening effects have been computed by Bühring[30]). A useful and practical formulation of the directional orientation coefficients $B_0^1(\beta)$, $B_2(\beta)$ and $B_0^3(\beta)$ for first - forbidden beta transition has been given by Simms[31]).

*Work supported by the U.S. Atomic Energy Commission

Acknowledgements

The author is greatly indebted to Mrs. K. Schatz for her skill and patience in typing this manuscript.

References

1. J.V. Dunworth, Rev. Scient. Instr. 11 (1940) 167.
2. D.R. Hamilton, Phys. Rev. 58 (1940) 122.
3. J.W. Gardner, Proc. Phys. Soc. (London) A62 (1949) 763.
4. G. Racah, Phys. Rev. 84 (1951) 910.
5. U. Fano, Nat'l Bureau of Standards Report 1214; Phys. Rev. 90 (1953) 577.
6. S.P. Lloyd, Phys. Rev. 85 (1952) 904.
7. K. Alder, Helv. Phys. Acta 25 (1952) 235.
8. F. Coester and J.M. Jauch, Helv. Phys. Acta 26 (1953) 3.
9. L.C. Biedenharn and M.E. Rose, Rev. Mod. Phys. 25 (1953) 729.
10. In this work the term 'angular distributions and correlations' includes the consideration of polarization phenomena. If only the directions of radiations are observed the term 'directional distribution' or 'directional correlation' is used.
11. A. Abragam and R.V. Pound, Phys. Rev. 92 (1953) 943.
12. K. Alder, H. Albers-Schönberg, E. Heer and T.B. Novey, Helv. Phys. Acta 26 (1953) 761.
13; R.M. Steffen and H. Frauenfelder, Chapter I of 'Perturbed Angular Correlations' Karlsson et al. Eds. North-Holland Publ. Company, Amsterdam 1964.
14. H. Gabriel, Phys. Rev. 181 (1969) 506.
15. E. Matthias, B. Olsen, D.A. Shirley, R.M. Steffen and J.E. Templeton, Phys. Rev. in press.
16. E.L. Brady and M. Deutsch, Phys. Rev. 72 (1947) 870;
    E.L. Brady and M. Deutsch, Phys. Rev. 74 (1948) 1541;
    E.L. Brady and M. Deutsch, Phys. Rev. 78 (1950) 558.
17. For a review see H. Frauenfelder and R.M. Steffen, Angular Correlations, Chapter XIX in Alpha- Beta- and Gamma-Ray Spectroscopy, K. Siegbahn, Ed. North-Holland Publishing Co. Amsterdam 1965.
18. For a review see R.M. Steffen, 'Angular Distributions of Conversion Electrons' in the Proceedings of the 'International Conference on Radioactivity in Nuclear Spectroscopy', Gordon-Breach, Science Publishers, Inc. New York 1971, pp. 795 - 879.
19. W.D. Hamilton, Progress in Nuclear Physics 10 (1968) 1;
    E.M. Henley, Ann. Rev. Nucl. Sc. 19 (1969) 367.
20. L. Grodzins, Ann. Rev. Nucl. Sc. 18 (1968) 291.
21. E. Matthias, D.A. Shirley, M.P. Klein and N. Edelstein, Phys. Rev. Letters 16 (1966) 974; E. Matthias and R.J. Holliday, Phys. Rev. Letters 17 (1966) 897; J.E. Templeton and D.A. Shirley, Phys. Rev. Letters 18 (1967) 240; K. Sugimoto, K. Nakai, K. Matuda and T. Minamisono, Phys. Letters 25B (1967) 130 and J. Phys. Soc. Japan 25 (1963) 1258; H.J. Besch, N. Kopf and E.W. Otten, Phys. Letters 25B (1967) 120; H.J. Besch, U. Köpf, E.W. Otten and Ch. v. Platen, Phys. Letters 26B (1968) 721; U. Köpf, H.J. Besch, E.W. Otten and Ch. v. Platen, Zeitschr. Physik 226 (1969) 297.
22. D.M. Brink and G.R. Satchler, Angular Momentum (Oxford Univ. Press, London, 1962).
23. R.J. Blin-Stoyle and M.A. Grace, Encycl. of Phys. Vol. XLII, p. 555 (Springer Verlag, Berlin 1957); T.P. Gray and G.R. Satchler, Proc. Phys. Soc. (London) A68 (1955) 349. Their orientation parameters B_λ (I) for λ = odd have the opposite sign as compared to the B_λ (I) defined in this work.
24. D. Dillenburg and Th.A. Maris, Nucl. Phys. 33 (1962) 208, 53 (1964) 159.
25. The interaction operator $H(\underline{k},p)_{op}$ is in general defined in terms of the absorption multipole operators $H_{LM}^{(\pi)}$ for radiation $LM\pi$. This is in accordance with the physical characterization of a spherical tensor operator $T_\mu^{(\lambda)}$ as the operator that transfers an angular momentum λ with the component μ to the state on which they act (see e.g. page 80 of ref. 28). The

hermitian conjugate $T^{\lambda\dagger}_{\mu}$ of a spherical tensor operator removes the angular momentum $\lambda\mu$ from the state on which it acts and hence $(T^{\lambda}_{\mu})^{\dagger}$ represents the emission operator. The hermitian adjoint $T^{\lambda\dagger}_{\mu}$ of a spherical tensor operator is related to T^{λ}_{μ} by $T^{\lambda\dagger}_{\mu} = (-1)^{\lambda+\mu+n}\, T^{\lambda}_{-\mu}$, where n is an integer that is independent of λ and μ.

26. A.R. Edmonds, Angular Momentum in Quantum Mechanics (Princeton Univ. Press, Princeton, 1957).

27. The matrix elements of U^{λ}_{q} (L'L) are here defined as $\langle L'M'|U^{\lambda}_{q}|LM\rangle = (-1)^{L+M}\langle L{-}ML'M'|\lambda q\rangle$.

28. A. Bohr and B.R. Mottelson, Nuclear Structure, Vol. 1 (W.A. Benjamin, Inc., New York, 1969), p. 381.

29. K. Alder et al., Rev. Mod. Phys. 28 (1956) 432.

30. W. Buhring, Nucl. Phys. 40 (1963) 472; W. Buhring, Nucl. Phys. 49 (1963) 190; W. Buhring, Nucl. Phys. 61 (1965) 110.

31. P.C. Simms, Phys. Rev. 138 (1965) 784.

MEANS OF INCREASING THE ACCURACY IN ANGULAR CORRELATION MEASUREMENTS

E. Karlsson
University of Uppsala, Institute of Physics, Uppsala, Sweden

Contents

1. Introduction

Like in most text-books on measurement we shall make a distinction between the terms 'precision' and 'accuracy'. In angular correlations 'statistical' errors are attached to the final result of A_2, A_4, $\omega\tau$ or other parameters determined. Sometimes they represent the reproducibility expected from the nature of radiative decay if repeated measurements were made by the same, completely stable instrument. Sometimes they come out of a fitting procedure, where data are compared to a theoretical function. Let us therefore start with some comments on systematic errors and, having these in mind, try to find out means to increase the precision in the collected data and the accuracy of the final result.

An ideal angular correlation measurement is made under the following premises
a) point source
b) point detectors
c) perfect centering
d) perfect energy selection when several γ - energies are present
e) no scattering in the source, surrounding material or between detectors
f) high intensity of radiation
g) with a perfect coincidence system having a δ - function type of resolution curve.
The practical situation deviates, of course, from the ideal one in all the respects mentioned. The optimum set - up and the ways to correct for finite source and detector, incomplete energy discrimination, etc., have been treated in standard texts (see for instance Steffen and Frauenfelder[1]). Progress has been made during the last few years on certain aspects of these problems. We will, without attempting completeness, take up such selected areas for discussion where improvement has been reported or can be expected. As for the uncertainty caused by the random nature of the decay (or the reaction events in the case of 'in - beam' - measurements) we note that increase in the number of collected counts is mainly dependent on advances in high - speed electronics, but also on the development of multi - detector systems.

In most aspects the following discussion is valid for accuracy problems in all directional correlation work. Certain cases of perturbed angular correlations (PAC) require separate error considerations which will be pointed out in the text.

2. Corrections for deviations from the ideal geometry

2.1 *Finite detectors*

The principles for treating the solid angle corrections were established long ago and applied successfully to sodium iodide detectors. With the refinements of Yates[2]) who realized the difference between photopeak (PI) and all intensity (AI) corrections and evaluated both by Monte Carlo techniques, such corrections can now be taken from tabulated data for most detector shapes. These computed attenuation factors Q_2 and Q_4 are generally considered reliable, but it should be noted that they apply to unshielded crystals, and individual shielding arrangements, absorbers, etc. often motivate an empirical determination as outlined by Feingold and Frankel[3]). Detector shielding as well as the presence of surrounding material (e.g. magnet pole tips) may increase the effective solid angle substantially.

The introduction of the Ge(Li)-detector into angular correlation work has been the subject of several recent investigations of solid angle effects. Drifted semi - conductor detectors do not always have as well - known properties as scintillation detectors. It is often difficult to give the exact shape and dimensions of the active volume, inhomogeneities with local dead zones may occur within this volume and these properties may change with time. Furthermore, Ge(Li) - detectors with non - cylindrical symmetry (e.g. trapezoidal shape) are sometimes used which makes computations quite complicated. The accurate knowledge of solid angle corrections for Ge(Li) - detectors becomes the more important when one observes a tendency among experimenters to use larger solid angles than with scintillation detectors in an attempt to compensate for lower detection efficiency.

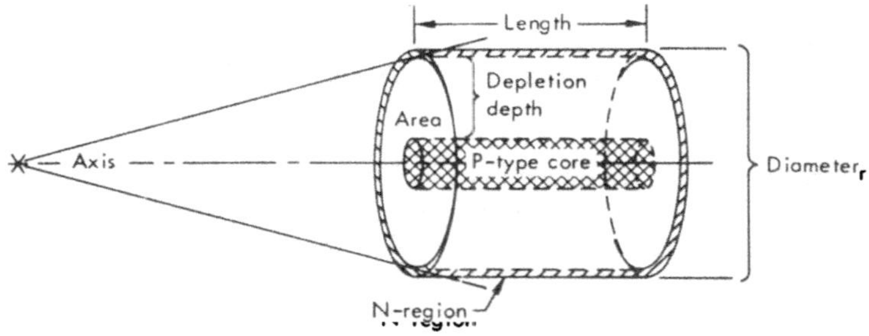

Fig. 1. Coaxial detector geometry (from Camp and van Lehn, Nucl. Instr. and Meth. 76 (1969) 192).

Among the recent treatments of these problems we find purely computational as well as empirical or semi - empirical approaches. Camp and van Lehn[4]) have calculated corrections for various cylindrically symmetric detectors of the plane or coaxial types as well as for trapezoidal shapes. Both theoretical absorption cross - sections and experimentally determined ones were used in the calculations but comparison showed

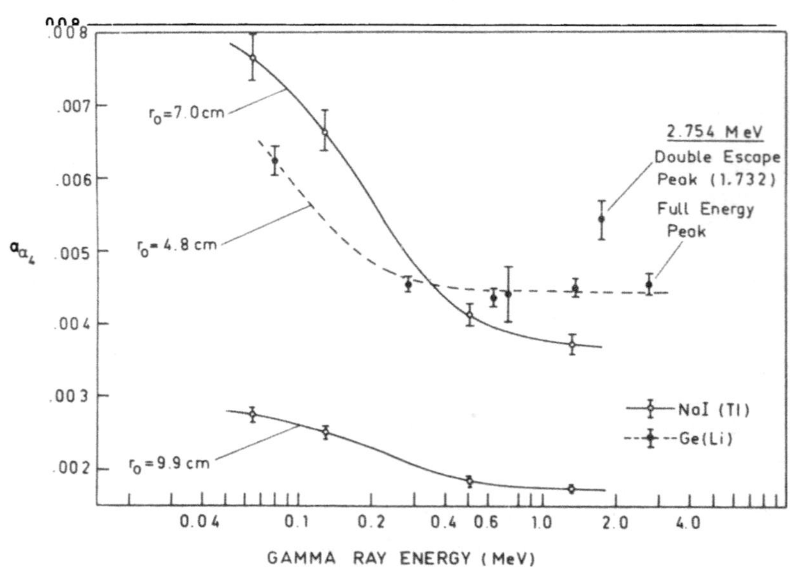

Fig. 2. Example of empirical Q_k - determinations (from Arns et al., Nucl. Instr. and Meth. 78 (1970) 295). The parameters a_K are related to Q_K by a series expansion.

that the corrections were not very sensitive to the use of slightly different cross - sections. PI and AI correction factors were obtained separately. Their modified Monte Carlo type of calculation (cf. Yates) showed identical results with the direct calculation for the AI factors but there was an unexplained difference between the two sets for the PI data at very short source - detector distances and higher γ - energies. Their results are found to be in good agreement with experimental values for one particular detector[5]), but it was later pointed out by Boddendijk[49]) that for cylindrical detectors one has to take into account intensity losses in the undepleted core and compton losses in the depleted region when calculating the efficiencies. This lowers Q_2 by 0.5 - 3% depending on photon energy. Camp and van Lehn also discuss finite source effects (see section 2.3) and recommend that directional correlation experiments be done at distances greater than 3.0 cm for a 6 cm^2 detector. With smaller distances it seems probable that other geometrical errors will exceed the errors in solid angle corrections.

Direct empirical determinations have been made by several other authors. Arns et al.[6]) use the method of Feingold and Frankel and obtain Q_2 and Q_4 with an error of 0.7% and 2.0% respectively. They point out that their detector was found to be slightly asymmetric and the information on the position and extension of the active

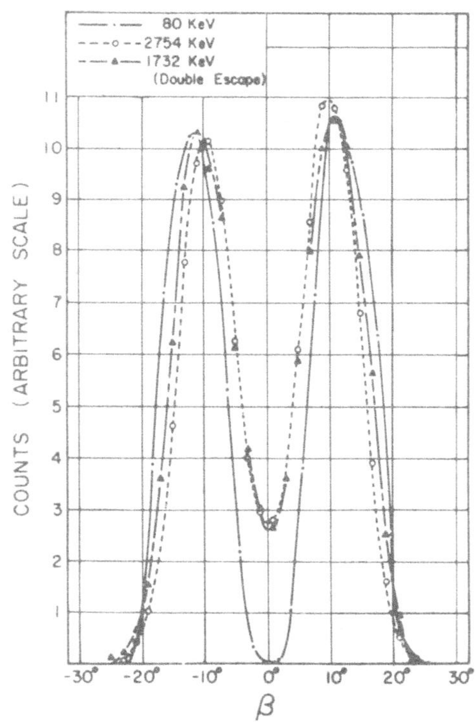

Fig. 3. Experimental efficiency function (from Arns et al., Nucl. Instr. and Meth. 78 (1970) 295).

volume, as given by the manufacturer, was found to be inaccurate (fig. 3).

An entirely different approach has been used by Winn and Sarantites[7]). It is based on the idea that time - consuming collimated beam experiments can be avoided by the use of an absorber which simulates the effects of solid angle in the following way: It can be shown that

$$Q_2 = \frac{3}{2}R_2 - \frac{1}{2} \quad \text{and} \quad Q_4 = \frac{35}{8}R_4 - \frac{30}{8}R_2 + \frac{3}{4}$$

where $$R_{2n} = \int_0^{\alpha} \epsilon(\beta)\cos^{2n}\beta \, d(\cos\beta) \Bigg/ \int_0^{\alpha} \epsilon(\beta)\, d(\cos\beta)$$

and the quantity R_{2n} can be determined by counting with and without an absorber of thickness

$$t(\beta) = (2n/\mu) \ln \sec\beta$$

and taking the ratio of counts at equal times (fig. 4). Examination of pulse height spectra showed that forward compton scattering did not introduce noticeable errors.

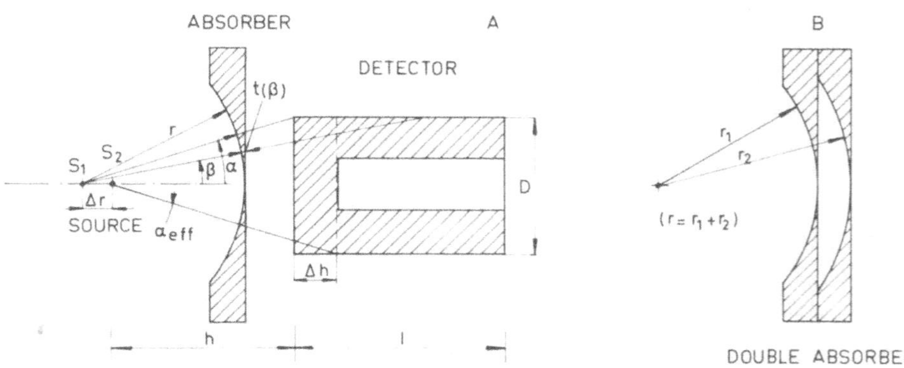

Fig. 4. Absorbers for solid angle corrections (from Winn and Sarantites, Nucl. Instr. and Meth. 66 (1968) 61).

The method was tested with good results for sodium iodide crystals (fig. 5) and then applied to 20 and 30 cm^3 coaxial type Ge(Li) detectors. The estimated errors in Q_2 and Q_4 were $\leqslant 0.5\%$ and $\leqslant 1.5\%$, respectively, for curves fitted to the data points. In a later paper[50]) the same authors make detailed investigations on two types of detectors including checks of symmetry and shape of the active volume. It was found for both cylindrical and trapezoidal detectors that the agreement between data from the absorber technique and the calculations was improved when the 'effective' geometry (obtained from scans with a collimated beam) was used in the calculations. The remaining trend towards higher calculated values might be explained in the way suggested by Boddendijk. It seems that great caution should be excersised in handling the

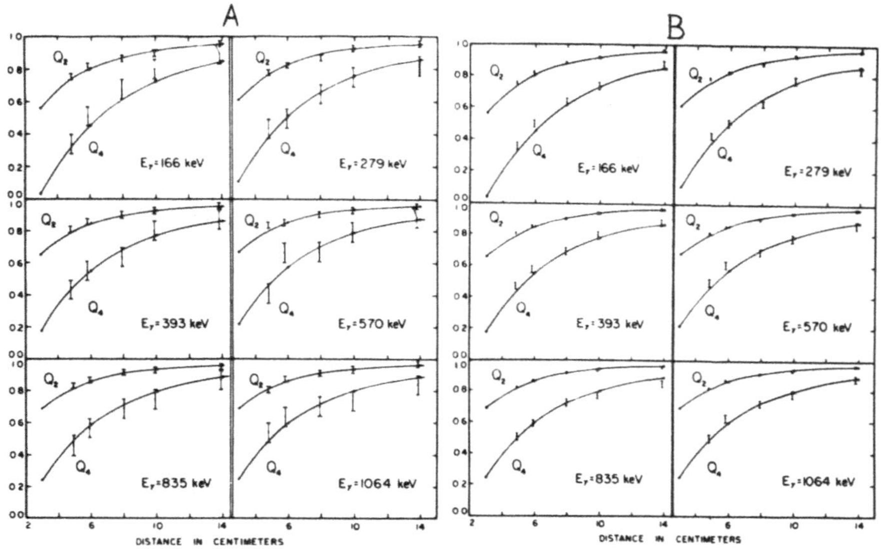

Fig. 5. Solid angle corrections for 3″ X 3″ NaI (Tl) - detectors by the absorber method (from Winn and Sarantites, Nucl. Instr. and Meth. 66 (1968) 61). Fig. B shows the results of fig. A after reanalysis by means of the relation $R_4 = (1+\sigma) R_2^2$.

solid angle correction for semiconductor detectors. Their individuality calls for an experimental determination if possible and all methods which can provide fast and safe corrections should be welcomed. Of course, the low counting rate at higher energies is a limiting factor on the accuracy and uncertainties in the 1% range is probably the practical limit at present.

For γ - rays above a few MeV the double escape peak (DEP) is the dominant peak in a Ge(Li) spectrum. It has been shown by Ingebrektsen[8]) that Q_2 and Q_4 are substantially higher when this peak is selected (fig. 6). The increase of DEP over AI factors is due to the fact that a double escape event is most likely to occur near the symmetry axis of the detector.

2.2 Decentering

The effects of a slight decentering of a point source on angular correlation coefficients can be computed accurately if the exact geometry of the detectors is known. As an example we show in fig. 7 the effect on the quantity $\omega\tau$ derived from a perturbed angular correlation experiment with a 4 - detector system[9]). Here, the A_2 and A_4 coefficients were determined in a separate experiment and $\omega\tau$ calculated from the measured value of

$$P = [\,W(157.5^\circ, +B)/W(157.5^\circ, -B)\,]^2 . [\,W(112.5^\circ, -B)/W(112.5^\circ, +B)]^2$$

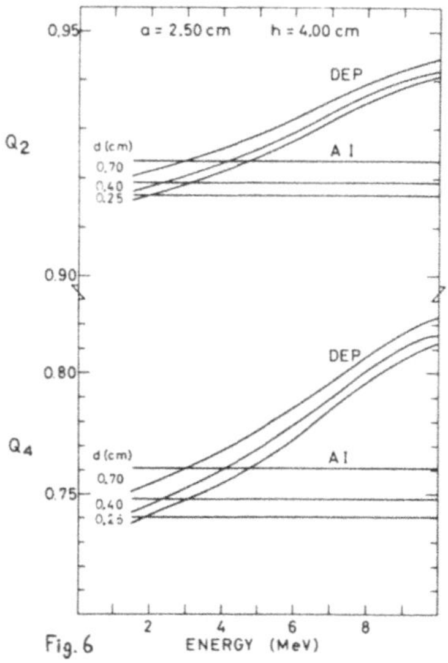

Fig. 6

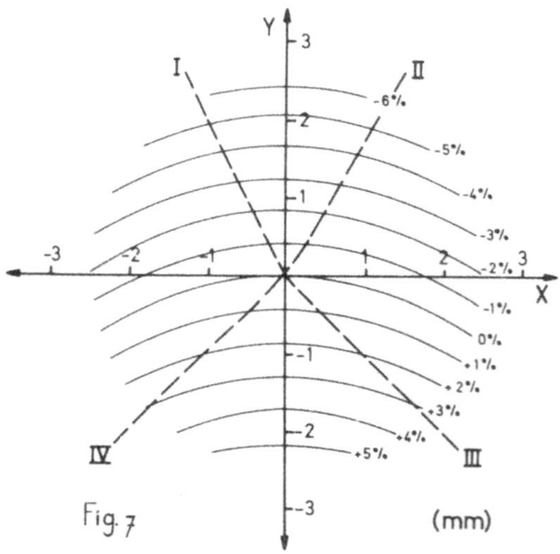

Fig. 7

Fig. 6. Comparison between 'double escape peak' and 'all intensity' correction factors (from Ingebrektsen, Nucl. Instr. and Meth. 60 (1968) 308).

Fig. 7. Effects of a slight decentering on a PAC - experiment (see text). 1% corresponds to $\Delta(\omega\tau) \approx 0.0002$.

of data taken with different sources. The specific angles were chosen at the steepest points on the correlation curve of the 613 - 576 keV cascade in ^{106}Pd and the perturbation was caused by the 620 kG hyperfine field for Pd in iron. For comparison, the attainable counting accuracy in this experiment corresponds to $\Delta P \approx 0.002$ or $\Delta(\omega\tau) \approx 0.0002$. When extended (plane) sources are used the finite source correction – not including scattering effects – can be calculated from fig. 7 by numerical integration.

Experimentally, the tedious source centering procedure is now computer - aided in some laboratories. Single counting rates at a few angles from an initial centering run are put into a desk - type computer which is programmed to decide in which direction and how far the source should be moved before taking the next series of single counts. Usually only a few such iterations are sufficient for a centering to 0.5% or better.

It is also possible to detect and correct for decentering effects by a careful analysis of the coincidence data obtained in the main angular correlation or distribution measurement. In section 5.1 it is discussed how improved curve fitting techniques allow the determination of a zero angle ϑ_0 of the angular scale, which occurs in beam experiments. The same method might be applied to γ - γ - coincidence set - ups to include the effect of a possible deviation of the source position from the mechanical center of the detector table.

2.3 *Finite source*
2.3.1 *Source extension effects*

The purely geometrical effects on the angular correlation of extended instead of point sources was first computed analytically for the case of a line source perpendicular to the detector plane[1]). The line source geometry has therefore always been selected by experimenters if possible. A recent example is given by Larsen et al.[10]), who use a continuous tape transport system for the study of very short - lived activities from the TRISTAN on - line separator. During the observation of parent activities the thin collector tape is running perpendicular to the detector table (cf. fig. 8) and

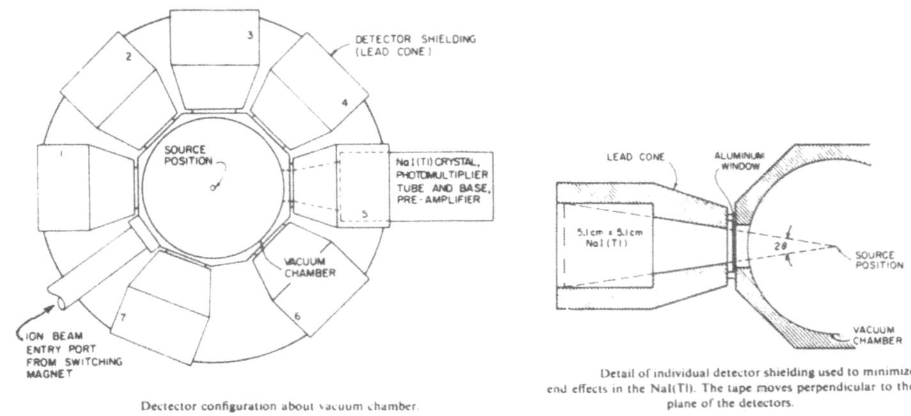

Fig. 8.Geometry of source and detector arrangement working on line with a mass separator (from Larsen et al., Nucl. Instr. and Meth. 69 (1969) 229).

interfering daughter products are carried away at a rate determined by the life - times of interest. With the recent extensive use of numerical computation one might expect at any time a complete list of corrections for various source shapes. So far, calculations have been performed for circular disc sources in combination with certain detector geometries[4]). The results are expressed as reduction factors P_k on the finite detector correction factors Q_k. As an example we quote the results for a cylindrical Ge(Li) - detector of 10 cm^2 area at a distance of 5 cm from a 6 mm dia source: Q_2 and Q_4 are reduced with 0.15% and 0.45% respectively. The same authors argue that essentially the same corrections would be valid for spherical sources of the same radius and a cylindrical source of diameter and height would behave approximately as a circular source whose area is d $\times$ h. For shorter distances than the one given in the example the sensitivity to the source extension and other geometrical uncertainties increases rapidly, which makes such geometries unsuitable for precision work. Cylindrical and rectangular sources have also been treated analytically by Verheul et al.[51]). They give closed expressions for source and detector corrections under the assumption of uniform density of activity and constant efficiency over the detector area. Then, the two corrections are independent up to second order in the parameter r/D, where

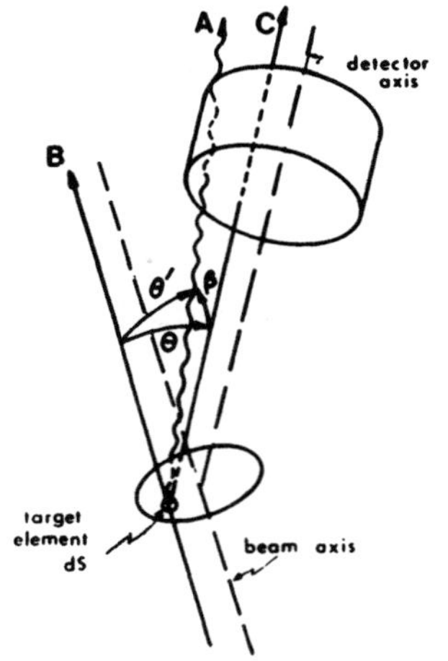

Geometry involved in the calculation of the effect of a finite size source on the attenuation of an angular distribution. A is the gamma-ray direction. B and C are parallels to the beam and detector axes respectively which pass through the target element *dS*.

Fig. 9. For calculation of finite source effects (from E.H. Heighway and J.D. MacArthur, Nucl. Instr. and Meth. 79 (1970) 224).

r is the radius of the circular source and D the source - detector distance. The total correction turns out to be relatively insensitive to the source extension.

An investigation of finite source effects in angular distribution measurements following nuclear reactions has been made by Heighway and Mac Arthur[11]). The calculation is valid for a circular target spot perpendicular to the beam axis. It was found that even with as big a target diameter as 1.0 cm the effect on Q_2 was only 0.9% and that on Q_4 2.3% for a 23 cm^2 germanium detector at 3.34 cm. Such extended targets may be necessary in cases of low yield where compensation by increased beam current presents heat dissipation problems. An interesting feature of their investigation was the empirical finding that the exact Q_k for a finite source and a finite detector was very well approximated by the product of the coefficient calculated for a point source and a finite detector with the coefficient for a finite source and a point detector. If a Rose type of calculation gives Q_k^1 for a point source another coefficient Q_k^2 of the form

$$Q_k{}^2 = \frac{\int P_k\ (\cos\beta)\ dS\ \sin\beta}{S\ \sin\beta}$$

for finite source and point detector is calculated, then $Q_k = Q_k{}^1 \cdot Q_k{}^2$

This separability is in agreement with the expressions given by Verheul et al. and facilitates the procedure in practical cases.

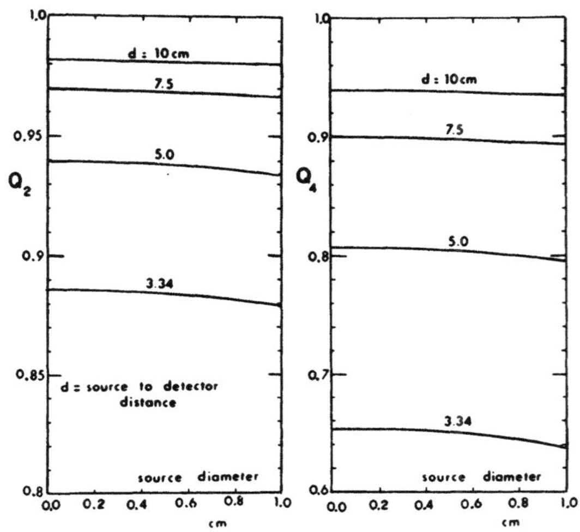

Fig. 10. Finite source (target) effects for circular sources (from Heighway and MacArthur, Nucl. Instr. and Meth. 79 (1970) 224).

2.3.2 Internal absorption and scattering

Internal effects in the sources have not, in spite of their general interest, been treated very much in the literature. The reason is probably that each source presents very individual problems, depending on shape, density, atomic number and the energies of the radiations involved. As pointed out by Frauenfelder and Steffen there is an absorption effect, which always decreases the counting rate more at 180° than at 90°, and a scattering or deflection effect caused by Compton events.

The scattering effect is fairly well under control in present - day experiments due to the careful photo - peak selection made possible by multi - channel analyzers. One should, however, be aware of the fact that, even if the selected cascade is relatively free from scattered events, other interfering coincident transitions which give rise to a background for the photopeak may contain 'in - source' scattered γ - rays and therefore differ from the 'standard' line shapes mostly used in stripping (see section 3.2).

A recent trend which tends to make the absorption (and scattering) problems more serious than before is the increased use of solid sources in unperturbed as well as perturbed angular correlations. It has been observed that activity dissolved in a regular matrix of cubic symmetry e.g. copper metal in most cases, gives far better environment conditions for avoiding perturbations than liquids do. In perturbed angular correlations (PAC) solid sources are used for the study of hyperfine interactions and a sufficient dilution sometimes requires fairly extended sources. For PAC - experiments with a Hf - single crystal an attenuation of the anisotropy by 15% was found to agree with calculations of internal absorption effects[12]). In a contribution to this conference Lopiparo and Rasera present results of calculations of scattering and absorption for cylindrical solid sources which show that these effects are serious with sources of intermediate Z and photon energies about 100 keV.

When counting statistics allow a high precision the well - known method of extrapolation to zero in a plot of A_2 and A_4 versus source diameter should be recommended. A careful study of the line shape in singles or coincidence spectra is also helpful in revealing the presence of such effects.

2.4 *Other angular dependent effects*

Serious errors may be introduced if the various angles are not equivalent from the point of view of the movable detector. Such errors are usually of a trivial nature but should be considered before any attempt at a precision measurement. Wagner and Forker[13]) have recently given examples of this type of systematic error. Their study was made on a time - differential system and showed that the results were extremely sensitive to small deviations from symmetry. In their case systematic effects were revealed clearly since they were working with a four - detector system (but with each detector pair working as an ordinary two - channel apparatus with one fixed and one moving detector) and the results of different pairs could be observed. The very discouraging results for A_2 (t) shown in fig. 11a were observed after the standard adjustment procedures. Calculations showed for the specific example of the 1276 - 123 keV cascade in ^{154}Gd that if $T_{1/2}$ = 1.15 ns, λ_2 = 0.26 ns^{-1} λ_4 = 0.32 ns^{-1} $A_2(0)$ = = $A_4(0)$ = 0.20 and 2τ = 2.25 ns, then a difference in time resolution at 90° and 180° of 2% is sufficient to change the attenuation parameters derived from the experiments by up to 60%. To be negligible in comparison with the statistical error the time resolution should be constant to within 0.5%. Only if $2\tau < 5T_{1/2}$ the deduced λ_2 and λ_4 become relatively insensitive to systematic changes in 2τ. Similar calculations showed that a drift in the zero point of the time scale of 30 ps changes the attenuation parameters by 100%.

Systematic variations of the shape and position of the time curve are caused by magnetic or temperature dependent effects. The zero - point changes can be characterized as an amplification effect (through the triggering of limiters) whereas time spectrum variations depend in a more complicated manner on the charge collection in the photomultipliers. The presence of iron in the detector table caused in the above-mentioned case differences inside the shielded detectors of approximately 2mOe,which could be reduced by demagnetization. The temperature dependence of the zero point was 15 ps / °C, which prompted the use of a separate ventilation in each detector box. After these precautions the time - differential spectra looked more convincing (fig. 11b). Time integral angular correlation results are of course not so sensitive to the above effects and a careful control of peak positions using modern (> 1000) - channel analyzers is sufficient as an isotropy test.

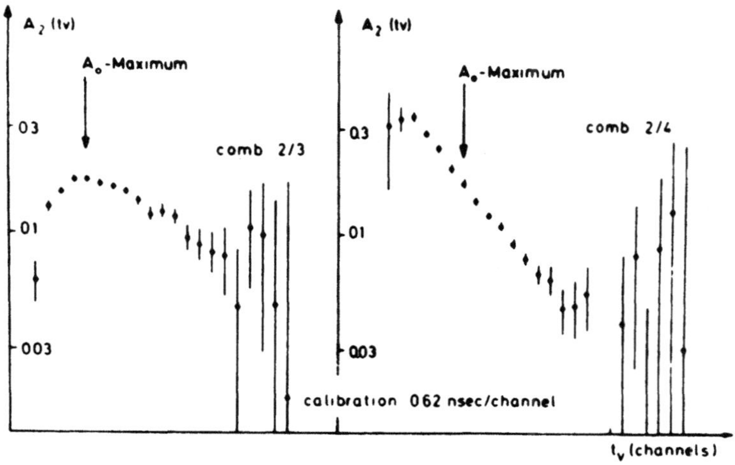

First measurement with ^{154}Gd.

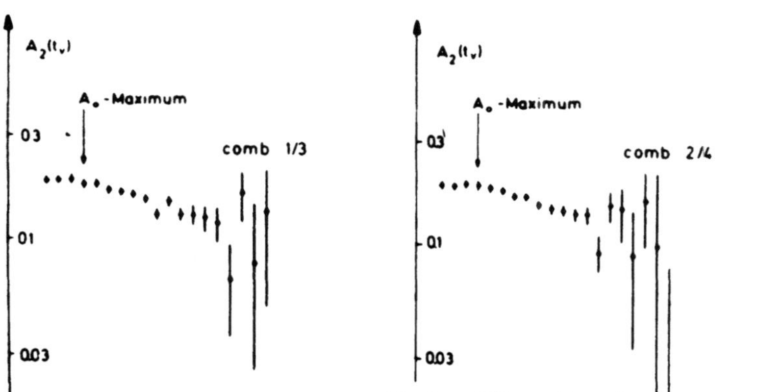

Measurement with ^{154}Gd after the discussed precautions.

Fig. 11. Example of systematic errors in time - differential angular correlations (from Wagner and Forker, Nucl. Instr. and Meth. 69 (1969) 197).

3. The selection of the desired quanta

Angular correlation techniques are nowadays applied to quite complicated γ - decays both in radioactivity and nuclear reactions. Much of the accuracy of the final result depends on how safely the transitions of interest can be picked out from the multitude of other coincident γ - transitions in the spectrum. Moreover, one must have means to handle bremsstrahlung and other backgrounds, scattered quanta, etc.

3.1 Detector progress

Since the perfect γ - detector has not yet been invented one can only in the case of the highest energy γ - transitions select the wanted photopeak in the coincidence spectrum free from admixtures of pulses from other transitions. The germanium detector has meant a revolution in the handling of complex spectra and has been introduced in angular correlation work in spite of its low detection efficiency, but it has not at all relieved the experimenters from the trouble of 'the photopeak sitting on the compton background'.

It was realized in the beginning of Ge(Li) - detector spectroscopy that the high resolution required a multichannel analyzer with an ADC of a few thousand channels. In angular correlation work one needs in principle a two - parameter analysis of the same spectrum taken at a series of angles and a subsequent treatment in a high capacity computer. This means that a tremendous amount of data should be handled and we shall see in the next section that such an ambitious procedure requires a substantial computer effort already for scintillation detectors.

Ge(Li) - detectors have, nevertheless, been used extensively in complex decays in situations where at least one photopeak stands out clearly from the background. An example of an experimental set - up is the arrangement used by the Bonn group[14]). By replacing one of the germanium detectors with a Si(Li) - detector one has in most cases sufficient resolution in the electron channel to study conversion electron - gamma angular correlations. A group at the Jülich[15]) reactor uses a Ge(Li) - Ge(Li) - set - up for the study of γ - γ correlations after thermal neutron capture where one finds on the average 100 to 150 γ - lines of measurable intensity in each decay. A two parameter pulse height analysis of the coincident pulses with two 4k - ADC's is used and the address pairs were put onto magnetic tape for subsequent computer analysis. To reduce the data handling to manageable proportions up to 16 digital windows set to contain the regions of interest were connected to each of the ADC's. In the Jülich case measurements have so far been carried out in the neutron captures of ^{167}Er, ^{45}Sc and ^{39}K, an example which shows how angular correlations techniques are presently being applied to cases of ever increasing complexity.

3.2 Stripping of γ - spectra

The stripping of γ - spectra from scintillation or semi - conductor detectors has been the subject of a great number of papers[16]). The application to angular correlation has been discussed by, e.g., Daniels[17]), who describes a system (fig. 12) used at Manchester including a PDP - 7 computer. In the data reduction of two - dimensional coincidence spectra from scintillation detectors he follows the procedure proposed by Slavinskas et al.[18]). The data are represented in the form of a matrix $M(E_2, E_1)$ where E_2 and E_1 are the energies left in the crystals and the stripping is first performed for each one - dimensional spectrum M_1 $(E_1) = M$ (E_2, E_1) corresponding to a certain energy E_2 left in crystal 2 which gives the contributions of each γ_i for each pulse height interval from detector 1. This procedure is repeated for all values of E_2 and one can now construct point by point a coincidence spectrum N_2 (γ_i, E_2) which is a one - dimensional spectrum of events from detector 2 which are in coincidence with the

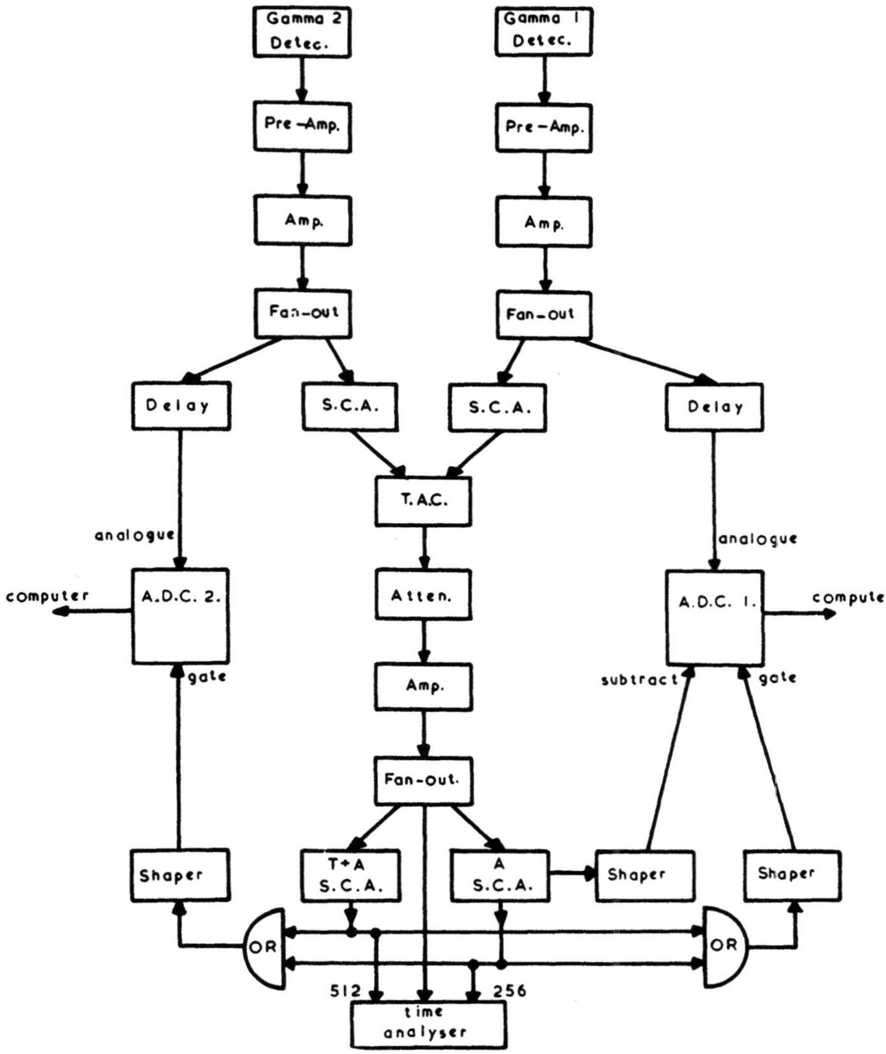

Fig.12. Electronic set-up for two-parameter analysis of coincidence spectra (from Nucl. Instr. and Meth.59(1968)214).

pure γ_i - transition. To find the required number of coincidences N_c (γ_i, γ_i) one has to perform a second stripping of the m (m is the number of transitions) spectra N_2 (γ_i, E_2) which gives a matrix containing m^2 numbers. This procedure is repeated for all angles necessary.

The actual stripping procedure used by Daniels starts with the best possible em-

Number of coincidences found from the normal data matrix.

E_{γ_2} \ E_{γ_1}	1.002	1.136	~1.4	1.805	2.138	~2.5	2.941	3.091
1.002	−12	1182*	−3	813*	7	−2	47*	−2
1.136	1229*	−41	512*	5540*	−15	21	−5	8
~1.4	−30	756*	66*	702*	−30	2	19*	−5
1.805	378*	6152*	376*	616*	85*	1106*	0	37*
2.138	−6	−9	−23	226*	−5	−1	0	1
~2.5	−26	37	0	1423*	−1	0	−2	−1
2.941	54*	10	10*	36	2	0	1	−1
3.091	−9	−10	−2	64*	−1	0	1	1

Number of coincidences found from the transpose of the data matrix.

E_{γ_2} \ E_{γ_1}	1.002	1.136	~1.4	1.805	2.138	~2.5	2.941	3.091
1.002	−82	1205*	−36	789*	7	20	39*	−5
1.136	1292*	−459	652*	5410*	−17	23	−21	14
~1.4	−30	749*	15	676*	−10	−8	9*	4
1.805	431*	5962*	457*	584*	185*	1137*	−12	37*
2.138	−1	38	30	215*	−1	−1	0	1
~2.5	−21	11	20	1390*	−2	2	2	1
2.941	46*	7	17*	15	5	0	0	0
3.091	11	5	−2	75*	−1	−1	−1	1

Fig. 13. Table of coincidences for different cascades derived from a two - dimensional analysis (from T. Daniels, Nucl. Instr. and Meth. 59 (1968) 214).

pirical line shapes S_i (E_1) of the m monochromatic γ - rays, and uses weighting and least-squares fitting. Least - squares fits are to be preferred because of the fact that the accumulative errors, which enter in the normal line - after - line stripping, are avoided. In the final evaluation of the angular correlation coefficients of all the coincident cascades one can make use of the following facts:

1. If, among the correlations, there is one which is completely known from other evidence (such as a 0-2-0 cascade) it can be used as a check on the normalization of all the other cascades.
2. In the case of levels reached in nuclear reactions all transitions proceeding from the same state contain information on the population parameters of this specific initial state and the ratio of unknown parameters to data points is reduced.

As an example, stripping using the least squares method took in the case of a 80 - 79 channel coincidence spectrum about 2 hours on a PDP - 7 computer when the number of γ - components was m = 8. The results for one angle are shown in fig. 13. The programs also allow the display of all intermediate steps in the data handling, stripped spectra, χ^2 - plots, etc. An indication of the reliability of the method can also be found if one treats the original data matrix and the transpose of the data matrix separately. It was found that the deviations between the tabulated final data from these two sets were roughly twice the square root of the mean entry. These deviations were used in assigning errors to the final result for A_2 and A_4. Further reduction of the systematical errors can probably be achieved by taking more care over the line shapes and including also some weak components.

In stripping coincidence spectra one should be aware of the possibility that the compton parts have shapes different from line shapes taken at the same distance with a single detector. Scattering often gives extra contributions with unpredictable line shapes and it is difficult to separate out the true compton background from each gamma transition for the stripping. Of course, all means to avoid such scattering should be tried, but are not always practically possible as examplified in section 3.4. Thus, the difficulty is stripping is that the angular dependent weighting factors a_i (θ) for the functions S_i (E) might be different for the photopeak region (true angular correlation) and for the compton region if it contains quanta scattered into the crystal from material surrounding the source (e.g. magnet poletips in PAC) or the crystal (such as detector shielding). If the photopeak sits on a flat background the safest correction procedure is probably the old - fashioned one: to determine the angular correlation coefficients of the background mixture for windows above and below the photopeak without attempting a complete stripping into components. The critical points in the unfolding of germanium detector spectra and their effect on angular correlations are also discussed in the proceedings of the Nashville conference on Radioactivity in Nuclear Spectroscopy[52]).

The demand for analyzer and memory capacity is large already for the stripping of scintillation spectra. In the example from Manchester mentioned above an 8K store PDP - 7 computer (0 - 17777 octal) was used with positions 0 - 3245 for program, 3246 - 17577 for storing data, and 17600 - 17777 for loaders. This means 6357 decimal locations for the store which allows a matrix of 79 X 80 channels. Another system constructed at Ames[10]) uses for a seven - detector array a 16384 - channel analyzer and the contents of the memory are put on magnetic tape for data processing in a large computer. It works with SCA's for selecting one photopeak region and a subtraction program for photopeak backgrounds in the resulting coincidence spectra.

To carry out complete stripping procedures for complicated spectra one must – when germanium detectors are involved – envisage two - dimensional spectra of at least the order of 1000 X 1000 channels. This would require memory extensions of a magnitude which is prohibitive from the economical point of view. Therefore, the direct storing on magnetic tape at the time of occurrence of the event (by writing a pair of addresses) presents itself as an attractive alternative. By using two small buffer memories alternatingly for storing the coincidence events before they are dumped onto the tape, the accumulation of data need not be interrupted by storing on the tape and an on - line computer can be utilized efficiently. As discussed by Der Matosian[19]) the cost of storage discs has now come down to levels which makes them acceptable as another alternative data storing device. Disc storage allows (in contrast to tape storage) display during accumulation since the scanning time is virtually eliminated, but the disc is slow if one requires random access to stored experimental data. The rate of storage can, however, be improved to about 1000 coincidence counts per second, which is sufficient for most experiments.

From the experimenter's point of view it is desirable to reduce the waiting period between the time when the data are taken and the delivery of the results from the computer centre. It is hoped that several calculations in angular correlation experiments, such as stripping and evaluation of pure coincidence numbers for transitions i, j can be soon carried out truly on - line, e.g., for one angle while data for the next angle are accumulating, Simple calculations can also be performed in intervals between counts if only coincident events are registered on-line. Murray and Macefield have given some general aspects on the conditions for foreground-background programming (ref. 57-59).This would bring the experimental situation back to the stage where the experimenter can react directly on preliminary results in order to optimize the experiment.

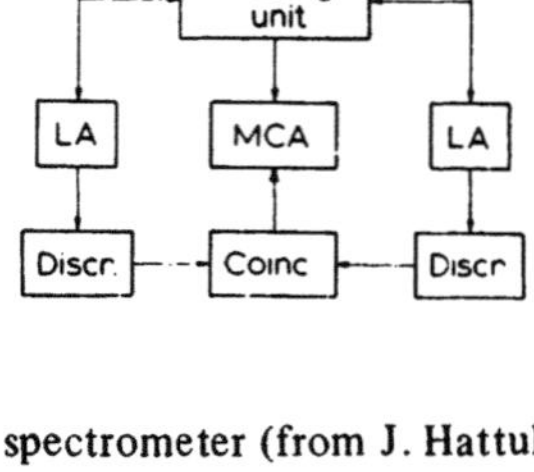

Fig. 14. A sum - coincidence spectrometer (from J. Hattula et al., Nucl. Instr. and Meth. 65 (1968) 77).

3.3 Sum - coincidence methods

The sum - coincidence technique, originated by Hoogenboom[20]) and used extensively for coincidence studies on decay schemes, has been applied also to angular correlations. A systematic study of a number of phenomena present in this application has been carried out by Hattula et al.[21]). Their spectrometer was of the integral bias summing type (fig. 14) and the method proved to be more insensitive to small electronic drifts than the conventional method where single channel windows are used. They considered also backscattering and annihilation effects and applied the apparatus to angular correlation studies in ^{106}Pd (fig. 15). The advantage of the method lies of course, in the reduction of the compton backgrounds in the selection of the photopeaks. The extension to Ge(Li) - detectors would give a similar improvement[60]) but their low efficiency does not make the method very attractive except in special cases.

Another application of the sum - coincidence technique has been discussed by Bozek et al.[22]). They report the measurements of g - factors for two cascades in ^{147}Sm (680 - 120 keV and 600 - 200 keV) observed in the sum - coincidence spectrum gated at a sum of 800 keV. In addition to the simplification of the spectrum a suitable combination of the data for γ_1 - γ_2 and γ_2 - γ_1 can be employed in order to cancel out effects of small electronic drifts and changes in photomultiplier amplification due to stray magnetic fields.

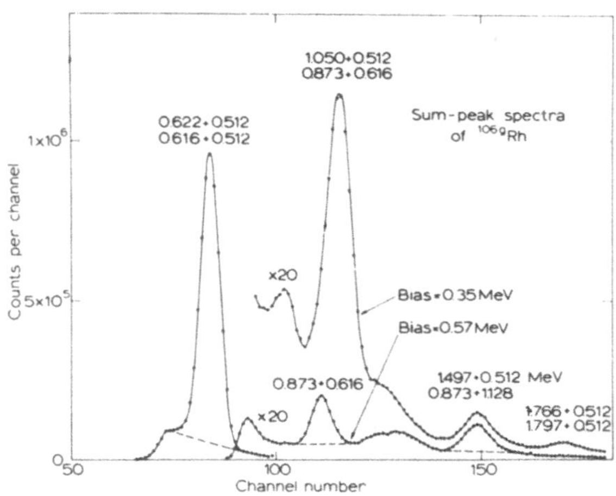

Fig. 15. Example of the use of sum peak spectra for angular correlations (from J. Hattula et al., Nucl. Instr. and Meth. 65 (1968) 77).

3.4 Scattered quanta

All quanta scattered before they reach the detector crystal must be corrected for if they give rise to pulses in the region to be analyzed. Scattering in the source and source - holder, other outer arrangements such as cryostats and magnet pole tips

or from detector shielding into the detector are in principle, detectable in a single - spectrum taken with good statistics. Scattering from one crystal to another is a much more serious problem since each events produces a false coincidence. Thus, even if the probability is small for reaching the second detector through lead shielding etc., it may become comparable to the true coincidence rate when weak cascades are studied. The effect is most pronounced at 180° which produces the so - called backscatter peak, but is present over the whole 90° - 180° range giving rise to varying pulse height distributions depending on angle, detector, geometry, etc. Such scattering may be especially annoying in multi - detector systems.

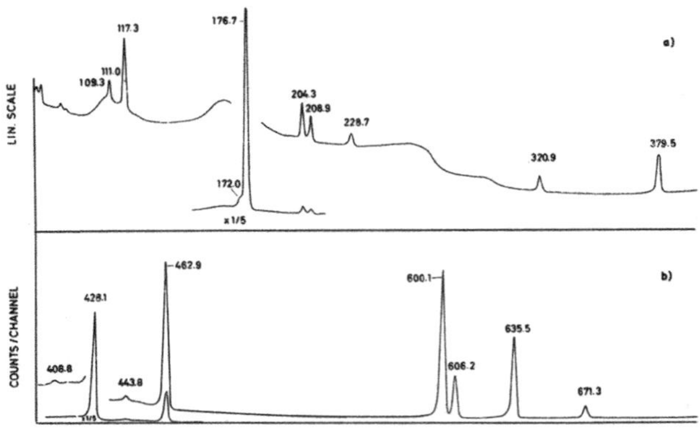

Fig. 16. Singles spectrum of ^{125}Te.

As an example from recent work at Uppsala[23]) we consider the cascades 321 - 177 and 204 - 177 keV in ^{125}Te (cf. fig. 16). The 321 and 204 keV lines are quite weak compared to the 428 line and the 600 - keV group. Gating on the 177 keV photopeak produced the coincidence spectra shown in fig. 17. (The 180° spectrum is not used because it contains a prohibitive amount of backscattering, although all usual precautions including conical lead shields were taken). The shape of the scatter contributions originating from the higher energy γ - rays were derived from coincidence spectra taken with the single line sources ^{198}Au (411 keV) and ^{137}Cs (661 keV) in the same geometry as the main experiment. With these shapes available the main spectra could be stripped for each angle and the relative numbers of all contributing coincidences determined. The experiment was part of a g - factor determination were an earlier measurement using the 204 - 177 keV cascade had given g = = 0.58 ± 0.06. By taking proper account of the scattering our A_2 and A_4 coefficients and $\omega\tau$ - values were substantially revised and the new g - factor came out as g = = 0.204 ± 0.007. A control measurement using the 321 - 177 cascade which is less sensitive to scattering but has a lower anisotropy gave g = 0.192 ± 0.016. The extreme difficulties encountered in this case are due to the weak cascade decay as compared to the strong cross - over transitions and to the energy relations involved.

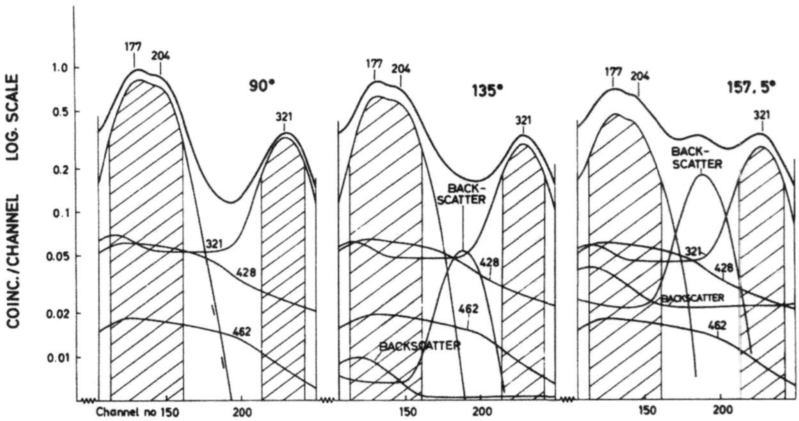

Fig. 17. Coincidence spectra with the 177 keV - region of ^{125}Te showing detector - detector scattering.

3.5 Annihilation, bremsstrahlung and X - rays

Other contributions which make trouble in the selection of the desired quanta from the total number of coincidences are due to bremsstrahlung - gamma (for high energy β - branches), annihilation - gamma and X - ray - gamma coincidences.

If annihilation quanta are present they may distort the angular correlation seriously when both energies of the cascade are below 511 keV. This is a well - known effect, but usually no great nuisance in spectrum stripping since the line shapes of the quanta are well - known. However, annihilation may also cause trouble when one of the γ - rays is of relatively high energy. As has been pointed out by Svensson et al.[24]) for the cases of ^{118}Te and ^{119m}Te experiments show two types of effects at high energy. Firstly, coincident summing of the type 511 - (511 $+\gamma$) gives a large contribution when the γ - ray is fed by a strong positron branch. Secondly, annihilation - in - flight (AIF) quanta give coincidences up to the end - point energy. Due to their energy distribution AIF quanta are most serious if one of the windows for the coincidence spectrum is set in the region just above 250 keV. When sources and source - holders are made thin in order to avoid γ - absorption and scattering the AIF takes place not only in the source but also in the air or in detector absorbers or crystals and the angular dependence will be quite complicated.

Bremsstrahlung contributions have to be considered in the stripping of the coincidence spectra (for an example see ref. 38). When the γ - rays are stopped by absorbers in front of the detectors β - γ anisotropies are reflected also in the bremsstrahlung - γ coincidences.

A couple of years ago there appeared a paper[25]) with the suggestion that coincidences between characteristic X - rays and gammas may in some cases show an anisotropy. Experimentally, it was later shown by Fechner et al.[26]) that all investigated cases were isotropic to within 0.4%, which is comforting for all workers in angular correlations who have had to correct for X - ray admixtures in low energy peaks.

4. The efficiency of data collection

After having discussed the most common types of systematic errors which occur in angular correlation measurements, we now turn to the precision achievable in determining the radiation intensity at the chosen angles. To bring down the statistical error in coincidence counting to the level where meaningful conclusions (about mixing ratios, for instance) can be made precision in the 0.1 - 1% range is usually required. Improvement in collection efficiency proceeds along two lines: (1) Electronic development in the handling of high counting rates, which allows the use of stronger sources and (2) The construction of multi - detector systems, which increase the useful solid angle and therefore can collect the same number of data more than one order of magnitude faster than a conventional system at the same source strength. Multi - detector systems are often used to detect weak activities or very short - lived activities where the pile - up, etc. is not the limiting factor. For PAC - measurements, where extremely small effects are looked for, one tries to realize a combination of the two approaches: a high counting rate, multi - detector system.

4.1 Problems with increased counting rates

Increase of the source strength leads to deterioration of the energy resolution in amplifiers, other pile up effects in various parts of the coincidence system and to a high proportion of accidental coincidences.

4.1.1 Amplifiers

Recent efforts in amplifier technology have concentrated on maintaining the energy resolution at high counting rates through the use of baseline restorers etc., mainly with germanium detector spectroscopy in mind.

Optimizing the energy resolution means a careful selection of time constants in the circuits, which leads to pulse - lengths of the order of microseconds and corresponding restrictions on the counting rate. A typical limit for existing amplifiers is 50,000 counts per second. With a sacrifice of energy resolution pulse - lengths can be cut down and some systems (using scintillation detectors) have been operated with detector pulses clipped to about 0.2 μs. Another device, called pile - up rejector[27]) works on the principle that a pulse closely following an earlier one has not the proper cross - over time in a bipolar pulse amplifier (fig. 18) and can be stopped in the gate system.

Among the highest published counting rates used in angular correlation we can find a system used at Caltech[28]) where the perturbation on the 722 - 558 keV cascade in ^{114}Cd was studied with 200,000 counts per second in each of the detectors.

4.1.2 Accidental coincidences

As is well - known the number of accidental coincidences increases when the source strength N_0 is increased. For selected windows with pulse rates N_1 and N_2 the

Fig. 18

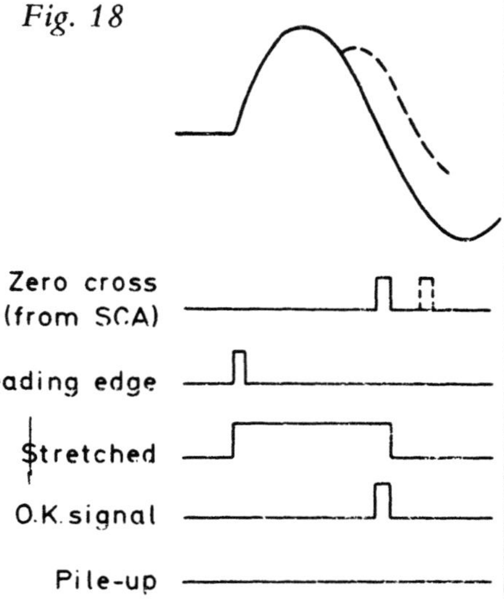

Fig. 19

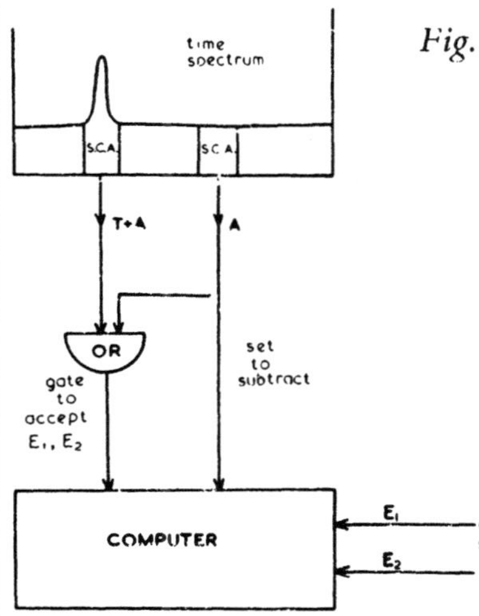

A schematic diagram illustrating the method of runtime random subtraction. $T+A$ indicates that the pulse may be generated by a true or accidental coincidence. A indicates that the pulse may be generated by an accidental coincidence only.

Fig. 18. Pile - up rejection principle.
Fig. 19. Coincidence analysis with subtraction of accidentals.

number of accidentals is to first order $2\tau N_1 N_2$. Since one prefers to work with a small accidental - to - true ratio the reduction of the coincidence resolving time 2τ is a main goal for workers in angular correlation.

Most modern angular correlation systems include a time - to - amplitude converter (TAC) for the display of the time-spectrum of the registered coincidences. For prompt coincidences it serves to separate the true coincidences in the peak from the accidental coincidences which form the background. Often two SCA's with the same window width are connected after the TAC,one set on the prompt peak for the counting of total coincidences and another on the background for the accidentals (fig. 19). In other versions the SCA set on background is connected to the main multi - channel analyzer in such a way that a pulse corresponding to the proper energy is subtracted from the total. Both versions involve a slight loss in accuracy compared to the method of accumulation on accidentals spectrum with good statistics which is later normalized and subtracted. When the numbers of accidental and true coincidences are about equal this loss of accuracy may be considerable, but when the ratio is as small as 1 : 10 the increase in the statistical error on the number of true counts is only about 4%[17]). Under favourable conditions the proportion of accidental coincidences can be brought down considerably. As an example we quote again the measurement[28]) of the Pasadena group on ^{114}Cd where a half - width of the prompt peak of 1.5 ns resulted in an accidental - to - true ratio of 1 : 10 at a singles counting rate of 200,000 per second (fig. 20).

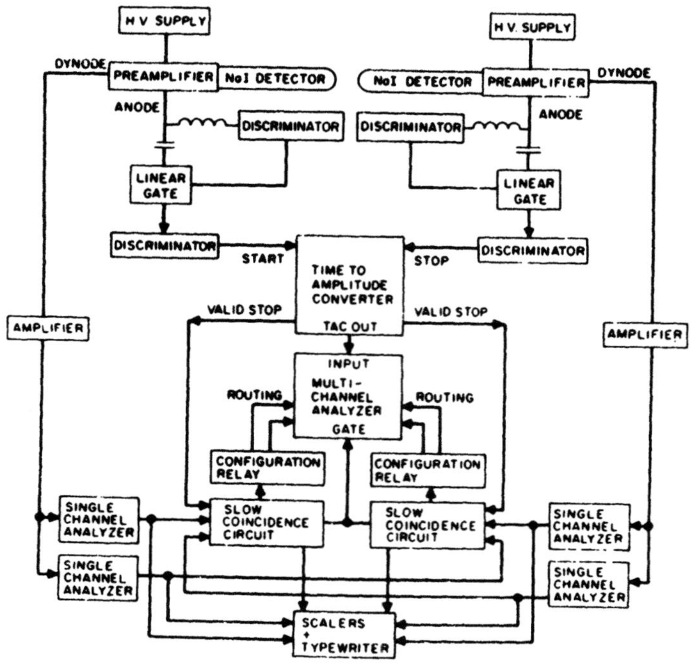

Fig. 20. Two - channel fast electronics used by the Pasadena group (courtesy M. Levanoni and F. Boehm).

The use of the TAC has brought in the possibility to get a detailed picture of the performance of the coincidence system. If the time spectrum can be analyzed carefully one can allow a higher background under the prompt peak and the source strengths can be increased. A very detailed study, such as the one carried out by Braunsfurth et al.[29]) reveals, however, that the situation is often not as simple as sketched so far. At high counting rates and especially when a weak coincidence pair is to be selected in the presence of other strong ones several higher order effects enter which makes the TAC - background sloping and unsymmetric. The shape is naturally dependent on instrumental parameters such as triggering of anode pulse formers, time resolution of the coincidence circuits, noise characteristics and the energy spectrum of the radiation. Investigation shows that accidental contributions of higher order than $2\tau N_1 N_2$ are noticeable already at normally used counting rates and that in the set - up used by Braunsfurth et al. they exceed the first order term at a total counting rate of 90,000 per second. This was the case although the authors were using a time overlap type of TAC which should be superior to the start - stop type when there are high counting rates in both channels[30]). (The coincidences with the inverted start - stop conditions were suppressed by a special connection in the TAC). When the anode pulses are amplified by a fast (200 MHz) linear amplifier and NaI(Tl) detectors are used, the time - spectrum will contain contributions of the form sketched in fig. 21. The asymmetry indicates that background coincidences are caused by time overlap of wanted pulses with noise and after - pulses and also photo electron packets from the tail of the scintillator pulse. Pile - up will cause a broadening of the prompt curve. Exact formulas for the dependence of the different contributions on the instrumental parameters could not be given but the following approximations are useful.

First order:

accidental coincidences $$Z_1 \approx N_1 N_2 \tau_{FS} \approx N_t^2 \tau_{FS}$$

2nd order:

true coincidences shifted on time scale by noise or after - pulses

$$Z_{21} \approx N_t^3 \tau_{FS}^2 (C_{21} \frac{\tau_{BG}}{\tau_{BL}} + 1)$$

accidental coincidences with noise

$$Z_{22} \approx N_t (R_1 + k_1 N_1)(R_2 + k_2 N_2)(C_{22} \frac{\tau_{BG}}{\tau_{BL}} + 1) \tau_{FS}^2$$

true coincidences time - shifted by pile - up

$$Z_{23} \approx N_t^3 \tau_{FS} \tau_{SL}$$

accidental coincidences with pile - up

$$Z_{24} \approx N_t^3 \tau_{FS} \tau_{SL}$$

3rd order:

true coincidences time - shifted by noise and pile - up (k = 1)

$$Z_{31} \approx N_t^2 (R_1 + k_1 N_1)(R_2 + k_2 N_2)(C_{3k} \frac{\tau_{BG}}{\tau_{BL}} + 1) \tau_{SL}^2 \tau_{FS}$$

and

accidental coincidences with noise and pile - up (k = 2)

$$\approx N_t^4 \tau_{SL}^2 \tau_{FS} (C_{3k} \frac{\tau_{BG}}{\tau_{BL}} + 1)$$

where N_t is the average rate of useful pulses, R_i noise pulses, N_i useful individual detector pulse rates, k_i the mean number per pulse of photo - electron packets from scintillator signal tails which trigger the TAC. The times involved are the fast - slow (τ_{FS}) and slow (τ_{SL}) resolving times, the blocking signal of the TAC (τ_{BL}) and the duration from the end of the coincidence signal up to the start of the blocking signal (τ_{BG}).

It is evident that contributions of type Z_{21}, Z_{22}, Z_{23}, Z_{31} and Z_{32} may cause trouble in the background determination of a time - spectrum when this type of anode pulse formers are used. When triggering circuits, e.g. tunnel diode discriminators

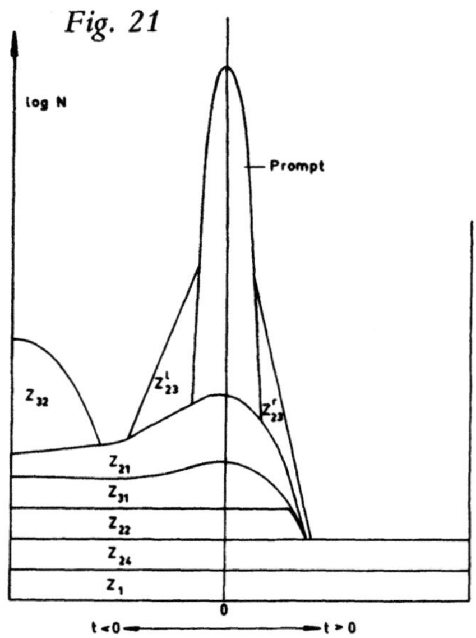

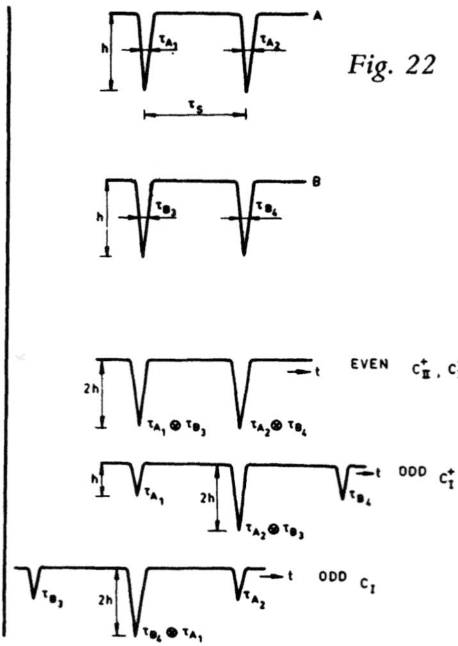

Fig. 21. Contributions to a TAC - spectrum. The Z_{ik} are defined in the text. The various contributions are not drawn to scale (from Braunsfurth et al., Nucl. Instr. and Meth. 67 (1969) 45).
Fig. 22. Double pulse technique for subtraction of accidental coincidences (from Z. Cho and T.R. Gerholm, Nucl. Instr. and Meth. 73 (1969) 67).

are used, the contributions Z_{21}, Z_{22}, Z_{31} and Z_{32} become approximately flat, which means a considerable increase of the background on the right hand side. It is also worth noticing that the higher order contributions increase as N_t^3 and N_t^4 respectively and soon exceed the first order accidentals. The slow coincidence resolution is critical as well as the fast one for the higher order terms and efforts to bring down τ_{SL} in angular correlation systems are motivated. Furthermore, the proportion of accidental coincidences increased faster than exponentially with relative amplitude C/A when the level C of the anode pulse shaper was lowered in the experiment of Braunsfurth et al. In a later paper[53]) Braunsfurth compares circuits with trigger pulse formers to those with limiters and shows the importance of the length of the noise pulses, τ_R. The quantity τ_R should be added to τ_{FS} in the expressions for Z_{21}, Z_{22}, Z_{31} and Z_{32}. In the commonly used combination of trigger plus start - stop type of TAC the noise pulses may be serious and it pays to minimize all the instrumental resolving times and dead times.

Some simple methods for the direct subtraction of accidentals without using a TAC have been suggested. A recent paper by Cho and Gerholm[31]) describes a system where a double pulse (with separation time τ_S) is produced in each channel. As shown in fig. 22 the coincidence circuit produces three types of pulses:

C_{II} for true and accidental coincidences proportional to $C_t + C_a$

C_I^+ for an accidental[delay time τ_S]

C_I^- for an accidental[delay time $-\tau_S$]

The true coincidences are then given by the expression

$$C_t = C_{II} - \tfrac{1}{2}(C_I^+ + C_I^-)$$

where all operations are carried out using the addition, subtraction and coincidence modes of the multi - channel analyzer. The method has been carefully tested and is accurate provided the shapes of the two double pulses in the two channels are equal. Later Cho et al.[54]) have described another subtraction circuit and its performance at high counting rates.

4.2 *Intensity correlations - a different approach*

Pulse counting has always been the cornerstone for detection of nuclear radiations. In the effort to increase source strengths and with an aim to avoid the problem

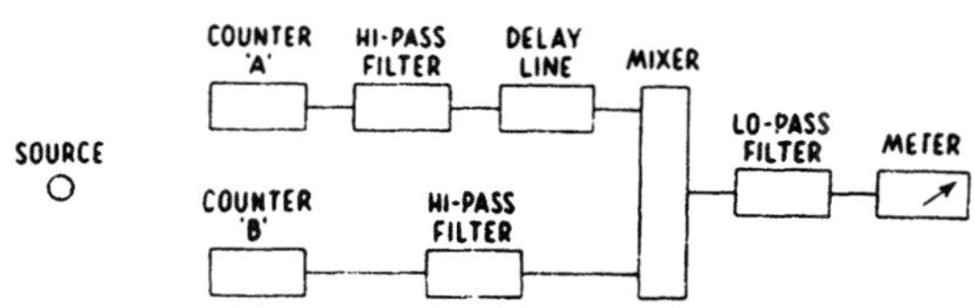

Block diagram of angular correlation apparatus for measuring time spectrum of delayed coincidences.

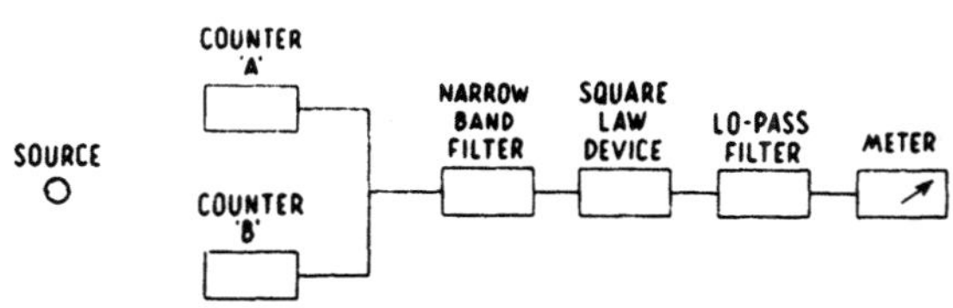

Block diagram of angular correlation apparatus for measuring frequency spectrum of delayed coincidences. Note that there is no delay line in the circuit.

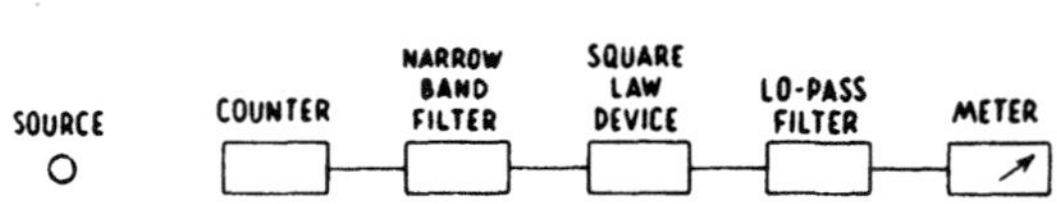

Block diagram of apparatus for measuring frequency spectrum of delayed coincidences of similar radiations.

Fig. 23. Block diagrams for intensity correlations (from R.R. Lewis, Nucl. Instr. and Meth. 75 (1969) 203).

of pulse pile - up the possibility to use continuous currents rather than pulses has been analyzed by Lewis[32]). These ideas were suggested by the single counter measurement by Lobashov[34]) in symmetry experiments, where counters were used as rate meters.

Contrary to one's intuition, the information inherent in currents can be used to determine true coincidence rates even when individual pulses overlap. Moreover, the statistical errors are the same. There is a correspondence between circuit elements; instead of a pulse shaping circuit, a filter; instead of a coincidence circuit, a mixer, etc.

In the paper by Lewis it is first shown that the current and pulse methods are equivalent for singles counting as far as the relative statistical error is concerned. For the coincidence circuit of fig. 23 the two currents I_a and I_b pass through a filter and into a mixer whose output (I_c) is proportional to the product of the two inputs. The output is passed through a narrow - band filter to a meter. The cross - correlation $< I_a(t)\, I_b(t) >$ of the two currents can be split up into two terms by introducing the deviations from the mean $I_a = < I_a > + i_a$; $I_b = < I_b > + i_a$. Then

$$< I_c(t) > \; = \; < I_a(t) I_b(t) > = < I_a > < I_b > + < i_a(t)\; i_b(t) >$$

The first term results from chance coincidences in the two counters and grows like the square of ν, the number of decays per second

$$< I_a > < I_b > \; = \; \nu^2\, \epsilon^2\, q^2$$

where q is the integrated charge of the pulse and ϵ is the counter and solid angle efficiency. The second term represents the true coincidences and contains the angular correlation W and the pulse width Δt

$$< i_a(t)\; i_b(t) > \; = \; \nu\epsilon^2\, q^2 W / \Delta\, t$$

The two terms can be measured separately by suitable filters; fig. 23 shows the filter choice for measuring accidentals. The relative statistical error is given by an expression similar to the one for pulse counting

$$\frac{\Delta I_c}{< I_c >} = \sqrt{[(1+2\nu R/W)/\nu\epsilon^2\, WT]}$$

where R is roughly the reciprocal of the bandwidth of the mixer or the counter, whichever is smaller.

Analysis in the frequency domain is natural in this type of arrangement. Lewis suggests a one - counter experiment for studying the spectrum of delayed coincidences (fig. 23, bottom) and Rama Reddy[33]) analyses in detail applications to perturbed angular correlations.

If intensity correlations are observed there is no limit on the pulse rate as long as the detector response is linear. It seems, however, to be limited to very simple decays since the energy information is lost. Applications to PAC in long - lived states $1\,\mu\text{sec} < \tau < 1$ sec might help to bridge the gap between other conventional techniques. No experimental results are available as yet.

4.3 *The use of multi - detector systems*

The use of several detectors on the same angular correlation table is an obvious way to increase the coincidence collection efficiency and has been used by several groups. Multi - detector systems have mainly been applied to two classes of experiments.

1. Directional correlation measurements on very short - lived isotopes, where the measuring time is limited by half - lives of the order of minutes or hours.
2. Perturbed angular correlation measurements, where the need for good counting statistics is stronger than in plain angular correlations since a second order effect is to be determined.

Furthermore, many 'in - beam' measurements of angular correlations and distributions are carried out with multi - detector[35]) arrangements for obvious reasons. Typical examples of the short - lived activity applications are the 8 - detector (formerly 12 - detector) system used at Kyoto[36]) and the 7 - detector system used in connection with the TRISTAN isotope separator project at Ames[10]). For PAC - measurements in magnetic fields a 3 - detector system in Hamburg[38]) and a 4 - detector system at Uppsala[37]) have been in constant use for almost 10 years. We will discuss below some of the experience gained with these systems and some points of interest for future construction.

The increase of the utilized solid angle compared to a 2 - detector system is, when all coincidence combinations are used and detector sizes and distances comparable, 3, 6, 21, 28 and 66 for 3, 4, 7, 8 and 12 detectors respectively. When both coincidence possibilities γ_1 (i)γ_2 (k) and γ_1 (k) γ_2 (i) can be selected for each detector pair i, k the efficiency increases by 2 for both two - and multi - detector systems. The optimum number of detectors when all are placed in the same plane has been considered by Okano et al.[36]) who find (for 3" NaI - crystals) that it was favourable to decrease the number of counters from the original 12 to 8 'as the merit from the decrease of source - detector distances was found to exceed the de - merit from the decrease of the number of counters'. A further increase of solid angle can be obtained with three - dimensional arrangements, for instance with twelve counters arranged perpendicular to each edge line of a regular octahedron, but difficulties in source handling and problems with detector - detector scattering make such approaches less attractive.

4.3.1 Fixed versus movable detectors

A system with several movable detectors arranged so that the angle between each pair of detectors can be varied over at least one quadrant can be considered as a simple extension of the conventional two - detector arrangement as far as adjustment, data treatment and technical requirements are concerned. The total result is obtained through a suitable averaging over coincidence versus angle data from each individual pair. Examples are found in the 3 - detector systems in Hamburg and Bonn[38,39]). When the number of detectors exceeds three, the desired angular range can be covered only with the use of more or less complicated mechanical devices for the permutation of detectors on the table (fig. 24).

Systems of 4 detectors or more are therefore most often operated in a fixed detector configuration. This type of arrangement brings in several new problems in the electronic adjustments and the analysis of data. The cause of the trouble is, of course, that one tries to compare the number of coincidence counts C_{ij} from different pairs. Now,

(1) coincidence efficiencies may differ appreciably from one, at least when the coincidence resolving times are shortened.

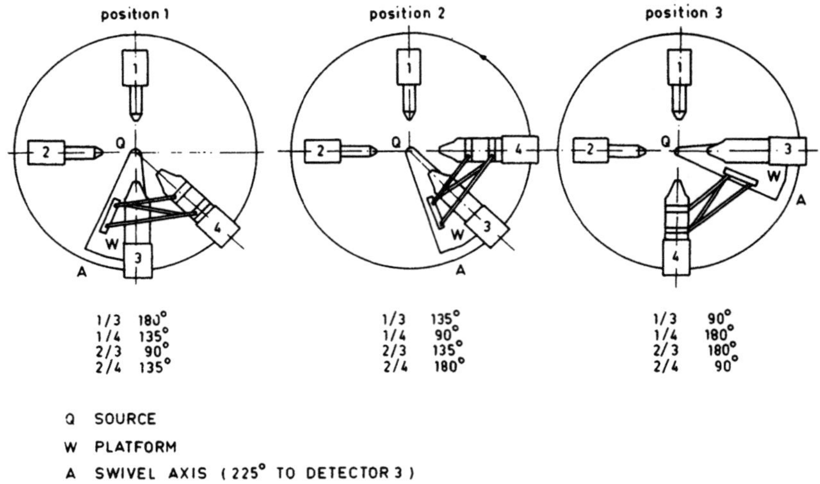

Fig. 24. 4 - detector system with mechanical arrangement for interchanging detectors 3 and 4 (from H.F. Wagner and M. Forker, Nucl. Instr. and Meth. 69 (1969) 197).

(2) the coincidence numbers are not proportional to the singles rate N_i and N_j in the pulse height region selected by the SCA's, because of the presence of pulses from other gammas under the selected peaks,
(3) two detectors are never equal in efficiency, photo - peak to compton ratio, energy resolution and noise characteristics,
(4) the centering control is difficult.

A 100% coincidence efficiency is approximated in the older arrangements only when much effort is put into the design and checking of the fast - slow system and even then at the sacrifice of time resolution. Present time - integral systems work at best with $2\tau \approx 20$ ns (above 100 keV)[10]) more often with 40 - 50 ns for NaI(Tl) - detectors and still higher when Ge(Li) - detectors are used. With TAC's for each coincidence pair and with an accurate subtraction routine for random background, one has in principle 100% coincidence efficiency at a time resolution corresponding to the width of the window (which should not be chosen too narrow considering the contributions Z_{23} of section 4.1.2) in the TAC spectrum. Recently developed multi - detector systems employ this technique[13,17,40]), which for a multi - detector set - up means a large increase in the cost of electronic equipment and data reduction.

Working with a fixed detector system, one soon realizes that it is virtually impossible to select detectors with equal characteristics and to set SCA windows, etc., such that coincidence pairs become comparable within the 1% range. Relative differences of 1 - 10% are more realistic. Most of the differences stem from different peak - to - background ratios in the windows and a normalization with single counts N_j works only when channels j select a photo - peak of a high energy γ - ray which is clean in the spectrum gated by channel i. Then, a series of angles for each i represented by C_{ij} can be compared in a reliable way.

In a method first suggested by Belyaevskii and Garýan[41]) and later also used by Okano et al.[42]) it is observed that if one writes

$$N_{iA} = \bar{N}_A\,(1 + X_{iA}) \quad \text{and} \quad N_{jB} = \bar{N}_{jB}\,(1 + X_{jB})$$

where $\bar{N}_A$ and $\bar{N}_B$ are the mean counting rates for the channels set on γ_A and γ_B respectively, then

$$C_{iAjB}(\vartheta) = W(\vartheta)K\,\bar{N}_A(1+X_{iA})\,\bar{N}_B(1+X_{jB})$$

for a clean spectrum and 100% coincidence efficiency. Taking the average of N detectors

$$C(\vartheta) = NW(\vartheta)K\,\bar{N}_A\bar{N}_B\,[1+\frac{1}{N}\Sigma' X_{iA}X_{jB}]$$

In such a simple case the differences can be corrected for. Belyaevskii and Garýan also treat the coincidence efficiencies in the same manner. However, the relation between $C_{iA\,jB}$, N_{iA} and N_{jB} is not as simple in a realistic case with compton backgrounds and Okano et al. made instead a statistical investigation of the effect of terms of the type $X_{iA}X_{jB}$. Random generation of numbers X_{iA} and X_{jB} from a normal distribution with certain standard deviations give the deviations of the mean counting rates according to table 1 for an 8 - detector system.

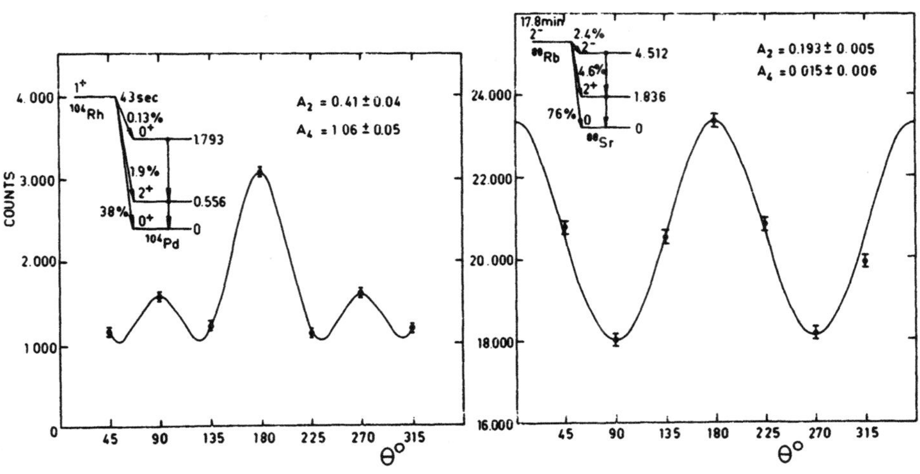

Fig. 25. Angular correlations of the 1.237 - 0.556 MeV cascade in ^{104}Pd from the decay of 43 sec ^{104}Rh and of the 2.676 - 1.836 MeV cascade in ^{88}Sr from the decay of 17.8 min ^{88}Rb determined with the 8 - detector system at Kyoto (from K. Okano et al., Nucl. Instr. and Meth. 75 (1969) 159).

Table 1

standard deviation of the distribution of N_i (or N_j) in %.	standard deviation of coincidence counting rate (%) due to spread in N_i and N_j.	number of coincidence counts required for the equivalent standard deviation
10	0.27	135.000
20	1.10	8,300
30	2.46	1.650

With still more detectors the cancelling will improve. but. as pointed out earlier, the gain in total solid angle is not very large, and the size of the electronic system will make it difficult to handle.

The above method seems to be a desperate solution of the fixed detector problem and a more detailed analysis of the spectral contributions would be preferred, but the last column of table 1 shows that. in cases where only a limited number of coincidences can be collected, the detector and channel setting inequalities may not be the main cause of error. Results from the study of very short - lived isotopes are shown in fig. 25, where an accuracy of a few per cent in the data is well sufficient to give meaningful results on nuclear transitions and states.

For a more accurate treatment of the background problem in a multi - detector system a large multi - channel analyzer is required. Larsen et al. use a 16 384 - channel analyzer with their 7 - detector system as mentioned in section 3.2. The

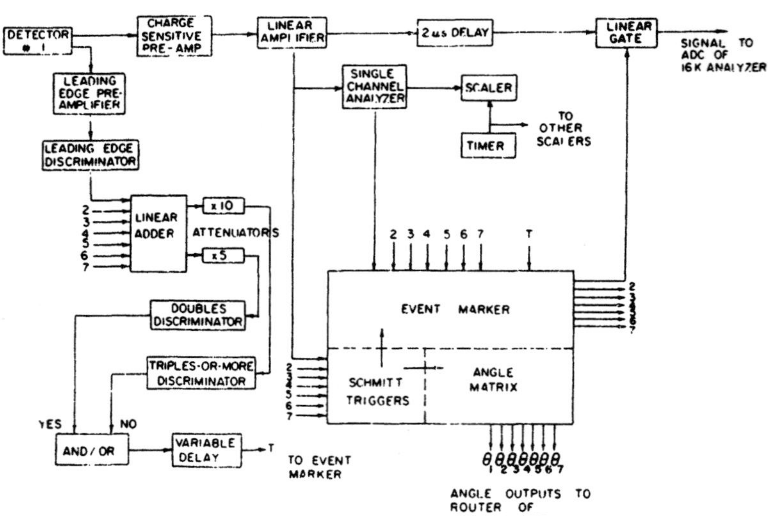

Fig. 26. Electronics for a 7 - detector system used by Larsen et al. (Nucl. Instr. and Meth. 69 (1969) 229).

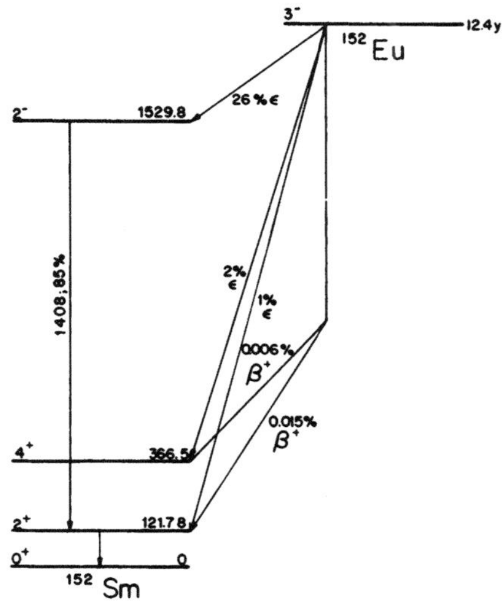

Fig. 27a. Decay scheme of ^{152}Eu.

block scheme of their electronics is shown in fig. 26. The gate is set on the photo - peak which has the least background of the coincident transitions of interest and the event marker routes the 7 coincidence spectra into seven out of 42 possible subsections of the memory, each containing 256 channels. The background is subtracted from the observed peaks in the coincidence spectra and normalized by dividing with the number of gate openings of the gating channel for each coincidence combination. Inequalities between different channels are then reduced to a large extent: if photo - peak efficiencies are the same for the 6 coincidence numbers $C_{ij}(\vartheta)\, j = 1, \ldots, i-1, i+1, \ldots\ldots 7$ with common gate i, they represent the angular correlations equally well as in a two - detector system, but comparability of the different sets i still requires equal peak and background contributions. An advantage of the normalization with gate signals (from channel i) over the earlier mentioned normalization with single rates is that the error introduced is restricted to the unequalities of peak - to - compton background of the coincidence spectrum within the gating window, which is considerably cleanerthan the singles spectrum. Test results for the ^{60}Co cascade indicated that remaining systematical errors were very small ($A_2 = 0.101 \pm 0.003$, $A_4 = 0.014 \pm 0.004$); other results for the 1408 - 122 keV cascade in ^{152}Sm, where one NaI - detector was replaced with a Ge(Li) - detector and 6 coincidence spectra were used, also showed very good agreement with results from two - channel instruments (fig. 27).

A complete two - parameter analysis with stripping of both spectra from each pair in the way described by Daniels (see section 3.2) would give the most reliable treatment of fixed - detector angular correlation data. No attempt in this direction

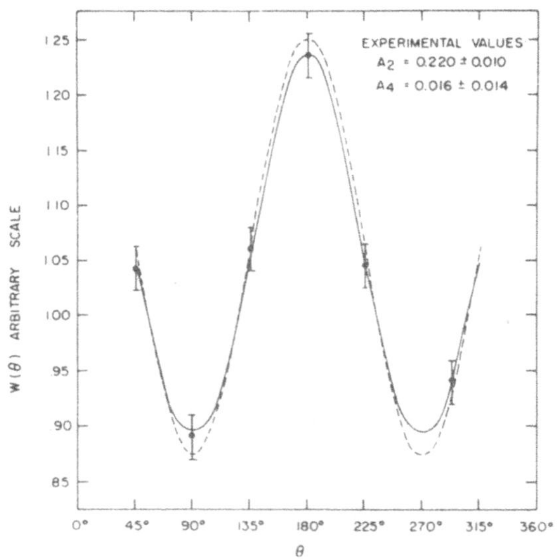

Fig.27b. Example of an angular correlation recorded by the 7 - detector system. Theoretical values represented by broken curve (from Larsen et al., Nucl. Instr. and Meth. 69 (1969) 229).

has been described as yet, probably because such an elaborate analysis would require a large storing capability: at least 4 k per combination and there are 6 - 28 combinations(for 4 - to 8 - detector systems) with NaI - detectors; in addition, a time - analysis is needed to have coincidence efficiencies under control. Anyhow, the advent and use of large MCA's and on - line computers has made possible a much more reliable use of multi - detector systems with a precision sufficient for most nuclear applications.

A major problem with multi - detector systems is detector - detector scattering. Placing thick absorbers between the crystals decreases the solid angle of the apparatus considerably. When the lower gamma energy is higher than about 300 keV the undesirable coincidences are small except at $\vartheta = 30^\circ$ and 60° [43]) and can be corrected for as described in section 2.4. For a 192 (A) - 544 (B) keV correlation in ^{199}Au Hayashi et al. have described an approximate method in which the result is compared with that of an auxiliary measurement with the source surrounded by an absorber of thickness T. Then the true coincidences decrease as $e^{-(\mu_A + \mu_B)T}$ whereas the scattered quanta decrease as $e^{-\mu_C T}$, where C is the energy of the quantum before scattering.

4.3.2 *The use of symmetries*

In some applications the use of a multi - detector arrangement does not only decrease the statistical error in the pulse counting, but can also reduce certain systematical errors present in conventional systems. The perturbation by a magnetic field on an angular correlation is (for nuclear lifetimes $\tau_N < 1$ ns) most often a very small shift of the correlation, with almost negligible attenuation of the pattern. In cases where the effect of the field,

$$\Delta C = C(+B) - C(-B)$$

is small compared to the anisotropy of the cascade, the effect is usually observed only at the most sensitive angles, i.e., 45°, 135°, 225° and 315° for dominant A_2 and approximately 67.5°, 112.5°, 157.5°, 202.5°, etc., when the A_4 - term is large. At Uppsala (ref. 37) a g - factor or hyperfine field determination is then always carried out in two steps:

(1) determination of the A_2 and A_4 coefficients for the particular source using two of the detectors of the main experiment and with the source placed in the same geometry (including magnet pole tips, cryostat etc.) as in the perturbation experiment; by using the conventional method the uncertainties involved with fixed detectors are avoided,

(2) measurement of the effect when the magnetic field is changed from +B to −B using 4 detectors placed at sensitive angles. (In cases where detector - detector scattering might be expected by the addition of two more detectors, coincidence spectra are compared and scattering contributions subtracted). The ratios W (135°, +B) / W (135°, −B), etc., can be treated separately but combination of the ratios in a suitable way is useful. For instance, when the detectors are placed in the following order I (90°) II (45°) III (90°) IV (135°) I, the following ratio is formed

$$P = \frac{C^+_{I\,III}}{C^+_{I\,IV}} \cdot \frac{C^+_{II\,IV}}{C^+_{II\,III}} \cdot \frac{C^-_{I\,IV}}{C^-_{I\,III}} \cdot \frac{C^-_{II\,III}}{C^-_{II\,IV}}$$

which in the absence of coincident admixtures is equal to $[W(135°, +B)/W(135°,-B)]^4$. One realizes that the coincidence efficiency problem is avoided and if only one cascade is present the expression P is insensitive to small drifts in the electronics caused by temperature variations or effects on PM - tubes of magnetic fields that remain after shielding, since changes in singles counting rates cancel out. When background coincidences are present this cancellation is not perfect but is still helpful when extremely small effects are looked for. The extraction of the quantity $\omega\tau$ from P, A_2 and A_4 is straight - forward for un - mixed cascades. With mixing, the appropriate formulas were worked out by Johansson et al.[44]).

The combined effect of increased counting efficiency (8 times in the PAC - application described here) and the use of symmetries has reduced the error limits of a W (+B) / W (−B) ratio to 0.0005 in favourable cases (10^6 coincidences collected per combination), corresponding to angular shifts of 4×10^{-3} rad or 0.2 degrees for the typical value of $A_2 = 0.20$. The absence of systematic effects can be checked by variation of the data treatment (section 5.2) and the measurement of zero - effect cases with very short level life - time.

5. Numerical methods in the determinations of angular correlation parameters

5.1 Fitting of angular correlation functions to measured data

Having a number of properly corrected coincidence data, the expected theoretical functions have to be fitted to the experimental points. The number of angles for which the coincidence rate is observed should exceed the number of parameters to be determined by at least a factor of two. Usually least - squares - fitting is carried out in the manner described by Rose[45]). Since the analytical form of γ - γ angular correlations is known, a χ^2 - test of the goodness of fit serves as a check if all systematical errors have been taken into account properly. One should like to see the results of such tests always when angular correlation data are presented.

5.1.1 Cases where parameters enter non - linearly

The conditions for the use of the least - squares method without approximation are (1) that Gaussian statistics apply to the experimental points and (2) that the fitting functions are linear in all the parameters to be determined from the data. The first condition is usually fulfilled since the number of counts is large (say greater than 100), but it is sometimes difficult to fulfill the second one. In a recent paper Orth et al.[46]) describe a numerical method for fitting functions of the type

$$Y = \Big(a_0P_0\{\cos(\vartheta+\vartheta_0)\} + a_1[P_1\{\cos(\vartheta+\vartheta_0)\} - (Q_3/Q_1)P_3\{\cos(\vartheta+\vartheta_0)\}] + a_2P_2\{\cos(\vartheta+\vartheta_0)\} + a_4P_4\{\cos(\vartheta+\vartheta_0)\}\Big)(1+k\epsilon)$$

which was applied to the d (p, γ)^{3}He reaction. The following parameters are involved:
ϑ_0 - the zero of the angular scale (could not be well determined by direct observation),
a_0 - a_4 coefficients of a Legendre polynomial series of the order of 4,
Q_1, Q_3 solid angle corrections
and k a normalization factor. Two sets of data were taken, with detectors A and B, for which one should put $\epsilon = 1$ and 0 respectively.

The above means that theoretically the angular distribution is expected to have the same value at 0° and 180°. The parameters a_0 - a_4, ϑ_0 and k should be determined from the fit.

This non - linear problem has to be handled with the more general maximum likelihood method and Orth et al. develop an iterative approach for maximizing the likelihood function. The convergence was fast enough to reach a situation in which all the parameters were simultaneously incremented by less than 0.1% on the sixth iteration. The fitted distribution function is shown in fig. 28 and the maximum likelihood function is plotted against parameter value in fig. 29. Differences from a Gaussian shape are small which means that the standard deviations obtained from the error matrix for the parameters are clearly acceptable.

The approach sketched here could be used also in other cases where centering is a problem or when data from different pairs of a multi - detector system have to be combined.

5.1.2 Determination of mixing ratios

A mixing ratio δ in a nuclear transition is usually determined by the following procedures: A_2 and A_4 are first obtained from a least - squares fit. Then, (1) the coefficients are plotted as functions of $Q = \delta^2/(1+\delta^2)$ for every possible spin sequence or (2) an elliptical plot in the A_2 - A_4 plane is plotted with the values of δ , or more conveniently, $\delta/(1+|\delta|)$ along the curve. Tables to facilitate the calculations have been given by Arns and Wiedenbeck[55]). Experimental values of $A_2 \pm \Delta A_2$ and

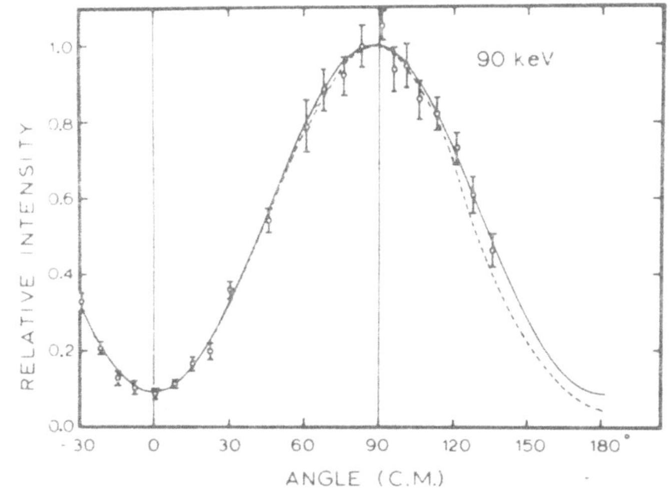

Angular distribution data obtained for the d(p,γ)³He reaction. The dotted line represents the initial fit to the data, the solid curve the "best" fit.

Results of the maximum likelihood analysis on the angular distribution data.

A_0	A_1	A_2	A_3	θ_0	k	χ^2/ν	Iteration number
2402.2 ± 356.4	173.6 ± 88.8	−2238.9 ± 485.9	61.66 ± 171.64	3.507 ± 1.486	−0.855 ± 0.025	26.82/16	0
2526.8 ± 246.5	102.4 ± 87.3	−2335.4 ± 329.6	121.56 ± 121.18	0.891 ± 1.175	−0.862 ± 0.015	10.29/16	1
2348.8 ± 226.8	125.2 ± 80.6	−2096.5 ± 301.7	47.05 ± 110.52	1.163 ± 1.202	−0.850 ± 0.016	9.84/16	2
2368.7 ± 227.4	122.3 ± 81.2	−2124.5 ± 302.2	57.28 ± 110.61	1.140 ± 1.198	−0.851 ± 0.016	9.70/16	3
2367.2 ± 227.2	122.6 ± 81.2	−2122.5 ± 302.0	56.62 ± 110.52	1.142 ± 1.198	−0.851 ± 0.016	9.70/16	4
2367.3 ± 227.2	125.5 ± 81.2	−2122.7 ± 302.0	56.69 ± 110.52	1.142 ± 1.198	−0.851 ± 0.016	9.70/16	5
2367.3 ± 227.2	125.5 ± 81.2	−2122.7 ± 302.0	56.69 ± 110.52	1.142 ± 1.198	−0.851 ± 0.016	9.70/16	6

Note: Iteration number 0 refers to the "initial" estimates. ν = degrees of freedom.

Fig. 28. Fit to an angular distribution with shift of angular scale (from P.H.R. Orth et al., Nucl. Instr. and Meth. 65 (1968) 307).

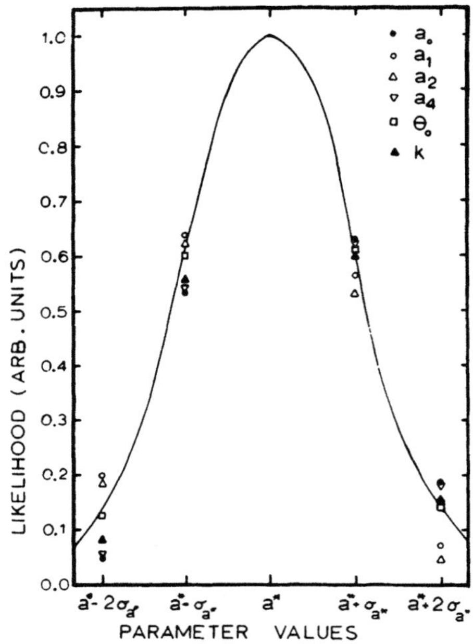

Likelihood as a function of parameter values. The solid curve represents a Gaussian with unit height and half width $(\sigma_a)^*$.

Fig. 29. Likelihood curves for the fit in fig. 28 (from Orth et al., Nucl. Instr. and Meth. 65 (1968) 307).

$A_4 \pm \Delta A_4$ are represented by rectangles in the A_2 - A_4 plane.

The errors ΔA_2 and ΔA_4 obtained from the fit are not independent however, and it would be more appropriate to replace the error rectangle by an ellipse within this rectangle[49]). Furthermore, there is usually other independent information about the multipolarity which should be taken into account in the evaluation of δ and its error limits. In order to put the evaluation on a more objective basis Boddendijk[49]) treated the angular correlation data $N_c\ (\vartheta)$ and the internal conversion coefficients $a_{K,exp}$ on the same basis. For each cascade and each value of δ one can evaluate

$$\chi^2 = \sum_i \left\{ \frac{N_{th}(\vartheta_i) - N_c(\vartheta_i)}{\Delta N_c(\vartheta_i)} \right\}^2 + \left\{ \frac{a_{K,\,th} - a_{K,\,exp}}{\Delta a_{K,\,exp}} \right\}^2$$

where the notations are self - evident. In this way one obtains a number of minima for χ^2 indicating possible solutions for the spin and parity sequence and δ. The error in δ corresponds to an increase of 1 above the value at the minimum[56]). The likelihood function

$$L(i) = \frac{e^{-x_i^2/2}}{\sum_j e^{-x_j^2/2}}$$

is evaluated for each possible solution and replaces the experimenter's judgement of mixing ratio plots by a quantitative measure. The treatment can of course be applied to cases where only angular correlation data exist. As an example Boddendijk evaluates for the $13/2^+$(M4+E5) $5/2^-$(E2) $1/2^-$ sequence in ^{207}Pb

δ	χ^2	L
0.007 ± 0.022	0.06	72%
-0.87 ± 0.04	1.92	28%

5.2 *Statistical considerations of counting sequences*

Another case where the final treatment of data is very important for the reliability of the result is found in the measurement of very small effects on the counting rate due to a periodical change of some external parameter. For the analysis of data taken with alternating signs of magnetic field in PAC - experiments Cruse and Norlin (ref. 47) have investigated different alternating sequences and extended the trend removal techniques earlier applied to parity experiments (determination of circular polarization) by Cruse and Hamilton[48]). Results of applications of several versions of moving averages, etc. to the same set of data {[$W(\theta; +B) / W(\theta; -B) - 1$] - values of the order of 1% }obtained in a g - factor experiment have increased the confidence in the derived quantities.

6. Concluding Remarks

Although the techniques of angular correlations are well developed there are still possibilities for improvement in several areas. Here, we have reviewed some recent work, mainly published in the period 1968 - 70, which seems to be of general interest. Some of the ideas and methods presented ought to be extended and applied more generally. The more extensive use of high capacity multichannel analyzers or on - line computers can increase the precision in the case of complicated decays and the data treatment can be improved. One should follow closely the advances in electronics for high counting rates and fast timing. Multi - detector arrangements are not the best solution for ordinary angular correlations when high accuracy is wanted and good statistics can be reached in a reasonable time with a conventional system, but they are valuable for short - lived or weak activities or 'in - beam' measurements and in particular, for PAC - measurements.

Several important areas have not been discussed. Angular correlations for reaction gammas have been mentioned only briefly and the specific problems in PAC only partly (analysis of time - differential PAC deserves a special chapter). It is also important for the extraction of nuclear or hyperfine field data to know if the source used shows the unperturbed correlation. The methods to find suitable environments for the nuclei are treated elsewhere[1]).

References

1. H. Frauenfelder and R.M. Steffen, in Alpha-, Beta- and Gamma-Ray Spectroscopy (ed. K. Siegbahn, North-Holland Publ. Co., Amsterdam 1965).

2. M.L.J. Yates, in Perturbed Angular Correlations (ed. E. Karlsson, E. Matthias and K. Siegbahn, North-Holland Publ. Co., Amsterdam, 1964).

3. A.M. Feingold and S. Frankel, Phys. Rev. 97 (1955) 1025.

4. D.C. Camp and A.L. van Lehn, Nucl. Instr. and Methods 76 (1969) 192.

5. D.E. Raeside and M.L. Wiedenbeck, Nucl. Instr. and Methods 78 (1970) 331.

6. R.G. Arns, S.E. Caldwell and W.G. Monahan, Nucl. Instr. and Meth. 78 (1970) 295.

7. W.G. Winn and D.G. Sarantites, Nucl. Instr. and Meth. 66 (1968) 61.

8. F. Ingebrektsen, Nucl. Instr. and Meth. 60 (1968) 308.

9. L.O. Norlin, private communication.

10. J.T. Larsen, W.C. Schick Jr., W.L. Talbert Jr., and D.I. Haddad, Nucl. Instr. and Meth. 69 (1969) 229.

11. E.A. Heighway and J.D. MacArthur, Nucl. Instr. and Meth. 79 (1970) 224.

12. M. Salomon, L. Boström, T. Lindqvist, M. Perez and M. Zwanziger, Arkiv Fysik 27 (1964) 97.

13. H.F. Wagner and M. Forker, Nucl. Instr. and Meth. 69 (1969) 197.

14. E. Bodenstedt, private communication.

15. W. Delang, P. Göttel and H. Seyfarth, private communication.

16. R.G. Helmer et al., Nucl. Instr. and Meth. 47 (1967) 305; F.C.P. Huang, C.H. Osman and T.R. Ophel, Nucl. Instr. and Meth. 68 (1969) 141.

17. T. Daniels, Nucl. Instr. and Meth. 59 (1968) 214.

18. D.D. Slavinskas, T.J. Kennett, W.V. Prestwich and G.L. Keech, Nucl. Instr. and Meth. 41 (1966) 1.

19. E. Der Matosian ,Nucl. Instr. and Meth. 73 (1969) 77.

20. A.M. Hoogenboom, thesis, Utrecht 1958.

21. J. Hattula, J. Kantele and A. Sarmanto, Nucl. Instr. and Meth. 65 (1968) 77.

22. E. Bozek, A.Z. Hrynkiewics, G. Zapalski, R. Kulessa and W. Walus, Nucl. Instr. and Meth. 58 (1968) 325.

23. D.C. Cruse. K. Johansson and E. Karlsson, submitted to Nuclear Physics.

24. A.G. Svensson, L.O. Norlin and P.N. Tandon, Arkiv Fysik 34 (1967) 113.

25. V.V. Perepelkin, JETP Letters 5 (1967) 81.

26. J. Fechner, C. Günther, H. Hübel, R.S. Raghavan, K.-H. Speidel, H. Toschinsky and B. Wolbeck, Phys. Letters 26B (1968) 374.

27. B. Herskind, private communication.

28. S.K. Bhattacherjee, J.D. Bowman and E.N. Kaufmann, Phys. Rev. Letters 18 (1967) 223.

29. J. Braunsfurth, D. Rüter and H. Winkler, Nucl. Instr. and Meth. 67 (1969) 45.

30. P.C. Simms, Rev. Sci. Instr. 32 (1961) 894.

31. Z.H. Cho and T.R. Gerholm, Nucl. Instr. and Meth. 73 (1969) 67.

32. R.R. Lewis Jr., Nucl. Instr. and Meth. 75 (1969) 203.

33. K. Rama Reddy, Nucl. Instr. and Meth. 81 (1970) 183.

34. V.M. Lobashov, V.A. Narenko, L.F. Saenko, L.M. Smotritskii and G.I. Kharkevich, JETP Letters 5 (1967) 59.

35. S. Gorodetzky, F.A. Beck, P. Engelstein, T. Byrski and A. Knipper, Nucl. Instr. and Meth. 69 (1969) 163.

36. K. Okano, Y. Kawese, S. Uehara and T. Hayashi, Nucl. Instr. and Meth. 75 (1969) 159.

37. B.E. Karlsson, Arkiv Fysik 22 (1962) 1; B.E. Karlsson, C.-A. Lerjefors and E. Matthias, Arkiv Fysik 22 (1962) 23; K. Johansson, A. Karlsson and J. Kozyczkowski, Arkiv Fysik 37 (1968) 251.

38. E. Bodenstedt, H.J. Körner, G. Strube, C. Günther, J. Radeloff and E. Gerdau, Z. Physik 163 (1961) 1.
39. E. Bodenstedt, private communication.
40. T.R. Gerholm, private communication.
41. A.I. Belyaevskii and Yu.A. Garýan, Bull. Acad. Sci. USSR 25 (1961) 1300.
42. T. Hayashi, K. Okano, K. Yuasa, Y. Kawese and S. Uehara, Nucl. Instr. and Meth. 53 (1967) 123.
43. K. Okano, Y. Kawese, S. Uehara and T. Hayashi, Nucl. Phys. A136 (1969) 321.
44. K. Johansson, S. Gustafsson and A.G. Svensson, Arkiv Fysik 34 (1967) 97.
45. M.E. Rose, Phys. Rev. 91 (1953) 610.
46. P.H.R. Orth and M.A. Olivo, Nucl. Instr. and Meth. 65 (1968) 307.
47. D.W. Cruse and L.O. Norlin, to be published.
48. D.W. Cruse and D. Hamilton, Nucl. Instr. and Meth. 57 (1968) 29.
49. H.G. Boddendijk, thesis, Univ. of Amsterdam, 1970.
50. W.G. Winn and D.G. Sarantites, Nucl. Instr. and Meth. 82 (1970) 230.
51. H. Verheul, J. Block, H.G. Boddendijk and B.G. Vonck, Physica 31 (1965) 565.
52. Radioactivity in Nuclear Spectroscopy (ed. J.H. Hamilton and J.C. Manthuruthil), Gordon and Breach Publ. Co., 1970).
53. J. Braunsfurth, Nucl. Instr. and Meth. (to be published).
54. Z.H. Cho, U. Bäverstam and R. Othaz, Nucl. Instr. and Meth. 84 (1970) 321.
55. R.G. Arns and M.L. Wiedenbeck, Phys. Rev. 111 (1958) 1631.
56 R.A. Arndt and M.H. MacGregor, in Methods of Computational Physics, Vol. 6 (1966) 256.
57. G.L. Murray and B. Macefield Nucl. Instr. and Meth. 62 (1968) 122.
58. G.L. Murray and B. Macefield, Nucl. Instr. and Meth. 57 (1967) 37.
59. B. Macefield, Proceedings of the Fall Joint Computer Conference, Dec. 1968.
60. J. Kantele and P. Souminen, Nucl. Instr. and Meth. 86 (1970) 65.

Contributed Paper

INTRASOURCE SCATTERING AND ABSORPTION EFFECTS ON $\gamma-\gamma$ ANGULAR CORRELATION MEASUREMENTS*

P.C. Lopiparo and R.L. Rasera
Department of Physics and Laboratory for Research on the Structure of Matter, University of Pennsylvania, Philadelphia, Pa. 19104, U.S.A.

Perturbed angular correlation (PAC) is a useful technique for the study of hyperfine interactions. However, its application to solids requires that effects associated with the use of dense sources be taken into consideration. In particular, time - integral PAC measurements, in which the magnitude of the anisotropy is often a measured quantity of interest, must be free of or corrected for directional effects from intrasource interactions of the cascade γ - rays after their emission. These effects are usually negligible in the aqueous solutions commonly used as sources in nuclear - spectroscopic applications, but they may become appreciable in extended solid samples of higher density and atomic number. In this paper we describe numerical calculations of the effects of γ - ray scattering and absorption on measured angular correlations. The results of this procedure may be used to choose favourable source configurations in order to minimize these distortions, as well as to correct the data for remaining attenuations.

If the true angular correlation is expressed in the form $W(\hat{k}_1, \hat{k}_2) = \Sigma_j A_j P_j (\hat{k}_1, \hat{k}_2)$, then the observed correlation function, modified by scattering and absorption, may be written

$$W_{obs}(\hat{e}_1,\hat{e}_2) = f_0(\hat{e}_1,\hat{e}_2)\Sigma_j A_j P_j(\hat{e}_1.\hat{e}_2)\,[f_j(\hat{e}_1,\hat{e}_2)/f_0(\hat{e}_1,\hat{e}_2)],$$

where

$$f_j(\hat{e}_1, \hat{e}_2) = \int_{\text{source}} \frac{d^3r}{V} \int d\Omega_{k1} \int d\Omega_{k2} [\,P_j(\hat{k}_1 . \hat{k}_2)/P_j(\hat{e}_1 . \hat{e}_2)]\, K(\hat{k}_1, \hat{e}_1 ; \hat{k}_2, \hat{e}_2 ; \underline{r})$$

In the last expression, $K(\hat{k}_1, \hat{e}_1 ; \hat{k}_2, \hat{e}_2 ; \underline{r})$ represents the probability / (steradian)^2 that γ_1 and γ_2, emitted from $\underline{r}$ in directions $\hat{k}_1$ and $\hat{k}_2$, will emerge from the source in direction neighbourhoods of $\hat{e}_1$ and $\hat{e}_2$, respectively. The evaluation of the f_j involves substituting a suitable expression for K and performing the multiple integrations using a high speed computer. The computations were performed for a variety of source configurations under the following assumptions:

1. The source is a uniform, right circular cylinder with axis perpendicular to the detector plane.
2. The distances between the source and the detectors greatly exceed any linear dimension of the source.
3. All angular correlation functions are probability densities/ (steradian)^2, so that the effects of finite solid angle subtended by the detectors at the source may be considered separately in the usual way.
4. Each detector responds to only one of the radiations.
5. Only photoelectric absorption, Compton scattering, or Rayleigh scattering can interfere with photon escape.
6. Either photon may be scattered at most once, and events involving simultaneous scattering of both photons are ignored.

Assumption (6) above was justified by the use of a Monte - Carlo routine which traced 200,000 simulated photon histories through 'unfavourable' sources until the photons either escaped, were degraded enough in energy to escape detection by typical NaI(Tl) systems, or were photoelectrically absorbed. Multiple scattering events never exceeded 3% of all detected single - scattering events. Under these assumptions, the functions f_j were decomposed into the form $f_j = f_j(\bar{1}\bar{2}) + f_j(1\bar{2}) + f_j(\bar{1}2)$, where $f_j(m\bar{n})$ is the contribution from cases where the mth photon was scattered but the nth was not. Published[1]) values were used for the photoelectric cross - sections, and the differential Compton cross - section was calculated using the Klein - Nishina formula. The Rayleigh cross - sections were synthesized from the work of Debye[2]), Franz[3]) and Moon[4]). The resulting 6 - fold integrals were evaluated by random sampling techniques on an IBM 360/75 computer. Three general observations may be made concerning the results of these calculations.

1. Rayleigh (coherent) scattering is completely negligible for the energies considered (> 60 keV). This is true, even at low energies, because of the sharp forward

peak in the differential cross - section for this type of scattering[2,3,4]).

2. Even in the absence of scattering, anisotropic photoelectric absorption can alter the observed correlation, in some cases markedly, by depressing the raw counting rate at 180° with respect to 90°. The cause of this distortion may be understood using a simple intuitive argument given by Frauenfelder and Steffen[5]). If one of the photons is strongly absorbed, the centroid of source points from which coincidences are likely to be detected will lie close to the corresponding detector, obliging the other photon to traverse an angle - dependent distance to escape the source. The resultant anisotropy shows up primarily in the coefficient $f_0(\hat{e}_1, \hat{e}_2)$, and varies roughly as cos ϑ. If an integral rotation is being measured, the rotation may be falsified by the artificially induced central tendency of the data about 180°. Representative values of the anisotropy of f_0 are presented in fig. 1.

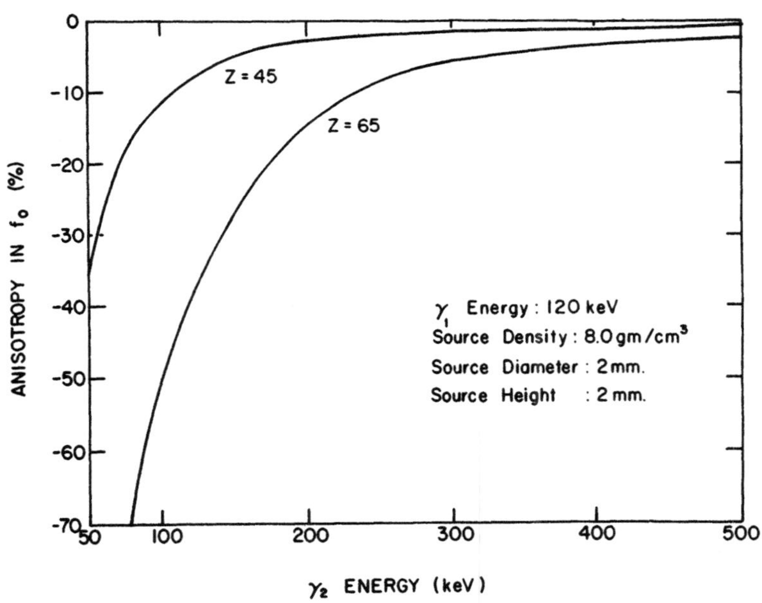

Fig. 1. Anisotropy [(f_0 (180) / f_0 (90)) − 1] due to photoelectric absorption as a function of the energy of one of the gamma rays for a typical case. These calculations are valid only above the K - absorption edge of the sample material.

3. Compton scattering is relatively a weaker effect, altering the coefficients A_j of the angular correlation by at most several per cent. The scattering effects are contained primarily in the factors $f_j(\hat{e}_1, \hat{e}_2) / f_0(\hat{e}_1, \hat{e}_2)$, which are nearly isotropic and may be treated as attenuation coefficients. It is interesting to note that photoelectric absorption moderates the scattering effects by determining the source regions from which coincidences are likely to be detected. Fig. 2 shows the energy dependence of the coefficients f_j/f_0. At the lowest energies, the larger scattering cross - sec-

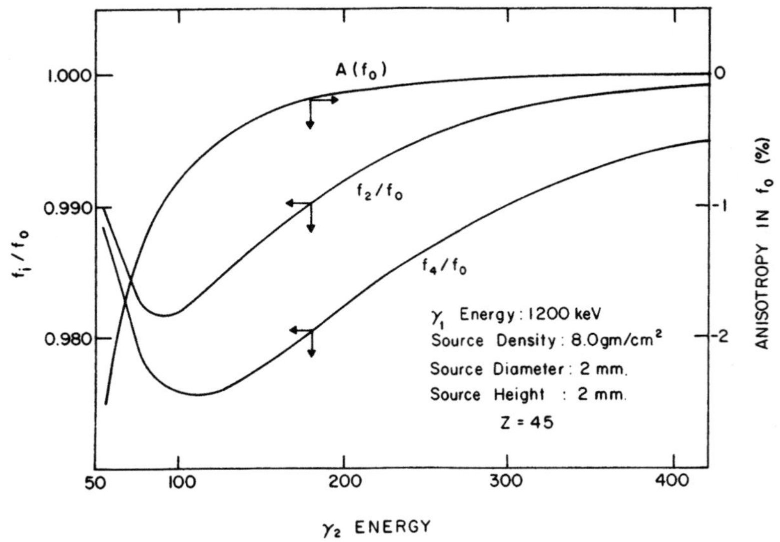

Fig. 2. Energy dependence of the scattering and absorption parameters for a typical case.

tions are compensated by the fact that photoelectric absorption restricts detected coincidences to originate in an outer region of the source. At highest energies, the photoelectric effect makes no restrictions on the origin of detected coincidences, but the scattering cross - sections are relatively smaller. The competition between these two effects causes a peak in the scattering coefficients f_j/f_0 at 100 - 120 keV, depending on Z. The same competition causes scattering effects to be worse in sources of lower Z, where the photoelectric effect is not particularly strong.

The above considerations suggest a procedure for correcting angular correlation data for scattering and anisotropic absorption. The raw data, after correction for random coincidences, is divided at each angle by f_0 $(\hat{e}_1, \hat{e}_2)$ before further analysis. Then, since the ratios f_j/f_0 are nearly isotropic, the angular correlation coefficients A_j derived from this data can be divided by these ratios to obtain the true A_j.

Because of the wide variety of possible conditions, the results presented here are representative rather than exhaustive. The computer program itself is straightforward and may be used to evaluate f_0 and f_j/f_0 for sources of arbitrary (cylindrical) dimensions, Z, density, and γ - γ cascade energies. A FORTRAN IV source listing of this program is available from the authors on request.

* Supported by the Advanced Research Projects Agency and the U.S. Army Research Office (Durham).

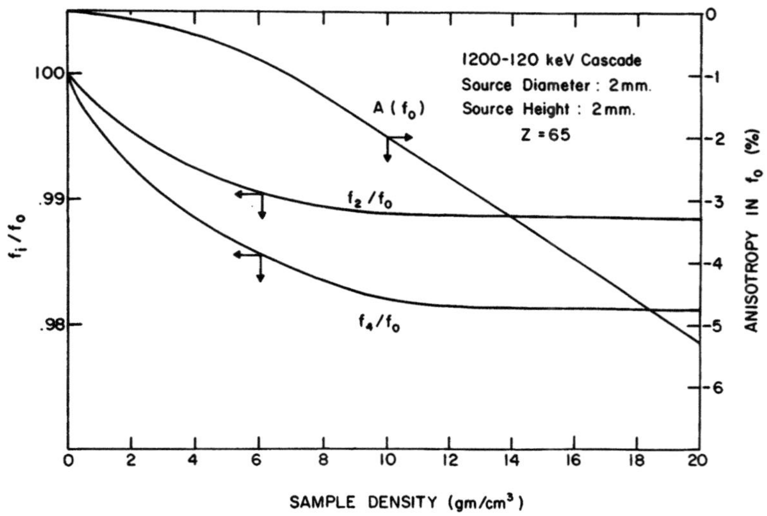

Fig. 3. Typical dependence of the parameters on sample density.

References

1. J.H. Hubbel, U.S. Natl. Bur. Stds. NSRDS - NBS 29 (August 1969).
2. P. Debye, Physik. Zeitschr., 31 (1930) 419.
3. W. Franz, Z. Phys. 98 (1935) 314.
4. P.B. Moon, Proc. Phys. Soc, A63 (1950) 80.
5. H. Frauenfelder and R.M. Steffen, in Alpha-, Beta- and Gamma-Ray Spectroscopy, ed. K. Siegbahn (North - Holland Publ. Co., Amsterdam 1965) Vol. 2, Ch. XIX.A.

Contributed Paper

A FIFTYSIX CHANNEL GONIOMETER

T.R. Gerholm, B.G. Pettersson, C. Bargholtz, Z.H. Cho, L. Eriksson and L. Gidefeldt

University of Stockholm, Institute of Physics, Stockholm, Sweden

In order to increase the efficiency in gamma - gamma correlation measurements a Multi - Channel - Goniometer (MCG) has been developed. The construction utilizes eight NaI (Tl) - scintillation detectors between which all possible coincidence combinations are simultaneously recorded with a fast - slow technique.

All the fiftysix coincidence channels are time analyzed in the same Time - to - Amplitude - Converter from which the time information is fed into a Multi - Channel - Analyzer where each time spectrum is displayed over sixtyfour channels. This enables us to make measurements in a time integral or differential mode.

A computer program has been developed for time integral measurements. Test experiments have been performed on ^{60}Co and ^{181}Ta.

In the near future the MCG will be equipped with facilities for solid state applications (magnet, cryostat). Due to the high efficiency the MCG is well suited for the series of measurements with changes in the nuclear environment involved in this kind of research.

At present the apparatus is used in nuclear structure studies measuring weak gamma - gamma transitions (^{59}Fe, ^{115m}Cd, ^{207}Bi). The complete MCG system utilizes a fast - slow technique. In the fast part of the system the coincident pairs of pulses of the observed gamma - gamma cascade are time analyzed and in the slow part energy analysis takes place. The short description of the various functions of the MCG system, given below closely follows the block diagram.

The anode pulses of the eight Na (Tl) - scintillation detectors are shaped in fast discriminators according to a low - level - timing technique in order to extract the best possible time information. The output signals from the fast discriminators are fed to the suppression unit and the fast interface. These two systems perform a 'majority coincidence' logical function. The fast interface works in a 'two or more out of eight' mode. It only transmits pairs of pulses coincident within a predetermined but easily variable time interval. In this way a time filtering function is achieved. The obvious reason for this is to protect the TAC from being exposed to the total single counting rates of the eight detectors. The fast interface also has a routing function, i.e., all possible coincidence combinations must be presented to the TAC in a unique way.

If higher order of coincident pulses are transmitted through the fast interface then the suppression unit, working in a 'three or more out of eight' mode, will send an error signal to the control system and the coincidence event will be cancelled.

The following TAC has two functions. Firstly, it time - to - amplitude converts the transmitted pairs of pulses and secondly, a 'yes' pulse is fed to the control system to signify the presence of a fast coincidence.

The analog time information is transferred to the ADC where it is temporary stored until a clearing (OK-) signal from the control system (see below) initiates the analog - to - digital conversion.

In the slow part of the system energy analysis takes place through eight dual differential discriminators. From each detector both gamma ray energies are accepted. The output pulses from the high energy acceptance interval we refer to as high and from the low energy acceptance interval as low.

The coding unit translates the various combinations of high and low signals into a six bit binary address code that is unique for each coincidence combination.

If the TAC has indicated the presence of a fast coincidence through the 'yes' signal and if no error signal from the suppression unit is present, then the control system will check that the accepted address code represents an appropriate high and low energy combination. If it is found to be correct then the control system initiates the

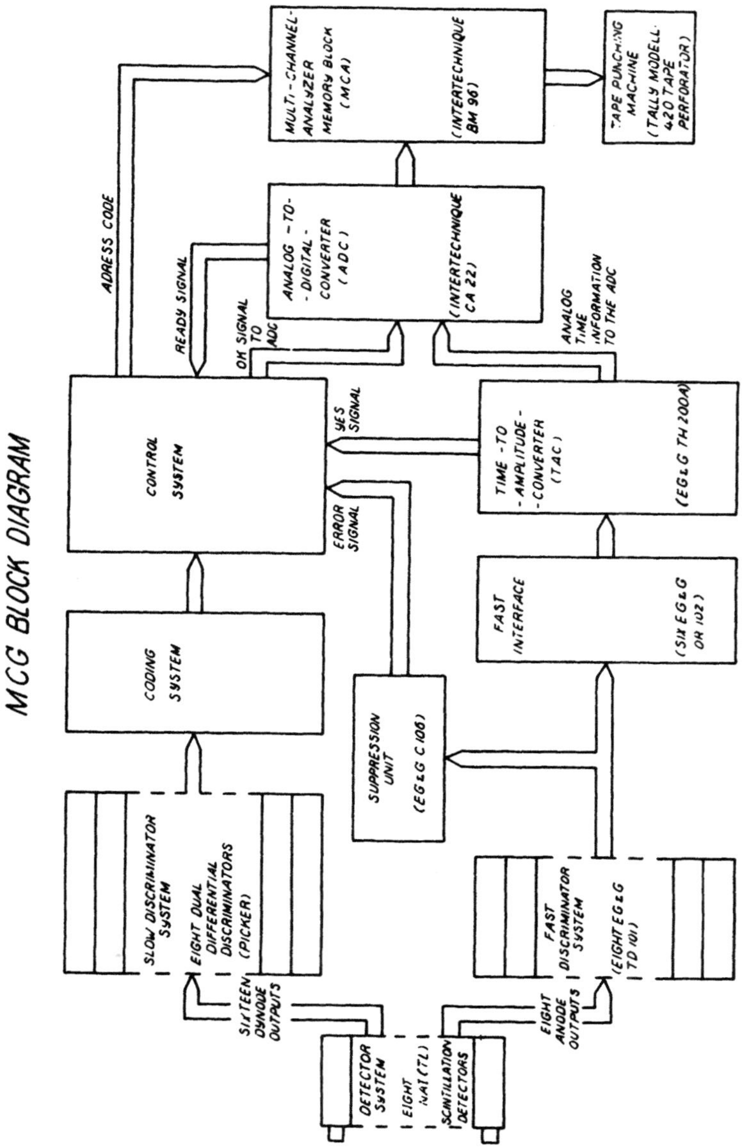

Fig. 1. MCG block diagram

analog - to - digital conversion of the time information in the ADC. The address code, temporarily stored in the control system is released by a clearing (ready-) signal from the ADC after the time period required for the analog - to - digital conversion. In this way, the binary address code and the digital time information are simultaneously presented to the MCA memory block which automatically places the time information at its appropriate position in a 64 X 64 angular - time matrix. Thus the MCA will show fiftysix time spectra of sixtyfour channels, each spectrum belonging to a definite coincidence combination.

The information from the MCA is taken out on punched paper tape. A computer program has been developed for time integral measurements. The IBM 360/75 computer of Stockholms Datamaskincentral is used. The analysis of data from time integral measurements is based on a least square fit to the equations

$$C_{ij} = \epsilon_i^H \epsilon_j^L [\, 1 + A_2P_2(\cos\theta) + A_4P_4(\cos\theta)\,] \qquad \begin{matrix} i = 1,8 \\ j = 1,8 \end{matrix} \quad i \neq j$$

where C_{ij} is the number of true coincidences between the detectors i and j with the high energy accepted in detector i and the low energy in detector j. A_2 and A_4 are the coefficients to be determined. ϵ_i^H and ϵ_j^L are the efficiencies of the coincidence channels i and j.

Several test runs on ^{60}Co with sources of different strengths and different detector set - up geometries show results in excellent agreement with theory.

In order to test time differential measurements (average time resolution is about 1.3 ns)*, the halflife of the 482 keV level in ^{181}Ta was determined. The result, 10.9(2) ns, is in good agreement with that obtained by other authors.

* measured with ^{60}Co.

References

1. T.R. Gerholm, T. Lindqvist and H. de Waard, Nucl. Instr. and Meth. 1 (1957) 107.
2. B. Erik Karlsson, Arkiv Fysik 22 (1962) 1.
3. B. Erik Karlsson, Arkiv Fysik 22 (1962) 27.
4. K. Fricke, F. Münnich and U. Wallner, Z. Instrument. 71 (1963) 247.
5. T. Hayashi, K. Okano, K. Yuasa, Y. Kawase and S. Uehara, Nucl. Instr. and Meth. 53 (1967) 123.
6. K. Okano, Y. Kawase, S. Uehara and T. Hayashi, Nucl. Instr. and Meth. 75 (1969) 159.
7. H.K. Walter and A. Weitsch, Nucl. Instr. and Meth. 62 (1968) 189.
8. H.F. Wagner and M. Forker, Nucl. Instr. and Meth. 69 (1969) 197.
9. J.T. Larsen, W.C. Schick,Jr., W.L. Talbert,Jr., and D.I. Haddad, Nucl. Instr. and Meth. 69 (1969) 229.
10. Table of Isotopes 6th ed. C.M. Lederer, J.M. Hollander and I. Perlman, eds. John Wiley & Sons, New York 1967.

Discussion

C. Broude: Speaking as an ex - Chalk River man I cannot resist pointing out that the multi - counter systems that are used with tandem accelerators are very much more complicated than the system that you just described. I do not really want to go into that, but I just want to mention that a LOTUS system at Chalk River has been in operation for several years and this system allows seven 5″ by 6″ NaI crystals to be moved both in and out of the plane. The data are taken of course in the form of energy spectra from every detector. The information goes through an on - line computer and then through two other computers before the information comes out. I do not really want to relate with that point. The point of interest to me, though, and I really do make this comment to someone who is not very familiar with source work, is the comment you made that you discard triple coincidences. I just like to make it clear that if in fact the pair of gamma - rays you are interested in, belong to a cascade that includes more than 2 gamma -rays then you cannot throw away triple coincidences. In fact if you do, as in the case of e.g. the sort of work that we have been doing recently, we have been exciting levels of spin around 7 and 8 which decay through a 6, a 4, a 2 and a 0, then, if you consider the efficiency for 7 or 8 detectors for observing a triple coincidence, this is very high indeed. In fact, if you throw away the triple coincidences, then what you are doing is subtracting out of your double correlation a triple correlation. It is as if you are putting holes into the space, because every time an uninteresting gamma - ray happens to go into a NaI detector instead of into empty space, you are throwing away that event. Now, in our case it is certainly not possible to throw these away without very strongly biasing the correlations. Would you care to comment on that?

Gerholm: I really did not say that we throw away triple coincidences but we have a possibility of doing so. We do not have to take advantage of it, but we have the possibility of testing the amount of triple coincidences and to see to what extent they effect. You are quite right that if you do have a triple correlation cascade, then this has to be taken into consideration. That is not necessarily always the case, but when it is the case, we have the possibility of doing so.

Contributed Paper

COMPUTER - CONTROLLED 70 - IN. CHAMBER*

J.L. Yntema

Argonne National Laboratory, Argonne, Illinois, U.S.A.

The 70 - in. chamber shown schematically in fig. 1 basically consists of four concentric ring gears. Each gear ring carried a radial arm which is accurately adjusted. The angular position of each arm can be read to better than 0.01° from the primary readout system, consisting of high - precision engraved quartz rings R. The secondary readout system, circular encoders which read from the gear teeth on the rings, is reliable to ± 0.02°. A detector carrier is attached to each radial arm through a worm gear which runs radially through each arm. This system allows the distance of the detectors from the center to be adjusted in 1 mm steps with an accuracy of ± 0.02 mm. The drives for the four gear rings and the four worm gears are interfaced with the computer. The computer can thus be used to position the detectors at the desired radii. Additional links to the computer permit the control of target position, selection of various standard sources, and control of electromagnetic fields. The computer arran-

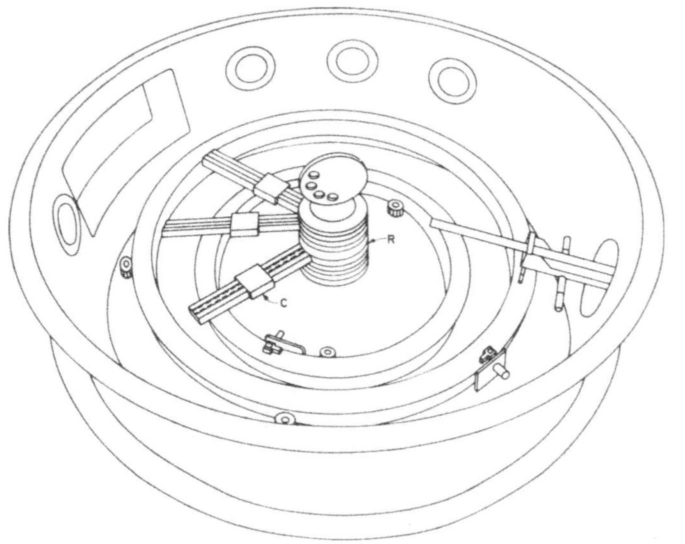

Fig. 1. Schematic view of 70 - in. chamber

gement and auxiliary equipment is shown schematically in fig. 2. The signals from the detectors are fed back to the computer through four dual ADC's and seven multichannel analyzers.

The computer can thus control the experimental apparatus, collects the data, and gives access to the data during accumulation. One obvious advantage is the use of the computer as bookkeeper in multidetector arrangements. Besides identifying all the components associated with a given data set, the computer can also prevent collisions between the movable and the fixed components of the chamber, and can select the best sequence of motions if a change in the configuration of the detector system is required. It will adjust true - to - accidental ratios, perform kinematic calculations, make energy calibrations, make normalizations, perform background corrections, and decide when a run has achieved the desired statistical accuracy, after which it will terminate the run, proceed to the next set of parameters and start up again.

In order to do this with a small computer, the use of the 96 k external memory is essential. This external memory has two banks (65.836 - and 327.68 - word capacity). The external memory can be used as a large - capacity pulse - height analyzer. One can operate it either as a two - parameter analyzer (up to 256 X 256 channels) or with up to eight one - parameter analyzers having 4000 channels each. It is also possible to transfer the contents of the seven multichannel analyzers nondestructively to the computer and subsequently store these temporarily in the external memory. The second bank of the memory is normally used for program storage. This allows

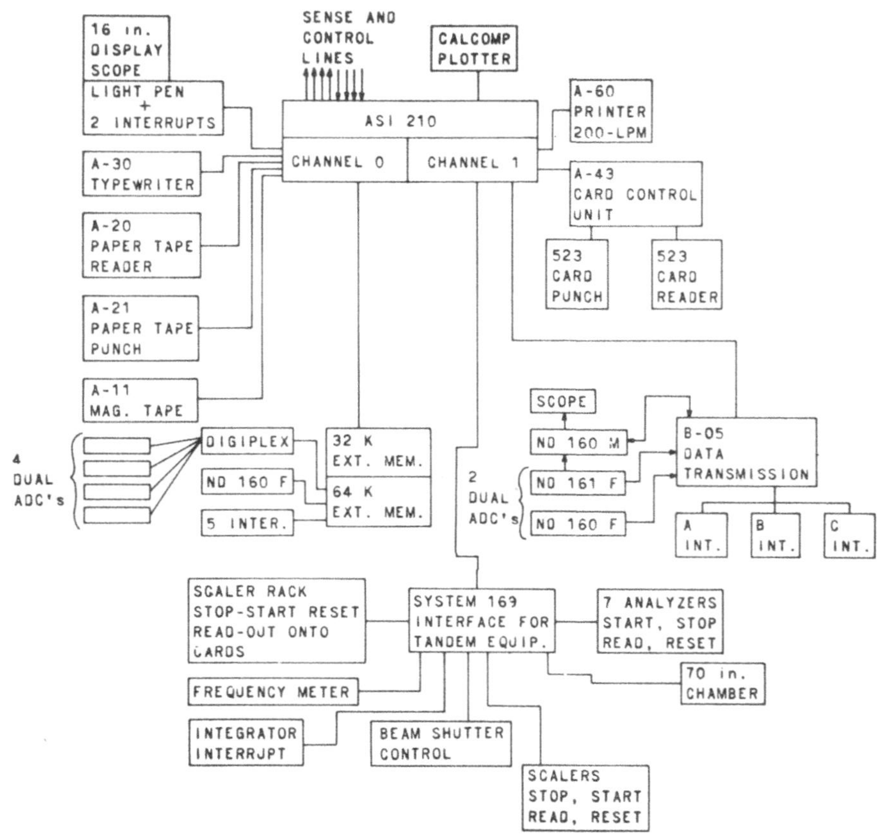

Fig. 2. Block diagram of computer system.

one to call up various programs, and after using a program to store it again and call the next program. The computer can thus be used to provide a continuing analysis of the data during accumulation.

To calibrate γ - ray spectra one can select any one of several known calibrated standard sources, bring it up to each detector, accumulate a spectrum, and display it on the oscilloscope. The lines to be used in the calibration are selected by pointing them out with the light pen. Since the computer has been programmed for these γ - lines, it will automatically fit a quadratic calibration funciton to the centroids of these peaks. During a run, it can determine the γ - lines of the accumulated spectrum and (by referring to the energy calibration for the detector used) can give the energies of the lines observed while the experiment is still in progress.

The 70 - in. chamber is on one of the beam lines of the 15 - MeV tandem Van de Graaff. It has been used extensively for multidetector experiments, especially angular correlation experiments with multidetector arrays – for which the required positions of the three slave arms with respect to the master arm were calculated by the computer. Such experiments would in many cases have been impractical without this system.

The projected use of the system for decay studies can be illustrated by the case of ^{72}Ga. It was recently observed that the 3^- ground state of ^{72}Ga is not excited in the ^{71}Ga (d, p) ^{72}Ga reaction, although the simple shell model leads one to expect a reasonably large $\ell = 4$ transfer. Studies of other reactions with a ^{71}Ga target have shown that this is not caused by an anomaly in the ^{71}Ga g.s. configuration. It appears therefore that the g.s. configuration of ^{72}Ga is more complex than one would have anticipated and that it may be of interest to investigate the γ decay of higher excited states of ^{72}Ga with the pulsed beam of the tandem on different targets.

The chamber was constructed in the Central Shops at Argonne National Laboratory. The control systems and the interface between the chamber and the computer were constructed by R. Pecina and J. Bicek. The computer system was devised by D. S. Gemmell[1]) and the programs used in the computer system were mostly written by M.R. Storm, J. Scherer, J.W. Tippie, R.C. Bearse, R.E. Holland, and D.S. Gemmell.

* Work performed under the auspices of the U.S. Atomic Energy Commission.

Reference

1. D.S. Gemmell, Nucl. Instr. and Methods 46 (1966) 1.

Discussion

Sújkowski: What kind of detectors you use for your gamma work?

Yntema: We have used NaI crystals. There are some difficulties with Ge(Li) - detectors in this system in order to provide adequate cooling while retaining mobility. We have made some progress in the solution of this problem and expect to have an operational Ge(Li) - system in a few months.

THE INFLUENCE OF AFTER EFFECTS ON ANGULAR CORRELATION MEASUREMENTS

J.E. Thun

University of Uppsala, Institute of Physics, Uppsala, Sweden

It is well - known that a radioactive decay may cause severe disturbances of the electron configuration of the decaying nucleus.

The term **after effects** is used to describe this process. The after effects may influence (or perturb) the directional correlation between successive radiations. However there exists a variety of other mechanisms which act on the nucleus to cause a perturbation. This makes in general the detection of the after effects difficult.

After effects, i.e., their influence on the angular correlation, has often been restricted to describe the special case of a time dependent magnetic hyperfine interaction, between the nuclear magnetic moment and the ionized atom created through α, β, electron capture and internal conversion decay.

The review will discuss the concept of after effects, and its relation to the hyperfine interactions in general.

The present knowledge about after effects in angular correlation will be reviewed and some recent results from alpha, electron capture and internal conversion decay, will be presented. Recent calculations will also be discussed, as an example of a more quantitative approach to the physical situation.

The environment of the 'decaying' atoms is known to play a decisive role for the manifestation, of the after effects, where e.g. the supply of conduction electrons in condensed media is known to be important. The equally important question of relaxation mechanisms in different media will be emphasized.

Introduction

Let me first remind you about the conference at Uppsala 1963[1]) when Frauenfelder used the following words to summarize the discussion of after effects held at the occasion. 'I am walking away from this discussion not much less confused than before'.

Frauenfelder and Steffen[2]) (who summarized the Uppsala conference, also mentioning the influence of after effects (AE)) have however helped to clear up the situation regarding AE influence.

Since 1963 also quite a number of experiments have been performed that have increased our knowledge about AE and their influence on angular correlations. I would also like to point at the great interest shown in AE by Mössbauer experimenters. Some of these results have been very useful for the understanding of after effects.

Before we proceed further, let us define what we are talking about.

A nuclear decay will in most cases result in a disturbance of the electron cloud of the decaying nucleus. The atomic processes set forth are named aftereffects.These after effects may of course have a strong influence of the correlation pattern of a cascade decay. The interaction between electrons and its nucleus is named hyperfine interaction. What we shall talk about today is therefore 'the influence on angular correlations of hf - interactions caused by after effects'. I have not found a proper short - hand notation. Let us see how this influence comes about, by looking at a K - capture decay as an example.

The K - electron is removed and through successive Auger effects a highly ionized atom will result. This ion will be neutralized if it is placed in a conducting surrounding. We call this process recovery. If the ion is placed in an insulator no conduction electrons are available but some recovery is still possible. The ion will also interact with the surrounding through relaxation. Relaxation will also take place in the metallic case. If the atom (ion) is free recovery is not possible since no electrons are available. A kind of relaxation may take place, through electronic deexcitation.

An angular correlation of a cascade following the K - capture may be perturbed through a hf - interaction between the ion and the nucleus **provided** the interaction strength and the time of the interaction are large enough and **provided** the relaxation time is large enough. Whether this perturbation should be called time dependent or static depends on the relation between relaxation time and observation time. The observation time is in general the nuclear lifetime.

When considering the possibility of a perturbation caused by the AE a number of conditions must be regarded. In general the ion created in the decay will be paramagnetic and the hf - interaction of magnetic nature. The environment of the ion will now determine, if a perturbation will appear.

As soon as the nucleus is placed in solid environment the interactions between the nuclear electric quadrupole moment and the field gradients of the lattice set in. This will in general make an observation of the influence of AE difficult, sometimes even impossible. In solids the strength of this interaction is determined by the degree of symmetry of the crystal structure. Of course the nuclear life time is of general importance. In liquids the field gradients change direction in a random manner. If the time between these changes — the correlation time — is short enough no perturbation occurs. If the time gets longer the correlation may be attenuated. It can even go below the hard core value. These interactions will in general be present, but in free atoms their influence should be negligible and thus a vacuum environment should be the most suitable for observing AE influence.

In solids (or liquids) the relaxation mechanism will be of decisive importance. This subject has been discussed by several authors; in 1953 by Abragam and Pound[3]) and more recently at the Asilomar conference by Blume[4]). The interaction between the electronic spin and the surroundings (other spins, lattice, etc.) will lead to a motion of the electronic spin which will be observed as a time dependence of the interaction. The time dependence is determined by the relaxation time.

To detect the presence of AE influence on angular correlations different ways have been used. Changes of environment and change of temperature are two frequent methods, where one must keep in mind that the relaxation time is in general temperature dependent. The application of a longitudinal magnetic field should in principle lead to a decoupling of the hf - interaction. For atoms in the solid state we are however no longer dealing with a free magnetic hf - interaction because of the crystalline fields and their interaction with the atomic spin. This interaction is in general anisotropic and is not possible to decouple in the usual way[3]). It will need a very high field to obtain decoupling.

Normally the observation of an attenuation below the hard core is regarded as evidence for AE influence. This should be expressed so that such an attenuation only can be caused by time dependent interactions or perturbations.

Let us now look at the experimental evidence for AE influence.

Experiments on the influence of after effects

K - capture decay

In gamma - gamma correlations following K - capture there will be an extra time lag due to the lifetime of the initial level of the cascade (if this is fed directly through the decay, otherwise lifetimes of higher levels should also be included). During this time recovery and relaxation processes are at work.

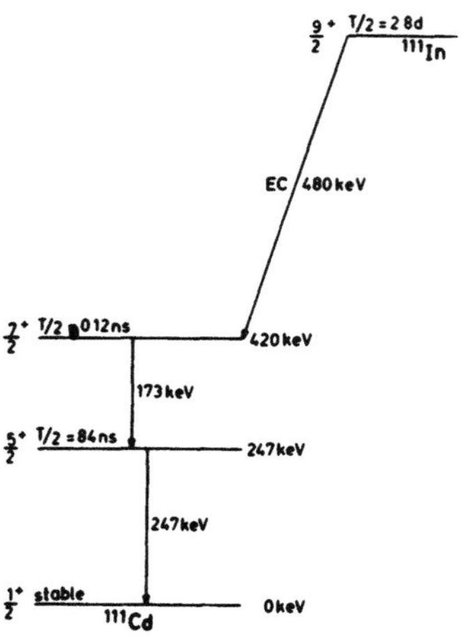

Fig.1.

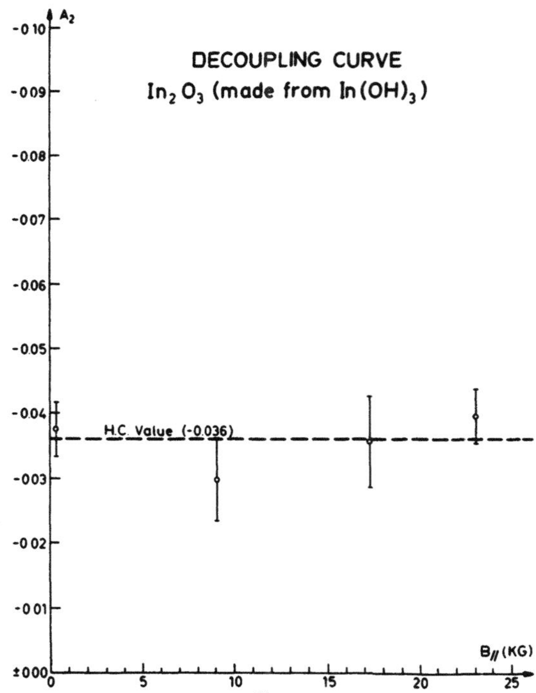

Fig 2

Fig. 1. The decay scheme of ^{111}In.

Fig. 2. The dependence of a longitudinal magnetic field of the 173 - 247 keV correlation in ^{111}Cd. In_2O_3 - source.

The classical nucleus studied in this respect is ^{111}Cd (fig. 1). It has a conveniently simple decay and a conveniently long half life of the intermediate state. The latest investigation of AE influence on the ^{111}Cd AC was performed by our group at Uppsala some years ago[5]). The group consists of Falk, Linnfors and Orre and myself. We studied several compounds of indium to see if we could find any AE - influence. We started out by looking at indium oxide In_2O_3. The correlation is perturbed to the HC value in this compound. We applied a parallel magnetic field to the source; the field was varied from 0 - 23 kilogauss (fig. 2). There is obviously no effect of the field on this correlation.

The same result is obtained if we use another compound, namely indium sulfide (fig. 3). Changing the temperature of the In_2O_3 source did not alter the correlation between 20 and 200° C (fig. 4). The parallel magnetic field was also applied at the higher temperature. No change was detected. The results for a metal - organic compound is interesting. We used In $(C_9H_7NO)_3$ (fig. 5). The figure shows two things: 1) the correlation is below the HC value at zero field, 2) the parallel magnetic field has no influence.

To summarize these results: no aftereffects of the K - capture decay are seen in

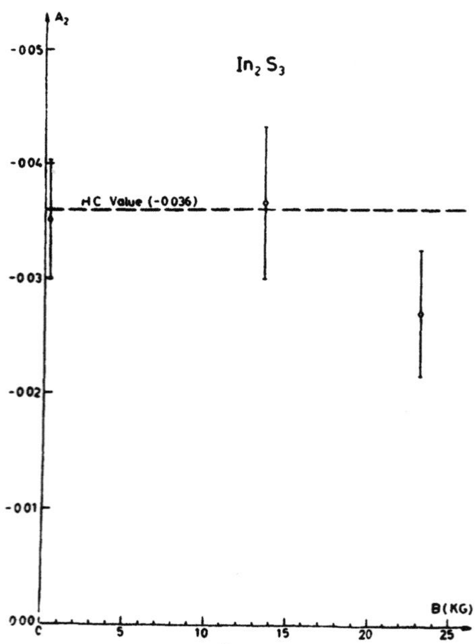

Fig.3.

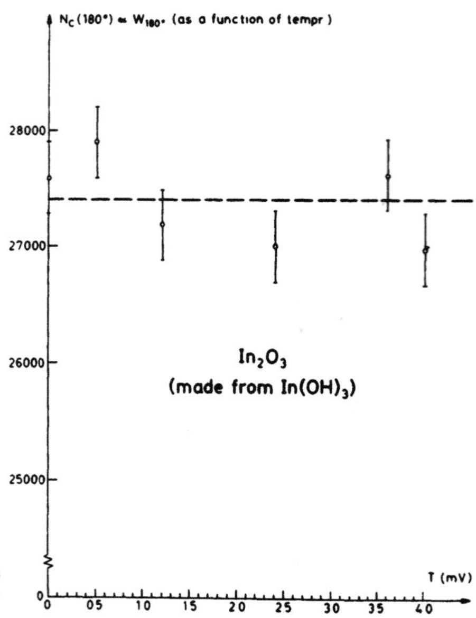

Fig.4.

Fig. 3. Same as fig. 2. In_2S_3 - source.

Fig. 4. Temperature dependence of the coincidence probability in 180°. In_2O_3 - source.

the insulating indium compounds. The result on $In(C_9H_7NO)_3$ shows a time-dependent (electric) interaction. This interaction might be the result of a break-up of the molecule, set forth by the decay. It is interesting to note that our investigation was partly initiated by the results of an earlier investigation[6]) which showed that the time - differential correlation was affected by a change of the temperature. These authors used indium oxide prepared by heating. We did a measurement of the same material at different parallel fields. (Fig. 6). The decoupling behaviour is due to the traces of **indium metal** present in the sample, in which the rather weak static **quadrupole** interaction is partly decoupled.

The experiment of Leisi et al.[7]) shows a dependence of the gamma - gamma correlation in ^{127}I on the pressure of the gaseous source. At very low pressures the correlation is slightly above the minimum hard core value and when the pressure is increased it goes even below this. At higher pressures it approaches the value of A_{22} in solid Al.

Internal conversion

If we now turn to an electron - gamma cascade the situation is somewhat different. The electron is ejected through internal conversion. This process will of course

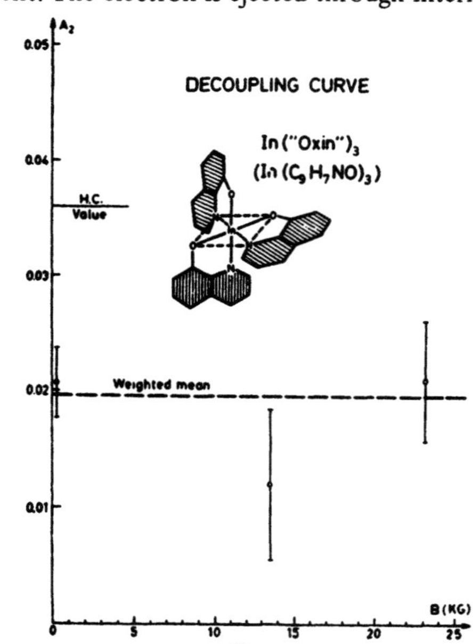

Fig.5.

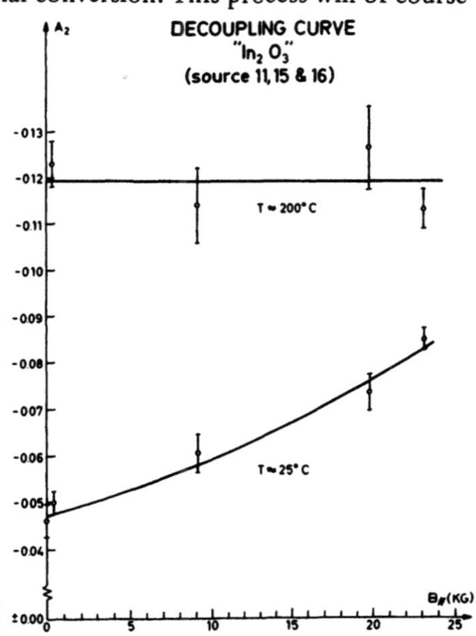

Fig.6

Fig. 5. Same as fig. 2. In $(C_9H_7NO)_3$ - source.

Fig. 6. Same as fig. 2. Source prepared by oxidizing metallic In in air.

lead to the same kind of aftereffect as in K - capture resulting in a highly ionized atom. Even in this case there is an almost classical nucleus – ^{197m}Hg. During the fifties several investigations were carried out. The results were however diverging as to the influence of aftereffects[8,9,10]). The active material was placed on insulating backing materials. In 1961 Pettersson, Gerholm and myself[11,12]) carried out a series of measurements on the same isotope. The correlation was strongly perturbed in insulating media. It was possible to show that – apart from the quadrupole interaction present in this as well as in most metallic environments – there is an extra attenuation present. This assumption was confirmed when a parallel magnetic field of 3 kilogauss was applied. The correlation was then partially restored. A few years ago Bäverstam and Gerholm[13]) repeated the experiment on ^{197m}Hg. The earlier result was confirmed. A time dependence of the attenuation was reported at this conference by Bäverstam. The sources in these investigations were evaporated carrier - free through a special distillation technique on the insulating organic backings. Recently Bäverstam[14]) investigated the electron - gamma correlation of the 121 - 280 keV cascade in ^{75}As. In this case he tried different methods to place the atoms in an insulator. When the active nuclei were shot into an aluminium oxide foil the correlation was reduced to

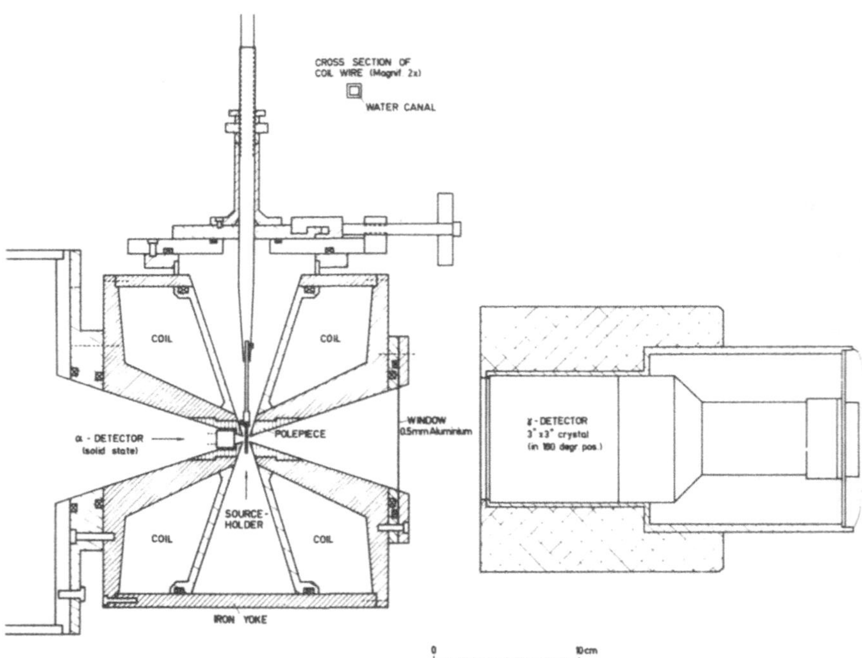

Fig. 7. Drawing of the apparatus used for longitudinal field measurements on alpna - gamma correlations, by the Uppsala group.

85 - 90% of its unperturbed value. This was not so in an aluminium backing. A 3 kilogauss parallel field restored the correlation to its unperturbed value.

In sodium selenide and in selenium chloride attenuations down to $G_{22} \approx 0.6$ were observed. A weak magnetic field did not restore the correlation. The gamma - gamma correlation was the same in aluminium and in aluminium oxide but not in the other compounds. The results in Al_2O_3 point to the same conclusion as for ^{197m}Hg. The perturbation in the other cases can be due to a quadrupole interaction.

Alpha decay

In alpha decay the experimental situation is in many ways different. The reason is that we are here dealing with much greater energies. The recoiling nucleus has an energy of around 100 keV and can be thrown away from the source and backing material. It can travel several hundred atomic layers. The large recoil energy will of course complicate the situation when the recoils enter the backing material. The influence of AE on alpha - gamma correlations is well established. If the recoils are emitted in-

Fig. 8. Photograph of the set - up of fig. 7.

to **vacuum**, severe perturbations are observed in many cases.

Siegbahn and Asaro [15,16]) studied these perturbations in ^{243}Am. The attenuation in vacuum was $G_{22} = 0.11$. This pointed to the presence of a time dependent perturbation. In a parallel magnetic field the perturbation was reduced with the increasing field. In 1965 the investigation was repeated with a field up to 23 kilogauss [17]) (twice the maximum field of Siegbahn and Asaro) by our group at Uppsala. We were then able to obtain the unperturbed correlation. No time dependence of the correlation was found, however. Since that time we have measured other alpha - gamma correlations to study the influence of after effects.

The apparatus [18]) we use is seen in the drawing of fig.7 and on the photograph of fig. 8. It is centered around the magnet, giving a maximum field of 23 kilogauss with the pole tips used for A.E. studies. The principle of the performed correlation measurements is seen in fig.9. ^{249}Cf presents an interesting case [19]). The decay scheme is shown in fig.10. The main cascade is the 5.81 MeV alpha-388 keV gamma. The behavior of this correlation in a parallel magnetic field is seen in fig.11. The most remarkable property is that already at comparably low fields the correlation is increased.

Another nucleus studied is ^{226}Ra [20]). Fig.12. The behavior of the correlation (measured as the coincidence rate at 180 °) is described by fig.13. Again a strong dependence on the field is seen, and as in californium occurring at fairly low field strengths. A perpendicular field also affects the correlation (fig.14).

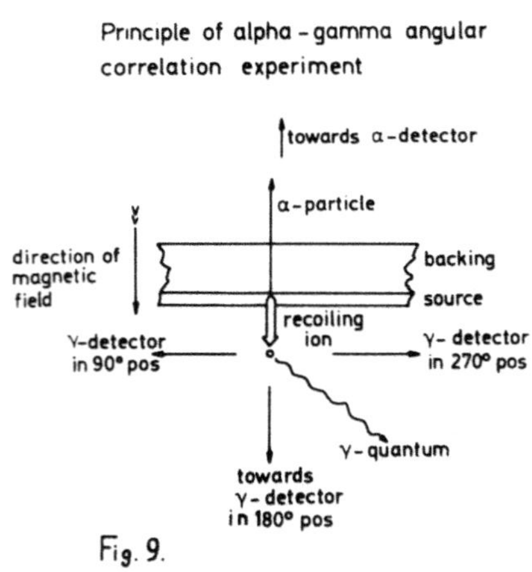

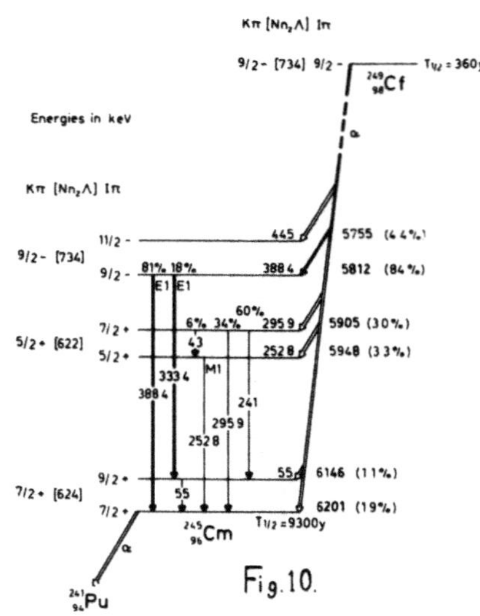

Fig. 9. Principle of an alpha - gamma angular correlation experiment. Recoils going into vacuum.

Fig. 10. The decay scheme of ^{249}Cf.

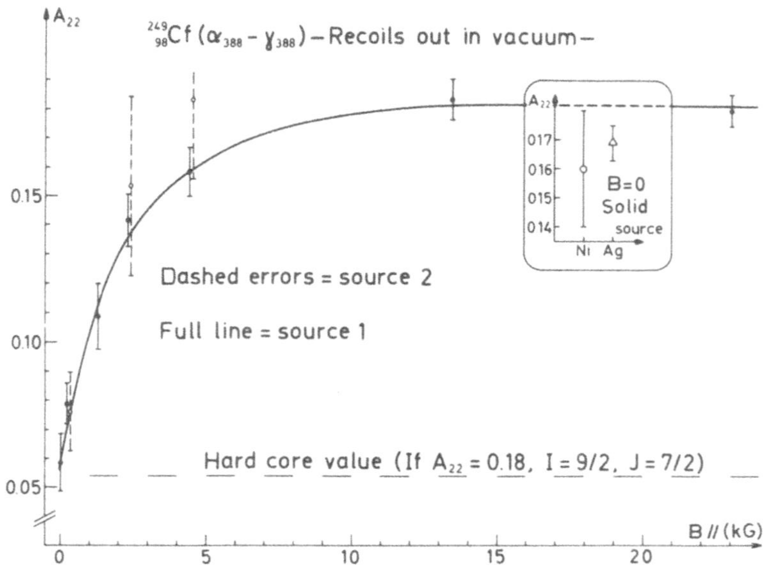

Fig. 11. Experimental A_{22} - value of the $\alpha_{388} - \gamma_{388}$ cascade in ^{249}Cf as a function of applied longitudinal field. Recoils into vacuum. Insert show result for recoils into Ag and Ni.

In all the cases that we measured we found no influence of the magnetic field when we had the recoils going into different insulating media. The correlation was definitely perturbed anyway. The results point at a quadrupole interaction in most cases. The ^{222}Rn recoils into mylar are affected by electric interactions when the recoils damage the complex organic structure. Fig. 15 shows no magnetic field influence.

Beta decay

Concerning beta decay AE and their influence on AC there exists very little evidence. The electron cloud is not so heavily affected in this type of decay as in the other types. Rather few direct investigations on beta decay AE are reported.

Discussion of experimental results and their interpretation

After having summarized some of the relevant experimental results I will now turn to the next part – the interpretation of these results.

I think it is wise to start with a few general remarks, that add to those of the introduction. First of all it is, I think, necessary to see what happens after the decay.

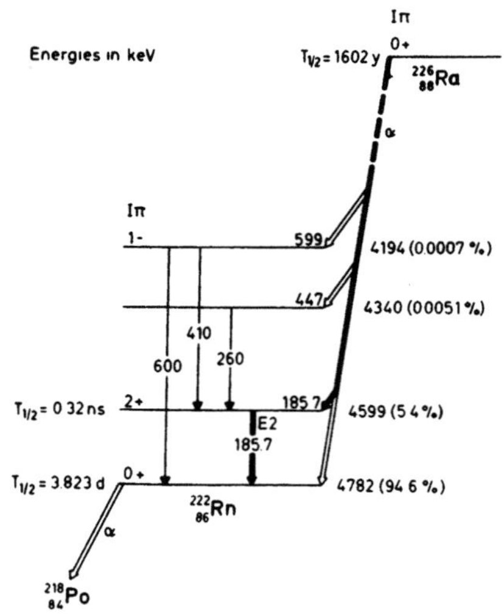

Fig. 12. The decay scheme of ^{226}Ra.

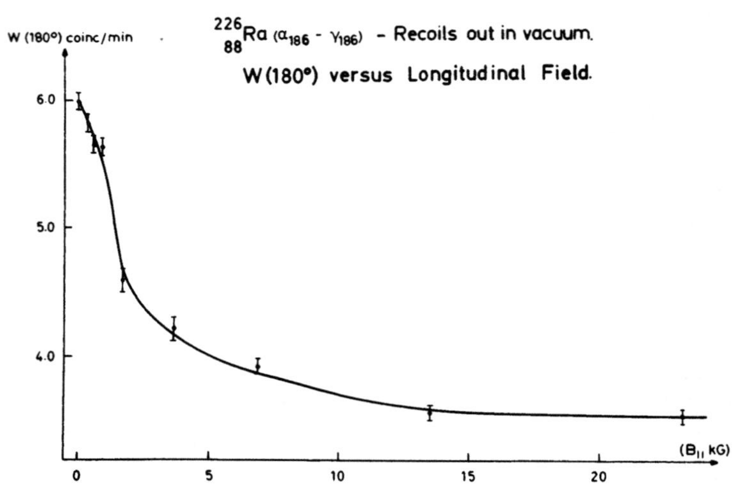

Fig. 13. The coincidence probability $(\alpha_{186} - \gamma_{186})$ in ^{226}Ra (in 180°) as a function of applied longitudinal magnetic field. Recoils into vacuum.

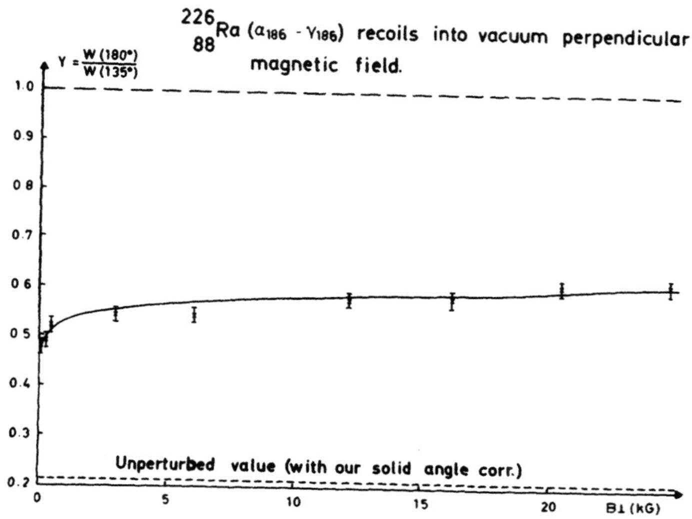

Fig. 14. The ratio of coincidences W (180°) to W (135°) for a perpendicular magnetic field. ^{226}Ra, recoils into vacuum.

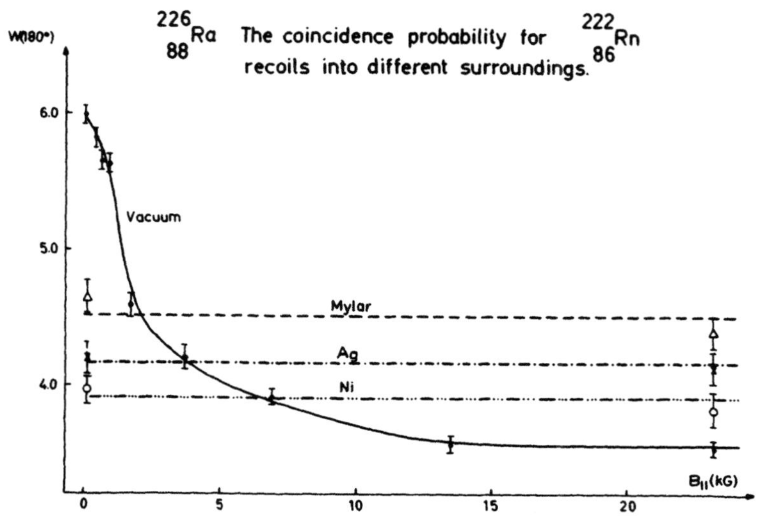

Fig. 15. Same as fig. 13. Recoils into vacuum and different solids.

In K - capture and K - conversion we are left with a hole in the K - shell. These states are very unstable. The estimated lifetime of a 1s state is of the order of 10^{-16} sec[21]). Even if the hyperfine field created by a 1s - electron is tremendous the short lifetime makes an influence of this field unlikely. In fact no such effects have been observed and one might add that this is a happy situation for the electron correlation people. When talking about AE we should therefore focus our attention on the outer electron shells. The lifetimes of these shells are determined by several factors. We will return to this later.

Another factor which is of interest in this context is the charge of the ions created in the decay. This is of interest mainly when these ions are not recovered through electron addition as for free atoms. After K - capture the average charge of the atom is considerable[2]). It can amount to 10 e in a normal case. Certain ions of course have still higher charges. In alpha decay a very small percentage with charges up to 8 e have been reported, but one has reason to believe that charges of 1 and 2 e are predominant (ref. 22 - 25). Around 25% of the recoils is supposed to be neutral. For free atoms these states are not likely to become neutralized during the nuclear lifetime.

One should keep in mind however that regardless of the charge state we are dealing with a number of different excited states of the atom.

K - Capture

The result of Leisi[7]) at low pressures points to an AE of the Xe EC decay. A magnetic hf - interaction – of time dependent nature – is at work. The relaxation of the hf - field is obviously the reason for the variation of the correlation with pressure at higher pressure. Filling of vacancies might also occur, and will depend on the relative probability of charge transfer in a collision compared to spin transfer.

The experiments on gamma - gamma correlations in different insulators show no evidence that AE in these cases cause any influence.

Internal conversion

No measurements on free atoms arc reported. The results from Uppsala 1961 and Stockholm 1967 on Hg is evidence for an AE influence on angular correlations where the activity is placed on insulator backings. The source conditions seem to be of decisive importance however. This is confirmed by the recent results from Stockholm on arsenic where the stochiometric compounds of selenium show no decoupling effect. It might be so that for Hg when only active atoms are deposited on the insulating backing these atoms are weakly coupled to the plastic backing material. The relaxation time might be fairly long in this case thereby allowing the hf - interaction to work for a reasonable period.

The A_{22} value without field is rather high. If we suppose that the relaxation time is determined by the strength of the coupling of J to the surrounding one might need a field that is strong enough to decouple this latter interaction. The situation is somewhat similar to that of the well - known Stiening and Deutsch experiment[26]) on magnetic hf - interaction in Gd^{3+} ions in an aqueous solution. There was a decoupling effect, but decoupling occurred at 3000 gauss which is more than expected for a free hf interaction. The relaxation time is in this case supposed to be unusually long which is the reason for the perturbation. On the other hand the Gd ions are not free but coupled to the surrounding which explains the fairly high field needed. Concerning the mercury results one is of course not able to state clearly what kind of surrounding the active nuclei have. Since a quadrupole interaction is said to be present the nuclei can hardly be regarded as free, lying on the insulating surface. They should therefore have diffused into the backing. The arsenic case indicates that the interstitial As ions interact rather weakly with the aluminium oxide host lattice. On the other hand the perturbation is fairly weak in this case, maybe because of the short life time.

Alpha decay

The results on alpha decay with recoils into vacuum present the most obvious examples of the influence of AE on AC.

The recoiling daughter ions do not recover during the normal lifetime of the intermediate cascade state. The relaxation time of the electronic state is normally also larger than the nuclear time. The hf - interaction may therefore be regarded as nearly static.

The I - J coupling is completely removed by the application of a fairly strong parallel magnetic field. The field at which decoupling is obtained is different for the three cases measured. A measure of this can be the field at which half decoupling is

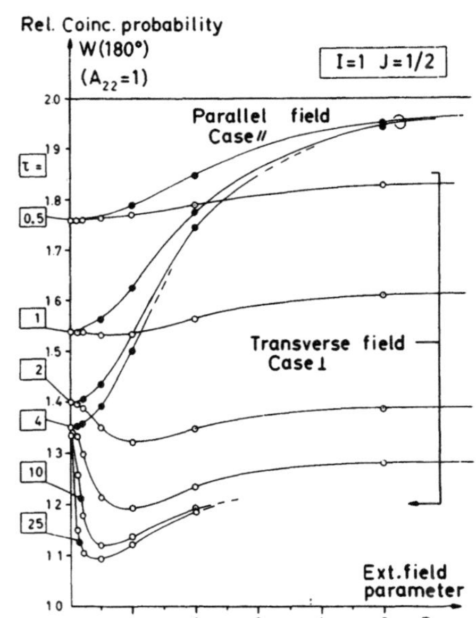

Fig. 16. Dependence of an external magnetic field. Detectors at 180°. $I = 1$, $J = 1/2$, $A_{22} = 1$. The parameter is the mean lifetime of the intermediate nuclear state (τ).

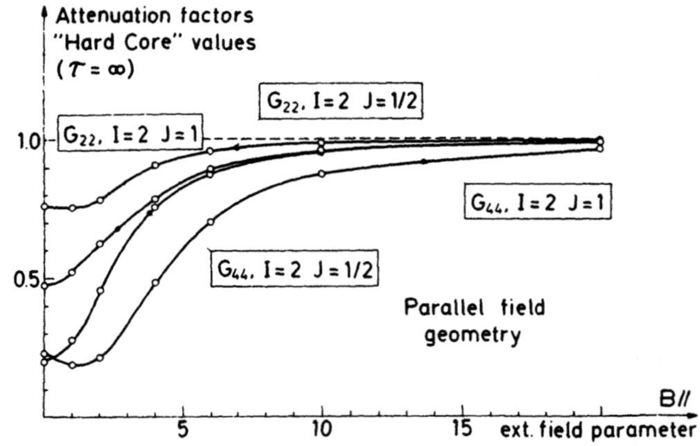

Fig. 17. Integral attenuation factors as a function of external longitudinal field.

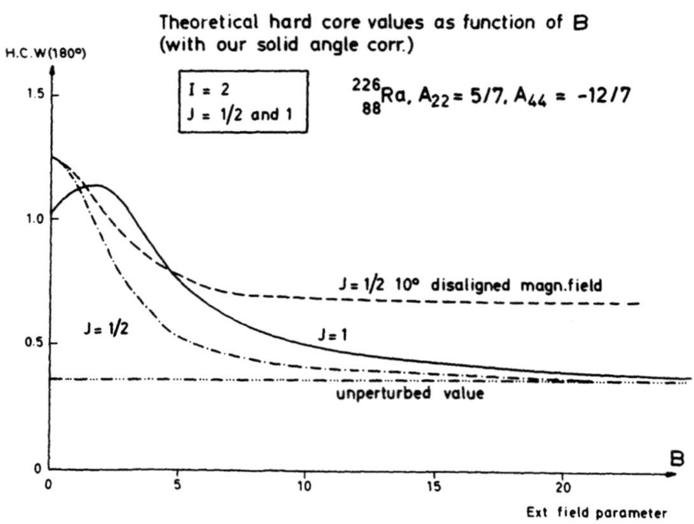

Fig. 18. Theoretical hard core values as function of a longitudinal magnetic field. Calculations applied to the case of ^{226}Ra (I = 2) and for J = 1/2 and 1.

attained. For ^{243}Am this is higher than in Cf and Ra. For these cascades half decoupling is reached at about 1.5 kilogauss.

It is illustrative to discuss a little more closely the cases of ^{249}Cf and ^{226}Ra. The I - J coupling in ^{249}Cf is very weak. If one assumes that most of the recoils consist of Cm^{1+} ions in their electronic ground state an unusually low internal field is expected because of the half - filled 5f - shell. According to Hund's rules the seven 5f - electrons form a $L = 0$, $S = 7/2$ term. This should give no hf - interaction because of the spherical symmetry. Because of admixtures and polarization effects there will nevertheless be a hf field. Furthermore Am^0 has small internal interactions. Cm^0 has one 6d - electron in addition to the seven 5f - electrons. This gives rise to rather weak hf fields.

In this context I would like to mention a few words about the calculations we have performed at Uppsala[27,28]). These calculations refer to a static magnetic hf interaction and are performed for a number of cases, both for longitudinal and transverse fields. Nuclear spins from 1 to 5/2 and atomic spins J less than 3 are considered. For the Cf case we did not try to fit the calculations to the experimental result. It is nevertheless interesting to have a look at some of the results at this point. Fig. 16 shows how the coincidence probability for detectors placed opposite to each other is affected by a magnetic field. The parameter is the mean life time. Fig. 17 shows the field dependence of the hard core values.

For ^{226}Ra the spin I of the first excited Rn state is 2. The experimental result for a parallel magnetic field agrees with the calculation for $J = 1/2$ at small fields but does not go down enough even when it is corrected for a disalignment of the field. Fig. 13 and 18.

The results on alpha - gamma correlations with recoils into vacuum show clear evidence for the influence of AE. They can furthermore be fairly well described quantitatively by a static hyperfine interaction of magnetic nature. The results with accelerated particles show also that the magnetic interaction dominates. High energy recoils into vacuum or into gasses (see e.g. ref. 29) give a G_2/G_4 ratio which is consistent with that of a magnetic hf - interaction. Here the interaction is time dependent, however.

I would also like to discuss the results on recoils into different solids. These results are of course not so simple to interpret in terms of AE. They should give some aspects on relaxation and on the origin of the interactions. The velocity of the recoil is rather high because of its large energy (around 100 keV). (This energy is of course much smaller than that of the ions implanted by accelerators). The heavy ion will cause severe damage of the host lattice. Fig. 19 and 20 indicate the situation after an implantation. The transient magnetic fields which occur during the stopping in ferromagnets are very large but are also short lived. They will certainly not influence the correlation.

In ^{249}Cf[19]) we used one insulating medium — ice. The correlation was severely perturbed but not below hard core. It was not changed by a 23 kilogauss longitudinal field. In silver and nickel very small perturbations occurred. The half - filled 5f - shell is probably the explanation of this result. Compare also the reversal of the hf - field in Fe and Ni for 4f - elements when going from an element with less than seven 4f electrons to one with more than seven.

In ^{226}Ra[20]) a number of solids were used for implantation. Perturbations up to 30% were observed for recoils into mylar. No field effect was registered, however.

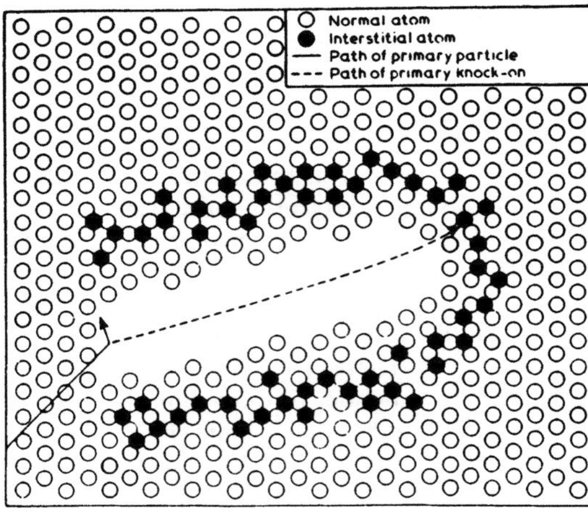

Fig. 19. Schematic representation of interstitial atoms around a multiple vacancy during production of a Brinkman displacement spike.

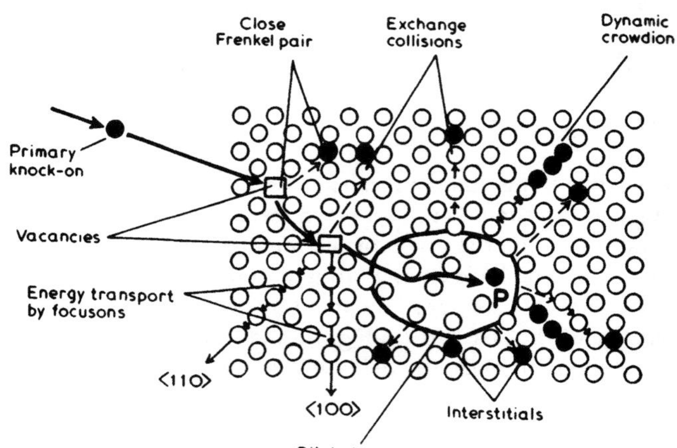

Fig. 20. Schematic two - dimensional representation of the Seeger depleted zone and its environment in a face - centered cubic crystal. A primary knock - on impinges from the left and comes to rest on the site P.

The interaction can in this case be explained as being of electric nature arising from the large field gradients created when the recoils damage the complex organic structure. It is interesting to note that by studying the G_2/G_4 ratio Ansaldo et al.[30]) found that Rn, Ra and Th recoils from alpha - decay show an electric quadrupole interaction when injected into copper. Our results on ^{241}Am[31]) indicate a time - dependent interaction. The temperature dependence also points to a spin lattice relaxation.

Final Remarks

I will now try to summarize the present situation described by the results presented and their interpretation. I might add here that I have not mentioned all experiments which have been made on the influence of AE on AC. Partly it is intentional, partly a matter of time and space. I hope the second category authors will accept my apologies. It is also easier to speak about ones own results.

First of all I think we should now be prepared to agree that after effects may influence angular correlation measurements in a significant way. Secondly we should also agree that the AE influence is not that frequent as perhaps was thought in the early days of angular correlations. Thirdly AE are of several kinds and their influence is not restricted to 'time - dependent magnetic hf - interaction'.

In alpha decay with recoils into vacuum there is clear cut evidence for a magnetic interaction, which is fairly well described as a static atomic state interacting with the nuclear spin. The K - capture decay in near vacuum gives rise to a free magnetic hf - interaction. In gases with some pressure collision times get shorter and relaxation phenomena set in. At high pressures the relaxation (and perhaps recovery) times are so short that the perturbation is very small.

When relaxation times are of the order of nuclear life times the most complicated situations arise and generally the most severe perturbations, if the life time is long enough. Relaxation times in solids are in general very short (10^{-12} sec) and this will prevent the development of a perturbation. For S - states this time will be longer, however – around 10^{-10} secs.

These short relaxation times are one of the reasons why AE do not affect angular correlations more seriously. The lifetimes of the excited electronic states in insulators are less than 10^{-9} secs but should lie between 10^{-10} and 10^{-11} secs. The electron - gamma results on solids show how sensitive the AE are to the environment of the ion. A suitable combination of experimental conditions might give influences of AE which can be removed by a longitudinal field. I would like to end by saying that the field of aftereffects is large, but by restricting it to those cases where they clearly influence the AC, this field certainly becomes smaller. The question if there is an aftereffect which causes a perturbation of an angular correlation should be put in each case since it cannot be answered generally. We know that there exist cases – alpha decay with recoils into vacuum – where we can predict the existence of an influence, but most other cases must be investigated individually before we can ascertain any influence. On the other hand, we also know many situations in which aftereffects do not exist or in which recovery and/or relaxation is so fast that they do not act for any time of significance.

References

1. Perturbed Angular Correlations (ed. E. Karlsson, E. Matthias and K. Siegbahn) North - Holland Publ. Co. Amsterdam 1964, p. 329.
2. H. Frauenfelder and R.M. Steffen, in Alpha-, Beta- and Gamma-Ray Spectroscopy (ed. K. Siegbahn), p. 1182, North - Holland Publ. Co. Amsterdam 1965.
3. A. Abragam and R.J. Pound, Phys. Rev. 92 (1953) 943.
4. M. Blume, in Hyperfine Structure and Nuclear Radiation, p. 911 (ed. E. Matthias), North - Holland Publ. Co. Amsterdam 1968.
5. F. Falk, A. Linnfors and J.E. Thun, UUIP - 573 (1968), Nucl. Phys. (to be published).
6. M. Salomon, Nuclear Physics 54 (1964) 171.
7. F.N. Gygax, J.Egger and H.J. Leisi, in Hyperfine Structure and Nuclear Radiation (ed. E. Matthias), p. 948, North - Holland Publ. Co. Amsterdam 1968.
8. H.H. Coburn, J.V. Kane and S. Frankel, Phys. Rev. 105 (1957) 1293.
9. F. Gimmi, E. Heer and P. Scherrer, Helv. Phys. Acta 29 (1956) 147.
10. E. Breitenberger, Proc. Phys. Soc. 69 (1956) 566.
11. B.G. Pettersson, J.E. Thun and T.R. Gerholm, Nucl. Phys. 24 (1961) 223.
12. B.G. Pettersson, T.R. Gerholm, J.E. Thun and K. Siegbahn, Phys. Rev. Letters 6 (1961) 14.
13. U. Bäverstam, A. Johansson and T.R. Gerholm, Arkiv Fysik 35 (1967) 451.
14. U. Bäverstam, Private communication.
15. K. Siegbahn and F. Asaro, Phys. Letters 2 (1962) 323.
16. F. Asaro and K. Siegbahn, Ref. 1, p. 233.
17. F. Falk, S. Törnkvist, J.E. Thun, H. Snellman, K. Siegbahn and F. Asaro, Zeitschr. f. Phys. 198 (1967) 106.
18. F. Falk, A. Linnfors, B. Orre and J.E. Thun, Nucl. Instr. and Methods, 83 (1970) 261.
19. F. Falk, A. Linnfors, B. Orre and J.E. Thun, Physica Scripta 1 (1970) 13.
20. B. Orre, A. Linnfors, F. Falk, J.E. Thun and L. Johansson, Nucl. Phys. A148 (1970) 516.
21. G. Günther and I. Lindgren, Ref. 1, p. 357.
22. G. Günther, F. Asaro and A.C. Holmholz, Phys. Rev. Lett. 16 (1966) 362.
23. W. de Wieclawik and N. Perrin, I de Physique Suppl. Cl 29 (1968) 104.
24. N. Perrin and W. de Wieclawik, Bull. Acad. Sci. USSR, 31 (1967) 86.
25. W. de Wieclawik, C.R. Acad. Sci, Paris 266 (1968) 577.
26. R. Stiening and M. Deutsch, Phys. Rev. 121 (1961) 1484.
27. F. Falk, A. Linnfors, B. Orre and J.E. Thun, Nucl. Instr. and Meth. 80 (1970) 129.
28. F. Falk, A. Linnfors, B. Orre and J.E. Thun, UUIP - 690 (1970).
29. I. Ben Zvi, P. Gilad, M. Goldberg, G. Goldring, A. Schwarzschild, A. Sprinzak and Z. Vager, Nucl. Phys. A121 (1968) 592.
30. E.J. Ansaldo, Private communication.
31. B. Orre, F. Falk, A. Linnfors and J.E. Thun (this conference). (To be published).

Discussion

Gerholm: I would like to comment on the relation between the recovery time constant and the strength of the hyperfine interaction caused by the aftereffects of internal conversion. On the basis of our experiments on ^{197}Hg in 1961 we concluded that the recovery process must be quite short indeed; less than 10^{-9} secs.

Now the main point in the study of ^{75}As recently made by Bäverstam et al. in Stockholm is that there the lifetime of the intermediate level is nearly two orders of magnitude shorter than in ^{197}Hg. It is in fact about 0.3×10^{-10} secs. Yet the aftereffects do show up also in ^{75}As when they are expected to show up, i.e. in insulating source environments. And this hyperfine coupling can be decoupled by a magnetic field of the order of a few thousand gauss. The attenuation is smaller in ^{75}As than it is in ^{197}Hg. But not very much smaller. About 15% rather than the 25 - 30% observed for ^{197}Hg. This seems to confirm our 1961 conclusion that the recovery time is quite short, and puts an upper limit to the recovery time constant. On the other hand it is also possible to place a lower limit. In the decay of ^{197}Hg there is a competing feeding to ^{197}Au where there is a cascade suitable for electron - gamma correlations. But in this case the lifetime of the intermediate level is much shorter or about 10^{-11} secs. Thus we have the possibility to compare, under identical source conditions, aftereffects attenuations for two widely different nuclear lifetimes. It turns out that the ^{197}Au cascade is unperturbed even in those cases where attenuations are observed in the ^{197}Hg cascade. This means that the ^{197}Au lifetime is too short to allow for any appreciable attenuation. Thus 10^{-11} secs. seems to represent a lower limit for the recovery time constant. The recovery therefore seems to take place at a rate of about 10^{-11} to 10^{-10} seconds. In order to give use to the observed attenuations the strength of the coupling must exceed certain limits. If one makes the simplifying assumption that the hyperfine coupling is purely magnetic, the effective magnetic field, as experienced by the nucleus during the recovery, appears to be of the order of 10 MG. This seems quite reasonable in view of what is known from Mössbauer studies of paramagnetic ions and similar experiments.

The fact that the aftereffects take place in such a short period of time may explain why they are so hard to detect in gamma - gamma correlations and Mössbauer experiments. I would like to add that all of what I have said applies to sources at room temperature. It is conceivable, and in fact quite probable, that the recovery times are temperature dependent and are longer the lower the temperature becomes. It would be interesting to investigate the aftereffects at very low temperatures. Experiments along these lines are now being planned at our institute.

Rasera: I would like, in this connection, to mention some preliminary results of an experiment we have just completed in which we looked for aftereffects following the β^- - decay of ^{160}Tb. We put the ^{160}Tb in the cubic compound $Eu_xTb_{1-x}S$, where by varying the proportion of terbium, one can vary the conduction - electron concentration in the lattice over eight orders of magnitude.

We observed the integral rotation of succeeding gamma - gamma cascades, with a view to observing the change in the atomic shells as reflected in the paramagnetic correction factor between our applied magnetic field and that seen at the nucleus. We also did the same experiment on the most tightly bound ionic material we could find, namely TbF_3. Our preliminary data shows no effects whatever that can be attributed to aftereffects, over a very wide range of conduction - electron densities.

Leisi: I would like to mention one aspect of aftereffects which has not been discussed. The very strong hyperfine fields present in free atoms after electron capture may offer the possibility to study electromagnetic moments of short - lived nuclear states. More information on hyperfine interactions induced by electron capture may be found in recent publications (H.J. Leisi, Phys. Rev. 1A (1970) 1654 and following paper).

Contributed Paper

A TIME - DIFFERENTIAL STUDY OF AFTER-EFFECTS IN THE DECAY OF ^{197m}Hg

U. Bäverstam and R. Othaz*

University of Stockholm, Institute of Physics, Stockholm, Sweden

Present knowledge of after - effects of hole-formation as a consequence of alpha- or beta - decay, a K - capture or a conversion - process has been discussed by Thun at this conference. In this paper we will therefore omit all theory.

Earlier experiments have shown the after-effects to exist, and it has also been possible to put some limits to the strength of the interactions involved, and to get some indirect information concerning the time constant of the interaction. This time constant has however never been directly measured.

We have in the present experiment measured this time constant in the most direct way, viz. by means of a time - differential angualr correlation measurement, using a TAC, MCA, etc. We have measured electron - gamma cascades by means of a magnetic spectrometer (Gerholm type) and NaI - crystal. Our time resolution (FWHM of the prompt time spectrum) was of the order of 2 ns.

Experiment

We have chosen to investigate the isotope ^{197m}Hg for the following reasons. After-effects have earlier been found in this isotope[1,2]). The cascade investigated, 165L - 134 gamma, has a fairly large anisotropy (A_2 a little larger than 20%). The intermediate level has a half - life of 7 ns, which permits time - differential measurements.

We have performed AC measurements with the active mercury deposited onto different backing materials; mica, which is a good insulator, and aluminum. The source was prepared by means of a special evaporation technique, described in reference 2. (The source preparation is of great importance, since no trace of metal should be found on the mica surface).

Four different angles have been used, and since the A_4 - term is known to be very small, we have in this way a good control of systematic errors. Several measurements on metallic and insulating backings have been performed, all with a total number of coincidences of about $5 \cdot 10^5$.

Result and calculations

For metallic backings the A_2 (t) slowly decreases by about 50% in three half -

lives (21 ns). This can well be accounted for by an electrostatic perturbation; the integral perturbation factor G_2, which is about 0.7, also agrees with such an interaction. No well - defined value can be given because of the relatively short half - life of the nuclear level.

For insulating backings the picture is quite different. Here the A_2 (t) decreases rapidly down to a value less than 50% of the original, and is then almost constant. In order to account for the shape and finite width of the prompt time curve, we have performed a numerical convolution of the prompt curve and a theoretical one, adjusting three different parameters for a best least squares fit to the experimental results. (In principle the convolution could have been performed by means of a numerical Fourier - analysis. In that case no a priori analytical shape would have to be given to the theoretical curve). Several different analytical expressions were tried, and it was found that the shape

$$A_2(t) = C_1.\exp(-t/T) + C_2$$

gave the best result. (More parameters would of course give a better fit). Our result from the calculation is:

$$A_2(0) = C_1 + C_2 = 0.205(10)$$

$$C_2 = 0.070(5)$$

$$T = 1.2(2) \text{ ns}$$

It should be pointed out, that no explanation seems possible on basis of static quadrupole interactions. (A comparison of the time integral G_2 and the quick slope of A_2 shows this).

Thus our results show that the after-effect of hole-formation in a case like ours is a fast and strong effect, capable of severely perturbing an angular correlation. Our result for A_2 seems to be the first published one.

* On leave from University of LaPlata, LaPlata, Rep. of Argentine.

References

1. B.G. Pettersson, J.E. Thun and T.R. Gerholm, Nucl. Phys. 24 (1961) 223.
2. U. Bäverstam, A. Johansson and T.R. Gerholm, Arkiv Fysik 35 (1967) 451.

Contributed Paper

EFFECTS OF ELECTRON SHELL RECOVERY AND RELAXATION MEASURED BY PAC WITH SHORT - LIVED NUCLEAR STATES OF THE RARE EARTH ISOTOPES

R. Wäppling, E. Karlsson, G. Carlsson and M.M. Bajaj*
University of Uppsala, Institute of Physics, Uppsala, Sweden

Introduction

In the investigations of after-effects in a radioactive decay, the study of hyperfine interactions usually is the only way to obtain information on the electron configuration. It is the purpose of this contribution to show that the time - integral method for observation of the perturbation of a γ - γ angular correlation can give important information on the electron shell recovery. The measured precession is for short observation times insensitive to quadrupole interactions and the only parameter of interest is $\omega_B \tau_N$ where ω_B is the Larmor precession angular frequency and τ_N is the observation time, i.e. the nuclear level life - time. For instance, in some of the rare earth elements strong magnetic hyperfine fields can be obtained which make studies in the 10^{-12} s region possible; we will present results for Dy - ions in $TbAl_2$. The measurements on Er^{3+} ions in $HoAl_2$ by the present authors have shown that crystal field effects are of no importance in this type of compounds.

After-effects and relaxation

In a radioactive decay (e.g. β - decay) there is a sudden change in the potential energy for the electrons of the ion under study. For the 4f - shells this leads to a small change of binding energy and a correspondingly larger change for inner shells. It is known for free ions that the ensuing shake - off and Auger processes give rise to a highly ionized daughter atom state. The refilling of inner shells is supposed to take place very fast, in less than 10^{-15} s, and recovery of the outer shells and the formation of the stable daughter atom state depend very much on the supply of electrons from the vicinity. The excess energy can be taken up by lattice vibrations which propagate through the crystal at the approximate speed of one atomic distance in 10^{-13} s.

The physical situations in PAC are classified as static or time - dependent cases. In the purely static case each nucleus experiences the full hyperfine field over the whole observation period, i.e. $\tau_{rel} \gg \tau_N$, where τ_{rel} is the relaxation time of the source of the field: in this case the 4f - electron state. In the opposite extreme, when $\tau_{rel} \ll \tau_N$, each nucleus reacts to a series of hyperfine fields with the full strength but varying in direction at the rate τ_{rel}. Then for the para- and ferromagnetic case the angular correlation pattern shows no net precession but will be attenuated according to

$$G_k = [1 + \tfrac{2}{3}\, \tau_{rel} \tau_N\, \omega_B^2\, I(I+1)(1-(2I+1)W(I\,1\,k\,I;II))]^{-1}$$

in the absence of external polarizing fields. In the presence of such a field the paramagnetic moments become partly polarized and there is a net component of the hyperfine field B_{hf} along the external field direction. Under the condition $\tau_{rel} \ll \tau_N$ relaxation is fast enough for each nucleus to feel the thermal equilibrium value of B_{hf} as determined by the Boltzmann distribution over electronic m_J - states.

Experimental results

The nucleus ^{160}Dy has two excited states suitable for perturbed angular correlation measurements namely the 87 keV and 966 keV states with lifetimes of 2.9 ns and 3.1 ps respectively.

The anisotropy of the angular correlation for the 1178 - 87 keV cascade was measured at six different temperatures between 4.2° K and room temperature. Dividing the full anisotropy by the values obtained (corrected for solid angle effects) gives the attenuation factor G_2 at the different temperatures from which the relaxation times were calculated using the formula given earlier. The results are summarized in fig. 1.

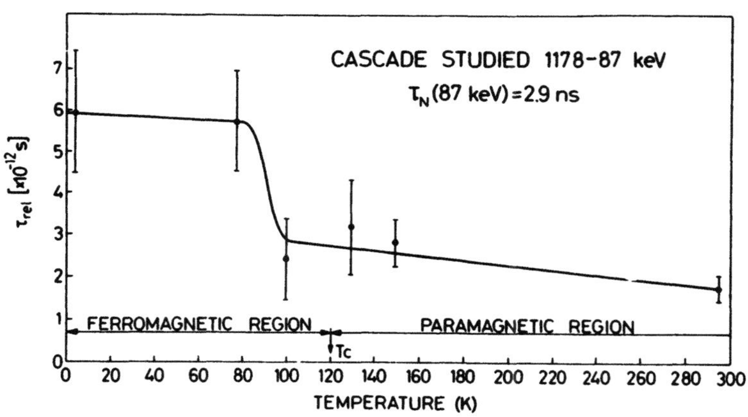

Fig. 1. Spin relaxation time as function of temperature.

The angular shift for a polarized sample was measured for the 299 - 966 keV cascade at seven temperatures between 1.5° K and 130° K and from the results the quantity $\omega_B \tau_N$ was calculated for each temperature. The result is summarized in fig. 2. The effective hyperfine field at the ^{160}Dy nucleus is proportional to $\omega_B \tau_N$ and can be calculated provided the lifetime and g-factor of the nuclear state are given and the effect of relaxation is known. If the relaxation time is constant the variation of $\omega_B \tau_N$ as function of temperature displays the variation of the effective hyperfine field.

It is evident from fig. 2 that the measured values of the precession are lower than the theoretical values.

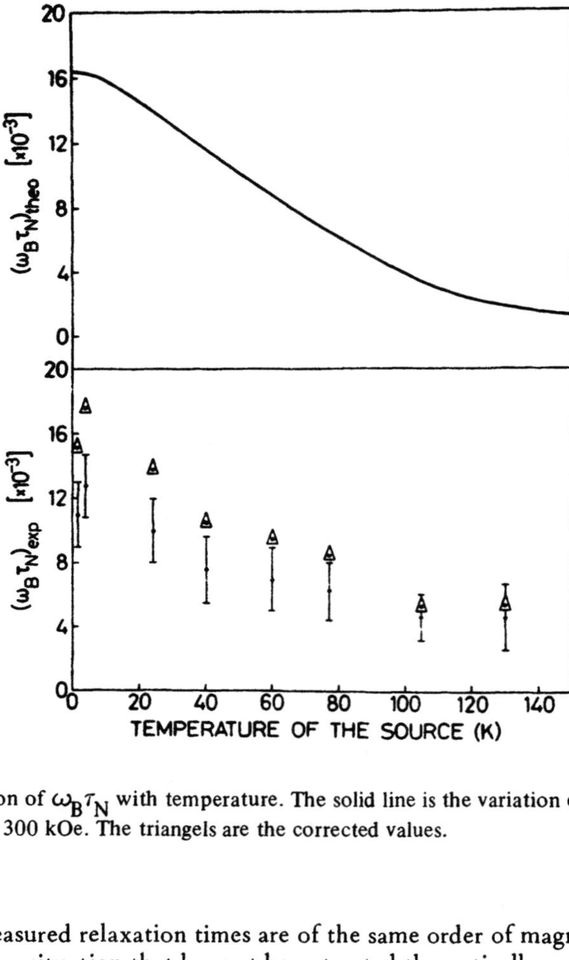

Fig. 2. Variation of $\omega_B\tau_N$ with temperature. The solid line is the variation expected for an exchange field of 300 kOe. The triangels are the corrected values.

The measured relaxation times are of the same order of magnitude as the nuclear level lifetime, a situation that has not been treated theoretically earlier and we will therefore give a simple model for the effect on a PAC measurement. Assuming that the approach to equilibrium in the magnetic field takes place from the completely disordered state, we obtain the effective population W_{m_J} of the magnetic sublevel m_J by folding the thermal equilibrium value $\overline{W}_{m_J}$ with the nuclear decay curve and the curve describing the approach to equilibrium, giving

$$W_{m_J} = \int_0^\infty \overline{W}_{m_J} (1 - e^{-t/\tau_{rel}})\, e^{-t/\tau_N}\, dt = \overline{W}_{m_J} \frac{\tau_N}{\tau_N + \tau_{rel}}$$

Using the approximation of a well defined spin temperature, the same relation is then obtained between the observed precession frequency and the thermal equilibrium value of the precession frequency. If the approach to thermal equilibrium is characterized by the same relaxation time as measured by the attenuation coefficients G_k it is possible from PAC - measurements to obtain the temperature dependence of the hyperfine field even in the presence of relaxation.

Using this simple theory the experimental values can be corrected as is indicated in fig. 2 and we can conclude that the observed deviations from the theoretical precessions can be satisfactorily explained through relaxation effects and that no after-effects are present. This is not very surprising since the compound used is a good conductor with presumably very short recovery times. Further work on insulators are in progress.

* On leave from University of Delhi, India.

PROBLEMS CONNECTED WITH THE THEORY AND MEASUREMENT OF α - γ - ANGULAR CORRELATION

H.J. Mang

Technische Universität, Physik - Department, Munich, W. Germany

This talk will consist of two parts. In a first part I shall discuss the information on nuclear properties which can be obtained from angular correlation measurements. The second part will deal with a few selected experiments. The first figure explains the notation

$$ {}^{Z}_{N}A \longrightarrow {}^{Z-2}_{N-2}A{-}4 + \alpha + \gamma $$

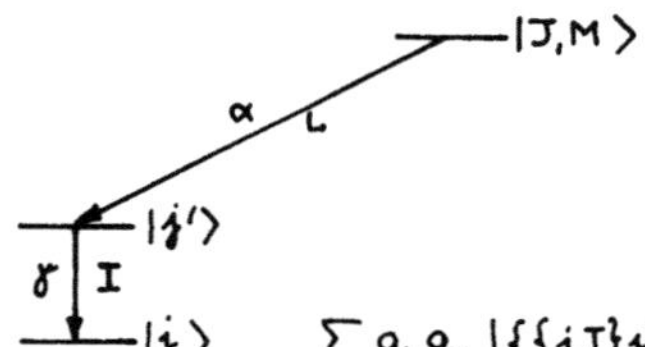

$$ \sum_{LI} a_L g_I \,|\{\{jI\}j'L\}J,M\rangle $$

$$ W(\vec{k}_\alpha, \vec{k}_\gamma, \vec{\xi}) = \sum_{M=-I}^{I} P_M \Big| \sum_{LI} a_L g_I \langle \{\{jI\}j'L\}J,M \,|\, \vec{k}_\alpha, \vec{k}_\gamma, \vec{\xi} \rangle \Big|^2 $$

Fig. 1.

J, L, j′, I, j denote the various angular momenta involved in the process. The short-hand notation $\{j,I\}j'$ means that the angular momenta j and I are coupled to give the resulting angular momentum j′. In writing the correlation $W(\underline{k}_\alpha, \underline{k}_\gamma, \underline{\xi})$ it has only been assumed that there is no perturbation acting on the intermediate state j′ [1]).
The symbols P_M, a_L, g_I, $\underline{k}_\alpha$, $\underline{k}_\gamma$, have the following meaning:

p_M is the occupation probability of magnetic substates in the parent nucleus.
a_L is the amplitude of the α - partial wave of angular momentum L.
g_I is the amplitude of the multipole radiaction.
$\underline{k}_\alpha$ is the wave vector of the α - particle.
$\underline{k}_\gamma$ is the wave vector of the γ - quantum.
$\underline{\xi}$ describes the polarisation of the γ - quantum.

The information on nuclear properties which can be obtained from a measurement of unperturbed α - γ angular correlations is contained in the amplitudes a_L and g_I. These are the quantities which must be extracted from the experimental data. At this point it should be emphasized that in general several partial waves of the emitted α - particle (amplitudes a_L) will contribute; indeed one has to expect all allowed partial waves ($|J - j'| \leqslant L\, |J + j'|$) to contribute significantly in most cases. This can make the analysis rather cumbersome because the correlation W depends on several continuous complex variables a_L. The situation is less complex for the γ - transition. Here only one or at most two multipoles (amplitudes g_I) will be important. Of course one can in principle also deduce the spins from the measured correlation. Clearly this complicates the analysis even further.

Now I shall concentrate on the quantities related to the α - particle, even though in some cases the multipole amplitudes g_I must also be deduced from the angular correlation. The amplitude a_L is given as[2])

$$a_L \propto G^J_{j'L} \cdot P_L^{\frac{1}{2}}(\epsilon_\alpha) \cdot e^{i\delta_L} \qquad (1)$$

with

$$G^J_{j'L} = \sqrt{\binom{N}{2}\binom{Z}{2}}\, R_o^{3/2} \int \Phi_J^{*M}\, \chi_\alpha \{Y_L \Psi_{j'}\}_J^M \, d\xi_\alpha\, d\xi_K \,|\, R = R_o$$

$$P_L(\epsilon_\alpha) = 2R_o\, q \exp\{-2 \int_{R_o}^{r_o} q(r)\, dr\}$$

$$q(r) = \sqrt{\{U(r) - \frac{\hbar^2 (L+{}^1\!/_2)^2}{2mr^2} - \epsilon_\alpha\} \frac{2m}{\hbar^2}} \qquad \text{WKB-approximation}$$

δ_L = phase shift of the partial wave L. Of course one is not forced to use the WKB - approximation. The coordinates ξ_α and ξ_k are the internal coordinates of α - particle and daughter nucleus respectively. R is the distance between α - particle and daughter nucleus. R_0 is defined as the smallest R at which one may still safely describe the interaction of the α - particle and the daughter nucleus by a potential U (R). On the other hand one must assume that a nuclear model (the shell model for instance) provides a satisfactory approximation for the wavefunction Φ_J^M of the parent nucleus in the vicinity of the point $R = R_0$. The justification for such an assumption is the success of the theory in explaining a great variety of experimental data. This cannot be considered accidental[2]).

Equation (1) is valid only for spherical nuclei or, to be more precise if the outgoing α - particle does not induce transitions in the daughter nucleus. Clearly if the daughter nucleus is deformed the large quadrupole moment produces a strong coupling of the α - particle to all the members of a rotational band. Interband transitions may be neglected unless there is considerable band mixing. Then, however, the situation becomes very complex.

The formulas for a well deformed nucleus are as follows

$$a_l \propto P_l^{\frac{1}{2}} e^{i\delta_l} \sum_L \left[B_l^{LK}{}^{K_f}_{i,j} G_{L\, K_i - K_f} + (-1)^{J-L-K_f} B_l^{L\, K_i K_f}{}_{j} G_{L\, K_i + K_f} \right] \qquad (2)$$

with

$$G_{L\, K_i - K_f} = \sqrt{\binom{N}{2}\binom{Z}{2}} R_o^{3/2} \int \chi^*_{K_i}\, \chi_\alpha\, Y_L^{K_i - K_f}\, \chi_{K_f}\, d\xi_\alpha\, d\xi_K \,|\, R = R_o$$

$$G_{L\, K_i + K_f} = \sqrt{\binom{N}{2}\binom{Z}{2}}\, R_o^{3/2}\, \pi_f \int \chi^*_{K_i}\, \chi_\alpha\, Y_L^{K_i + K_f}\, \chi_{K_i - K_f}\, d\xi_\alpha\, d\xi_K \,|\, R = R_o$$

$$B_{l\ j'}^{L\ K_i-K_f} = (-1)^{l+K_i-K_f}\ C(Jlj';K_i\ K_f-K_i)\ k_{lL}^{K_i-K_f}$$

$$B_{l\ j'}^{L\ K_i+K_f} = (-1)^{l+K_i+K_f}\ C(Jlj';-K_i\ K_f+K_i)\ k_{lL}^{K_i+K_f}$$

π_f = parity of the final state $\qquad k_{lL}^m = \int Y_l^{*m}\ e^{BP_2}\ Y_L^m$

The expressions for $B_{lj'}^{Lm}$ have been derived by Fröman[3]) with the help of a three - dimensional WKB - approximation.

Summarizing one may say that the theory has been developed so far as to allow a calculation of the amplitudes a_L and hence a direct comparison between theory and experiment is possible. The problems are the approximations that go into the formulation of α - decay - theory. These approximations are:

a) The use of the WKB - approximation
b) The introduction of the'separation distance R_0'
c) The use of nuclear model - wavefunctions.

Point a) is not really a problem because of the possibility to solve the Schrödinger - Equation for an α - particle in a potential U (R) exactly. All one needs is a reasonably big computer. Problems b) and c) are intimately related. The better the nuclear model the less critical is the assumption of the existence of a separation distance R_0 which fulfills the criteria given before. Indeed if one could use exact solutions of the nuclear Schrödinger - Equation the final result would be completely independent of R_0. Besides these general problems there remains a special one. In the transition region between spherical and well deformed nuclei no microscopic nuclear model is available since neither the wavefunctions of the spherical shell model nor those of the Nilsson model provide a reasonable approximation to the exact nuclear wavefunction. There would be no need to bother about this fact if there were not quite a number of well studied α - emitters in the transition region around A = 220.

Before I can begin part II of this talk – a discussion of a few selected experiments – a few words must be said about perturbations. In general electromagnetic fields at the site of the nucleus will interact with the nuclear moments which means that an intermediate state $|j'm'\rangle$ is changed according to

$$\Lambda(t)\ |j'm'\rangle = \sum_{m''} \langle j'm''\ |\Lambda|\ j'm'\rangle j'm''\rangle \qquad (3)$$

t = 0 corresponds to the time at which the α - particle is emitted: $\Lambda(0) = 1$.
For a static perturbation V the operator Λ has the form

$$\Lambda(t) = e^{-i/\hbar\ Vt} \qquad (4)$$

This change of the states $|j'm'\rangle$ must be taken into account when the wavefunction of the final state ($\alpha + \gamma +$ daughter nucleus) is written down and consequently the correlation W ($\underline{k}_\alpha, \underline{k}_\gamma, \xi$) is altered. There is only one case in which one may disregard the possibility of a perturbation, namely, if the lifetime of the intermediate state $|j'\rangle$ is very short, then the perturbation has no time to act.

In general the situation in α - γ - angular correlations is more complex than in β - γ - and γ - γ - angular correlations and this for two reasons

a) In α - decay the daughter nucleus gets an appreciable recoil (80 - 90 keV in typical cases)

b) The nuclear charge is changed by two units and the electron cloud rearranges itself accordingly.

Hence one knows generally less about the fields acting in the intermediate state than in the other cases. If, however, the fields are known then one can measure nuclear moments with the help of perturbed α - γ - angular correlations[4]). There are now two types of experiments. First one can determine the unperturbed correlation and deduce the amplitudes a_L and g_I[5,6]). The second possibility is to use a perturbed correlation with known a_L and g_I and deduce moments of the intermediate state $|j'\rangle$[4]). Examples of both types of experiments will now be briefly discussed.

a) The region around ^{208}Pb

α - γ - angular correlations in the decays

$$^{211}\text{Po} \rightarrow {}^{207}\text{Pb} + \alpha + \gamma \qquad\qquad ^{211}\text{Bi} \rightarrow {}^{207}\text{Tl} + \alpha + \gamma$$

The decay schemes are given in fig. 2. It is important to note that in the case of the ^{211}Bi decay also a polarization correlation measurement has been made.

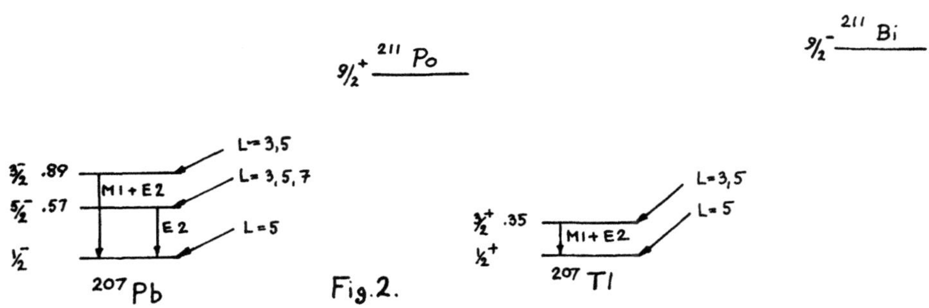

Fig. 2.

Results in terms of quantities a_L and g_I are given in table 1 [6]). In the case of the ^{211}Po decay only α - γ - angular correlations have been measured[5]). Therefore one can only deduce a value for a_5/a_3 if one makes an assumption about the M1 - E2 mixing of the transition $3/2^- \rightarrow 1/2^-$ in ^{207}Pb. Two possibilities for this mixing are given in table 1 and are marked by Exp$^{(1)}$ and Exp$^{(2)}$. For the transition $^{211}\text{Bi} \rightarrow {}^{207}\text{Tl}$ the amplitude a_5 relative to a_3 can be deduced unambiguously from the measurements. In this case also the polarization correlation has been measured[6]). Finally it should be mentioned that the decay

$$^{210}\text{Po}\,(0^+) \overset{\alpha}{\rightarrow} {}^{206}\text{Pb}\,(2^+) \overset{\gamma}{\rightarrow} {}^{206}\text{Pb}\,(0^+)$$

was the first example of an unperturbed correlation[7]). The reason is the fast E2 transition. ($2^+ \rightarrow 0^+$ $E_\gamma = 0.8$ MeV). For the same reason one can also expect that the correlations which have just been discussed are essentially unperturbed.

b) The actinide region.

The situation is more complicated for α - emitters in the actinide region. Gamma - energies are smaller and hence lifetimes of intermediate states are generally longer. This means one has to deal with perturbed correlations in most cases[8]). Of course one might think of using decays of even nuclei $0^+ \rightarrow 2^+ \rightarrow 0^+$ to study perturbations or to measure the g - factor of a 2^+ state with the help of a perturbed α - γ - angu-

Table 1

Transition	Angular Momentum L	Amplitude $\bar{a}_L$ $(=P_L^{-1/2} e^{-i\delta_L} a_L)$	Multipolarity	Amplitude g_I	Remarks
$^{211}Po_{9/2^+} \rightarrow {}^{207}Pb_{5/2^-}$	L = 3	1.0			Theory
		1.0			Exp.
	L = 5	-.90			Theory
		-.96			Exp.
	L = 7	.61			Theory
		-.55			Exp.
$^{207}Pb_{5/2^-} \rightarrow {}^{207}Pb_{1/2^-}$			E2 (I=2)	$g_2 = 1.0$	
$^{211}Po_{9/2^+} \rightarrow {}^{207}Pb_{3/2^-}$	L = 3	1.0			
		1.0			
	L = 5	-.35			Theory
		-.14			Exp. (1)
		-.35			Exp. (2)
$^{207}Pb_{3/2^-} \rightarrow {}^{207}Pb_{1/2^-}$			M1 (I=1)	$g_1 = 1.0$	Exp. (1)
				$g_1 = 1.0$	Exp. (2)
			E2 (I=2)	$g_2 = 0$	Exp. (1)
				$g_2 = 0.13$	Exp. (2)
$^{211}Bi_{9/2^-} \rightarrow {}^{207}Tl_{3/2^+}$	L = 3	1.0			Theory
		1.0			Exp.
	L = 5	-.35			Theory
		-.34			Exp.
$^{207}Tl_{3/2^+} \rightarrow {}^{207}Tl_{1/2^+}$			M1 (I=1)	$g_1 = 1.0$	Exp.
			E2 (I=2)	$g_2 = 0.27$	Exp.

lar correlation[4]). The α - emitter was ^{226}Ra and the g - factor of the 2^+ state of ^{222}Rn turned out to be 0.45 ± 0.07. This value is larger than Z/A and clearly disagrees with all theoretical values which have been obtained on the basis of theories of collective motion[4]).

Finally I wish to discuss the determination of α - amplitudes a_L in the decay of strongly deformed nuclei. In a few favoured decay cases of odd A nuclei the mixtures of l - values including relative phases have been tested by angular correlation experiments[8,9]). The experiments confirm the theoretical prediction[2,10]) that l = 0 and l = 2 amplitudes have the same sign. The limited experimental information bearing on the l = 4 amplitudes indicates[11]) that the phase of l = 2 and l = 4 is the same in ^{233}U and opposite in ^{253}Es which is in agreement with theory[2,10]). No angular correlation experiments have yet tested the l - mixture predictions[10]) for hindered decay. Theory in this case predicts that considerable admixture occurs and that l - values other than the minimum may be dominant. Clearly a systematic comparison of experimentally determined amplitudes a_l with the great number of theoretical values which are already available[10]) would be extremely helpful in our understanding of single particle states in the actinide region.

References

1. H. Frauenfelder and R.M. Steffen, Angular Correlations, Ch. XIX, vol. 2, in Alpha-, Beta- and Gamma-Ray Spectroscopy, ed. K. Siegbahn (North - Holland, Amsterdam 1965).
2. H.J. Mang, Alpha Decay, Ann. Rev. Nucl. Sci. 14 (1964) 1.
3. P.O. Fröman, Kgl. Danske Vidensk. Selskab. Mat. - Fys. Skr. 1, No. 3 (1957).
4. B. Orre, A. Linnfors, J.E. Thun and L. Johansson, Nucl. Phys. A148 (1970) 516.
5. S. Singes (Doctoral thesis, University of Illinois, Urbana, III, 1957, unpublished).
6. S. Gorodetzky, A. Gallmann, A. Knipper and R. Armbruster, Compt. Rend., 237 (1953) 245. S. Gorodetzky, F. Beck, A. Knipper, R. Manquenouille and R. Richert, Compt. Rend., 254 (1962) 2319.
7. S. de Benedetti and G.H. Minton, Phys. Rev. 85 (1952) 944.
8. K. Siegbahn and F. Asaro, Phys. Lett. 2 (1962) 323. F. Falk, S. Törnkvist, J.E. Thun, H. Snellman, K. Siegbahn and F. Asaro, Z.Phys. 198 (1967) 106. F. Falk, A. Linnfors, B. Orre and J.E. Thun, UUIP - 651; Physica Scripta 1 (1970) 13. E. Flamm and F. Asaro, Phys. Rev. 129 (1963) 290.
9. V.E. Krohn, T.B. Novey and S. Raboy, Phys. Rev. 105 (1951) 234.
10. J.K. Poggenburg, H.J. Mang and J.O. Rasmussen, Phys. Rev. 181 (1969) 1697.
11. See discussion by J.O. Rasmussen in Alpha-, Beta- and Gamma-Ray Spectroscopy, vol. I, p. 742, ed. K. Siegbahn (North - Holland, Amsterdam 1965).

Discussion

Alder: I would like to ask if the alpha - gamma correlations depend on the pairing of the nucleons? As far as I remember the absolute lifetime depends somewhat on the pairing and it might be that alpha - gamma correlations do not depend so strongly on these pairing properties.

Mang: I would say that in a certain way, they depend somewhat on the pairing because the pairing decides how many Nilsson levels, for instance in a favoured decay, will contribute to making up the alpha amplitude, but since one averages, the dependence will not be as strong as the absolute rate; the absolute rate gets a tremendous enhancement by the pairing. Of course, any partial wave gets this enhancement factor. This is a common factor everywhere and so, the alpha - gamma correlation, the relative amplitudes for l equals 2, 4 and so on, do not depend very strongly on the pairing. This has been checked in calculations where the pairing strength and the number of states have varied.

Contributed Paper

PERTURBED ALPHA - GAMMA ANGULAR CORRELATIONS IN THE DECAY OF ^{241,243}Am

B. Orre, F. Falk, A. Linnfors and J.E. Thun

University of Uppsala, Institute of Physics, Uppsala, Sweden

Introduction

Angular correlations involving α - particles differ from those involving or following β - decay and γ - transitions. They may be strongly perturbed even at lifetimes as short as 10^{-10} sec.

The most severe perturbation generally occurs with the recoils going into a vacuum. The resulting hyperfine interaction can be decoupled by a longitudinal magnetic field and the unperturbed correlation can be obtained. This method was first successfully used by Siegbahn and Asaro[1]) and later by Falk et al.[2,3]).

With the recoils into a solid the perturbations have in some cases appeared to be of electrical character[4,5]) but with paramagnetic atoms one has also to take into consideration the possibility of a magnetic interaction. The α - γ correlation in the ^{226}Ra decay has been examined in regard to its dependence on a parallel magnetic field with the recoils occurring into many different environments[6]). Except for vacuum we found no clear field effects. The perturbations were in general small because of the stable radon core and the rather short half - life (0.3 ns).

Among the transuraniums the perturbations are more severe. These heavy nuclei have large quadrupole moments and the 5 - f shell generally creates large internal fields at the nucleus.

Flamm and Asaro have carefully examined the perturbations of the α - γ correlations in the decay of ^{241}Am, ^{243}Am and ^{243}Cm by varying the environment of the recoiling ion[7]). The attenuations in metals were less than in insulators. For ^{241}Am (τ = 91 ns), the attenuation coefficients G_{22} were below the hard - core value for a static perturbation. In this paper we describe a study of these α - γ perturbations for the two americium isotopes ^{241}Am and ^{243}Am.

Experiments and results

A partial decay scheme of the two isotopes is shown in fig. 1. For both cascades approximately the same A_{22} - value = - 0.36 has been reported[7]).

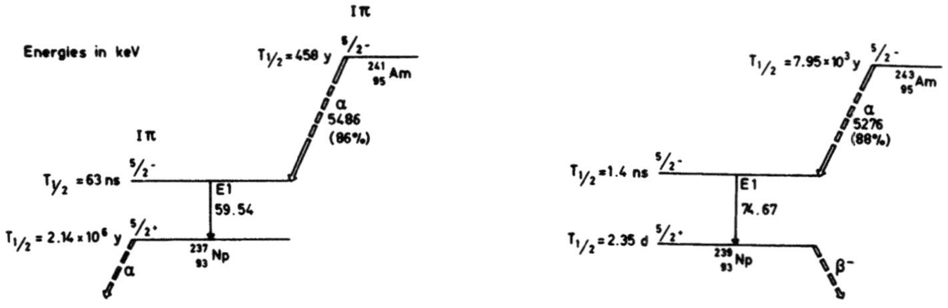

Fig. 1. The decay scheme (taken from C.M. Lederer, J.M. Hollander and I. Perlman, Table of Isotopes, sixth edition).

In the present experiment the α - particles were detected in a surface barrier detector thus ensuring that the daughter nuclei recoil into the backing with an energy of about 95 keV. The gamma radiation was detected in a movable NaI(Tl) detector. The resolving time in the time - differential measurements was about 7 ns. The decoupling magnet had a maximum field strength of 23 kG parallel to the α - detector. The equipment has been described in ref. 2 and 8. Both time - integral and - differential measurements have been made. The results of the time - integral measurements (carried out at room temperature) are summarized in tables 1 and 2.

Table 1

^{241}Am $(\alpha_{60} \cdot \gamma_{60})$ angular correlation results. A_{22} (unperturbed) = - 0.36 τ = 91 ns.

Environment of the recoiling ion	A_{22} (B = 0)	G_{22}	Does the correlation change in a parallel magnetic field? 0 - 23 kG.
nickel	-0.08 ± 0.01	0.22	yes
iron	-0.03 ± 0.01	0.08	yes
cobolt	-0.04 ± 0.01	0.11	yes
lead	-0.03 ± 0.01	0.08	no
gadolinium	-0.03 ± 0.01	0.08	no
uraniumoxide	-0.03 ± 0.01	0.08	no
silver	-0.04 ± 0.01	0.11	no
gold	-0.02 ± 0.01	0.06	no
aluminium	-0.04 ± 0.01	0.11	no
indium	-0.01 ± 0.01	0.03	no
mylar	-0.03 ± 0.01	0.08	no
air	-0.04 ± 0.01	0.11	no

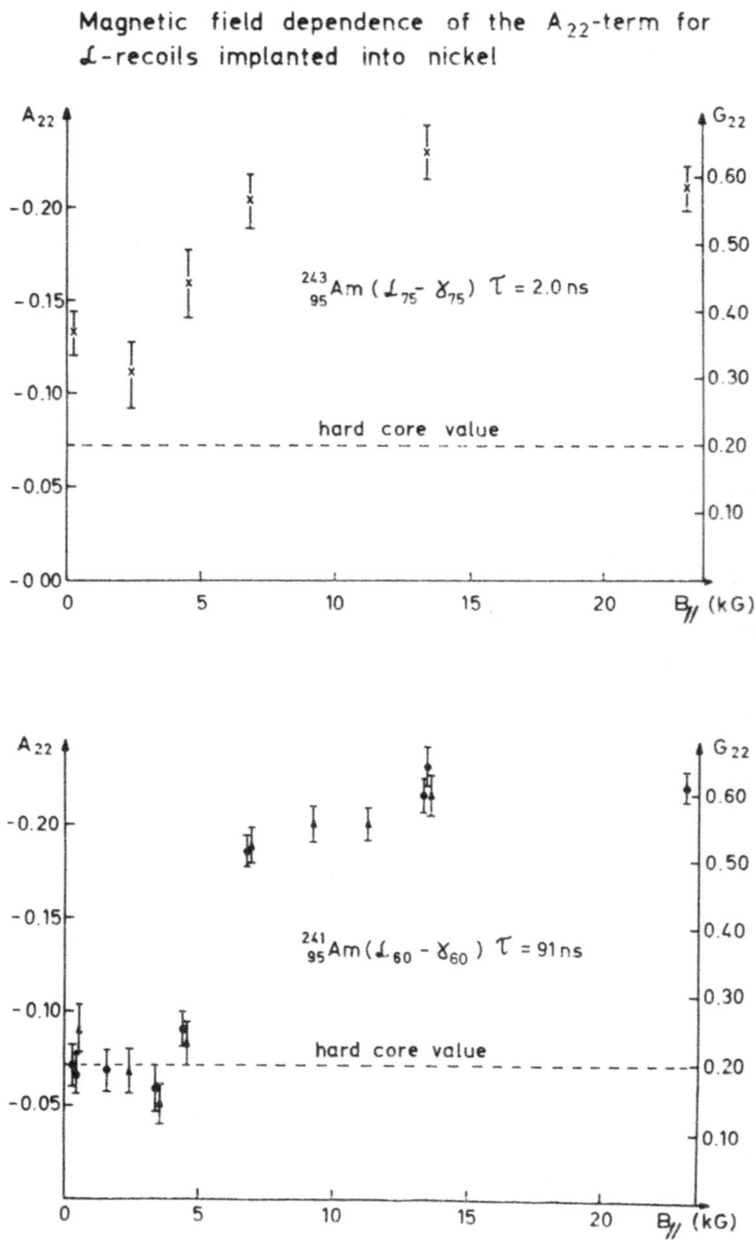

Fig. 2. The longitudinal field effect in nickel at room temperature. In the first figure the mean values for two different $^{243}_{95}Am$ sources are indicated, while in the second the results for three $^{241}_{95}Am$ sources are shown.

Table 2

^{243}Am ($\alpha_{75} \cdot \gamma_{75}$) angular correlation results. A_{22} (unperturbed) = - 0.36τ = 2.0 ns.

Environment of the recoiling ion	A_{22} (B = 0)	G_{22}	Does the correlation change in a parallel magnetic field? 0 - 23 kG.
nickel	-0.13 ± 0.01	0.36	yes
silver	0.16 ± 0.02	0.44	no
gadolinium	0.17 ± 0.03	0.47	no

The results are in agreement with Flamm and Asaro's measurements. Only in the ferromagnetic elements there was an influence of the external field on the correl-

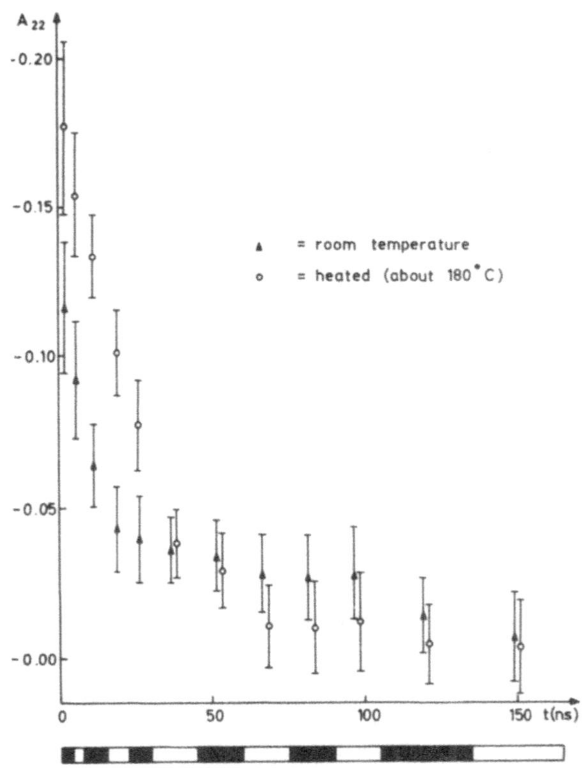

Fig. 3. A time - differential measurement for $^{241}_{95}$Am with recoils into aluminium. (Not corrected for admixtures and the finite resolving time).

ation. Nickel has been extensively examined in this respect. The 'field effect' curves for the two americium isotopes are shown in fig. 2. This longitudinal effect is almost identical in the two cases in spite of the difference in lifetimes. The correlation cannot be restored to its unperturbed value although the external and the aligned internal field are both in the longitudinal direction and should therefore in principle not perturb the correlation.

At present we can only point out some hypotheses to explain this discrepancy. One possibility is that the α - recoil damages the lattice structure so that large angles arise between the local direction of the magnetization and that of the applied magnetic field, see Aharoni, ref. 9. These radiation damages may also give large quadrupole (perhaps time - dependent) perturbations. In addition, magnetic or possibly electric relaxation phenomena may be important in this context.

We have studied this nickel effect from room temperature up to 180° C but the changes were insignificant except for the change of the magnetization in nickel with the temperature. The fall of the curves at low fields must be attributed to some kind of transverse effect of magnetization. It disappeared when the thin foil (800 $\mu g/cm^2$) with the source made an angle of 45 degrees with respect to the field direction. The field effect otherwise appeared to be the same.

The amount of the hyperfine field in nickel can be estimated from the strength of the perturbation according to the theoretical work of Matthias et al.[10]). A crude estimate gives $|H_{int}(Ni)| > 260$ kG.

The field dependence has also been studied for ^{241}Am in iron and cobalt, but changes of the correlation occur at field strengths of about 20 kG and higher.

The perturbation in some non - ferromagnetic metals (Al, Au and In) have been examined by time - differential technique. The results for aluminium (see fig. 3) indicate the existence of a time - dependent interaction. The slower decay at a higher temperature points to magnetic (atomic) spin - lattice relaxation.

References

1. K. Siegbahn and F. Asaro, Phys. Lett. 2 (1962) 323.
2. F. Falk, S. Törnkvist, J.E. Thun, H. Snellman, K. Siegbahn and F. Asaro, Z. Phys. 198 (1967) 106.
3. F. Falk, A. Linnfors, B. Orre and J.E. Thun, Physica Scripta 1 (1970) 13.
4. J.C.D. Milton and J.S. Fraser, Phys. Rev. 85 (1952) 944.
5. E.J. Ansaldo, L. Grodzins and R. Kalish, Phys. Lett. 30B (1969) 538.
6. B. Orre, A. Linnfors, F. Falk, J.E. Thun and L. Johansson, Nucl. Phys. A148 (1970) 516.
7. E. Flamm and F. Asaro, Phys. Rev. 129 (1963) 290.
8. F. Falk, A. Linnfors, B. Orre, J.E. Thun, Nucl. Instr. and Meth. 83 (1970) 261.
9. A. Aharoni, Phys. Rev. Lett. 22 (1969) 856.
10. E. Matthias, S.S. Rosenblum and D.A. Shirley, Phys. Rev. Lett. 14 (1965) 46.

Discussion

Deutch: A question and a comment. What are the thicknesses of your sources and what are the thicknesses of the oxide layers of your foils?

Orre: The thickness of our source is about 200 Å and the backing is about 800 μg/cm^2. For the source in the radium application we used a very thin foil. These new thin foils were only 80 μg/cm^2. By measuring the energy loss of alpha particles we determined the source thickness. We found thicknesses between 500 and 600 μg/cm^2 and we still have recoils coming out of them.

Deutch: Your radium sources were 500 μg/cm^2?

Orre: Yes. We were surprised that we could use radium sources that thick.

Deutch: I would venture to say that if the source was that thick, it would never get into an iron foil. If you look at the range distribution even with a thickness as small as 10 μg/cm^2, you will have a great deal of your alpha activity in the oxide layer on the foils.

Orre: In what oxide layer?

Deutch: And I would believe that much of this analysis is done with not only material implants in magnetic material but also in the source and the oxide layer.

Orre: Yes.

Dar: Did you observe also the recoil into vacuum? What did you get for the G_{22}?

Orre: Yes, we have made this measurement before, for ^{243}Am. The value was slightly above 'hard core'.

Thun: I would just like to comment on the statement of Deutch a little. Our results on recoils into vacuum show that most alpha particles enter the vacuum environment. These measurements were made on the same sources as used in the implantation experiments. The hyperfine interaction in vacuum was fully restored, which is evidence that the recoils really enter the implantation material. This material is in fact evaporated on top of the active layer.

De Waard: I think that also, if every atom penetrates into the nickel or iron, the results you got are more or less what one would expect. There are two things which I would like to ask. You are implanting in nickel and iron and these are both ferromagnets. Now you talk about decoupling and I wonder if you maybe really mean magnetization of the foils, because, do I understand correctly that you magnetize perpendicular to a thin foil?

Orre: Yes.

De Waard: Then of course you have to overcome the demagnetization field first and it means, as you see in the picture, that your iron results start to change only at 20 kGauss, which you would expect, and the nickel result starts to change at 6 or 5 or so which you would also expect because you have to magnetize the foil. Now then, you reach with the nickel foil complete magnetization and you find G_{22} equal to about 0.6, I think, while you would like to have it 1, isn't it? I do not know about the actinides, but the lanthanides implanted in iron, and properly implanted I daresay, give us exactly the same picture. And the way we understand it is that about half of the nuclei or maybe 40% are not in a good site. That is to say they are associated with one or more vacancies. And these vacancies set up a strong quadrupole field and actually in those sites you just cannot magnetize the sites at whatever field you want, well maybe at a million Gauss, but not at these ......

Orre: So you have a direction and angles?

De Waard: That is right. Random angles for part of the implanted atoms. Even if they are properly implanted, because after implantation a fraction, varying between a maximum of 75% to a minimum of 30%, is always associated with vacancies, which set up a strong quadrupole coupling which makes your exchange field incapable of having a good influence. That is what we think happens.

Orre: This quadrupole interaction could be both static and time - dependent.

Deutch: I would like to reply to Thun with reference to recoil in vacuum. We will present evidence later where we have actually measured the oxide layers with a 900 keV ^{16}O (d, p) ^{17}O reaction and have calculated the distribution of a 100 keV recoil and we will show how many of these alpha's actually do not end up inside the material.

Contributed Paper

ALPHA - GAMMA LINEAR POLARIZATION MEASUREMENTS IN THE DECAY OF $^{223}Ra \rightarrow ^{219}Rn$

M.J. Canty and K. Krien

Institut für Strahlen und Kernphysik der Universität Bonn, W. Germany

Recent investigations (ref. 1 - 4) of the decay of $^{223}Ra \rightarrow ^{219}Rn$ lead to spin assignments for 7 low lying levels. The results of one[1]) of these investigations are summarized in fig. 1. A ground state spin of I = 3/2 for ^{219}Rn has been deduced assuming a ground state spin for ^{223}Ra of $I_i = 1/2$. This assignment is in contradiction with our investigation[5]) of the decay of $^{219}Rn \rightarrow ^{215}Po$ in which we proposed a spin of I = 7/2 for the ground state of ^{219}Rn. In order to clear up this discrepancy we have measured the α - γ linear polarization correlations of the 269 keV and 445

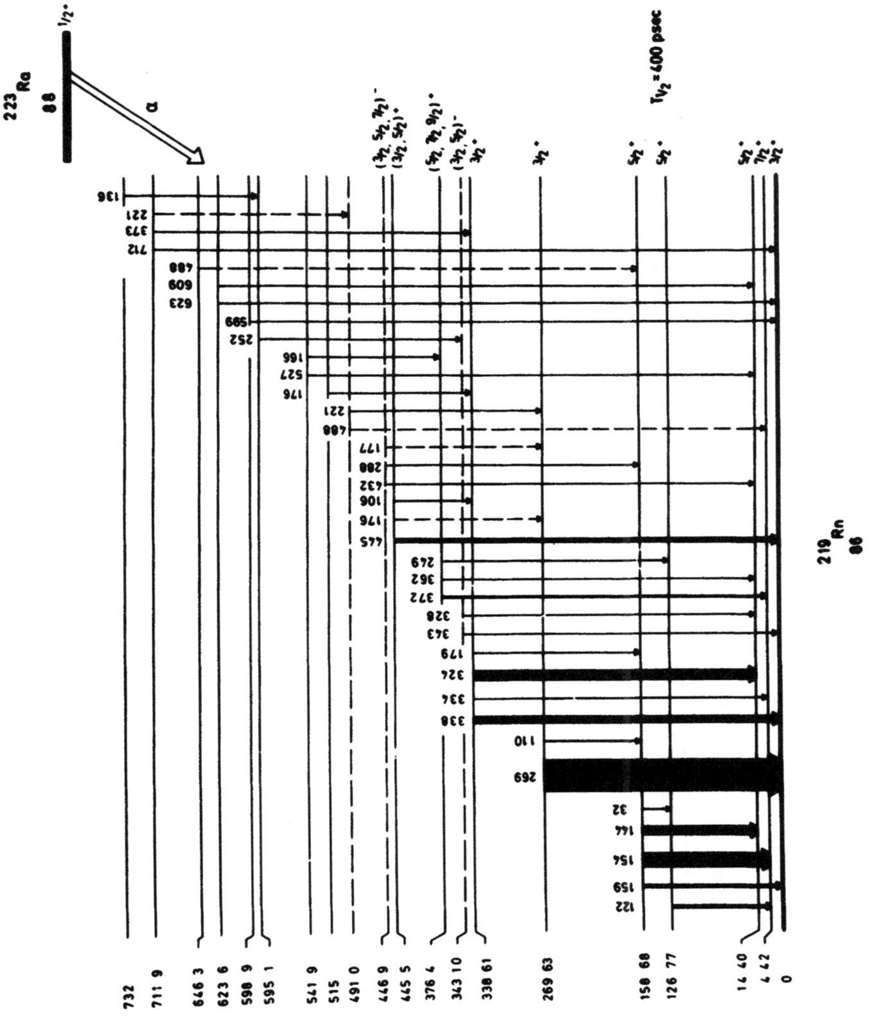

Fig. 1. Decay scheme of $^{223}Ra \rightarrow ^{219}Rn$ (from ref. 1 modified).

keV transitions. They were measured with our Compton effect polarimeter at Bonn (ref. 6). Instead of the four NaI (Tl) detectors outside the polarimeter a single 100 μ thick silicon α - detector of about 50 keV FWHM resolution was used. The detector was mounted together with the source in a vacuum chamber of 5 cm diameter. Sources of ^{223}Ra and ^{227}Ac for the polarizations involving the 269 keV and 445 keV transitions respectively were prepared by an evaporation technique (ref. 1) and mounted such that nuclei belonging to the detected α - particles recoiled into the aluminum backing, in order to minimize attenuations[7]) by free hyperfine fields. We measured the linear polarization at different angles between the direction of emission of the α - and γ - radiations in separate runs with the vacuum chamber moved to the appropriate positions. In both the α - 269 keV and α - 445 keV linear polarization correlations, corrections with respect to background were unnecessary because it was

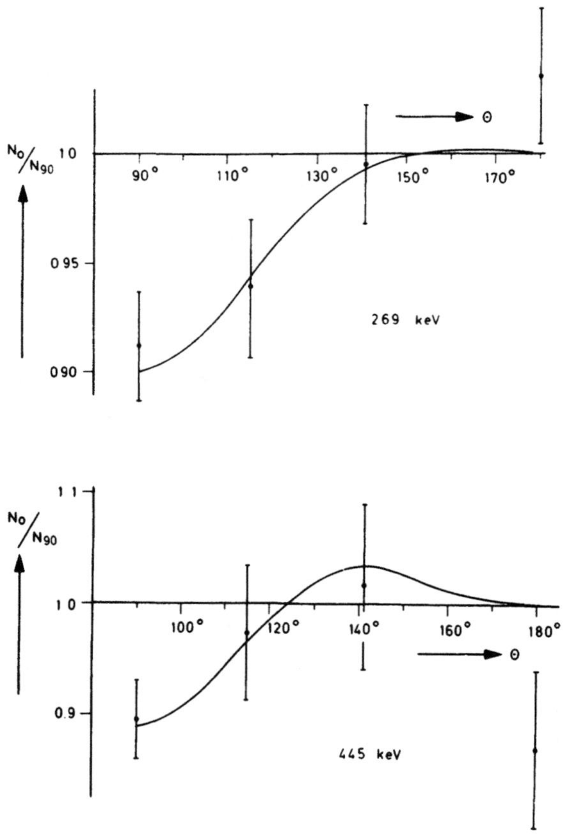

Fig. 2. α · γ linear polarization correlation of the 269 keV and 445 keV transitions.

possible to select the appropriate α - line by a single channel analyzer. The linear polarization P (θ) is defined by

$$P(\theta) = I_0(\theta) / I_{90}(\theta) \tag{1}$$

where I_0 (θ) and I_{90} (θ) are the intensities of the components of the γ - radiation polarized parallel and antiparallel to the θ - plane respectively. The theory[8]) of the linear polarization correlation yields

$$P(\theta) = \frac{W(\theta) + B_2 P_2^2 (\cos\theta) + B_4 P_4^2 (\cos\theta)}{W(\theta) - B_2 P_2^2 (\cos\theta) - B_4 P_4^2 (\cos\theta)} \tag{2}$$

where W (θ) is the angular correlation function $P_\nu^{\ 2}$ are the associated Legendre polynomials, and B_ν are parameters.

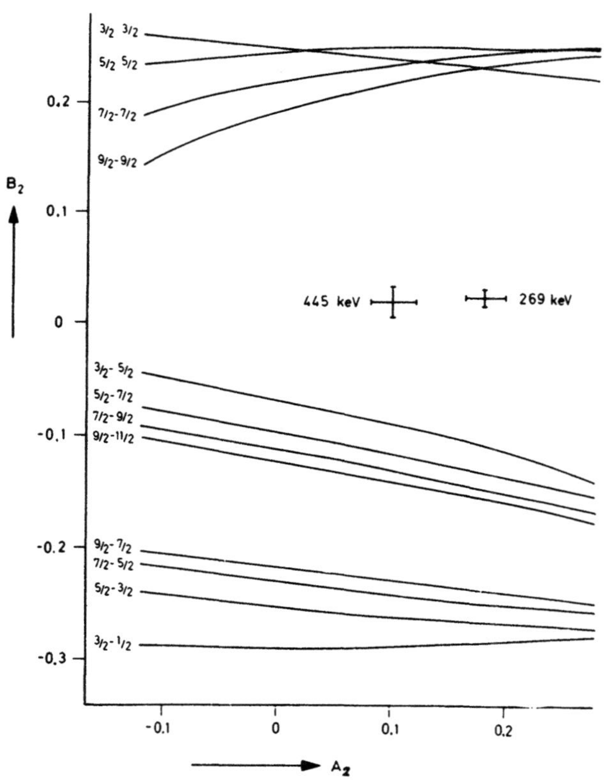

Fig. 3. Comparison of theoretical and experimental values of A_2 and B_2 for the α- 269 keV γ- and α- 445 keV γ- cascades, assuming the ground state spin of ^{223}Ra to be 1/2.

In figure 2 a plot of the triple coincidence counting rate $N_0(\theta) / N_{90}(\theta)$, a quantity which is related to $I_0(\theta) / I_{90}(\theta)$ by the sensitivity[6]) of the polarimeter, is shown. The solid curve represents a least squares fit of equation (2), taking the sensitivity correction into account, to the measured points. In this fit only the parameters B_ν were allowed to vary, while the parameters of the angular correlation function were taken from the work of Krien et al.[1]). The results are:

269 keV: $B_2 = 0.0234 \pm 0.0070$ $B_4 = -0.0023 \pm 0.0027$

445 keV: $B_2 = 0.020 \pm 0.014$ $B_4 = -0.0067 \pm 0.0068$

The multipolarities of both γ - transitions have been determined from conversion data (ref. 1 - 4) to be mainly M1. In figure 3 we plot theoretical curves of B_2 as a function of A_2 where A_2 varies with the E2/M1 mixing ratio within a range greatly exceeding the limits of errors in the multipolarities. The L waves must be pure, since in this plot a ground state spin of $I_i = 1/2$ for ^{223}Ra has been assumed. This is the only severe restriction while the intermediate and final spins range from 3/2 to 9/2 and 1/2 to 11/2 respectively. The experimental values are indicated by horizontal and vertical bars for A_2 and B_2 respectively where the A_2 has been taken from the literature[1]). It is obvious from this figure that no solution is in agreement with theoretical predictions. This means that the assumption of $I_i = 1/2$ cannot be maintained. Since all spin assignments in previous investigations, except our work[5]) on ^{215}Po, were based on the assumption that the nucleus ^{223}Ra has a ground state spin $I_i = 1/2$, it is clear why they are in disagreement with this last investigation. Although one discrepancy has been removed, now we have the difficulty that our measurements are in contradiction with previous spin assignments[9]) for the ground state of ^{223}Ra. On the other hand, these assignments are based mainly on α - and γ - transition probabilities and energy spacings fitting nicely a K = 1/2 band but they have not been sufficiently verified. Recently measured[10]) α - γ angular correlations in the decay of $^{227}Th \to$ $\to {}^{223}Ra$ are also in contradiction with a spin assignment of I = 1/2 for the ground state of ^{223}Ra. Unfortunately, from our data we cannot yet definitely assign new spins to the levels involved, but the work is still in progress.

References

1. K. Krien, C. Günther, J.D. Bowman and B. Klemme, Nucl. Phys. A141 (1970) 75.
2. W.F. Davidson and R.D. Connor, Nucl. Phys., in press.
3. Ch. Briançon, C.F. Leang and R. Walen, Compt. Rend. 266 (1968) 1533.
4. Ch. Briançon and C.F. Leang, J. de Phys. 29 (1968) 61.
5. K. Krien, M.J. Canty and P. Herzog, Nucl. Phys., to be published.
6. P. Weigt, H. Hübel, P. Göttel, P. Herzog and E. Bodenstedt, Nucl. Inst. Meth. 57 (1967) 295.
7. E. Flamm and F. Asaro, Phys. Rev. 129 (1963) 290.
8. L.W. Fagg and S.S. Hanna, Rev. Mod. Phys. 31 (1959) 711.
9. Ch. Briançon, thesis, Orsay and references quoted.
10. R. Walen, private communication.

CONVERSION ELECTRON CORRELATIONS IN NUCLEAR STRUCTURE RESEARCH

T.R. Gerholm and L. Holmberg
University of Stockholm, Institute of Physics, Stockholm, Sweden

The theory for angular correlations involving conversion electrons is briefly discussed. The expressions for the analysis of conversion electron correlations are summarized. The experimental techniques are reviewed, with an emphasis on recent developments. Some results from electron correlation experiments are presented. E0, E1, M1 and E2 transitions are treated separately with comments on the evidence for the effects of nuclear penetration and 'anomalous' effects. Finally, a few references are given to applications in neighbouring fields such as perturbed angular correlations, aftereffects of hole formation, and conversion electrons following nuclear reactions. The main conclusion is that the theory of conversion electron correlations now appears to be well understood and in good agreement with the experiments. The conversion electron correlations therefore present a valuable tool for the study of nuclear structure, regardless of whether nuclear penetration effects show up in the internal conversion process. One may anticipate a shift of emphasis towards studies directly inspired by problems in nuclear structure. The inherent potentialities in the electron correlation method, when applied to conversion electrons following nuclear reactions, represents a challenge for the future.

1. Introduction

This paper presents an attempt to define the present state - of - art in the field of electron - gamma angular correlations†. It seems reasonable, however, to start with a brief historical account of its twenty years of existence.

The pioneering experiments, performed by use of GM - counters, were carried out around 1950. From the beginning, the Zürich group was one of the most active in the field[2]).

It soon became evident that angular correlations involving conversion electrons call for an energy resolution in the electron channel sufficient to separate at least the K conversion electrons from those of the L subshells. Magnetic momentum selection presented an obvious solution to this problem. Already in 1951, Sevenson and Deutsch developed an instrument for electron - gamma angular correlation measurements including a magnetic lens spectrometer to single out the electron lines[3]). Several other instruments of a similar kind were soon to be constructed[4,5,6,7]).

Parallelling this development of experimental techniques the theory of conversion electron angular correlations was explored by Biedenharn and Rose, who managed to cast the complex theory into surprisingly neat and simple formulae[8,9]). They showed that, by means of a set of particle parameters, the expression already given

† Electron polarization is not discussed. There have been rather few experimental studies of the polarization of conversion electrons, cf. ref. 1. Thus, we confine ourselves to what is properly called directional correlations.

for gamma - gamma angular correlations easily could be transformed into corresponding expressions involving conversion electrons. Thus, conversion electron correlations presented an attractive alternative to the already by then widely used gamma - gamma angular correlation technique. It could be used in those cases where high conversion probability makes gamma - gamma correlations difficult to perform, as is the case for transitions of low energy, or of high multipolarity. It was attractive also because the resolution offered by magnetic spectrometers is much superior to what is possible to achieve with the sodium - iodide scintillation detector. The particle parameters also turned out to depend on the multipolarity as well as the character of the transitions. Consequently, conversion electron correlations carry the additional information of the parity change. Furthermore, angular correlations involving conversion electrons are sometimes considerably more sensitive to minute multipolarity admixtures than are the gamma -gamma correlations.

At this time, the internal conversion process itself came into the focus of interest. To match the accurate techniques developed for the measurement of conversion coefficients, and particle parameters, it became necessary to perform considerably more careful theoretical calculations. The rise of computer technology made such a heavy undertaking feasible, and two independent sets of fairly complete tables of conversion coefficients appeared[10,11]). Underlying these theoretical achievements was a much better understanding of the internal conversion process. This turned out to be of great importance for the conversion electron correlation theory. The effect of the finite size of the nucleus, discovered by Sliv[12]), caused Church and Weneser to consider the conversion process taking place within the nuclear volume[13]). Their study led to an understanding of E0 transitions, and to the soon confirmed prediction that such transitions could take place in competition with E2 and M1 transitions (ref. 14). As is now well known, the presence of E0 conversion electron is solely due to the existence of penetration matrix elements, a concept introduced by Church and Weneser and soon extended to apply to penetration effects of higher multipolarities such as E1 or M1 transitions[15]). Sizeable penetration contributions were identified and determined. These provide new and independent information. Hence, conversion electron correlation studies became a subject of considerable interest.

2. Theory

The foundation for the theory of conversion electron angular correlations was laid by Rose, Biedenharn and Arfken in 1952[8,9]). They published the first tabulation of parameters necessary for the interpretation of the experimental data. Since then several other theoretical investigations and tables of the particle parameters have been published[16,17,18,19,20]).

A major point in the theoretical considerations is the observation that the nuclear charge distribution has a vanishingly small geometrical extension in the atomic scale. The atomic electron - nuclear interaction in the interior of the nucleus may then be neglected. The conversion process solely depends on the electromagnetic field outside the nucleus and the wave functions of the electrons. The conversion interaction amplitude may be factorized into two parts, nuclear and atomic respectively. As the gamma - ray emission is a measure of the electromagnetic field created by the nucleus, the nuclear part cancels in the comparison of conversion electron and gamma - ray

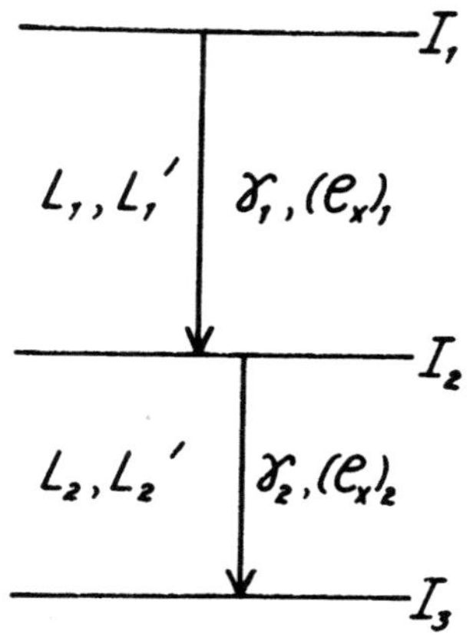

emission.

This line of arguments leads to a very comfortable conclusion. The theory of conversion electron correlations may be considered as composed of two independent parts. These are the theory of the angular distribution of gamma - rays, which automatically takes care of the nuclear part of the problem, and the theory of the internal conversion process, which entirely deals with extra - nuclear problems.

In practice this means that the expression for gamma - gamma angular correlations is simply converted into an analogous expression for conversion electron correlations by the introduction of mutliplicative factors, b_k. These particle parameters are defined as the ratio between the radiation parameters for conversion electrons and for gamma radiation. That means e.g. that the coefficient $A_k\ (L, L + 1, I_3, I_2; \delta; \gamma)$ entering

$$W(\vartheta) = \sum_{k\ \text{even}} a_k P_k(\cos\vartheta),$$

where

$$a_k = A_k\ (L_1, L_1 + 1, I_1, I_2; \delta_1; \gamma)\ A_k\ (L_2, L_2 + 1, I_3, I_2; \delta_2; \gamma)$$

and expressed as

$$A_k(L, L + 1, I_3, I_2; \delta; \gamma) = [F_k(L, L, I_3, I_2) + 2\delta F_k(L, L + 1, I_3, I_2) + \delta^2 F_k(L + 1, L + 1, I_3, I_2)]\ (1 + \delta^2)^{-1}$$

is changed into

$$A_k(L, L+1, I_3, I_2; \delta_e; e_X) = [b_k(\pi L, \pi L; X)\, F_k(L, L, I_3, I_2)$$
$$+ 2\delta_e b_k(\pi L, \pi' L+1; X)\, F_k(L, L+1, I_3, I_2)$$
$$+ \delta_e^2 b_k(\pi' L+1, \pi' L+1; X)\, F_k(L+1, L+1, I_3, I_2)]\,(1+\delta_e^2)^{-1}$$

where

$$\delta_e = \delta \sqrt{\frac{a(\pi' L+1; X)}{a(\pi L; X)}}$$

The particle parameters used in this expression have been normalized in order to keep the coefficient for P_0 (cos θ) equal to unity. The normalized particle parameters are related to the corresponding unnormalized quantities $\bar{b}_k$ by

$$b_k\,(LL') = \bar{b}_k\,(LL')\,[\bar{b}_0\,(LL)\,\bar{b}_0\,(L'L')]^{-1/2}$$

where

$$\bar{b}_0\,(LL) = a\,(L),$$

i.e. the corresponding conversion coefficient.

General, more or less equivalent, expressions for the particle parameters have been published by several authors. For instance, Hager and Seltzer in a recent publication[20]) have given the following formula

$$\bar{b}_k\,(LL') = 4\pi^2 a K \sum_{\kappa,\hat{\kappa}} \sqrt{\frac{(2j+1)(2\hat{j}+1)(2\ell+1)(2\hat{\ell}+1)}{(2L+1)(2L'+1)}}$$

$$\times\,(-1)^{j+\hat{j}+j'-1/2+(\hat{\ell}-\ell)/2} \begin{Bmatrix} \ell & \hat{\ell} & k \\ \hat{j} & j & 1/2 \end{Bmatrix} \begin{Bmatrix} L & L' & k \\ \hat{j} & j & j' \end{Bmatrix} \begin{pmatrix} \ell & \hat{\ell} & k \\ 0 & 0 & 0 \end{pmatrix} \Big/ \begin{pmatrix} L & L' & k \\ -1 & 1 & 0 \end{pmatrix}$$

$$\times\; e^{i(\delta_\kappa - \delta_{\hat{\kappa}})} \langle \kappa \| \underline{j}\cdot\underline{B}_L \| \kappa_i \rangle \langle \hat{\kappa} \| \underline{j}\cdot\underline{B}_{L'} \| \kappa_i \rangle^*$$

where K denotes the photon energy and where

$$\langle \kappa \| \underline{j}\cdot\underline{B}_L(e) \| \kappa_i \rangle = \frac{i^L}{\sqrt{L(L+1)}} \langle \kappa \| Y_L \| \kappa_i \rangle R^{EL}_{\kappa\kappa_i},$$

$$\langle \kappa \| \underline{j}\cdot\underline{B}_L(m) \| \kappa_i \rangle = i^{L-1}\; \frac{(\kappa+\kappa_i)}{\sqrt{L\,(L+1)}} \langle -\kappa \| Y_L \| \kappa_i \rangle R^{ML}_{\kappa\kappa_i}$$

The B_L are outgoing - wave electromagnetic potentials and the $R_{\kappa\kappa_i}$ denote electron radial integrals. The initial electron state is denoted by κ_i and the final states by

κ and $\hat{\kappa}$. The relation of κ_i to spectroscopic and X - ray notation is given in table 1, together with the selection rules for the conversion of a given multipole (p.136).

The crucial factors are the radial integrals. The techniques used to calculate these ultimately limit the accuracy that may be claimed for the theoretical particle parameters. The radial integrals are the same as those entering the expression for the conversion coefficient. However, there the sum of squares of matrix elements appear. As seen, the product of the matrix elements enters the particle parameter, which thus may be more sensitive to minor effects in the smaller of these quantities.

The most recent tabulations of particle parameters are those by Pauli[19]) and by Hager and Seltzer[20]). All calculations are based on lowest - order perturbation theory. Hager and Seltzer use a Fermi distribution for the nuclear charge, and the electron wave functions are obtained by relativistic selfconsistent - field calculations. Vacuum polarization is taken into account, as well as the effect of the hole of the emitted electron both in the direct and exchange potentials. Experimental values for the binding energies are used.

In the theoretical calculations published so far, higher order effects involving two electrons and radiative corrections have not been taken into consideration. They might turn out to be of some importance. The same remark possibly applies to the effects of nuclear electromagnetic moments. For outer shells the chemical state of the atom certainly enters the picture.

Let us return to the fundamental physical assumption underlying these theoretical calculations. To neglect the part of the internal conversion process that takes part in the interior of the nucleus is generally justified. There are interesting exceptions, however, as was first pointed out by Church and Weneser[13,15]).Whenever the gamma ray emission is hindered because of specific nuclear selection rules, the effects of electron penetration into the nuclear matter may become important. This applies when the selection rules do not impose similar restrictions on the interior part of the conversion process. In sect. IV we give several examples. In some cases the penetration effects give the by far dominating contribution to the internal conversion process. Particularly striking are of course the E0 conversion electrons, which originate exclusively from the interior of the nucleus.

The effects of nuclear penetration cannot be taken into account in a model - independent way. However, the penetration effect is in principle described by the internal conversion theory. Beside the in general dominating term in the interaction Hamiltonian, which is exactly factorable into nuclear and electron variables, there appear terms corresponding to the penetration effects. In these the integration in electron variables is taken over the nuclear volume only. For convenience the gamma transition amplitude is still taken out as a common factor. The parameter describing the penetration contribution is thus expressed as the ratio between penetration and gamma ray matrix elements. Contrary to the radial matrix elements previously discussed, the effects of nuclear penetration depend on the actual form of nuclear current and charges.

The dependence on the details of nuclear structure can be included in the expression for the particle parameters in a very convenient way, as shown in ref. 21 and 22. An anomaly factor is defined, which is a polynomial in the penetration parameters.

Whereas the tabulated b_k are normalized, the anomaly factors, Q_k, are relevant for the unnormalized quantities. Thus, when penetration effects are included, the particle parameters have to be renormalized by including the anomaly factor for the conversion coefficient, Q_0. As this factor appears also in the electron mixing ratio. the resulting expression for an angular correlation coefficient with penetration effects included is

$$A_k(L, L+1, I_3, I_2) = [b_k(\pi L, \pi L)\, Q_k(\lambda_i)\, F_k(L, L, I_3, I_2)$$

$$+ 2\delta_e b_k(\pi L, \pi' L+1)\, Q'_k(\lambda_i)\, F_k(L, L+1, I_3, I_2)$$

$$+ \delta_e^2 b_k(\pi' L+1, \pi' L+1)\, F_k(L+1, L+1, I_3, I_2)]\,(Q_0(\lambda_i) + \delta_e^2)^{-1}$$

This expression adheres to the treatment by Hager and Seltzer, where penetration effects are included only for the lowest multipole.

The penetration matrix element in the numerator of the penetration parameter has a similar structure as the gamma - ray matrix element except for a radial weighting factor. This function is expanded in powers of r/R. Different authors treat the series of penetration matrix elements in different ways. For the comparison of experimental results with nuclear model predictions, it has been suggested to keep only the first term, or the two first, or to include higher order terms in an approximate way[23,24]. Hager and Seltzer find the use of lowest order terms sufficient, except for E1 transitions, where a second term is included. The explicit form for their lowest order penetration parameters is given below. The gamma - ray matrix element in the denominator is expanded and the next important term beyond the long - wave approximation is kept.

$$\lambda = \frac{\langle I_f \| \underline{J}_n \cdot (\underline{L}\, Y_L)\, (\frac{r}{R})^{L+2} \| I_i \rangle}{\langle I_f \| \underline{J}_n \cdot (\underline{L} Y_L)\, (\frac{r}{R})^{L} \| I_i \rangle + C \langle I_f \| \underline{J}_n \cdot (\underline{L} Y_L)\, (\frac{r}{R})^{L+2} \| I_i \rangle} \equiv \bar{\lambda}\,(1 + C\bar{\lambda})^{-1},$$

where

$$C = -\tfrac{1}{2}(KR)^2/(2L+3)$$

for magnetic conversion and

$$\lambda_1 = \frac{-K \langle I_f \| \underline{J}_n \cdot \frac{\underline{r}}{r} (\frac{r}{R})^{L+1} Y_L \| I_i \rangle}{\langle I_f \| \underline{J}_n \cdot (\nabla (\frac{r}{R})^L\, Y_L) \| I_i \rangle - CK \langle I_f \| \underline{J}_n \cdot \frac{\underline{r}}{r} (\frac{r}{R})^{L+1}\, Y_L \| I_i \rangle}$$

$$\equiv \bar{\lambda}_1\,(1 + C\bar{\lambda}_1)^{-1}$$

and

$$\lambda_2 = \frac{\langle I_f \| \underline{J}_n \cdot (\nabla (\frac{r}{R})^{L+2}\, Y_L) \| I_i \rangle}{\langle I_f \| \underline{J}_n \cdot (\nabla (\frac{r}{R})^L Y_L) \| I_i \rangle - CK \langle I_f \| \underline{J}_n \cdot \frac{\underline{r}}{r} (\frac{r}{R})^{L+1} Y_L \| I_i \rangle}$$

$$\equiv \bar{\lambda}_2 (1 + C\bar{\lambda}_1)^{-1}$$

where

$$C = -KR / (L + 1)$$

for electric conversion. (Numerically the difference between λ_i and $\overline{\lambda}_i$ is often negligible compared to experimental uncertainties).

The influence of the penetration parameters on the internal conversion process is determined by the electron - wave - function dependent weighting factors. These penetration coefficients have been tabulated, most extensively by Hager and Seltzer†.

For 'normal' transitions, where the penetration effects are negligible, the electron correlation depends on dynamical effects of nuclear structure solely through the mixing ratio. As the determination of mixing ratios may turn out as the most fruitful application of conversion electron correlations in the near future, it seems worth - while to mention the dependence of the expression for the angular correlation function on the phase convention used. Especially so, as there have recently been theoretical predictions of the magnitude, as well as the sign, for the mixing ratio in several nuclear transitions.

The confusing situation with various conventions in current use, often not well defined, was pointed out by Huby[26])††. The topic has recently been treated in considerable detail by Rose and Brink who derive the pertinent formulae from first principles.

The Biedenharn - Rose theory[9]) uses symmetric formulae for both transitions in a two - component cascade, that is coefficients of the form

$$A_k(L, L', I, I_2) = [F_k(L, L, I, I_2) + 2\delta F_k(L, L', I, I_2) + \delta^2 F_k(L', L', I, I_2)] (1+\delta^2)^{-1}$$

This symmetry is arrived at only by use of different phase conventions for the mixing ratios of the first and the second transitions. Thus

$$\delta \equiv \frac{\langle I_i || L+1 || I_2 \rangle}{\langle I_i || L || I_2 \rangle}$$

where i = 1 or i = 3, and the intermediate state is always to the right. This means that the result for the mixing ratio depends on whether the transition happens to be the first, or second, member of a studied cascade. For a more complicated situation, for instance a triple correlation, this specification is not practicable and Rose and Brink choose to consistently use absorption matrix elements. Hager and Seltzer[20]) and Steffen[29]) instead use emission matrix elements

$$\delta_1 \equiv \frac{\langle I_2 || L_1+1 || I_1 \rangle}{\langle I_2 || L_1 || I_1 \rangle} , \qquad \delta_2 \equiv \frac{\langle I_3 || L_2+1 || I_2 \rangle}{\langle I_3 || L_2 || I_2 \rangle}$$

† Beside circulating a comprehensive table[19]), H.-C. Pauli has made available his program for the calculation of conversion coefficients, particle parameters and penetration coefficients[25]).

†† Ofer discussed this from an experimental point of view[27]).

This consistent phase definition implies a change in the expression for the angular correlation function. A factor $(-1)^{L'-L}$ appears in front of the interference term in the coefficient for the first transition. For the usual case of (L + 1)/L interference the appropriate expression is

$$A_k(L, L+1, I_1, I_2; \delta_e; e_X) = [\, b_k(\pi L, \pi L; X)\, F_k(L, L, I_1, I_2)$$
$$- 2\delta_e b_k(\pi L, \pi' L+1; X)\, F_k(L, L+1, I_1, I_2)$$
$$+ \delta_e{}^2 b_k(\pi' L+1, \pi' L+1; X)\, F_k(L+1, L+1, I_1, I_2)]\,(1+\delta_e{}^2)^{-1}$$

3. Experimental techniques
3.1 Introduction

From the theoretical point of view the following looks like a fairly simple problem: An isolated nucleus emits cascading nuclear radiations of which at least one involves conversion electrons. The object is to measure the transition probability as a function of the angle between the directions of emission. Because of the existing symmetry the experiment can be confined to a plane.

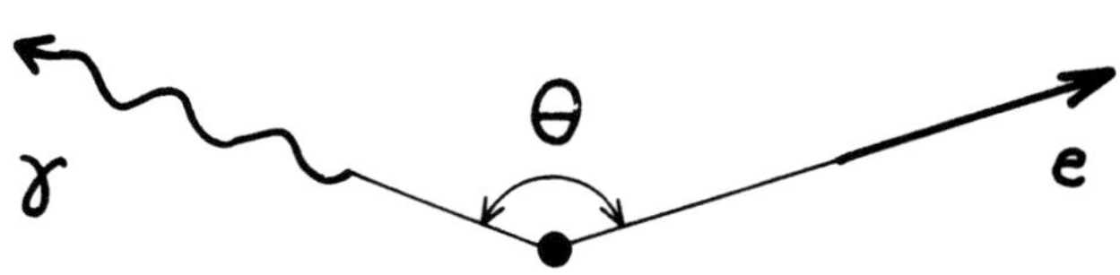

When expressed in the language of experimental physics, however, the problem is not that simple. Rather than the theoretical concept of an isolated atom, the experimentalist is forced to deal with an assembly of nuclei, closely packed together to form a radioactive source. For most cases of practical interest, the source has to be solid, which means that the possible influence of the environment on the angular correlation must be considered. It is well known that at least for cascades where the intermediate level has a lifetime in the nanosecond region, substantial attenuations occur.

Thus, the gross structure of the source is a parameter that has to be considered, and it may change in the course of measurements, due to oxidation or hygroscopical effects. These effects may show up in the experiments as lack of consistency and reproducibility. Thus, the experimentalist frequently has to take special measures to avoid what is sometimes referred to as 'the source - ageing effect'.

Moreover, electrons from the interior of the source pass through solid material to reach the surface, which results in problems of energy degradation, and scattering, affecting the angular correlation. Further, electrons scattered from the source backing and the material of the source holder may give a contribution to the coincidence counting rate.

The random coincidences must be corrected for, and this puts an upper limit on the source strength.

The source should be placed in the centre of rotation of the movable detector (when a two - detector system is used). Imperfect centering is generally corrected for by a normalization procedure. This may be dangerous when the set - up contains bulk material, causing a highly anisotropic background of scattered radiation.

For intensity reasons the radiations have to be accepted within finite apertures, which means that solid angle corrections must be applied. There are convenient formulae for these corrections. In order to apply these formulae, however, it is necessary to know the aperture angles, and the detection efficiency as a function of the angle, quite precisely. This is true in particular for measurements where spectrometers of the magnetic lens type are used in the electron channel. Here the correction for the range of angles of acceptance may be as large as 30 per cent (for the P_2 distribution, and correspondingly larger for P_4). To keep the uncertainties in the solid angle corrections a minor fraction of the statistical accuracy achieved, calls for a precision method.

In addition to these experimental difficulties there is a series of special problems related to the possibility of closely spaced lines in the spectra, as well as unavoidable radioactive contaminations. The energy resolution in the two channels is one of the most important aspects to be considered.

The development of the last decade has to a great extent aimed at methods to overcome these experimental limitations.

3.2 *Multidetector systems*

Since they are based on the accumulation of coincidences, rather than singles counting rates, angular correlation measurements are necessarily tedious to perform. For gamma - gamma correlations various multidetector systems have been devised to enhance the data acquisition rate. The most efficient technique is the multichannel goniometer (MCG), where one records all possible coincidence combinations between any two of a number of simultaneously operated gamma detectors. Thus, in the MCG the efficiency increases as N (N − 1), where N is the number of detectors used. Systems with up to eight detectors have been developed[30]).

A necessary prerequisite for an MCG mode of operation is the simultaneous registration in all detectors of both cascading transitions. This automatically excludes electron - gamma correlations. An MCG adapted for electron - electron correlations, however, is conceivable, in particular if silicon detectors are used.

More moderate advantages may be obtained by the use of several simultaneously operating detectors in various kinds of coincidence arrangements. Here, the efficiency gain, as well as the complexity of the experimental equipment, becomes roughly proportional to the number of detectors used. The major difficulties are, in a set - up of this kind, associated with the lack of symmetry. By a permutation of the position of the detectors, one may, however, to a certain extent overcome these difficulties.

An elaborate system including up to four magnetic spectrometers has been developed by Siegbahn et al.[31]).

There are certain advantages with a three detector system for simultaneous measurements of electron - gamma and gamma - gamma correlations, as will be discussed in the following, in the connection with attenuation effects. In the use of such systems it is also possible to eliminate, or at least substantially reduce, some possible sources of errors, arising from an erroneous normalization for an imperfect centering of the radioactive source with respect to the centre of movement of the gamma detector. For positive particle parameters the fixed detector should be placed at 180°, with respect to the spectrometer axis, whereas for negative particle parameters the 90° position is to be preferred. With this arrangement, normalization errors present in both of the measured correlations tend to cancel, when the ratio between electron and gamma correlation coefficients is looked for, rather than the absolute values of the coefficients[32]).

3.3 Storage and display of coincidence spectra

For the analysis of electron - gamma correlations it is sufficient to define the accepted energies in both channels. In practice, however, the finite acceptance intervals necessarily include admixtures of neighbouring lines and Compton distributions for transitions of higher energy. These have to be corrected for, and the same applies to scattered background radiation which, as has already been mentioned, may give a highly anisotropic contribution. For these reasons it is preferrable to record the complete gamma spectrum in coincidence with the conversion electrons. With suitable programming the memory of a multichannel analyzer is split into a set of independent memories used to store coincidence spectra recorded at different angular positions.

These spectra represent the total coincidence counting rate, that is the sum of genuine and random coincidences. There are different techniques for automatic subtraction of the random coincidences[33,34,35]). A convenient method takes advantage of the subtraction function of modern multichannel analyzers[33,36]). The simultaneous random - coincidence subtraction facilitates the use of stronger sources. It is also of particular advantage in the case of shortlived activities where the ratio of true to random coincidences varies during the course of measurement. The same applies when activities are growing in, as a consequence of a preceding radioactive decay.

It may also be noted that the fast electronics developed during recent years has made the problems connected with random coincidences less severe. A time resolution of a few nanoseconds is now easily achieved. For most practical cases this is sufficient to give quite acceptable signal - to - noise ratios.

3.4 Static attenuations

The fact that essentially all electron - gamma correlations are made with solid sources does not imply that one is always faced with more or less strongly attenuated correlations. On the contrary, according to our experience, attenuations only occur in a limited number of cases. Generally, one anticipates under what conditions the angular correlation is affected.

However, one cannot always avoid to treat this problem. Conversion electron correlations often aim at determinations of the particle parameters b_k rather than A_k coefficients. The b_k parameters may be absolutely determined, regardless of any attenuations. The attenuation coefficients cancel in the ratio

$$b_k = \frac{G_k \, a_k(e\gamma)}{G_k \, a_k(\gamma\gamma)} \quad \dagger$$

The comparison method has an additional advantage. It makes the experimental results independent of competing admixtures in the common gamma channel, since $A_k(\gamma_2)$ as well as G_k cancels in the comparison. The coefficients $a_k(e_1 \gamma_2)$ and $a_k(\gamma_1 \gamma_2)$ may be taken as average values of various competing gamma transitions.

To take full advantage of these possibilities, systems with three detectors have been introduced. These may comprise spatially fixed electron and gamma channels, simultaneously operated in coincidence with a movable gamma detector[32]). Since the same detector is used for both correlations, one eliminates errors due to different acceptance intervals of the gamma spectrum.

A second line of approach implies the determination of and correction for the effects of static attenuation. For this purpose time - differential angular correlations may be used. Present techniques facilitate time resolutions for the energy range of interest of the order of 1 or 2 nanoseconds, when sodium - iodide crystals are used. More important than this half - width of the resolution curve is the shortest time interval that can be separated by unfolding the differential time spectrum.

Thus, one may study the initial fraction of the decay. This is about an order of magnitude shorter than the full - width at half maximum of the time - resolution curve, which means that the effects of static quadrupole interaction may be considered negligibly small. Moreover, the complete time - differential measurements may be used to determine the strength of the quadrupole interaction, such that proper corrections for these effects can be applied.

3.5 *Scattering and energy degradation*

The problems of scattering and energy degradation of the electrons are very difficult to treat theoretically. Frankel[37]) showed how the scattering effects caused by source thickness could be treated theoretically when the scattering function is assumed to be known. However, a nomogram by Gimmi et al.[38]), based on Frankel's theoretical treatment, is now known to give erroneous results. Most experimentalists have preferred to restrict themselves to isotopes available with such a high specific activity that the scattering attenuation becomes negligible. To overcome this limitation Stefansson et al.[39]) worked out a semiempirical method. The relation between energy

† This argument presupposes that the nucleus will experience its environment in precisely the same way in the two measurements. This is true, of course, for gamma - gamma and gamma - electron correlations, but not necessarily for electron - gamma correlations, since a preceding conversion process sometimes gives rise to a specific time - dependent hyperfine - interaction (this so called K - hole effect is briefly discussed in Sect.V), and possibly also recoil effects.

degradation and scattering attenuation was used for the construction of a set of curves for different Z - values. The experimentally determined energy degradation is used as an input parameter to find the relevant scattering attenuation coefficient. By this method, it has become possible to extend the range of applicability to isotopes available only in fairly low specific activities, including many interesting reactor - produced isotopes.

Backing and source holder scattering is also associated with energy degradation. This may be used both to detect the amount of scattering present and to discriminate against unwanted scattered electrons. However, in the presence of very high energy transitions, scattered electrons may be accepted at lower energy settings. Their contribution to the coincidence counting rate requires that the high energy transition is in coincidence with the cascading gamma. The magnitude of such admixtures is reduced with increased resolution. The energy spectrum of scattered electrons in general varies smoothly and an offline setting of the electron spectrometer is then sufficient for a determination of the background correction. An extensive investigation of scattering effects in a lens spectrometer has been published by Kleinheinz et al.[40]).

3.6 The electron channel

We proceed to deal with the detection of the cascading radiations, considered as two independent problems, and will first discuss the electron channel.

The optimum choice of angular acceptance interval depends of course on the choice of the electron detection system and of the angular resolution required, and in addition, for magnetic spectrometers, on the resolving power required. For lens type spectrometers, where the central cone is screened off, it may be shown that optimum figures of merit are obtained for an average take - off angle of $a \approx 21^\circ$ for P_2 measurements, and $a \approx 12^\circ$ for P_4 measurements[41]). This implies fairly large solid angle corrections, which, however, may be accurately determined by simple experimental methods[41,42]). Theoretical solid angle correction factors for various kinds of magnetic spectrometers have been given by Vos et al.[43]).

The very successful development of the silicon detector has presented the experimentalist with a tempting alternative to magnetic spectrometers. As far as the energy resolution is concerned, for the energy range of interest in most conversion electron correlations, the silicon detector now competes favourably.

However, experimentalists in the field generally agree that the problem has more dimensions than just that of energy resolution. The multichannel mode of operation, which is the outstanding feature of the silicon detector in all singles counting applications, is not necessarily an advantage in a coincidence experiment. Here the data collection time required is ultimately limited by the gathering of coincidences corresponding to the transition of weakest intensity of the cascades under investigation. The simultaneous measurement of more intense cascades present in the decay scheme frequently amounts to a non - sensical accumulation of excess information.

From this particular point of view, electron correlation studies with the single channel magnetic spectrometer may in fact turn out to be less time consuming. The simple reason for this paradoxical statement is the fact that in the magnetic spectrometer the radiation under investigation is selected before detection. Thus one discriminates against all unwanted electrons, singling out the events of interest. A weak feeding may then be compensated for by the use of a correspondingly stronger source.

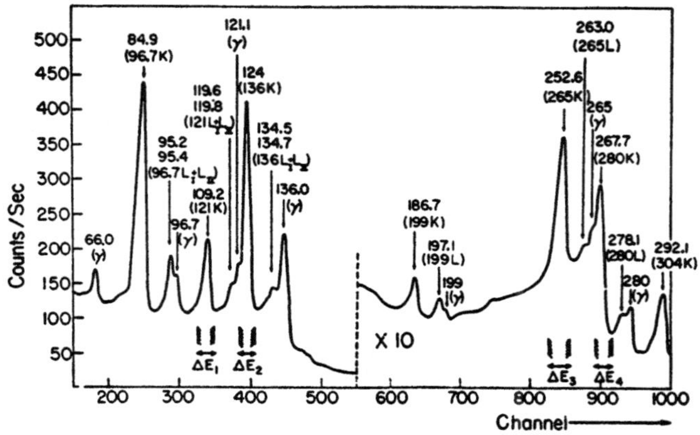

Fig. 3. The decay of ^{75}Se observed with a Si (Li) detector in an electron - gamma angular correlation setup. (Becker and Steffen[29]).

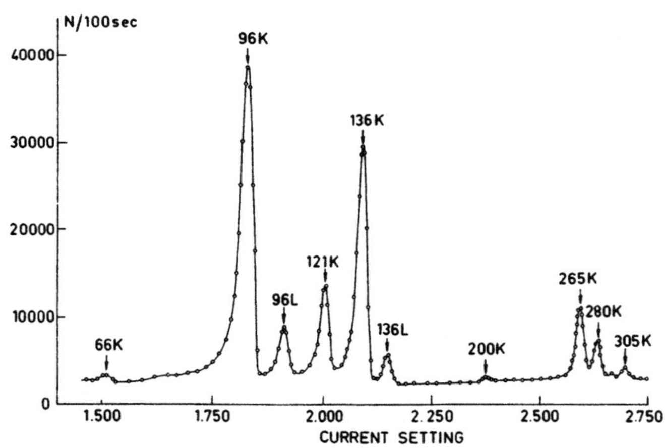

Fig. 4. The internal conversion electron spectrum of ^{75}Se registered by use of a magnetic lens spectrometer for angular correlation studies. (Vignau et al.[43a]).

The spectral response of the magnetic spectrometer is certainly superior to that of the silicon detector. This is true particularly in the low energy region where the resolution is better, and where the silicon spectra are disturbed by the presence of x - ray lines.

The silicon detector is certainly nevertheless very useful for electron - gamma angular correlations, as has been demonstrated by Coppage and Steffen[44]), Speidel et al.[45]), Kawakami et al.[46]), and others (e.g. ref. 47, 47a, 48).

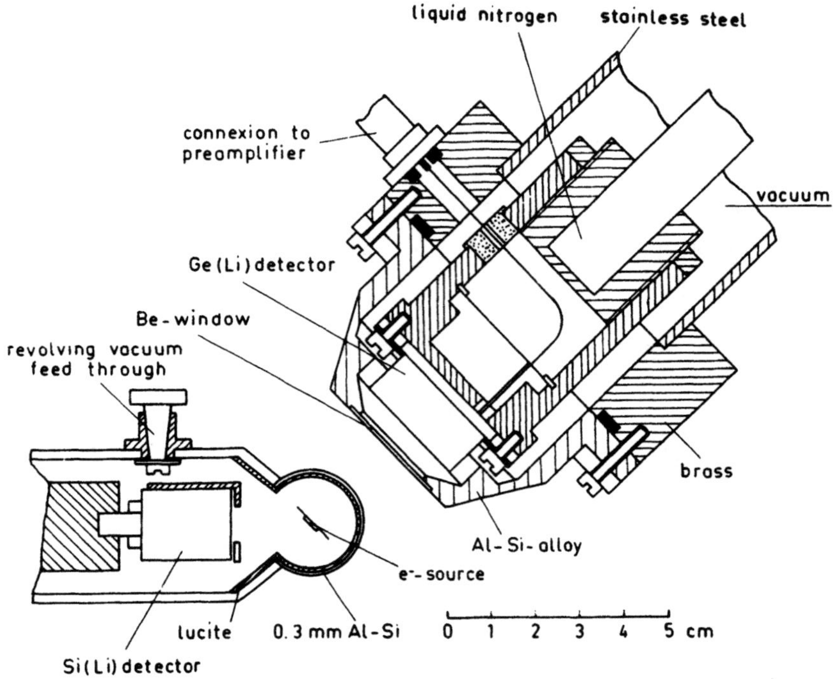

Fig. 5. A Si (Li) - Ge (Li) detector combination for electron - gamma angular correlations, due to K.-H. Speidel et al.

It seems likely, however, that it will find its most fruitful applications in combinations with magnetic spectrometers. So far, there have been fairly few experimental studies of electron - electron correlations. There are a few instruments particularly developed for this kind of measurements, using magnetic momentum selection in both channels, either lens- or sector-field spectrometers[31,49]). The silicon detector, in combination with a single magnetic spectrometer, here represents an attractive alternative for electron - electron correlations.

Another combination of great potential value is magnetic spectrometers equipped with a silicon detector. This arrangement combines the virtues of the two analysing systems.

So far, however, most experiments have been performed with lens type spectrometers encapsuled in iron[41,50]), and with sector field focussing instruments[51,52]). The figures of merit of these are comparable. For high - resolution studies a few double - focussing instruments have been adapted for electron - gamma correlations[53]).

Some specific advantages of the electron - gamma technique, as compared to gamma - gamma correlations, seem to be emphasized when magnetic spectrometers are used. First, there is no contribution from Compton - scattered gamma - radiation under the line of interest. Second, the acceptance of coincident radiation scattered between the two detectors should be absent.

In connection with the problem of solid angle corrections for lens spectrometers there has been a great deal of discussion concerning the deviation from rotational

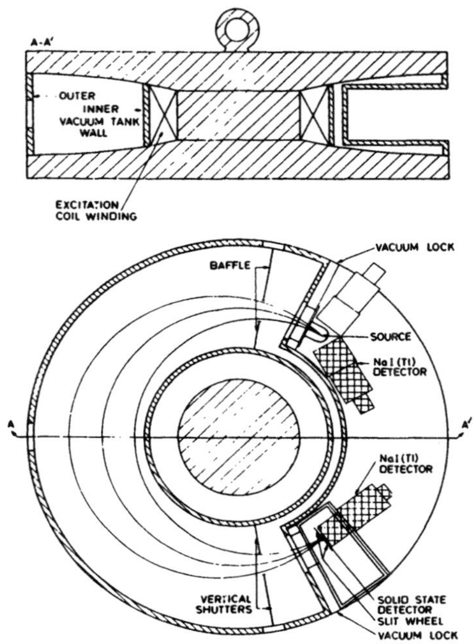

Fig. 6. The 50 cm radius iron - core double focussing beta - ray spectrometer in Uppsala adapted for electron - gamma angular correlations with high resolution in the electron channel. (From S. Antman et al.[53]).

symmetry. In a recent study it was shown that minute rotational asymmetries easily can be detected and, if necessary, corrected for[54]). This is achieved simply by comparing the coincidence counting rates at the 135° and 225° angular settings.

3.7 *The gamma channel*

With very few exceptions electron - gamma correlations so far have been studied with sodium - iodide crystals as detectors in the gamma channel. It is somewhat surprising that the germanium detector, with its superior energy resolution properties has been so slow in finding a more general acceptance. The reason has probably been that germanium detectors have a smaller detection efficiency per unit volume. Since germanium detectors of large volume only recently have become commercially available, experimentalists have considered them too inefficient for correlation experiments. In addition there have been technical problems, related to the time response characteristics of the germanium detectors. The more sofisticated preamplifier electronics, pile - up rejection, pole - zero cancellation and base - line restoration has substantially improved their low capacity. Integrated counting rates of the order of 50 k are now feasible without appreciably reduced energy resolution. The germanium detector, therefore, now presents the superior alternative and should be introduced as a replacement for the sodium - iodide crystals. With its superior energy resolution properties, the germanium detector facilitates the study of many interesting decay schemes of great complexity, so far avoided by the experimentalists in the field.

4. Experimental results

4.1 *Introduction*

As discussed in the foregoing, the conversion electron correlation depends partly on the internal conversion process as such, and partly on the nuclear structure. In principle the internal conversion process may be considered perfectly accounted for by quantum electrodynamics. It is a complex problem however, and a great deal of efforts have been directed towards the presentation of the internal conversion process in terms of physical observables. In general, the agreement between the present theory and the experimental results is quite satisfactory. Some E2 transitions have been disturbing, since they seemed to indicate the presence of theoretical shortcomings. This problem, however, now seems to be clarified as will be discussed below.

In the following, magnetic dipole, and electric quadrupole, dipole and monopole transitions are treated separately. As remarked by Steffen[55]), there are few transitions of higher multipoles suitable for electron - correlation experiments. An exception is M4. Experimental and theoretical particle parameters for the M4 transitions are in good agreement[55]).

It seems yet to be an unsettled question how accurate the theoretical particle parameters are. New anomalies are likely to be claimed by the experimentalists in the future, who thus will urge the theorists to reconsider the description underlying the tabulations of particle parameters. However, it seems that for this one needs more refined experimental techniques, still to be developed.

The rather comforting conclusion is that at present'anomalies' only appear when they are to be excepted due to penetration effects[15,21,56,57]). Hence, one may safely proceed to the study of problems directly related to nuclear structure. The parameters determined in the study of penetration effects are purely nuclear. They provide the same kind of information as life - time measurements. Thus, for the interpretation of their physical significance a diversified set of models of the nuclear structure is invoked. However, since different kinds of problems in nuclear structure in a fairly natural way are related to different transition multipolarities, the subdivision into transition multipolarity and character nevertheless seems appropriate for the following summary of experimental results.

4.2 *E2 transitions*

There are at least two different continuum states available for an emitted electron cf. table 1. The partial waves of lowest κ are most affected by details of the nuclear charge distribution, due to the larger amplitude of the corresponding wave function in the vicinity of the nucleus. However, for K shell conversion of electric transitions, a cancellation occurs among the radial matrix element for a transition to this state. This is the reason that the internal conversion coefficients obtained in the point nucleus approximation do not differ much from those calculated by use of more realistic nuclear models.

Table 1

Selection rules for κ relevant for magnetic multipoles. For electric transitions final states with the opposite sign of κ are allowed.

Initial state			Final state ML transitions
κ_i	Spectroscopic notation	x-ray notation	
-1	$s_{1/2}$	$K, L_1, M_1 \ldots$	$-L, L+1$
1	$p_{1/2}$	$L_2, M_2 \ldots$	$L, -(L+1)$
-2	$p_{3/2}$	$L_3, M_3 \ldots$	$-(L-1), L, -(L+1), L+2$
2	$d_{3/2}$	$M_4 \ldots$	$L-1, -L, L+1, -(L+2)$
-3	$d_{5/2}$	$M_5 \ldots$	$-(L-2), L-1, -L, L+1, -(L+2), L+3$

It has been suggested that for strongly deformed nuclei the large nuclear quadrupole moment might cause a static deformation of the electron wave functions. This might show up as anomalies in conversion coefficients and particle parameters. Theoretical studies of this effect, however, furnish little support for this hypothesis (ref. 57 and 58).

For paramagnetic atoms it is conceivable that electron wave functions could be affected to such an extent that deviations from the normal internal conversion coefficients and particle parameters would show up in the experimental results. This problem has not been theoretically investigated, and so far there seems to be no experimental evidence for the existence of such an effect.

The chemical state of the atom, however, is known to affect the internal conversion process, but this applies only the outer shell electrons and is therefore of no concern in the present context[59,60]).

Recent precision measurements of the K conversion coefficients are in excellent agreement with the theoretical predictions. There are, however, certain problems related to the L_1 subshell E2 conversion coefficients. Experimental results for L_1/L_2 and L_1/L_3 subshell ratios differ significantly from the theoretical values, and the reason for this is not completely understood[61]). In a very recent investigation Band et al. showed that the inner part of the atom is of major importance for the internal conversion coefficients[60a]). However, for $\kappa_i = -1$ conversion of electric transitions it was found that comparatively large distances have to be considered in order to obtain the asymptotic values for the conversion coefficients. It is suggested that this may be the reason for the anomalies in the intensity ratios where L_1 electrons are involved.

As far as the particle parameters are concerned some disturbing deviations have been reported. In his recent review Steffen calls attention to a discrepancy of about 15 per cent between experimental and theoretical K and L particle parameters for some fast E2 transitions[55]). In table 2 the pertinent experimental information is brought up to date. Our conclusion is that there is no remaining evidence for significant deviations from the theory.

Table 2

Particle parameters $b_2(E2; e_K)$ for some E2 transitions for which the existence of anomalies has been suggested.

^{152}Sm		^{154}Gd	^{160}Dy	
122 keV	245 keV	123 keV	87 keV	Reference
1.65 ± 0.17	1.05 ± 0.16	1.41 ± 0.07		65
1.90 ± 0.07	1.76 ± 0.07*	1.88 ± 0.06		66
1.60 ± 0.05	1.74 ± 0.07*	1.37 ± 0.07		67
1.82 ± 0.06*	1.68 ± 0.07*	1.89 ± 0.08		68
			1.72 ± 0.08	69
1.82 ± 0.05*			1.90 ± 0.07*	70
		1.87 ± 0.04		63
		1.77 ± 0.13		71
1.77 ± 0.03				72
1.93 ± 0.04*		1.93 ± 0.08*		73
1.97 ± 0.11	2.11 ± 0.19			64
1.89	1.72	1.88	1.92	Theory

*) The b_4 method has been applied.

In our opinion, the deviations reported are due to experimental difficulties encountered for the particular cascades in ^{152}Sm and ^{154}Gd. For instance, Van Krugten and Koopmann obtained irreproducible results when their europium - chloride activity was irregularily exposed to air[63]). This was explained by hygroscopic effects which influence the gross structure of the source, and thereby change the strength of the quadrupole coupling. In addition, there are some difficulties related to the gamma-gamma correlations where a proper background correction presents certain problems (ref. 64).

The b_4 method, successfully applied in a number of conversion electron correlations, offers unique possibilities here. It results in reduced systematic errors and renders it possible to check the internal consistency of the results obtained†.

The transitions surveyed in table 2 are of collective nature and the transition rate is enhanced relative to the single - particle estimate. Thus, no effects of nuclear penetration are expected.

4.3 M1 transitions

The K shell conversion of magnetic multipoles is dominated by the electron transition most sensitive to details of the nuclear - charge and current distribution, $\kappa_i = -1$ to $\kappa = -L$. (This means for instance that M1 conversion is dominated by the

† One makes use of the fact that particle parameters of different ranks are related by recurrence formulae. For quadrupole conversion of $|\kappa_i| = 1$ transitions the relevant expression is $b_2 = 1.4 - 0.4\, b_4$.

transition to $\kappa = -1$. Consequently, the anisotropy of the electron distribution is small). However, even when the angular momentum transfer is as small as possible, i.e. in M1 transitions, the particle parameters are only moderately influenced by static effects of the nuclear size. The M1 transitions are often retarded. The study of M1 particle parameters has thus concentrated on the determination of penetration effects.

Table 3. Experimental results for λ(M1) from angular correlation measurements.[†]

Isotope	Transition energy(keV)	Retardation factor	λ	Remark	Reference
^{121}Te	212	30	-7 ± 1	$\Delta l = 2$	86
^{123}Te	159		-2 ± 2		87
^{131}Cs	92		23^{+7}_{-11}		83
^{133}Cs	303	97	$-2.0^{+3.3}_{-3.1}$	$\Delta l = 0$	82
	223	4000	15^{+4}_{-3}		80a
	161	230	41 ± 6		82
			40^{+12}_{-15}		80a
			41 ± 7		81
	81	420	2.8 ± 3.3	$\Delta l = 2$	82
^{181}Ta	482˙	3×10^6	210 ± 30	$\Delta\Lambda = 2$	78,79
^{201}Tl	361		-0.5 ± 2	$\Delta l = 0$	79,89
	330		5 ± 4	$\Delta l = 2$	79,89
^{203}Tl	400		1.3 ± 0.8	$\Delta l = 0$	79,88
	279		7.5 ± 1.5	$\Delta l = 2$	79,88
			9 ± 1		89a
^{207}Tl	350		$0.5 < \lambda < 1.2$	$\Delta l = 2$	90

[†]) Cf. also table 4.

Expressions for the penetration parameters for magnetic transitions are given in section 2. By use of an explicit expression for the nuclear current the following expression for the penetration parameter is obtained, in the formalism of Hager and Seltzer:

$$\overline{\lambda} = [-2\langle I_f \| g_l(\nabla\cdot(\tfrac{r}{R})^{L+2}Y_L)\cdot \boldsymbol{\ell} \| I_i\rangle$$

$$+ \tfrac{L}{R}\langle I_f \| g_s\underline{\sigma}\cdot\tfrac{\mathbf{r}}{R}(\tfrac{r}{R})^L Y_L \| I_i\rangle - \tfrac{L+3}{2}\langle I_f \| (\tfrac{r}{R})^2 g_s\underline{\sigma}\cdot(\nabla(\tfrac{r}{R})^L Y_L) \| I_i\rangle]$$

$$\times [-2(L+1)\langle I_f \| (\nabla\cdot(\tfrac{r}{R})^L Y_L)\cdot(\tfrac{g_l \boldsymbol{\ell}}{L+1} + \tfrac{g_s}{4}\underline{\sigma}) \| I_i\rangle]^{-1}$$

(Summation over all nucleons is understood). The interesting aspects of this expression is the appearance of the radial weighting factor $(r/R)^2$ in the numerator, and the additional (second) term of form $\underline{\sigma} . \underline{r}$. Owing to the presence of these factors, there might be observable nuclear structure dependent effects in the M1 internal conversion process.

In table 3 are summarized some M1 transitions studied by the electron correlation method, and the deduced values for the penetration parameter†. The transitions denoted by a change of the quantum number l may be accounted for in the shell-model picture. The M1 gamma ray transition is retarded by orders of magnitude when the $\Delta l \neq 0$ selection rule is effective. However, due to the $\underline{\sigma} . \underline{r}$ term in the penetration matrix element, nuclear states differing by two units of orbital angular momentum are connected. Using an experimental value for the forbidden gamma matrix elements one may make shell model predictions for the corresponding λ values.

An interesting example is ^{203}Tl where the single particle model predicts $\lambda \approx 10$, to be compared with the theoretical result $\lambda = 7.5 \pm 1.5$. R.J. Lombard[76]) and J. Le Tourneux[77]) have studied how the value for λ is changed when one allows for a certain amount of core - excitation.

The 482 keV transition in ^{181}Ta represents the largest λ - value so far reported. Here, the sizeable effects of nuclear penetration may be understood as an extremely strong retardation of the M1 gamma - ray transition. This is in accordance with the selection rules for the quantum numbers in the Nilsson model. The change in the quantum numbers I K π [N n_z Λ] is 5/2 5/2 + [4 0 2] → 7/2 7/2 + [4 0 4]. This selection rule, however, does not apply to the penetration matrix element. The experimental value is in reasonable agreement with the theoretical estimate.

This particular M1 transition offers the opportunity to search for an anomalous E1 contribution. A minute parity - violating E1 admixture is to be expected because of the weak internucleon interaction. Estimates of its magnitude have been made, based on the current - current hypothesis. If the amplitude of the part of the wave - function of irregular parity is denoted by F, that is $\Psi = \Psi_{reg} + F \Psi_{irreg.}$, the ratio of parity - irregular to regular transition amplitudes becomes F R, where R is a nuclear structure dependent factor. Wahlborn introduces the expression

$$FR = \lambda U(\widetilde{E}1) \; / \; U(M1; \text{pen.})$$

where U denotes the corresponding transition amplitude[80]). By use of the experimental value for λ it is possible to obtain the magnitude and sign of F R. There is qualitative agreement between this result and experiments. This is an exciting aspect of nuclear structure research, where one may expect a great deal in the near future.

Recently some l - forbidden M1 transitions in cesium nuclei have been the subject of several experimental investigations[80a,81,82,83]). The structure of these nuclei has been discussed in terms of the pairing - plus - quadrupole model by Kisslinger and Sorensen. However, no detailed comparison between experimental results for the

† Before 1964 erroneous values for the interference particle parameter were used. The sign for b_2 (M1, E2) in the Biedenharn - Rose paper[9]) was corrected by Church et al.[74]). The consequences of the reinterpretation of experimental results is discussed e.g. in ref. 75.

penetration parameters and predictions of the model has so far been carried out. Penetration effects in several other nuclei in this region were studied in a recent theoretical investigation by use of a unified model with particle - phonon coupling[84]).

There have been several studies of transitions between vibrational levels. These are supposed to be forbidden for both conversion electrons and gamma rays. However, to what extent the degree of forbiddenness results in a more or less perfect cancellation is difficult to estimate. λ - values differing from zero could therefore be possible. Some results are given in the following in connection with E0 - transitions.

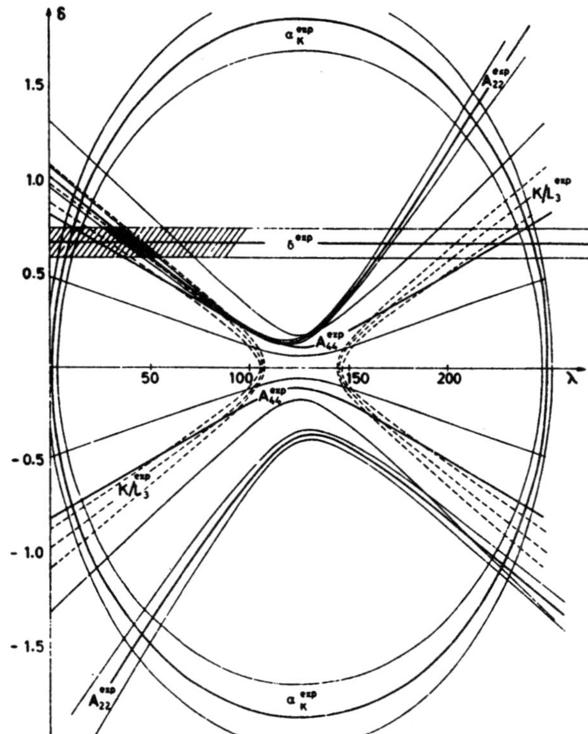

Fig. 7. Analysis of the 161 keV M1 - E2 transitions in ^{133}Cs. The following observables are plotted in the λδ plane: a_2 (276γ - 161K), a_4 (276γ - 161K), α_K (161), and K/L_3 (161). The value for δ is obtained from the (276K - 161γ) angular correlation. (from ref. 80a).

Reiner, in an early theoretical study, investigated M1 transitions between members of the same rotational band[85]). Since the matrix element M_γ (M1) is proportional to $g_\Omega - g_R$, an accidental cancellation may result in a vanishing M1 gamma-ray contribution. Reiner concluded that appreciable penetration effects might occur due to the spin - dependent term in the matrix element. Experimental verification of this prediction is hampered by the fact that the E2 component is dominating in transitions with a highly retarded M1 strength.

4.4 *E1 transitions*

Also the E1 transitions present a fertile ground for the study of the effects of nuclear penetration in the internal conversion process[24,57,91,92]. It is well known that strongly retarded E1 transitions are frequent in the regions of deformed nuclei. Sizeable 'anomalies' were discovered by Asaro et al. who made a systematic investigation of E1 L - subshell ratios[93]. Since then several experiments of this kind have been reported. Conversion electron correlations, however, have proved to be an alternative of great potential value.

The description of penetration in electric dipole transitions is more complex than in the case of magnetic transitions. The expanded form of the electric vector potential contains both a gradient and a radial term. Thus, two distinctly different kinds of penetration matrix elements appear, as mentioned in section 2. By use of the explicit form of the nuclear current the following expressions are obtained for the penetration parameters of nuclear current and nuclear charge type:

$$\bar{\lambda}_1 = -\frac{1}{2MR}\Big[2 < I_f \| g_\ell \left(\tfrac{r}{R}\right)^L Y_L \, \underline{r} \cdot \nabla \| I_i >$$

$$+ \; (3+L) < I_f \| g_\ell \left(\tfrac{r}{R}\right)^L Y_L \| I_i > + < I_f \| \tfrac{g_s}{2} \, \underline{\sigma} \cdot \underline{\ell} \left(\tfrac{r}{R}\right)^2 Y_L \| I_i > \Big]$$

$$\times \; \left[< I_f \| g_\ell \left(\tfrac{r}{R}\right)^L Y_L \| I_i > \right]^{-1}$$

$$\bar{\lambda}_2 = < I_f \| \left(\tfrac{r}{R}\right)^{L+2} g_\ell Y_L \| I_i > / < I_f \| \left(\tfrac{r}{R}\right)^L g_\ell \| I_i >$$

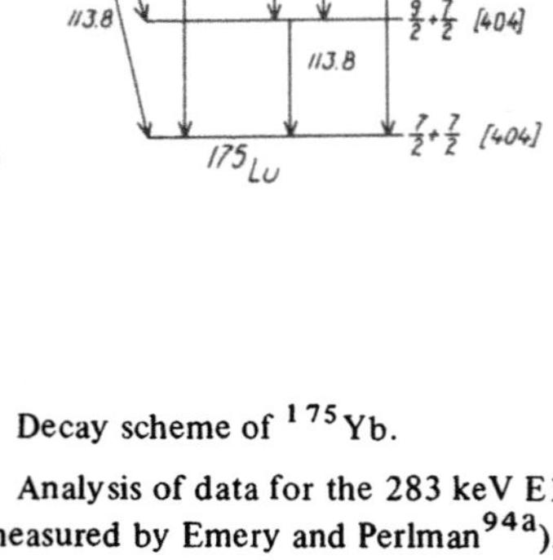

Fig 8

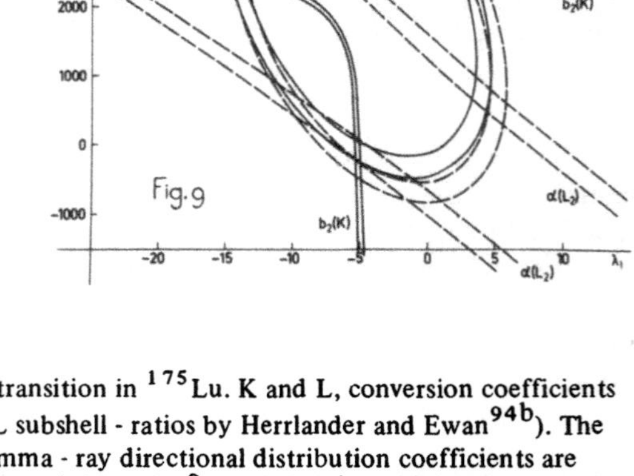

Fig. 8. Decay scheme of ^{175}Yb.

Fig. 9. Analysis of data for the 283 keV E1 transition in ^{175}Lu. K and L, conversion coefficients were measured by Emery and Perlman[94a], L subshell - ratios by Herrlander and Ewan[94b]. The ratio between K conversion - electron and gamma - ray directional distribution coefficients are taken from L. Holmberg et al.[94] who used the mixing ratio $\delta = -0.053 \pm 0.010$ for the analysis.

In general, penetration effects are expected to be relatively more important for the electron transition corresponding to the final - state partial wave of lowest $|\kappa|$. However, as first pointed out by Church and Weneser, for E1 transitions there occurs a cancellation in the $\kappa = +1$ weighting function for the nuclear charge penetration term of first order (λ_2). In comparison, this cancellation makes the nuclear current contribution (λ_1) much more important than otherwise expected.

The 283 keV transition in ^{175}Lu demonstrates the power of the electron correlation method[94, 95]. Here the E1 transition is retarded by a factor of 10^6, as compared to the Weisskopf estimate. The analysis of the conversion electron data gives for the penetration parameters (cf. fig. 9)

$$\lambda_1 = -5.0 \pm 0.2, \quad \lambda_2 = -150 \pm 150$$

These results are accounted for by the Nilsson model. Selection rules for the asymptotic quantum numbers exclude the E1 gamma ray matrix element as well as the nuclear charge matrix element, whereas the nuclear current matrix element is expected to have its nominal value.

As shown in fig. 8 there are three competing E1 transitions originating from the 396 keV level. To all three of these the same arguments apply, and the experimental results show that the relative values obtained for the nuclear current matrix elements are in accordance with what is to be expected for allowed transitions.

A measurement of the 208 keV transition in ^{177}Hf is discussed in a following contribution.

H.-C. Pauli has called attention to the interesting case of the 84.2 keV transition in ^{231}Pa which is retarded[24] by a factor $2\ 10^6$. The L - subshell ratios verify the existence of large penetration effects. However, the value for the nuclear charge penetration parameter λ_1 cannot be uniquely determined by these experimental data. The data seem to favour a positive sign which is hard to reconcile with the negative sign obtained for λ_1 in the analogous transition in ^{233}Pa. Törnkvist et al. measured relative gamma - L subshell electron correlations[96]. For the quantity

$$X = a_2(\gamma - (L_1+L_2))/\, a_2(\gamma - L_3)$$

they obtain

$$X(\text{exp.})/X(\text{theor.}) = -0.07 \pm 0.07.$$

This result shows the value of λ_1 to be negative, $\overline{\lambda}_1 < 4$.

4.5 *E0 transitions*

The E0 interaction Hamiltonian reduces to

$$H(L{=}0) = -e \int_0^\infty \rho_n \langle e_f |\, \frac{1}{r_n} - \frac{1}{r_e} \,| e_i \rangle\, d\tau_n$$

and the nuclear strength parameter is defined as

$$\rho = \int \varphi_f^* \left[\left(\frac{r}{R}\right)^2 - \sigma \left(\frac{r}{R}\right)^4 + \ldots \right] \varphi_i \, d\tau_n ,$$

summation over protons understood. This interaction gives rise to the emission of E0 electrons. The E0 deexcitation mode is of special importance in the $0^+ \to 0^+$ transitions where, of course, no gamma radiation can be emitted. However, as pointed out by Church and Weneser, E0 conversion is expected in any $\Delta I = 0$, no,transitions, i.e. in direct competition with components of higher multipolarity[13]).

As there is no corresponding gamma radiation, it is not meaningful to define internal conversion coefficients or particle parameters for E0 conversion processes. When E0 conversion occurs together with E2, there is an anisotropic E0 - E2 interference term. The analysis becomes somewhat complex since both M1 and E2 admixtures have to be considered.

The existence of the E0 component in competition with M1 and E2 was demonstrated by a measurement of the electron - gamma angular correlation for the $2^+ \to 2^+ \to 0^+$ cascade in ^{196}Pt. As mentioned in section 4.3 a penetration contribution to the M1 component cannot be excluded in this kind of vibrational transitions. For the electron - gamma correlation the relevant expression then is

$$
\begin{aligned}
W(\vartheta) = 1 + [&F_2(1,1,2,2)\, b_2(M1,M1)\, Q_2(\lambda) - 2\delta_e F_2(1,2,2,2) \\
&\times b_2(M1,E2) Q_2'(\lambda) + 2\delta_e \bar{\delta}_e b_2(E0,E2) + \delta_e^2 F_2(2,2,2,2) b_2(E2,E2)] \\
&\times [Q_0(\lambda) + \delta_e^2 + \bar{\delta}_e^2]^{-1} F_2(2,2,0,2) P_2(\cos\vartheta) + \delta_e^2 F_4(2,2,2,2) \\
&\times b_4(E2,E2)\, [Q_0(\lambda) + \delta_e^2 + \bar{\delta}_e^2]^{-1} F_4(2,2,0,2)\, P_4(\cos\vartheta)
\end{aligned}
$$

where the mixing ratios are expressed in emission matrix elements

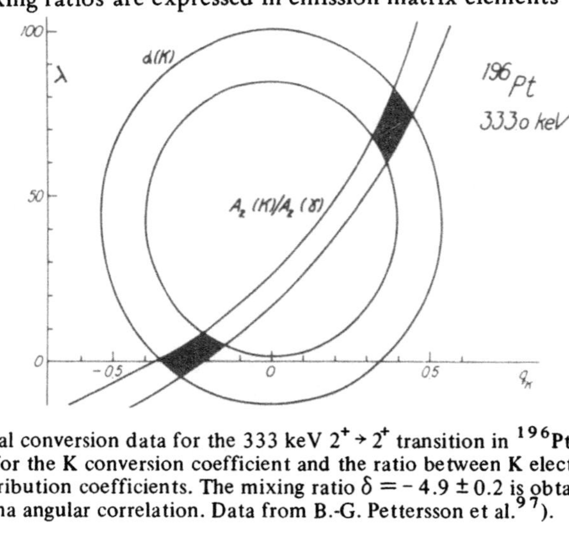

Fig. 10. Internal conversion data for the 333 keV $2^+ \to 2^+$ transition in ^{196}Pt. Shown are experimental results for the K conversion coefficient and the ratio between K electron and gamma directional distribution coefficients. The mixing ratio $\delta = -4.9 \pm 0.2$ is obtained from the electron gamma angular correlation. Data from B.-G. Pettersson et al.[97]).

$$\delta_e = \frac{\langle I_2 \| \underline{j}_n \cdot \underline{A}_2(e) \| I_1 \rangle}{\langle I_2 \| \underline{j}_n \cdot \underline{A}_1(m) \| I_1 \rangle} \sqrt{\frac{\alpha(E2)}{\alpha(M1)}} ,$$

$$\bar{\delta}_e = \frac{\langle I_2 \| (\frac{r}{R})^2 \| I_1 \rangle}{\langle I_2 \| \underline{j}_n \cdot \underline{A}_1(m) \| I_1 \rangle} \sqrt{\frac{A(E0)}{\alpha(M1)}} .$$

The anomaly factors for the particle parameters (Q_2, Q_2') and for the conversion coefficient (Q_0) are polynomials in the M1 penetration parameter. The electronic factor A (E0) and the penetration coefficients are tabulated by Hager and Seltzer[22]).

Electron - gamma and gamma - electron correlation measurements as well as internal conversion coefficient data are used for a determination of the parameter $q_K = \bar{\delta}_e / \delta_e$, cf. fig. 10.

Several other $2^+ \to 2^+$ transitions in the W - Hg region have been studied. An appreciable effect was found for instance in the 676 keV transition in ^{198}Hg, investigated by Sakai et al.[98]). The value of q is in general rather small for these nuclei, which requires the electron correlation method. Even if the E0 intensity may seem vanishingly small, the correlation can be appreciably affected due to the E0 - E2 interference.

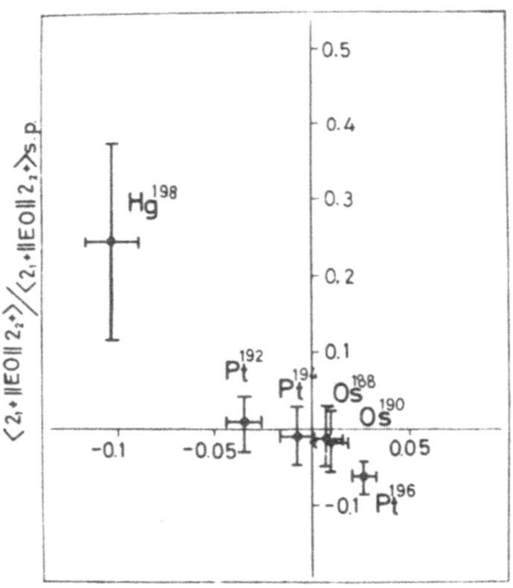

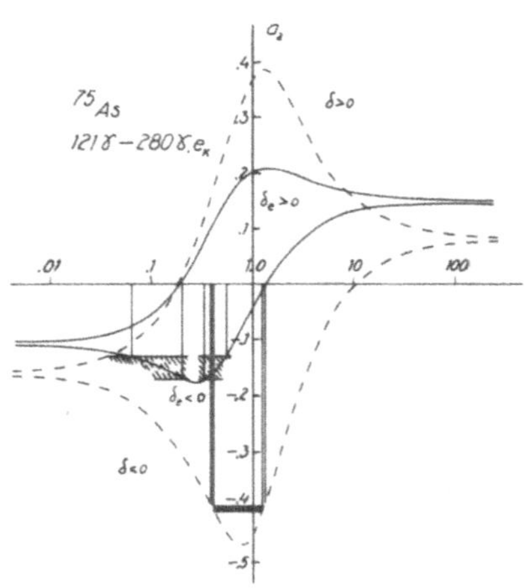

Fig.11 ⟨2,+‖M1‖2,+⟩/⟨2,+‖M1‖2,+⟩s.p. Fig.12

Fig. 11. Deduced values for E0 and M1 reduced matrix elements for transitions in Os, Pt, and Hg nuclei. The matrix elements are in single particle units and the figure shows that they are strongly hindered and that there is a correlation between the two kinds of matrix elements. (From Ikegami et al.[102]).

Fig. 12. Results for the electron - gamma and gamma - gamma angular correlation of the 121 keV - 280 keV cascade in ^{75}As obtained by Becker and Steffen[29]). As seen from the plot a unique value for δ is obtained from correlation data alone.

The experimental technique of electron - electron angular correlations has been used in some of the studies, for instance in the investigation of the 371 keV transition in ^{190}Os by Samuelsson et al.[99]). The mixing ratio δ (E2/M1) is determined from the 2^+ (γ) 2^+ (e) 0^+ correlation, and the E0 component from the corresponding electron - electron correlation and conversion coefficient measurements.†

The experimental results for the electric monopole strength are of considerable theoretical interest. From q (E0/E2) and B (E2) a value for the nuclear strength parameter ρ is obtained. In general ρ is of the order of 1 per cent in the transitions discussed so far. The model - description of nuclei in this transition region is not very certain. Experimental results for ρ have been used to distinguish between predictions from different collective models and are also compared with the results of the extensive calculations by Baranger and Kumar[101]). In fig. 11 the E0 strength for some transitions in this region is compiled and compared with the strength of the M1 component[102]).

An investigation of E0 strengths in ^{152}Sm and ^{154}Gd was recently carried out by Riedinger et al.[103]) by use of intensity measurements. Although the situation is more clear - cut in this end of the deformed region the predictions of the Bohr - Mottelson model are not very well fulfilled. As expected, large E0 components were found in $I_\beta \to I_{gr.}$ but not in $I_\gamma \to I_{gr.}$ transitions. However, even for the transitions from the beta - band the E0 strengths are appreciably smaller than predicted by theory. An electron - gamma angular correlation measurement of the $2_\beta \to 2_{gr.}$ transition of 689 keV in ^{152}Sm is discussed in a following paper.

Table 4. Experimental results for q_K (E0/E2) and λ (M1) in $2^+ \to 2^+$ transitions

Isotope	Transition energy (keV)	λ (M1)	q_K (E0/E2)	Reference
^{152}Sm	689	-	2.5 ± 0.2	104
^{184}W	793	$-82 < \lambda < 196$	0.18 ± 0.19	99a
		13^{+80}_{-40}	$0.30^{+0.08}_{-0.3}$	105
		-77^{+40}_{-36}	$-0.16^{+0.4}_{-0.12}$	
^{190}Os	371	-	-0.08 ± 0.17	99
^{192}Pt	296	-	0.19 ± 0.11	106, 107
		$-2 < \lambda < 15$	$0.01^{+0.15}_{-0.14}$	108
^{194}Pt	293	-	-0.25 ± 0.13	107
^{196}Pt	333	2 ± 5	-0.25 ± 0.07	97
		71 ± 8	0.39 ± 0.06	
^{198}Hg	676	$5 < \lambda < 26$	$0.8 < q < 1.6$	98
		$31 < \lambda < 52$	$-1.6 < q < -1.1$	

† It has recently been remarked that E0 transitions can be studied purely by gamma - gamma techniques, that is by measurement of a γ_1 - γ_3 cascade[100]). However, in this case no E0 - E2 interference occurs.

Some of the $2^+ \rightarrow 2^+$ E0 transitions studied by angular correlation techniques are summarized in table 4.

4.6 General applications to nuclear structure research

In the preceding sections we have dealt with experiments aimed at the study of 'anomalous' conversion processes. We would like to stress, however, that conversion electron correlations have a much wider range of applicability in nuclear structure research. In fact one may foresee a gradual shift of emphasis towards a more general application of the method. The studies inspired by problems in the theory of internal conversion as such are giving way to investigations proceeding directly from nuclear structure research.

Conversion electron correlations should be considered a supplement to gamma - gamma directional correlations, and the measurements of internal conversion coefficients. However, electron correlations sometimes offer particular advantages. Contrary to gamma - gamma correlations they distinguish between electric and magnetic transitions of the same multipole order. They may therefore be used to determine not only the spins but also the parities of the nuclear energy levels.

Gamma - gamma correlations are frequently used to determine the L + 1 / L mixing ratio. Conversion electron correlations are even more sensitive to minute admixtures of multipole order L + 1, owing to the higher weight given to this transition in the electron channel. This is true in particular in the case of parity - favoured transitions, ML + 1 / EL.

L - subshell conversion correlations offer some interesting possibilities. B.I. Deutch et al.[109]) and Törnkvist et al.[110]) have demonstrated the power of this technique for determinations of mixing ratios and particle parameters.

Finally, conversion electron correlations combined with gamma - gamma angular correlations frequently offer a unique determination of the spin and parity sequence, as well as the mixing ratio of the transition under investigation. Instructive examples may be found in the literature. See for instance the paper by Becker and Steffen concerning transitions in ^{75}As[29]) (cf. fig. 12) or the study of the 133.5 keV transition in ^{131}Cs by Fechner et al.[111]).

5. Applications to neighbouring fields and current developments

We have in the preceding survey restricted ourselves to the use of conversion electron correlations in nuclear structure research. We conclude that the experimental technique, as well as the underlying theory, now seems to be sufficiently developed to make electron correlations attractive when compared to alternative techniques. There are some applications where angular correlations involving conversion electrons offer special advantages, such as the study of nuclear penetration effects in internal conversion and the determination of minute multipolarity admixtures.

We like to conclude, however, by some very brief references to the use of conversion electron correlations in neighbouring fields, only indirectly related to nuclear structure research. The study of hyperfine interactions by means of perturbed angular correlations is a field of increasing importance, as witnessed for instance by the following (Sept. 1970) international conference in Rehovoth, Israel.

So far, most of these experiments have been confined to gamma radiation, with the interesting exception, however, of the several studies of aftereffects of electron - hole formation. Since these aftereffects are dealt with in a separate contribution to this conference, we restrict ourselves to the observation that these experiments have contributed to the understanding of atomic recovery processes.

Conversion electrons are deflected by externally applied magnetic fields. They are therefore less suited than gamma rays for g - factor measurements and similar studies. This limitation may be at least partly overcome by using the localized magnetic fields, as has been shown by B.I. Deutch (cf. separate contribution to this conference). Along these lines further progress is to be expected.

For technical reasons electron - correlation experiments have been confined to the medium and higher ranges of energy. The low - energy region has so far been left almost unexplored. The preacceleration technique, very successfully used in magnetic spectrometers for delayed coincidence experiments, seems to be less suited to angular correlations, owing to the difficulties associated with the angular definition in the presence of the combined electric and magnetic fields. However, some interesting results were recently obtained by Bäverstam et al. who replaced the scintillation detector with an electron multiplier[112]). The relative halfwidth of the magnetic spectrometer is of course independent of the energy, or rather the momentum, of the focussed electrons. Provided proper precautions are taken, the noise level in the extreme low - energy range may be substantially suppressed. Electrons of energies below 1 keV have been recorded with an excellent signal - to - noise ratio. Conversion electrons following Môssbauer excitation have been recorded, and this technique offers interesting possibilities for the study of the conversion process, as well as of surface physics effects.

Closely related to directional correlations is the study of the spatial distribution of conversion electrons from oriented nuclei. This technique was first applied by Frankel et al. who measured the particle parameter for the 255 keV M4 transition in ^{137}Ce [113]). The technique has some attractive features. One here measures singles counting rates rather than coincidences. Provided a sufficient amount of nuclear orientation can be achieved, the experiments of this kind are quite expedient and of course not restricted to cascading events. An interesting application would be the study of crossover E1 transitions in deformed nuclei (e.g. the 396 keV transition in ^{175}Lu (cf. fig. 8). Conversion coefficients demonstrate the presence of large nuclear penetration effects but they are not sufficient for a complete analysis. Coincidence experiments are generally excluded and nuclear orientation seems to be the only alternative.

So far, there have been fairly few experiments on the angular distribution of conversion electrons following nuclear reactions. There are considerable experimental difficulties to be overcome related both to the target itself and the electron detection system. Some pioneering experiments were performed already in 1966 by Sakai and his coworkers[114]). They studied the directional distribution of electrons from states excited in (p, 2n) reactions. A device for the study of directional distributions of electrons emitted in nuclear reactions has recently been constructed by Sawa and Aström[115]). They also suggest the measurement of electron - gamma correlations in beam - off periods. When the measurements are confined to 'isomeric' transitions several experimental difficulties can be avoided.

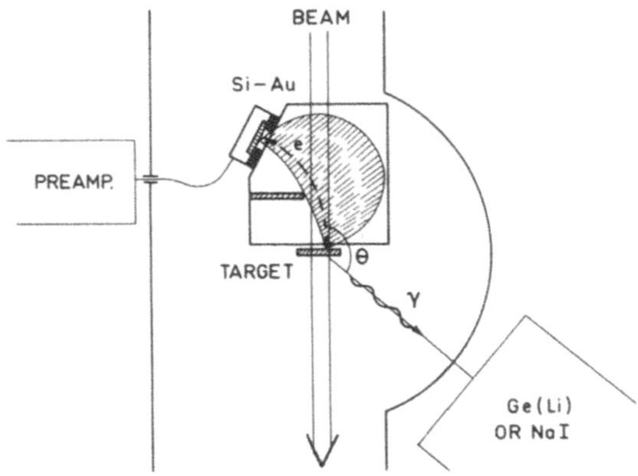

Fig. 13. A spectrometer for studying the directional distribution of conversion electrons from nuclear reactions, due to Sawa and Åström[115]). A possible application would be the measurement of electron - gamma correlations in beam - off periods.

The successful studies of the spatial distribution of gamma rays following nuclear reactions are very encouraging. However, it should be born in mind that the replacement of gamma rays with conversion electrons implies a substantial geometrical extension of the recoiling system to be considered. Moreover, contrary to gamma rays, conversion electrons originate from the loosely bound atomic - electron configuration. Theoretical as well as experimental physicists here find themselves at the frontier of our knowledge, but the increasing use of tandem van de Graffs and isochronous cyclotrons offers possibilities that represent a challenge for the future.

Acknowledgement

We express our gratitude for preprints sent to us by Professors I.M. Band and L.A. Sliv, Z.W. Grabowski, W.D. Hamilton, R.M. Steffen and J.-E. Thun. We are indebted to Dr. Z. Sawa for communications on work in progress. Our coworkers, in particular Prof. B.-G. Pettersson, have contributed by many illuminating discussions.

General references

Role of Atomic Electrons in Nuclear Transformations, Proceedings of International Conference (Nuclear Energy Information Center, Warsaw, 1965).

Internal Conversion Processes, Including the Proceedings of the International Conference on the Internal Conversion Process, edited by J.H. Hamilton (Academic Press, New York, 1966).

Proceedings of the International Conference on Radioactivity in Nuclear Spectroscopy, Nashville, 1969, to be published.

Theory

L.C. Biedenharn and M.E. Rose, Theory of Angular Correlation of Nuclear Radiations, Rev. Modern Phys. 25 (1953) 729.

H. Frauenfelder and R.M. Steffen, Angular Correlations, in Alpha-, Beta- and Gamma-Ray Spectroscopy, edited by K. Siegbahn (North - Holland Publishing Company, Amsterdam, 1965).

R.M. Steffen, Angular Distributions of Conversion Electrons, Department of Physics, Purdue University Report, to be published in Proceedings of the International Conference on Radioactivity in Nuclear Spectroscopy, Nashville, 1969.

Experimental aspects

T.R. Gerholm and B.-G. Pettersson, Remarks on Conversion Electron - Gamma Directional Correlations in Internal Conversion Processes, op. cit.

Penetration effects

E.L. Church and J. Weneser, Nuclear Structure Effects in Internal Conversion, Ann. Rev. Nuclear Sci. 10 (1960) 193.

G. Kramer and S.G. Nilsson, The Penetration Effect in Electric Dipole Internal Conversion, Nuclear Physics 35 (1962) 273.

T.R. Gerholm and B.-G. Pettersson, Some Consequences of the Finite Nuclear Size, in Alpha-, Beta- and Gamma-Ray Spectroscopy, op. cit.

H.-C. Pauli, Finite Nuclear Size Effects in Internal Conversion, Helv. Phys. Acta 40 (1967) 713.

Tables

M.E. Rose, Internal Conversion Coefficients (North - Holland Publishing Company, Amsterdam, 1958).

I.M. Band, M.A. Listengarten and L.A. Sliv, Tables of the Conversion Matrix Elements and Phases, Particle Parameters for Angular Correlations of Conversion Electrons, in Alpha-, Beta- and Gamma-Ray Spectroscopy, op. cit.; also in Internal Conversion Processes, op. cit.

E.L. Church, Tables of Electron Wavefunctions and Associated Data, Brookhaven National Laboratory Report (BNL 50002, Upton, 1966).

H.-C. Pauli, Tables of Internal Conversion Coefficients and Particle Parameters, Department of Physics, Purdue University Report (1967).

H.-C. Pauli, A Computer Program for Internal Conversion Coefficients and Particle Parameters, The Niels Bohr Institute Report (Copenhagen), to be published.

R.S. Hager and E.C. Seltzer, Internal Conversion Tables. Part I: K-, L-, M-Shell Conversion Coefficients for Z = 30 to Z = 103, Nuclear Data A4 (1968) 1; Part II: Directional and Polarization Particle Parameters for Z = 30 to Z = 103, Nuclear Data A4 (1968) 397; Part III: Coefficients for the Analysis of Penetration Effects in InternalConversion, Nuclear Data A6 (1969) 1.

Compilations of Experimental Particle Parameters

B.I. Deutch and P. Hornshøj, Particle Parameters Measured in Pure Transitions, in Internal Conversion Processes, op. cit.; also Nuclear Structureless Internal Conversion Particle Parameters in the K- and L-Shells, Nuclear Physics A95 (1967) 65.

R.M. Steffen, Angular Distributions of Conversion Electrons, op.cit.

References

1. R.M. Steffen and H. Frauenfelder, in Alpha-, Beta- and Gamma-Ray Spectroscopy, ed. by K. Siegbahn (North - Holland Publishing Co., Amsterdam, 1965).
2. H. Frauenfelder, M. Walter and W. Zünti, Phys. Rev. 77 (1950) 557.
3. D.T. Stevenson and M. Deutsch, Phys. Rev. 83 (1951) 1202.
4. K. Siegbahn, Arkiv för Fysik 4 (1952) 223.
5. F. Gimmi, E. Heer and P. Scherrer, Z. Physik 138 (1954) 394.

6. N. Goldberg and S. Frankel, Phys. Rev. 93 (1954) 1425.
7. T.R. Gerholm and B.-G. Pettersson, Nucl. Instr. and Meth. 4 (1959) 107.
8. M.E. Rose, L.C. Biedenharn and G.B. Arfken, Phys. Rev. 85 (1952) 5.
9. L.C. Biedenharn and M.E. Rose, Rev. Modern Physics 25 (1953) 729.
10. M.E. Rose, Internal Conversion Coefficients (North - Holland Publishing Co., Amsterdam, 1958).
11. L.A. Sliv and I.M. Band, Coefficients of Internal Conversion of Gamma Radiation, Part I: K - shell (Physico - Technical Institute, Academy of Science, Leningrad, 1956); Tables of Gamma - ray Conversion Coefficients. Part II: L - shell (Physico - Technical Institute, Academy of Science, Leningrad, 1958).
12. L.A. Sliv, Zhur. Eksptl. i Teoret. Fiz. 21 (1951) 770.
13. E.L. Church and J. Weneser, Phys. Rev. 100 (1955) 943.
14. T.R. Gerholm and B.-G. Pettersson, Phys. Rev. 110 (1958) 1119.
15. E.L. Church and J. Weneser, Phys. Rev. 104 (1956) 1382.
16. I.M. Band, M.A. Listengarten, L.A. Sliv and J.E. Thun, in Alpha-, Beta- and Gamma-Ray Spectroscopy, op. cit.
17. M.A. Listengarten, I.M. Band, E.F. Zganjar and J.H. Hamilton, in Internal Conversion Processes, ed. by J.H. Hamilton (Academic Press, New York, 1966).
18. A. Miranda, P. Hornshøj and B.I. Deutch, in Internal Conversion Processes, op. cit.
19. H.-C. Pauli, Tables of Internal Conversion Coefficients and Particle Parameters, Department of Physics, Purdue University Report (1967).
20. R.S. Hager and E.C. Seltzer, Nuclear Data A4 (1968) 397.
21. H.-C. Pauli, Helv. Phys. Acta 40 (1967) 713.
22. R.S. Hager and E.C. Seltzer, Nuclear Data A6 (1969) 1.
23. E.L. Church, Tables of Electron Wavefunctions and Associated Data, Brookhaven National Laboratory Report (BNL 50 002, Upton, 1966).
24. H.-C. Pauli and K. Alder, Z. Physik 202 (1967) 255.
25. H.-C. Pauli, A Computer Program for Internal Conversion Coefficients and Particle Parameters, The Niels Bohr Institute Report (Copenhagen), to be published.
26. R. Huby, in Proceedings of the International Conference on Nuclear Physics, July 1958 (Crosby, Lockwood and Son, Ltd., London, 1958).
27. S. Ofer, Phys. Rev. 114 (1959) 870.
28. H.J. Rose and D.M. Brink, Rev. Modern Physics 39 (1967) 306.
29. A.J. Becker and R.M. Steffen, Phys. Rev. 180 (1969) 1043.
30. T.R. Gerholm, B.–G. Pettersson, C. Bargholtz, Z.H. Cho, L. Eriksson and L. Gidefeldt, this conference.
31. K. Siegbahn, Arkiv för Fysik 22 (1963) 435; P. Kleinheinz, L. Samuelsson and K. Siegbahn, Nucl. Instr. and Meth. 32 (1965) 1.
32. L. Holmberg and Z.H. Cho, to be published.
33. A.E. Blaugrund and Z. Vager, Nucl. Instr. and Meth. 29 (1964) 131.
34. Z.H. Cho and T.R. Gerholm, Nucl. Instr. and Meth. 73 (1969) 67.
35. Z.H. Cho, U. Bäverstam and R. Othaz, Nucl. Instr. and Meth. 84 (1970) 321.
36. J. Becker and L. Holmberg, to be published.
37. S. Frankel, Phys. Rev. 83 (1951) 673.

38. F. Gimmi, E. Heer and P. Scherrer, Helv. Phys. Acta 29 (1956) 147.

39. V. Stefánsson, L. Holmberg. B.-G. Pettersson and B. Fogelberg, in Proceedings of the International Conference on Radioactivity in Nuclear Spectroscopy, Nashville, 1969, to be published.

40. P. Kleinheinz, R. Vukanović, L. Samuelsson, M. Zupancić, D. Krmpotić, C. Besev and K. Siegbahn, Nucl. Instr. and Meth. 57 (1967) 181.

41. T.R. Gerholm, R. Othaz and M.S. El-Nesr, Arkiv för Fysik, 21 (1962) 253.

42. B.-G. Pettersson, L. Holmberg and T.R. Gerholm, Nucl Phys. 65 (1965) 454.

43. J.W.E. Vos, C.W.E. van Eijk, W.H.G. Lewin and F. Schutte, Nucl. Instr. and Meth. 60 (1968) 285.

43a. H. Vignau, A. Mocorva, R. Othaz and N. Baade, Nucl. Phys. A110 (1968) 56.

44. D.L. Coppage and R.M. Steffen, to be published.

45. K.-H. Speidel, D. Kolb, K. Mittag, J. Voss, B. Wolbeck, G. Dammertz and K.-G. Plingen, Nucl. Phys. A115 (1968) 421.

46. H. Kawakami, K. Manabe and K. Hisatake, Japan. Journ. of Appl. Physics 7 (1968) 1397.

47. J.S. Geiger, Physics Letters 7 (1963) 48.

47a. R.S. Stone, T. Yamazaki, D.A. Landis and J.M. Hollander, quoted by J.M. Hollander in Nuclear Chemistry, ed. by L. Yaffe (Academic Press, New York, 1968).

48. H.E. Bosch, L.F. Gattó, M. Behar and G.J. García, Nucl. Instr. and Meth. 68 (1969) 88.

49. M. Sakai, H. Ikegami and T. Yamazaki, Nucl. Instr. and Meth. 25 (1964) 328.

50. F. Falk, S. Törnkvist, H. Snellman and J.E. Thun, Nucl. Instr. and Meth. 49 (1967) 240.

51. K.M. Bisgard, Nucl. Instr. and Meth. 29 (1964) 213.

52. M. Sakai, H. Ikegami and T. Yamazaki, Nucl. Instr. and Meth. 9 (1960) 154.

53. S. Antman, Y. Grunditz, A. Johansson, B. Nyman, H. Pettersson, B. Svahn and K. Siegbahn, Nucl. Instr. and Meth. 82 (1970) 13.

54. L. Holmberg, N. de Sousa, V. Stefánsson, B.-G. Pettersson and T.R. Gerholm, in Proceedings of the International Conference on Radioactivity in Nuclear Spectroscopy, Nashville, 1969, to be published.

55. R.M. Steffen, Angular Distributions of Conversion Electrons, Department of Physics, Purdue University Report, to be published in Proceedings of the International Conference on Radioactivity in Nuclear Spectroscopy, Nashville, 1969.

56. T.A. Green and M.A. Rose, Phys. Rev. 110 (1958) 105.

57. E.L. Church and J. Weneser, Ann. Rev. Nucl. Sci. 10 (1960) 193.

58. J.J. Matese, Phys. Rev. 173 (1968) 1165.

59. J.P. Bocquet, Y.Y. Chu, O.C. Kistner, M.L. Perlman and G.T. Emery, Phys. Rev. Lett. 17 (1966) 809.

60. T.A. Carlson, P. Erman and K. Fransson, Nucl. Phys. A111 (1958) 371.

61. P. Erman, G.T. Emery and M.L. Perlman, Phys. Rev. 147 (1966) 858.

62. I.M. Band, L.A. Sliv and M.B. Trzhaskovskaya, Formation Region of Internal Conversion Coefficients (Academy of Sciences of the USSR, A.F. Ioffe Physico - Technical Institute, Leningrad, 1970).

63. E.W. Koopmann and H. van Krugten, Physica 42 (1969) 262.

64. H.A. Doubt and W.D. Hamilton, following paper.

65. J.H. Hamilton, E.F. Zganjar, T.M. George and W.H. Hibbits, Phys. Rev. Lett. 14 (1965) 567; E.F. Zganjar, T.M. George and J.H. Hamilton, Nuclear Physics A114 (1968) 609.

66. L. Holmberg, V. Stefánsson and B.G. Pettersson, Nucl. Phys. A96 (1967) 33.

67. H.M. Nasir, Z. W. Grabowski and R.M. Steffen, Phys. Rev. 162 (1967) 1118.

68. K.B. Nielsen, N. Rud, J.C. Overgard and H.E. Sørensen, Nucl. Phys. A124 (1969) 445.

69. J. Voss, Thesis (Mathem. - naturw. Fakultät der Universität Bonn, Bonn, 1968).

70. Y.K. Agarwal, C.V.K. Baba and N.C. Mukhopadhyay, Nucl. Phys. A127 (1969) 661.

71. A.W. Sunyar and B.I. Deutch, quoted in B.I. Deutch, Proc. Roy. Soc. A. 311 (1969) 151.

72. H. Blumberg, J. Fechner, A. Hammesfahr, P. Herzog, A. Kluge and J. Voss, Z. Physik 228 (1969) 391.

73. L. Holmberg, V. Stefánsson, L. Eriksson and B.-G. Pettersson, in Proceedings of the International Conference on Radioactivity in Nuclear Spectroscopy, Nashville, 1969, to be published.

74. E.L. Church, A. Schwarzschild and J. Weneser, Phys. Rev. 133 (1964) B35; L.C. Biedenharn and M.E. Rose, Phys. Rev. 134 (1964) B8.

75. T.R. Gerholm and B.-G. Pettersson, in Alpha-, Beta-, and Gamma-Ray Spectroscopy, op.cit.

76. R.J. Lombard, Phys. Letters 9 (1964) 254.

77. J. LeTourneux, Nucl. Phys. 74 (1965) 172.

78. Z. Grabowski, B.-G. Pettersson, T.R. Gerholm and J.E. Thun, Nucl. Phys. 24 (1961) 251.

79. T.R. Gerholm, B.-G. Pettersson and Z. Grabowski, Nucl. Phys. 65 (1965) 441.

80. S. Wahlborn, Phys. Rev. 138B (1965) 530.

80a. S. Törnkvist, L. Hasselgren, S. Ström, J.-E. Thun and S. Antman, Nucl. Phys. A142 (1970) 238.

81. M.R. Meder, Phys. Rev. C 1 (1970) 1085.

82. F.T. Avignone, III, G.D. Frey and L.D. Hendrick, Phys. Rev. C 1 (1970) 635.

83. L. Hasselgren, S. Antman, H.S. Sahota and J.-E. Thun, following paper.

84. G. Vanden Berghe and K. Heyde, Nucl. Phys. A144 (1970) 558.

85. A.S. Reiner, Nucl. Phys. 5 (1958) 544.

86. A. Marelius, H. Pettersson, S. Törnkvist, S.-E. Hägglund and R. Dumitrescu, Arkiv för Fysik 37 (1968) 435.

87. S. Törnkvist, S. Ström and L. Hasselgren, Nucl. Phys. A130 (1969) 3.

88. T.R. Gerholm, B.-G. Pettersson, B. van Nooijen and Z. Grabowski, Nucl. Phys. 24 (1961) 177.

89. B.-G. Pettersson, T.R. Gerholm, Z. Grabowski and B. van Nooijen, Nucl. Phys. 24 (1961) 196.

89a. J.-E. Thun, S. Törnkvist, F. Falk, H. Snellman and F. Kropff, Phys. Letters 8 (1964) 326.

90. S. Gorodetzky, F. Beck and A. Knipper, Nucl. Phys. 82 (1966) 275.

91. E.L. Church and J. Weneser, Nucl. Phys. 28 (1961) 602.

92. G. Kramer and S.G. Nilsson, Nucl. Phys. 35 (1962) 273.

93. F. Asaro, F.S. Stephens, J.M. Hollander and I. Perlman, Phys. Rev. 117 (1960) 492.

94. L. Holmberg, L. Gidefeldt, M. Gunnerhed and B.-G. Pettersson, Nucl. Phys. A96 (1967) 2.

94a. G.T. Emery and M.L. Perlman, Phys. Rev. 151 (1966) 984.

94b. C.J. Herrlander and G.T. Ewan, Role of Atomic Electrons in Nuclear Transformations, Proc. International Conference (Warsaw, 1963).

95. J.-E. Thun, Nucl. Phys. 91 (1967) 653.

96. S. Törnkvist, L. Hasselgren, J.-E. Thun and A. Marelius, Nucl. Phys. A133 (1969) 3.

97. B.-G. Pettersson, L. Holmberg and T.R. Gerholm, Nucl. Phys. 65 (1965) 466.

98. M. Sakai, M. Nozawa, H. Ikegami, T. Yamazaki, Nucl. Phys. 53 (1964) 529.

99. L. Samuelsson, R. Vukanović, M. Migahed, M. Zupancić, L.O. Edvardson and L. Westerberg, Nucl. Phys. A135 (1969) 3.

99a. M. Zupancić, R. Vukanović and L. Samuelsson, Arkiv för Fysik 39 (1969) 313.

100. I. Anicin, Dj. Krmpotić, A. Kukec and R. Vukanović, Nucl. Instr. and Meth. 83 (1970) 293.

101. K. Kumar and M. Baranger, Phys. Rev. Lett. 17 (1966) 1146; Nucl. Phys. A122 (1968) 273.

102. H. Ikegami, T. Hirose, M. Sakai, T. Yamazaki and K. Sugigama, in Proceedings of the International Conference on Nuclear Structure (Physical Society Japan, Tokyo, 1968).

103. Riedinger, Johnson and Hamilton, Phys. Rev. 179 (1969) 1214.

104. V. Stefánsson, V. Sergeev, J. Becker and L. Holmberg, following paper.

105. H.A. Doubt, W.D. Hamilton, K.E. Davis and Z.W. Grabowski, to be published.

106. D.K. Butt and B.C. Dutta, Nucl. Phys. 39 (1962) 517.

107. S. Alwyn and D.K. Butt, Nucl. Phys. A100 (1967) 177.

108. L. Marinkov, I. Anicin, I. Bikit and R. Stepić, Nucl. Phys. A131 (1969) 601.

109. B.I. Deutch and P. Hornshøj, Nucl. Phys. 53 (1964) 497.

110. S. Törnkvist, J.-E. Thun, S. Ström, R.G. Albridge and V. Schmidt, Z. Physik 214 (1968) 214.

111. J. Fechner, A. Hammesfahr, A. Kluge, S.K. Sen, H. Toschinski, J. Voss, P. Weight and B. Martin, Nucl. Phys. A130 (1969) 545.

112. U. Bäverstam, private communication.

113. R.B. Frankel, D.A. Shirley and N.J. Stone, Phys. Rev. 136B (1964) 577.

114. H. Ejiri, M. Ishihara, M. Sakai, T. Inamura and K. Katori, Nucl. Phys. 89 (1966) 641.

115. Z. Sawa and B. Aström, in Annual Report 1969, Research Institute of Physics, Stockholm, and private communication.

Discussion

Manthuruthil: I will make a small comment regarding the value of lambda for the 161 keV transition in ^{133}Cs that you showed there. There are 3 measurements that you showed which are all in excellent agreement, but there has been a recent gamma-gamma correlation measurement published in Physical Review. In this measurement two Ge(Li) - detectors are used to measure the correlation and now they find that there is really no penetration effect at all for this particular transition. So I am wondering whether the use of the Ge(Li) - detectors for gamma - gamma correlations may change the results that we have available up to now.

Gerholm: Well I am not aware of the experiment that you refer to. I do not think that the replacement of one detector system with another will change the results obtained so far, if the results obtained so far are correct. That is about all I can say. But I would like to add that in the tables I have shown we have tried to summarize the most recent and accurate results. But it is quite possible that we may have overlooked some contributions which should be incorporated in those tables. And I would like to ask you if you are aware of any of such experiments to let us know about this so that the tables published in the proceedings of this conference will be as complete as possible.

J.H. Hamilton: Just a brief comment. Since Steffen set some records straight historically last night. I discovered about 4 years ago in reading the literature that penetration effects were first pointed out in 1930 and not in 1956 by Church and Weneser, both for E0 transitions and 2 - 2 cases. And for some reason nobody has bothered to read those papers, they were not very explicit, but at least people compared conversion coefficients and calculated the penetration parameter in 1930.

Gerholm: They were clever in those days, were not they?

W.D. Hamilton: As one of the people who at one time saw an almost penetration particle parameter I would like to show some slides if I may of the techniques which Doubt has been using at Sussex using an electron microscope to examine the electron source.And one canget an exceptionally high magnification with such an instrument and one sees beautiful details of the positive source and I think such a technique is almost essential if one is going to claim a rather maximum estimate about scattering corrections within the source.

The following series of photographs of electroplated sources were taken using a scanning electron microscope. The source backing was a stainless steel foil 2.2 μm thick and the photographs were taken after the source had decayed. The sources were used in a series of e - γ correlation measurements by Andrew Doubt and myself, and these photographs illustrate the variation in the quality of the sources that may be obtained.

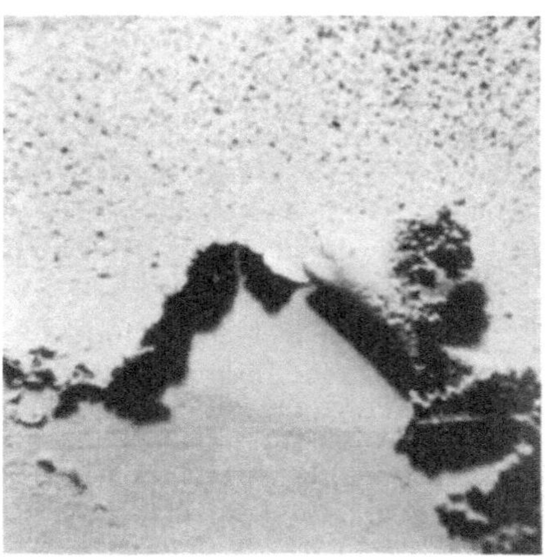

A preliminary $^{194,196}Au$ source having a mean thickness of 0.6 μm. The edge of the source has been purposely torn. The magnification is 680.

This shows the same source at a magnification of 6500 and it is obvious that the uniformity is poor and the effective thickness is much greater.

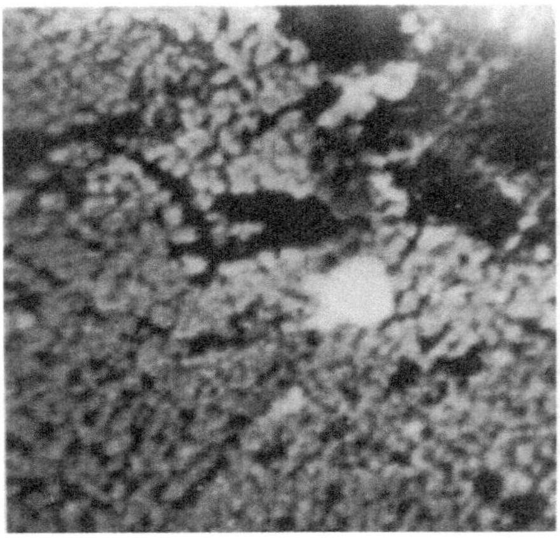

This indicates the improvement in uniformity that may be achieved by reducing the plating current. The source is 0.2 μm thick and the magnification is 13000. The same stock solution was used as in the preparation of the preliminary source.

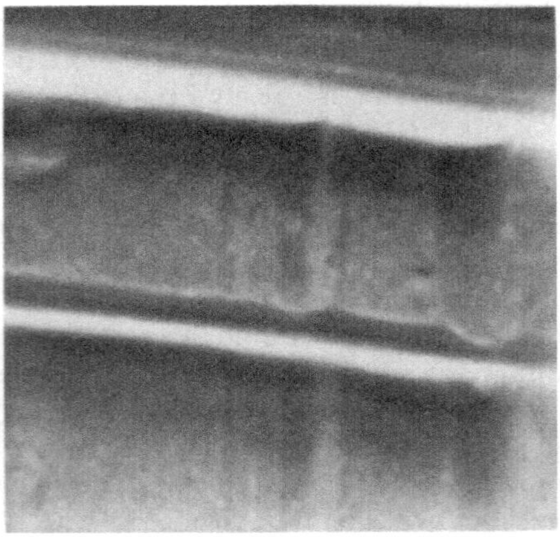

A ^{192}Ir source prepared by plating for a period of 10 hrs. using a carrier - free hexachloridic acid in dilute HCl. The scratches across the surface were made in order to estimate the thickness when a shallow electron scattering angle of 25° was used. The source is 0.2 μm thick and the magnification is 6300.

Gerholm: Here in Holland the Dutch scientist Van Leeuwenhoek first started to look at the drinking water through a microscope and saw all sorts of ugly things like small worms and eels and everything. But still I think it is possible to obtain rather reliable attenuation correction coefficients by the technique developed by Stefansson. You can check it directly experimentally and it really works out. Of course it is not as precise as if you are to work with high specific activities, I agree.

Sastry: Speaking of these corrections for electron scattering in the source there seems to be some difference of experience. The results that you are referring to as I recall were presented last year at the Nashville conference and the report was made that the attenuations of the angular correlations were not as much as predicted by the so - called Frankel theory which has been in work for quite a while. We have made a lot of measurements on these. Not with the sources of the kind that Dr. Hamilton pointed out, but using a thin source and placing a uniform absorber right against it to verify the theory. In a contribution to this conference we present the results which seem to verify quite well the predictions of Frankel's theory. We also verified this in the case of the beta - gamma correlations for ^{86}Rb. Our results are still tentative at the moment.

Gerholm: I am not going to comment in detail on this very complicated problem. I think it will be better to refer to Stefansson's publication which will soon appear. But it is certainly true that you can get agreement with the theory if you want to. You can also get disagreement. It all depends on how you arrange your experiment. It is very sensitive to the setting of the peak in the electron channel. And this has been very widely studied by Stefansson. So, I think one has to very careful in claiming that Frankel's theory is correct or is not correct; it all depends on how you do the experiment. It would take much more time than I am allowed to here to go into all details and I am not really equiped for it either.

Sújkowski: At the end of your talk you mentioned the very low energy electron - spectroscopy. Was it really in connection with angular correlations? Do you really believe the results of correlations with electrons of that energy and would you have a theory to account for it?

Gerholm: No, it has not been used for angular correlations. Bäverstam is here present, so you can ask him directly, but so far it has only been used to study conversion electrons following Mössbauer excitation. It is perfectly true that one probably cannot do angular correlation work and compare the results with the theory at these very low energies. But what one can do, is to perform relative measurements under different source conditions and to study the chemical effects on the angular correlations and this is one possible application of the technique. It may not have so much to learn us about nuclear physics but it may have a lot to do with solid state physics and particularly with surface physics.

EFFECT OF MULTIPLE SCATTERING OF ELECTRONS ON $e^- - \gamma$ DIRECTIONAL CORRELATIONS*

N.K. Gupta† and K.S.R. Sastry
University of Massachusetts, Amherst, Massachusetts, U.S.A.

In directional correlation studies involving conversion electrons or β - radiations, an important systematic effect arises due to scattering of the particles in the source material. These effects are often rendered negligible by the use of 'thin' sources. However, there are several practical situations requiring use of liquid or implanted sources, or sources of low specific activity for which quantitative corrections need to be applied to obtain the true correlations. In such cases, the effective directional distribution function W (θ) is given by

$$W(\theta) \propto 1 + \Sigma\, C_K A_K P_K(\theta), \quad (K \geqslant 2, \text{even})$$

where the coefficients A_K are the true correlation coefficients and C_K represents the effect of scattering. According to Frankel[1]), C_K is given by $(1 - e^{-a_K t_K}) / a_K t$ for a uniform disc source of thickness t, and for a thin source with an external scattering foil, we have $C_K = e^{-a_K t}$. The coefficients a_K depend on the details of the scattering. Assuming elastic multiple scattering as the most dominant process, Golden and Frankel[2]) gave an explicit expression for a_K using the theoretical results of Goudsmit and Saunderson[3]). According to them, a_K is essentially proportional to Z^2/A of the scatterer and varies roughly as $(1/\epsilon)^2$, ϵ being the kinetic energy of electrons in mc^2 units. It is customary to evaluate the influence of scattering on measured correlations using nomograms[4]) prepared on the basis of these expectations. However, few systematic experimental studies are reported in the literature[2,5]). The purpose of this paper is to present preliminary results of an investigation of the validity of the above theoretical expectations. Specifically, the a_2 coefficients are reported for internal conversion electrons of average kinetic energies of 500 and 990 keV from ^{207}Pb for three elements: aluminium (Z = 13), niobium (Z = 41) and gold (Z = 79).

The experiment consisted in measuring the attenuation of $e^- - \gamma$ and $\gamma - e^-$ correlations associated with the 1063 keV (M4) and 569 keV (E2) cascade transitions in ^{207}Pb for different scattering foils placed at a distance of 0.8 mm in front of a thin ^{207}Bi source using a multichannel β - γ coincidence spectrometer[6]). Measurements were made with three foils of Al (thickness 3.4 to 13.7 mg/cm^2), three foils of Nb (thickness 4.3 to 15.2 mg/cm^2) and two foils of Au (thickness 9.5 and 18.9 mg/cm^2). The foil thicknesses were measured experimentally by at least two of the following three methods: α - particle range, low energy X - ray absorption, and weighing. The thicknesses are believed to be accurate to better than 10%. For the thinnest foils of each element, the effective A_2 and A_4 coefficients were obtained from measurements at four angles: 90°, 120°, 150° and 180°. For thicker scatterers, the effective A_2 values were obtained from the anisotropy of the correlation from measurements at 90° and 180° only in view of the small A_4 coefficient and large

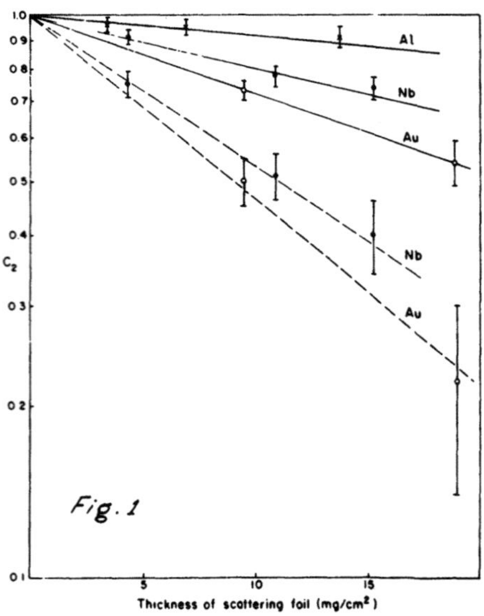

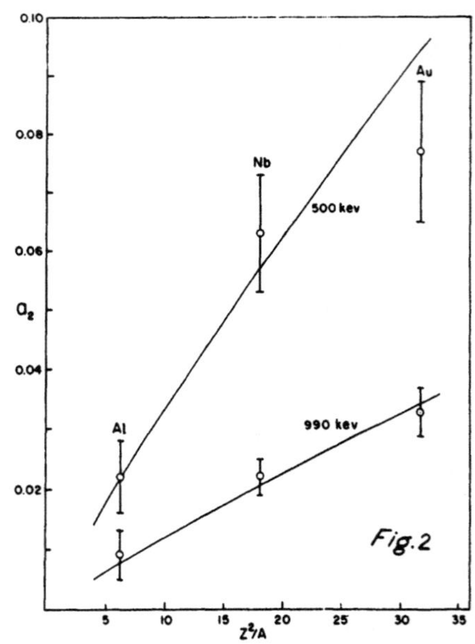

Fig. 1. Plot of experimental C_2 values versus thickness of scattering foil for Al (crosses), Nb (solid circles) and Au (open circles). The straight lines are best fits to the data, the solid ones for electrons of average energy 990 keV and the dashed ones for 500 keV electrons.

Fig. 2. Plot of a_2 values versus Z^2/A. The solid curves represent the predictions of Frankel's theory.

attenuation effects expected. In the absence of any scatterer, our directional correlation results are:

$$A_2 = 0.227 \pm 0.003 \qquad A_4 = -0.024 \pm 0.006 \qquad (1069e^- - 569\gamma)$$

$$A_2 = 0.273 \pm 0.009 \qquad A_4 = -0.015 \pm 0.012 \qquad (1069\gamma - 569e^-)$$

in good agreement with predictions for a $13/2^+ \rightarrow 5/2^- \rightarrow 1/2^-$ M4 - E2 cascade.

The experimental attenuation coefficients C_K, obtained as ratios of A_2^{eff} (with foil) to A_2 (without foil), are plotted in fig. 1 as a function of the thickness of the scatterer. The straight lines (solid ones for 990 keV, dashed ones for 500 keV) are the best fits to the data in each case. These results substantiate the expectation of exponential decrease of C_2 with increasing foil thickness in the range of thicknesses used. The results for Al at 500 keV are not shown as they closely overlap those for Nb at 990 keV. The a_2 coefficients given by the slopes of these straight lines are plotted in fig. 2. The predicted a_2 values are shown as a function of Z^2/A by the solid curves. Our results are in good agreement with Frankel's theory although some inadequacy might exist for high Z scatterers at the lower energy. Further work is in progress to extend the results to more Z - values over a wider energy range.

In conclusion, we observe that the present measurements along with the work of Golden and Frankel[2]), and Mellema and van der Veen[5]), who studied variation of C_2 with source thickness, support the theory of Frankel and show that the predominant mechanism involved here is elastic multiple scattering of electrons although somewhat different results have been recently reported[7]).

The assistance of Mr. Dandamudi V. Rao is deeply appreciated.

* Work supported in part by U.S. Army Research Office, Durham, North Carolina.

† Present address: The Harshaw Chemical Company, Cleveland, Ohio, U.S.A.

References

1. S. Frankel, Phys. Rev. 83 (1951) 673.
2. R.K. Golden and S. Frankel, Phys. Rev. 102 (1956) 1053.
3. S. Goudsmit and J.L. Saunderson, Phys. Rev. 57 (1940) 24.
4. F. Gimmi et al., Helv. Phys. Acta 29 (1956) 147.
5. J. Mellema and P.H. van der Veen, Nucl. Phys. A117 (1968) 376.
6. R.A. Amadori et al., Nucl. Phys. A118 (1968) 33.
7. V. Stéfansson et al., Proceedings of International Conference on Radioactivity in Nuclear Spectroscopy, Nashville, Tennessee, August, 1969.

Contributed Paper

DIRECTIONAL CORRELATION MEASUREMENTS ON THE DECAY OF ^{176}Ta

H.G. Boddendijk, S. Idzenga, G. Timmerman and H. Verheul
Natuurkundig Laboratorium der Vrije Universiteit, Amsterdam, The Netherlands

The level scheme of ^{176}Hf was established by recent investigations[1,2,3]), making directional correlation measurements desirable. We performed these measurements on all cascades involving strong gamma rays or converted transitions.

The source material was deposited on a 125 $\mu g/cm^2$ Zapon foil, the actual source thickness was $< 20\ \mu g/cm^2$ of organic material from the ion - exchange column used in the chemical separation procedure. Sources were circular and had a diameter of about 6 mm.

For the directional correlation measurements an automatic apparatus was used, described elsewhere[3]). Geometrical attenuation coefficients for the solid state gamma-ray and electron detectors were calculated, the calculations were tested with the aid of the $13/2^+$ (M4) $5/2^-$ (E2) $1/2^-$ cascade occurring in the decay of ^{207}Bi [3]).

In all cascades of interest only one of the radiations was mixed. So the directional correlation data could be analyzed with the aid of parametric plots. To avoid the disadvantages of this method we compared the various possible spin sequences by carrying out a χ^2 - analysis of the experimental data, to which we added the experimental value of the conversion coefficient of the mixed transition. We wrote a computer program carrying out the following procedure: for all spin - and parity sequences of interest the value of the mixing ratio δ^1 of the mixed transition is varied from -1 to $+1$ in steps of 0.002. At each value of $\delta^1 = \delta/1 + |\delta|$ the value of χ^2 is calculated. Minima in χ^2 indicate possible solutions for the spin and parity sequence and the value of δ^1. The error in δ^1 corresponds to an increase of 1 above the value at the minimum. The i - th solution now has a likelihood L equal to

$$L = \frac{e^{-\chi_i^2/2}}{\sum_j e^{-\chi_j^2/2}}$$

The results of the analysis of the experimental data on the 0.2010γ - 0.0881γ cascade agree with the known 4^+ (E2) 2^+ (E2) 0^+ character of this cascade (cf. table 1), indicating that effects of extra - nuclear fields were absent.

To be able to perform the analysis of the experimental data of the $e^- - \gamma$ directional correlation measurements we first fitted the data of the 0.2010γ - ce_L 0.0881 and the ce_K 0.2010 - 0.0881γ cascades to the form of a 4^+ (E2) 2^+ (E2) 0^+ cascade in order to obtain the values of the electron particle parameters. The experimental values of these parameters thus obtained are b_2 (ce_L, 0.0881) =

TABLE 2. Analysis of the directional correlation data.

Cascade	Solutions with L > 1%	δ_1 (γ)	χ^2	L
*1.1588 - 0.0881	$1^-(E1+M2)2^+(E2)0^+$	0.19 ± 0.01	6.00	23%
	$2^-(E1+M2)2^+(E2)0^+$	-0.37 ± 0.015	3.60	77%
1.5530 - 0.0881	$1^-(E1+M2)2^+(E2)0^+$	0.19 ± 0.035	3.45	95%
	$2^-(E1+M2)2^+(E2)0^+$	-0.25 ± 0.035	8.70	5%
1.5812 - 0.0881	$1^-(E1+M2)2^+(E2)0^+$	0.30 ± 0.035	1.15	23%
	$2^-(E1+M2)2^+(E2)0^+$	-0.20 ± 0.04	0.10	39%
	$3^-(E1+M2)2^+(E2)0^+$	0.24 ± 0.04	0.15	38%
*1.8251 - 0.0881	$2^+(M1+E2)2^+(E2)0^+$	<-17.5;>14	3.60	17%
	$2^-(E1+M2)2^+(E2)0^+$	0.45 ± 0.06	4.30	12%
	$3^+(M1+E2)2^+(E2)0^+$	-2.3 ± 0.5	2.70	27%
	$3^-(E1+M2)2^+(E2)0^+$	-0.31 ± 0.09	4.60	10%
	$4^+(E2+M3)2^+(E2)0^+$	0.0 ± 0.08	2.30	34%
*2.8315 - 0.0881	$1^-(E1+M2)2^+(E2)0^+$	0.16 ± 0.025	3.40	90%
	$2^-(E1+M2)2^+(E2)0^+$	-0.40 ± 0.045	7.80	10%
*0.7102 - 1.1588 - 0.0881	$1^-(M1+E2)2^-(E1+M2)2^+(E2)0^+$	0.26 ± 0.05	1.35	31%
	$2^-(M1+E2)2^-(E1+M2)2^+(E2)0^+$	-0.27 ± 0.08	1.10	36%
	$3^-(M1+E2)2^-(E1+M2)2^+(E2)0^+$	-0.16 ± 0.08	1.25	33%
1.0228 - 0.2010 - 0.0881	$3^-(E1+M2)4^+(E2)2^+(E2)0^+$	0.10 ± 0.035	1.25	100%
1.6948 - 1.1588 - 0.0881	$1^+(E1+M2)2^-(E1+M2)2^+(E2)0^+$	0.37 ± 0.07	2.30	21%
	$1^-(M1+E2)2^-(E1+M2)2^+(E2)0^+$	<-7.6	4.45	7%
	$1^-(M1+E2)2^-(E1+M2)2^+(E2)0^+$	0.34 ± 0.07	5.30	4.5%
	$2^+(E1+M2)2^-(E1+M2)2^+(E2)0^+$	-0.27 ± 0.10	4.60	6.5%
	$2^-(M1+E2)2^-(E1+M2)2^+(E2)0^+$	$3.6^{+2.0}_{-1.0}$	0.20	61%

Table 1

possible solutions	$\delta_1 (\gamma)$	χ^2	L
4^+ (E2 + M3) 2^+ (E2) 0^+	0.013 ± 0.035	0.45	35%
3^- (E1 + M2) 2^+ (E2) 0^+	-0.23 ± 0.02	0.16	40%
2^- (E1 + M2) 2^+ (E2) 0^+	-0.23 ± 0.02	1.05	25%

= 1.22 ± 0.10 and b_2 (ce_K, 0.2010) = $1.65 {}^{+0.45}_{-0.29}$. These values agree with the values 1.24 and 1.72 respectively as calculated by Hager and Seltzer[4]).

Inserting the experimental value of b_2 (ce_L, 0.0881) all other direct cascades involving the ce_L 0.0881 MeV transition were analyzed. For the analysis of the triple cascades the experimental results on the 1.1588 MeV - 0.0881 MeV cascade were used (cf. table 2 on page 160). The results of the analysis are given in table 2. For cascades marked with a star the possible solutions follow from the conversion coefficient of the mixed transition (1), the $e^- - \gamma(\theta)$ data (2) and the $\gamma - \gamma(\theta)$ data (3). The solutions for the other cascades follow from the data (1) and (2).

From the results summarized in table 2 and the results of our earlier investigation on the decay of ^{176}Ta [2]) it was possible to determine the quantum number K for a number of ^{176}Hf levels using the Alaga branching rules. They are listed in table 3.

Table 3

K - values of excited states of ^{176}Hf

E (MeV)	I^π	K	Possible level character
1.1499	0^+	0	β - vibration
1.2483	2^-	2	octupole - vibration
1.2932	0^+	0	β - vibration
1.3133	3^-	(2)	
1.3410	2^+	2	γ - vibration
1.6414	1^-	0	octupole - vibration
1.6693	$1^-, 2^-, 3^-$	0 or 1	octupole - vibration
1.7207	1^-	0 or 1	octupole - vibration
1.8615	2^+	2	γ - vibration
1.9132	2, 3, 4	I = 3 with K = 2	
2.0438	1^+	1	particle excitation
2.9196	1^-	1	particle excitation
2.9400	2^-	1	

Recently, calculations were performed on the ^{176}Hf levels by Soloviev [5]), Malov and Soloviev [6]) and Neergård [7]). Experimental counterparts can be found for most of the quadrupole - vibrational and octupole - vibrational levels predicted by Soloviev; a more detailed comparison can be found in ref.3. Most of the pairing - vibrational 0^+ levels calculated by Mikoshiba[8]) are not yet found experimentally. The earlier results of Bés[9]) and Marshalek[10]) seem to describe the quadrupole - vibrational levels fairly well.

References

1. F.M. Bernthal, UCRL report 18651 (1969).
2. H.G. Boddendijk et al., Nucl. Phys. A134 (1969) 442.
3. H.G. Boddendijk, Thesis, Free University, Amsterdam (1970).
4. R.S. Hager and E.C. Seltzer, Nuclear Data, Sect. A, Vol. 4, no. 2 (1968).
5. V.G. Soloviev, Private communication.
6. L.A. Malov, V.G. Soloviev and U.M. Fainer, Conf. Int. Symp. Nucl. Str. Dubna (1968) p. 78.
7. K. Neergård and P. Vogel, Nuclear Phys. A145 (1970) 33.
8. O. Mikoshiba, R.K. Sheline and T. Udagawa, Nucl. Phys. A101 (1967) 202.
9. D.R. Bès, Nucl. Phys. 49 (1963) 544.
   D.R. Bès, P. Federman, E. Magueda and A. Zuker, Nucl. Phys. 65 (1965) 1.
10. E.R. Marshalek, UCRL report 10046 (1962).

Contributed Paper

SOME ASPECTS OF DIRECTIONAL CORRELATION MEASUREMENTS IN ^{152}Sm

H.A. Doubt and W.D. Hamilton
School of Mathematical and Physical Sciences, University of Sussex, Brighton, England

The recent extensive efforts to determine whether anomalous internal conversion particle parameters occur has emphasized that not only are γ - e directional correlations difficult to measure but, more significantly, the equivalent $\gamma \cdot \gamma$ correlation may give widely differing results. To date there exist 7 independent sets of data aimed at checking the particle parameters of the converted 122 keV E2 transition in ^{152}Sm by examining the 4^+ (245) 2^+ (122) 0^+ [1,2]) and 2^- (1408) 2^+ (122) 0^+ (ref. 1, 3, 4, 5 and 6) cascades. These data are presented in tables 1 and 2. The outstanding feature of the data is the lack of agreement between the different measurements which reach similar conclusions.

We have remeasured these correlations and have investigated possible mechanisms to explain the disagreement in the 1408 - 122 $\gamma \cdot \gamma$ correlation measurements.

All the solid - source measurements used a source prepared by irradiation of

Table 1

Directional correlation results for the 4^+(245 keV)2^+(122 keV)0^+ cascade in ^{152}Sm. Both γ-γ and γ-e results are shown

	γ-γ: G_2A_{22}	G_4A_{44}	G_2	γ-e: G_2A_{22}	G_4A_{44}	G_2	b_2^K(E2; 122)
Theory	+0.1020	+0.0091	1	+0.1931	−0.0112	1	1.89
Solid Ref. 2	0.086 ± 0.008	0.013 ± 0.016	0.84 ± 0.08	0.142 ± 0.006	−0.028 ± 0.011	0.74 ± 0.03	1.65 ± 0.17
Ref. 1	0.087 ± 0.008	0.002 ± 0.010	0.85 ± 0.08	0.140 ± 0.003	−0.018 ± 0.009	0.73 ± 0.02	1.61 ± 0.13
Present work	0.081 ± 0.006	0.006 ± 0.013	0.80 ± 0.06	0.171 ± 0.009	−0.037 ± 0.016	0.88 ± 0.05	2.11 ± 0.19

mass - separated ^{151}Eu in a neutron flux of 1.5×10^{14} for two weeks.† As the thickness was only 1 μg·cm^{-2}, no corrections were applied for multiple scattering of electrons. Conventional angular correlation systems were used, with special care taken to minimize drifting of gates (when the coincident spectra were not displayed in a multichannel analyser). Singles and coincidence rates were constantly monitored, and random rates and relative timing were regularly checked. Corrections were applied

Table 2

Results for γ-γ and γ-e directional correlation measurements of the 2^-(1408)2^+(122)0^+ cascade in ^{152}Sm

	γ-γ: G_2A_{22}	G_4A_{44}	γ-e: G_2A_{22}	G_4A_{44}	b_2^K(E2; 122)
Theory: Pure E1	+0.2500	0	+0.4733	0	1.89
Solid Ref. 3	0.144 ± 0.005	0.011 ± 0.007	0.272 ± 0.007	0.017 ± 0.022	1.90 ± 0.07
1	0.172 ± 0.004	0.005 ± 0.007	0.275 ± 0.005	0.010 ± 0.015	1.60 ± 0.05
4	0.158 ± 0.008	0.008 ± 0.011	0.294 ± 0.011	0.015 ± 0.021	1.86 ± 0.13
5	0.168 ± 0.004	0.001 ± 0.006	0.305 ± 0.010	0.002 ± 0.013	1.82 ± 0.08
6	0.138 ± 0.003[a)] 0.129 ± 0.003[b)]	<0.01	0.249 ± 0.003[c)]	<0.01	1.77 ± 0.06[d)]
I [*)]	0.134 ± 0.007	0.022 ± 0.014	0.264 ± 0.007	0.024 ± 0.014	1.97 ± 0.11
II [*)]	0.167 ± 0.003	0.011 ± 0.008			
III [*)]	0.150 ± 0.003	0.007 ± 0.008			

*) Our measurements. Roman numerals are explained in the text.

a) Average of a series of 3.8 × 5.1 cm NaI(Tl) results from Ref. 6.

b) Measurements with 122 keV γ-ray in Ge(Li) diode, also taken from Ref. 6.

c) Measurements with 122-K electron in lens spectrometer, also taken from Ref. 6.

d) This is a typical value of the particle parameter obtained in Ref. 6. There were six other similar values.

† The isotope separated source was kindly supplied by Prof. Z.W. Grabowski and was prepared as described in ref. 1.

in the usual way for source - miscentring and the finite solid angles of the detectors. (Q - values were either taken from already published works[7,8]), or computed using a Monte Carlo routine; in the case of the beta spectrometer, the Q - values were measured under the experimental conditions[9]).

The higher energy was always selected in a 7.6 X 7.6 cm NaI (Tl) detector. The 122 keV transition was selected on different occasions in 7.6 X 7.6 cm or 0.6 X 3.8 cm NaI (Tl) detectors or in a Ge(Li) 0.8 X 2.6 cm diode. For the 245 - 122 keV γ - γ cascade, this introduced no significant difference in the correlation coefficients, but for the 1408 - 122 keV cascade the dependence on detector size is quite marked. This may be seen in table 2 and in fig. 1.

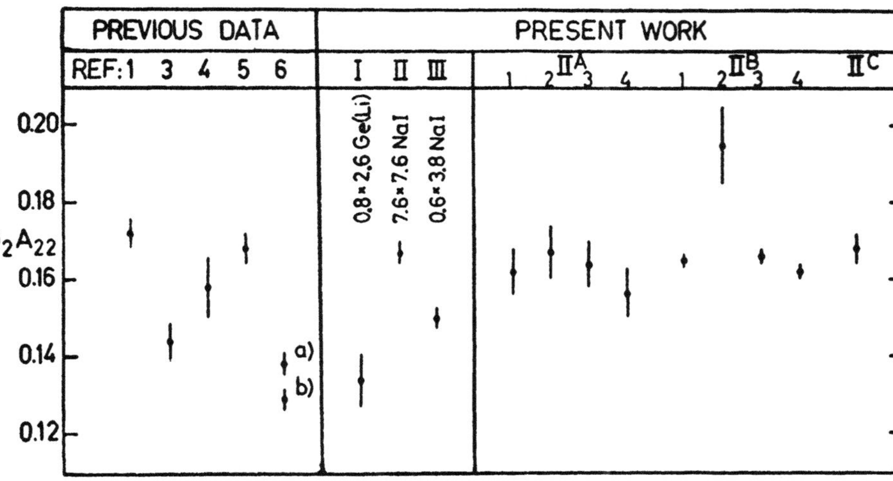

Fig. 1. The G_2A_{22} directional correlation coefficients for the 2^- (1408 keV) 2^+ (122 keV) 0^+ γ - γ cascade.

At no time was any dependence of the correlation coefficients on source - detector distances noted. Investigation of the effect of choice of angle was performed by separate analysis of the same measurement for different combinations of the angular positions used. An example of this, for two 7.6 X 7.6 cm NaI detectors (combination II) is shown in fig. 1 and table 3. This type of analysis demonstrates that it is possible (Result II B.2) to choose angles which give very poor coefficients, but that it is not too difficult to recognize such an effect. Such an analysis also indicates that the possible presence of annihilation radiation around the 180° position is not significant.

A background scattering effect could be a possible explanation for the variation of the correlation with the size of the 122 keV detector. Simple crystal - crystal scattering of the correlation γ- rays is not energetically possible, but scattering of the 344 keV γ - ray and coincident summing with the 1298 γ - ray in ^{152}Gd could certainly occur. Because this scattering is not markedly angular - dependent, it is difficult to establish it unambiguously, although it might be expected to show up in the kind of analysis presented in table 3. It will, however, depend on detector size.

It would seem, therefore, that the NaI - Ge(Li) measurements (I) are the most realistic. Our measurements (II) with the large detector are compatible with those of ref. 1, in which the same size detectors were used. The results of references 3 and 4 are not inconsistent with our analysis, and only ref. 5 is in actual disagreement. We have also averaged some unpublished measurements[6]) for a variety of source forms and detector sizes. Here too, the Ge(Li) detector results are lower than the others, and consistent with our own.

Table 3

An example of measurements of the γ-γ directional correlation $2^-(1408)2^+(122)0^+$ using two 7.6 × 7.6 cm NaI detectors at 13.8 and 10.5 cm respectively. Each block contains the results obtained by processing the same data for different combinations of the angles used.

	G_2A_{22}	G_4A_{44}	Angles used (in degrees)	Effective angles (in degrees)
II. A.1	0.162 ± 0.006	0.001 ± 0.013	75,120,165,210,255	105(2),120,150,165
2	0.167 ± 0.007	0.020 ± 0.016		105(2),120,150
3	0.164 ± 0.006	-0.013 ± 0.017		105(2), 150,165
4	0.156 ± 0.006	0.011 ± 0.013		105(2),120, 165
II. B.1	0.165 ± 0.002	0.001 ± 0.004	85,130,175,220,265	95(2),130,140,175
2	0.195 ± 0.010	0.043 ± 0,015		95(2),130,140
3	0.166 ± 0.002	-0.007 ± 0.005		95(2), 140,175
4	0.162 ± 0.002	0.008 ± 0.005		95(2),130 175
II. C	0.168 ± 0.004	-0.001 ± 0.006	90,135,180,225,270	90(2),135(2), 180

Taking the Ge(Li) measurements to be correct, there is no evidence in either correlation of any attenuation in the b_2^K (E2; 122) particle parameter (see tables 1 and 2). In fact it is slightly overestimated. Assuming a normal particle parameter, the perturbation of the 122 keV level calculated from our 4 - 2 - 0 γ - γ and γ - e measurements is:

$$G_2(2^+) = 0.85 \pm 0.05$$

This is in good agreement with the 4 - 2 - 0 γ - γ results of references 2 and 3 and suggests that their γ - e measurements are low. This value also gives an unperturbed directional correlation coefficient for the 1408 - 122 γ - γ cascade of

$$A_{22} = +0.157 \pm 0.012 \,,$$

and implies an M2 admixture in the $2^- - 2^+$ transition with a mixing ratio of

$$\delta(M2 : E1) = -(0.13 {}^{+0.02}_{-0.04})$$

Such an M2 admixture gives a theoretical K - conversion coefficient of

$$a_K = 0.00053 {}^{+(2)}_{-(3)}$$

which is compatible with 0.00047 ± (4), the experimental value[10,11]). Former measurements for this transition, using liquid sources[4,12]), were also made with large detectors and so will be subject to errors similar to those discussed above.

Thus we have shown that it may often be difficult to perform γ - γ measurements that have both precision and accuracy, without a detailed investigation of the experimental conditions.

This work was supported by a research grant from the Ministry of Technology to whom H.A.D. acknowledges receipt of a studentship.

References

1. H.M. Nasir, Z.W. Grabowski and R.M. Steffen, Phys. Rev. 162 (1967) 1118.
2. E.F. Zganjar, T.M. George and J.H. Hamilton, Nuclear Phys. A114 (1968) 609.
3. L. Holmberg, V. Stefansson and B.G. Pettersson, Nucl. Phys. A96 (1967) 33.
4. K. Bonde Nielsen, N. Rud, J.Chr. Overgård and H.E. Sørensen, Nucl. Phys. A124 (1969) 445.
5. Y.K. Agarwal, C.V.K. Baba and N.C. Mukhopahyay, Nucl. Phys. A127 (1969) 661.
6. H. Blumberg, J. Fechner, A. Hammesfahr, P. Herzog, A. Kluge and J. Voss, Report: Institut für Strahlen- und Kernphysik, Bonn (1969).
7. M.J.L. Yates, Nucl. Instr. Methods 23 (1969) 152.
8. D.C. Camp and A.L. van Lehn, Nucl. Instr. Meth. 76 (1969) 192.
9. P. Kleinheinz, L. Sameulsson, R. Vukanovic and K. Siegbahn, Nucl. Instr. Meth. 32 (1965) 1.
10. G. Malmsten, O. Nilsson and I. Andersson, Ark. Fys. 33 (1967) 361.
11. L. Varnell, J.D. Bowman and J. Trischuk, Nucl. Phys. A127 (1969) 270.
12. P. Debruner and W. Kundig, Helv. Phys. Acta 33 (1960) 395.

Contributed Paper

E0 - CONTRIBUTION TO THE 689K - 122γ ANGULAR CORRELATION IN ^{152}Sm

V. Stefánsson, V. Sergeev, J. Becker and L. Holmberg
University of Stockholm, Institute of Physics, Stockholm, Sweden

In some recent experiments (ref. 1 - 9) beta - vibrational states in ^{150}Nd,

^{152}Sm, and $^{154,156}Gd$ were investigated. The results were found to disagree with the collective - model description of these states. The relative magnitude of the E2 matrix elements for transitions connecting the beta and ground state bands cannot be described by introducing only one mixing parameter.

In their extensive investigation of transitions in ^{152}Sm and ^{154}Gd Riedinger et al.[2]) determined the ratios between the E0 and E2 intensities. They obtained values 2.5 to 5 times smaller than predicted by the models used for comparison with the experimental data.

This work aims at a determination also of the sign of the mixing parameter q_K (E0/E2) for one of the transitions in ^{152}Sm. The angular correlation between K electrons emitted in the 689 keV $2_\beta \rightarrow 2_{gr}$ transition and the 122 keV gamma radiation is studied.

The conversion electrons emitted between bands with the same spin and parity are in general a mixture of E2, M1, and E0 components. The method used here is based on the fact that the second order angular correlation coefficient contains a term proportional to q. For an electron - gamma correlation Church, Rose and Weneser[10]) give the expression

$$A_2(E0, M1, E2) = \frac{(1+p^2)\, A_2(M1, E2) + b_0 q}{1+p^2 + q^2}$$

where p (M1/E2) is the mixing ratio for electrons and b_0 is an interference parameter.

The angular correlation was measured by means of a Gerholm magnetic lens spectrometer[11]) and a movable NaI (Tl) detector. Coincidences were collected at four different positions of the gamma detector and collected in the memory of a 4 X 1000 channel analyser. The coincidence counting rate was normalized to the single counting rate; background and accidental coincidences were subtracted. The A_2 and A_4 coefficients were determined and corrected for the finite solid angles of the detectors and the source thickness. An isotope - separated source was used[12]).

The major difficulty is caused by the strong background of β - γ coincidences. The beta particles are coincident with the 344 keV gamma transition in ^{152}Gd and the angular correlation has a fairly large negative anisotropy. The contribution of coincident gamma radiation under the 122 keV peak was determined by separate measurements with an off - line setting of the spectrometer current.

From the data collected so far the following result is derived

$$G_2 A_2 = 0.05 \pm 0.07$$

and $G_4 A_4$ is equal to zero within the limits of error.

The attenuation factor was determined in a recent investigation[12]). $G_2 A_2$ of the 1408 keV - 122 keV gamma - gamma cascade was measured for the same solid source, and a liquid ^{152}Eu source. Applying the result for G_2 to the 689K - 122γ correlation yields

$$A_2 = 0.08 \pm 0.11$$

The M1 admixture in the 689 keV transition is expected to be small. This has recently been confirmed by Coulomb excitation experiments[5,13]). Fraser et al.[5]) report δ (E2/M1) $= -67^{+58}_{-\infty}$.

This value was used in our analysis together with theoretical conversion coefficients and particle parameters tabulated by Hager and Seltzer[14]). Neglecting penetration contributions to the weak M1 component we obtain

$$q_K \geqslant 0.3$$

The value by Riedinger et al. is

$$|q_K| = 2.5 \pm 0.2$$

The result of the present work is thus consistent with the positive value.

References

1. Y-t Liu, O.B. Nielsen, P. Salling and O. Skilbreid, Bull. Acad. Sci. USSR, Phys. Ser. 31 (1967) 69.
2. L.L. Riedinger, N.R. Johnson and J.H. Hamilton, Phys. Rev. 179 (1969) 1214.
3. R.A. Meyer, Phys. Rev. 170 (1968) 1089.
4. L. Varnell, J.D. Bowman and J. Trischuk, Nucl. Phys. A127 (1969) 270.
5. I.A. Fraser, J.S. Greenberg, S.H. Sie, R.G. Stokstad, G.A. Burginyon and D.A. Bromley, Phys. Rev. Letters 23 (1969) 1047.
6. B. Elbek, Proceedings International Conference on Properties of Nuclear States, Montréal 1969, p. 63, Les Presses de l'Université de Montréal, 1969.
7. J.S. Greenberg, Proceedings op. cit. p. 83.
8. G.T. Ewan and G.I. Andersson, Contributions Nucl. Structure Conf. Tokyo 1967.
9. J.H. Hamilton, D.J. McMillan, J.C. Manthuruthil, S.R. Amtey and J.J. Pinajian, Contribution International Conference on Properties of Nuclear States Montréal 1969.
10. E.L. Church, M.E. Rose and J. Weneser, Phys. Rev. 109 (1958) 1299.
11. T.R. Gerholm, R. Othaz and M.S. El - Nesr, Arkiv. Fysik 21 (1962) 253.
12. L. Holmberg, V. Stefánsson, L. Eriksson and B.-G. Pettersson, Contribution International Conference on Radioactivity in Nuclear Spectroscopy, Nashville 1969.
13. F.K. McGowan, R.O. Sayer, P.H. Stelson, R.L. Robinson and W.T. Milner, Bull. Am. Phys. Soc. 13 (1968) 895.
14. R.S. Hager and E.C. Seltzer, Nucl. Data A4 (1968) 1, 397, Nucl. Data A6 (1969) 1.

Contributed Paper

ELECTRON - GAMMA ANGULAR CORRELATIONS OF THE 208 - 113 keV CASCADE IN ^{177}Hf

L. Holmberg and V. Stefánsson

University of Stockholm, Institute of Physics, Stockholm, Sweden

1. Introduction

The beta decay of ^{177}Lu is well known; spins and parities of the levels in ^{177}Hf

have been determined and are well described by the unified model. In this work electron angular correlations of the 208 keV - 113 keV cascade were carried out. Possible deviations from the 'normal' values for the conversion electron particle parameters were studied.

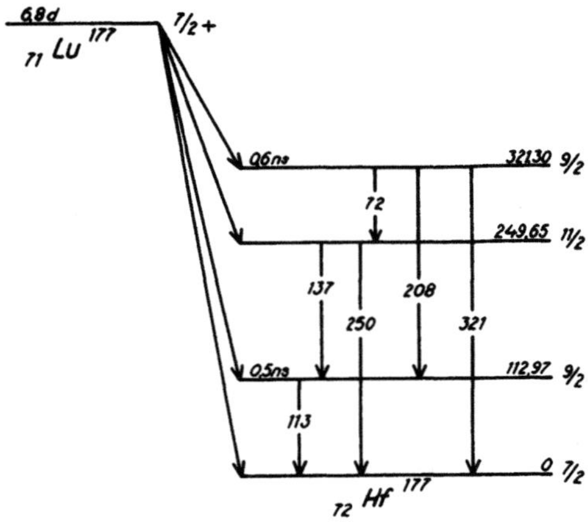

Fig. 1. Decay scheme of ^{177}Lu.

The electron - gamma and gamma - electron correlations were measured. A Gerholm spectrometer[1]) was used for the energy selection of the electrons and the gamma radiation was detected in a movable NaI (Tl) detector. The corresponding gamma - gamma correlation was measured both in a two - detector set - up and in a multichannel goniometer using eight NaI (Tl) detectors. Solid as well as liquid sources were used for the gamma - gamma correlations.

The extensive tabulations by Hager and Seltzer[2]) were used for the analysis of the conversion processes.

2. The 208 keV E1 transition

The $9/2^+ \to 9/2^-$ transition of 208 keV is of electric dipole type and retarded. The hindrance factor[3]) relative to the Weisskopf estimate is 3.1×10^4.

Thun et al.[4]) obtained a value for the K - electron particle parameter in reasonable agreement with the theoretical value. The conversion coefficient and subshell measurements by Hager and Seltzer[5]) also indicate an almost normal internal conversion process.

The magnitude of the penetration effect expected for this transition is consistent with any of the two experiments. However, the electron correlation and conversion coefficient data indicate a different sign of the penetration parameter.

The present study was carried out by use of sources obtained by evaporation of lutetium oxide onto aluminium backings and subsequent neutron irradiation. The K - electron - gamma and gamma - gamma correlations were measured simultaneously. Effects of static attenuations and possible normalization errors then cancel in the determination of the particle parameter.

The contribution of magnetic quadrupole radiation has to be known for an analysis of the conversion process. The M2 admixture was estimated from the theoretical relation between δ and the penetration parameter[6,7]). It was found to be negligibly small.

The experimental result for the particle parameter is

$$b_2 = -1.66 \pm 0.04 ,$$

to be compared with the theoretical value of b_2 (E1) = -1.68 and the result by Thun et al.[4]), $b_2 = -1.40 \pm 0.24$.

3. The 113 keV M1 + E2 transition

The conversion process of the rotational 113 keV $9/2^- \rightarrow 7/2^-$ transition of mixed M1 + E2 type is not well understood. The L - subshell ratios by Novakov and Hollander[14]) do not allow an unambiguous determination of the mixing ratio. An extensive internal conversion study by Törnkvist et al.[8]) involving several gamma - electron correlations gave for the mixing ratio the value $\delta = -3.1 \pm 0.6$. This corresponds for the gamma - gamma correlation to the value (if the 208 keV transition is pure E1) $A_2 = -0.245^{+0.4}_{-0.5}$, which is appreciably smaller than the measured value.

In this work the gamma - K electron and gamma - L electron correlations were measured. The experimental results are given in table 1; A_2 $(\gamma\,\gamma)$ obtained with a source produced by evaporation of Lu - metal onto aluminium backing is also listed.

Table 1

Experimental results for the angular correlation coefficients A_2 of the 208 keV - 113 keV cascade.

	Törnkvist et al.[8])	Present work
γ–K	0.022 ± 0.004	0.025 ± 0.004
γ - L_2	–0.019 ± 0.004	
γ - L_3	–0.004 ± 0.003	
γ - L		–0.012 ± 0.003
γ - γ	–0.132 ± 0.003	–0.128 ± 0.004

The result for the gamma - gamma correlation in a liquid source is

$$A_2(\gamma\gamma) = -0.157 \pm 0.001.$$

The moderate attenuation as compared with the result for a solid source, $G_2 = 0.82 \pm 0.03$, does not indicate any large remaining attenuation.

From the result for A_2 $(\gamma\,\gamma)$ we derive $\delta = -4.9 \pm 0.1$. If this mixing ratio is used for analysing the electron correlations, no significant deviation from the normal values is found. We obtain

$$A_2(\gamma K)/A_2(\gamma\gamma) = -0.195 \pm 0.031,$$
$$A_2(\gamma L)/A_2(\gamma\gamma) = 0.094 \pm 0.023.$$

The corresponding theoretical values, for $\delta = -4.9$, are -0.234 and 0.064.

4. Discussion

Penetration contributions to the conversion process have been found for many retarded E1 transitions in deformed nuclei. This is for instance true for transitions depopulating the 396 keV level in ^{177}Lu [6,7]).

The 208 keV transition in ^{177}Hf is similar to the 282 keV transition in ^{175}Lu. Both are spin - flip transitions. In ^{175}Lu it was found that the penetration contributions of nuclear current type obey the branching rules for allowed transitions. If the same rules are assumed to be valid for the decay of the 321 keV state in ^{177}Hf, the effect in the 208 keV transition may be found from λ (321), determined from conversion coefficient data. One obtains for the 208 keV transition $|\lambda| = 0.4$.

The subshell ratios obtained by Hager and Seltzer yield $\lambda = 0.4 \pm 0.4$. (Here a possible contribution from the nuclear charge type penetration has been neglected). Our result for the particle parameter is consistent with $\lambda = -0.1 \pm 0.3$. Thus it may be concluded that a positive value for the penetration parameter is favoured by the experimental data.

The magnitude of the gamma radiation matrix element is experimentally determined from the life - time of the 321 keV level. As the penetration matrix element is unhindered its value may be directly calculated from the Nilsson model. It is then possible, by use of the experimental value for λ, to obtain also the sign of the gamma radiation matrix element, which is of interest for comparison with more detailed nuclear structure calculations.

The 272 keV transition in ^{173}Yb was investigated by Hornshøj and Deutch[9]) who found a normal value for the particle parameter, cf. table 2. The hindrance factor for this transition, 5.3×10^4, is somewhat larger than for the 208 keV transition. However, it does not seem likely that even a higher statistical accuracy would reveal

Table 2. K-conversion data for some retarded E1 transitions

	^{173}Yb	^{175}Lu	^{177}Hf
Asymptotic quantum numbers			
Initial	633↑	514↑	624↑
Final	512↑	404↓	514↓
I'→ I + 1 transitions			
Energy (keV)	272	282	208
Hindrance factor	5.3×10^4	9.7×10^5	3.1×10^4
b_2^{exp}/b_2^0 (E1)	1.03 ± 0.07[9])	-0.16 ± 0.10[6])	0.99 ± 0.03
I'→ I transitions			
Energy (keV)	351	396	321
Hindrance factor	6.6×10^6	1.7×10^6	6.2×10^6
α^{exp}/α^0 (E1)	1.24 ± 0.17[12])	4.68 ± 0.07[7])	5.5 ± 0.3[5])

The hindrance factors are taken from ref.3.

any penetration contribution in this case. The penetration contribution to the E1 conversion processes in ^{175}Lu and ^{177}Hf proceeds through the interaction of the spin - part of the nuclear current. This transition is allowed by the selection rules for asymptotic Nilsson quantum numbers[3,10,11]). In ^{173}Yb the change in asymptotic Nilsson quantum numbers is different. The nuclear current gives no contribution in lowest order, instead the nuclear charge part of penetration is allowed. Due to a smaller dimensional value of the corresponding matrix element this effect is expected to be of less importance. This is confirmed by the experimental conversion coefficients for the cross - over transitions in the three nuclei surveyed in table 2.

The 113 keV transition occurs between two levels of the same rotational band. Reiner[13]) pointed out that even for transitions of this kind observable penetration effects may exist, due to an accidental cancellation of the collective and particle gyromagnetic ratios. For the 113 keV transition $g_K - g_R = -0.034 \pm 0.001$.

The results obtained in this work do not furnish support for the existence of a penetration contribution. In fact, the K and L-particle parameters indicate different signs of λ, if an effect should exist in the M1 component. As Törnkvist et al. concluded the conversion data alone indicate a somewhat larger value for δ than the one used in this analysis.

Experimental and theoretical values for the particle parameters are generally in good agreement. This has been verified in particular for E2 conversion processes. In the 113 keV transition also the M1 - E2 interference term has a decisive influence. However, due to the large E2 component the correlations are too small to be suitable for studies of the conversion process since high accuracy is wanted.

We are grateful to Prof. T.R. Gerholm for the excellent research facilities put at our disposal, and for his continuous interest in this work. We are indebted to M. Gunnerhed who participated in the measurements, to L. Eriksson and L. Gidefeldt for carrying out the MCG runs, and to V. Sergeev who called our attention to the work by Balalaev et al.[12]).

This work is sponsored by Statens Rad för Atomforskning.

References

1. T.R. Gerholm, R. Othaz and M.S. El - Nesr, Arkiv Fysik 21 (1962) 253.
2. R.S. Hager and E.C. Seltzer, Nuclear Data A4 (1968) 1, 397, Nuclear Data A6 (1969) 1.
3. W.B. Ewbank, in Internal Conversion Processes, ed. J.H. Hamilton, Academic Press, New York (1966).
4. J.E. Thun, Z. Grabowski, W.D. Hamilton and M.S. El - Nesr, Nuclear Physics 29 (1962) 13.
5. R. Hager and E.C. Seltzer, in Internal Conversion Processes, op. cit.
6. L. Holmberg, L. Gidefeldt, M. Gunnerhed and B.-G. Pettersson, Nuclear Physics A96 (1967) 305.
7. G.T. Emery and M.L. Perlman, Phys. Rev. 151 (1966) 984.
8. S. Törnkvist, S. Ström, J.E. Thun, V. Schmidt and N. Baade, Nuclear Physics, A117 (1968) 336.
9. P. Hornshøj and B.I. Deutch, Nuclear Physics 67 (1965) 342.
10. E.L. Church and J. Weneser, Nuclear Physics 28 (1961) 602.
11. G. Kramer and S.G. Nilsson, Nuclear Physics 35 (1962) 273.

12. V.A. Balalaev et al., in Program of XX. Soviet Nuclear Spectroscopy Conference, Leningrad, 1970 (in Russian).

13. A.S. Reiner, Nuclear Physics 5 (1958) 544.

14. T. Novakov and J.M. Hollander, Nuclear Physics 60 (1964) 593.

Contributed Paper

NUCLEAR PENETRATION EFFECTS IN THE 92 keV TRANSITION IN ^{131}Cs

L. Hasselgren, S. Antman, H.S. Sahota and J.E. Thun
University of Uppsala, Institute of Physics, Uppsala, Sweden

1. Introduction

The level scheme of ^{131}Cs has been troublesome to interpret both theoretically and experimentally[1,2,3,4]). The angular correlation measurements of Fechner et al.[4]), however, provided unique spin assignments for some of the lower lying states for which previously reported results had been contradictory. Recent measurements of

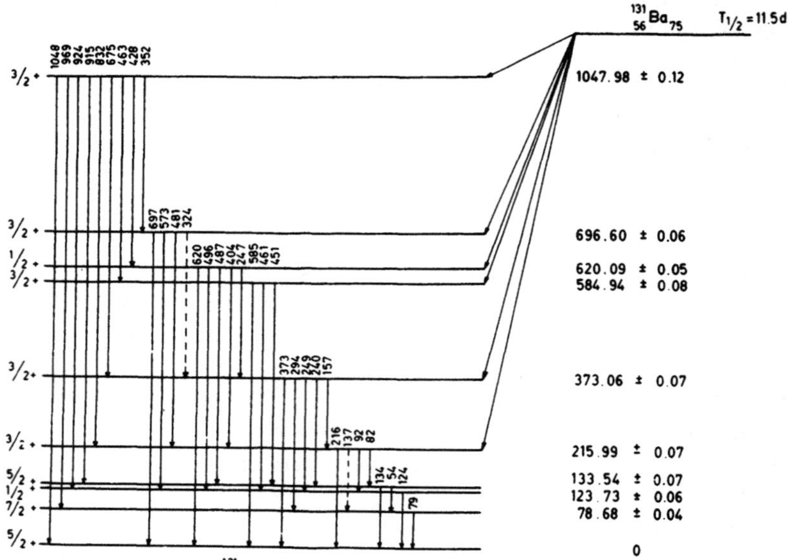

Fig. 1. The decay scheme of ^{131}Ba.

internal conversion and gamma ray intensities by our group in Uppsala[5]) strengthened the results obtained by Fechner et al.[4]). The decay scheme of ^{131}Cs is shown in fig. 1.

2. Measurements and results

The barium activity was produced by irradiating ultra - pure ceasium targets with 30 MeV protons at the Gustav Werner Institute synchro-cyclotron. The directional correlation measurements were performed using the spectrometer described by Falk et al.[6]).

Special care was exercised to obtain very thin sources thereby avoiding source scattering effects. The line shapes indicated that no corrections for such attenuations should be necessary and the consistent results from runs using different source thicknesses were another indication that these effects were negligible. Special attention was also paid to the settings of the discriminators of the gamma channel to reduce possible interference from unresolved gamma rays detected by the NaI detectors. Coincidence spectra showed that such settings were possible and no corrections for unwanted gamma rays were found to be necessary.

The results obtained after proper corrections for finite solid angles are:

$$A_{22}(404\gamma - 216K) = -0.040 \pm 0.004;$$

$$A_{22}(404\gamma - 92K) = -0.017 \pm 0.008.$$

The quoted errors represent one statistical standard deviation unit.

3. Analysis

The 92 keV transition might have a highly retarded M1 component, possibly as much as 10^3 times, compared to the Weisskopf estimate. Since penetration effects in the internal conversion process might then be present, the analysis of the directional correlation results was performed with this possibility in mind.

The A_{22} - value of the (404γ - 92K) angular correlation of -0.017 ± 0.008 reported above is used in this analysis; knowledge of the multipolarity mixture of the 404 keV transition is required. Since the 404 keV transition is rather weak, angular correlation measurements, in which the 404 keV K - conversion line is detected in the electron channel of the apparatus, were not possible. However, the internal conversion and L - subshell measurements on the 216 keV transition (see ref. 5) resulted in a deduced multipolarity mixing of $< 2.4\%$ E2 for this M1 (+E2) transition. Using this value for the 216 keV transition in A_{22} (404γ - 216K) (= −0.040 ± 0.004), we obtain: $-1 \leqslant A_2(404\gamma) \leqslant -0.41$ and $\delta > 0$ for the 216 keV transition. The sign of δ is determined according to the phase convention of Ferentz and Rosenzweig[7]). These limits of A_2 (404γ) and our value of A_{22} for the (404γ - 92K) - directional correlation then give the following limits for A_2 (92K) : $0.009 \leqslant A_2(92K) \leqslant 0.061$.

These values for A_2 (92K) are plotted in the $\lambda \cdot \delta$ - plane in fig. 2 for the two spin sequences 3/2 — 3/2 and 3/2 — 1/2 for the 216 and 124 keV states respectively. In fig. 2 we also included the K- and M_1-conversion coefficients of ref. 5.

From fig. 2 it can be seen that overlapping solutions are obtained only for δ - values around 0 and λ - values around 25 or 225 and for the spin sequences 3/2 — 1/2

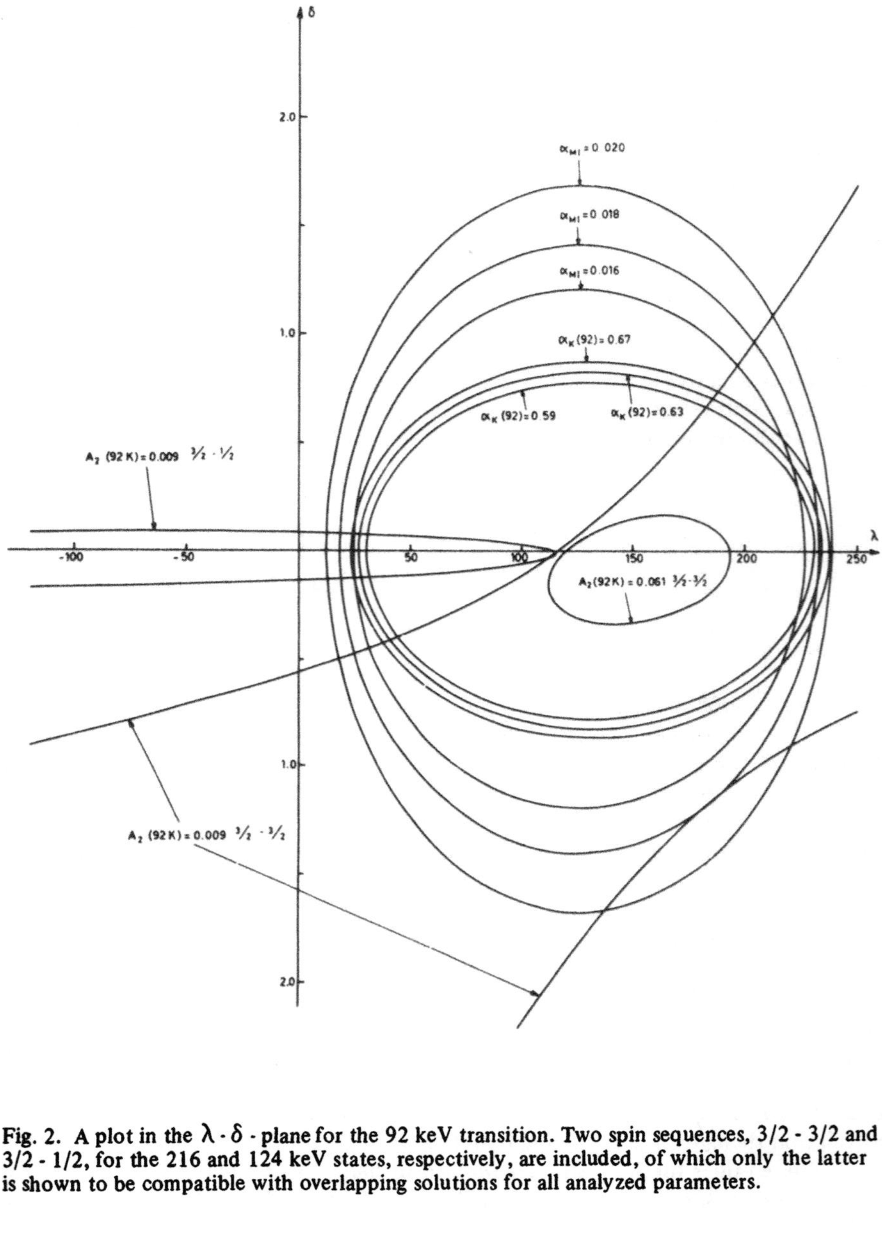

Fig. 2. A plot in the λ - δ - plane for the 92 keV transition. Two spin sequences, 3/2 - 3/2 and 3/2 - 1/2, for the 216 and 124 keV states, respectively, are included, of which only the latter is shown to be compatible with overlapping solutions for all analyzed parameters.

and 3/2 – 3/2, respectively. The L_1 - analysis shows overlapping solutions around $\lambda = 25$ but not around $\lambda = 225$, thus excluding the latter unreasonably high λ - value. From the figure it can clearly be seen that the conversion coefficient curves diverge in the region where the spin sequence 3/2 – 3/2 is valid, thus strongly supporting the 3/2 – 1/2 assignment. The value deduced for the penetration parameter λ can be judged directly from fig. 2. In the quotation below the value is given with errors which overlap both the K- and M_1 -limits of error, giving: $\lambda = 23^{+7}_{-11}$. This penetration parameter λ was determined using the theory by Pauli[8,9]).

The multipolarity for the 92 keV transition is then determined to be M1 + < 1.7% E2 and for the 404 keV transition to be E2 + < 20% M1.

References

1. L.S. Kisslinger and R.A. Sorensen, Rev. Mod. Phys. 35 (1963) 853.
2. N. Freed and W. Miles, Nuovo Cim. 60B (1969) 301.
3. D.J. Horen, J.M. Hollander and R.L. Graham, Phys. Rev. 135 (1964) B301.
4. J. Fechner, A. Hammesfahr, A. Kluge, S.K. Sen, H. Toschinski, J. Voss, P. Weight and B. Martin, Nucl. Phys. A130 (1969) 545.
5. L. Hasselgren, S. Antman, H.S. Sahota and J.E. Thun, UUIP - 678 (1970).
6. F. Falk, S. Törnkvist, H. Snellman and J.E. Thun, Nucl. Instr. and Meth. 49 (1967) 240.
7. M. Ferentz and N. Rosenzweig, Alpha-, Beta- and Gamma-Ray Spectroscopy, ed. K. Siegbahn (North - Holland, Amsterdam, 1965), p. 1687.
8. H.C. Pauli and K. Alder, Z. Phys. 202 (1967) 255.
9. H.C. Pauli, PROGRAM MONICA, private communication.

Contributed Paper

THE INFLUENCE OF NUCLEAR STRUCTURE ON THE PARTICLE PARAMETERS FOR ANGULAR CORRELATIONS INVOLVING CONVERSION ELECTRONS AND ON INTERNAL CONVERSION COEFFICIENTS

I.M. Band, M.A. Listengarten, M.B. Trzhaskovskaya

A.F. Ioffe Physico - Technical Institute, A.A. Zhdanov Leningrad State University, Leningrad, U.S.S.R.

Modern development of the experimental techniques makes it possible to study the high energy gamma and conversion transitions in nuclei (5 - 10 MeV and higher). This problem is treated in some of the recent experimental[1]) and theoretical[2]) investigations.

It was supposed earlier that for the high energy transitions the penetration effects might be important even for the unhindered transitions. The greatest penetration effect was expected to occur for the M1 transitions.

The present work is to show that in the energy region of about 15 MeV in heavy nuclei, the sensitivity of the internal conversion coefficients (ICC) to the penetration effects is small for unhindered M1 - transitions. At the same time the sensitivity of the b_2's (i.e. particle parameters) to these effects is still high for angular correlations involving conversion electrons.

We calculated the ICC's and b_2's for the K - shell in the 'surface current' model (ref. 3) and in the 'no - penetration' model[4]) for the atomic numbers Z = 62, 92 and 102, and for the transition energy k up to 50 mc^2. The wave functions are found numerically. The nucleus was taken as a uniformly charged sphere. The ICC of multipolarity τL (EL or ML) in the q - shell, $a_q^{(\tau L)}$, is equal to:

$$a_q^{(\tau L)} = \Sigma_\kappa |M_\kappa|^2 , \qquad M_\kappa = \int_0^\infty dM_\kappa$$

The quantum number $\kappa = (l - j)(2j + 1)$ determines the state of the free electron. And

$$b_2^{(\tau L)} = 1 + |\Sigma_\kappa a_\kappa M_\kappa e^{i\delta_\kappa}|^2 / D\Sigma_\kappa |M_\kappa|^2 \qquad (1)$$

where δ_κ is the phase of the free electron in the infinity, a_κ and D are constants.

The relative contribution from the nuclear region to the ICC may be characterized by the quantity

$$\Omega(R_0) = \sqrt{\Sigma_\kappa |S_\kappa(R_0)|^2 / \Sigma_\kappa |M_\kappa|^2}$$

where $S_\kappa(R_0) = \int_0^{R_0} dM_\kappa$ and R_0 is the nuclear radius. At high energies (i.e. $10 \leqslant k \leqslant 50$) this contribution is equal to $\Omega(R_0) = 8\%$ (Z = 92) and $\Omega(R_0) = 13\%$ (Z = 102) for the surface current model or $\Omega(R_0) = 20\%$ (Z = 92) and $\Omega(R_0) = 32\%$ (Z = 102) for the no - penetration model (for K - shell, M1).

However, in spite of the great value of the contribution from the nuclear region to the ICC (i.e., great $\Omega(R_0)$), and in spite of the great difference in the values of these contributions $\Omega(R_0)$ in the two models considered, the relative difference between the ICC's calculated for these two models is small (see table 1). It increases only up to k = 10, when this difference drops and passes through the zero (k = 36, Z = 92, M1).

A decrease of the difference between the ICC's calculated for the two models is caused by the fact that the difference in the values of the arguments of the complex quantities M_κ and $S_\kappa(R_0)$ amounts to $\frac{\pi}{2}$ (for both $\kappa = -1$ and $\kappa = +2$) at the transition energy $k = 30 \div 40$ (Z = 92, 102). Since the radial electron wave functions were taken to be real, the only complex quantity contained in the matrix element M_κ is the Hankel function. This function has a great imagninary part at the small values of kr. For this reason the contribution from the nuclear region to the ICC (i.e. $S_\kappa(R_0)$) remains almost purely imaginary up to transition energies for which $kR_0 \approx 1$. At the

Table 1

M1, K-shell ICC's α and nuclear region contribution Ω for surface current and no-penetration models. In brackets is the power of ten by which the preceding number must be multiplied.

k (mc^2)	Z = 92				Z = 102				Z = 92	Z = 102
	Surf. current		No-penetr.		Surf. current		No-penetr.		$\frac{\alpha_K^{M1}(n.-p.)-\alpha_K^{M1}(s.c.)}{\alpha_K^{M1}(s.c.)}$	
	α_K^{M1}	Ω%	α_K^{M1}	Ω%	α_K^{M1}	Ω%	α_K^{M1}	Ω%		
0.5	1.25(0)	2.23	1.32(0)	5.02	3.01(0)	3.16	3.27(0)	6.86	5.6%	8.56%
1.0	1.88(-1)	2.46	2.00(-1)	5.49	4.44(-1)	3.44	4.85(-1)	7.70	6.3%	9.40%
5.0	2.56(-3)	3.97	2.83(-3)	8.75	5.20(-3)	5.54	5.98(-3)	12.1	10.5%	15.0%
10.0	3.79(-4)	5.64	4.27(-4)	12.3	6.45(-4)	8.13	7.76(-4)	17.4	12.7%	20.4%
20.0	6.72(-5)	7.84	7.36(-5)	17.2	8.32(-5)	12.5	9.96(-5)	26.6	9.5%	19.6%
30.0	3.40(-5)	8.80	3.47(-5)	20.0	3.75(-5)	13.9	3.88(-5)	31.2	+2.1%	+3.5%
40.0	2.44(-5)	8.80	2.36(-5)	20.0	2.72(-5)	13.8	2.53(-5)	32.2	-3.3%	-7.2%
50.0	1.93(-5)	8.82	1.78(-5)	20.2	2.29(-5)	13.5	1.99(-5)	32.0	-7.7%	-13%

same time the phase of M_κ changes quickly with the energy (the phase depends approximately linearly on ln 2p) and at a certain energy it passes through zero. The difference $\frac{\pi}{2}$ between the phases of M_κ and S_κ (R_0), however, is at the same time the condition of the weak dependence of M_κ on S_κ (R_0) $(d\,|M_\kappa|^2\,/\,dS_\kappa\,(R_0) \approx 0$ if the phase difference equals $\frac{\pi}{2}$). This is the reason of the ICC's weak dependence on the nuclear structure in the region of the above transition energies.

On the other hand, in the process of calculating the $b_2^{(\tau L)}$, the phase δ_κ must be added to the phase of M_κ. Essentially this addition changes the relation between the arguments of the complex numbers considered so that the effect of the small sensitivity does not occur in $b_2^{(\tau L)}$ in this energy region.

If the nuclear structure is taken into account the expression for M_κ is

$$M_\kappa = M_\kappa^{(0)} + \Sigma_n C_{n\kappa} (\lambda_n - 1)$$

Here, $M_\kappa^{(0)}$ is calculated for the surface current model, λ_n are the parameters characterizing the penetration effects in ICC[5]), $C_{n\kappa}$ are the coefficients for the estimation of the nuclear structure influence on ICC and are calculated with the aid of electron wave functions; $\lambda = 0$ in the no - penetration model and $\lambda = 1$ in the surface current model. The same M_κ has to be used in equation (1) to include the penetration effects in $b_2^{(\tau L)}$.

For allowed gamma transitions the influence of the structural dynamical effects is of the order of the difference between b_2 (or between ICC's) with $\lambda = 0$ and with $\lambda = 1$.

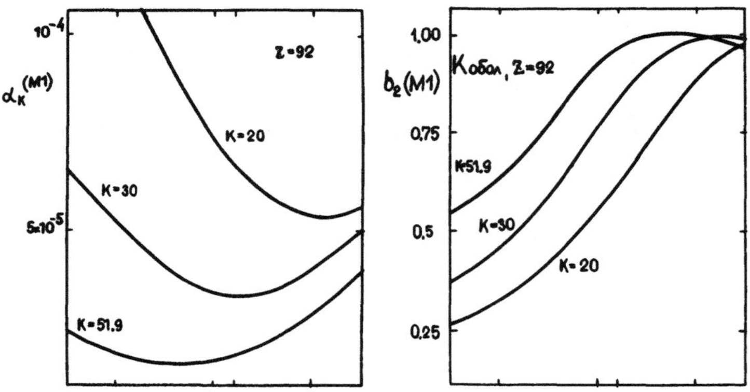

Fig. 1. Dependence of $a_K^{(M1)}$ (the internal conversion coefficients (ICC)) and b_2 (the particle parameters for angular correlations of conversion electrons) on the nuclear penetration parameter λ (the data are given for K - shell conversion and for M1 - transitions); $\lambda = 1$ in the surface current model and $\lambda = 0$ in the no - penetration model. The transition energy k is in $mc^2 = 511$ keV units, atomic number $Z = 92$.

ICC's and b_2's for different λ's and for a transition energy k equal to 20, 30 and 50 (i.e. 10 - 25 MeV) are presented in figs. 1 and 2. Fig. 1 shows that the dependence of $a_K^{(M1)}$ on λ in the region of $\lambda \sim 0 \div 1$ is weaker at a high transition energy ($k \sim 30$) than at the smaller energies, while the dependence of $b_2^{(M1)}$ (K) on λ at the transition energy $k \sim 20 \div 30$ and $\lambda \sim 0 \div 1$ is seen in fig. 2 to be very considerable. At $k \sim 20$, for example, b_2 changes by 13% when λ varies from 0 to 1 ($Z = 92$).

Thus, a situation is possible (M1 - transitions, K - shell, large Z) in which b_2 at a high transition energy is still far from its asymptotic limit ($b_2 = 1$ for M1) and depends strongly on the nuclear structure. In the case of allowed transitions b_2 depends even stronger on the nuclear structure than the ICC's at the same energies.

References

1. R.K. Smither, Phys. Lett. 25B (1967) 128.
2. T. v. Egidy, W. Kaiser, Z. Phys. 201 (1967) 378.
   B.C. Dutta et al., Z. Phys. 207 (1967) 153.
   V.A. Balalajev, B.S. Dzhelepov, A.I. Medvedev, V.E. Ternersesjan, S.A. Shestopalova, Izvestiya Acad. Sci. USSR, ser. fiz. 32 (1968) 730.
   Yu. N. Panin, V.I. Pelehov, preprint of the Inst. of Atomic Energy, No. 1832, Moscow 1969 and the Programm and Theses of Report of XX Annual Conference on Nuclear Spectroscopy and Structure of Atomic Nucleus, part I, page 107, 'Nauka', Leningrad, 1970.
   J. Moragues et al., Phys. Lett. 27B (1968) 441.

2. C.O. Carroll, R.F. O'Connell, Nucl. Phys. 80 (1966) 500; A125 (1969) 637.
   E.L. Church, Phys. Lett. 28B (1968) 168.
   V.F. Trusov, Jadernaya Fizika 9 (1969) 740; 11 (1970) 312.
   O. Dragoun, Nucl. Phys. A132 (1969) 374.

3. L.A. Sliv, JETP 21 (1951) 770.
   L.A. Sliv, I.M. Band, 'Tables of the Internal Conversion Coefficients of the Gamma-Rays', Acad. Sci. USSR, Moscow - Leningrad, 1956, 1958; the books 'Gamma - Lutchi', Acad. Sci. USSR, M. - L., 1961, 'Alpha-, Beta- and Gamma-Ray Spectroscopy', North - Holland Publ. Co., Amsterdam, 1965.

4. M.E. Rose, 'Internal Conversion Coefficients', North - Holland Publ. Co. Amsterdam, 1958.
   R.S. Hager, E.C. Seltzer, Nucl. Data A4 (1968) 1.

5. E.L. Church, J. Weneser, Phys. Rev. 104 (1956) 1382.

E2 - M1 ADMIXTURES IN TRANSITIONS IN DEFORMED AND TRANSITIONAL EVEN - EVEN NUCLEI

J.H. Hamilton
Vanderbilt University, Physics Department†*, Nashville, Tennessee 37203, U.S.A.*

1. Introduction

When the nuclear spin - parity selection rules allow more than one multipole radiation in a gamma transition, only the two lowest multipole orders have been observed to compete. By far the most prevalent admixtures observed are E2/M1. As has been pointed out[1]), measurements of E2/M1 admixtures long have provided important tests of nuclear models in several different mass regions. This point was emphasized in earlier surveys[2-10]) of E2/M1 multipole mixing ratios in which the available experimental data were compared with the predictions of various models. The most recent comparison of theory and experiment for E2/M1 admixed transitions from all classes of nuclei has been made by Grechukhin[11]) who surveyed the experimental data to January 1, 1968.

Experimental techniques for the measurements of E2/M1 mixing ratios have improved markedly in the last three years,however, so that accurate E2/M1 admixtures can be attained in very complex decays for even weak transitions[1]). Accurate experimental measurements of the M1 admixtures in $\Delta I = 0,1$ transitions in deformed and transitional, even - even nuclei are providing significant new insight into the structure of nuclei as they provide sensitive tests of the predictions of nuclear models.

In 1969, surveys were made of E2/M1 admixtures in transitions from particular classes of nuclei or levels. Kumar[12]) surveyed the data in the osmium and platinum isotopes, Bodenstedt[13]) the data in the $2+2_\gamma \to 0+2_g$ transitions in deformed nuclei, and Hamilton[1]) the data on the $0+2_\beta \to 0+2_g$ transitions in deformed nuclei. This last paper[1]) also discussed the techniques and applicability of different systems for gamma - gamma directional correlation studies with Ge(Li) detectors.

The use of Ge(Li) detectors in γ - γ (θ) studies and angular distribution of gamma rays following Coulomb excitation has revolutionized the quality and quantity of data on E2/M1 admixtures. Within the year since the 1969 surveys[1,12,13]) many significant new results have been obtained. The present paper presents a survey of the available data to August, 1970 on E2/M1 admixtures in transitions from nuclei with $A \geqslant 100$.

The theoretical work of Kumar and Baranger[12]) have shown that both the sign and magnitude of the multipole mixing parameter δ are important in a critical comparison of theory and experiment. Because such detailed comparison with theory is now possible, particular attention has been given to the problem of the accuracy of experimental results and to the use of a consistent convention for the sign of δ. This latter problem will be considered first so that all data can be presented in a consistent fashion. Then some recent results will be presented and interpreted along with some discussion of experimental techniques. Finally the survey itself will be presented and discussed.

2. Definition of the multipole mixing ratio δ

While, in general, the physics community has long had a practice of adapting a single uniform definition for a physical observable, this has not been the case for the multipole mixing ratio δ. Several different definitions exist in the literature. Biedenharn and Rose[14]) were among the first to introduce a definition of δ. Their notation was to define δ in such a way that an absorption matrix element was used for the first member of a cascade and a emission matrix element for the second. This definition describes the reaction process as the prototype. Tables[15]) have appeared for the analysis of data based on their definition[14]). Ferguson[16]) has reviewed the arguments of Biedenharn and Rose[17]) for the continued adoption of their notation. Their arguments[17]) include simplicity under time reversal, a sign anomaly when applied to reactions and the long usage of the notation. On the other hand when one measures δ in two γ - $\gamma(\theta)$ experiments on the same transition where the transition is the first transition in experiment 1 and the second in experiment 2, a different sign is obtained with their definitions[14]). Similar ambiguities occur in triple cascade and skipped correlation measurements. To eliminate this, Ferguson[16]) adopted the notation of Devons and Goldfarb[18]) with the choice of positive sign as recommended by Huby[19]) where all processes are emission type in contrast to the Biedenharn and Rose[14]) definition.

Steffen[20]) in his Handbook of Angular Correlations adopted a notation for δ in terms of emission processes and made a sign choice such that his δ had the same sign as Biedenharn and Rose[14]) for the first member of a cascade but the opposite sign for the second member. Rose and Brink[21]) proposed a definition of δ in which one wrote the initial state on the left as in adsorption processes. Through a choice of sign, their δ's have the same sign as those of Steffen[20]). There is also the question of whether you define δ as the ratio of the L + 1 to L multipole radiation matrix elements or the L to L + 1 radiation. All of the above definitions used the (L + 1) / L ratio of matrix elements. In their discussion of directional correlations that involve E0 + M1 + E2 admixtures, Church, Rose and Weneser[22]) introduced an inverse definition of δ where δ^2 was the ratio of the M1 to the E2 gamma - ray transition probability where otherwise the convention of Biedenharn and Rose[14]) was followed. This choice of δ has been much used in studies of E0 + M1 + E2 admixtures from e - $\gamma(\theta)$ and γ - $\gamma(\theta)$ measurements.

The question of the definition of δ has been carefully studied most recently by Krane and Steffen[23]). Their δ is carefully defined in terms of emission matrix elements and it is the recommendation of this paper that this be the adopted notation for all future work. Their notation is adopted in the present survey. Since their work has not yet been published the authors graceously consented to a reproduction of their definition here for completeness.

The transition matrix element for the emission of an electromagnetic multipole radiation πL ($\pi = E$ or $\pi = M$ for electric or magnetic multipole radiation, respectively) from an initial nuclear state $|I_i m_i\rangle$ to a final nuclear state $|I_f m_f\rangle$ is

$\langle I_f m_f | \underline{j}_N \underline{A}^{\pi *}_{LM} | I_i m_i \rangle$ where $\underline{j}_N$ is the nuclear current operator and $\underline{A}^{\pi}_{LM}$ are the multipole fields in the form: (in units $\hbar = m = c = 1$)

$$\underline{A}^{E}_{LM}(k, \underline{r}) = \frac{i^L}{k} \frac{\underline{\nabla} \times \underline{L}}{[L(L+1)]^{1/2}} \; j_L(kr)\, Y_{LM}(\hat{r}) \tag{1a}$$

$$\underline{A}^{M}_{LM}(k, \underline{r}) = i^L \frac{\underline{L}}{[L(L+1)]^{1/2}} \; j_L(kr)\, Y_{LM}(\hat{r}) \tag{1b}$$

The j_L (kr) are spherical Bessel functions and k is the transition energy (in units mc^2). The vector fields (1) satisfy the equation:

$$\underline{A}^{\pi *}_{LM} = (-1)^{L-M+1} \underline{A}^{\pi}_{L-M} \tag{2}$$

Hence the interaction operator $\underline{j}_N \underline{A}^{\pi}_{LM}$ is related to its Hermitian adjoint by $(\underline{j}_N A^{\pi}_{LM})^{\dagger} = (-1)^{L-M+1} (\underline{j}_N \underline{A}^{\pi}_{L-M})$. For local nuclear forces the nuclear current operator $\underline{j}_N$ is proportional to the nuclear momentum and spin operators $\underline{p}$ and $\underline{\sigma}$, respectively, which transform under the time - reversal operation T as $T \underline{p} T^{-1} = -\underline{p}$ and $T \underline{\sigma} T^{-1} = -\underline{\sigma}$. Hence the interaction operators $\underline{j}_N \underline{A}^{\pi}_{LM}$ transform under time - reversal according to $T (\underline{j}_N \underline{A}^{\pi}_{LM}) T^{-1} = (-1)^{L-M} (\underline{j}_N \underline{A}^{\pi}_{L-M})$. If the nuclear eigenstates $|Im\rangle$ are chosen with their phases such that $T|Im\rangle = (-1)^{I-m}|I{-}m\rangle$, which is always possible, the matrix elements $\langle I_f m_f | \underline{j}_N \underline{A}^{\pi}_{LM} | I_i m_i \rangle$ are real.

The reduced gamma emission matrix elements $\langle I_f || \underline{j}_N \underline{A}^{\pi}_{L} || I_i \rangle$ are defined by the Wigner - Eckart theorem:

$$\langle I_f m_f | \underline{j}_N \underline{A}^{\pi *}_{LM} | I_i m_i \rangle = (-1)^{L-M+1} \langle I_f m_f | \underline{j}_N \underline{A}^{\pi}_{L-M} | I_i m_i \rangle$$

$$= (-1)^{L-M+1} (-1)^{I_f - m_f} \begin{pmatrix} I_f & L & I_i \\ -m_f & -M & m_i \end{pmatrix} \langle I_f || \underline{j}_N \underline{A}^{\pi}_{L} || I_i \rangle \tag{3}$$

where $\begin{pmatrix} I_f & L & I_i \\ -m_f & -M & m_i \end{pmatrix}$ is the Wigner 3 - j symbol. Note that the initial states of a transition are always written on the right side in the matrix elements. Equation (3) refers to the emission of electromagnetic radiation, i.e. $E_i > E_f$.

The (real) reduced matrix elements defined in eq. (3) are related to the reduced matrix elements of Bohr and Mottelson[24]) and of Alder et al.[25]) by

$$\langle I_f || \underline{j}_N \underline{A}^{E}_{L} || I_i \rangle = \frac{k^L}{(2L+1)!!} \left(\frac{L+1}{L}\right)^{1/2} \langle I_f || i^L \mathscr{M}(EL) || I_i \rangle_{BM} \tag{4a}$$

$$\langle I_f \| \underline{j}_N \underline{A}_L^M \| I_i \rangle = -\frac{k^L}{(2L+1)!!}\left(\frac{L+1}{L}\right)^{1/2} \langle I_f \| i^{L-1} \mathcal{M}(ML) \| I_i \rangle_{BM} \qquad (4b)$$

The reduced matrix elements $\langle I_f \| \underline{j}_N \underline{A}_L^\pi \| I_i \rangle$ are related to those of Rose and Brink[21]) by:

$$\langle I_f \| \underline{j}_N \underline{A}_L^\pi \| I_i \rangle = -(-1)^{I_1 - I_2 + L + 1} (2\pi(2L+1))^{-1/2} (2I_i+1)^{1/2} \langle I_i \| T_L^\pi \| I_f \rangle_{RB} \qquad (5)$$

Note that in Rose and Brink's paper[21]) the initial state for emission processes is always written on the left side of the interaction operator ($E_i > E_f$).

The directional correlation of two mixed multipole gamma radiations, emitted according to the scheme $I_1 \xrightarrow{\gamma_1} I_2 \xrightarrow{\gamma_2} I_3$, is given by

$$W(\vartheta) = \sum_\Lambda B_\Lambda(\gamma_1) A_\Lambda(\gamma_2) P_\Lambda(\cos\vartheta) \qquad (6)$$

where the normalized orientation coefficient $B_\Lambda(\gamma_1)$ is given by:

$$B_\Lambda(\gamma_1) = \frac{\sum\limits_{L_1\pi_1 L_1'\pi_1'} F_\Lambda(L_1 L_1' I_1 I_2)(-1)^{L_1 - L_1'} \langle I_2 \| \underline{j}_N \underline{A}_{L_1}^{\pi_1} \| I_1 \rangle \langle I_2 \| \underline{j}_N \underline{A}_{L_1'}^{\pi_1'} \| I_1 \rangle}{\sum\limits_{L_1\pi_1} \langle I_2 \| \underline{j}_N \underline{A}_{L_1}^{\pi_1} \| I_1 \rangle^2} \qquad (7)$$

and the directional distribution coefficient $A_\Lambda(\gamma_2)$ is given by

$$A_\Lambda(\gamma_2) = \frac{\sum\limits_{L_2\pi_2 L_2'\pi_2'} F_\Lambda(L_2 L_2' I_3 I_2) \langle I_3 \| \underline{j}_N \underline{A}_{L_2}^{\pi_2} \| I_2 \rangle \langle I_3 \| \underline{j}_N \underline{A}_{L_2'}^{\pi_2'} \| I_2 \rangle}{\sum\limits_{L_2\pi_2} \langle I_3 \| \underline{j}_N \underline{A}_{L_2}^{\pi_2} \| I_2 \rangle^2} \qquad (8)$$

The F - coefficients $F_\Lambda(LL'I'I)$ are defined and tabulated in reference 26. It is useful to distinguish between the orientation coefficients $B_\Lambda(\gamma_1)$ which are characteristic of the (axially symmetric) orientation of the intermediate state I_2 and the directional distribution coefficients $A_\Lambda(\gamma_2)$ which characterize the directional distribution of the gamma radiation γ_2 with respect to the orientation axis $\underline{z}$ of the state I_2 ($\underline{z}$ = = propagation direction of γ_1).

Most mixed multipole transitions are of the type E2 + M1. It is then convenient to introduce the 'mixing ratio'

$$\delta(\gamma_n) = \frac{\langle I_{n+1} \| \underline{j}_N A_2^E \| I_n \rangle}{\langle I_{n+1} \| \underline{j}_N \underline{A}_1^M \| I_n \rangle} = k_n \frac{\sqrt{3}}{10} \frac{\langle I_{n+1} \| \mathcal{M}(E2) \| I_n \rangle}{\langle I_{n+1} \| \mathcal{M}(M1) \| I_n \rangle} \tag{9}$$

where the initial state of the transition is written on the right side in the reduced matrix elements ($E_n > E_{n+1}$). The matrix elements and k_n in eq. (9) are expressed in units $\hbar = m = c = 1$.

The orientation parameter is then simply

$$B_\Lambda(\gamma_1) = (1 + \delta^2(\gamma_1))^{-1} (F_\Lambda(11\, I_1 I_2) - 2\,\delta(\gamma_1)\, F_\Lambda(12\, I_1 I_2) + \delta^2(\gamma_1)\, F_\Lambda(22\, I_1 I_2)) \tag{10}$$

and the directional distribution coefficient is

$$A_\Lambda(\gamma_2) = (1 + \delta^2(\gamma_2))^{-1} (F_\Lambda(11\, I_3 I_2) + 2\,\delta(\gamma_2)\, F_\Lambda(12\, I_3 I_2) + \delta^2(\gamma_2)\, F_\Lambda(22\, I_3 I_2)) \tag{11}$$

The mixing ratios defined in eq. (9) are related to those of Biedenharn and Rose[14]) by

$$\delta(\gamma_1) = -\,\delta_{\text{Biedenharn-Rose}} \tag{12a}$$

$$\delta(\gamma_2) = \delta_{\text{Biedenharn-Rose}} \tag{12b}$$

The directional correlation of γ_1 with a gamma radiation γ_N, that follows some unobserved radiations $\gamma_2, \gamma_3, \ldots\ldots \gamma_{N-1}$ which are emitted in cascade from the oriented state I_2 is given by

$$W(\vartheta) = \sum_\Lambda B_\Lambda(\gamma_1)\, U_\Lambda(\gamma_2)\, U_\Lambda(\gamma_3) \ldots\ldots U_\Lambda(\gamma_{N-1})\, A_\Lambda(\gamma_N)\, P_\Lambda(\cos\vartheta) \tag{13}$$

where the reorientation parameters $U_\Lambda(\gamma_n)$ for the nth transition depend on I_n, I_{n+1} and on the multipole intensities L_n (but not the interference terms) of the unobserved radiation:

$$U_\Lambda(\gamma_n) = \frac{\sum_{L_n \pi_n} (-1)^{L_n} \begin{Bmatrix} I_n & I_n & \Lambda \\ I_{n+1} & I_{n+1} & L_n \end{Bmatrix} \langle I_{n+1} \| \underline{j}_N \underline{A}_{L_n}^\pi \| I_n \rangle^2}{\sum_{L_n \pi_n} (-1)^{L_n} \begin{Bmatrix} I_n & I_n & 0 \\ I_{n+1} & I_{n+1} & L_n \end{Bmatrix} \langle I_{n+1} \| \underline{j}_N \underline{A}_{L_n}^\pi \| I_n \rangle^2} \tag{14}$$

or for an unobserved E2 + M1 transition:

$$U_\Lambda(\gamma_n) = (-1)^{I_n+I_{n+1}}[(2I_n+1)(2I_{n+1}+1)]^{\frac{1}{2}}(1+\delta(\gamma_n)^2)^{-1} \times$$

$$[\delta^2(\gamma_n)\begin{Bmatrix} I_n & I_n & \Lambda \\ I_{n+1} & I_{n+1} & 2 \end{Bmatrix} - \begin{Bmatrix} I_n & I_n & \Lambda \\ I_{n+1} & I_{n+1} & 1 \end{Bmatrix}] \qquad (15)$$

The reorientation coefficients of eqs. (14) and (15) are normalized to unity, i.e. $U_0(\gamma_n) = 1$.

In the present investigation only dipole and qaudrupole radiations are involved. Hence the directional correlations are of the form:

$$W(\vartheta) = 1 + A_{22}P_2(\cos\vartheta) + A_{44}P_4(\cos\vartheta) \qquad (16)$$

where the directional correlation coefficients $A_{\Lambda\Lambda}$ are given by

$$A_{\Lambda\Lambda} = B_\Lambda(\gamma_1)U_\Lambda(\gamma_2)\ ......\ U_\Lambda(\gamma_{N-1})A_\Lambda(\gamma_N) \qquad (17)$$

Due to the finite size of source and detectors the experimentally observed directional correlation function is given by

$$W(\vartheta)' = A_{00}' + A_{22}'P_2(\cos\vartheta) + A_{44}'P_4(\cos\vartheta) \qquad (18)$$

from which the theoretical correlation coefficients $A_{\Lambda\Lambda}$ can be extracted on the basis of

$$A_{\Lambda\Lambda} = \frac{A'_{\Lambda\Lambda}}{A'_{00}}\ \frac{1}{\bar{Q}_\Lambda(\gamma_1)\bar{A}_\Lambda(\gamma_2)} \qquad (19)$$

where $\bar{Q}_\Lambda(\gamma_i) = Q_\Lambda(\gamma_i)/Q_0(\gamma_i)$ are the normalized geometrical correction coefficients for the detector that observes γ_i.

Finally one observes that the Krane and Steffen[23]) $\delta(\gamma_1)$ and $\delta(\gamma_2)$ have the opposite sign from those of the older Steffen[20]) and they have the opposite sign from those of Rose and Brink[21]).

3. Recent experimental results

In the last year significant new measurements of E2/M1 admixtures in transitions in both the deformed, rare earth region and the transitional region of the osmium and platinum isotopes have been made. Some of these measurements are presented in detail before the general survey of all measurements.

3.1 ^{178}*Hf*

We have recently completed careful studies of the level structure of ^{178}Hf from the radioactive decay[27]) of ^{178}Ta and from Coulomb excitation[28]) and as well carried out more expensive γ - γ (θ) studies[29]) since our first report[1]) at the Vanderbilt Conference in 1969. This work will show the power of present γ - γ (θ) systems to accurately measure mixing ratios and illustrate the importance of such accurate data to nuclear models. Because of the significance of the results in ^{178}Hf for the development of the Bohr - Mottelson model, this work will be discussed in detail. The level structure of ^{178}Hf as obtained from our recent studies of the decay of ^{178}Ta is shown in fig. 1. This scheme includes 5 new levels and 13 transitions unplaced or unknown in earlier work[30,31]).

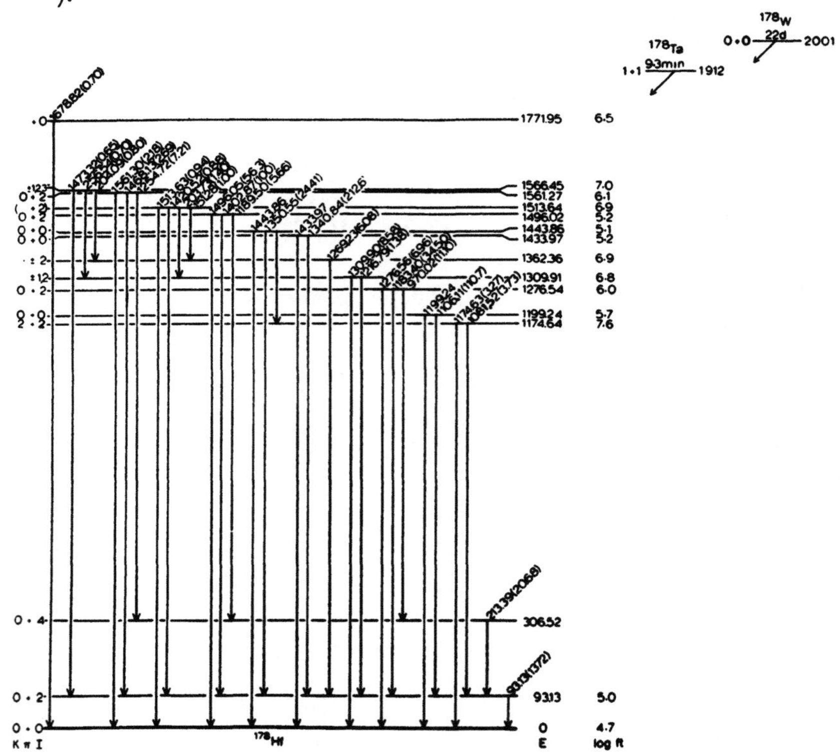

Fig. 1. The revised decay scheme[27]) of ^{178}Hf.

In 1967 three groups[32-34]) reported anomalous gamma - ray branching ratios from the 2^+ members of β - vibrational bands in ^{152}Sm, ^{154}Gd and ^{156}Gd, when the experimental data were compared with the predictions of the Bohr - Mottelson model which included mixing of the beta and ground state bands[35]). An early suggestion that would bring theory and experiment into agreement was that there existed a very large M1 component in the $2_\beta^+ \rightarrow 2_g^+$ transitions of these three nuclei despite the long standing prediction of essentially no M1 admixtures in the quadrupole beta vibrations[36]). At the International Conference on Nuclear Structure in Tokyo, however, Mottelson[35]) pointed out how the theory could be modified to include sizeable M1 admixtures and stressed the crucial importance of observing these M1 admixtures in order to preserve the application of rotational relationships to these nuclei. Our γ - $\gamma(\theta)$ measurements in ^{154}Gd have shown[37]), however, that the 692 - keV, $2_\beta^+ \rightarrow 2_g^+$ transition is $99.2^{+0.3}_{-0.7}$ %E2 which clearly excludes the 50%M1 component required[32,33]) to obtain a consistent band mixing parameter, Z_β, from the reduced transition probabilities in ^{154}Gd.

On the other hand, Nielsen, Nielsen, and Rud[31]) carried out γ - $\gamma(\theta)$ measurements in ^{178}Hf on the 1183 - 93 keV cascade. The 1276 - keV, 2$^+$ level where this cascade originates has been assigned as a beta vibrational state[30]). An M1 admixture of (85.6 ± 10)%was found in the 1183 - keV transition[31]). There is evidentally an error in their paper on the percent M1. They report[31]) (83 ± 10)% but their δ value yields (85.6 ± 10)%. When this large M1 admixture was considered, they obtained[31]) a consistent band mixing parameter for the 1276 - keV level as a beta vibrational state.

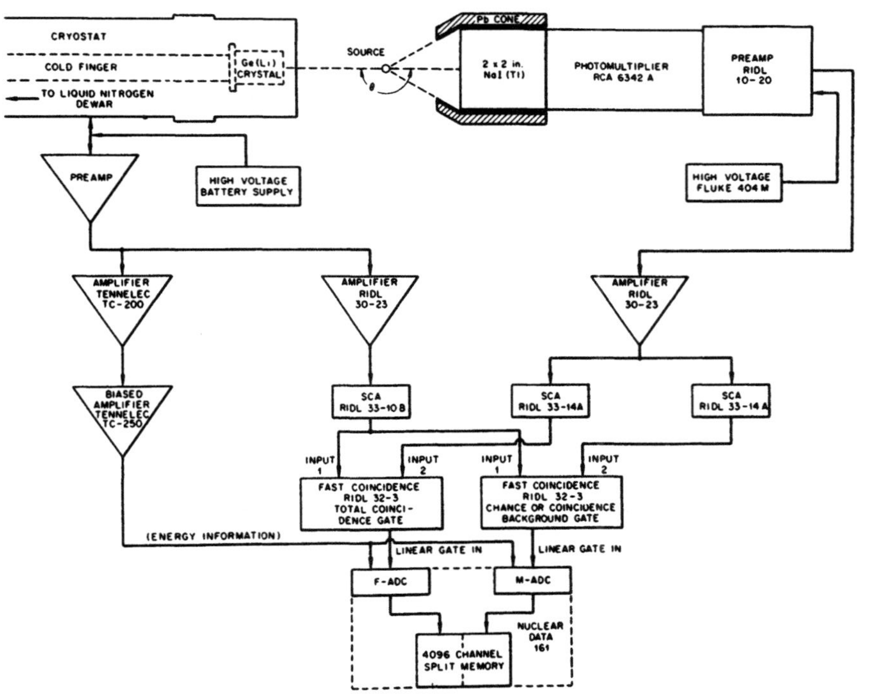

Fig. 2. A schematic diagram of the electronic for γ - $\gamma(\theta)$ studies at Vanderbilt.

Recent Coulomb excitation work[38,39]) also has shown a very low (<2%) M1 admixture in the $2^+_\beta \to 2^+_g$ in ^{152}Sm. It now appears[38,40]) that instead of the $2^+_\beta \to 2^+_g$ transition being responsible for the inconsistent branching ratios, the problem is in the $2^+_\beta \to 4^+_g$ transition. Thus, it is surprising to find reported[31]) in ^{178}Hf that it is the $2^+ \to 2^+_g$ transition that is involved in the alteration of the B (E2) branching ratios. It is, of course, possible that both experiments are correct and that the differences arise from the transitional character of ^{152}Sm and ^{154}Gd which have softly deformed nuclei in comparison to the well deformed ^{178}Hf nucleus.

Nielsen et al.[31]) employed two NaI detectors in their $\gamma \cdot \gamma(\theta)$ studies. The 0.11% 1183 - keV transition of interest is seen as a weak shoulder on the 0.44% 1106 - keV photopeak in the NaI spectrum. Their studies were also complicated by a 0.02%, 1189 - keV transition (1189 - (213) - 93 keV cascade) that is very difficult to resolve even by computer. It was this cascade that led them to increase their error limits on δ for possible systematic effects. Also, we have discovered an unreported[30]) 1174 - keV transition with 0.015% intensity that was not considered in the work of Nielsen et al.[31]). It is important for the development of the theory to establish that it is indeed an M1 admixture in the $2^+ \to 2^+_g$ transition in ^{178}Hf that gives rise to inconsistent Z_β values in contrast to ^{152}Sm where it appears that the $2^+_\beta \to 4^+_g$ transition[38,40]) is responsible. Thus, we repeated[29]) the ^{178}Hf $\gamma \cdot \gamma(\theta)$ measurements with our directional correlation arrangement which has one Ge (Li) detector so that the 1183 - keV transition is well resolved in the spectrum. The intensities of the gamma transit-

Table 1. Directional Correlation Coefficients of Cascades in ^{178}Hf.

(Results of second experiment ref. 29)

Energies (keV) Spins	$Q_2G_2A_2$	$Q_4G_4A_4$	A_2	A_4
1106-93 (0-2-0)	0.245 ± 0.011	0.577 ± 0.010		
1341-93 (0-2-0)	0.253 ± 0.006	0.572 ± 0.007		
1351-93 (0-2-0)	0.252 ± 0.006	0.580 ± 0.006		
Average of Above Three	0.250 ± 0.004[a]	0.577 ± 0.005[a]		
1183-93 (2-2-0)	-0.045 ± 0.020	0.032 ± 0.022	-0.064 ± 0.028	0.063 ± 0.043
			-0.056 ± 0.051[b]	-0.025 ± 0.057[b]
1403-93 (2-2-0)	0.346 ± 0.009	0.072 ± 0.010	0.494 ± 0.015	0.143 ± 0.020
			0.415 ± 0.035[b]	0.122 ± 0.040[b]

[a]Coupled with the theoretical values of 0.357 and 1.14 for A_2 and A_4 these numbers give $Q_2G_2 = 0.700 \pm 0.011$ and $Q_4G_4 = 0.505 \pm 0.004$.

[b]Data from ref.31. The errors include statistical ones only. When systematic effects were taken into account[31]), the errors on delta are increased a factor of two over that obtained from the statistical errors.

ions in ^{178}Hf were carefully measured[27]) with a much larger Ge(Li) detector in order to have accurate values for computation of relative B(E2) values. From these data, one can ascertain what value of an M1 admixture is needed to obtain a consistent Z_β.

The apparatus for the ^{178}Hf γ - γ (θ) measurements[29]) was the same as used previously[37]) except that new detectors were used. Figure 2 (see page 188) is a diagram of the system. The ^{178}Hf experiment was carried out first with a Ge(Li) detector which had a resolution of 2.8 keV FWHM at 1.33 MeV, and at the same energy an efficiency of 3.8% relative to that of a 3 in. X 3 in. NaI detector. Then it was repeated with a much larger Ge(Li) detector which had a 10% efficiency and 2.5 keV resolution at 1.33 MeV. With our two gate system, the correlations of all gamma - rays coincident with the 93 - keV, $2_g^+ \rightarrow 0_g^+$ transition and with the Compton background just above the 93 - keV peak were simultaneously measured. The latter correction was found to be negligible. A lead - cadmium - copper absorber was placed over the Ge(Li) detector to attenuate the low energy portion of the spectrum, and a 0.5 mm cadmium absorber over the NaI detector to reduce the strong ^{178}Ta and ^{178}Hf K X-rays in that detector. A biased amplifier was used to cut out the Ge(Li) spectrum below 500 keV. The gamma - ray singles spectrum was also measured with the Ge(Li) detector with the larger efficiency. The source was ^{178}W ($T_{1/2}$ = 22d) in the form of H_2WO_4 dissolved in dilute NaOH.

The correlations of all gamma - rays in coincidence with the 93 - keV peak and with the Compton background just above it were measured simultaneously. The first of the two separate experiments was carried out for 14 cycles and the second for 7 cycles. Each cycle was for three days and consisted of a 23h run at 90°, 135°, 180°, 180°, 135° and 90° compensated for the source decay to better than 1%. The NaI gate counts recorded for each 23 h period were constant to better than 1%. These counts were used to normalize the coincidence results, and coupled with the measured 2τ and the Ge(Li) singles spectra taken between each run, were also used to calculate chance corrections for both the 93 -keV and background correlations. The true to chance ratio varied from 9 to 25. Since the 93 - keV state has a relatively long half-life ($T_{1/2}$ = 1.50 ns), correlations through this state are attenuated. The attenuation corrections Q_2G_2 and Q_4G_4 for the Ge(Li) and NaI detectors were obtained from known correlations in the decay.

There are three previously known 0^+ states[30,41]) with strong transitions to the 93 - keV 2_g^+ level. Cascades with $0 \rightarrow 2 \rightarrow 0$ spins have unique correlation coefficients $A_2 = 0.357$ and $A_4 = 1.14$. The Q_2G_2 and Q_4G_4 for each run were obtained from a comparison of these values and the experimental $Q_2G_2A_2$ and $Q_4G_4A_4$ coefficients of these cascades. The results of the second measurements with the large detector are given in table 1 (see page 189). Note the excellent consistency in table 1 of the $Q_2G_2A_2$ and $Q_4G_4A_4$ values for the $0 \rightarrow 2 \rightarrow 0$ cascades. The attenuation corrections for each run were applied to their respective 1183 - 93 and 1403 - 93 keV, $2^+ \rightarrow 2^+ \rightarrow 0^+$ cascade data.

Figure 3 (cf. p. 191) shows the 1183 - 93 keV cascade data. These data show that even with very large Ge(Li) detectors one must carefully analyze the Compton backgrounds as has been previously emphasized[1]). The 1186 keV Compton edge seen in fig. 3 could have been overlooked if long regions of the background on both sides of the peak had not been plotted. The corrected A_2 and A_4 values are given in columns 6 and 7 of table 1. Within the limits of error, the same results were obtained

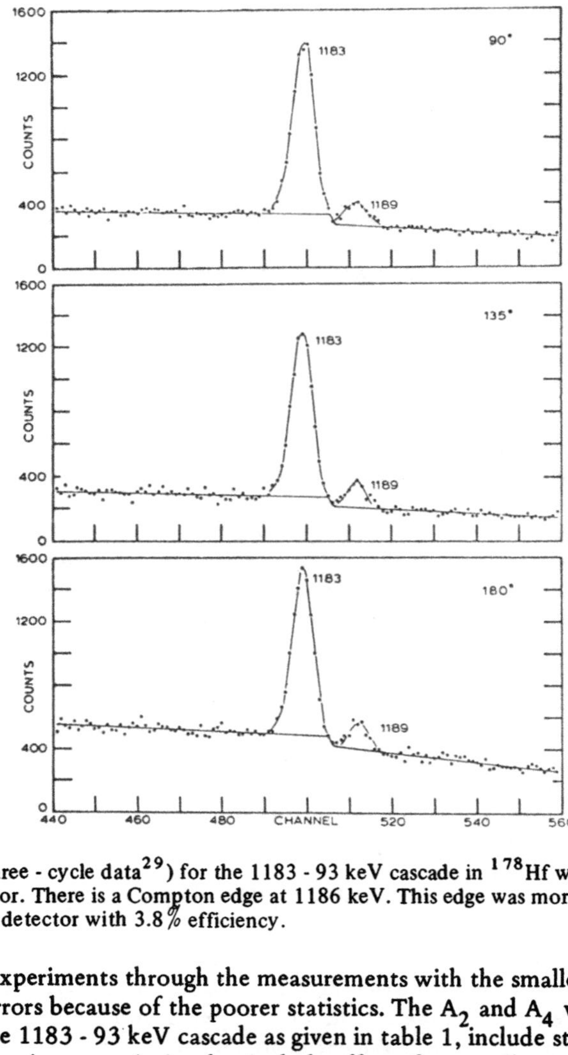

Fig. 3. The three - cycle data[29]) for the 1183 - 93 keV cascade in ^{178}Hf with a 10% efficient Ge (Li) detector. There is a Compton edge at 1186 keV. This edge was more pronounced in the data from the detector with 3.8% efficiency.

in our two experiments through the measurements with the smaller Ge(Li) detector have large errors because of the poorer statistics. The A_2 and A_4 values of Nielsen et al.[31]) for the 1183 - 93 keV cascade as given in table 1, include statistical errors only. When systematic uncertainties that include effects from weak unresolved lines with unknown correlations were taken into account, they found[31]) $\delta = -(0.41 \pm 0.17)$ and an M1 admixture of (85.6 ± 10)% for the 1183 keV transition. From a weighted average of our data ($A_2 = -0.054 \pm 0.023$ and $A_4 = 0.041 \pm 0.033$) as shown in fig. 4, we find $\delta = -(0.410 \pm 0.036)$ which corresponds to an $(85.6^{+1.1}_{-2.1})$% M1 admixture. Because of our good resolution, there are no known systematic effects from unresolved gamma - rays and Compton backgrounds to distort these data. Thus, a large M1 admixture in the 1183 - keV transition is firmly established.

The relative intensities of the transitions from the 1276 keV, 2^+ state as obtained with our more efficient detector, are given in table 2 (see page 193) along with

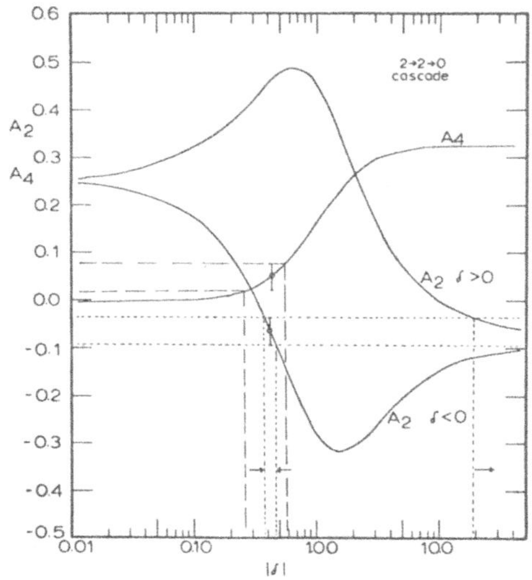

Fig. 4. A plot[29]) of the A_2 and A_4 coefficients as theory for 2 - 2 - 0, 1183 - 93 keV cascade in ^{178}Hf. Here the definition of δ is that of ref. 14.

the previous results[31,42]). From our branching ratio data, an M1 admixture of $(73.0 \pm 1.3)\%$ in the $2^+ \to 2^+$, 1183 - keV transition would yield a consistent Z_β for the various branching ratios from the 1276 - keV, 2^+ state. Siddiqi et al.[42]) recently obtained from their branching ratio data an M1 admixture of $(70.5 \pm 1.5)\%$ in agreement with our results. When one subtracts the 85.6% M1 admixture in this $2^+ \to 2^+$ transition found from our γ - γ (θ) work, it is impossible to find a consistent Z_β value. It takes 6 standard deviations in the γ - γ (θ) work or 10 standard deviations in our branching ratio data for the M1 admixtures obtained in these two ways to agree. Thus we are forced to conclude that while the M1 admixture is large from this state in ^{178}Hf the branching ratio data are in no better agreement with theory than in ^{154}Gd. Thus this work does not confirm the expectation of Mottelson[35]) as earlier reported (ref. 31) and removes the one beta band case that appeared to fit the suggested[35]) new theoretical description with large M1 admixtures.

A similar analysis was carried out for the next 2^+ level at 1496 - keV. From the γ - γ (θ) results, we obtain $\delta = -0.8^{+0.2}_{-0.3}$ while Nielsen et al.[31]) reported $\delta = -0.2$ to -1.3 for the 1403 keV transition. It is interesting to note that δ has the opposite sign than that for the 1183 keV transition. From our more accurate δ, an M1 admixture of $(61.0^{+0.7}_{-1.2})\%$ was found in the 1403 - keV $2^+ \to 2^+$ transition. This admixture is somewhat smaller than in the 1183 - keV transition. If one assumes that this is the 2^+ beta vibrational state, then a 61% M1 admixture leads to consistent branching ratios for a single Z_β value. Since the branching ratios agree with a $K = 0$ assignment after subtracting a 61% M1 admixture in the $2^+ \to 2^+$ transition, one is tempted to identify this as a beta vibrational state. However, to assign one of the known 0^+ states

Table 2. Relative Intensities and Branching Ratios from the 1276.6 and 1496.1 keV 2^+ Levels in ^{178}Hf Populated in the Decay of ^{178}Ta.

Energy keV	Transitions	Gamma Intensity			Present Work Experimental Relative B(E2) Values[a]			Theoretical Relative B(E2) Values for Z_β Values[b] of:			
		Ref. 42	Ref. 31	This work[27])	M1 = 0	M1 = 73%	$M1=85.6^{+1.1\%}_{-2.1\%}$	0	0.020	0.026	0.040
970.0	$2^+ \to 4^+$	3.66 ± 0.21	3.57 ± 1.09	3.54 ± 0.18	86.9 ± 4.4	322 ± 16	$603\ ^{+35}_{-32}$	180	295	323	437
1183.4	$2^+ \to 2^+$	11.00	11.00	11.00	100	100	100	100	100	100	100
1276.6	$2^+ \to 0^+$	2.52 ± 0.20	2.07 ± 0.64	2.22 ± 0.10	13.8 ± 0.6	51.1 ± 2.2	$95.8\ ^{+4.7}_{-4.3}$	70	54	51.2	40
		Ref. 30		This work[27])	M1 = 0	M1 = 62.5%	$M1=61^{+0.7\%}_{-1.2\%}$	0	-0.020	-0.040	-0.060
1189.5	$2^+ \to 4^+$	4.65		5.66 ± 0.28	12.9 ± 0.6	34.4 ± 1.6	33.1 ± 1.6	180	93.3	34.8	4.6
1402.9	$2^+ \to 2^+$	100		100.0	100	100	100	100	100	100	100
1496.1	$2^+ \to 0^+$	65.9		56.3 ± 1.9	40.8 ± 1.4	109 ± 3.7	$104.6^{+3.8}_{-3.6}$	70	87.4	108	129

[a]These values are for various M1 components in the 1183.4 keV and 1402.9 keV $2^+ \to 2^+$ transitions as indicated.

[b]The notation here is the same as ref. 32. The lower half of the table is for the 1496.1 keV level and the Z_β values are negative.

as the ground state member of such a beta vibrational band would yield a moment of inertia quite different from that of the ground state band. Nielsen et al.[31]) already have pointed out that the electron capture feeding to the 1493 - keV level is a factor of two higher than predicted by Alaga's rules for a rotational band built on either the 1434 or 1444 - keV, 0^+ states. This branching problem would only be worse for an undetected 0^+ state with an energy spacing similar to that of the ground state band. There is also a problem in the E0 strength of the 1403 - keV $2^+ \rightarrow 2^+$ transition. Our electron data and that of earlier work[30]) indicate that the E0 strength of the 1403 - keV transition is about 10 times weaker than that of the 1183 - keV transition.

While the K = 0 character of the 1276 keV level is established by the strong E0 component in the 1183 - keV transition[30]) there was no evidence in the earlier work (refs. 30 and 31) that the 0^+ 1199 and 2^+ 1276 keV levels assigned as a beta band have the collective character of such a band. Because of the importance to the further development of the Bohr - Mottelson model of knowing how the beta vibrational bands behave in a well deformed nucleus, the interpretation of the 1276 - keV level as a beta vibration must be demonstrated more clearly.

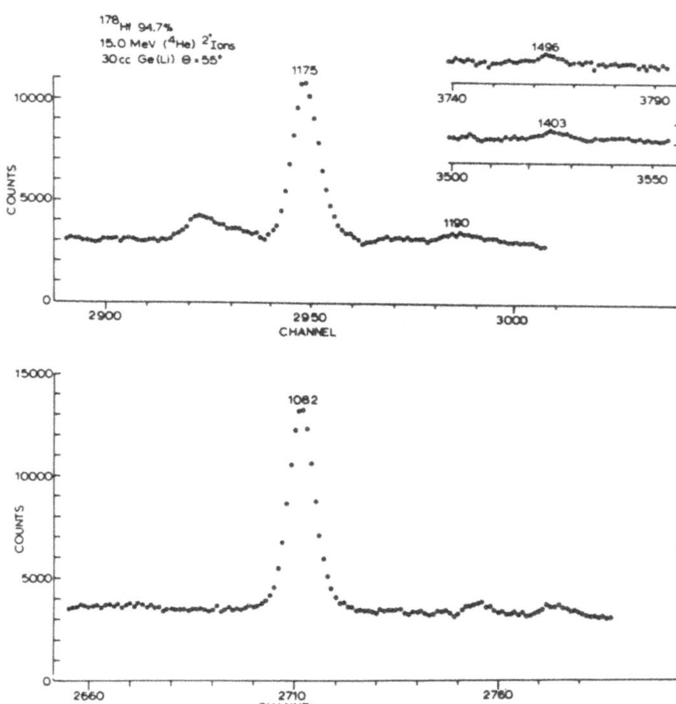

Fig. 5. The gamma - ray spectrum of ^{178}Hf from Coulomb excitation with α particles[28]). The 1175 keV 2_γ - 0_g and 1082 keV 2_γ - 2_g transitions are very strong in comparison to the similar transitions from the 2^+, 1496 keV level.

To determine whether any of the levels observed in radioactive decay to ^{178}Hf (fig. 1) are indeed the collective beta and gamma bands, Coulomb excitation experiments were performed with the alpha particle beam of the ORNL Tandem Van de Graaff accelerator[28]). The thick metal target was enriched to 95% in ^{178}Hf. Gamma - ray yield measurements were taken at energies of 13 and 15 MeV. The de-excitation gamma - rays were observed with a 30 cm^3 Ge(Li) coaxial detector. The results were analyzed with the programs developed by the Oak Ridge Coulomb excitation group[43]). Gamma - ray angular correlation measurements were taken at 0° and 90° with an alpha - particle energy of 15 MeV. The correlation from the $2^+ \rightarrow 0^+$, 1175 keV transition was used to determine the alignment corrections for the $2^+ \rightarrow 2^+$, 1082 keV gamma - ray correlation.

A portion of the gamma - ray spectrum is shown in fig. 5 (cf. p. 194). The level at 1175 keV can be identified as the 2^+ band head of the gamma band from the branching ratios and its collective nature. This band is also reported[44]) in subsequent work on the basis of an observed sequence of states up to 5^+. The measured excitation probability of the 1175 keV level is $B(E2)_{exc} = (10.0 \pm 0.8) \times 10^{-50}$ e^2 cm^4. The ratio of this value to the single - particle estimate is 3.2 ± 0.3. Thus this level is confirmed as a gamma vibrational state by its collective character.

The results of the angular correlation experiments yield two values of the mixing parameter δ for the 1082 keV transition. Because the a_4 particle parameter is very small, it is not possible to determine a unique value of δ. The allowed lower value, $\delta = 0.6$, is ruled out from $\gamma \cdot \gamma (\theta)$ results on the 1082 - 93 keV cascade from radioactive decay (additional results obtained from ref. 29). The Coulomb excitation data yield $\delta = -32$ where limits on δ are $\delta > +32$ or $\delta < -11$ which yields an M1 admixture of < 1 %.

Upper limits can be placed on the excitation strengths of the previously known 2^+ levels at 1276 and 1496 keV. The former was proposed as a beta vibrational level. From the gamma - ray yields, these values are $B(E2)_{exc} \leqslant 2 \times 10^{-51}$ e^{2}cm^4 for the 1276 - keV level, and $B(E2)_{exc} \leqslant 10^{-50}$ e^2 cm^4 for the 1496 keV level. The ratio of these values to the single particle estimates are $\leqslant 0.07$ and $\leqslant 0.34$, respectively. Since these values are considerably less than the single particle estimate, the levels do not appear to be collective. If the three 0^+ excited levels in ^{178}Hf as well as the 2^+ levels at 1276 and 1496 keV are all interpreted as two - quasi - particle states, the beta decay feedings[27]) agree well with the predictions of Soloviev[45]). If these states are two - quasi - particle, then M1 admixtures can be expected, since M1 transitions to the ground state band are not hindered in general nor the E2's enhanced as for quadrupole vibrational states. The M1 transitions could be K forbidden but such is not the case in ^{178}Hf as here $\Delta K = 0$.

Thus, in conclusion, we have shown that $K\pi I = 0 + 2 \rightarrow 0 + 2$ transitions from the 1276 and 1496 keV levels in ^{178}Hf do indeed have large M1 components but that these levels are not collective beta vibrational states.

3.2 ^{156}Eu

It is important to the development of a theoretical understanding of the beta type vibrational levels in deformed nuclei to have accurate data on a well deformed nucleus. Unfortunately, there are few nuclei with well established beta vibrational bands. The 0^+, 2^+ and 4^+ members of the beta vibrational band in ^{156}Gd have been well established[34,46]) from the decays of ^{156}Eu and ^{156}Tb. While ^{156}Gd is not a strongly deformed nucleus, it is further into the well deformed region than either ^{152}Sm or ^{154}Gd.

Thus studies of the 2^+_β level in ^{156}Gd would provide tests of the model. It is already known [34,46,47]) that it would take a (62.5 ± 5)% M1 admixture in the $2^+_\beta \to 2^+_g$ transition to obtain branching ratios consistent with a one parameter Z_β, fit to the data.

We have carried out γ - γ (θ) measurements in ^{156}Gd populated by ^{156}Eu ($T_{1/2}$ = 15d) to measure the M1 component in the $2^+_\beta \to 2^+_g$ transition[48]).

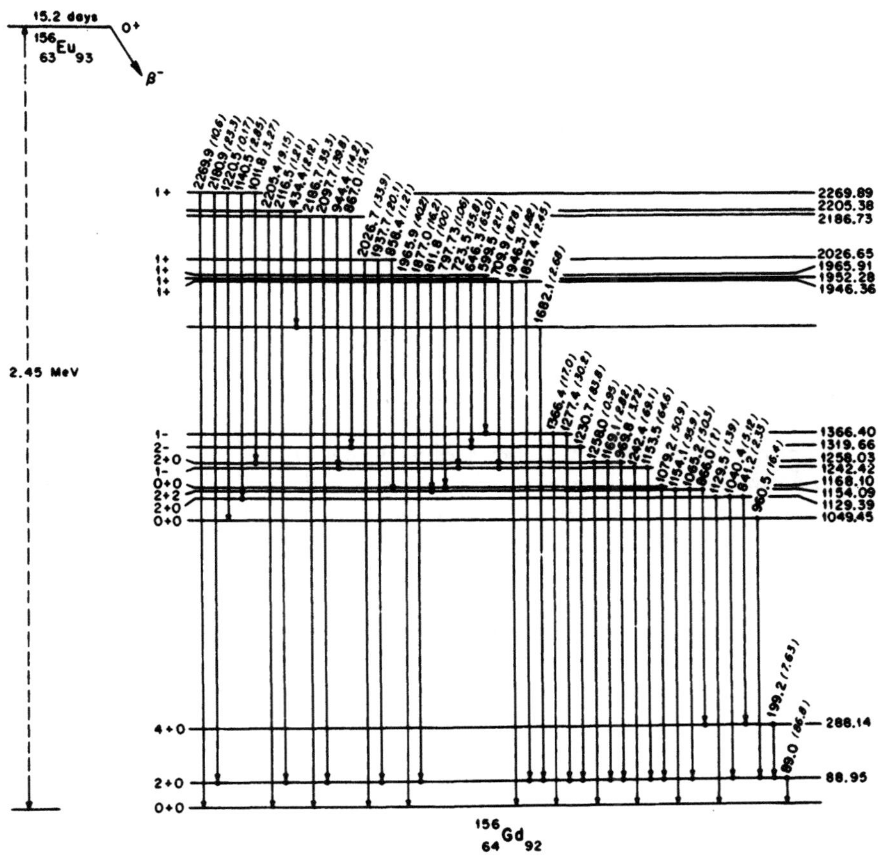

Fig. 6. The primary levels in ^{156}Gd populated by ^{156}Eu from the recent work of Kluk, Johnson and Hamilton[47]).

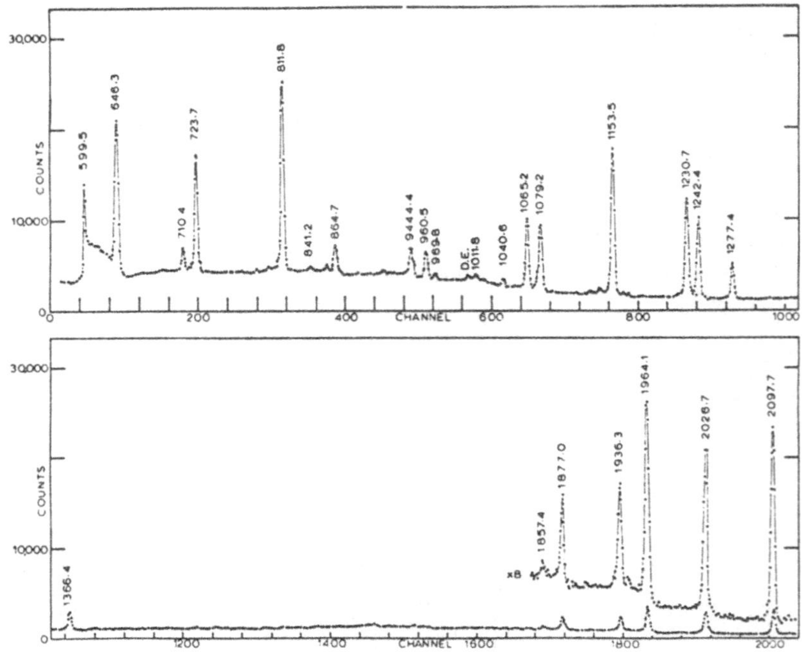

Fig. 7. A selected region[47]) of the gamma ray spectrum from the decay of ^{156}Eu. The 1040.6 keV transition is the $2^+_\beta \rightarrow 2^+_g$ transition of particular interest in the γ - $\gamma(\theta)$ work.

The primary levels in ^{156}Gd populated in the decay of ^{156}Eu are given in fig. 6. A selected region of the gamma - ray spectrum is shown in fig. 7. The 1040 - 89 keV, $2^+_\beta \rightarrow 2^+_g$ cascade is the primary one of interest. One observes that the 1040 keV transition is quite weak in the spectrum, so this 1040 - 89 keV correlation serves to illustrate the power of present γ - $\gamma(\theta)$ systems to obtain significant results under very unfavourable circumstances.

The same experimental γ - $\gamma(\theta)$ set - up was used as for the first ^{178}Hf experiment described earlier. A total of twenty 3 - day cycles were taken in a manner similar to the ^{178}Hf work. The source strength was increased from time to time after 6 and 12 cycles. Once again one must account for attenuation of correlations across the 89 keV level which has a half - life of 2.2 ns. There was one 0^+ state already known in ^{156}Gd and a second was discovered in the course of this work[49]). These $0 \rightarrow 2 \rightarrow 0$ cascades were used to obtain the attenuation coefficients for both solid angle and perturbation effects. Here too care must be exercised as the 1079 keV, $0 \rightarrow 2$ transition has the double escape peak of the 2098 keV transition at 1076 keV under its low energy side and it is not obvious in fig. 8. This escape peak if neglected would introduce an error of a few per cent in $Q_2G_2A_2$. Fig. 8 also illustrates the problem of how Compton edges can be hidden under peaks and can be seen only from a careful analysis of the background which includes a calculation of where all Compton edges and

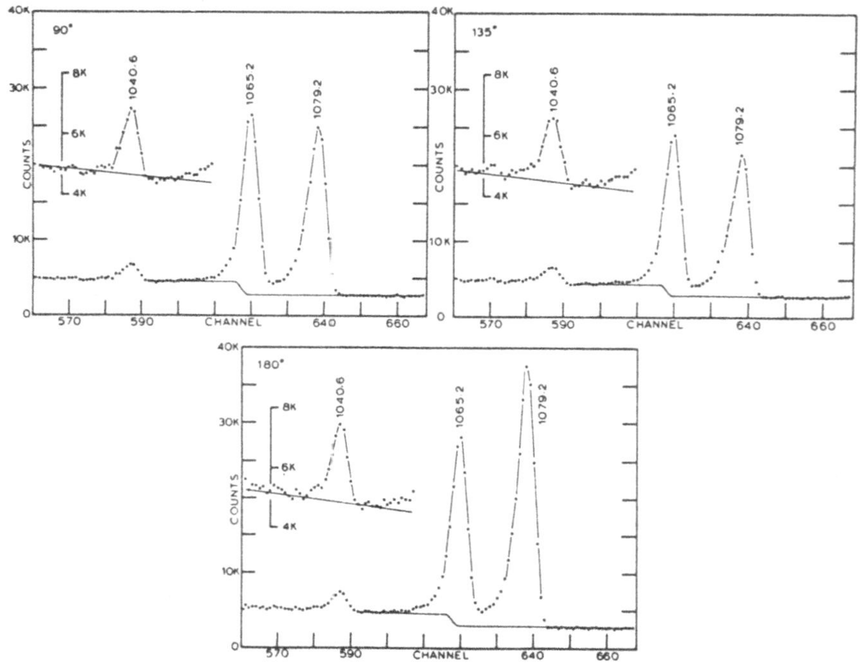

Fig. 8. Three - angle data for the 1040.6, 1065.2, and 1079.2 - 89 keV correlations in ^{156}Gd.

escape peaks occur in the decay.

The three angle data for the 1040.6, 1065.7 and 1079.7 - 89 keV cascades in ^{156}Gd are shown in fig. 8. The attenuation correlated results are given in table 3. We also measured the multipole admixtures of several transitions in 1 → 3 skipped correlations[50]). We find the following results: the $2^+_\gamma \to 2^+_g$ transition is nearly pure E2 (97.7 ± 0.1)% as is true in other nearby nuclei. The $2^+_\beta \to 2^+_g$ transition also is essentially pure E2, $(97^{+1.7}_{-2.2})$%, as is found in ^{152}Sm and ^{154}Gd. This value is in clear disagreement with the 62% M1 needed to obtain a consistent band mixing parameter for the beta band from the branching ratios. Thus as one moves more into the deformed region, one does not find any increase in the M1 admixture in the 2 → 2 transitions from either the beta or gamma band. There is still no agreement between theory and experiment for the gamma ray branching ratios from the beta vibrational band in any nucleus. In table 3 one sees that we have 4 cascades from odd parity states which should decay by nearby pure E1 radiation to the 2^+ level, and such is found.

What is most interesting in this experiment is the M1 admixtures in the $1^+ \to 2^+_g$ transitions from the high excited states at 1966, 2027 and 2187 keV. All three of these transitions show very large M1 admixtures to the 2^+_g state. From a skipped

Table 3. The Results of a Directional Correlation Study of the Consecutive Cascades in ^{156}Gd. (The gating transition was the 89 keV $2^+ \rightarrow 0^+$ ground state transition; ref. 48)

Energy(keV) (Transition)	Spin Sequence ($x \rightarrow 2_g^+ \rightarrow 0_g^+$)	A_2[a,b]	A_4[a,b]	δ	Admixture (%)
961	0^+	0.357	1.14	--	E2 100
1041	2_β^+	0.054 (43)	0.25 (8)	$5.8\ ^{+3.1}_{-1.5}$	E2 97 $^{+1.7}_{-2.2}$
1065	2_γ^+	-0.035 (06)	0.337 (12)	$19\ ^{+4}_{-3}$	E2 97.7 ± 0.1
1079	0^+	0.357	1.14	--	E2 100
1153.5	1^-	-0.254 (10)	0.004 (09)	$\lvert\delta\rvert < 0.023$	E1 >99.95
1231	2^-	0.260 (10)	0.002 (09)	$0.018\ ^{+0.010}_{-0.018}$	E1 99.97 $^{+0.03}_{-0.05}$
1277	1^-	-0.237 (12)	-0.016 (15)	$\lvert\delta\rvert < 0.031$	E1 >99.90
1857.4	1	-0.22 (9)	-0.12 (16)	0.024 ± 0.008	E1 99.94 $^{+0.03}_{-0.04}$
1877	1^+	-0.588 (28)	-0.126 (34)	$-0.43\ ^{+0.09}_{-0.06}$	E2 15.6 $^{+5.7}_{-3.6}$
1937	1^+	0.321 (21)	-0.198 (27)	$0.57\ ^{+0.03}_{-0.01}$	E2 24.5 $^{+2.0}_{-0.6}$
2098	1^+	0.581 (22)	-0.410 (20)	1.14 ± 0.03	E2 56.5 ± 1.3

[a]The last digit of the error, in parenthesis, occurs in the third decimal place.
[b]$G_2\Omega_2 = 0.624 \pm 0.021$, $G_4\Omega_4 = 0.385 \pm 0.008$.

correlation[50]), we find that the 812 keV, $1^+ \rightarrow 2_\gamma^+$ transition from the 1966 keV level to the 2_γ^+ level is essentially pure M1.

These 1^+ states are rather strongly populated in the decay of 0^+ ^{156}Eu and evidently arise from quasi - particle excitations with K = 1. The large M1 admixtures in these K allowed transitions from quasi - particle states support the interpretation in ^{178}Hf that it is the K allowed quasi - particle excitations that have large M1 admixtures and not the collective excitations. The K forbidden transitions in ^{156}Gd and ^{172}Yb (see table 11 in the last section) do not have as large M1 admixtures as seen in the K allowed ones from the quasi - particle states on ^{156}Eu and ^{178}Hf.

3.3 Tungsten - Osmium - Platinum Isotopes

Within the last year several groups have reported new measurements of E2/M1 admixtures in the tungsten, osmium and platinum isotopes. This region has taken on new significance with the theoretical calculations of the sign and magnitude of δ for transitions in these nuclei by Kumar and Baranger[12]). By and large the agreement in the experimental δ values obtained by different groups is exceptionally good. There are, however, a few exceptions which dictate a word of caution. An analysis of these data points out both the accuracy of the present techniques and some of the difficulties associated with the measurements.

Table 4. Angular Correlation Coefficients of the 308-612 keV, $3^+ \rightarrow 2^{+'} \rightarrow 0$ and 604-316 keV, $3^+ \rightarrow 2^+ \rightarrow 0^+$ Cascades in ^{192}Pt and the δ Values Obtained from these Coefficients and Nuclear Orientation Studies.

Cascade	A_2	A_4	δ	Method	Ref.
308-612	-0.124 ± 0.008	-0.080 ± 0.013	9.9 ± 1.0[a]	γ-$\gamma(\theta)$Na-Ge	Grabowski[51]
	-0.094 ± 0.015	-0.023 ± 0.020	$7.3^{+1.0}_{-0.8}$[b]	γ-$\gamma(\theta)$Na-Ge	Hamilton & Davies[52]
	-0.120 ± 0.020	-0.027 ± 0.020	9.4 ± 1.5[c]	γ-$\gamma(\theta)$Na-Ge	Kenyon et al.[53]
	-0.084 ± 0.010	-0.075 ± 0.012	$6.3^{+0.6}_{-0.5}$[b]	γ-$\gamma(\theta)$Na-Ge	Beraud et al.[54]
	(-0.094 for δ = 7.3)		7.3 ± 0.2	Nucl.Orient.	Reid et al.[55]
	-0.10 ± 0.02	-0.07 ± 0.04	8 ± 2 [d]	γ-$\gamma(\theta)$Na-Ge	Khan et al.[56]
604-316	-0.492 ± 0.015	-0.051 ± 0.020	-2.0 ± 0.2	γ-$\gamma(\theta)$Na-Ge	Grabowski[51]
	-0.480 ± 0.016	+0.34 ± 2.23	$-2.1^{+0.2}_{-0.3}$	γ-$\gamma(\theta)$Na-Ge	Hamilton & Davies[52]
	(-0.524 for δ = -1.5)		-1.5 ± 0.1	Nucl.Orient.	Reid et al.[55]
	-0.450 ± 0.010	-0.065 ± 0.013	-2.5 ± 0.2	γ-$\gamma(\theta)$Na-Ge	Beraud et al.[54]
	-0.410 ± 0.015	-0.070 ± 0.022	-3 ± 1	γ-$\gamma(\theta)$Na-Ge	Kenyon et al.[56]

[a] The 308(Ge)-612(NaI) keV cascade was corrected for 308-(296+316) keV coincidences. This correction increased δ by 11% and was attributed[51] as a major difference between the results of Grabowski and Hamilton & Davies.

[b] No mention is made for the correction mentioned in (a) and the same Na-Ge gates as used by Grabowski were employed.

[c] Summing effects as discussed in (a) were included.

[d] Here the 612 keV was selected in the Ge(Li) detector.

In table 4 the A_2 and A_4 coefficients and δ values for the 308 keV, $3^+ \rightarrow 2'^+$ and 604, $3^+ \rightarrow 2^+$ transitions in ^{192}Pt as obtained by several different groups are compared. The results do not show the wide spread in values that was typical in earlier surveys (refs. 2 - 9), when different groups' results for the same transition were compared. On the other hand, there are observable differences seen so that some A_2, A_4 or δ values do not overlap within one standard deviation. It must be remembered that for six measurements one can expect that two of the measurements should deviate from the average by more than one standard deviation. Still the accuracies quoted on the A_2 and A_4 coefficients given in table 4 seem to be somewhat better than the spread in

the coefficients would indicate. However, the δ values do not change much so now one is dealing with small refinements in the values of δ. The results of the nuclear orientation work and the γ - γ (θ) studies in the 604 keV transition could indicate a small systematic effect in one or the other of the types of experiments in that case. Grabowski[51]) has shown in the ^{192}Pt that even with NaI - Ge(Li) systems, one may have observable corrections from systematic effects such as gamma summing in the NaI gate. In an earlier survey[1]) it was shown that while γ - γ (θ) studies with NaI - Ge(Li) systems are capable of quite accurate results, one must take care to analyze carefully the backgrounds under all photopeaks.

Table 5. Angular Correlation Coefficients of the 630-137 keV, $2^+ \rightarrow 2^+ \rightarrow 0^+$ Cascade in ^{186}Os and the δ Values Obtained from these Coefficients and from Coulomb Excitation Studies.

A_2	A_4	δ	Method	Source	Ref.
-0.06 ±0.01	0.28 ±0.03	<-20	γ-γ(θ) 2 Na	Re in HNO_3	Lerjefors et al.[57]
-0.073±0.010	0.310±0.014	$-150^{+95}_{-\infty}$ [b]	γ-γ(θ) 2 Na	$HReO_4$ in H_2O	Bodenstedt et al[58]
-0.070±0.024	0.305±0.038	δ<-15		Re_2O_7 in H_2O	Vervier[59a]
-0.058±0.021	0.281±0.033			Re_2O_7 in H_2O	Vervier[59a]
			γ-γ(θ) 2 Na	+ glycerine	Vervier[59a]
-0.093±0.026	0.313±0.041	δ>10		Re metal	Vervier[59a]
-0.104±0.032	0.249±0.047			Re_2O_7 powder	Vervier[59a]
-0.129±0.017	0.332±0.011	$+13^{+5.5}_{-3.5}$	γ-γ(θ) 2 Na	^{186}Re liquid	Schölzig et al.[60]
-0.001±0.052	0.302±0.049	-10^{+4}_{-15}	γ-γ(θ) 2 Ge	^{186}Ir liquid	Krane & Steffen[61]
-0.065±0.026	0.292±0.031	$-60^{+40}_{-\infty}$ [b]	γ-γ(θ) 2 Ge	^{186}Re liquid	Lange[62]
		$\|\delta\|>25$	Coul. Exc.		Casten et al.[63]

[a]While differences were obtained for the different sources, the errors are so large that it is not clear that a true effect is observed except for the A_4 term in the powder source. The half life is 0.8 ns.

[b]When the A_2 and A_4 coefficients allow ∞ as an error limit, then large values of δ with the opposite sign are allowed within 2σ limits.

Now look at the results in table 5 on the 630 keV, $2'^+ \rightarrow 2^+$ transition in ^{186}Os. There is a disagreement in the A_2 and A_4 values from γ - γ (θ) results. The differences are not too large but they are sufficient to yield different signs for the δ values obtained by the different groups. There is also disagreement between γ - γ (θ) and Coulomb excitation results in the magnitude of δ. The sign difference in δ from the γ - γ (θ) studies is a serious problem since theories can predict the sign in this case. The sign difference arises from the differences in the A_2 values since the A_4 values are reasonably consistent for the errors assigned. As A_2 goes from about -0.07 to -0.10 with $A_4 \cong$ constant, δ changes sign. This fact is easily seen from the theoretical curves in fig. 4 as pointed out by Schötzig et al.[60]) for this particular case, when 2 NaI detectors are used, there are problems with chance coincidences and impurities of

^{188}Re in ^{186}Re sources. They[60]) used enriched ^{185}Re to make ^{186}Re to minimize this latter effect. The chance coincidences should not have had a serious effect in the work with 2 Ge(Li) detectors[61]) nor probably in the NaI - Ge(Li) studies[62]) which disagree in sign with the work of Schötzig et al.[60]). Also the work of Krane and Steffen[61]) was with ^{186}Ir though ^{188}Ir was a problem. It is not clear, at present, which of these results are right though there are more data to support the negative value of δ. It appears that either the statistical or systematic errors or both have been underestimated by some groups. Though here one must be careful for there is a few per cent probability that one can get an A_2 coefficient of -0.129 ± 0.017 when the true value is -0.075.

It should be pointed out that it is possible that one standard deviation in the error limits may yield a δ with one sign only, but the opposite sign is allowed for 2 σ variation when A_2 is between -0.05 to -0.12 in a $2 \to 2 \to 0$ cascade. Very good statistical accuracy must be obtained when A_2 is in this range to obtain a unique sign of δ as can be seen from fig. 4.

A similar situation arises in Coulomb excitation work so that when δ is large one cannot definitively establish the sign of δ. Thus several values in this mass region in the later tables will only be presented as absolute values. It can be stated as a general rule that when δ is large, $\gtrsim 25$ or has error limits that extend to infinity that one cannot definitively establish the sign of δ. One must also remember that because of the variation of the correlation coefficients with δ that 2σ limits on δ may not be obtained from a doubling of the 1σ limits. More particularly, one cannot double one σ limits on the % E2 to get 2σ limits. A word of caution must be said about upper and lower limits. Some people quote upper limits for one standard deviation in a correlation coefficient. Certainly one should not quote upper limits unless two standard deviations have been taken in the statistical errors.

Finally, these data also serve to affirm that great care must be taken in cases where weak cascades are involved for a single measurement may yield wrong conclusions. As will be pointed out later in the survey, there are a few other such cases where different signs are obtained for δ from different experiments in $2 \to 2 \to 0$ cascades when A_2 is in the above range.

By and large, as the survey in the next section will show, however, the accuracy of present experiments is generally good so that one can now make reliable comparisons of experiments with theoretical calculations, nucleus by nucleus.

4. E2/M1 Admixtures from Transitions in Nuclei $A \geqslant 100$

Since this paper is a review of the information on E2/M1 admixtures in transitional and deformed nuclei, the survey was limited to nuclei with $A \geqslant 100$. The survey includes data through August 10, 1970. In several cases there are included data privately communicated prior to publication. The sign convention for δ is that of Krane and Steffen[23]) as discussed in the second section. Special effort was made to put all older results into this convention and the original published data, A_2 and A_4 coefficients in γ - γ (θ) studies, were checked for the sign and magnitude of δ. In fig. 9 is shown the energy spectra predicted by the pure vibrational model for near spherical nuclei and by the Bohr - Mottelson model for deformed nuclei. Only the transitions where E2/M1 admixtures may occur are shown. In the survey to follow the data are presented in

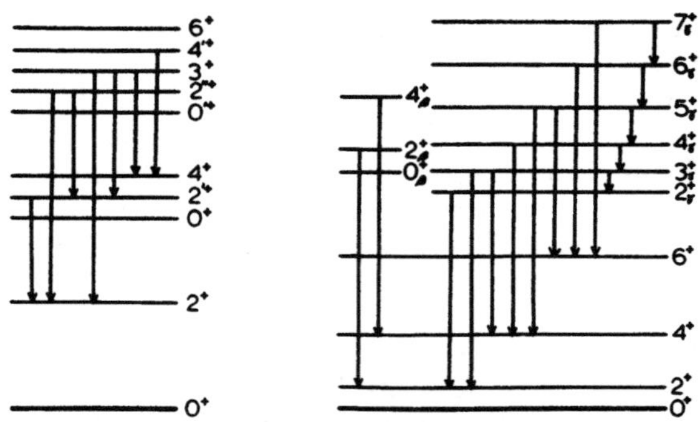

Fig. 9. The energy spectra predicted by the pure vibrational model for near spherical nuclei, and the Bohr - Mottelson model for deformed nuclei. Only the transitions where E2/M1 admixtures may occur are shown.

tables according to the various classes of transitions that one sees in fig. 9. There is obviously some overlap of assignment in the transitional region from one type excitation spectra to the other as occurs in the tungsten, osmium and platinum region.

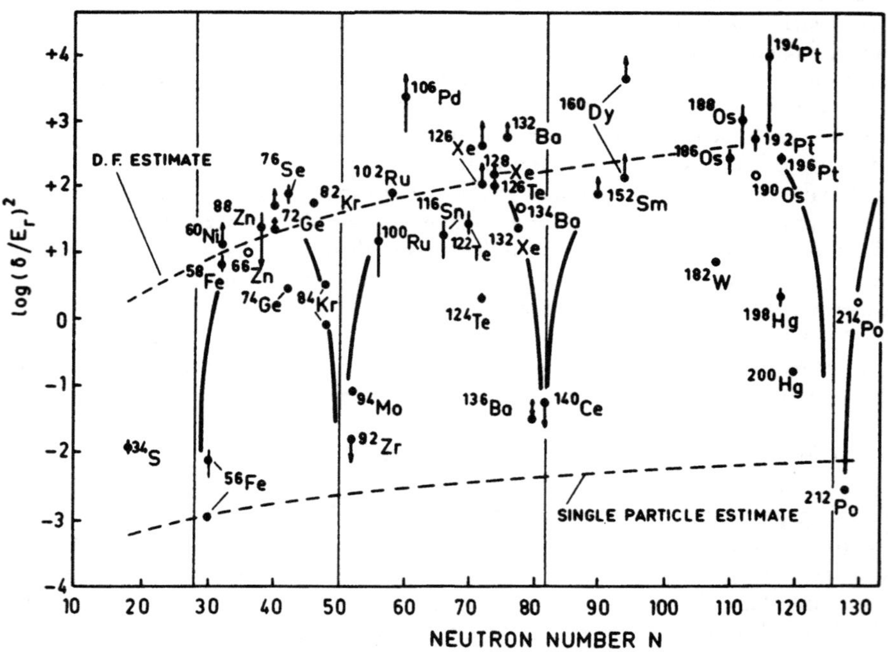

Fig. 10. The log of the reduced mixing ratio $(\delta/E_{\gamma})^2$ as a function of neutron number. This figure is taken from the survey of Potnis and Rao[9]). The predictions of Davydov and Filippov (ref. 64) and the single particle estimates are shown.

There have been several theoretical efforts to predict the values of δ or δ^2. Davydov and Filippov[64]) have worked out in the framework of their model of axially asymmetric nuclei for $2'^+ \to 2^+$ transitions, $\delta^2 = 8.1 \times 10^{-11} E_\gamma^2 Z^2 A^{4/3}$. The choice of the nuclei radius and other such constants make a small perturbation on the constant in this equation. This equation was taken from ref. 5. As can be seen in some of the earlier surveys[3,5,9]), the agreement of the δ^2 predicted by this model with experiment is reasonably good for most cases as to the general magnitude of δ^2 for $2'^+ \to 2^+$ transitions in even - even nuclei as shown in fig. 10 taken from ref. 9. But even in the earlier surveys[3,5,9]), there were notable exceptions to the rather smooth variation of $(\delta/E_\gamma)^2$. In particular the most glaring exceptions seemed to fall around the closed neutron shells. Actually the plots were of the log $(\delta/E_\gamma)^2$ so that only the general magnitudes were compared and any finer details were lost. Of course, the experimental accuracy at the time of these survery[3,5,9]) did not justify more than such a general comparison. The situation has now changed so that accurate results are available. An inspection of the following tables will show variations in δ from nucleus to nucleus that are related to the structure of their respective nuclei not experimental errors. Only a microscopic theory can explain these variations. Thus further comparison of experimental δ's with the general trends of the Davydov and Filippov[64]) model is not enlightening. *

Tamura and Yoshida[7]) studied the relative phase of the M1 and E2 radiations with a model in which quasi - particle excitations were coupled to originally pure collective states. The relative phase was found to be determined primarily by the sign of the magnetic moments of the quasi - particle states. Except in rare cases they expected to get the same sign of δ (negative in the notation of ref. 23) in the $2'^+ \to 2^+$ transitions. Their calculations fitted the then available data rather well.

In 1966 Greiner[10]) put forth an explanation of the mixing ratios based on different pairing forces for the neutrons and protons. This difference has the interesting consequence[10]) that protons and neutrons have different deformations. The magnitude of δ is predicted reasonably well in general by this theory as shown by Greiner (ref. 10) and in more recent work[60]). The major drawback of Greiner's theory however, is that all $2'^+ \to 2^+$ transitions are predicted to have the same sign for δ.

Finally in terms of general trends, Grechukhin[11]) has derived an expression for δ in terms of a phenomenological model of collective excitations. This model provides a general test of the collective character of the levels from a comparison with experimental δ values. Grechukhin[11]) concludes that only in the region of $150 < A < 190$ and $A > 230$ one can apply phenomenological collective models with reasonable success and that it is doubtful for the $50 \leqslant A \leqslant 150$ region. He qualifies his conclusion, however, because of the limited accuracy of the experimental data available even in January, 1968.

What was said earlier that a comparison of accurate experimental data with the more general prediction of Davydov and Filippov[64]) did not bring further enlightenment seems true for the predictions of Greiner[10]) and Grechukhin[11]). These models have given us insight into the behaviour and possible origin of the M1 admixtures observed in competition with E2 radiations in the decays of collective levels. Present experimental data are sufficiently accurate, however, that one must move on to a microscopic description of the problem. Kumar and Baranger[12]) have followed such a microscopic approach in the tungsten, osmium and platinum region with surprising-

*Bès et al [98]) have studied theoretically the properties of γ-vibrational levels. They compared the available data to a parameter proportional to δ^2 and found reasonable agreement with 4 cases reported in nuclei between ^{152}Sm and ^{168}Er but poor agreement for the tungsten isotopes.

ly good success as will be shown later in table 13.

Before going on to the survey, a final word should be said about the interpretation of earlier surveys that there were strong effects observed at the closure of neutron shells at N = 28, 50, 82 and 126. In fig. 10 are shown the results of the survey of Potnis and Rao[9]). A similar plot with the same general features has been given more recently by Grechukhin[11]). The data certainly seem to show sharp peaks toward the single particle estimates near the magic numbers for neutrons. But note there is no effect observed on the closing of the proton shells. A careful analysis of the data that yield δ values near the single particle estimates in both surveys[9,11]) reveal that several of the values are not correct. The ^{94}Mo and ^{136}Ba results are now known to be $4 \to 2$, not $2 \to 2$ transitions. The $\delta^2 \leqslant 1$ reported from a private communication in ^{214}Po in ref. 9 is now reported as $\delta^2 = 18.5^{+55}_{-12}$. The ^{200}Hg results of fig. 10 involve mixed cascades that were not accounted for and so this result[65]) was not included in the present survey. The ^{212}Po results[66]) were analyzed only for an A_2 term under the assumption that the transition in question was nearly pure M1. The ^{92}Zr and ^{140}Ce results are the work of Coleman[2,67]). The ^{92}Zr work[2]) involves a 900 (1.9%) – 930 (98.1%) keV cascade. Though the results[2]) are not necessarily wrong, we have shown[68]) that under such circumstances erroneous results were obtained in work with NaI detectors in ^{58}Fe. There is nothing obviously questionable about the ^{140}Ce work. This leaves two cases, ^{140}Ce and ^{56}Fe, or perhaps three to support the conclusion about the large M1 admixtures at the closed neutron shells. Thus at present it seems that the earlier conclusion has not been demonstrated very definitively if at all.

4.1 $2'^{+} \to 2^{+}$ *Transitions,* $100 \leqslant A \leqslant 150$

In table 6 are presented the mixing ratios for transitions from the second excited 2^+ state, designated $2'^{+}$, to the first excited 2^+ state. The year is given before each reference. In this table there are a few cases where Coulomb excitation studies do not yield a unique δ and both choices are given. In Coulomb excitation studies where only the angular distribution of the gamma rays is observed with respect to the beam direction, the a_4 coefficient is generally too small to assign a unique δ. Other data must be used or one can look at the angular distribution of gamma rays in coincidence with backscattered particles where a_4 need not be small. From a study of all the measured δ values in the region, one would favour the larger value of δ in the Coulomb excitation work but in table 7 as discussed next, one sees that such an interpretation may be in error there so one cannot follow this rule blindly.

All of these δ values are far from the single particle estimates in this region so that collective effects are dominating. One does observe that there is a minimum in the δ values around the closed shell at Z = 50 though the minimum is actually in the cadmium (Z = 48) isotopes. One also observes that all the δ's have negative signs between ^{100}Ru and ^{128}Xe.

There are several cases in table 6 where more than one measurement has been made and in every case there is consistency in the sign of δ. In only one case is there a disagreement outside one standard deviation in the magnitude, and that is ^{106}Pd where the errors in two groups are large and the results overlap within 2σ. Thus these data indicate that the accuracy of experimental measurements is good and justifies a theoretical microscopic approach to this region.

Table 6. E2/M1 Mixing Ratios in $+2' \to +2$ Transition for $100 \leqslant A \leqslant 150$

Isotope	Level keV	Transition Energy	$(\pi I)_i$	$(\pi I)_f$	Mixing ratio, δ	% E2	Method	Ref.
$^{100}_{44}$Ru	1366	826	+2′	+2	$+3.1^{+1.3}_{-0.5}$	$90.6^{+4.5}_{-0.6}$	$\gamma \cdot \gamma(\theta)$ 2NaI	64. Koi[73]
$^{102}_{44}$Ru	1103	628	+2′	+2	$-5.5^{+0.5}_{-0.7}$	$96.8^{+0.7}_{-1.4}$	$\gamma \cdot \gamma(\theta)$ 2NaI	61. His[74]
$^{104}_{44}$Ru	893	535	+2′	+2	$-8.5^{+1.5}_{-2.5}$ or -0.6	$98.6^{+0.4}_{-1.3}$ / 26.5	Coul Exc	68. McG[75]
$^{106}_{46}$Pd	1127	615	+2′	+2	-12^{+5}_{-15}	$99.3^{+0.6}_{-1.3}$	Coul Exc	69. Rob[76]
					$-30^{+15}_{-\infty}$	$99.9^{+0.1}_{-0.3}$	$\gamma \cdot \gamma(\theta)$ 2NaI	60. Rob[77]
					-4.35 ± 0.80	$95.0^{+1.4}_{-2.4}$	$\gamma \cdot \gamma(\theta)$ 2NaI	67. Koc[68]
$^{106}_{48}$Cd	1718	1085	+2′	+2	$-8.4^{+4.7}_{-100}$ or -0.6	$98.6^{+1.3}_{-5.4}$ / $26.5^{+12.7}_{-9.7}$	Coul Exc	69. Mil[79]
$^{108}_{46}$Pd	931	497	+2′	+2	$-5.2^{+1.4}_{-2.5}$ or -0.73 ± 0.10	$96.4^{+1.9}_{-2.0}$ / $34.6^{+5.9}_{-6.5}$	Coul Exc	69. Rob[76]
$^{108}_{48}$Cd	1603	970	+2′	+2	$-1.5^{+0.6}_{-1.5}$	69^{+21}_{-24}	Coul Exc	69. Mil[79]
$^{110}_{46}$Pd	814	440	+2′	+2	$-4.6^{+1.2}_{-1.9}$	$95.5^{+2.2}_{-3.3}$	Coul Exc	69. Rob[76]
$^{110}_{48}$Cd	1476	818	+2′	+2	-1.20 ± 0.15	$59.0^{+5.5}_{-12.1}$	$\gamma \cdot \gamma(\theta)$ 2Ge	70. Kra[23]
					$-1.5^{+0.4}_{-0.9}$	69^{+16}_{-14}	Coul Exc	64. Mil[79]
$^{112}_{48}$Cd	1312	695	+2′	+2	$-3.5^{+0.8}_{-0.9}$ or -0.87 ± 0.10	$92.4^{+2.7}_{-4.5}$ / $43.1^{+5.4}$ [illegible]	Coul Exc	69. Mil[79]

^{114}Cd	1208	650	$+2'$	$+2$	$-$	$1.5^{+0.5}_{-0.9}$	69^{+16}_{-19}	Coul Exc	69. Mil[79]
$^{116}_{48}Cd$	1212	699	$+2'$	$+2$	$-$	$1.5^{+0.4}_{-0.9}$	69^{+16}_{-14}	Coul Exc	69. Mil[79]
$^{116}_{50}Sn$	2109	819	$+2'$	$+2$	$-$	3.5 ± 1.5	$92.4^{+3.7}_{-12.4}$	$\gamma \cdot \gamma(\theta)$ 2NaI	56. Sch[80]
$^{122}_{52}Te$	1170	686	$+2'$	$+2$	$-$	3.7 ± 0.3	$93.2^{+0.9}_{-1.2}$	$\gamma \cdot \gamma(\theta)$ 2NaI	61. Asp[81]
					$-$	3.49 ± 0.07	92.4 ± 0.2	$\gamma \cdot \gamma(\theta)$ 2NaI	67. Koc[78]
					$-$	3.4 ± 0.5	$92.0^{+1.8}_{-2.6}$	$\gamma \cdot \gamma(\theta)$ 2NaI'	57. Lin[82]
$^{124}_{52}Te$	1326	722	$+2'$	$+2$	$-$	3.4 ± 0.6	$92.0^{+2.1}_{-3.3}$	$\gamma \cdot \gamma(\theta)$ Na-Ge	67. Ste[83]
					$-$	3.3 ± 0.1	$91.6^{+0.4}_{-0.5}$	$\gamma \cdot \gamma(\theta)$ Na-Ge	70. Kra[84]
					$-$	3.7 ± 0.3	$93.2^{+0.9}_{-1.2}$	$\gamma \cdot \gamma(\theta)$ 2Ge	70. Kra[84]
					$\pm$	1.00 ± 0.08	$50.0^{+3.8}_{-4.2}$	$\gamma \cdot \gamma(\theta)$ 2NaI	57. Lin[82]
					$-$	3.6 ± 0.3	$92.9^{+0.7}_{-1.3}$	$\gamma \cdot \gamma(\theta)$ Na-Ge	70. Bak[85]
$^{126}_{52}Te$	1420	750	$+2'$	$+2$	$-$	$8.8^{+1.7}_{-2.6}$	$98.7^{+0.5}_{-0.7}$	$\gamma \cdot \gamma(\theta)$ 2NaI	61. Asp[81]
					$-$	5.5 ± 0.5	$96.8^{+0.5}_{-0.7}$	$\gamma \cdot \gamma(\theta)$ 2NaI	59. Sak[86]
$^{126}_{54}Xe$	870	380	$+2'$	$+2$		$\lvert\delta\rvert > 5$	> 96.1	$\gamma \cdot \gamma(\theta)$	61. Asp[81]
					$-$	$5^{+0.5}_{-1.4}$	$96.1^{+1.5}_{-0.8}$	$\gamma \cdot \gamma(\theta)$ 2NaI	59. Sak[86]
$^{128}_{54}Xe$	969	528	$+2'$	$+2$	$+$	6.4 ± 1.5	$97.6^{+0.8}_{-1.6}$	$\gamma \cdot \gamma(\theta)$ 2NaI	61. Asp[81]
					$+$	$6.4^{+1.2}_{-2.6}$	$97.6^{+1.2}_{-1.2}$	$\gamma \cdot \gamma(\theta)$ 2NaI	61. Sun[87]
$^{132}_{54}Xe$	1298	630	$+2'$	$+2$	$+$	3.2 ± 0.8	$91.1^{+3.0}_{-5.9}$	$\gamma \cdot \gamma(\theta)$ 2NaI	62. Rob[88]
$^{132}_{56}Ba$	1031	567	$+2'$	$+2$	$-$	$100^{+84}_{-\infty}$	$99.9^{+0.1}_{-0.3}$	$\gamma \cdot \gamma(\theta)$ 2NaI	62. Rob[88]
					or	> 20	99.7		

Table 7. E2/M1 Mixing Ratios in Transitions from Higher Spin States for $100 \leq A \leq 150$

ISOTOPE	LEVEL keV	TRANSITION ENERGY	$(\pi I)_i \to$	$(\pi I)_f$	MIXING RATIO, δ	% E2	METHOD	REF.
$^{110}_{48}Cd$	2162	1505	+3	+2	-0.55 ± 0.10	23.3 ± 6.3	γ-$\gamma(\theta)$ 2 Ge	70. Kra[23]
	2162	687	+3	+2'	$-1.1^{+0.8}_{-0.4}$	55^{+14}_{-28}	γ-$\gamma(\theta)$ 2 Ge	
	2162	620	+3	+4	-0.80 ± 0.50	39^{+24}_{-31}	γ-$\gamma(\theta)$ 2 Ge	
	2219	677	+4'	+4	-0.25 ± 0.20	$5.9^{+10.8}_{-5.6}$	γ-$\gamma(\theta)$ 2 Ge	
	2926	447	+5	+6	-0.45 ± 0.20	$16.7^{+13.0}_{-10.8}$	γ-$\gamma(\theta)$ 2 Ge	
	2926	1384	+5	+4	-0.37 ± 0.03	$12.0^{+1.8}_{-3.4}$	γ-$\gamma(\theta)$ 2 Ge	
	2926	707	+5	+4'	-1.0 ± 0.3	$50.0^{+12.8}_{-17.1}$	γ-$\gamma(\theta)$ 2 Ge	
$^{112}_{48}Cd$	1468	851	+2"	+2	$2.0^{+0.7}_{-0.5}$ or 0.05	$80.0^{+7.9}_{-10.8}$ 0.25	Coul Exc	69. Mil[79]
$^{114}_{48}Cd$	1363	804	+2"	+2	$1.5^{+0.7}_{-0.4}$ or 0.15 ± 0.15	$69.2^{+13.7}_{-14.5}$ $2.2^{+6.0}_{-0.0}$	Coul Exc	69. Mil[79]
$^{132}_{54}Xe$	1803	1136	+3	+2	$+0.30 \pm 0.05$	$8.2^{+2.7}_{-2.3}$	γ-$\gamma(\theta)$ 2 NaI	62. Rob[88]
	1985	1317	+2"	+2	-0.077 ± 0.025	$0.6^{+0.4}_{-0.3}$	γ-$\gamma(\theta)$ 2 NaI	62. Rob[88]

Transitions from $2''^{+}$, 3^{+}, $4'^{+}$ and 5^{+} states, $100 \leqslant A \leqslant 150$

In table 7 are presented the results for transitions from third excited 2^{+} ($2'''^{+}$) states and higher spin 3^{+}, $4'^{+}$ and 5^{+} states. Here one has the reverse case from table 6 in a comparison of the two values of δ obtained from Coulomb excitation work. On the basis of systematics, one would favour the lower value of δ here. But as clearly stated this can be dangerous as indicated by the differences in tables 6 and 7. One sees that in all cases in table 7 where a unique δ is established, there is a large M1 component from 45% to 99%. This is, of course, expected in terms of the vibrational model since the $2''^{+}$ and 3^{+} states would be associated with the three phonon triplet (fig. 9) and E2 transition to the first 2^{+} states would be highly hindered over decays to the 2 phonon states. In ^{110}Cd the $3^{+} \to 2'^{+}$ decay does indeed have a larger E2 component than the $3^{+} \to 2^{+}$ decay. The vibrational model works even better in ^{132}Xe. There the very small E2 admixtures in the $3^{+} \to 2^{+}$ and $2''^{+} \to 2^{+}$ transitions more strongly indicate that the 3^{+} and $2''^{+}$ states are three phonon states as already proposed from gamma - ray branching ratios[69]).

4.2 $2 + 2_{\gamma} \to 0 + 2_{g}$ *Transitions,* $150 \leqslant A \leqslant 180$, $A > 225$

As is well known, the region of deformed nuclei sets in rapidly in nuclei around A = 150 with N = 90 in contrast to the gradual transition in the osmium - platinum region. At present there are no accurate data on E2/M1 admixtures in nuclei just outside the start of the deformed region at A = 150. In table 8 are presented the data on $K\pi I$, $2{+}2_{\gamma} \to 0{+}2_{g}$ transitions between the 2^{+} gamma vibrational band head and the 2^{+} member of the ground band..These transitions may be thought of as analogous to the $2'^{+} \to 2^{+}$ transitions in table 6. One sees from a comparison of tables 6 and 8 that while the M1 admixtures are reasonably small in the $2'^{+} \to 2^{+}$ transition, $A < 150$, they are very, very small in the $2{+}2_{\gamma} \to 0{+}2_{g}$ transitions. These very small M1 admixtures have been one of the successes of the early predictions of the Bohr - Mottelson model (refs. 36 and 70) that transitions from the quadrupole vibrational bands of K = 2, gamma and K = 0, beta were essentially pure E2.

There is complete agreement as to the sign of δ in all cases where more than one result is available on the same transition. Here one must recall the earlier warning, however, that in cases where δ is large, the opposite sign of δ is not excluded within 2σ limits.

There is one case, ^{152}Sm, where there is a small disagreement in the magnitude of δ from different experiments. Kumar[71]) has made a preliminary calculation of δ for ^{152}Sm and finds for the $2{+}2_{\gamma} \to 0{+}2_{g}$, $\delta = -24.3$. There is agreement in the sign of δ but the magnitude is larger than most of the experimental values. The agreement among the different experiments in ^{154}Gd and ^{160}Dy is remarkably good. One observes that in table 8 there are changes in the sign of δ with 5 cases with negative sign and 5 with positive. The more well deformed nuclei have positive signs but ^{172}Yb is an exception. These δ values will certainly provide a good test for any microscopic theory.

4.3 $0 + 2 \to 0 + 2$ *Transitions,* $150 \leqslant A \leqslant 180$, $A > 225$

Results on transitions from $K = 0, 2^{+}$ levels, primarily of the beta vibrational

Table 8. E2/M1 Admixtures in 2^+_γ-2^+_g Transitions in Deformed Nuclei

Isotope	Level keV	Transition Energy	$(K\pi I)_i-(K\pi I)_f$	Mixing Ratio, δ	% E2	Method	Ref.
$^{152}_{62}Sm$	1086	964	$2+2_\gamma$ -0+2	$-27\ ^{+11}_{-55}$	$99.86\ ^{+0.12}_{-0.25}$	Coul Exc	69. Fra[89]
				$-7>\delta>-11$	98.0<E2<99.2	γ-$\gamma(\theta)$ 2 Na	60. Deb[90]
				$-6>\delta>-13$	97.3<E2<99.4	γ-$\gamma(\theta)$ 2 Na	59. Lid[91]
				$-13.7\ ^{+2.4}_{-3.4}$	$99.47\ ^{+0.18}_{-0.25}$	γ-$\gamma(\theta)$ 2 Na	68. Kes[92]
				-3.8 ± 1.8	$93.5\ ^{+3.4}_{-13.5}$	Coul Exc	68. McG[93]
				-11.0 ± 1.0	$99.18\ ^{+0.13}_{-0.18}$	γ-$\gamma(\theta)$ 2 Ge	70. Lan[62]
$^{154}_{64}Gd$	996	873	$2+2_\gamma$ $0+2_g$	$-10.1\ ^{+1.1}_{-1.9}$	$99.02\ ^{+0.29}_{-0.24}$	γ-$\gamma(\theta)$ 2 Na	68. Kes[92]
				-10.2 ± 1.0	$99.04\ ^{+0.16}_{-0.21}$	γ-$\gamma(\theta)$ 2 Ge	70. Lan[62]
				-10.2 ± 1.8	$99.04\ ^{+0.27}_{-0.44}$	γ-$\gamma(\theta)$ 2 Na	69. Var[94]
				$-9.7\ ^{+1.5}_{-2.5}$	$98.94\ ^{+0.39}_{-0.41}$	γ-$\gamma(\theta)$ Na-Ge	69. Ham[95]
$^{156}_{64}Gd$	1154	1065	$2+2_\gamma$ $-0+2_g$	$-19\ ^{+3}_{-4}$	99.7 ± 0.1	γ-$\gamma(\theta)$ Na-Ge	69. Ham[48]
$^{160}_{66}Dy$	966	879	$2+2_\gamma$ $-2+2_g$	$-17.7\ ^{+2.8}_{-5.2}$	$99.68\ ^{+0.13}_{-0.13}$	γ-$\gamma(\theta)$ 2 Na	65. Gün[96]
				$-11\ ^{+3}_{-13}$	$99.18\ ^{+0.64}_{-1.36}$	γ-$\gamma(\theta)$ Na-Ge	67. Jak[97]
				-13.5 ± 3.0	$99.45\ ^{+0.18}_{-0.35}$		70. Kra[61]
$^{162}_{66}Dy$	890	811	$2+2_\gamma$ $0+2_g$	$+7\ ^{+\infty}_{-6}$ or $0.65\ ^{+1.3}_{-0.3}$	$98.0\ ^{+2.0}_{-48}$ / $28\ ^{+50}_{-19}$	Coul Exc	70. Eng[99]
$^{164}_{66}Dy$	762	689	$2+2_\gamma$ $0+2_g$	$+8\ ^{+\infty}_{-6}$ or $+0.6\ ^{+1.5}_{-0.2}$	$98.0\ ^{+2.0}_{-18}$ / $26\ ^{+55}_{-13}$	Coul Exc	70. Eng[99]
$^{166}_{68}Er$	787	706	$2+2_\gamma$ $0+2_g$	$\lvert\delta\rvert \geq 21$	≥ 99.77	Coul Exc	68. McG[93]
$^{168}_{68}Er$	822	742	$2+2_\gamma$ $0+2_g$	$\lvert\delta\rvert \geq 70$	≥ 99.97	Coul Exc	68. McG[93]
				> 6.3	>97.54	γ-$\gamma(\theta)$ 2 Na	64. Rei[100]
				$65\ ^{+135}_{-26}$	$99.97\ ^{+0.02}_{-0.04}$	γ-$\gamma(\theta)$ Na-Ge	70. Lan[101]
$^{170}_{68}Er$	933	855	$2+2_\gamma$ $0+2_g$	≥ 10	≥ 99.00	Coul Exc	68. McG[93]
$^{172}_{70}Yb$	1466	1387	$2+2_\gamma$ $0+2_g$	$-9.3\ ^{+2.5}_{-5}$	$98.85\ ^{+0.66}_{-0.97}$	γ-$\gamma(\theta)$ 2 Ge	70. Lan[62]
	1477	1398	2+2' $0+2_g$	$0.4 \leq \delta \leq 1.25$ or $2.2 \leq \delta \leq \infty$	14 - 56 / 83 - 100	γ-$\gamma(\theta)$ 2 Ge	70. Lan[62]
$^{178}_{72}Hf$	1174	1082	$2+2_\gamma$ $2+2_g$	$\delta < -11$ or $\delta > 32$	$>99\%$	Coul Exc	70. Var[28]
				$\lvert\delta\rvert > 1.3$	$>63\%$	γ-$\gamma(\theta)$ Na-Ge	70. Ham[29]
$^{232}_{90}Th$	786	736	$2+2_\gamma$ $0+2_g$	23 ± 10	$99.81\ ^{+0.09}_{-0.40}$	Coul Exc	70. McG[102]

type, are given in table 9. Here there are 4 cases that are established to be beta vibrational in character and two in ^{178}Hf that are not beta vibrational. Where accurate data are available, the $0+2_\beta \rightarrow 0+2_g$ transitions are essentially pure E2 as predicted in the first theoretical formulation[36,70]). There is no evidence to support the (50 - 60)% M1 admixtures that were sought to bring about agreement between theory and experiment for the gamma ray branching ratios from these beta bands (see, for example, ref. 72 for a discussion of the branching ratios). Here experimental measurements of E2/M1 admixtures are playing a crucial role in the further development of the Bohr - Mottelson model. At present, one is left without a theoretical description of the beta vibrational states and the first major crisis in the application of the rotational relationships to deformed nuclei as pointed out by Mottelson[35]).*

Table 9. E2/M1 Admixtures in 0+2-0+2 Transitions in Deformed Nuclei

Isotope	Level keV	Transition Energy	$(K\pi I)_i$ -	$(K\pi I)_f$	Mixing Ratio, δ	% E2	Method	Ref.
$^{152}_{62}$Sm	810	689	$0+2_\beta$	$-0+2_g$	$+67^{+\infty}_{-54}$	$99.97^{+0.03}_{-0.56}$	Coul Exc	69. Fra[89]
					13^{+28}_{-5}	$99.41^{+0.53}_{-0.95}$	Coul Exc	68. McG[93]
$^{154}_{64}$Gd	815	692	$0+2_\beta$	$-0+2_g$	$+11\pm3$	$99.18^{+0.31}_{-0.72}$	γ-$\gamma(\theta)$ Na-Ge	69. Ham[37]
$^{156}_{64}$Gd	1129	1040	$0+2_\beta$	$-0+2_g$	$-5.8^{+1.5}_{-3.1}$	$97.0^{+1.7}_{-2.2}$	γ-$\gamma(\theta)$ Na-Ge	70. Ham[48]
$^{178}_{72}$Hf	1276	1183	0+2	$0+2_g$	$+0.410\pm0.036$	$14.4^{+2.1}_{-1.1}$	γ-$\gamma(\theta)$ Na-Ge	70. Ham[29]
					$+0.41\pm0.17$	14 ± 10	γ-$\gamma(\theta)$ 2 Na	68. Nie[31]
	1496	1403	0+2'	$0+2_g$	$-0.8^{+0.2}_{-0.3}$	$39^{+1.7}_{-1.2}$	γ-$\gamma(\theta)$ Na-Ge	70. Ham[29]
					-0.2 to -1.3	30 - 65	γ-$\gamma(\theta)$ 2 Na	68. Nie[31]
$^{232}_{90}$Th	775	725	$0+2_\beta$	$0+2_g$	-(0.8 to 4.3)	39 to 95	Coul Exc	70. McG[102]

It is interesting to note that the δ values for the $0+2_\beta \rightarrow 0+2_g$ transitions have the opposite sign from the $2+2_\gamma \rightarrow 0+2_g$ transitions in every case, though for ^{152}Sm one must recall that for these large values there could be a change in sign so that the sign is not definitely established there. Kumar[71]) has preliminarily calculated a value of 10.8 for the $0+2_\beta \rightarrow 0+2_g$ transition in ^{152}Sm. For the reason just discussed, one cannot say there is a disagreement here, however.

The two 0+2 levels in ^{178}Hf have already been discussed. They are interpreted as two quasi - particle excitations and as such the M1 transitions are not expected to be hindered as for the beta vibrational states. It is interesting that these two $0+2 \rightarrow$ $\rightarrow 0+2$ transitions have the opposite sign for δ. Thus they too provide a sensitive test of any microscopic theory.

* Agreement with Bohr-Mottelson model for a 2^+_β state was recently found in ^{174}Hf, H. Ejiri and G.B. Hagemann, Nucl. Phys. A161(1971) 449.

Table 10. E2/M1 Admixtures in Transitions from States with KπI=2+3 and 2+4 in Def. Nucl.

Isotope	Level keV	Transition Energy	$(K\pi I)_i$	$-(K\pi I)_f$	Mixing Ratio, δ	% E2	Method	Ref.
$^{152}_{62}Sm$	1235	1113	2+3	0+2	$-9.1^{+2.1}_{-3.9}$	$98.80^{+0.61}_{-0.80}$	γ-γ(θ) 2 Na	68. Kes[92]
					-8.9 ± 1.0	$98.75^{+0.24}_{-0.33}$	γ-γ(θ) 2 Ge	70. Lan[62]
					23^{+47}_{-9}	$99.81^{+0.13}_{-0.30}$	γ-γ(θ) 2 Na	59. Lid[91]
$^{154}_{64}Gd$	1128	1005	2+3	0+2	-22^{+7}_{-24}	$99.78^{+0.20}_{-0.23}$	γ-γ(θ) Na-Ge	70. Ham[103]
					$\lvert\delta\rvert>15$	>99.55	γ-γ(θ) 2 Na	69. Var[94]
					-9.1^{+3}_{-5}	$98.80^{+0.69}_{-1.42}$	γ-γ(θ)	68. Kes[92]
					-6.6 ± 0.7	$97.75^{+0.41}_{-0.55}$	γ-γ(θ) 2 Ge	70. Lan[62]
$^{160}_{66}Dy$	1049	962	2+3	0+2	$-9.0^{+1.7}_{-2.8}$	$98.78^{+0.50}_{-0.62}$	γ-γ(θ) 2 Na	70. Lan[62]
					-18.0 ± 5	$99.69^{+0.12}_{-0.28}$	γ-γ(θ) 2 Ge	70. Kra[61]
					$7.9^{+6.1}_{-2.4}$	$98.42^{+1.07}_{-1.62}$	γ-γ(θ) Na-Ge	67. Jak[97]
^{172}Yb	1549	1470	2+3	0+2	-5.6^{+2}_{-3}	$96.90^{+1.76}_{-4.07}$	γ-γ(θ) 2 Ge	70. Lan[62]
$^{152}_{62}Sm$	1235	869	2+3	0+4	-5.0 ± 0.7	$96.2^{+0.9}_{-1.3}$	γ-γ(θ) Na-Ge	70. Lan[62]
$^{154}_{64}Gd$	1005	757	2+3	0+4	-5.7 ± 0.2	$97.01^{+0.19}_{-0.21}$	γ-γ(θ) Na-Ge	69. Ham[95]
					-4.9 ± 0.3	$96.00^{+0.43}_{-0.52}$	γ-γ(θ) 2 Na	69. Var[94]
$^{160}_{66}Dy$	1049	765	2+3	0+4	$-4.7^{+1.7}_{-5.0}$	$95.7^{+3.3}_{-5.7}$	γ-γ(θ) 2 Ge	70. Lan[62]
$^{152}_{62}Sm$	1373	1006	2+4	0+4	-6^{+3}_{-19}	$97.3^{+4.6}_{-7.3}$	γ-γ(θ) 2 Ge	70. Lan[62]
					$-12.9^{+7.6}_{-\infty}$	$99.4^{+0.6}_{-2.8}$	γ-γ(θ) 2 Na	69. Fra[89]
$^{154}_{64}Gd$	1141	893	2+4	0+4	-4.4 ± 0.5	$95.1^{+0.9}_{-1.2}$	γ-γ(θ) Na-Ge	69. Ham[95]
					-3.0 ± 1.0	90^{+4}_{-10}	γ-γ(θ) 2 Na	69. Var[94]

Transitions from $2+3_\gamma$, and $2+4_\gamma$ levels $150 \leqslant A \leqslant 180$

As for the $2+2_\gamma \rightarrow 0+2_g$ transitions, the $2+3_\gamma \rightarrow 0+2_g$ transitions given in table 10 all show very, very small M1 admixtures as early predicted by Bohr and Mottelson (ref. 70). There are two cases where the problem discussed earlier for 2 - 2 - 0 cascades applies to 3 - 2 - 0 cascades. In ^{152}Sm the large positive δ value in the $2+3_\gamma \rightarrow 0+2_g$ transition disagrees in sign with the other two measurements but within 2σ limits the results overlap. The same type problem occurs in ^{160}Dy where differences in the A_2 coefficient yield differences in the sign. In 3 - 2 - 0 cascades, one must have very good statistics in order to uniquely determine the sign of δ.

The $2+3_\gamma \rightarrow 0+4_g$ and $2+4_\gamma \rightarrow 0+4_g$ transitions all have smaller δ values than for the $(2+2_\gamma$ and $2+3_\gamma) \rightarrow 0+2_g$ transitions, though the M1 admixtures are still small. As already pointed out in the work of Hamilton et al.[103]) on ^{154}Gd, the accurate results in this nucleus show a definite increase in the M1 component as one goes to transitions to the $+4_g$ state over the $+2_g$ state and as one goes from $+2_\gamma$ to $+3_\gamma$ to $+4_\gamma$ for the initial level. Kumar[71]) has preliminarily calculated a $\delta = -9.6$ for the $2+4_\gamma \rightarrow 0+4_g$ transition. This value agrees on the sign and is in agreement with the magnitude of the experimental δ.

4.4 *Transitions from 0+4, 4+4, 1+1, 3+3 and 3+4 States, $150 \leqslant A \leqslant 180$*

The $0+4_\beta \rightarrow 0+4_g$ transitions from the beta bands are given in table 11. The M1 admixtures of $\gtrapprox 50\%$ needed to explain the gamma - ray branching ratios[72]) from the $0+4_\beta$ levels in ^{152}Sm and ^{154}Gd are ruled out. This is all part of the same problem discussed under the $0+2_\beta \rightarrow 0+2_g$ transitions. Kumar's preliminary calculations (see ref. 71) yield a $\delta = 4.1$ for the $0+4_\beta \rightarrow 0+4_g$ transition which has the same sign as experiment and reasonable agreement on the magnitude. The ^{154}Gd $0+4_\beta \rightarrow 0+4_g$ transition is an example of the big change in δ that can occur if you allow two standard deviations on the A_2 and A_4 coefficients in some cases. For 1σ on A_2 and A_4 E2 = (97 ± 2)% but for 2σ limits, $\delta > 2$ and E2 > 80%. It is often the case the the % E2 allowed by 2σ limits on A_2 and A_4 cannot be obtained from the % E2 and its error as obtained from the 1σ limits on A_2 and A_4.

The 4+4 band in ^{156}Gd could be a two K = 2 phonon band or arise from quasi - particle excitations. In any case the highly K forbidden, $\Delta K = 4$, $4+4 \rightarrow 0+4_g$ transition does not show a very large M1 component as in ^{178}Hf.

The 1^+ states in ^{156}Gd are evidently K = 1 levels of quasi - particle character. Thus the large M1 admixtures can be understood since M1 transitions would not be hindered as from quadrupole vibrational states. The very small E2 component in the $1^+ \rightarrow 2+2_\gamma$ transition is surprising. Again as in ^{178}Hf there is a difference in the sign of δ for one transition so that these levels offer very sensitive tests of any microscopic model.

It is interesting to note that the K forbidden, $\Delta K = 3$ decays of the 3+3 quasi-particle states in ^{172}Yb have strong E2 components to the ground state band. There is a decrease in the E2 admixture in the $3+3 \rightarrow 0+4_g$ transition over that in the $3+3 \rightarrow 0+2_g$ transition but an increase in the $3+4 \rightarrow 0+4_g$ transition as well as a change of sign for δ.

Three transitions within the K = 2 gamma band in ^{166}Er are reported in table

Table 11. E2/M1 Admixtures in Transitions from States with KπI=0+4,4+4,+1,3+3 and 3+4

ISOTOPE	LEVEL keV	TRANSITION ENERGY	$(K\pi I)_i$	$-(K\pi I)_f$	MIXING RATIO, δ	% E2	METHOD	REF.
$^{152}_{62}Sm$	1023	675	0+4	0+4	$2.94^{+1.86}_{-0.85}$	$89.6^{+6.2}_{-8.3}$	Coul Exc	69. Fra[89]
$^{154}_{64}Gd$	925	677	0+4	0+4	+6 ±2	$97.3^{+1.2}_{-3.2}$	γ-γ(θ) Na-Ge	69. Ham[95]
$^{156}_{64}Gd$	1511	1222	4+4	0+4	2.3±0.3	$84.1^{+3.0}_{-4.1}$	γ-γ(θ) 2 Na	59. Ofe[104]
					$2.12^{+0.41}_{-0.37}$	$81.8^{+4.7}_{-6.4}$	γ-γ(θ) Na-Ge	67. Ken[105]
$^{156}_{64}Gd$	1946	1857	(+)1	0+2	-0.024±0.008	$0.06^{+0.04}_{-0.03}$	γ-γ(θ) Na-Ge	70. Ham[48]
	1966	1877	+1	0+2	$+0.43^{+0.06}_{-0.09}$	$15.6^{+5.7}_{-3.6}$	γ-γ(θ) Na-Ge	
	1937	2027	+1	0+2	$-0.57^{+0.01}_{-0.03}$	$24.5^{+2.0}_{-0.6}$	γ-γ(θ) Na-Ge	
	2187	2098	+1	0+2	-1.14±0.03	56.5±1.3	γ-γ(θ) Na-Ge	
	1966	812	+1	2+2	-0.045±0.02	0.2 ±0.1	γ-γ(θ) Na-Ge	
$^{172}_{70}Yb$	1174	1095	3+3	0+2	-3.6±0.4	$92.8^{+1.3}_{-1.7}$	γ-γ(θ) 2 Ge	70. Lan[62]
					-3.3±0.4	$91.6^{+1.6}_{-2.2}$	e-γ(θ)	69. Vuk[106]
					-3.2±0.2	$91.1^{+0.9}_{-1.0}$	γ-γ(θ) Na-Ge	67. Blu[107]
	1172	912	3+3	0+4	-1.7±0.2	$74.3^{+4.0}_{-5.1}$	e-γ(θ)	69. Vuk[106]
					-1.8±0.8	$76.4^{+10.7}_{-16.4}$	γ-γ(θ) Na-Ge	67. Blu[107]
						82 ±12	L Subshell	68. Kay[108]
	1263	1003	3+4	0+4	>13	>99.4	e-γ(θ)	69. Vuk[107]
					> 8	>98.5	γ-γ(θ) Na-Ge	67. Blu[107]
						88 ±11	L Subshell	68. Kay[108]

12. The measurements are somewhat indirect and only δ^2 values are obtained. The M1 admixtures are larger than for transitions to the ground state band but small by single particle estimates. The same is true for the K = 3 band in ^{172}Yb.*

4.5 *Transitions in the Transitional Region 180 ⩽ A ⩽ 214*

The transition from deformed to near spherical nuclei takes place slowly as one moves through the tungsten, osmium and platinum region. To cover this transformation data are given in table 13 from nuclei where K = 2 gamma bands seem well established as in ^{182}W to more clearly vibrational nuclei like ^{198}Hg. It is just this region where Kumar and Baranger[12]) have carried out their microscopic calculation based on pairing + quadrupole forces. As emphasized by Kumar[12]), the E2/M1 mixing ratio provides a rather sensitive test of nuclear wave functions.

By and large the agreement among the various experimental results in this

*More accurately values on M1/E2 mixing ratios for transitions from beta- and gamma vibrational levels in ^{152}Sm: see contributed paper on page 247.

Table 12. E2/M1 Admixtures in Transitions between States within the same K=2 band in ^{166}Er, and within the same K=3 band in ^{172}Yb.

Isotope	Level keV	Transition Energy	$(K\pi I)_i$ -	$(K\pi I)_f$	Mixing Ratio, δ^2	% E2	Method	Ref.
$^{166}_{68}$Er	1075	119	2+5	2+4	$3.2^{+1.0}_{-0.6}$	$76.2^{+4.6}_{-4.0}$		69. deB[109]
					$4.43^{+3.36}_{-1.33}$	$81.6^{+7.0}_{-6.0}$	I_γ Ratios	69. Sun[110]
	1215	140	2+6	2+5	$2.5^{+5.0}_{-1.0}$	$71.4^{+16.8}_{-11.4}$		69. deB[109]
					≥ 1.6	≥ 61.5	I_γ Ratios	69. Sun[110]
	1375	160	2+7	2+6	$2.0^{+0.9}_{-0.5}$	$66.7^{+7.7}_{-6.7}$		69. deB[109]
					$1.22^{+0.47}_{-0.27}$	$55.0^{+7.9}_{-6.2}$	I_γ Ratios	69. Sun[110]
$^{172}_{70}$Yb	1262.9	90.6	3+4	3+3	$2.88^{+0.47}_{-0.44}$	74 ±3	L Subshell	68. Kay[108]
	1375.7	112.8	3+5	3+4	$2.03^{+0.55}_{-0.40}$	67 ±5	L Subshell	68. Kay[108]
	1510.0	134.3	3+6	3+5	$2.0^{+1.8}_{-0.8}$	67 ±12	L Subshell	68. Kay[108]

region is good as pointed out in an earlier discussion. But as also pointed out here, there are some disagreements. The ^{186}Os case was discussed earlier as an example. Other examples of sign differences between different results are ^{182}W and ^{190}Os. The ^{190}Os work of Yamazaki[122]) has a different sign from first reported[117]) because of a sign error in the theoretical conversion electron particle parameter for mixed cascades. While Yamazaki[122]) claims that this reverses the sign of δ for 3 - 2 - 0 cascades, in fact he changed only the ^{190}Os sign. This could imply that his work on ^{188}Os on the $3^+ \rightarrow 2^+$ transition has the wrong sign in table 13. In any case the analysis of these e - γ results are difficult because of possible penetration effects in the M1 radiation so a more careful analysis of these two cases is called for. The results in ^{182}W may simply arise from a systematic error in one or more of the 3 experiments. The results of Schiff and Metzger[132]) on ^{198}Hg in table 13 are the results of an analysis of the data by this author. The ^{188}Os $2'^+$ - 2^+ transition is a case in point where the Coulomb excitation work cannot really determine the sign.

The calculations of Kumar[12]) for these isotopes agree well with the experimental results in both sign and magnitude of δ from ^{186}Os to ^{194}Pt.

The three cases in ^{192}Pt are in particularly nice agreement. Since there is a sign change within these three cases in the same nucleus, this illustrates the point of the sensitivity of tests of nuclear wave functions provided by δ values. The largest discrepancy in magnitude is in ^{196}Pt and the only one in sign is ^{184}W. Kumar[12]) has pointed out the two most probable approximations to influence the results: 1) since the matrix elements are evaluated by summing over a large number of terms that correspond to different points of a β - γ mesh[12]), the numerical accuracy of nearby forbidden transition matrix elements is comparatively poor; and 2) the

Table 13. E2/M1 Admixtures in Transitions in the Transitional Region 182 ≤ A ≤ 198

Iso-tope	Level keV	Transition Energy	$(K\pi I)_i \rightarrow$	$(K\pi I)_f$	Mixing ratio, δ	Kumar	% E2 (exp)	Method	Ref.
$^{182}_{74}$W	1222	1122	2+2	0+2	17^{+36}_{-7}	16.8	$99.6^{+0.3}_{-0.6}$	Coul Exc	68. Mil[111]
					6^{+6}_{-2}		97^{+2}_{-3}	γ-γ(θ) 2 NaI	60. Hic[112]
					$-2.6^{+1.9}_{-3.0}$		87^{+10}_{-54}	γ-γ(θ) 2 NaI	56. Wil[113]
$^{184}_{74}$W	904	793	2+2	0+2	-19^{+6}_{-14}	45.4	$99.72^{+0.09}_{-0.31}$	Coul Exc	68. Mil[111]
					-20^{+5}_{-8}		$99.75^{+0.12}_{-0.20}$	e - γ(θ)	69. Zup[114]
					$-15^{+0.9}_{-1.4}$		$99.55^{+0.06}_{-0.06}$	γ-γ(θ) 2 NaI	64. Kör[115]
$^{186}_{74}$W	730	610	2+2	0+2	$-13.5^{+2.5}_{-3.5}$	-216	$99.47^{+0.20}_{-0.27}$	Coul Exc	68. Mil[111]
					-30^{+25}_{-10}		$99.89^{+0.05}_{-3.74}$	Coul Exc	61. McG[116]
$^{186}_{76}$Os	767	630	+2'	+2	$-150^{+95}_{-\infty}$	- 14.7	$99.99^{+0.01}_{-0.03}$	γ-γ(θ) 2 Na	61. Bod[58]
					<-15 > 10			γ-γ(θ) 2 Na	63. Ver[59]
					|δ| > 25		>99.84	Coul Exc	69. Cas[63]
					-45^{+16}_{-65}		$99.95^{+0.04}_{-0.07}$	γ-γ(θ) 2 Na	61. Ler[57]
					$13.0^{+3.0}_{-5.5}$		$99.41^{+0.20}_{-1.16}$	γ-γ(θ) 2 Na	69. Sch[60]
					-10^{+4}_{-15}		$99.00^{+0.75}_{-1.00}$	γ-γ(θ) 2 Ge	70. Kra[61]
					$-60^{+40}_{-\infty}$		$99.97^{+0.03}_{-0.22}$	γ-γ(θ) 2 Ge	70. Lan[62]
	910	773	+3	+2	-13^{+6}_{-9}	- 13.5	$99.31^{+0.44}_{-0.85}$	γ-γ(θ) Na-Ge	70. Kra[61]
$^{188}_{76}$Os	633	478	+2'	+2	$-14.3^{+4.3}_{-10.7}$	- 9.6	$99.51^{+0.10}_{-0.51}$	e - γ(θ)	63. Yam[117]
					- 15.7 ± 7.0		$99.59^{+0.21}_{-0.90}$	γ-γ(θ) 2 Na	56. Pot[118]
					|δ| > 25		>99.8	Coul Exc	70. Mil[119]
					- 12.3 ± 2.8		$99.34^{+0.22}_{-0.44}$	γ-γ(θ) 2 Ge	70. Kra[61]
					|δ| > 30		>99.89	Coul Exc	69. Cas[63]
	790	635	+3	+2	- 6.9 ± 3.2	- 10.5	$97.9^{+1.1}_{-4.8}$	γ-γ(θ) Na-Ge	70. Kra[61]
					$-8.3^{+3.0}_{-1.2}$		$98.5^{+1.2}_{-2.0}$	e - γ(θ)	63. Yam[117]
$^{190}_{76}$Os	557	371	+2'	+2	$-14.3^{+4.3}_{-10.7}$	- 7.6	$99.51^{+0.33}_{-0.51}$	γ - e(θ)	63. Yam[117]
					- 11 ± 3		$99.18^{+0.31}_{-0.71}$	Coul Exc	69. Rob[120]
					- 6.5 ± 1.3		$97.68^{+0.70}_{-1.25}$	γ - e(θ)	69. Sam[121]
					- 8.5 ± 1.0		$98.63^{+0.27}_{-0.38}$	γ-γ(θ) 2 Ge	70. Kra[61]
					-11^{+6}_{-4}		$99.18^{+0.37}_{-0.03}$	Coul Exc	69. Cas[63]
	756	569	+3	+2	- 9.0 ± 1.5	- 9.9	$98.78^{+0.32}_{-0.53}$	γ-γ(θ) 2 Ge	70. Kra[61]
					$+14.3^{+\infty}_{-7}$		$99.51^{+0.49}_{-1.4}$	e - γ(θ)	65. Yam[122]
$^{192}_{76}$Os	489	283	+2'	+2	$-4.7^{+0.7}_{-0.6}$	- 5.2	$95.67^{+0.89}_{-1.56}$	Coul Exc	69. Rob[120]
					$-5.2^{+1.6}_{-1.1}$		$96.43^{+1.11}_{-3.60}$	Coul Exc	69. Cas[63]
$^{192}_{76}$Os	690	484	+3	+2	$-10.9^{+1.5}_{-2.1}$	- 4.8	$99.16^{+0.25}_{-0.28}$	γ-γ(θ) Na-Ge	69. Gra[51]
					$-7.6^{+1.3}_{-2.4}$		$98.30^{+0.70}_{-0.76}$	γ-γ(θ) Na-Ge	70. Ber[54]
					-10^{+3}_{-10}		$99.00^{+0.75}_{-1.0}$	γ-γ(θ) Na-Ge	69. Kha[56]

Table 13 (continued)

Isotope	Level keV	Transition Energy	$(K\pi I)_i$	$\rightarrow(K\pi I)_f$	Mixing ratio, δ	Kumar	% E2 (exp)	Method	Ref.
$^{192}_{78}Pt$	612	296	+2'	+2	10 ± 1	14.8	$99.00^{+0.18}_{-0.22}$	Lin Polar γ-γ(θ) NaGe	68. Ham[52]
					$9.1^{+2.7}_{-1.7}$		$98.80^{+0.49}_{-0.60}$	γ-γ(θ) 2 Ge	69. Gra[51]
					$7.2^{+1.6}_{-2.6}$		98.1 ± 0.62	γ - e(θ)	67. Kuk[123]
					6.5 ± 1.5		$97.68^{+0.78}_{-1.53}$	γ-γ(θ) 2 Na	57. Mra[124]
					6.5 ± 0.5		$97.68^{+0.32}_{-0.28}$	γ-γ(θ) 2 Na	67. Koc[78]
	921	308	+3	+2'	$7.3^{+1.0}_{-0.8}$	9.2	$98.16^{+0.41}_{-0.48}$	γ-γ(θ) NaGe γ Lin Polar	68. Ham[52]
					9.9 ± 1.0		$98.99^{+0.17}_{-0.24}$	γ-γ(θ) NaGe	69. Gra[51]
					9.4 ± 1.5		$98.88^{+0.28}_{-0.46}$	γ-γ(θ)	69. Ken[53]
					$6.3^{+0.6}_{-0.5}$		$97.54^{+0.40}_{-0.43}$	γ-γ(θ) NaGe	70. Ber[54]
					7.3 ± 0.2		$98.16^{+0.09}_{-0.11}$	Nucl Orient	69. Rei[55]
					8 ± 2		$98.46^{+0.54}_{-1.16}$	γ-γ(θ) NaGe	69. Kha[56]
$^{192}_{78}Pt$	921	604	+3	+2	$-2.1^{+0.2}_{-0.3}$	- 2.2	$81.5^{+3.7}_{-6.9}$	γ-γ(θ) NaGe γ Lin Polar	68. Ham[52]
					- 2.0 ± 0.2		$80.0^{+2.9}_{-3.6}$	γ-γ(θ) NaGe	69. Gra[51]
					- 1.5 ± 0.1		$69.2^{+2.7}_{-3.0}$	Nucl Orient	69. Rei[55]
					- 2.5 ± 0.2		$86.2^{+1.7}_{-2.1}$	γ-γ(θ) NaGe	70. Ber[54]
					- 3 ± 1		90^{+4}_{-10}	γ-γ(θ) NaGe	69. Ken[53]
$^{194}_{78}Pt$	622	294	+2'	+2	30^{+39}_{-11}	19.9	$99.89^{+0.09}_{-0.17}$	γ-γ(θ) NaGe	69. Ham[125]
					14.3 ± 2.1		$99.51^{+0.11}_{-0.18}$	γ-γ(θ) 2 Ge	70. Kra[61]
					19		99.7	e - γ(θ)	68. Krm[126]
$^{196}_{78}Pt$	688	333	+2'	+2	- 4.03± 0.12	-101.4	$94.20^{+0.31}_{-0.34}$	γ-γ(θ) NaGe	69. Ham[123]
					- 5.7 ± 0.3		$97.00^{+0.20}_{-0.22}$	γ-γ(θ) 2 Ge	70. Kra[61]
					- 5.0 ± 0.5		96.2 ± 0.8	γ-γ(θ) 2 Na	63. Ike[128]
					- 5.4 ± 0.5		$96.68^{+0.53}_{-0.68}$	γ-γ(θ) 2 Na	56. Pot[127]
					- 5.4 ± 1.0		96.2 ± 2.1	γ-γ(θ) 2 Na	53. Ste[129]
$^{198}_{80}Hg$	1088	676	+2'	+2	1.1 ± 0.1		$54.8^{+4.3}_{-4.8}$	γ-γ(θ) 2 Na	66. Pet[130]
					0.96± 0.10		$48.0^{+5.0}_{-5.4}$	γ-γ(θ) 2 Na	53. Sch[131]
					1.7 ± 0.3		$74.3^{+5.7}_{-8.1}$	γ-γ(θ) 2 Na	53. Sch[132]
					1.37± 0.09		$64.2^{+3.0}_{-3.3}$	γ-γ(θ) 2 Na	67. Koc[78]
$^{214}_{84}Po$	1378	769	+2'	+2	$4.3^{+4.5}_{-1.8}$		$94.9^{+3.9}_{-8.7}$	γ-γ(θ) Na	61. Tay[133]

calculated M1 transition matrix elements (which vanish in the simple collective model where the gyromagnetic ratio is independent of deformation and the M1 operator is just a constant times the nuclear angular momentum operator) are reduced by one to two orders of magnitude compared to the diagonal matrix element. Attempts are being made to improve the numerical accuracy. Deviations from the adiabatic assumption of the present, microscopic treatment of collective states may be particularly serious for M1 matrix elements, which in contrast to E2 matrix elements are only mildly affected by nuclear collectivity or deformation.

5. Summary

In conclusion this survey has shown that currently experimental techniques are sufficiently well developed to obtain reliable mixing ratios both in sign and magnitude except for very large δ values where the sign may be difficult to obtain. Nevertheless, even with Ge(Li) detectors care must be taken to eliminate systematic errors. Careful attention should be given to the experimental errors and what happens when two standard deviations are allowed in the experimental A_2 and A_4 coefficients. Upper limits should only be quoted when two standard deviations in the statistical errors are taken. The present experimental results are sufficiently accurate over a wide mass range to test microscopic calculations and the nuclear wave function used in such calculations.

† Work supported in part by a grant from the National Science Foundation.

Acknowledgements

I would like to thank Prof. R. Steffen for his very kind permission to reproduce the work of Dr. Krane and himself on their definition of δ. Also they along with Prof. Z. Grabowski, Dr. J. Lange, Dr. W.T. Milner, Prof. A.V. Ramayya, Dr. R.L. Robinson and Dr. D.M. van Patter are thanked for their private communication of data prior to publication and their assistance in the tabulation of these results. Conversations with Prof. Steffen, Dr. K. Kumar and Prof. Ramayya have been particularly helpful. The assistance of my colleagues and students at Vanderbilt in the tabulation of data and preparation of figures and slides is most appreciated. These include Dr. S. Brahmavar with help from Dr. P.E. Little, Mr. K. Baker, Mr. D. Dunn and Mr. G. Phillips and the secretarial assistance of P. Smith, Karen Taylor and K. Cravens.

References

1. J.H. Hamilton, Radioactivity in Nuclear Spectroscopy, ed. J.H. Hamilton, J.C. Manthuruthil, Gordon and Breach (in press).
2. C.F. Coleman, Nucl. Phys. 7 (1958) 1488.
3. S.S. Malik, V.R. Potnis and C.E. Mandeville, Nucl. Phys. 11 (1959) 691.
4. D.M. van Patter, Nucl. Phys. 14 (1959) 42.
5. E.P. Grigoriev and M.P. Avotina, Nucl. Phys. 19 (1960) 248.
6. C.A. Mallmann, Nucl. Phys. 24 (1961) 535.
7. T. Tamura and H. Yoshida, Nucl. Phys. 30 (1962) 579 and M. Sakai, INS Report 37.

8. D.P. Grechokhin, Nucl. Phys. 40 (1963) 422.

9. V.R. Potnis and G.N. Rao, Nucl. Phys. 42 (1963) 620.

10. W. Greiner, Nucl. Physics, 80 (1966) 417.

11. D.G. Grechokhin, Yad. Fiz. 10 (1969) 94; Trans. Soviet J. Nucl. Phys. 10 (1970) 55.

12a. K. Kumar, Physics Letters 29B (1969) 25;
12b. K. Kumar and M. Baranger, Phys. Rev. Letters 17 (1966) 1146; Nucl. Phys. A92 (1967) 608 and Nucl. Phys. A122 (1968) 273.

13. E. Bodenstedt, Radioactivity in Nuclear Spectroscopy, ed. J.H. Hamilton and J.C. Manthuruthil, Gordon and Breach (in press).

14. L.C. Biedenharn and M.E. Rose, Revs. Mod. Phys. 25 (1953) 729.

15. H.W. Taylor and R. McPherson, γ - γ Directional Correlation Coefficients A_2 and A_4 as Functions of the Mixing Parameter δ. (Queen's University, Kingston, Canada, 1960, with supplement of 1961).

16. A.J. Ferguson, Angular Correlations Methods in Gamma - Ray Spectroscopy, North - Holland Publishing Co., Amsterdam, 1965, p. 6.

17. L.C. Biedenharn and M.E. Rose, private communication to A.J. Ferguson.

18. S. Devons and L.J.B. Goldfarb, Handbuch der Physik 42 (1957) 362. (Springer, Berlin).

19. R. Huby, Proc. Inter. Conf. Nucl. Phys. Paris, Crosby, Lockwood and Son, Ltd., London, 1958, p. 895.

20. R.M. Steffen, Handbook for Angular Correlations, Purdue University, Lafayette, Indiana, 1965.

21. H.J. Rose and D.M. Brink, Rev. Mod. Phys. 39 (1967) 306.

22. E.L. Church, M.E. Rose and J. Weneser, Phys. Rev. 109 (1958) 1299.

23. K.S. Krane and R.M. Steffen, Phys. Rev. C2 (1970) 724.

24. A. Bohr and B.R. Mottelson, Nuclear Structure, Vol. I (W.A. Benjamin, Inc. New York, 1969) p. 381.

25. K. Alder, H. Bohr, T. Huus, B. Mottelson and A. Winter, Rev. Mod. Phys. 28 (1956) 432.

26. H. Frauenfelder and R.M. Steffen, Alpha-, Beta- and Gamma-Ray Spectroscopy, ed. K. Siegbahn, (North - Holland Publishing Co., Amsterdam, 1965) p. 947.

27. P.E. Little, J.H. Hamilton, A.V. Ramayya and N.R. Johnson, to be published.

28. L. Varnell, J.H. Hamilton and R.L. Robinson, to be published.

29. J.H. Hamilton, A.V. Ramayya, P.E. Little and N.R. Johnson, to be published; P.E. Little, Ph. D. Thesis, August 1970, Vanderbilt University.

30. H.L. Nielsen, K. Wilsky, Y. Zylicz and G. Sorensen, Nucl. Phys. A93 (1967) 385.

31. H.L. Nielsen, K. Bonde Nielsen and N. Rud, Phys. Letters 27B (1968) 150.

32. L.L. Riedinger, N.R. Johnson and J.H. Hamilton, Phys. Rev. Letters 19 (1967) 1243; Phys. Rev. 179 (1969) 1214.

33. Y.T. Liu, O.B. Nielsen, P. Salling and O.S. Skilbried, Izv. Akad. Nauk SSSR, Ser. Fiz. 31 (1967) 63 (translation, Phys. Ser. 31 (1967) 69).

34. G.T. Ewan and J. Bower, as quoted by R.L. Graham, Bull. Acad. Sci. USSR (Columbia Tech. Transl.) Phys. Ser. 31 (1967) 1.

35. B.R. Mottelson, International Conference on Nuclear Structure, Supp. J. Phys. Soc., Japan 24 (1968) 87.

36. A. Bohr and B.R. Mottelson, Beta- and Gamma-Ray Spectroscopy, ed. K. Siegbahn, North - Holland Publishing Co., Amsterdam, Netherlands, 1955, p. 482.

37. J.H. Hamilton, A.V. Ramayya and L.C. Whitlock, Phys. Rev. Letters 22 (1969) 65.

38. J.S. Greenberg, International Conference on Radioactivity in Nuclear Spectroscopy: Modern Techniques and Applications, (Gordon and Breach) 1970; I.A. Fraser, J.S. Greenberg, S.H. Sie, R.G. Stokstad and D.A. Bromley, Phys. Rev. Letters 23 (1969) 1047.

39. F.K. McGowan, International Conference on Radioactivity in Nuclear Spectroscopy: Modern Techniques and Applications, (Gordon and Breach) 1970; F.K. McGowan, R.O. Sayer, P.H. Stelson, R.L. Robinson and W.T. Milner, Bull. Am. Phys. Soc. 13 (1968) 845.

40. L.L. Riedinger, G.B. Hagemann and E. Eichler, private communication: to be published in Proceedings of the Heavy Ion Conference, Heidelberg, 1969.

41. H. Huber, W. Kesternich, P. Weigt and E. Bodenstedt, Nucl. Phys. A107 (1968) 106.

42. T.A. Siddiqi, R.C. Carlson and G.T. Emery, Bull. Am. Phys. Soc. 14 (1969) 1174.

43. W.T. Milner, Ph. D. Thesis, University of Tennessee (1968).

44. M.M. Minor, R.K. Sheline and E.T. Jurney, Bull. Amer. Phys. Soc. 15 (1970) 523.

45. V.G. Soloviev, Nucl. Phys. 49 (1965) 1.

46. D.J. McMillan, J.H. Hamilton, J.C. Manthuruthil and J.J. Pinajian, Z. Phys., 241(1971)103.

47. A.F. Kluk, N.R. Johnson and J.H. Hamilton, to be published.

48. J.H. Hamilton, P.E. Little, A.V. Ramayya, E. Collins and N.R. Johnson, to be published; P.E. Little, Ph. D. Thesis, Vanderbilt University, August 1970.

49. E. Collins, M.S. thesis, Vanderbilt Univ., May 1970 and E. Collins, J.H. Hamilton, P.E. Little, A.V. Ramayya, M.U. Kim, A.F. Kluk and N.R. Johnson, contribution to this Conference.

50. A.V. Ramayya, J.H. Hamilton, P.E. Little, E. Collins and J.J. Pinajian, contribution to this Conference.

51. Z. Grabowski, Phys. Rev. 183 (1969) 1019.

52. W.D. Hamilton and K.E. Davies, Nucl. Phys. A122 (1968) 165.

53. D.B. Kenyon, L. Keszthelyi and J.A. Cameron, Can. J. Phys. 47 (1969) 2395.

54. R. Beraud, I. Berkes, R. Chéry, R. Haroutunian, M. Lévy, G. Marguier, G. Marert and R. Rougny, Phys. Rev. C1 (1970) 302.

55. P.G.E. Reid, M. Sott and N.J. Stone, Nucl. Phys. A129 (1969) 273.

56. M.Y. Khan, L.D. Wyly, C.H. Braden and E.T. Patronis, Jr., Phys. Rev. 182 (1969) 1259.

57. C.A. Lerjefors, E. Matthias and E. Karlsson, Nucl. Phys. 25 (1961) 404.

58. E. Bodenstedt, H.J. Körner, G. Strube, C. Günther, J. Radehoff and E. Gerdon, Z. Physik 163 (1961) 1.

59. J. Vervier, J. Physique, 24 (1963) 763.

60. U. Schötzig, H. Schrader, R. Stipplen and F. Münnich, Z. Physik 222 (1969) 479.

61. K.S. Krane and R.M. Steffen, Phys. Rev. 2C(1970) 724.

62. J. Lange, Thesis, University of Bonn (1970).

63. R.F. Casten, J.S. Greenberg, S.H. Sie, G.A. Burginyon and D.A. Bromley, Phys. Rev. 187 (1969) 1532.

64. A.S. Davydov and G.P. Filippov, Nucl. Phys. 8 (1958) 237.

65. T. Lindquist, Ark. Fys. 12 (1957) 495.

66. Giannini, Nuovo. Cim. 22 (1961) 31.

67. C.F. Coleman, Phil. Maz. 46 (1955) 1132.

68. N.C. Singhal, A.V. Ramayya, J.H. Hamilton and S. Raman, contribution to this conference.

69. J.H. Hamilton, H.K. Carter and J.J. Pinajian, Phys. Rev. C1 (1970) 666.

70. A. Bohr and B.R. Mottelson, Kgl. Danske Viedenskol Selskal, Mat. - Fys. Medd. 27 (1953) 16.

71. K. Kumar, private communication, August 1970.

72. L.L. Riedinger, N.R. Johnson and J.H. Hamilton, Phys. Rev. 179 (1969) 1214.

73. M. Koike, K. Histake, N. Oho and K. Takahaski, Nucl. Phys. 54 (1964) 129.

74. K. Histake, J. Phys. Soc. Japan 16 (1961) 1280.

75. F.K. McGowan, R.L. Robinson, P.H. Stelson and W.T. Milner, Nucl. Phys. A113 (1968) 529.

76. R.L. Robinson, F.K. McGowan, P.H. Stelson, W.T. Milner and R.O. Sayer, Nucl. Phys. A124 (1969) 553.

77. R.L. Robinson, F.K. McGowan, and W.G. Smith, Phys. Rev. 119 (1960) 1692.

78. J. Koch, F. Münnich and U. Schötzig, Nucl. Phys. A103 (1967) 300.

79. W.T. Milner, F.K. McGowan, P.H. Stelson, R.L. Robinson and R.O. Sayer, Nucl. Phys. A129 (1969) 687.

80. R.P. Scharenberg, M.G. Stewart and M.L. Wiedenbeck, Phys. Rev. 101 (1956) 689.

81. I. Asplund, L.G. Stromberg, and T. Wiedling, Ark. Fys. 18 (1961) 65.

82. T. Lindquist and I. Marklund, Nucl. Phys. 4 (1957) 189.

83. P.H. Stelson, Phys. Rev. 157 (1967) 1098.

84. K.S. Krane and Z.W. Grabowski, private communication.

85. K. Baker, G. Highland, A.V. Ramayya and J.H. Hamilton, Bull.Am.Phys.Soc.15(1970)1358.

86. M. Sakai, H. Ikegami, T. Yamazaki and K. Sugiyama, J. Phys. Soc. Japan 14 (1959) 938.

87. R.E. Sund, R.G. Arns, L.I. Yin and M.L. Wiedenbeck, Bull. Am. Phys. Soc. 6 (1961) 239.

88. R.L. Robinson, E. Eichler and N.R. Johnson, Phys. Rev. 128 (1962) 252.

89. I.A. Fraser, J.S. Greenberg, S.H. Sie, R.G. Stokstad, G.A. Barginyon and D.A. Bromley, Phys. Rev. Letters 23 (1969) 1047.

90. P. Debrunner and W. Kundig, Hel. Phys. Acta 33 (1960) 395.

91. R.W. Lide and M.L. Wiedenbeck, Phys. Rev. 113 (1959) 840.

92. W.K. Kesternich, J. Lange, R.L. Rasera, W. Schaffner, Verk. d. Deutsch. Phys. Gesellschaft 4 (1968) 335; W. Schaffner, Diplomarbeit Universität Bonn (1969).

93. F.K. McGowan, R.O. Sayer, P.H. Stelson, R.L. Robinson and W.T. Milner, Bull. Am. Phys. Soc. 13 (1968) 895.

94. L. Varnell, J.D. Bowman and J. Trischek, Nucl. Phys. A127 (1969) 270.

95. J.H. Hamilton, A.V. Ramayya and L.C. Whitlock, Phys. Rev. Letters 23 (1969) 1178; Phys. Rev., C3 (1971) 313.

96. C. Günther, G. Strube, U. Wehmann, W. Engels, H. Blumberg, H. Luig, R.M. Lieder, E. Bodenstedt, and H.J. Körner, Z. Phys. 183 (1965) 472.

97. J.M. Jaklevic, E.G. Funk and J.W. Mihelich, Nucl. Phys. A99 (1967) 83.

98. D.R. Bès, P. Federman, E. Maqueda and A. Zuker, Nucl. Phys. 65 (1965) 1.

99. G. Engler and S.A. Lane, Bull. Am. Phys. Soc. 15 (1970) 100.

100. J.J. Reidy, E.G. Funk and J.W. Mihelich, Phys. Rev. 133 (1964) B556.

101. J. Lange, R.L. Rasera and W. Schaffner, to be published.

102. F.K. McGowan, Radioactivity in Nuclear Spectroscopy, ed. J.H. Hamilton and J.C. Manthuruthil, Gordon and Breach, in press.

103. J.H. Hamilton, A.V. Ramayya and L.C. Whitlock, Phys. Rev. Letters 23 (1969) 1178. Data revised to include new measurement.

104. S. Ofer, Phys. Rev. 115 (1959) 412.

105. P.F. Kenealy, E.G. Funk and J.W. Mihelich, Nucl. Phys. A105 (1967) 522.

106. R. Vukanovic, M. Zupančic and H.A. Lindstrom, Ark. Fys. 39 (1969) 113.

107. N. Blumberg, K.-H. Speidel, H. Schlenz, P. Weigt, H. Hübel, P. Göttel, H.-F. Wagner and E. Bodenstedt, Nucl. Phys. A90 (1967) 65.

108. G. Kaye, Nucl. Phys. A108 (1968) 625.

109. D.R. De Boisblane, R.M. Bragger and R.L. Heath, IN - 1218, TID 4500, Idaho Nuclear.

110. A.W. Sunyar, Phys. Rev. 186 (1969) 1227.

111. W.T. Milner, Ph. D. Thesis, Univ. of Tennessee, 1968, ORNL - TM - 2121.

112. G.D. Hickman and M.L. Wiedenbeck, Phys. Rev. 118 (1960) 1049.

113. R.C. Williams, K.I. Roulston, Can. J. Phys. 34 (1956) 1087.

114. M. Zapančić, R. Vukanović and L. Samuelsson, Ark. Fys. 39 (1969) 313.

115. H.J. Körner, E. Gerdau, J. Heisenberg and J. Braunsfurth, ed. Karlsson, Mathis and Siegbahn, Perturbed Angular Correlations, North - Holland Publ. Co., Amsterdam, p. 200 (1964).

116. F.K. McGowan and P.H. Stelson, Phys. Rev. 123 (1961) 1274.

117. T. Yamazaki, Nucl. Phys. 44 (1963) 353.

118. V.R. Potnis, V.S. Dubey and C.E. Mandeville, Phys. Rev. 102 (1956) 459.

119. W.T. Milner, private communication, August 1970.

120. R.L. Robinson, F.K. McGowan, P.H. Stelson, W.T. Milner and R.O. Sayer, Nucl. Phys. A123 (1969) 193.

121. L. Samuelsson, R. Vukanović, M. Migahed, M. Zupančić, L.O. Edwardson and L. Westerberg, Nucl. Phys. A135 (1969) 657.

122. T. Yamazaki, Nucl. Phys. 61 (1965) 497.

123. A.H. Kukoć, L. Marinkov, D.M. Krmpotic and R. Vukanović, Bull. Boris Kidric Inst. 18 (1967) 9.

124. J. Mraz, Nucl. Phys. 4 (1957) 457.

125. W.D. Hamilton, Nucl. Phys. A136 (1969) 257.

126. D.M. Krmpotic, Ph. D. Thesis, Univ. of Belgrade (1968).

127. V.R. Potnis, Indian J. Phys. 30 (1956) 375; S.S. Malik, V.R. Potnis and C.E. Mandeville, Nucl. Phys. 11 (1959) 691.

128. H. Ikegami, K. Sugiyama, T. Yamazaki and M. Sakai, Nucl. Phys. 91 (1963) 130.

129. R.M. Steffen, Phys. Rev. 89 (1953) 665.

130. B.G. Pettersson, L. Holmberg, and T.R. Gerholm, Internal Conversion Processes, ed. J.H. Hamilton, Academic Press (1966) p. 473.

131. C.D. Schrader, Phys. Rev. 92 (1953) 928.

132. D. Schiff and F.R. Metzger, Phys. Rev. 90 (1953) 849.

133. H.W. Taylor and R. McPherson, Can. J. Phys. 39 (1961) 1235.

Discussion

Bloom: I have what I believe is half a query and half a comment. In analyzing angular distributions of gamma - rays in other kinds of experiments I have had the experience that what we do is use the angular distribution directly and get a χ^2 fit versus delta, the value of A_2 and the value of A_4. I am quite well aware that it is common, or used to be at least, to derive the value of A_2 and the value of A_4 from the angular distribution and then from that deduce a value of delta. These two procedures are by no means equivalent. I should like to know, if in your survey you have determined what fraction of the data was analyzed in the last mentioned way. Because I believe it is the first mentioned way which is the correct one.

J.H. Hamilton: In the analysis of the data for the A_2 and the A_4 coefficients?

Bloom: No, it is the analysis of the angular distribution as a function of delta where you plot the χ^2 versus delta very simply.

J.H. Hamilton: I think generally the measurements are such that one analyzes the data by fitting to A_2 and A_4 coefficients. This is not the procedure you are recommending.

Steffen: In view of the remark made by Dr. S. Bloom I would like to make the following comments. We always analyze our data in terms of a non - linear least square fit as a function of delta and in terms of A_{kk} (δ) from which then a value of delta is derived. In general, the extracted values of delta agree quite well. In some cases, however, the first method gives definitely a more reliable value of δ. It is quite clear, that the method of evaluating directional correlation data directly in terms of δ is to be preferred.

Glaudemans: Which conclusion can one draw from a theoretically calculated sign of an E2/M1 mixing ratio and is there any systematics involved in these signs?

J.H. Hamilton: The relative phase of the M1, E2 matrix elements.

Sakai: I would like to comment on the ratio E2/M1. About ten years ago, I did the same sort of compilation of E2/M1 in $2^+ \rightarrow 2^+$ transitions as Dr. Hamilton showed several minutes ago in his slides. It stimulated theoretical physicists. T. Tamura and H. Yoshida (published in Nuclear Physics) tried to understand the physical meaning of the sign and magnitude of δ. The calculation was based on the coupling of the second 2^+ state with low - lying two quasi - particle states. It turned out that the sign mainly depends on the sign of the magnetic moment in the orbit j.

Contributed Paper

DETERMINATION OF MIXING RATIOS IN ^{27}Mg BY γ - γ ANGULAR CORRELATION MEASUREMENTS WITH A MULTI - COUNTER GONIOMETER

G. Costa and F.A. Beck

Laboratoire des Basses Energies respectively Laboratoire de Physique Nucléaire et Physique des Particules, C.R.N., Strasbourg 3, France

Since 1967, ^{27}Mg has received a relatively high amount of attention both theoretically and experimentally. Most of the main results on branching ratios, multipole mixing parameters of the gamma - ray transitions and spins of levels in ^{27}Mg have been obtained by using method II of Litherland and Ferguson, through the ^{26}Mg (d, p) reaction[1,2,3,4]) but the mixing ratio parameter for the 984 keV γ-ray from the first excited state could not be deduced. Its determination should provide a fairly sensitive test of the collective model[5]). Lacambra et al.[5]) by a P.W.B.A. analysis and Wagner[4]) by a D.W.B.A. analysis, found two different values for this mixing - ratio parameter. Anyway these kind of calculations are poor approximations. Therefore, it was our purpose to obtain the value of this mixing - ratio by measuring γ - γ angular correlations with a multi - counter goniometer and γ angular distributions with a Ge(Li) detector.

The multi - counter goniometer[6]) was set in a spatial configuration of five NaI (Tl) 3″ X 3″ crystals and one Ge(Li) (27 c.c.) detector. Three NaI counters plus the Ge(Li) were in the horizontal plane and two NaI counters in the vertical plane. Each NaI detector was at a distance of 14 cm from the target, while the front of the Ge(Li) was at 4 cm from the target. The experimental arrangement of the five counters made the A1, A2 and D geometries as defined by Ferguson[7]) possible. In fact, the use of the multi - counter goniometer gave us the opportunity to analyze the triple angular correlation function W $(\theta_1, \theta_2, \phi)$ in terms of θ_1, θ_2 and ϕ. The efficiency of the different coincidence circuits was the same and so, the normalization for all the counters was obtained from the counting rate of the single gamma - rays. This was checked by measuring the angular correlation of the 1.17 and 1.33 MeV γ - rays from ^{60}Co source placed at the target position. The Hoogenboom method was used. A window was set in the Ge(Li) and five NaI spectrometers around the sum energy. The ^{26}Mg (d, p) ^{27}Mg reaction has been used at a bombarding energy of 2 MeV deuterons (10nA); the beam was produced in a 3 MeV Van de Graaff accelerator. The target was a foil of ^{26}Mg of 3 mg/cm^2 thickness.

The angular distributions of the gamma - rays emitted in the ^{26}Mg (d, p) ^{27}Mg reaction have been first measured with a moving Ge(Li) detector (27 c.c.) and a fixed one (30 c.c.). Assuming that the gamma transition depopulating the 1939 keV state to the ground state is a pure E2 (in agreement with our lifetime measurements), we were able to extract the statistical tensors of the aligned state from the angular distribution of the 1939 keV transition. Therefore, the values of the angular correlation of the 955 keV - 984 keV radiations are used, together with the angular distribution of the 955 keV γ - ray, to determine the values of the mixing - ratio parameters, using the statistical tensors of the 1939 keV state previously determined.

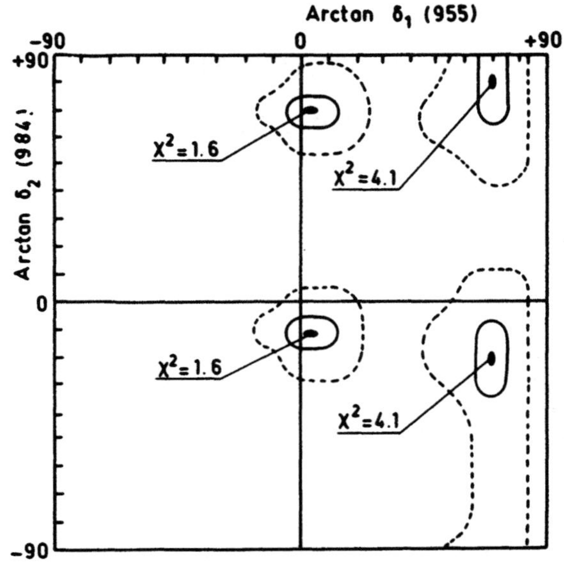

Fig. 1. The dashed and solid curves represent the isobaric lines for χ^2 values of 500 and 100 respectively. Regions outside the area defined by the dashed curves are those of χ^2 values higher than 500. The points in the plots are the regions of χ^2 minimum.

A χ^2 analysis versus the two multipole mixing parameters δ_1 (955 keV) and δ_2 (984 keV) show four regions of minima for the spin sequence $5/2^+ \rightarrow 3/2^+ \rightarrow 1/2^+$ (fig. 1). Only two of them give an acceptable probability; they correspond to the following values of the mixing ratio:

$$\delta_1(955 \text{ keV}) = +0.05 \pm 0.04$$

$$\delta_2(984 \text{ keV}) = -0.22 \pm 0.02 \quad \text{or} \quad 3.15 \begin{array}{l} +0.20 \\ -0.15 \end{array}$$

Our lifetime measurements[8]) allow us to discriminate between the two values of the 984 keV mixing ratio. The lowest and highest value of the mixing ratio would correspond to an E2 enhancement of 5 and 109, respectively. Only the value of $\delta = -0.22 \pm 0.02$ seems to give an acceptable E2 transition enhancement factor (ref. 9). Detailed results and discussion will be published elsewhere[8]).

We are indebted to dr. Gallmann for putting the 27 cc Ge(Li) detector at our disposal.

References

1. P.M. Endt and C. van der Leun, Nucl. Phys. A105 (1967).
2. M.A. Eswaran, M. Ismail and N.L. Ragoowansi, Phys. Rev. 185 (1969) 1458.
3. D.H. Sykes, W.C. Olsen, D.A. Hutcheon, C.W. Vermette, D.M. Sheppard, B.D. Sowerby and P.J. Twin, Nucl. Phys. A135 (1969) 335.
4. P. Wagner, Thesis, University of Strasbourg (1968).
5. J.M. Lacambra, D.R. Tilley and N.R. Roberson, Nucl. Phys. A92 (1967) 30.
6. S. Gorodetzky, F.A. Beck, P. Engelstein, T. Byrski and A. Knipper, Nucl. Inst. Meth. 69 (1969) 169.
7. A.J. Ferguson, Angular Correlation Methods in Gamma - Ray Spectroscopy, ed. North - Holland Publishing Company, 1965.
8. G. Costa and F.A. Beck, to be published.
9. S.J. Skorka, J. Hertel and T.W. Retz - Schmidt, Nucl. Data 2 (1966).

Contributed Paper

STUDIES OF THE PROPERTIES OF ^{29}P LEVELS

F.A. Beck, T. Byrski, P. Engelstein, A. Knipper and S. Gorodetzky
Laboratoire de Physique Nucléaire et de Physique des Particules –CRN – Strasbourg 3, France

The ^{29}P nucleus has been studied by several authors[1]) in different reactions. But many properties of the electromagnetic transitions and even the spins of many levels have not been determined. Such measurements are interesting in order to check the different nuclear models. In a first step the ^{29}P nucleus has been considered as a spherical ^{28}Si inert core plus a single proton. Other models, more interesting, have been developed in the weak coupling representation by assuming a deformation of the ^{28}Si core[2]). Strong coupling models[3,4]) are also successfully applied in this region. Therefore, new experimental data must give some information on these theoretical considerations.

The ^{28}Si (p, γ) ^{29}P reaction has been used to produce the ^{29}P levels like in our previous experiments[5]). Radiative proton capture by even - even target nuclei is a powerful tool for the measurement of nuclear spins and of γ - ray mixing ratios, due to the high degree of alignment that is achieved; usually, the measurement of the angular distribution of the γ - rays is sufficient to determine these parameters, especially with detectors having good energy resolution. In our experiments, solid state Ge(Li) detectors of 27 cc, 46 cc and 85 cc have been used. It can happen that one or more angular distributions of the γ - rays cannot be measured because of target contaminations or because of a weak transition being obscured by a strong one. In such cases γ - γ angular correlation measurements may solve the problem. To do this we use a multi - counter goniometer (six NaI detectors 3" X 3") permitting the simultaneous measurement at five geometries of Litherland and Ferguson or 22 angles of the triple angular correlation function W (θ_1, θ_2, ϕ). Another reason for measuring γ - γ angular correlations with a multi - counter goniometer can be found in the fact that, for the same measuring time, the quantity of data one gets is much greater than in a measurement of γ - ray angular distributions with a single Ge(Li) detector. Energy and intensity measurements of γ - rays, which are of particular importance, have been made at each resonance. The reaction is also well suited for lifetime measurements with the Doppler Shift Attenuation Method (D.S.A.M.). Isotopic ^{28}Si targets of 50 μg, 100 μg and 300 μg, furnished by A.E.R.E. Harwell, have been used. The bombarding beam, 3 to 30 μA, was obtained from Van de Graaff accelerators (2 and 3 MeV).

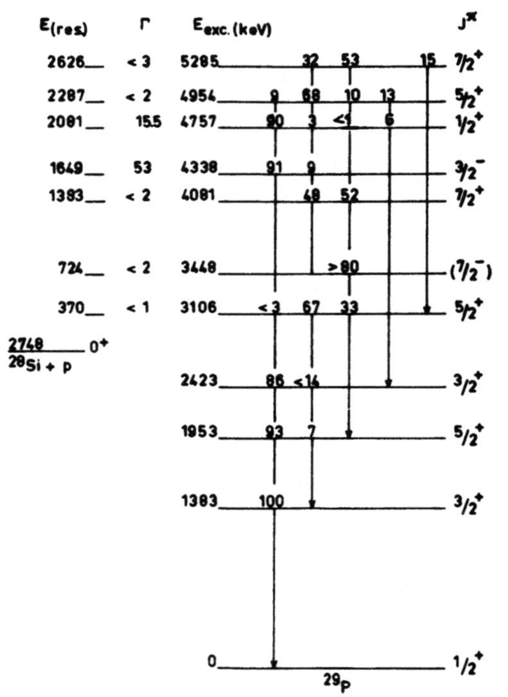

Fig. 1. Disintegration level scheme of ^{29}P nucleus obtained in the ^{28}Si (p, γ) reaction.

The results obtained at the different resonances are summarized in the fig. 1.

Resonance at E_p = 2081 keV

Angular distribution measurements in the ^{28}Si (p, γ) reaction, have made it possible to assign $J^{\pi} = 1/2^+$ to the 4757 keV level. This resonance is strongly perturbed and could not be analyzed[6]). A weak transition has been found to the level at 2423.5 ± 0.3 keV, whose lifetime has been determined to be $\tau = 44^{+7}_{-8}$ fs.

Resonance at E_p = 2287 keV

This resonance previously indicated at E_p = 2300 ± 15 keV has been found at E_p = 2287 ± 2 keV. The resonance has been checked as being well isolated. The γ - ray angular distributions and γ - γ angular correlations determine uniquely the spin $J^{\pi} = 5/2$ of the level at 4954 keV. This result and the lifetime measurement, $\tau = 24^{+5}_{-4}$ fs, show that the 4954 keV level decays essentially to the first three excited levels by M1 transitions and to the ground state by an E2 transition. The value of the spin of the 2423.5 keV level has also been uniquely determined to be 3/2 and the mixing ratio of the γ - transition decaying from this level to the ground state is found to be $\delta = -0.28 \pm 0.05$. These results are in agreement with those recently obtained from the ^{32}S (p, α) ^{29}P reaction[8]).

Resonance at E_p = 2627 keV

This resonance previously given at E_p = 2640 keV has been determined to be at E_p = 2627 ± 2 keV. This resonance ($\Gamma < 2$ keV) has been extracted from the large resonance at E_p = 2880 keV (Γ = 425 keV) by using thinner targets. Measurements at the resonance and at each side of it show that the 5285 keV resonant level decays essentially to the 1953 keV, 1383 keV and the 3106 keV levels. The lifetime τ = 18 ± 6 fs and the feeding to the 1383 keV level indicate a positive parity for the 5285 keV level.

Resonance at E_p = 724 keV

A level at an excitation energy of 3446 ± 16 keV has been observed in ^{28}Si (d, n) and ^{32}S (p, α) reactions but has not previously been observed in the ^{28}Si (p, γ) reaction. With a Ge(Li) detector of 85 cc at 1.5 cm from the target and a beam of 30 μA we measured this resonance at E_p = 724 ± 2 keV. The level of 3447 ± 2 keV is found to decay mainly (> 80%) to the 1953 keV level. These results involve the probability of a spin value J = 7/2 for this level. It would be the isobaric analogue of the $J = 7/2^-$ 3623 keV level in ^{29}Si.

Measurements in the ^{28}Si (d, n) ^{29}P reaction are in progress in order to measure directly the lifetimes of the low lying excited states. The results of all these measurements will give us more information on the structure of the ^{29}P nucleus.

The authors are very grateful to Professor D. Magnac for putting the 85 cc Ge(Li) detector at their disposal.

References

1. P.M. Endt and C. van der Leun, Nuclear Physics A105 (1967) 1.
2. B. Castel, K.W.C. Stewart and M. Harvey, Intermediate - coupling Structure of Odd - A Nuclei in the 2s - 1d shell, University of Kingston - Canada.

3. H. Ejiri, Nucl. Phys. 52 (1964) 578.
4. F.B. Malik and W. Scholtz, p. 34 in Nuclear Structure (1967), ed. North - Holland Publ. Co., Amsterdam.
5. S. Gorodetzky, F.A. Beck, P. Engelstein, T. Byrski and A. Knipper, Comptes Rendus Acad. Sc. 270 (1970) 169.
6. S. Gorodetzky, F.A. Beck, P. Engelstein, T. Byrski and A. Knipper, Nucl. Inst. Meth. 69 (1969) 169.
7. J.O. Newton, Nucl. Phys. 21 (1960) 529.
8. J.S. Forster, C. Broude and W.G. Davies, C.R.N.L., Chalk River, private communication.

Contributed Paper

DETERMINATION OF THE MIXING RATIO IN THE DECAY OF THE LONG - LIVED 1.612 MeV LEVEL IN ^{37}Ar†

P. Taras, J. Matas*, A.Turcotte and R. Vaillancourt
Laboratoire de Physique Nucléaire, Université de Montréal, Canada

Rosner and Schneid[1]) have shown that the ^{36}Ar (d, p) ^{37}Ar reaction populates the 1.612 MeV level in ^{37}Ar via l_n = 3 stripping. This measurement implies a spin and parity of $J^\pi = 5/2^-$ or $7/2^-$ for this level. They assigned $J^\pi = 7/2^-$ using shell - model arguments. Goosman and Kavanagh[2]) have measured the mean lifetime of this level to be $\tau = 5.15 \pm 0.7$ ns, using a recoil - Doppler - shift method. They concluded that the 1.612 MeV level has $J^\pi = 7/2^-$ on the basis of transition probability arguments. In this paper, we report on the determination of the mixing ratio of the gamma - ray transition from the 1.612 MeV level to the $3/2^+$ ground state of ^{37}Ar.

Because of the long lifetime of the 1.612 MeV level, the angular distribution or correlation of its decay gamma - rays can be expected to be perturbed. Consequently, the common angular correlation techniques, such as method II of Litherland and Ferguson, cannot be used to yield useful information on the nature of the gamma - ray transitions. However, both the angular distribution and the linear polarization of the gamma - rays can be expressed in terms of statistical tensors (see. e.g., McCallum[3])). Thus, when the angular distribution and the linear polarization are measured simultaneously, the polarization parameter p can be expressed directly in terms of the angular distribution coefficients, thereby eliminating the statistical tensors and the effect of perturbation, if any.

We have done such measurements at incident proton energies of 3.98 and 4.38 MeV, using the ^{37}Cl (p, n) ^{37}Ar reaction and an enriched (99.3%) AgCl target. The angular distribution measurement was performed with a 30 cc Ge(Li) detector and the linear polarization was measured with a NaI (Tl) Compton polarimeter[4]). The results of the measurements on the 1.612 $(7/2^-) \rightarrow 0$ $(3/2^+)$ MeV transition are shown in table 1.

Table 1

Results of angular distribution and linear polarization measurements

E_p (MeV)	observed distribution coefficients		observed polarization	mixing ratio
	$a_2 \pm \Delta a_2$	$a_4 \pm \Delta a_4$	$p = J_0/J_{90}$	δ (M2E3)
3.98	0.12±0.01	-0.02±0.01	0.49 ± 0.04	-0.21 ± 0.02
4.38	0.16±0.01	-0.02±0.01	0.44 ± 0.02	-0.21 ± 0.02

The values of the mixing ratios δ, given in table 1, were obtained from the overlap of the experimental value of p and the theoretical value of p calculated as a function of δ using the observed angular distribution coefficients. Fig. 1 shows such a plot for a $7/2^-$ to $3/2^+$ transition, obtained at E_p = 3.98 MeV.

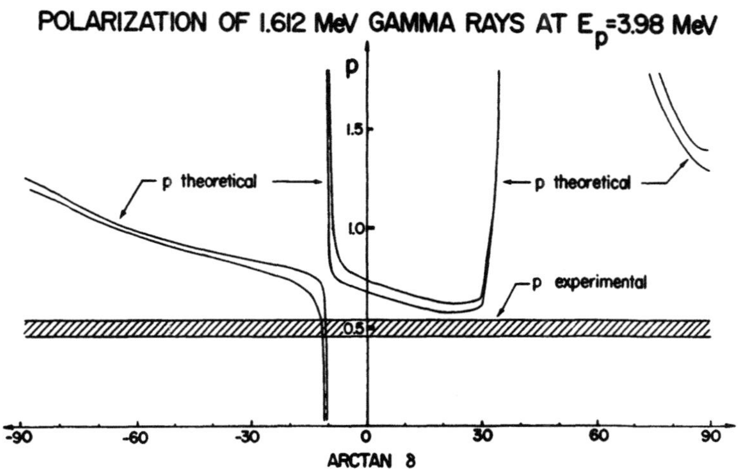

Fig. 1. Polarization parameter p as a function of the mixing ratio δ for the spin sequence $7/2 \rightarrow 3/2$ and M2E3 radiation.

From the gamma - ray spectrum obtained with the Ge(Li) detector, the energy of the level was found to be 1611.7 ± 0.8 keV, in agreement with the value of 1611 ± 1 keV found by Goosman and Kavanagh[2]). A possible transition to the $1/2^+$, 1.410 MeV level is estimated to be less than 3% of the transition to the ground state.

The properties of the 1.612 MeV level in ^{37}Ar are quite similar to the properties of the 3.162 MeV level[4]) in ^{35}Cl and the 3.105 MeV level[5]) in ^{37}Cl. These are given in table 2. All three levels have $J^{\pi} = 7/2^{-}$ and seem to arise from a $lf_{7/2}$ single - particle excitation[6]).

Table 2

Properties of $lf_{7/2}$ levels in ^{35}Cl, ^{37}Cl and ^{37}Ar

E_x (MeV)	lifetime τ (nsec)	mixing ratio δ	transition strengths (Wu) M2	E3
^{35}Cl 3.162	0.060 ± 0.007	-0.25 ± 0.04	0.19	6.5
^{37}Cl 3.105	$>$ 0.007[a)]	-0.18 ± 0.01	$<$ 1.9	$<$ 34
^{37}Ar 1.612	5.15 ± 0.7	-0.21 ± 0.02	0.07	6.7

a) Duncan et al., Phys. Rev. 185 (1969) 1515.

† Partially supported by the National Research Council of Canada.

* Permanent address: Nuclear Research Institute, CSAV, Rez, Czechoslovakia.

References

1. B. Rosner and E. Schneid, Phys. Rev. 139 (1965) B66.
2. D.R. Goosman and R.W. Kavanagh, Phys. Lett. 24B (1967) 507.
3. G.J. McCallum, Phys. Rev. 123 (1961) 568.
4. P. Taras and J. Matas, Can. J. Phys. 48 (1970) 603.
5. Hyder et al., Phys. Rev. 169 (1968) 899.
6. D.D. Watson and F.D. Lee, Phys. Lett. 25B (1967) 472.

Contributed Paper

863.6 - 810.6 keV γ - γ DIRECTIONAL CORRELATION IN ^{58}Fe

N.C. Singhal, A.V. Ramayya and J.H. Hamilton
Vanderbilt University, Physics Department†, Nashville, Tenn., U.S.A.
and
S. Raman
Oak Ridge National Laboratory, Oak Ridge, Tenn., U.S.A.*

1. Introduction

The decay scheme of ^{58}Co is particularly simple[1]) (fig. 1). The $\gamma\gamma$ directional correlation of the 863.6 - 810.6 keV cascade from the decay of ^{58}Co was measured by several authors[2-6]) with NaI (Tl) detectors. What should be, in principle, straight forward measurements are rendered difficult by the low intensity of the 863.6 keV γ - ray and by the proximity of the 863.6 keV γ - ray to the very intense 810.6 keV γ - ray. Nevertheless, reasonably consistent results have been obtained in five different measurements with ^{58}Co sources as shown in table 1.

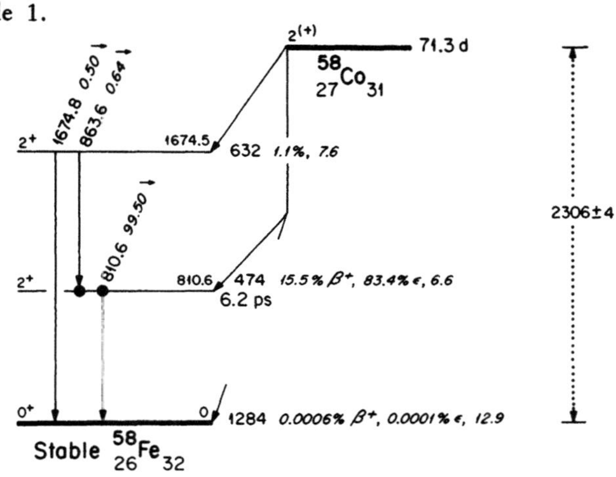

Fig.1. Decay scheme of ^{58}Co.

The 863.6 - 810.6 keV $\gamma\gamma$ - directional correlation has also been studied by Fanger et al.[7]) as part of a comprehensive $\gamma\gamma$ (θ) study following thermal neutron capture by ^{57}Fe. The A_2 and A_4 coefficients obtained by Fanger et al.[7]) (see table 1) differ markedly from the values obtained by the other groups (refs. 2 - 6). This discrepancy prompted the present study.

Unlike the situation in the ^{58}Co decay, the 863.6 keV γ - ray is strong [I (863.6) / I (810.6) = 0.27] in the ^{57}Fe (n, γ) reaction[7]). This advantage is partly offset by the fact that the 863.6 keV γ - ray observed in the (n, γ) studies can be placed at more than one location in the complex level structure of ^{58}Fe. However, Fanger et al.[7]) have shown through coincidence measurements that the 863.6 keV γ - ray located between the 1674.5 and 810.6 keV levels accounts for 99% of this γ - ray. As a further check, Schmidt et al.[8]) have studied the 8368 keV - (863.6 + 810.6) keV $\gamma\gamma$ angular correlation where the 8368 keV γ - ray is the primary capture γ - ray to the 1674.5 keV level. These results are consistent with the earlier work[7]) from (n, γ) studies. To resolve this discrepancy, we measured the 810.6 - 863.6 keV $\gamma\gamma$ (θ) from the decay of ^{58}Co with a NaI - Ge(Li) detector system.

TABLE 1

Source	A_2	A_4	Reference
^{58}Co decay	+0.39±0.07	+0.23±0.10	MacArthur et al.[2]
	+0.38±0.04	+0.24±0.05	Malmskog[3]
	+0.35±0.08	+0.23±0.05	Rama Mohan et al.[4]
	+0.37±0.015	+0.24±0.025	Vikram Singh et al.[5]
	+0.42±0.03	+0.17±0.025	Schotzig et al.[6]
$^{57}Fe(n,\gamma)$	+0.479±0.009	+0.081±0.013	Fanger et al.[7]
	810.6-863.6 keV		
^{58}Co decay	+0.51±0.03	+0.09±0.04	Present Work
	863.6-810.6 keV		
^{58}Co decay	+0.36±0.06	+0.14±0.07	Present Work

2. Experimental Procedure

The gamma - gamma directional correlation was measured with a 50.4 cm^3 Ge(Li) detector (at 1.33 MeV FWHM = 2.4 keV; peak to Compton ratio 30/1; efficiency of 10% of a 3 in.X 3 in. NaI(Tl) detector) and a 2 in.X 2 in. NaI(Tl) crystal, which was set on the composite 0.810 - 0.860 MeV gamma - rays and was rotated through a cycle of 90°, 135°, 180°, 225° and 270°. The coincidence spectrum and the chance coincidence spectrum at the above angles were stored simultaneously in different portions (2048 channels for each) of the system memory. An external fast coincidence unit (resolving time = 60 ns) opened the linear gate of the analyzer. Two cycles of data were taken. The source of ^{58}Co, in liquid form, was placed in a plastic vial and the rotating detector centered about it. The distances of the NaI detector and the Ge(Li) detector were 7 cm and 4.5 cm, respectively.

3. Data Reduction

The data from corresponding angles in a cycle and the data of the two cycles were summed, corrected for chance coincidences and normalized to the number of gating events in the NaI detector. In each of the cycles,data were recorded at each angle for a 23 - hr period. The data were fitted to the correlation function $W(\theta) = 1 + A_2P_2(\cos\theta) + A_4P_4(\cos\theta)$ and solid angle corrections[9,10]) were applied. Coefficients were obtained in two ways; one case from the 860 keV transition as seen in the Ge(Li) spectrum in coincidence with the 810 keV transition in the gate and then from the 860 (gate)-810 (spectrum) data. The results from both analyses are given in table 1.

4. Results and Discussion

The coefficients from the 810 (gate) - 860 (spectrum) analysis are in very good agreement with the results from the (n, γ) studies while the coefficients from 810 (spectrum) - 860 (gate) data are lower. The errors in the latter case, however, are sufficiently large to be consistent with either the (n, γ) or the other ^{58}Co results. The errors are large because approximately 80% of the counts in 810 keV peak in the Ge(Li) spectrum are from chance events. Our 810 keV gate 860 keV γ - ray correlation data are consistent with a δ value of 0.57 ± 0.07 whereas the older correlation data obtained[2-7]) with radioactive sources yield an average δ value of 1.5 ± 0.2.

The good agreement of the coefficients from our accurate results with those from the (n, γ) work indicate that the older results obtained with two NaI detectors have systematic errors that are not present in our 810 (gate) - 860 (spectrum) or the (n, γ) work. At present one can only speculate on such effects. For example, there may be a second 810 keV transition in coincidence with the $2^+ \to 0^+$, 810 keV transition or it may be that summing with annihilation radiation from the positron feeding to the 810 keV level alters the correlation data. A measurement of accidentals would not determine the latter which would depend on real coincidences of 511 - 810 keV with an accidental sum with the 511 keV photon. In any case these data point up the fact that even in simple decays some older results obtained with two NaI detectors may be in error.

† Research supported in part by a grant from the National Science Foundation.

* Research sponsored by the U.S. Atomic Energy Commission under contract with Union Carbide Corporation.

References

1. Nuclear Data Sheets, Vol. 3, Nos. 3 - 4 (1970).
2. D. MacArthur, R. Goodman, A. Artna, M.W. Jones, Nucl. Phys. 38 (1962) 106.
3. S. Malmskog, Nucl. Physics 51 (1964) 690.
4. R.V. Ramamohan, K.V. Reddy, B.B.V. Raju, Swami Jnanananda, Indian J. Pure Appl. Phys. 4: 420 - 2 (Nov. 1966).
5. V. Singh, P.N. Tandon, S.H. Devare, H.G. Devare, Nucl. Phys. A137 (1969) 278.
6. U. Schötzig, H. Schrader, R. Stippler, F. Munnich, Z. Physik 222 (1969) 479.
7. U. Fanger, W. Michaelis, H. Schmidt and H. Ottmar, Nucl. Phys. A128 (1969) 641.
8. H. Schmidt, W. Michaelis, U. Fanger, Nucl. Phys. A136 (1969) 122.
9. M.J.L. Yates, Alpha-, beta- and gamma-ray Spectroscopy, ed. K. Siegbahn, North - Holland Publ. Co., Vol. 2 (1965) 1691.

10. D.C. Camp, private communication.

Contributed Paper

ANGULAR DISTRIBUTION MEASUREMENTS IN THE (γ, γ') REACTION ON VARIOUS ISOTOPES

R. Moreh, A. Nof and O. Shahal
Nuclear Research Centre - Negev, Beer Sheva, Israel

The technique of employing neutron capture γ - rays for photoexciting high - energy nuclear levels (5 to 9 MeV) to study the spins and parities of resonance and low - lying levels by the (γ, γ') reaction is relatively new[1,2]).

The basic idea[3]) in this work is that a resonance event is obtained only when a random overlap exists between at least one of the incident γ - lines and one level in any isotope of the element studied. This condition places a severe limitation on the choice of the target; nevertheless it was possible to study about 15 isotopes using the present technique. In the present work, two examples, namely ^{74}Ge and ^{139}La, are discussed. Because of the thermal motion of the nuclei of the γ - source and the target, the shapes of the incident γ - lines and the resonance levels are Doppler broadened; the resonance scattering cross section is proportional to the overlap integral of the two lines.

The experimental assembly[1]) (fig. 1) consisted of an (n, γ) source containing several kilograms of iron or titanium discs, placed along a tangential beam tube of the IRR - 2 research reactor at a thermal neutron flux of 2 x 10^{13} n/cm^2.s. The resulting γ - beam was collimated (angular spread of $\sim 1°$), neutron filtered and allowed to hit an 11 - cm diameter metallic target (10 g/cm^2 thick). Typical γ - beam

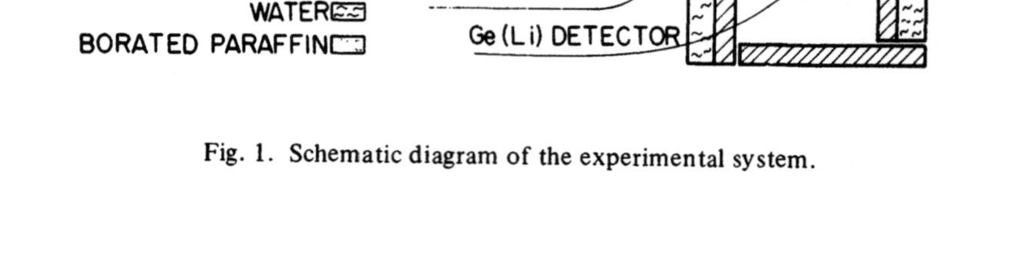

Fig. 1. Schematic diagram of the experimental system.

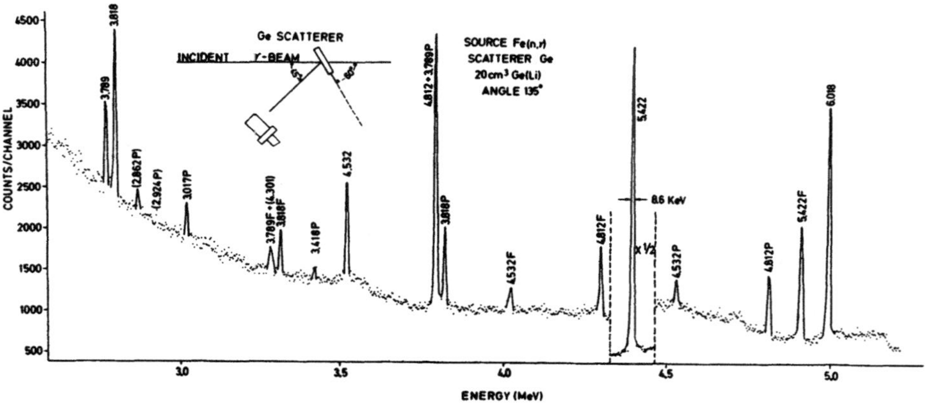

Fig. 2. Scattered radiation spectrum from a Ge target. P and F refer to photopeaks and first escape peaks; other lines refer to double escape peaks.

intensities are of the order of 10^8 monoenergetic photons / cm^2 .s on the target. The width of a γ - line of 7 MeV from an Fe (n, γ) reaction is ~ 20 eV, at a source temperature of 380° C.

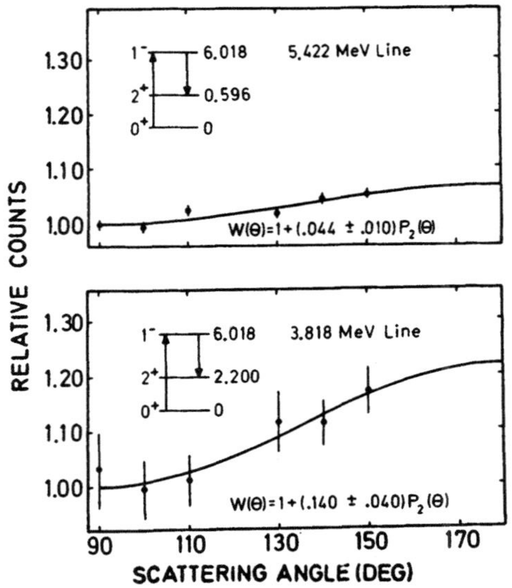

Fig. 3a. Angular distribution of the 3.818 and 5.422 MeV lines in ^{74}Ge measured by a 20 cc Ge(Li)-detector.

Fig. 2 shows a resonantly scattered spectrum from a Ge target as measured by a 20 cc Ge(Li) detector using an Fe (n, γ) source. Besides the 6018 keV line, the figure shows other lines corresponding to inelastic transitions to low - lying levels in one of the Ge isotopes. In this case the scattering isotope was identified and found to be ^{74}Ge.

For angular distribution measurements, the detector was mounted on a horizontally rotating arm at a distance of 50 cm from the target. Corrections for the nuclear and atomic absorption of the incident beam and the atomic absorption of the scattered beam in the target were introduced. In an elastic resonance scattering, one deals with a cascade of the form $J_0 \rightarrow J_r \rightarrow J_f$ where J_0, J_r and J_f are the spins of the ground-state resonance level and low - lying levels, respectively. Assuming that the first component of the cascade is pure and that the other component is either a pure or a mixed dipole - quadrupole transition, the theoretical distribution is given by

$$W(\vartheta) = A_0 + A_2 A_2' P_2(\cos\vartheta) \propto 1 + AP_2(\cos\vartheta)$$

where $A_0 = 1 + x^2$; $A_2 = a + 2bx + cx^2$; x is the amplitude of quadrupole - dipole mixing ratio, and a, b, c and A_2' are the corresponding F - coefficients.

Some experimental distributions of the elastic and of some strong inelastic scattered lines in ^{74}Ge and ^{139}La are shown in fig. 3 and fig. 4; the curves are least square fits of the form $W(\theta) = 1 + AP_2(\cos\theta)$. By comparing these values of A with those of theoretical distributions, the spins of the corresponding levels were deduced. In addition, the quadrupole - dipole mixing amplitudes x were obtained from a plot

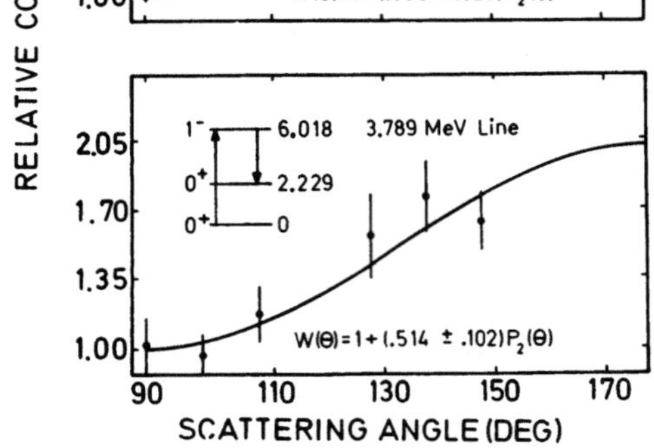

Fig. 3b. Angular distribution of the 3.789 and 6.018 MeV lines in ^{74}Ge measured by a 20 cc Ge(Li)-detector.

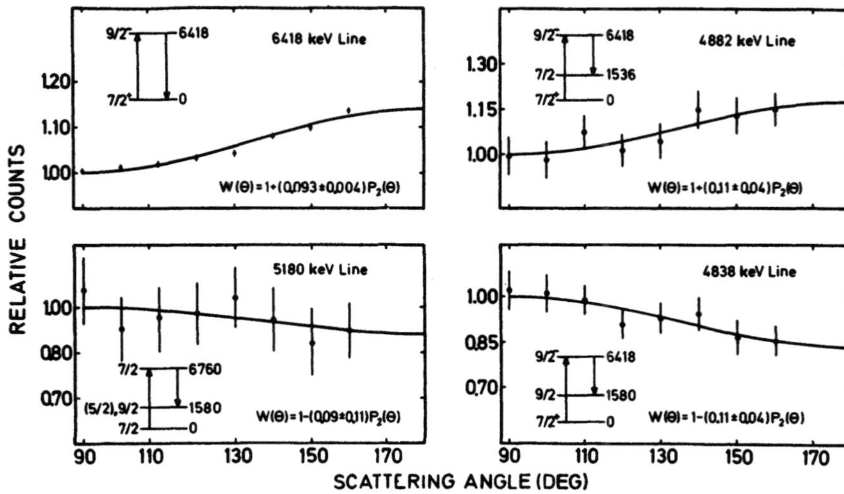

Fig. 4. Angular distribution of some scattered γ lines from a La target. The γ source is obtained from Ti (n, γ).

of the theoretical values of A as a function of x, after comparing with the experimental values of A.

The polarization of the elastic and inelastic scattered radiation was measured using a Compton polarimeter[4]). In this way the parities of some levels were determined. The following spin and parity determinations in ^{74}Ge and ^{139}La were made:

^{139}La: 0.166 ($5/2^+$), 1.384 (7/2), 1.536 (7/2), 1.580 (9/2), 6.018 ($7/2^-$), 6.418 ($9/2^-$), 6.760 (7/2).

^{74}Ge: 0.596 (2^+), 1.206 (2^+), 1.486 (0^+), 2.200 (2^+), 2.229 (0^+), 6.018 (1^-). In ^{74}Ge, the parities of only the 0.596 and 6.018 MeV levels were directly measured; the parities of the other levels were derived from the radiation strengths of the transitions[5]).

In the case of ^{139}La, the M2/E1 mixing ratio in the 6418 keV transition was found to be $x^2 < 2.5 \times 10^{-5}$; the results of other transitions were less precise and it was not possible to obtain x - values with good accuracy. In ^{74}Ge, a high value of the M2/E1 mixing ratio in the 3.818 MeV transition leading to the 2.200 MeV level was obtained: $x^2 \sim 1.7 \times 10^{-2}$. The enhancement factor of the M2 strength, over the single - particle Moszkowski estimate, was found to be 125. The absolute E1 strength (ref. 5) of the 3.818 MeV transition was found to have a normal value for this region of relatively high - energy transitions; the E1 strength was obtained by measuring the total radiative width of the 6.018 MeV resonance level[1]).

In conclusion, it appears that precise angular distribution measurements can provide valuable information concerning M2 transition strengths.

References

1. R. Moreh and A. Wolf, Phys. Rev. 182 (1969) 1236.
2. N. Shikazono and Y. Kawarasaki, Nucl. Phys. A118 (1968) 114.
3. G. Ben - David et al., Phys. Rev. 146 (1966) 852.
4. R. Moreh and M. Friedman, Phys. Letters 26B (1968) 579.
5. G.A. Bartholomew, Ann. Rev. Nucl. Sci. 11 (1961) 259.

Contributed Paper

GAMMA - GAMMA DIRECTIONAL CORRELATION MEASUREMENTS IN ^{124}Te†

Z.W. Grabowski, K.S. Krane and R.M. Steffen
Purdue University, Lafayette, Indiana, U.S.A.

The levels of ^{124}Te have recently been reinvestigated by Auer et al.[1]) and Meyer et al.[2]) from the decay of ^{124}Sb and by Ragaini et al.[3]) from the decay of ^{124}I. The low excited states of ^{124}Te exhibit a 'vibrational' pattern which could be expected for this nucleus (proton number 52 and neutron number differing by only 10 from the closed neutron shell). The measurements of Warner and Draper[4]) show, however, that the 4^+ level at 1249 keV could also be considered as a member of the ground state quasi - rotational band. The 723 keV transition proceeds between the first and the second excited 2^+ levels which would correspond to one and two phonon excitation respectively in the vibrational picture. The results of the measurements of the mixing ratio, δ_{723}, for this transition are conflicting.* From our previous measurements[6]) of β - γ directional correlation in the decay of ^{124}Sb we were able to calculate the mixing ratio of the 723 keV gamma transition. This measurement gave $-2.8 \leqslant \delta_{723} \leqslant -1.0$. The β - γ directional correlation measurement by Paul[7]) was not precise enough to distinguish between δ_{723} - values in the region $-5.0 \leqslant \delta_{723} \leqslant -0.7$. There is a mistake in the interpretation of the 1368γ - (723γ) - 603γ directional correlation result obtained by Lindquist and Marklund[8]). Their experimental value, $A_{22} = -0.027 \pm 0.008$, measured for this cascade gives $0.45 \leqslant |\delta_{723}| \leqslant 0.70$ and not $|\delta| \approx 1$, as stated in their paper. The above mentioned results are less precise than the value $\delta_{723} = -4.1 \pm 0.6$ obtained by Glaubman and Oberholtzer[9]) from the β - γ - γ directional correlation measurement. The γ - γ measurements by Dorikens - Vanpraet et al.[10]) performed with scintillation detectors involve large corrections due to unresolved cascades. They favour $\delta_{723} \approx -2.4$. The measurement of Stelson[11]) which

* All the δ values ($\delta = \frac{\langle \| L+1 \| \rangle}{\langle \| \ L \ \| \rangle}$) are defined in agreement with the notation proposed by Krane and Steffen[5]).

was performed using a Ge(Li) - NaI (Tl) detector combination yielded $\delta_{723} = -3.4 \pm 0.6$.

Our measurements of the 723γ - 603γ directional correlation were performed using a combination of coaxial Ge(Li) - NaI (Tl) and two coaxial Ge(Li) detectors. Our aim was to resolve the discrepancies among the different measurements and to improve the accuracy of the measured mixing ratio of the 723 keV transition.

The coincidence rate was determined at seven different angles, corrected for chance coincidences and for coincidences caused by competing cascades. The directional correlation coefficients A_{22} and A_{44} were calculated by fitting the experimental data to the directional correlation function in the form $W(\theta) = 1 + A_{22} P_2 (\cos\theta) + A_{44} P_4 (\cos\theta)$ with $A_{kk} = A_k(1) A_k(2)$, where 1 and 2 stand for the first and second transitions in a cascade. Correction for the finite solid angle of the detectors was subsequently applied[12]). The energy resolution of the Ge(Li) detectors was better than 3.5 keV and allowed a complete separation of the 723 keV and 714 keV lines (see figure). A least - squares fit of the Ge(Li) - NaI (Tl) data for the 723 keV - 603 keV cascade gave $W(\theta) = 1 + (0.152 \pm 0.008) P_2 (\cos\theta) + (0.304 \pm 0.012) P_4 (\cos\theta)$, and the result of the Ge(Li) - Ge(Li) measurement for the same cascade was $W(\theta) = 1 + (0.130 \pm 0.015) P_2 (\cos\theta) + (0.295 \pm 0.020) P_4 (\cos\theta)$. The weighted average for the E2/M1 mixing ratio obtained from these two measurements is

$$\delta_{723} = -3.4 \pm 0.1$$

This value is in good agreement with the value obtained by Stelson[11]).

Using our value for δ_{723} and Sliv's conversion coefficients[15]) one can calculate the K - conversion coefficient for the 723 keV transition in ^{124}Te as $\alpha_k(723) = 2.7 \times 10^{-3}$. The agreement between this calculated value and the measurements reported in refs. 3 and 16 is excellent. It is also interesting to compare the value of δ_{723} with the mixing ratios of the corresponding $2'^{+} \rightarrow 2^{+}$ transitions in ^{122}Te ($\delta_{564} = -3.6 \pm 0.3$) [8,13]) and ^{126}Te ($\delta_{750} = -6.0 \pm 0.4$) [13,14]) which differ from the

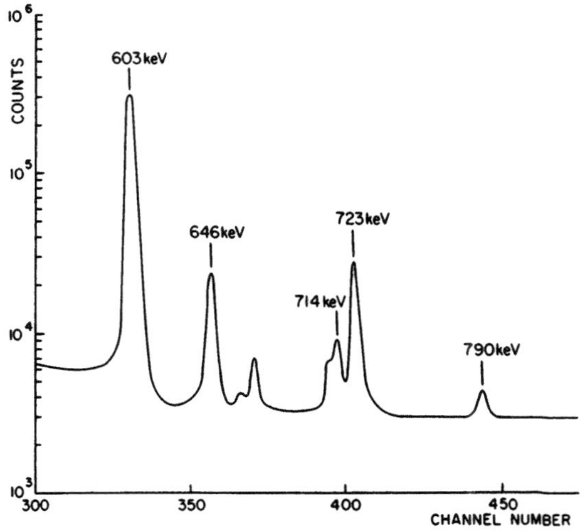

^{124}Te nucleus only by a pair of neutrons.

In addition, the 646γ - 603γ, 1691γ - 603γ and 2091γ - 603γ directional correlations in ^{124}Te have been measured. The results are summarized in the table together with the results of other authors (quoted here for comparison). It is worth mentioning that the results of directional correlation measurements alone cannot distinguish between spins I = 4 (646 keV transition pure quadrupole) and I = 3 ($\delta_{646} \approx 0.2$) for the 1249 keV level. On the basis of the systematics of the excited states of even - even nuclei in this mass region and in view of the absence of direct feeding to this level from the ^{124}I ($I = 2^-$) decay, the I = 4 assignment is preferred. The evidence from the K - conversion coefficient measurements is contradictory: the values of the conversion coefficients measured from the decay of ^{124}Sb give pure E2 character for the 646 keV transition[3,16]). The value α_k (646) = 4.8 X 10^{-3} obtained in ref. 3 from the decay of ^{124}I is, on the other hand, in agreement with pure M1 character for this transition. The 1691γ - 603γ and 2091γ - 603γ directional correlation measurements when combined with the measured values of the K - conversion coefficients[3,16])give unique 3^- spin assignments to the 2294 keV and the 2694 keV levels, with the 1691 keV and 2091 keV transition being essentially of pure E1 character ($|\delta| < 2 \times 10^{-2}$).

Cascade	$A_{22} \pm \Delta A_{22}$	$A_{44} \pm \Delta A_{44}$	Reference
645 keV - 603 keV	0.12 ± 0.04	0.0 ± 0.05	11
	0.099 ± 0.005	0.007 ± 0.008	present, Ge - NaI
	0.121 ± 0.025	0.010 ± 0.030	present, Ge - Ge
1691 keV - 603 keV	-0.063 ± 0.004	-0.009 ± 0.006	10
	-0.072 ± 0.005	-0.007 ± 0.005	17
	-0.099 ± 0.026	+ 0.03 ± 0.03	11
	-0.060 ± 0.006	0.0 ± 0.008	present, Ge - NaI
	-0.064 ± 0.020	-0.005 ± 0.025	present, Ge - Ge
2091 keV - 603 keV	-0.058 ± 0.005	0.001 ± 0.005	18
	-0.052 ± 0.012	0.037 ± 0.018	10
	-0.069 ± 0.015	0.004 ± 0.020	present

† Supported by the U.S. Atomic Energy Commission.

References

1. I.P. Auer, J.J. Reidy and M.L. Wiedenbeck, Nucl. Phys. A124 (1969) 199.

2. R.A. Meyer, W.B. Walters and R.C. Ragaini, Nucl. Phys. A127 (1969) 595.
3. R.C. Ragaini, W.B. Walters, A.A. Noyes and R.A. Meyer, Phys. Rev. 187 (1969) 1721.
4. R.A. Warner and J.E. Draper, Phys. Rev. C, 1 (1970) 1069.
5. K.S. Krane and R.M. Steffen, Phys. Rev. (in press).
6. R.S. Raghavan, Z.W. Grabowski and R.M. Steffen, Phys. Rev. 139 (1965) 1.
7. H. Paul, Phys. Rev. 121 (1961) 1175.
8. T. Lindquist and I. Marklund, Nucl. Phys. 4 (1957) 189.
9. M.J. Glaubman and J.D. Oberholtzer, Phys. Rev. 135 (1964) B1313.
10. L. Dorikens - Vanpraet, J. Demuynck and M. Dorikens, Nucl. Phys. 73 (1965) 539.
11. P.H. Stelson, Phys. Rev. 157 (1967) 1098.
12. D.C. Camp and A.L. van Lehn, Nucl. Instr. and Meth. 76 (1969) 192; M.J.L. Yates, Alpha-, Beta- and Gamma-Ray Spectroscopy (ed. K. Siegbahn, North - Holland Publishing Company, Amsterdam, 1965) p. 1691.
13. I. Asplund, L.G. Strömberg and T. Wiedling, Ark. Fysik 18 (1961) 65.
14. M. Sakai, H. Ikegami, T. Yamazaki and K. Sugiyama, J. Phys. Soc. Japan 14 (1959) 983.
15. L.A. Sliv and I.M. Band, in Alpha-, Beta- and Gamma-Ray Spectroscopy (ed. K. Siegbahn, North - Holland Publishing Company, Amsterdam, 1965) p. 1639.
16. E.P. Grigorev, A.V. Zolotavin, V.O. Sergeev and M.I. Sovcov, Izv. Acad. Nauk. SSSR, Ser. Fiz. 32 (1968) 773.
17. M.R. Meder and F.E. Durham, Nucl. Phys. A97 (1967) 78.
18. C. Weitkamp, Nucl. Phys. 43 (1963) 57.

Contributed Paper

DETERMINATION OF E2/M1 - MIXING PARAMETERS FOR COLLECTIVE TRANSITIONS IN DEFORMED NUCLEI

Jürgen Lange
Institut für Strahlen- und Kernphysik, Universität Bonn, Bonn, W. Germany

A survey of measurements of E2/M1 - mixing parameters for transitions between the γ - band head $I^{\pi}K = 2^{+}2$ and the first excited state $2^{+}0$ of the ground state rotational band has been given at the Nashville Conference in August 1969 [1]). In the meantime we have performed new angular correlation experiments with Ge(Li) - detectors especially with transitions between higher excited states in these bands. These measurements may show the effects of the admixture of other states on the M1 - matrix elements; these effects are well established for the E2 - transition probabilities (ref. 2). Furthermore, the measurements may serve as a test of the different models predicting mixing parameters[3-5]).

The following results were obtained: Measurements of coincidence spectra with the 122 keV line in ^{152}Sm led to the determination of the mixing parameter δ_{32} for the 1113 keV transition and also to a reinvestigation of δ_{22}. The final results, corrected for the interference of other cascades, the attenuation of the angular correlation in the $2^{+}0$ state, and the finite solid angles of the detectors, are:

$$\delta_{22}(965\ \mathrm{keV}) = 11.0\ (1.0) \quad \text{and}$$

$$\delta_{32}(1113\ \mathrm{keV}) = 8.9\ (1.0)$$

with the sign convention:

$$\operatorname{sign} \delta = \operatorname{sign} \frac{(I_i \| E2 \| I_f)}{(I_i \| M1 \| I_f)}$$

Coincidence measurements with the 245 keV transition gave

$$\delta_{34}(869\ \mathrm{keV}) = 5.0\ (0.7) \quad \text{and}$$

$$\delta_{44}(1006\ \mathrm{keV}) = 6^{+19}_{-3}$$

In ^{154}Gd the results

$$\delta_{22}(876\ \mathrm{keV}) = 10.2\ (1.0)^{\dagger} \quad \text{and}$$

$$\delta_{32}(1008\ \mathrm{keV}) = 6.6\ (0.7)$$

were obtained.
For ^{160}Dy coincidence measurements with the 215 keV transition gave:

$$\delta_{32}(765\ \mathrm{keV}) = 9.0^{+2.8}_{-1.7}$$

$$\delta_{34}(962\ \mathrm{keV}) = 4.7^{+5.0}_{-1.7}$$

In ^{172}Yb the $2^{+}2$ state at 1466 keV has a B (E2) - enhancement factor of about 1.5, whereas for the other $2^{+}2$ states this factor is about 0.5 or smaller[6]). By means of coincidence measurements with the 79 keV line the following mixing parameters were obtained:

$$\delta_{22}(1387\ \mathrm{keV}) = 9.3^{+5.0}_{-2.5} \quad ,$$

$$\delta_{32}(1470\ \mathrm{keV}) = 5.6^{+3.0}_{-2.0} \quad ,$$

† weighted average of the results of a NaI (Tl) - and a Ge(Li) - measurement[1]).

$$\delta_{22}(1530\ \mathrm{keV}) = -8.0^{-2.0}_{+1.5} \ ,$$

and for the 1398 keV transition:

$$-1.25 \leqslant \delta_{22} \leqslant -0.4 \quad \text{or}$$

$$\delta_{22} \leqslant -2.2,$$

where the second solution is favoured considering the B (E2) - transition ratios[7]).

For a comparison of the observed mixing parameters $\delta^{exp}_{II'}$ for different transitions between the γ - band and the ground state band of the same isotope we have evaluated 'reduced' mixing parameters Δ (see table 1), defined by

$$\Delta = \delta^{exp}_{II'} \ C(I, I') \ / \ E_{II'} \ f(z_2, I, I') \ ;$$

$E_{II'}$ is the transition energy, f (z_2, I, I') the band mixing correction for the E2 - transition probability [2,7,8]) and C (I, I′) stands for the vector coupling coefficients[3]). These parameters Δ should be constant in first order approximation.

Table 1

transition $I_\gamma - I_{gr}$	Δ (^{152}Sm)	Δ (^{154}Gd)	Δ (^{160}Dy)	Δ (^{172}Yb)
2 - 2	26.1 (2.4)	26.7 (2.7)	$49.5^{+13.6}_{-8.0}$ [1])	$16.9^{+9.0}_{-4.6}$
4 - 4	26^{+84}_{-13}	14.7 (4.8) [8])		
3 - 2	24.5 (2.7)	20.1 (2.1)	$27.6^{+8.4}_{-5.2}$	11^{+6}_{-4}
3 - 4	15.2 (2.1)	17.0 (1.0) [8])	19^{+10}_{-5}	

Although the errors are large an I - dependence is indicated. This I - dependence must be determined by the structure of the M1 - operator and the nature of the admixed states. There are specific I - dependent corrections due to the tensorial connection between the magnetic moment operator and the angular momentum operator[4,9]). The microscopic theory[4]) gives different corrections resulting from M1 - transitions between the admixed K = 1 states.

Although there is an indication for a better fit in the first case, the experimental values are not precise enough to reveal the true origin of the I - dependence.

References

1. E. Bodenstedt, International Conference, Nashville, August 1969.
2. C. Günther, D.R. Parsignault, Phys. Rev. 153 (1967) 1296.
3. W. Greiner, Nucl. Phys. 80 (1966) 417.
4. D.R. Bès et al., Nucl. Phys. 65 (1965) 1.
5. K. Kumar, Phys. Lett. 29B (1969) 25.
6. D.G. Burke, B. Elbeck, Dan.Mat. Fys. Medd. 36 (1967) no. 6.
7. K. Wien, Bericht der Technischen Hochschule, Darmstadt, 1968.
8. L. Varnell et al., Report Calt. - 63 - 95.
9. D.P. Grechukhin, JETP Lett. 3 (1966) 275.J. Nucl. Phys. USSR 4 (1966) 691.

Contributed Paper

γ - γ DIRECTIONAL CORRELATIONS OF THE 1079-89 AND 2089-89 keV CASCADES IN ^{156}Gd

E. Collins, J.H. Hamilton, P.E. Little, A.V. Ramayya and M.U. Kim†
Vanderbilt University, Physics Department, Nashville, Tenn., U.S.A.*

A.F. Kluk†† and N.R. Johnson
*Oak Ridge National Laboratory**, Oak Ridge, Tenn., U.S.A.*

1. Introduction

The spin of the 1.168 MeV level has been previously assigned[1]) as 0, 1, 2 on the basis of the branching to the 0^+ and 2^+ levels in the ground state rotational band. With the development of NaI - Ge(Li) detectory systems for measurement of $\gamma\gamma$ (θ) it has become very easy to measure $\gamma\gamma$ directional correlations on weak transitions without systematic errors. In the course of our experiments on the measurement of E2/M1 admixtures in the transitions from the β - band to the ground state rotational band in ^{156}Gd, we measured tne directional correlations of the 1079-89 and 2089-89 keV cascades. The directional correlation on the first cascade firmly establishes the spin of the 1.168 MeV level. Our results on the latter cascade deviates considerably from the older work of Bauer and Deutsch[2]). In this paper our results on these two cascades are presented.

2. Experimental Procedure

The directional correlation system used for these measurements is described in ref. 3. A 0.02 cm Cd absorber was used on the NaI detector to suppress the 92 - keV sum peak which results from the 43- and 49-keV X - rays. The 5 cm X 5 cm NaI detector was set 7 cm from the source center. The other detector was a 35 cm^3 active volume Ge(Li) detector with a photopeak efficiency of 3.5% relative to that of a 7.6 cm X 7.6 cm NaI detector for a transition energy of 1.3 MeV and resolution of 2.8 keV FWHM with a peak - to - Compton ratio of 16 : 1 at this same energy. The Ge(Li) detector was placed 5 cm from the source center and shielded by a graded lead - cadmium - copper absorber to decrease pile - up and reduce the count rate to improve the resolution.

With our two - gate system, we simultaneously stored the coincidence spectrum from the 89 - keV, $2^+ \to 0^+$ transition and the Compton background under this transition in a split 4k memory. The Compton background gate was set just above the photopeak. Data for each coincidence run at any angle were collected for 23 hr. The following sequence of angles, 90°, 135°, 180°, 180°, 225°, and 270° were used in each cycle. The data were collected for a total of 20 days at each angle (90 (90 + 270); 135 (135 + 225); and 180°).

A computer code was written to correct the raw coincidence data for chance events and Compton background coincidence events. The corrected coincidence spectra at each angle were summed for the final determination of the A_k coefficients.

It is well known, because of the 2.2 ns life - time of the 2^+ level at 89 keV, that the correlations through this level are attenuated. The product, G_kQ_k, of the source attenuation and solid angle correction factors were determined from normalizing the measured 961-89 keV $0^+ \to 2^+ \to 0^+$ correlation coefficients to the unique theoretical values $A_2 = 0.357$ and $A_4 = 1.14$.

3. Results and Discussion

1079-89 keV Correlation

A known 0^+ state occurs at 1050 keV and depopulates via a 961 keV transition in a $0^+ \to 2^+ \to 0^+$ cascade. Using the correction coefficients, G_kQ_k, obtained from this correlation the raw A_k coefficients of the 1079-89 keV correlation were corrected to obtain $A_2 = 0.348 \pm 0.007$ and $A_4 = 1.06 \pm 0.03$. The only spin sequence that admits such a large A_4 term is a $0 \to 2 \to 0$. Hence, the 1168 keV level is assigned a spin of 0. Since we now have two spin zero states, a weighted average of the correction coefficients as obtained from the two $0 \to 2 \to 0$ correlations gives $G_2Q_2 = 0.605 \pm$ $\pm\ 0.018$ and $G_4Q_4 = 0.362 \pm 0.010$. Since this work was completed, this spin has been confirmed in other work[4]), though the inaccuracy there was large and just sufficient to assign a 0 spin.

2098-89 keV Correlation

The 2098-89 keV correlation has a spin sequence of $1^+ \to 2^+ \to 0^+$. The anisotropy coefficients obtained before correction for solid angle and source attenuation are $A_2 = 0.362 \pm 0.006$ and $A_4 = -0.158 \pm 0.007$.

The corrected anisotropy coefficients obtained for this correlation are $A_2 =$ $= 0.599 \pm 0.021$ and $A_4 = -0.436 \pm 0.023$ which limit the mixing ratio to $1.24 \leqslant \delta_{2098} \leqslant 1.50$ when two standard deviations are allowed on the A_2 term. This

δ - value permits an admixture of 60.5% $\leqslant$ E2 $\leqslant$ 69.3%. Our A_k coefficients differ considerably from those of Bauer and Deutsch[2]), who obtain $A_2 = 0.27 \pm 0.02$ and $A_4 = -0.15 \pm 0.02$. Their coefficients are consistent with a δ - value of 0.5 ± 0.1 which gives a 20% E2 admixture. This latter work[2]) was carried out with two NaI detectors. If their gate was too wide, this could explain their low value in comparison to our result. This work points up the fact that some older results obtained in complex decays with NaI detectors are possibly wrong.

† NSF Summer Research Participant, 1969, from Lycoming College, Williamsport, Pa.

†† ORAU, Predoctoral Fellow from Vanderbilt University.

* Work supported in part by a grant from the National Science Foundation.

** Research sponsored by the U.S. Atomic Energy Commission under contract with the U.S. Atomic Energy Commission.

References

1. C.M. Lederer, J.M. Hollander and I. Perlman, Table of Isotopes, (John Wiley and Sons, New York, 1967).
2. R.W. Bauer and M. Deutsch, Phys. Rev., 128 (1962) 751.
3. J.H. Hamilton, A.V. Ramayya and L.C. Whitlock, Phys. Rev. Lett. 22 (1969) 65.
4. H.L. Nielsen, N. Rud and K. Wilsky, Phys. Letters 30B (1969) 169.

Contributed Paper

M1 - E2 ADMIXTURES IN TRANSITIONS FROM BETA- AND GAMMA LEVELS IN ^{152}Sm

A.V. Ramayya†, W. Lourens, B. van Nooijen and H. van Krugten
Laboratorium voor Technische Fysica, Delft University of Technology, Delft, The Netherlands

1. Introduction

There has been much interest in recent years[1-4]) in the detailed properties of beta- and gamma levels in intermediate nuclei since they exhibit large deviations from both the harmonic vibrator model and the rigid rotator model. A perturbation expansion around the equilibrium shape is therefore not allowed in these cases. Special theoretical techniques have been developed for the study of intermediate nuclei. For a number of nuclei, Kumar[4]) has recently calculated the M1 - E2 - mixing occurring in transitions from beta- and gamma vibrational levels. We decided to check his results experimentally on ^{152}Sm since that might contribute to showing the potentiality of the applied theoretical techniques. It is true that measurements of the M1 - E2 mixing ratios in transitions occurring in ^{152}Sm have been performed by other groups (ref. 5); however, we felt that for a proper judgement of the theoretical calculations of the small M1 - admixtures the experimental accuracy had to be improved considerably. For instance, the amplitude mixing ratio δ for the 689 keV $2_\beta \rightarrow 2$ transition has been measured by two groups applying Coulomb excitation techniques. They report

$$\delta = 67^{+\infty}_{-54} \quad \text{(ref. 3)} \quad \text{and} \quad \delta = 13^{+28}_{-5} \quad \text{(ref. 6)}.$$

2. Experimental Techniques and Measurements

The M1 - E2 admixtures can in general be measured by various types of directional correlation techniques. In these cases, the A_2 value alone determines the magnitude and sign of δ because A_4 is a very insensitive function of δ. Since our aim is to achieve a high degree of experimental accuracy on A_2 we decided to do electron - gamma angular correlation measurements with a system consisting of a magnetic β - ray spectrometer and a high efficiency, high resolution Ge (Li) detector. This system has the obvious advantage of having high resolution in both channels. The disadvantage of low counting rate is not a serious problem, because, firstly, the particle parameter b_2 which enters into the correlation has a value much larger than 1 which enhances the anisotropy and, secondly, prolonged measurements are feasible when the system has an extremely stable behaviour as a function of time.

The β - ray spectrometer is of the sector focussing type and has been described elsewhere[7]). The spectrometer was operated at a resolution of 5% and transmission of 2.5% in the case of the measurements on cascades involving the 122 keV level, and 1.5% and 1.8% respectively for the correlation measurements including the 245 keV transition. The Ge(Li) detector has an efficiency of 11.8% as compared to a 3 in. X 3 in. NaI crystal and has a resolution of 2.45 keV at 1.33 MeV. The peak to Compton ratio for this detector is 33/1 at 1.33 MeV. A constant fraction timing device, an external routing system and a 4096 channel analyzer were employed to simultaneously store the single γ - spectra and the true and chance coincidence spectra in separate groups of 1024 channels.

A thin carrier free europium chloride source of about 5 mm diameter was prepared by liquid drop method on an aluminium backing of 0.54 mg/cm^2 thickness covered with a zapon film. The movable Ge (Li) detector was centered with respect to the source with an accuracy of better than 1%. The gamma detector was placed at a distance of 7 cm from the centre of the source. The current of the β - ray spectrometer was adjusted to the top of the 122 keV K - conversion line in the first experiment and to the top of the 245 keV K - conversion line in the second experiment. In front of the Ge (Li) detector a Pb (1 mm) – Cu (1 mm) absorber was placed in order to reduce the counting rate in the low energy part of the γ - ray spectrum. This reduction improves the detector resolution because of lower overall counting rate. The time resolution of the coincidence system was 40 ns resulting in a true to chance ratio of approximately 5.

The singles γ - ray spectrum, the true and the chance coincidence spectra were simultaneously measured at 90°, 135°, 180° (twice), 225°, and 270° in a cycle. At each angle a measurement was carried out for 23 hr. During such a run the total number of electron counts, accidental and true coincidence counts were printed out every 30 min. as a check on the stability.

A separate scaler registered the total number of electron counts collected in 23 h. The attenuation correction which arises from the life - time of the 122 keV state and the solid angle correction factor of the Ge (Li) detector were determined from the known correlations in ^{152}Sm. The solid angle correction factors for the β - spectrometer were calculated from the transmission curve of the spectrometer[7]). Bauer and Deutsch[8]) have carefully measured the 1408 - 122 keV γ - γ directional correlation. Our measured 1408 - 122 keV correlation was normalized to their value to obtain the attenuation and solid angle correlation factors including the particle parameter, b_2.

This $b_2\Omega_2G_2$ - factor was then applied to the 964 - 122 keV correlation, thus giving a corrected A_2 - value. This value could subsequently be used to obtain the corrected A_4 - value since the error in A_2 is not important because of the very weak dependence of A_4 on A_2. From this A_4 - value the factor $b_4\Omega_4G_4$ could be determined by comparing it with the measured A_4 - value.

The results of our measurements on cascades involving the 122 keV transition are shown in table 1.

TABLE 1

e_K^--$\gamma(\theta)$	$\Omega_2G_2b_2A_2$	$\Omega_4G_4b_4A_4$	A_2	A_4	δ (expt)	δ (theory) ref.4	δ (theory) ref.9
1408(2^-)	0.315 ± 0.003	0.010 ± 0.002	0.225 ± 0.007[a]	−0.016 ± 0.003[d]			
964(2^+_γ)	0.009 ± 0.002	−0.203 ± 0.002	0.006 ± 0.002[b]	0.322 ± 0.005[c]	$-9.2^{+0.2}_{-0.2}$	−36.7	−24.3
1112(3^+_γ)	−0.349 ± 0.002	0.057 ± 0.003	−0.250 ± 0.008[b]	−0.091 ± 0.005[d]	$-16.8^{+3.6}_{-2.5}$		−26.6
689(2^+_β)	−0.165 ± 0.013	−0.205 ± 0.015	−0.118 ± 0.010[b]	0.325 ± 0.024[d]	$+17^{+5}_{-3}$	−108.7	10.8
869(3^+_γ)	0.250 ± 0.009	0.117 ± 0.009	0.151 ± 0.006[e]	−0.167 ± 0.013[f]	$-6.1^{+0.3}_{-0.3}$		−16.5
1006(4^+_γ)	0.018 ± 0.019	−0.07 ± 0.02	0.011 ± 0.012[e]	0.10 ± 0.03[f]	$-2.8^{+0.2}_{-0.3}$	−18.3	−9.6

γ-$\gamma(\theta)$	$\Omega_2\Omega_2'A_2$	$\Omega_4\Omega_4'A_4$	A_2	A_4	δ (expt)	δ (theory) ref.4	δ (theory) ref.9
656(4^+_β)	−0.178 ± 0.026[g]	0.133 ± 0.039[g]	−0.196 ± 0.029[g]	0.181 ± 0.053	$3.1^{+1.5}_{-1.4}$	−59.3	4.1

a) A_2 value of Bauer and Deutsch[8])
b) corrected with $b_2\Omega_2G_2$ = 1.400 ± 0.046 obtained as ratio of 0.315 ± 0.003/0.225 ± 0.007
c) A_4 (theory) compatible with A_2 (experimental)
d) corrected with $\Omega_4b_4G_4$ = −0.630 ± 0.007 obtained as ratio of -0.203 ± 0.002/0.322 ± 0.001
e) corrected with Ω_2b_2 = 1.653 ± 0.005
f) corrected with Ω_4b_4 = −0.700 ± 0.002
g) these values are obtained from a γ-γ (θ) experiment[11]) with NaI (4.3 cm. × 5 cm.) and Ge (Li) detector system, because of the statistical inaccuracy obtained for this case in the second experiment described above.

A second series of $e_K^- - \gamma(\theta)$ - measurements were carried out with the current of the β - ray spectrometer set to the top of the 245 keV K - conversion line. These results are also shown in table 1. We should point out that in these measurements, theoretical b_2 and b_4 - values and calculated solid angle correction factors were used in order to arrive at the correct anisotropy coefficients. The G_2 and G_4 factors were assumed to be equal to one because of the short half - life of the 367 keV level.

As a check on the results obtained for 869 - 245 keV cascade, we also calculated δ from the 869 - (245) - 123 keV (1 - 3) - correlation. The result $\delta = 6.3 \pm 0.2$ is in excellent agreement with the value quoted in table 1.

3. Discussion

The results listed in table 1 show that the experimental results deviate considerably from the values published by Kumar in ref. 4. The most serious discrepancy occurs in the case of the 689 (2_β^+) keV transition where even the sign of δ [10]) disagrees. Recently Kumar recalculated his numbers using a new method for the renormalization of the inertial functions and his new results[9]) are listed in column 6 of the table. It appears that there is now agreement with respect to the sign of δ (689 keV), and also the magnitude of δ is in reasonable agreement with Kumar's new results. However, the absolute values of δ are on the whole smaller than the theoretical values thus showing that the theoretical calculations can still be improved.

It is interesting to point out that in this case also the M1 admixtures in the transitions from the β and γ - levels to the 4^+ member of the ground state band are larger than the M1 admixtures in the transitions to the 2^+ member of the ground state band as observed in case of ^{154}Gd.

† On leave of absence from Vanderbilt University, Nashville, Tennessee, U.S.A.

References

1. J.H. Hamilton, A.V. Ramayya and L.C. Whitlock, Phys. Rev. Lett. 22 (1969) 65.
2. L.L. Riedinger, N.R. Johnson and J.H. Hamilton, Phys. Rev. Lett. 19 (1967) 1243.
3. I.A. Fraser et al., Phys. Rev. Lett. 23 (1969) 1047, 1051.
4. K. Kumar, in Proc. of the Conf. on Properties of Nuclei far from the region of β - stability, Leysin, Switzerland (1970).
5. J.H. Hamilton, in Proc. of Int. Conf. on Angular Correlations in Nuclear Disintegrations, Delft, The Netherlands, August 1970, p. 181. (see the references cited there for the individual measurements).
6. F.K. McGowan et al., Bull. Am. Phys. Soc. 13 (1968) 895; R.O. Sayer et al., Phys. Rev. C1 (1970) 1525.
7. H. van Krugten and E.W. Koopmann, Nucl. Instr. and Methods 71 (1969) 280.
8. R.W. Bauer, M. Deutsch, Phys. Rev. 128 (1962) 1783.
9. K. Kumar, Phys. Rev. Lett. 26 (1971) 269; private communication of results to be published.
10. K.S. Krane and R.M. Steffen, Phys. Rev. C7 (1970) 724.
11. To be published in Nuclear Physics.

Contributed Paper

THE M2/E1 MIXING RATIO FOR THE ANOMALOUS $\Delta K = 1$ 71.6 keV TRANSITION IN ^{177}Hf

B. Hrastnik, I. Basar, M. Dikšić, K. Ilakovac, V. Kos and A. Ljubičić
Institute 'Rudjer Bošković', Zagreb, Yugoslavia

In the region of deformed nuclei, the weak M2 transitions compete with the E1 transitions which are considerably slowed down. The M2 admixtures are mostly deduced from internal conversion coefficients. However, due to the penetration effect the deduced M2 values are somewhat uncertain. In these cases, the directional gamma - gamma correlations are expected to give more reliable M2/E1 mixing ratios.

The gamma - gamma directional correlation for the weak 71.6 - 249.7 keV cascade transition in ^{177}Hf has been measured. The energies, spins and parities of the low - lying ^{177}Hf levels have been determined and successfully interpreted in the Nilsson scheme[1]). The ground state and the two lowest excited states at 112.9 keV and 249.7 keV are members of the $7/2^-$ [514] ↓ rotational band, while the third excited state at 321.4 keV is the ground state of the $9/2^+$ [624] ↑ rotational band. The ($\Delta K=1$) 321.4 keV, 208.4 keV and 71.6 keV E1 transitions between these two rotational bands are delayed[2]), and have a small M2 admixture[3,4]). A possible small M2 admixture in the 71.6 keV transition was not previously determined with certainty[5]). The 249.7 keV is a pure E2 transition[1]) and therefore the angular correlation coefficients depend only on the M2/E1 mixing ratio of the 71.6 keV transition.

The angular correlation apparatus consisted of two coaxial Ge(Li) detectors of 15 cm^3 and 20 cm^3, respectively, in connection with a fast - slow coincidence system and a two - parameter time - energy 128 X 128 channel data analyzing and recording system. The source of ^{177g}Lu (6.8 d), obtained by a four days irradiation of lutetium oxide in the reactor at a flux of 10^{13} neutrons/$cm^2 \cdot s$, was prepared in the form of lutetium nitrate in a diluted aqueous solution. It was shown[5,6]) that for such sources one obtains the unperturbed correlation. The 71.6 - 249.7 keV integral correlation was measured at four angles: 90°, 135°, 180° and 225°. A total number of 1.12×10^5 of true coincidences was collected. From the internal conversion data[1,7]) the upper limit for the A_{44} coefficient was deduced to be $|A_{44}| \leqslant 0.001$. Neglecting A_{44}, we obtained, after the finite solid angle corrections,

$$A_{22} = -0.105 \pm 0.05$$

From this value of A_{22} the M2/E1 mixing ratio for the 71.6 keV transition was deduced,

$$\delta = -0.0173 \pm 0.0065$$

From the known partial width of the 71.6 keV gamma transition[2]) we deduced the partial E1 and M2 widths of 4.6×10^{-9} eV and 1.4×10^{-13} eV, respectively. The Weisskopf single particle estimates are well known[8])

$$\Gamma_{\gamma W}(E1) = 6.748 \times 10^{-2} A^{2/3} E_\gamma^3$$

$$\Gamma_{\gamma W}(M2) = 1.472 \times 10^{-8} A^{2/3} E_\gamma^5$$

where the width $\Gamma_{\gamma W}$ is in (eV) when energy E_γ is in MeV. The Nilsson estimates were calculated from the expressions (29), (35) and (36) of reference 9 taking the deformation parameter $\delta = 0.3$ for the ^{177}Hf nucleus and eigenfunctions for the deformed field. The hindrance factors

$$F_a(\sigma\lambda) = \frac{\Gamma_{\gamma a}(\sigma\lambda)}{\Gamma_{\gamma\ \exp}(\sigma\lambda)} \quad , \quad a = W, N$$

for E1 and M2 71.6 keV transitions were calculated. The results are:

$$F_W(E1) = (7.1 \pm 1.4) \times 10^4 \ , \qquad F_N(E1) = 0.68 \pm 0.14$$

and

$$F_W(M2) = 2.6\,^{-1.5}_{+3.9} \ , \ F_N(M2)_{g.s.\ (free)} = 0.19\,^{-0.11}_{+0.28}$$

Comparing with fig. 8 in reference 2, where F_N (M2) are plotted versus mass number, one can find a good agreement between the experimental values and the theoretical single particle estimates of Weisskopf and of Nilsson. Our values for F_W (M2) and F_N (M2) are therefore in good agreement with the general trend of M2 transitions in this region of nuclei.

References

1. Nuclear Data Sheets, as available in June 1965.
2. K.E.G. Löbner and S.G. Malmskog, Nucl. Phys. 80 (1966) 505.
3. B. Harmatz, T.H. Handley and J.W. Mihelich, Phys. Rev. 119 (1960) 1345.
4. E.D. Klema, Phys. Rev. 109 (1958) 1652.
5. B. Hrastnik et al., Fizika 1 (1969) 117.
6. B. Hrastnik et al., to be published.
7. R.Hager and E. Seltzer, in the 'Internal Conversion Processes', ed. by J.H. Hamilton, Academic Press New York and London, 1966.
8. S.J. Skorka, J. Hertel and T.W. Retz - Schmidt, Nuclear Data A2, no. 4, Nov. 1966, p. 347.
9. S.G. Nilsson, Mat. Fys. Medd. Vid. Selsk. 29 (1955) no. 16.

Contributed Paper

MEASUREMENT OF THE E2/M1 AMPLITUDE RATIOS OF GAMMA TRANSITIONS IN THE EVEN - EVEN ISOTOPES OF Os AND Pt[†]

K.S. Krane and R.M. Steffen

Purdue University, Lafayette, Indiana, U.S.A.

The nuclei of the osmium and platinum isotopes are situated in a transition region between deformed and spherical nuclei. Recent experimental evidence has indicated that these nuclei follow neither the rotational nor the vibrational descriptions of nuclear structure. Even though ground state rotational bands, in some cases up to a 10^+ state, have been discovered in these nuclei, the energy spacing deviates considerably from the simple I (I + 1) rule for rotational excitations, and the addition of a term proportional to $[I\ (I + 1)]^2$ does not improve the fit. An interpretation in terms of vibrational excitations is made difficult due to the large experimental quadrupole moments which have been measured, pure vibrational states having vanishing quadrupole moments.

In an attempt to resolve these difficulties, Kumar and Baranger[1]) have proposed a model, based on a microscopic description of the nuclei in the region $182 < A < 196$

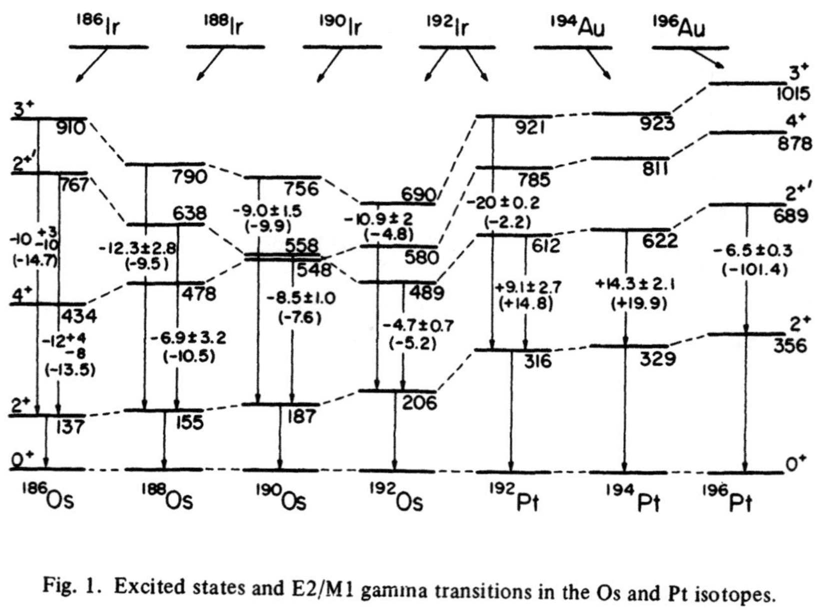

Fig. 1. Excited states and E2/M1 gamma transitions in the Os and Pt isotopes.

TABLE 1. Experimental and Theoretical Values of the E2/M1 Mixing Amplitudes for Gamma Transitions in Even-Even Os and Pt Isotopes

Isotopes	Transition	E(keV)	δ Experimental	δ Kumar - Baranger
^{186}Os	$2^{+'} - 2^{+}$	630	-10^{+3}_{-10}	-14.7
	$3^{+} - 2^{+}$	773	-12^{+4}_{-8}	-13.5
^{188}Os	$2^{+'} - 2^{+}$	478	-12.3 ± 2.8	-9.5
	$3^{+} - 2^{+}$	635	-6.9 ± 3.2	-10.5
^{190}Os	$2^{+'} - 2^{+}$	371	-8.5 ± 1.0	-7.6
	$3^{+} - 2^{+}$	569	-9.0 ± 1.5	-9.9
^{192}Os	$2^{+'} - 2^{+}$	283	$-4.7^{+0.6}_{-0.7}$ [a]	-5.2
	$3^{+} - 2^{+}$	485	$-10.9^{+2.1}_{-1.5}$ [b]	-4.8
^{192}Pt	$2^{+'} - 2^{+}$	296	$+9.1^{+1.7}_{-2.7}$ [b]	+14.8
	$3^{+} - 2^{+}$	604	-2.0 ± 0.2 [b]	-2.2
^{194}Pt	$2^{+'} - 2^{+}$	293	$+14.3 \pm 2.1$	+19.9
^{196}Pt	$2^{+'} - 2^{+}$	333	-6.5 ± 0.3	-101.4

[a] Reference 6 [b] Reference 7

in terms of the pairing and quadrupole interaction. A numerical solution of Bohr's collective Hamiltonian then enabled the calculation of energy levels, and static and dynamic electromagnetic multipole moments of even - even nuclei in this range of mass numbers.

In terms of the Bohr - Mottelson[2]) electromagnetic multipole operators, Kumar (ref. 3) has calculated the magnitude and the sign of the M1 matrix element in even - even nuclei in the region of A = 190. The present work describes an experimental investigation of the relative M1 and E2 transition amplitudes in this region.

The relative contribution of the M1 component, measured in terms of the E2/M1 mixing ratio δ, have been determined for gamma transitions in $^{186,188,190}Os$ and $^{194,196}Pt$ by measuring γ - γ directional correlations using high resolution Ge(Li) detectors and multichannel coincidence and routing techniques.

The γ - γ directional correlation data were corrected for chance coincidences, effects of competing cascades and Compton - scattered radiation, finite source and detector geometries, and perturbation effects (due to the electric quadrupole interaction in the liquid sources used). Details of the experimental equipment and of the data analysis are discussed in a recent paper[4]).

The excited states of the Os and Pt nuclei, whose gamma transitions were studied were populated by the radioactive decay of the isotopes of Pt and Ir, respectively, The sources of $^{186,188,190}Ir$ were produced by (α, xn) reactions on Re; the isotopes $^{194,196}Au$ were obtained by means of (d, xn) reactions on Pt.

A representation of the states involved in these measurements is given in fig. 1 (see page 253); the states of ^{192}Os and ^{192}Pt are shown for completeness.

From the γ - γ directional correlation results the amplitude mixing ratios

$$\delta = \frac{\langle I_f || \underline{j}_N \underline{A}_2^E || I_i \rangle}{\langle I_f || \underline{j}_N \underline{A}_1^M || I_i \rangle}$$

were extracted using the methods described in ref. 4. The multipole operators $\underline{A}_L^{(\pi)}$ have been explicitly defined in the work of Krane and Steffen[4]). The results of δ are represented in table 1 and in fig. 1. They represent the weighted averages of many individual measurements. Also included in table 1 and in fig. 1 (in brackets) are the theoretical mixing amplitudes as computed by Kumar[3]).

The agreement is seen to be reasonably good, with the sign of δ predicted correctly in every case. This is perhaps surprising, in view of the poor numerical accuracy of the calculation of the small M1 admixtures. The large discrepancy in the result for ^{196}Pt has been interpreted by Kumar[5]) as an accidental cancellation of terms in the calculation. It is possible that future higher - order calculations will improve the accuracy of the forbidden M1 matrix elements, and it is hoped that the apparent success of this type of calculation will result in its extension to nuclei in other mass regions of interest.

† Work supported by the U.S. Atomic Energy Commission.

References

1. K. Kumar and M. Baranger, Nucl. Phys. 92 (1967) 608; M. Baranger and K. Kumar, ibid. A110 (1968) 490; K. Kumar and M. Baranger, ibid. A110 (1968) 529; M. Baranger and

K. Kumar, ibid. A122 (1968) 241; K. Kumar and M. Baranger, ibid. A122, (1968) 273.
2. A. Bohr and B.R. Mottelson, Nuclear Structure (W.A. Benjamin, Inc., New York, 1969).
3. K. Kumar, Phys. Letters 27B (1969) 25.
4. K.S. Krane and R.M. Steffen, Phys. Rev. C (Aug. 1970).
5. K. Kumar, private communication.
6. R.L. Robinson, F.K. McGowan, P.H. Stelson, W.T. Milner and R.O. Sayer, Nucl. Phys. A123 (1969) 193.
7. Z.W. Grabowski, Phys. Rev. 183 (1969) 1019.

Contributed Paper

LOW - LYING LEVEL STRUCTURE OF ^{198}Hg AND ^{200}Hg

R. Béraud, I. Berkes, R. Haroutunian, G. Marest, Michèle Meyer - Lévy,

R. Rougny and A. Troncy

Institut de Physique Nucléaire, Université de Lyon - I, 69 - Villeurbanne, France

Recent interest in the structure of transition nuclei[1,2]) prompted us to re-investigate in detail the ^{198}Hg and ^{200}Hg level schemes from the disintegrations of ^{198m,198,200}Tl and to determine spin assignments and transition probabilities in these two nuclei. Energy spectra and transition rates of the low - lying levels as well as theoretical predictions[3,4]), suggested a strong vibrational nature and led us to search the members of the two phonon - triplet at about 1 MeV excitation energy.

The ^{198m,198,200}Tl activities were produced by irradiating a gold sandwich target composed of 100 mg/cm^2 foils with 56 - MeV α - particles in the Institute's synchrocyclotron. According to the energy cross section relationship[5]), we used the foil in which 35 MeV $\leqslant E_\alpha \leqslant$ 42 MeV and the one in which 20 MeV $\leqslant E_\alpha \leqslant$ 30 MeV to study the decays of ^{198m,198}Tl and ^{200}Tl respectively.

Angular correlation (A.C.) measurements were performed using a conventional fast - slow system with an overall time - resolution of 12 ns. The spectrum of a Ge(Li) detector in coincidence with the photopeak of another detector was accumulated in a 4000 - channel analyzer. An electronic clock controlled at each angle the counting time, the position of the moving counter and the read out of the coincidence spectra. Data were handled by an 1130 IBM computer. The γ - lines involved in these A.C. measurements are shown in fig. 1. Fitted A.C. coefficients (corrected for finite angular resolution) are presented in table 1.

Lifetimes of the two negative parity levels in ^{198}Hg [1,6]) have been measured by means of the delayed - coincidence method. For the 5^- state the time correlation of the 48 keV L-conversion electrons (detected by a thin plastic scintillator) and 587 keV γ - rays was compared with the β^- - γ prompt curve obtained with a ^{60}Co source.

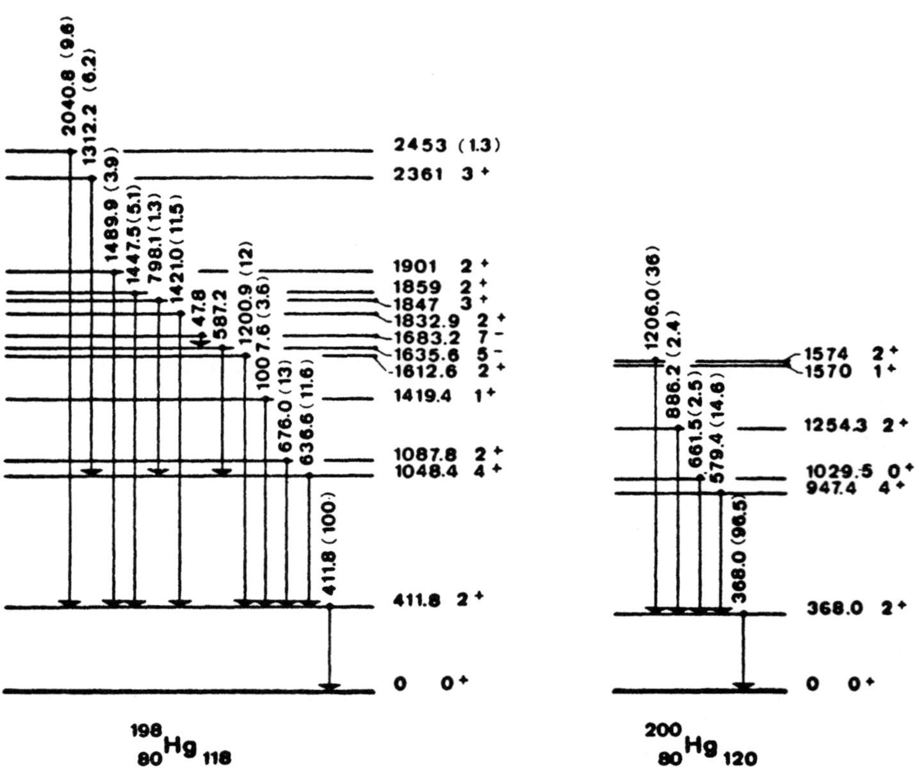

Fig. 1. Simplified level schemes. γ intensities are given between brackets. They are due to the g.s. decays.

The displacement between the centroids yielded: $\tau_{5^-} = 89 \pm 15$ ps. The mean life of the 1683.4 keV state was determined (from the slope of the time correlation curve by measuring coincidences between X - rays and 587 keV γ - rays) to be: $\tau = 10.7 \pm 0.5$ ns. (Since the lifetime of the 5^- state is comparatively very small, it has no effect on this measurement). This result agrees with a recent unpublished value[7]).

Spins of excited states obtained from our A.C. measurements are summarized in table 1. For the two first levels in ^{198}Hg and for the 947.4 keV and 1574.0 keV levels in ^{200}Hg our values confirm literature ones. δ^2 mixing parameters have been derived from graphical analysis (fig. 2). The transition rates deduced from our measurements are: B (E2) = 0.19 ± 0.02 $(eb)^2$ for the 48 keV transition and B (E1) = = (3.5 ± 0.6) 10^{-7} e^2b for the 587 keV transition. It is worth mentioning that while the $7^- \rightarrow 5^-$ reduced transition probability is comparable to that of the collective transition $2_1^+ \rightarrow 0_1^+$ (B (E2) = 0.22), the $5^- \rightarrow 4^+$ E1 transition is strongly forbidden (2.10^{-5} Weisskopf Units).

Table 1. Experimental A.C. coefficients for cascades studied in ^{198,200}Hg. (*) indicates $\delta^2\,(\frac{M2}{E1})$. An internal conversion coefficient measurement allowed us to derive $I^\pi = 1^+$ for the 1419.4 keV level.

Isotope	Level (keV)	Cascade (keV)	Experimental results		$\delta^2\,(\frac{E2}{M1})$	I^π
			A_2	A_4		
^{198}Hg	1048.4	637 - 412	$+0.091 \pm 0.009$	$+0.002 \pm 0.015$	∞	4^+
	1087.8	676 - 412	-0.284 ± 0.010	$+0.201 \pm 0.017$	1 2 ± 0.2	2^+
	1419.4	1008 - 412	$+0.21 \pm 0.04$	-0.08 ± 0.06	0.17 ± 0.03	1^+
	1612.6	1201 - 412	$+0.423 \pm 0.018$	$+0.025 \pm 0.027$	0.09 ± 0.03	2^+
	1635.6	587 - 637	-0.058 ± 0.015	$+0.033 \pm 0.020$	$< 5.7\ 10^{-4}$ *	5^-
	1832.9	1421 - 412	$+0.364 \pm 0.020$	$+0.025 \pm 0.029$	0.03 ± 0.01	2^+
	1847	798-(637)-412	$+0.29 \pm 0.07$	-0.2 ± 0.1	$4 \leq \delta^2 \leq 11$	3^+
	1859	1448 - 412	$+0.38 \pm 0.03$	$+0.04 \pm 0.05$	0.05 ± 0.03	2^+
	1901	1490 - 412	$+0.35 \pm 0.05$	$+0.09 \pm 0.07$	≈ 0.02	2^+
	2361	1312-(637)-412	-0.067 ± 0.026	$+0.02 \pm 0.04$	$(6^{+4}_{-3})10^{-3}$	3^+
		1312 - 637	-0.066 ± 0.022	-0.03 ± 0.03		
	2453	2041 - 412	-0.214 ± 0.025	-0.027 ± 0.038	0.029 ± 0.016	(1)
					0.19 ± 0.02	(3)
^{200}Hg	947.4	579 - 368	$+0.07 \pm 0.02$	$+0.04 \pm 0.03$	∞	4^+
	1029.5	662 - 368	$+0.36 \pm 0.10$	$+0.90 \pm 0.16$	∞	0^+
	1574.0	1206 - 368	$+0.10 \pm 0.04$	-0.03 ± 0.06	≈ 0.04	2^+

In order to explain the experimental low - lying levels we have applied the boson - approximation of nucleon pairs (with only s and d bosons) and a residual Q - Q interaction[8]) between the bosons. The two proton holes are assumed to be in the $(s_{1/2})^{-2}$ configuration. We take into account 3 bosons (6 neutron - holes in ^{200}Hg) and 4 bosons (8 neutron - holes in ^{198}Hg) without any interaction between core and holes. Results are reported in fig. 3. Our experimental and theoretical spectra agree with each other for the low levels in ^{200}Hg. To obtain negative parity levels[1,6,9]) we must include excitation states of single proton - holes: $(s_{1/2})^{-1}\ (h_{11/2})^{-1}$ and $(d_{3/2})^{-1}\ (h_{11/2})^{-1}$. The large B(E2) value of the transition between these states could then be explained by the following configurations: $|7^-\rangle = |s_{1/2}\ h_{11/2},\ 2_{core};\ 7^-\rangle$ and $|5^-\rangle = |s_{1/2}\ h_{11/2},\ 0_{core};\ 7^-\rangle$. The inclusion of a core - hole interaction is necessary to split the degenerate hole levels as in ref. 3 and 4. Such calculations are now in progress.

It has to be mentioned that two differences between ^{198}Hg and ^{200}Hg are yet not understood. Comparing their level schemes[2,10]) we note that the density of levels fed by the g.s. decay of ^{198,200}Tl in the range 1500 - 2400 keV is about two times higher in ^{200}Hg than in ^{198}Hg. On the other hand no 0_2^+ level, analogous to that of ^{200}Hg, could be identified in ^{198}Hg at about 1 MeV.

References

1. R.F. Petry, R.A. Naumann, J.S. Evans, Phys. Rev. 174 (1968) 1441.

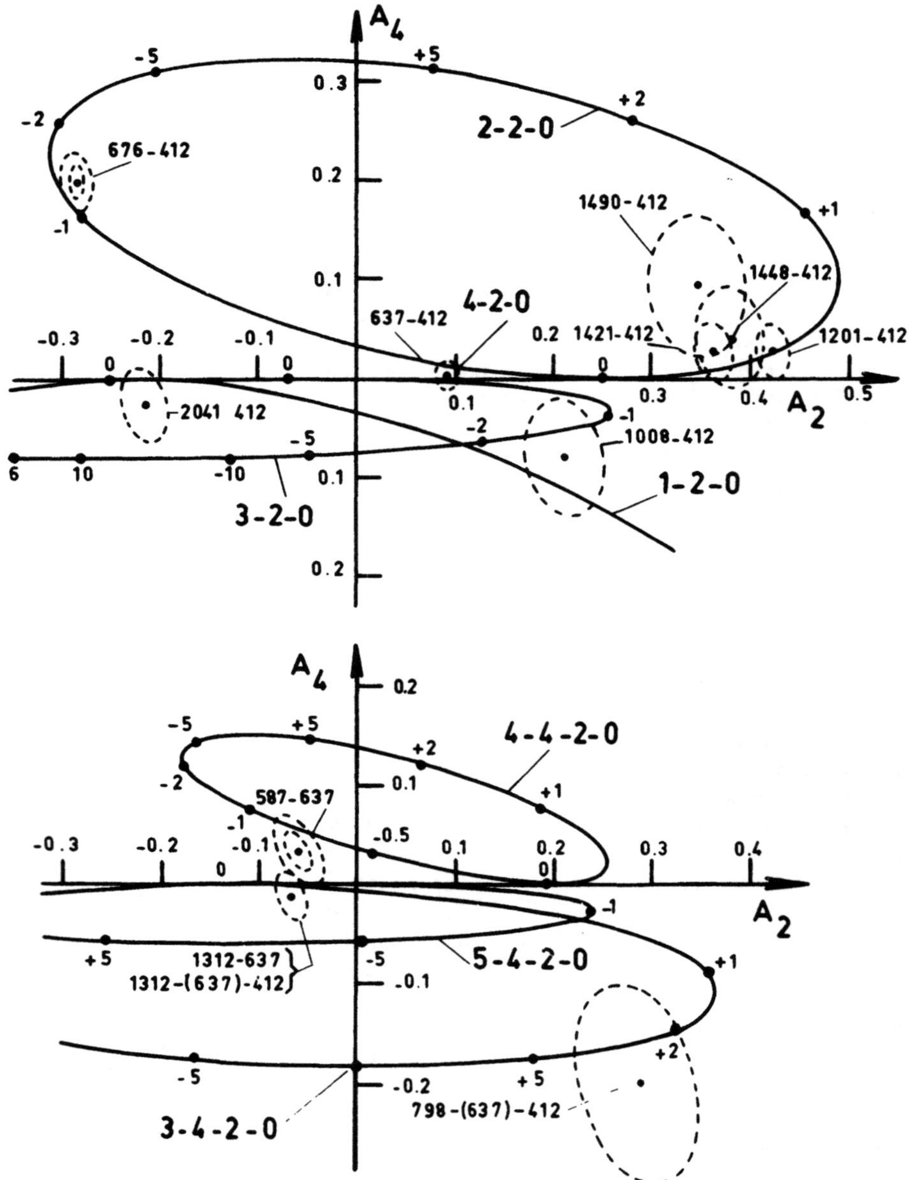

Fig. 2. ($A_2 - A_4$) plots (full lines). Some particular values of δ are reported. The dashed ellipses are the error ellipses. In two cases we give the confidence ellipse corresponding to $P_F(\chi^2) = 0.2$.

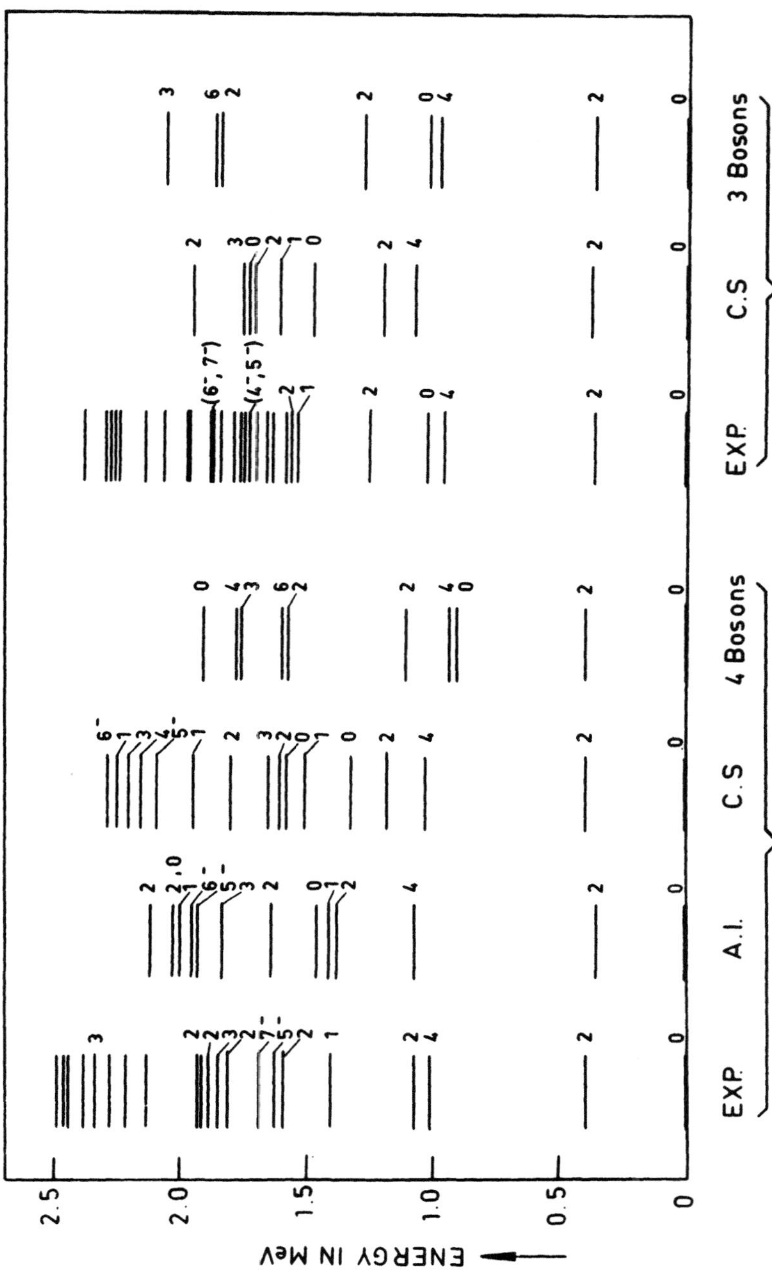

Fig. 3. Comparison of calculated and experimental level schemes. A.I. refer to calculations of ref. 3 and C.S. to those of ref. 4.

2. R. Béraud, I. Berkes, J. Danière, R. Haroutunian, M. Lévy, G. Marest, R. Rougny, Phys. Rev. 188 (1969) 1958.
3. G. Alaga, G. Ialongo, Phys. Lett. 22 (1966) 619.
4. A. Covello, G. Sartoris, Nucl. Phys. A104 (1967) 189.
5. D. Vinciguerra, K. Kotajima and R.E. Van der Vijver, Nucl. Phys. 77 (1966) 337.
6. R. Béraud, I. Berkes, R. Haroutunian, G. Marest and R. Rougny, Intern. Conf. Nashville, August 1969.
7. G.H. Dulfer, H. Ton, M.E.J. Wigmans and H. Verheul, to be published.
8. K. Taruishi, Progr. of Theor. Phys. 39 (1968) 53.
9. H. Ton, G.H. Dulfer, J. Brasz, R. Kroondijk and J. Block, to be published.
10. R. Béraud, I. Berkes, R. Haroutunian, G. Marest, R. Rougny, Rapport Annuel, Janvier 1970, Lycen 7002.

Contributed Paper

TABLES AND GRAPHS FOR THE ANALYSIS OF (1 - 3) $\gamma \cdot \gamma$ DIRECTIONAL CORRELATIONS

A.V. Ramayya, J.H. Hamilton, P.E. Little, and E. Collins
Vanderbilt University, Physics Department, Nashville, Tenn., U.S.A.*

J.J. Pinajian**
Oak Ridge National Laboratory, Oak Ridge, Tenn., U.S.A.

Introduction

Arfken, Biedenharn and Rose[1]) have pointed out as early as 1952 that the $\gamma \cdot \gamma$ directional correlations on the (1 - 3) cascade with the intermediate transition unobserved can be used to obtain multipolarities of transitions and level spins. Until now, because of poor resolution of the NaI detectors used in $\gamma \cdot \gamma\ (\theta)$ experiments, the analysis of the (1 - 3) correlations has been very complicated. With the recent development of NaI - Ge(Li) and Ge(Li) - Ge(Li) detector systems for $\gamma \cdot \gamma\ (\theta)$ experiments, the analysis of skip correlation data has become as feasible and profitable as consecutive cascades. The correlation coefficients for a skip cascade depend also on the spins of the intermediate levels and the multipolarity of the unobserved transition as well as the spins of the upper and lower states and multipolarities of the first and third gamma rays[2]) (see fig. 1).

Although convenient tables for consecutive cascades have been available[3]) for some time, similar tables have not been available for the 1 - 3 cascades. We have generated a set of tables and graphs for several of the most useful cases of spins and multipolarities for the (1 - 3) cascades.

THE CORRELATION COEFFICIENTS FOR THE (1-3) DIRECTIONAL CORRELATIONS ARE :

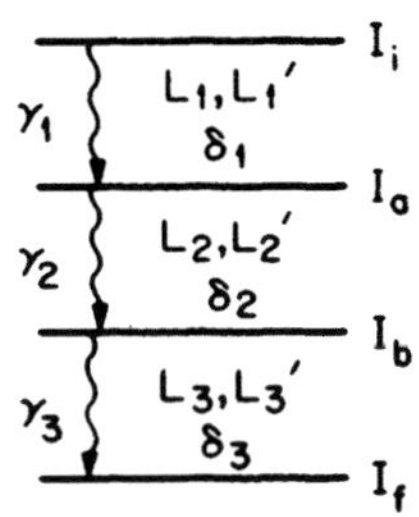

$$A_{kk} = A_k(\gamma_1)\, U_{kk}(\gamma_2)\, A_k(\gamma_3)$$

WITH $U_{00}(\gamma_2) = 1$ (NORMALIZED)

$$A_k(\gamma_1) = \frac{F_k(L_1 L_1 I_i I_a) - 2\delta_{em}(\gamma_1)\, F_k(L_1 L_1' I_i I_a) + \delta_{em}^2(\gamma_1)\, F_k(L_1' L_1' I_i I_a)}{1 + \delta_{em}^2(\gamma_1)}$$

$$A_k(\gamma_3) = \frac{F_k(L_3 L_3 I_f I_b) + 2\delta_{em}(\gamma_3)\, F_k(L_3 L_3' I_f I_b) + \delta_{em}^2(\gamma_3)\, F_k(L_3' L_3' I_f I_b)}{1 + \delta_{em}^2(\gamma_3)}$$

$$U_{kk}(\gamma_2) = \frac{(-1)^{L_2} \begin{Bmatrix} I_a & I_a & k \\ I_b & I_b & L_2 \end{Bmatrix} + \delta_2^2(\gamma_2)\, (-1)^{L_2'} \begin{Bmatrix} I_a & I_a & k \\ I_b & I_b & L_2' \end{Bmatrix}}{1 + \delta_2^2(\gamma_2)} (-1)^{I_a + I_b} \left[(2I_a + 1)(2I_b + 1)\right]^{1/2}$$

FOR $k \neq 0$

$$F_k(L\, L'\, I'\, I) = (-1)^{I' + I - 1} \left[(2L+1)(2L'+1)(2I+1)(2k+1)\right]^{1/2} \begin{pmatrix} L & L' & k \\ 1 & -1 & 0 \end{pmatrix} \begin{Bmatrix} L & L' & k \\ I & I & I' \end{Bmatrix}$$

WITH $F_0(L\, L'\, I'\, I) = \delta_{LL'}$

$$k \leq \mathrm{MIN}\,(L_1 + L_1',\ L_3 + L_3',\ 2I_a,\ 2I_b)$$

Fig. 1. (1 - 3) Directional Correlations

Tables and graphs

The relevant equations[2]) are given in fig. 1. Under the assumption that γ_3 is a pure multipole, each table is generated as a function of δ_1 for a given value of δ_2 and spin sequence. These conditions are useful for even - even nuclei which have 0^+ and 2^+ ground and first excited states and are often populated by energetic decays which have many triple cascades. For each spin sequence the values of δ_2 used are $\delta_2 = \infty$, 7.0, 4.36 and 0. Out of the 150 cases generated, graphs of 54 cases were constructed for easy analysis of the experimental data and will be available for the interested workers. The usefulness of such graphs is illustrated in figures 2, 3 and 4.

599 - (1277.5) - 89 Correlation in ^{156}Gd

The spin sequence of this triple cascade is $1^+ \to 1^- \to 2^+ \to 0^+$. Our A_K coefficients for this cascade are $A_2 = 0.154 \pm 0.037$ and $A_4 = -0.005 \pm 0.071$. For this spin sequence the theoretical A_4 value for a 1 - 3 correlation vanishes identically for all values of δ_1. The two values of δ_1 determined from these coefficients (fig. 2) are ~ 23 and 0.043 ± 0.043. Since very large M2 admixtures are not expected in competition with E1 admixtures, the δ_1 value of 23 is excluded. The δ_1 value of 0.043 ± 0.043 corresponds to an E1 component of $99.80^{+0.20}_{-0.54}$%.

646 - (1230.7) - 89 Correlation in ^{156}Gd

This correlation has the spin sequence $1^+ \to 2^- \to 2^+ \to 0^+$. We obtain $A_2 =$ $= -0.136 \pm 0.007$ and $A_4 = 0.003 \pm 0.012$. Fig. 3 illustrates the δ_1 determination from which $\delta_1 = 0.022 \pm 0.014$ results. This permits a $99.95^{+0.04}_{-0.08}$ % E1 admixture in the 646 keV transition.

812 - (1065) - 89 Correlation in ^{156}Gd

This correlation arises from a $1^+ \to 2^+ \to 2^+ \to 0^+$ spin sequence. With $A_2 =$ $= 0.045 \pm 0.007$ and $A_4 = -0.012 \pm 0.013$, the permitted mixing ratio value is $0 \leqslant \delta \leqslant 0.07$. Hence, $0\% \leqslant E2 \leqslant 5\%$ for the 812 keV transition (see fig. 4).

Conclusions

These tables and graphs are most useful for even - even nuclei and where the unobserved transition is most often a pure multipole and the third transition is a $2^+ \to 0^+$. This situation occurs often in deformed even - even nuclei and in spherical nuclei where the vibrational model works very well. These graphs (42 cases) are going to be published as an AEC Report. If the reader is interested in cases that are not published in the report, the generating program is readily available from the authors, and other tables can be generated by the user. Besides, a number of tables are given in Appendix I of the Proceedings.

* Work supported in part by a grant from the National Science Foundation.

** Research sponsored by the U.S. Atomic Energy Commission under contract with the Union Carbide Corporation.

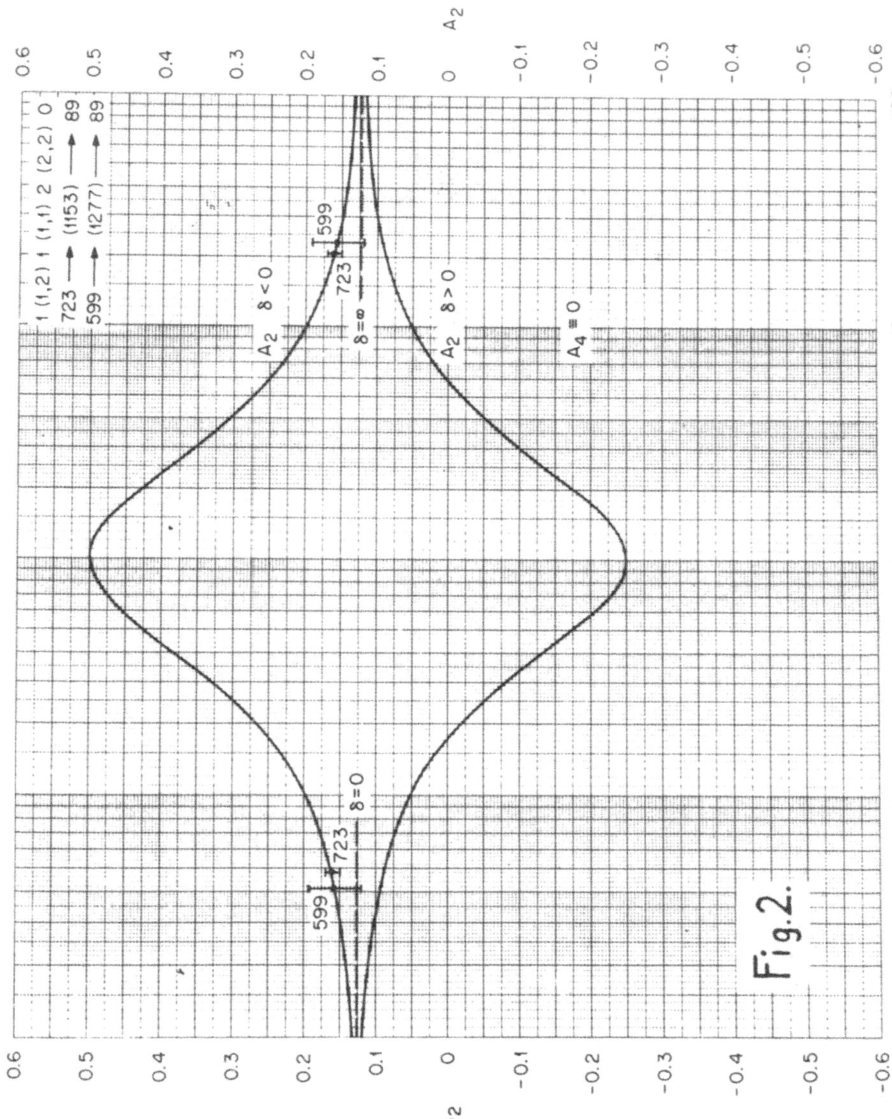
1 (1,2) 1 (1,1) 2 (2,2) 0
723 → (1153) → 89
599 → (1277) → 89
A2 δ < 0
A2 δ > 0
A4 ≡ 0
δ = ∞
δ = 0
599
723
A2
|δ|

Fig.2.

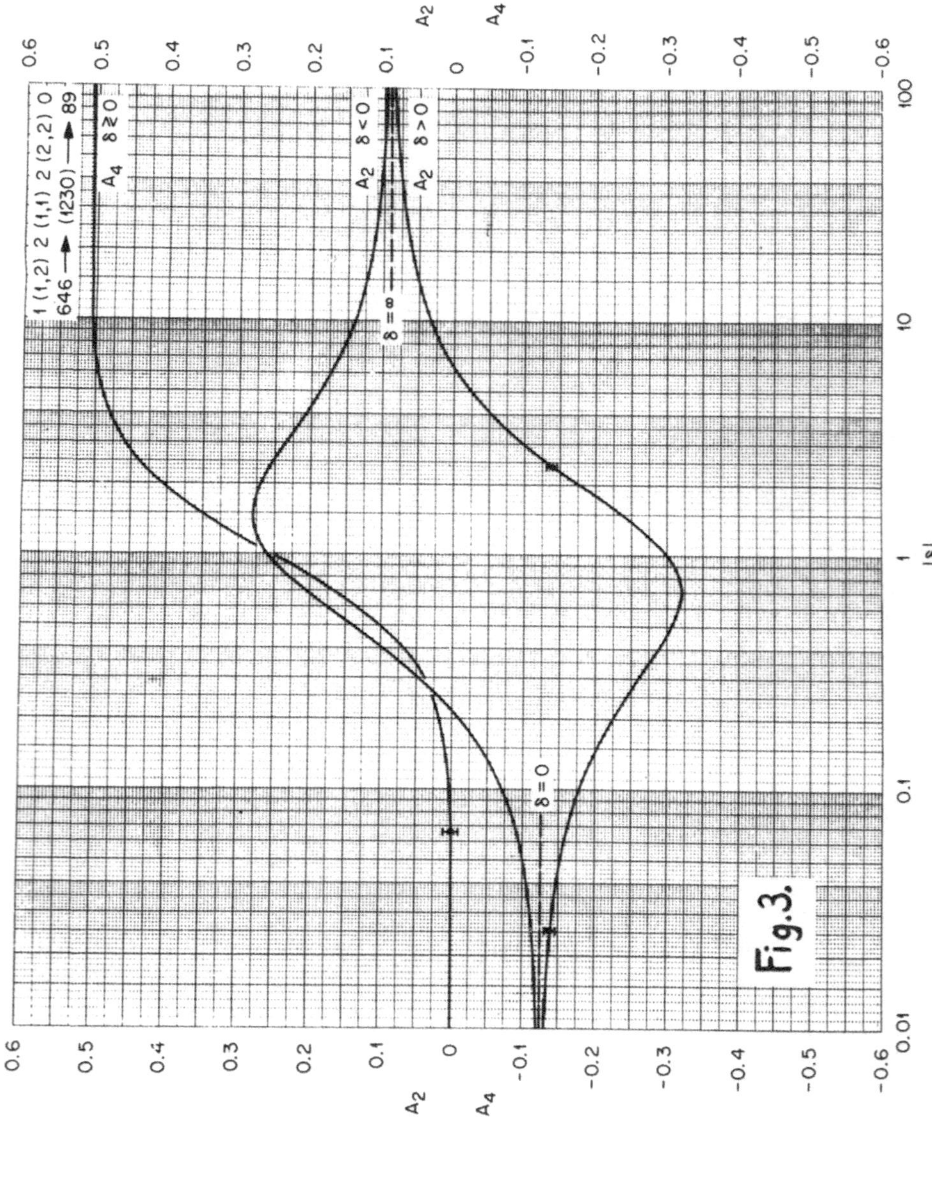

1 (1,2) 2 (1,1) 2 (2,2) 0
646 → (1230) → 89
A4 δ ≷ 0
A2 δ < 0
A2 δ > 0
δ = ∞
δ = 0
A2
A4
|δ|
0.01
0.1
1
10
100

Fig.3.

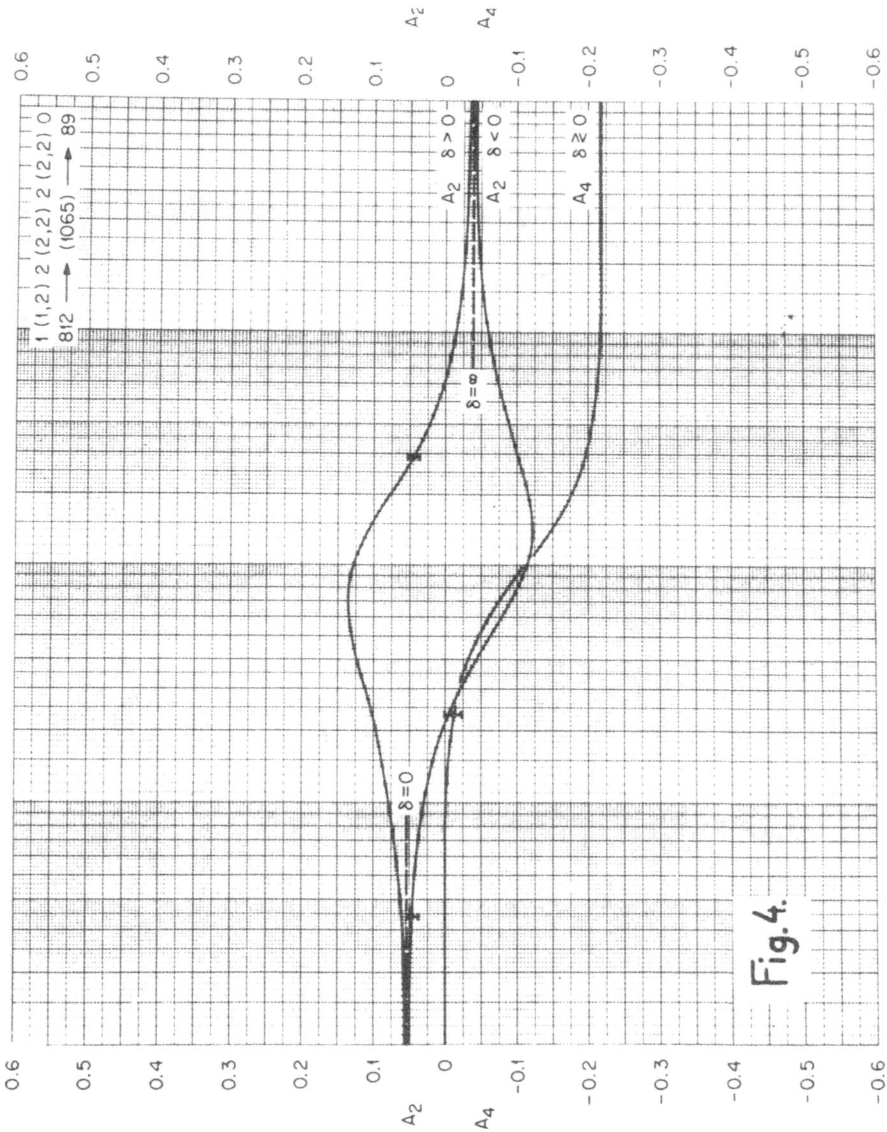

Fig. 4.

References

1. G.B. Arfken, L.C. Biedenharn and M.E. Rose, Phys. Rev. 86 (1952) 761.
2. R.M. Steffen, Angular Correlation Handbook, Purdue University, Lafayette, Indiana.
3. H.W. Taylor and R. McPherson, γ - γ Directional Correlation Coefficients A_2 and A_4 for Function of the Mixing Ratio. Physics Department, Queen's University, Kingston, Canada.

Contributed Paper

K X - RAY - γ - DIRECTIONAL CORRELATIONS IN THE DECAY OF ^{54}Mn, ^{65}Zn, ^{75}Se AND ^{85}Sr†

H.E. Bosch, M.A. Fariolli, M. Behar, and G. García Bermudez
Laboratorio de Radiaciones, IIAE, and Departamento de Física, Facultad de Ciencias Exactas y Naturales, Universidad de Buenos Aires, Buenos Aires, Argentina

1. Introduction

Correlations between circularly polarized X - rays and succeeding gamma rays, emitted after electron capture decay, have been theoretically described by Dolginov[1]). This author also predicted anisotropic X - γ - directional correlations if the electron capture occurs in the L_{III} subshell. It is not expected that any anisotropic X - γ - directional correlation will occur for other electron captures, namely in the K - shell.

Nevertheless, a somewhat contradictory result (2% anisotropy) has been reported by Perepelkin[2]) in the case of the K - electron capture decay of ^{54}Mn. Further work on K X - γ - directional correlations was done by Ramaswamy[3]) and by Mc Donnell and Ramaswamy[4]). These authors did not find, within experimental errors, any anisotropy in the correlation between K X - rays and γ - rays following the decay of ^{65}Zn, ^{85}Sr, ^{113}Sn, and ^{133}Ba. Therefore, re-study of the ^{54}Mn case as well as a study of other isotopes, which decay by electron capture, seems useful to probe any effect that might be in disagreement with the theoretically predicted isotropic K X - γ - directional correlations.

The present work deals with measurements on K X - γ - directional correlations occurring in the decay of ^{54}Mn, ^{65}Zn, ^{75}Se, and ^{85}Sr; this work is performed with a conventional scintillation set - up, a description of which is given in section 2. Results and discussions are presented in section 3 and conclusions are drawn in section 4.

2. Experiments

The radioactive sources were deposited as a spot of 2 to 3 mm diameter on a mylar film and were sprayed over. A thin plastic annulus was used as a frame.

The X - ray detector consisted of a NaI (Tl) crystal of 2.5 cm diameter and 2 mm thickness, covered with a beryllium window and coupled to a photomultiplier in an integrated unit.

Three other scintillation counters were set at 90°, 180° and 270° with respect to the source - to - X - ray detector direction. Two different sets of NaI (Tl) crystals were used, one of 5 cm diameter and of 5 cm height and the other of 7.5 cm diameter and of 7.5 cm height, both subtending a solid angle of about 1% steradian. Interchange of counters was made in order to avoid systematic errors.

Conventional fast - slow 'Cosmic Radiation' coincidence units, amplifiers, and single channel analyzers were employed. The same gamma - ray energy was discriminated in three single channel analyzers, and the corresponding pulses were coincided with the ones sorted by the X - ray detector. The coincidence pulses were sent into a logic unit which delivered a pulse to open the gate of a 512 - channel analyzer. The pulses sorted by the X - ray detector were fed into the input of the multichannel analyzer. From the logic circuit, three pulses were delivered to open three 128 - channel submemories.

Several runs of 25,000 counts each were taken. The areas of the X - rays peaks were normalized with respect to the single counting rate of the gamma rays detected in the corresponding counter. Corrections for chance coincidences were made and a χ^2 - test was applied. An anisotropy

$$A = \frac{C(180^o) - C(90^o)}{C(90^o)}$$

(C means normalized coincidences) was determined after adding all coincidences at each angle.

3. Results and Discussions

The ^{54}Mn decay

The ^{54}Mn decays by an allowed transition to the 845 - keV level of ^{54}Cr. Then a 5.4 - keV K X - ray and a 845 - keV gamma ray are emitted in cascade. The value of the measured anisotropy is shown in table 1, indicating an isotropic directional correlation within an accuracy of better than 1%. This result is in disagreement with the one previously reported by Perepelkin[2]).

The ^{65}Zn decay

The allowed decay of ^{65}Zn to the 1115 - keV level of ^{65}Cu was studied. The directional correlation between the 8.05 - keV K X - ray and the 1115 - keV gamma ray was measured and the anisotropy is quoted in table 1. This value is in agreement with the one obtained by Ramaswamy[3]), within experimental errors.

The ^{75}Se decay

Among the partial decays of ^{75}Se to the excited levels of ^{75}As, the allowed

Table 1. Values of the anisotropies reported from K X-γ-directional correlations measurements following the decay of ^{54}Mn, ^{65}Zn, ^{75}Se, and ^{85}Sr.

Parent nucleus	Electron capture transition	Electron capture branching %	K X-ray energy (keV)	γ-ray energy (keV)	Anisotropy A		Ref.
					Present results	Previous results	
^{54}Mn	$3^+ \rightarrow 2^+$	100	5.4	835	0.002 ± 0.003	0.025 ± 0.004	2
^{65}Zn	$\frac{5}{2}^- \rightarrow \frac{5}{2}^-$	49	8.0	1115	0.008 ± 0.014	0.006 ± 0.011	3
^{75}Se	$\frac{5}{2}^+ \rightarrow \frac{5}{2}^+$	94	10.5	401	0.003 ± 0.008		
^{85}Sr	$\frac{9}{2}^+ \rightarrow \frac{9}{2}^+$	99	13.4	514	0.03 ± 0.05	0.011 ± 0.016	3

decay (94%) to the 401 - keV level was chosen. The cascading 10.5 - keV K X - ray and the 401 - keV gamma ray did not show any anisotropy (see table 1).

The ^{85}Sr decay

The ^{85}Sr decays with an intensity of 99% to the 514 - keV isomeric state (~ 1 μs mean life) of ^{85}Rb. The directional correlation involving the 13.4 - keV K X-ray and the 514 - keV gamma ray was measured (see table 1). There is agreement, within experimental errors, with the value reported by Ramaswamy[3]).

4. Conclusions

The present measurements were carried out in a region of the periodic table between A = 54 and A = 85 and show no evidence of even a small anisotropy in K X - ray - γ - directional correlations. If any anisotropy would actually exists, it ought to be about 1% or less. The case of ^{85}Sr is rather trivial because it is expected to show an isotropic correlation owing to the long half - life of the intermediate state.

Further experimental studies on medium - weight and heavy elements are of interest to check whether the present results are also valid in other mass regions.

† Work sponsored by FORGE, INC., USA, and CNICYT, Argentina.

References

1. A.Z. Dolginov, Soviet Physics JETP 7 (1958) 644.

2. V.V. Perepelkin, Soviet Physics JETP Letters 5 (1967) 81.
3. M.K. Ramaswamy, Physics Letters 27B (1968) 215.
4. M. Mc Donnell and M.K. Ramaswamy, Phys. Rev. 171 (1968) 1278.

Contributed Paper

POTENTIAL OF $\gamma \cdot \gamma \cdot \gamma$ TRIPLE DIRECTIONAL CORRELATION STUDIES IN RADIOACTIVE DECAY

F.E. Coffman, J.H. Hamilton, A.V. Ramayya and M.U. Kim*
Vanderbilt University, Physics Department†, Nashville, Tenn., U.S.A.

R.L. Robinson
Oak Ridge National Laboratory††, Oak Ridge, Tenn., U.S.A.

1. Introduction

The theory for both double and triple directional correlations has been well known for years[1]). Gamma - gamma directional correlations have been long used for the determination of nuclear spins and the multiple admixture of transitions. While $\gamma \cdot \gamma \cdot \gamma$ directional correlations also are sensitive to these same parameters, this technique has been little used. However, triple angular correlations in nuclear reactions have been an important tool for determining spins and gamma - ray radiative admixtures. In these one or two of the three events are gamma rays. The success of this tool led to the recent suggestion[2]) that the triple angular correlation between three gamma rays from radioactive nuclei should be explored as a method for providing information on spins and radiative admixtures where other methods have failed.

We have carried out a feasibility study of $\gamma \cdot \gamma \cdot \gamma$ directional correlation measurements to test the usefulness of this method in radioactive decay. There are several different experimental configurations that can be used and each has its own theoretical expression for the angular dependence, W (θ) of the correlation.

It is well known[1-3]) that when two of the detectors of a triple correlation system are either parallel or perpendicular, a considerable simplification in the form of the directional correlation function results. Robinson[2]) has expressed the form of W (θ) for the geometry in which the first (γ_0) and third (γ_2) cascade gamma rays are counted at 180° relative to each other (fig. 1). This correlation function was derived from the more general form used by Ferguson[4]). The phase convention is that of Biedenharn and Rose[1]). The second gamma ray (δ_2) is counted at various angles relative to the direction of γ_0 or γ_2. The Clebsch - Gordon coefficients and 9 - j coefficients obey certain vector addition rules which yield selection rules for the

$$W(\theta) \propto a_0 + a_2 P_2(\cos\theta) + a_4 P_4(\cos\theta)$$

where

$$a_{k_1} = \sum_{k_0 k_2} C_{0\ 0\ 0}^{k_0 k_1 k_2}\ g_{k_0} g_{k_1} g_{k_2} \sqrt{(2k_0+1)(2k_1+1)}\,[F_{k_0}(11ab) - 2F_{k_0}(12ab)\delta_0 + F_{k_0}(22ab)\delta_0^2]$$

$$\cdot \left[3\,C_{1-10}^{1\ 1k_1}\begin{Bmatrix} b & 1 & c \\ b & 1 & c \\ k_0 & k_1 & k_2 \end{Bmatrix} - 2\sqrt{15}\,C_{1-10}^{1\ 2k_1}\begin{Bmatrix} b & 1 & c \\ b & 2 & c \\ k_0 & k_1 & k_2 \end{Bmatrix}\delta_1 - 5C_{1-10}^{2\ 2k_1}\begin{Bmatrix} b & 2 & c \\ b & 2 & c \\ k_0 & k_1 & k_2 \end{Bmatrix}\delta_1^2\right]$$

$$\cdot \left[F_{k_2}(11dc) + 2F_{k_2}(12dc)\,\delta_2 + F_{k_2}(22dc)\,\delta_2^2\right],$$

$$\delta_0 = \frac{\langle b||T(Q)||a\rangle}{\langle b||T(D)||a\rangle},\quad \delta_1 = \frac{\langle c||T(Q)||b\rangle}{\langle c||T(D)||b\rangle} \quad\text{and}\quad \delta_2 = \frac{\langle d||T(Q)||c\rangle}{\langle d||T(D)||c\rangle}$$

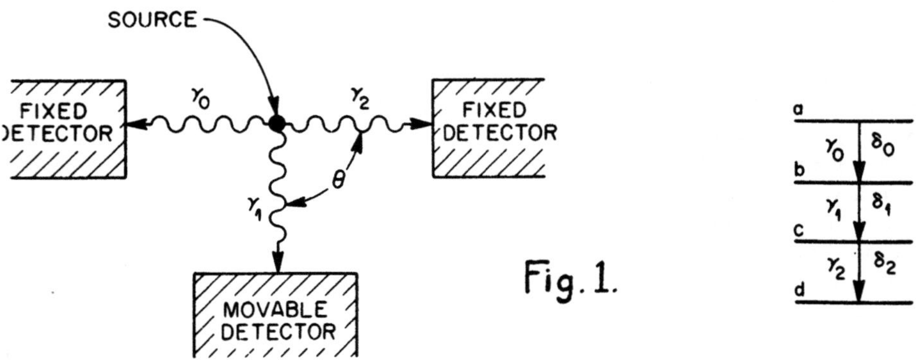

Fig. 1.

form taken by W (θ). A discussion of the selection rules applicable to other geometries is contained in the work of Biedenharn and Rose[1]).

2. Experimental Procedures and Results

To evaluate a triple directional correlation system, an isotope which contains a triple gamma - ray cascade with strong feeding through the cascade was desired. In ^{154}Eu, it is known that the relatively intense 591.8, 1004.8 and 123.1 keV gamma rays ($2^- \to 3^+ \to 2^+ \to 0$) feed - through strongly and each transition is essentially a pure multipole[5]).

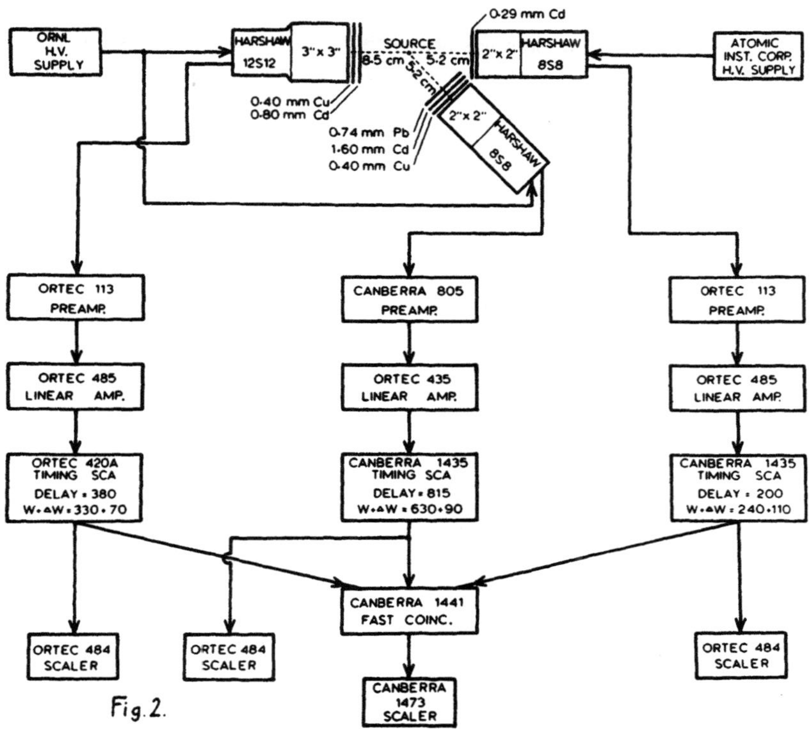
ORNL H.V. SUPPLY
HARSHAW 12S12
3" x 3"
SOURCE
8.5 cm
5.2 cm
5.2 cm
0.29 mm Cd
2" x 2"
HARSHAW 8S8
ATOMIC INST. CORP. H.V. SUPPLY
0.40 mm Cu
0.80 mm Cd
0.74 mm Pb
1.60 mm Cd
0.40 mm Cu
2" x 2"
HARSHAW 8S8
ORTEC 113 PREAMP.
CANBERRA 805 PREAMP.
ORTEC 113 PREAMP.
ORTEC 485 LINEAR AMP.
ORTEC 435 LINEAR AMP.
ORTEC 485 LINEAR AMP.
ORTEC 420A TIMING SCA DELAY = 380 W+ΔW = 330+70
CANBERRA 1435 TIMING SCA DELAY = 815 W+ΔW = 630+90
CANBERRA 1435 TIMING SCA DELAY = 200 W+ΔW = 240+110
CANBERRA 1441 FAST COINC.
ORTEC 484 SCALER
ORTEC 484 SCALER
ORTEC 484 SCALER
CANBERRA 1473 SCALER
Fig. 2.

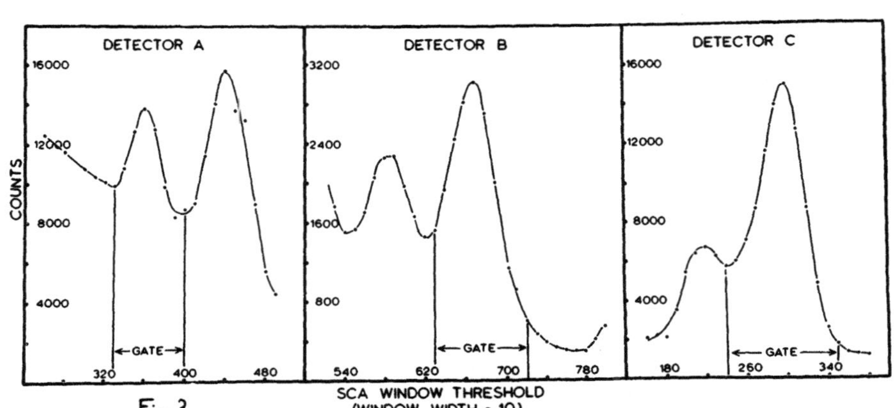
DETECTOR A
DETECTOR B
DETECTOR C
COUNTS
16000
12000
8000
4000
3200
2400
1600
800
16000
12000
8000
4000
GATE
GATE
GATE
320
400
480
540
620
700
780
180
260
340
SCA WINDOW THRESHOLD
(WINDOW WIDTH = 10)
Fig. 3.

The experimental arrangement and gate settings are shown in figures 2 and 3. Three double coincidence delay curves for each of the three possible double coincidence modes of the three input channels were run. Tests showed that, for the equipment being used (fig. 2) severe compromises had to be made. Many combinations of geometry, source strength, resolving time and absorber were tried in an effort to obtain a reasonable true triple coincidence count rate with an acceptable true - to - chance ratio and reasonable solid angle corrections. Finally, the geometry and absorber configuration of fig. 2 were selected. For this geometry, the source strength selected and the resolving time of 30 nanoseconds, a true - to - chance ratio of 1 : 1 was obtained with a total triple coincidence count rate of about 20 counts/hr.

Twenty - four hour total triple coincidence runs were taken with detector B at angles of 40°, 60° and 90° relative to detector C. The contributions to chance coincidences can be accounted for as follows:

$$N^{chance}_{total} = 2\tau \{N_{12\ tot.} N_{3\ singles} + N_{13\ tot.} N_{2\ singles} + N_{23\ tot.} N_{1\ singles} + \ldots . \}$$

where N_{ij} represents the double coincidence counts between detectors i and j and N_k is the singles gate count rate of detector k from the photopeak in that gate. The contribution from triple random events is omitted from the equation because it is negligible.

The first three terms were measured by introducing suitable delays. A sequence of seven 24 hour total triple coincidence runs were taken for each angle over a counting time of 21 days. The total triple coincidence counts were corrected for chance to yield the true triple coincidence counts at each angle, W (θ). These values were then substituted into the equation

$$W(\vartheta) = a + bP_2(\vartheta) + cP_4(\vartheta)$$

The data were analyzed in the standard manner with the solid angle corrections obtained from tables[6]). The attenuation effect from the lifetime of the 123 keV state is small[7]).

The experimentally measured angular correlation coefficients corrected for solid angle attenuation are: $A_2 = -(0.37 \pm 0.12)$, $A_4 = -(0.6 \pm 0.4)$. These values were compared to the theoretical coefficients for different values of δ_0 and for δ_1 and $\delta_2 = 10^4$. This comparison yielded choices of $-0.25 \leqslant \delta_0 \leqslant 0$ (5% dipole), or $1.5 \leqslant \delta_0 \leqslant 4.0$ (70 - 94% quadrupole). While the results are consistent with the known dipole character of the 591 keV transition, the data also allow quadrupole radiation.

As can be judged from the length of time and results obtained in this ideal case of ^{154}Gd, the technique has at best limited applicability from an experimental point of view, even with faster electronics.

A more careful examination of the triple correlation expression as given above reveals a more fundamental difficulty. If the ambiguities of the double correlation are in spin a or spin d, or in the admixutre δ_0 or δ_2, the triple correlation will have the same ambiguities. It can only resolve an ambiguity in spins b and c and in the admixture δ_1. This is also true for the general triple correlation in which the three detectors are placed at any angle in three dimensional space.

* NSF Summer Research Participant, 1969, from Lycoming College, Williamsport, Penn.

† Work supported by a grant from the National Science Foundation.

†† Research sponsored by the U.S. Atomic Energy Commission under contract with Union Carbide Corporation.

References

1. L.C. Biedenharn and M.E. Rose, Rev. Mod. Phys. 25 (1953) 729.
2. R.L. Robinson, Radioactivity in Nuclear Spectroscopy, eds. J.H. Hamilton and J.C. Manthuruthil (Gordon and Breach, New York, in press).
3. R.L. Robinson, F.K. McGowan, P.H. Stelson, W.T. Milner and R.O. Sayer, Nucl. Phys. A123 (1969) 193.
4. A.J. Ferguson, Angular Correlation Methods in Gamma - Ray Spectroscopy (North - Holland Publishing Co., Amsterdam, 1965).
5. R.A. Meyer, Phys. Rev. 170 (1968) 1089;

   L.L. Riedinger, Noah R. Johnson and J.H. Hamilton, Phys. Rev., in press.

Discussion

J.H. Hamilton: The part left out in the talk will be included in the paper. When you look at a triple correlation and look at the function more carefully you really discover that when a double cascade has given you an ambiguity in the delta, the triple cascade is going to carry in many cases, not in all cases, the same ambiguity and that even if you did the triple cascade you would end up with the same ambiguity that you got out of the double cascade. In some cases it will work, but the number of cases are few on both theoretical as well as experimental grounds. That we have not realized until we got into it more.

Broude: I just want to ask you if the choice of different geometries will effect your conclusion?

J.H. Hamilton: Yes, as far as we can determine, it will. There are obviously many geometries that one can try. We went through a number of configurations and came to the same conclusion, that even though you could vary the geometry considerably, this could not help in resolving the ambiguities. I will point out that there still remains a useful technique in particle - gamma - gamma correlations as the Oak Ridge people have already shown. There are many cases in particle - gamma - gamma correlations from where you can get information and resolve ambiguities.

DETERMINATION OF FIRST FORBIDDEN BETA MATRIX ELEMENTS FROM ANGULAR CORRELATION DATA†

P.C. Simms

Purdue University, Department of Physics, Lafayette, Indiana, U.S.A.

1. Introduction

First forbidden beta decay is interesting to study because a large number of nuclear matrix elements can be observed in a single transition. The large number of unknowns also makes it difficult to analyze the experimental data. Nevertheless, at the present time the experimental values of nuclear - matrix elements for several first forbidden transitions are more accurate than the nuclear model calculations which are used to understand the matrix elements. Thus these nuclear matrix elements provide a stringent test for the accuracy of nuclear model calculations.

An interesting relationship between isobaric analogue states and first forbidden beta transitions is illustrated in fig. 1. The state $|a\rangle$ is the isobaric analogue to the initial state $|i\rangle$ of the beta transition. There is a great deal of systematic experimental data which shows that the difference in energy between $|a\rangle$ and $|i\rangle$ is given primarily by two terms:

$2.4\,\frac{\alpha Z}{2\rho}$	the difference in coulomb energy caused by the addition of a proton
$\Delta E_m = 2.5\ m_e c^2$	the difference in the rest mass energy of a neutron and a proton.

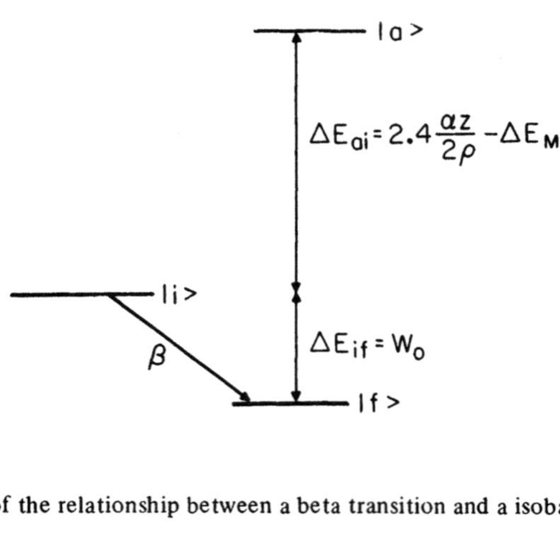

Fig. 1. Illustration of the relationship between a beta transition and a isobaric analogue state.

The total energy difference between states $|a\rangle$ and $|f\rangle$ is

$$E_{af} = 2.4\frac{\alpha Z}{2\rho} - 2.5\, m_e c^2 + W_0 \qquad (1)$$

The conserved vector current (CVC) theory of beta decay can be used to calculate the ratio Λ of two of the first forbidden matrix elements,

$$\Lambda \equiv \frac{\langle f|\underline{\alpha}|i\rangle}{\langle f|i\underline{r}|i\rangle}$$

One way to obtain the relationship for Λ is by forming the commutator $[H, i\underline{r}]$ of the nuclear hamiltonian and the coordinate operator $\underline{r}$. The nuclear hamiltonian is expressed as a sum of terms which represent kinetic energy H_K, coulomb energy H_C, rest mass energy H_M, and energy arising from nucleon - nucleon forces H_N.

$$H = H_K + H_C + H_M + H_N$$

Matrix elements are taken between the initial and final states for the total hamiltonian and for each of the separate terms.

$$\langle f|[H, i\underline{r}]|i\rangle = -W_0\langle f|i\underline{r}|i\rangle \qquad (2a)$$

$$\langle f|[H_K, i\underline{r}]|i\rangle = -\hbar c\langle f|\underline{\alpha}|i\rangle \qquad (2b)$$

$$\langle f|[H_M, i\underline{r}]|i\rangle = -(M_n - M_p)\, c^2 \langle f|i\underline{r}|i\rangle \qquad (2c)$$

The expressions for H_N and H_C are not as obvious. CVC theory can be used to extend the Seigert Theorem to vector first - forbidden beta decay (suggested independently by Fujita[1]) and Eichler[2])). Therefore it is expected that

$$[H_N, i\underline{r}] = 0 \qquad (2d)$$

Finally we consider the coulomb term which is most important for nuclear structure and isobaric analogue states.

$$\langle f|[H_C, i\underline{r}]|i\rangle = [\langle f|H_C|f\rangle - \langle i|H_C|i\rangle]\langle f|i\underline{r}|i\rangle + \\ + \{\sum_{f'\neq f}\langle f|H_C|f'\rangle\langle f'|i\underline{r}|i\rangle - \sum_{i'\neq i}\langle f|i\underline{r}|i'\rangle\langle i'|H_C|i\rangle\} \qquad (2e)$$

The term in square brackets is the difference in the coulomb energies of the initial and final states. As in the case of isobaric analogue states, this energy difference is caused by the addition of one proton, so the same estimate for the energy is used,

$$\Delta E_c = 2.4\frac{\alpha Z}{2\rho}$$

Z is the charge of the daughter nucleus, $\alpha = 1/137$, and ρ is the nuclear radius.

Ahrens and Feenberg[3]) suggested that the term in curly brackets can be neglected because the off diagonal elements of the coulomb hamiltonian ($\langle f|H_C|f'\rangle$, $\langle i'|H_C|i\rangle$) are small compared to the diagonal elements ($\langle f|H_C|f\rangle$, $\langle i|H_C|i\rangle$). This suggestion is supported by the existence of isobaric analogue states. The analogue state is quite similar to the parent state in spite of the fact that an extra proton has been added. If the off diagonal matrix elements of H_C were large, there would be a great deal of mixing in the analogue state.

When the individual terms in equations (2) are collected, and the Ahrens - Feenberg approximation is used, the expression for Λ can be written in the following form:

$$\Lambda = \frac{\langle f|\underline{\alpha}|i\rangle}{\langle f|i\underline{r}|i\rangle} = 2.4\frac{\alpha Z}{2\rho} - 2.5 + W_0 \tag{3}$$

Natural units where $m_e = \hbar = c = 1$ have been used. When eq. (3) and (1) are compared, it is seen that the matrix element ratio is equal to the excitation energy of the analogue state above the final state

$$\Lambda = E_{af} \tag{4}$$

(E_{af} in MeV is given by $0.511\,\Lambda$).

This simple relation has been used by a number of people to obtain information about beta decay transitions from measurements on isobaric analogue states. However, there is now clear evidence that the simple expression for Λ given by eq. (3) is **not** generally valid. The analysis of ^{140}La which will be discussed in detail later shows that Λ is quite different from the value predicted by eq. (3) In fact the relation $\Lambda = E_{af}$ would require that the energy of the isobaric analogue states would be several MeV below the systematic trend for isobaric analogue states. It is possible that the off diagonal elements of H_C are so large for ^{140}La that the analogue state is highly mixed – perhaps even looses its identity. However, this does not seem very likely since a great many isobaric analogue states have been found which do not deviate from the systematic line by more than 20 keV.

A more likely explanation for deviations from the simple expression for Λ has been proposed by Damgaard and Winther[4]). Even though the off diagonal elements of H_C are small, the correction term in eq. (2) could be large if the additional elements of the operator $\underline{r}$ ($\langle f'|i\underline{r}|i\rangle$, $\langle f|i\underline{r}|i'\rangle$) are larger than $\langle f|i\underline{r}|i\rangle$. Damgaard and Winther suggested that the matrix elements of the commutator be calculated using a realistic form[4]) for H_C. The result is,

$$\langle f|H_C, i\underline{r}|i\rangle = \frac{\alpha Z}{2\rho}(3-\lambda)\langle f|i\underline{r}|i\rangle$$

The parameter λ is the ratio of a higher order matrix element to the first order matrix element,

$$\lambda = \frac{<f|\,i\underline{r}\,(r/\rho)^2\,|\,i>}{<f|\,i\underline{r}\,|\,i>}$$

When $\lambda = 0.6$ the realistic calculation gives the same numerical value for Λ as the Ahrens - Feenberg approximation. In this more complete form, the expression for Λ is

$$\Lambda = \frac{\alpha Z}{2\rho}\,(3-\lambda) - 2.5 + W_0 \qquad (5)$$

It is important to realize that the beta decay matrix element ratio may be more sensitive to the off diagonal elements of H_C than the isobaric analogue state. In the expression for Λ, the effect of $<f\,|\,H_C\,|\,f'>$ can be enhanced by $<f'\,|i\underline{r}|\,i>$. When an unusual value of Λ is observed in beta decay, the effect on the energy of the isobaric analogue state need not be as dramatic as that suggested by the approximate expression for Λ (eq. 3)).

The treatment given here for the matrix element ratio Λ also provides a valuable way of getting additional nuclear structure information. The parameter λ is treated as an unknown, and it is varied until the experimental limits on Λ agree with the theoretical prediction given by eq. (5).

2. Beta Decay Matrix Element Extraction

The most important parameters that we wish to obtain from the experimental data are listed in table 1. A shorthand notation will be used to represent the matrix elements $<0> = <f\,|0|\,i>$. The experimental observables are primarily controlled by certain combinations of matrix elements. Therefore it is usually necessary to use the results of a number of different experiments in order to be able to set limits on the individual matrix elements.

Table 1

Matrix elements to be extracted from the experimental data

matrix elements	ΔJ	$\Delta\pi$
$<\gamma_5>$	0	yes
$<i\underline{\sigma}\cdot\underline{r}>$	0	yes
$<i\underline{r}>$	1	yes
$<\underline{\sigma}\times\underline{r}>$	1	yes
$<\underline{\alpha}>$	1	yes
$<iB_{ij}>$	2	yes

$\Lambda = <\underline{\alpha}>/<i\underline{r}>$ $\qquad\qquad$ $\lambda = <\underline{r}\,(r/\rho)^2>/<\underline{r}>$

The computer program[5,6]) which extracts the matrix elements can use the following experimental data: energy dependence of the $\beta \cdot \gamma$ directional correlation, energy and angular dependence of $\beta \cdot \gamma$ circular polarization, beta spectrum shape correction factor, angular distribution of gamma rays following the beta decay of oriented nuclei, longitudinal polarization of beta particles, and K - capture to positron ratio. It is not always necessary to have data available from all of these experiments in order to set reasonable limits on the matrix elements. Usually at least three different experiments are required. The most difficult situation occurs when the combinations of matrix elements referred to above dominate the transition. This frequently happens because the matrix element combinations are enhanced by the coulomb field of the nucleus. Unless a large variety of accurate experimental data is available, it is very difficult to set limits on the individual matrix elements.

The theoretical formulas which were used to calculate the experimental observables were derived by Bühring[7]). The formulas have been rearranged[5,6]) to make it easier to evaluate the contribution of the higher order matrix elements. The higher order matrix elements which are most likely to be important are included in the analysis as additional unknowns (nine higher order matrix elements are considered). The radial electron wave functions of Bhalla and Rose[8]) and the atomic screening factors of Bühring[9]) are used.

The computer program uses a random search for the matrix elements. In order to save time, the most sensitive characteristics of the experimental data are used to guide the search. Nevertheless, the more difficult isotopes require many hours of high speed computer time.

A second computer program is used to investigate which isotopes are most promising for further study. All of the available experimental data are used to find different sets of matrix elements which are consistent with the existing data. Then the unmeasured observables are calculated and arranged to show which experiments would be most useful in order to distinguish among the sets of matrix elements.

3. Results for the Matrix Elements

Of the many isotopes which have been studied in the course of this work, six provide good limits on the matrix elements. Most of the other isotopes will require additional experimental data before unambiguous solutions can be determined.

The size of the matrix elements and the limits of error are given in the references quoted for each isotope[6,10,11]). In order to provide a concise summary of the most important results, a different notation is used for the matrix elements in table 2. The notation $< \Delta J >$ $(\Delta J = 0, 1, 2)$ is used to represent the average value of the first order matrix elements which cause an angular momentum change of ΔJ. (That is, $<0>$ represents $< i\,\underline{\sigma} \cdot \underline{r}/\rho >$ and $< \gamma_5 >$). All of the matrix elements which contain the coordinate operator $\underline{r}$ have been divided by the nuclear radius ρ so they are normalized to have a maximum possible size of $\sqrt{2}$.

3.1 ^{140}La

The 2.2 MeV beta transition[6]) of ^{140}La provides a clear example of the need for a general treatment of the matrix element ratio Λ. When the off diagonal elements of H_C were neglected in the analysis of the experimental data (λ set equal to 0.6),

Table 2. Summary of the most significant features of the nuclear matrix elements.

Isotope	<0>	<1>	<2>	λ
$^{140}_{57}La_{83}$ $3^- \rightarrow 2^+$	0	0.09	0.04	2.45 ± 0.2
$^{86}_{37}Rb_{49}$ $2^- \rightarrow 2^+$	0.04	0.15	0.65	-0.6 → +0.8
$^{122}_{51}Sb_{71}$ $2^- \rightarrow 2^+$	0.06	0.13	0.09	-0.5 → +1.3
$^{124}_{51}Sb_{73}$ $3^- \rightarrow 2^+$	0	0.002	0.02	*
$^{76}_{33}As_{43}$ $2^- \rightarrow 2^+$	0.04	0.05	0.07	-10 → +0.8
$^{198}_{79}Au_{119}$ $2^- \rightarrow 2^+$	0.04	0.05	0.03 or 0.005	0.4 → 2.2

* No value of λ was found for which the experimental and theoretical value of Λ were in agreement.

the following results were obtained:

$$\Lambda \text{ (theory)} = 35$$
$$\Lambda \text{ (experimental)} = 21 \pm 3$$

The systematic trend for isobaric analogue states would require that the energy difference E_{af} would be 17.9 MeV. The experimental value of Λ (neglecting off diagonal elements of H_C) would predict that E_{af} is approximately 10.8 MeV. Unless the effect of the off diagonal elements of H_C on Λ are enhanced by the radial matrix elements, the effect of H_C on the energy of the isobaric analogue state would be dramatic! Fig. 2 shows the variation in the theoretical value and the experimental limits of Λ as a function of λ, that is, the contribution of the off - diagonal elements of H_C is treated as an unknown. The experimental limits on Λ change because an increase in λ also requires that high order matrix elements like $\langle f \mid \underline{r}\ (r/\rho)^2 \mid i \rangle$ make a larger contribution to the beta transition. For λ = 2.45 the experimental and theoretical value for Λ are in good agreement. Again it should be stressed that this result does not

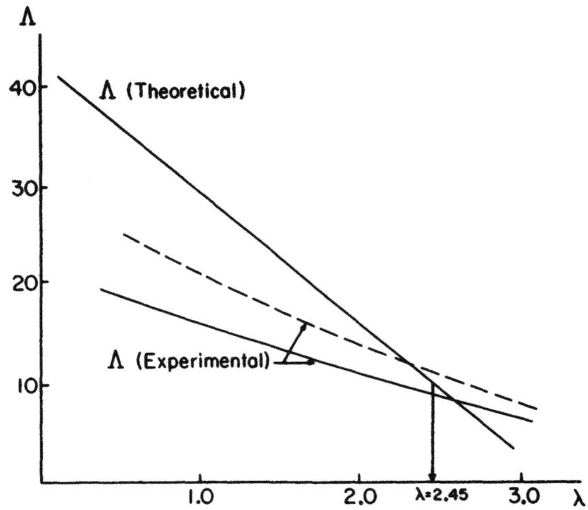

Fig. 2. Variation of the theoretical value and the experimental limits for the matrix element Λ as a function of the contribution of the off diagonal elements of the coulomb hamiltonian.

necessarily mean that there must be a large amount of mixing in the isobaric analogue state. Before the off diagonal elements of H_C can be evaluated, it will be necessary to estimate the size of the off diagonal radial matrix elements, and this must be done with a nuclear model which is constrained by the value of λ and the value of the other matrix elements which have been determined for this transition.

All of the matrix elements for this transition are reduced by an order of magnitude from the size that would occur for perfect overlap of the wave functions. This lack of overlap of the wave functions is not surprising since the last neutron in ^{140}La and the last proton in ^{140}Ce are in different major shells. The similarity in the size of the <1> and <2> matrix elements is also easy to understand because the shell model predicts that the transforming nucleon will have the same total angular momentum (7/2) before and after the transition. Thus it is equally probable for the lepton field to carry away 1 or 2 units of angular momentum. (<0> is identically zero for this $3^- \rightarrow 2^+$ transition). Since the matrix element $< f \mid i\underline{r} / \rho \mid i >$ is reduced by an order of magnitude, it is possible that the additional matrix elements ($< f' \mid i\underline{r} / \rho \mid i >, < f \mid i\underline{r} / \rho \mid i' >$) are large enough to account for the experimental value of Λ even if the off diagonal elements of H_C are not large.

The large size that is required for λ means that the higher order matrix element $< i\underline{r}\ (r/\rho)^2 >$ must be approximately 2.5 times larger than the first order matrix element $< i\underline{r} >$. This is an unusual constraint that a nuclear model calculation for this transition would have to meet. (To avoid confusion it should be noted that a higher order matrix element does not make a major contribution to the transition even if it is several times larger than the corresponding first order matrix element. In the formula for the transition probability, the higher order matrix elements have coefficients which are small[5,6]).

3.2 ^{86}Rb

The off diagonal elements of H_C do not make a significant contribution to the 710 keV transition[11]). The range of λ is limited and it contains the value 0.6 which corresponds to the Ahrens - Feenberg approximation. The computer program showed that a more accurate measurement of the energy dependence of the β - γ circular polarization would be most useful in setting better limits of error on λ. In fact if the present experimental data[11]) was confirmed and the limits of error were reduced by a factor of 3, the results for λ would be 0.5 ± 0.3.

The shell model gives a quite reasonable understanding of the order of magnitude of the matrix elements for ^{86}Rb. The primary component of the transition is expected to be a 1 $g_{9/2}$ neutron transforming into a 1 $f_{5/2}$ proton. Since these two particles are in the same major shell and $\Delta J = 2$ is allowed, the large size of $\langle 2 \rangle$ is to be expected. Configuration mixing must be responsible for the $\langle 0 \rangle$ and $\langle 1 \rangle$ matrix elements with $\Delta J = 0$ less likely than $\Delta J = 1$. This prediction is confirmed by the small size of $\langle 1 \rangle$ and still smaller size of $\langle 0 \rangle$.

3.3 ^{122}Sb

The contribution of the off diagonal elements of H_C can not be large for the 1.4 MeV transition of ^{122}Sb, and they may be zero[10]). In this respect the transition is similar to ^{86}Rb. The simple shell model predictions are also similar. The primary term in the transition would be expected to be a 1 $h_{11/2}$ neutron transforming into a 1 $g_{7/2}$ proton. Since both particles are within the same major shell, one would expect $\langle 2 \rangle$ to dominate the transition and be near to its maximum physical size ($\sqrt{2}$). However the experimental results do not confirm this prediction because the matrix elements for all angular momentum changes are reduced by an order of magnitude and they are approximately equal. In spite of the fact that there is only one proton above the major 50 shell, the shell model does not provide even a qualitative description of the transition.

3.4 ^{124}Sb

The 2.31 MeV transition of ^{124}Sb provides more dramatic evidence[10]) of the failure of the shell model to provide a basis for understanding these two nuclei. The shell model would predict that the primary component would be a 1 $h_{11/2}$ to 1 $g_{7/2}$ transition. The $\langle 2 \rangle$ would be quite large. However the experiments show that $\langle 2 \rangle$ is reduced by two orders of magnitude and $\langle 1 \rangle$ is reduced by three orders of magnitude.

^{124}Sb is the only isotope that we have investigated where it is impossible to obtain agreement between the experimental value of Λ and the theoretical value of Λ in its general form. This does not cast doubt on the validity of the general expression for Λ, but rather it is caused by the extreme suppression of the first order matrix elements. The most likely explanation[10]) is that the higher order matrix element $\langle \underline{a}\ (r/\rho)^2 \rangle$ is not suppressed as badly as the first order matrix element. If $\langle \underline{a}(r/\rho)^2 \rangle$

is approximately equal to 0.027 the experimental and theoretical value of Λ would agree for a wide range of λ.

3.5 ^{76}As

A very large contribution from the off diagonal elements of H_C is possible for ^{76}As, but the results are also consistent with a zero contribution[10]). Since neither the neutron or proton number is close to a major shell for ^{76}As, it is not reasonable to expect that the shell model would provide much insight into the nucleus. (In any case the same prediction would be made. The <2> would be large and should dominate the transition). A much more detailed model will be required to explain why all of the matrix elements are suppressed by more than an order of magnitude and are approximately equal in size.

3.6 ^{198}Au

The <2> is not nearly as well defined in the 959 keV transition of ^{198}Au as it is in the other isotopes. The combinations of matrix elements which are enhanced by the coulomb field of the nucleus dominate the transition, so it is difficult to extract the individual matrix elements. All of the applicable experiments have been performed on this nucleus with a high degree of accuracy, so it was possible to set good limits on <0> and <1>.

The contribution of the off diagonal elements of H_C need not be large, but a significant contribution is also possible. Again no simple single particle picture provides insight into this transition. All of the matrix elements are suppressed by more than an order of magnitude. This is probably caused by the transforming nucleons not being in the same major shell. The <0> and <1> are approximately equal in size, and the <2> may be the same size or an order of magnitude smaller.

3.7 Summary

This type of investigation provides very detailed information about nuclear structure and proposes a challenge for nuclear model calculations. We plan to continue experimental measurements and analysis with the hope of producing enough systematic information to guide the development of nuclear models. It should also be possible to use the information obtained about the off diagonal matrix elements of the coulomb part of the hamiltonian to obtain insight into particular isobaric analogue states.

4. Beta Energy Measurements at High Counting Rates

In beta - gamma circular polarization measurements the rate at which data can be collected is primarily limited by the counting rate that the beta detector can accept. Since the gamma detector only accepts scattered gamma rays, the gamma counting rate is usually small. The true to chance ratio need not limit the source strength if true and chance coincidences are recorded simultaneously.

We have developed[12]) a beta detector system which gives reasonable energy resolution and is very stable even at counting rates in excess of 10^6 counts per second (see fig. 3). A well shaped plastic scintillator is used to minimize back scattering of the beta particles. The scintillator is coupled to the photomultiplier with a short (1 inch) light pipe. A small (1/4 X 1/4 inch) incapsulated NaI (Tl) crystal which contains ^{241}Am is mounted in the light pipe. Alpha particles from the ^{241}Am cause the NaI crystal to emit well defined light pulses at an equivalent beta energy of greater than 10 MeV. The light pipe provides enough space so that the NaI light pulser uniformally illuminates the photocathode of the photomultiplier.

A bialkali high quantum efficiency photomultiplier is used. This type of tube provides approximately 30% better energy resolution than tubes with Cs - Sb photocathodes. The energy resolution is degraded by the shape of the plastic scintillator and the light pipe which contains the NaI crystal. Energy resolution of 12% was obtained for 1 MeV conversion electrons.

The linear output of the detector goes to a pair of double delay line clipped linear amplifiers (one with 100 ns clipping lines and the other with 750 ns clipping lines). The spectrum coming from the 750 ns amplifier shows a sharp peak from the light pulser well above intense 'noise' from the high beta counting rate. This peak is viewed by a single channel analyzer which controls the high voltage for the beta detector. The feedback systems works to keep the gain of the photomultiplier constant. The beta detector can be calibrated with a conversion electron source and the gain shifts by less than 2% when the high rate beta source is inserted in the system.

The short pulses from the 100 ns amplifier are used to measure the energy of the beta particles. (The amplifier is designed to recover rapidly from the large overload pulse produced by the NaI light pulser). A fast single channel analyzer is used to monitor the stability of the beta single counting rate. The pulses from the 100 ns linear amplifier also go to a gated pulse stretcher. When a β - γ coincidence occurs the beta pulse is stretched to 1 μs and fed into a conventional multichannel analyzer. Pile-up suppression is included in the beta channel by requiring a coincidence between a leading edge trigger and a zero crossing trigger.

† Work supported by the U.S. Atomic Energy Commission.

References

1. J.I. Fujita, Phys. Rev. 126 (1962) 202.
2. J. Eichler, Z. Phys. 171 (1962) 463.
3. T. Ahrens and E. Feenberg, Phys. Rev. 86 (1952) 86.
4. J. Damgaard and A. Winther, Phys. Letters 23 (1966) 345.
5. P.C. Simms, Phys. Rev. 138B (1965) 784.
6. H.A. Smith and P.C. Simms, Phys. Rev. 1C (1970) 1809.
7. W. Bühring, Nucl. Phys. 40 (1963) 472.
8. C.P. Bhalla and M.E. Rose, Oak Ridge National Laboratory, Report No. ORNL - 3207.
9. W. Bühring, Nucl. Phys. 61 (1965) 110.
10. H.A. Smith and P.C. Simms, to be published.

11. J.J. Bosken, D.E. Ohlms, and P.C. Simms, Phys. Rev. 3C (1971) 1168.

12. D.E. Ohlms, J.J. Bosken, and P.C. Simms, Phys. Rev. 1C (1970) 1803.

Discussion

Sastry: I have a comment and a question. First the comment. The parameter Λ connecting the vector matrix elements is very important. In the nuclei you talked about the final states are collective. On the otherhand, Fujita's estimate of Λ is particularly valid when the nuclear states involved are of single particle nature. In fact for odd-A nuclei (with shell model states) and ground - state to ground - state transitions in the spherical region, the comparison of Λ with Fujita's estimate seems to be good. Analog resonance work of Ejiri on ^{141}Pr is in agreement with this. This is also supported by β - decay work on ^{141}Ce (at Groningen, I suppose). My question is: The 2^+ first excited state of ^{140}Ce is very collective. Do you have a theoretical estimate of λ based on model dependent wave functions such as those given by ρ? If not, such a calculation would be very interesting.

Simms: Well, what we tried to show here, is that out of the 6 nuclei which have been worked out in detail, five are in agreement with Fujita - Eichler. But, they also can be in disagreement and there is one very clear example which is ^{140}La which clearly does not. So all I am saying is that in fact assuming the ratio of these two matrix elements is a very good place to start an extraction, it is a very good place to start your thinking. But you cannot depend upon it.

The λ is experimentally determined; the value for λ came off that graph, because you just move the λ until the experimental and theoretical values agree, so this we consider being a measured value of the ratio of those two matrix elements. I agree that a calculation of λ as you suggest would be very useful.

Manthuruthil: I wish to make a comment concerning the use of lambda. Recently, a paper has appeared in Physics Letters by two Russian authors concerning this ratio lambda. They point out that the use of the ratio lambda according to the procedure suggested by Damgaard and Winther is incorrect because they neglect the interaction of the quasi - particles of the nucleus, namely the residual force. They have calculated this ratio for many different nuclei including the residual force, using the Damgaard - Winther procedure, and using Fujita's relation for comparison. They find that for the cases they investigated, their ratio is in better agreement with Fujita than with Damgaard and Winther. Therefore, there is a lot of confusion as to what lambda one should use. These things do not affect the results presented by you, but is intended for those people who might want to use the ratio lambda to eliminate one parameter when they try to determine matrix elements from experimental data.

Coussement: In the CVC relation of Fujita you use Coulomb forces approximated by a single nuclear potential. In this case a relation between the second forbidden matrix element and the Fermi - matrix element can be found. We measured this relation on ^{66}Ga and found a serious disagreement. This means that one has to be careful when using the single nucleon approximation for the Coulomb forces.

Simms: This is an area that dr. Bloom will get into in the next paper, so I won't comment. Dr. Manthuruthil and I talked about this Russian paper and unfortunately I have not seen it. And perhaps I should leave it there; however my impression is the following. Damgaard and Winther calculated on the basis of nuclear models what the lambda should come out to be. We made no use of that whatsoever. As far as we are concerned we do not know nuclear wave functions, we do not attempt to calculate them. What we do is measure the overlaps experimentally. So I think –although I have not seen the paper and I am going to look at it at my first opportunity– that the Russian results will perhaps change the meaning of lambda but will not affect our results at all.

Manthuruthil: The only reason why I mentioned the Russian paper was that a lot of people use these ratios to eliminate the number of parameters and then determine the matrix elements. That is what I was actually referring to.

Simms: We avoid doing it.

Coussement: If you have some lambda which is not zero, you suddenly have to use a sort of Bühring approximation not the Bhalla approximation or such.

Simms: We always use the Bühring approximation. As I said the reason why the theoretical line in my graph goes down is obvious; the reason why the experimental line goes down is just what you are stating now. That is, when you do have the off-diagonal contributions or when you know that the higher order matrix elements are significant you have to bring them in a consistent way; that is precisely why those lines go down.

Sujkowski: Is the set of the experimental data you have presented a complete one? It seems to me that it would be very interesting to obtain information about β - transitions to two pure rotational states of the same band in a well deformed nucleus. This could give you a sort of cross - check for any possible interpretation.

Simms: The data I have presented is a complete set only in the following sense, it is all that we have finished including the lambda analysis. Now, presented at this conference is a result of Manthuruthil on ^{152}Eu. But most of the work has been around the single particle area.

Alder: I like to make a comment on the structure of the λ - parameter, which is very similar to the penetration parameter in E1 internal conversion. If the E1 transition is retarded, the λ - parameter is often larger than one. A similar situation might also occur here.

Simms: Thank you.

β - γ (CP) CORRELATION MEASUREMENTS IN ALLOWED TRANSITIONS†

S.D. Bloom

Lawrence Radiation Laboratory, Livermore, California, U.S.A.

1. Introduction

In the following I will review the results which have been achieved by applying the technique of β - γ (CP) correlation to allowed β - decays. My emphasis will be almost entirely on the kind of nuclear information which can be elicited from these measurements. The working assumption will be that all questions regarding the nature of the β - interaction are understood well within present - day experimental precision or theoretical understanding, insofar as any significant effects pertaining to nuclear structure are concerned. I will review briefly the basis for these statements later in this section.

I will explicitly not review the experimental technique per se because several excellent reviews have been available for some years now[1,2,3]). I will, however, begin by giving the formula for the correlation in a somewhat more general form than usual, the purpose being to include cases wherein the β and γ quanta whose correlation is

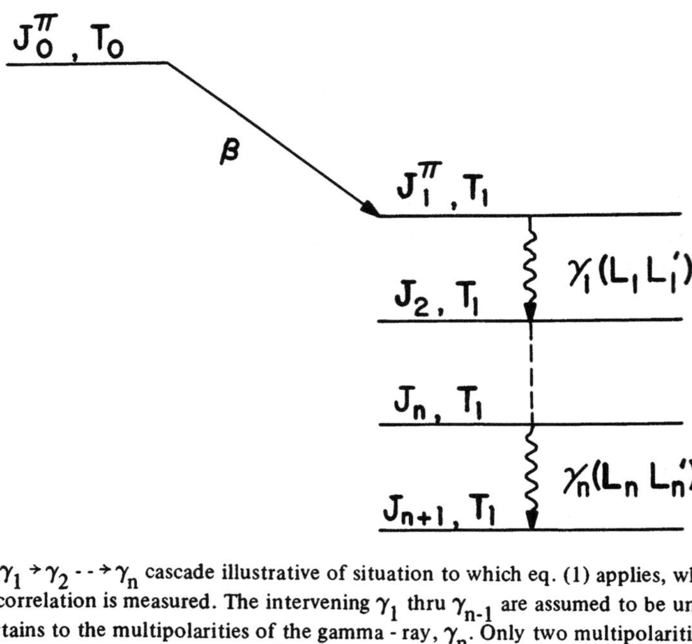

Fig. 1a. $\beta \to \gamma_1 \to \gamma_2 \cdots \to \gamma_n$ cascade illustrative of situation to which eq. (1) applies, wherein the $\beta \to \gamma_n$ (CP) correlation is measured. The intervening γ_1 thru γ_{n-1} are assumed to be unobserved. (L_n', L_n) pertains to the multipolarities of the gamma - ray, γ_n. Only two multipolarities (at most) are assumed to be significant and $L_n' \equiv L_n + 1$. The levels are labeled by spin - parity (J^π) and the 'ostensible' isospin, $T \equiv (N - Z)/2$.

being measured are not necessarily immediately sequential. I do this partly because of the possible bearing on new cases for investigation by this technique (see section 4) and partly to exhibit the convenience of the formulation (see below).

Fig. 1 illustrates the situation in general. The formula is as follows:

$$W(\vartheta_n, \tau_n) = \frac{1}{4\pi}(1 + A\tau_n \frac{v}{n} \cos\vartheta_n) \tag{1a}$$

$$A = B(J_0 J_1 y) \times \left\{ \prod_{p=1}^{n-1} U_p(J_p J_{p+1} \delta_p) \right\} \times F(J_n J_{n+1} \delta_n) \tag{1b}$$

where v is the velocity of the detected β - particle, c the velocity of light, θ_n the angle between the direction of the β - particle and γ_n (see fig. 1), τ_n the sense of circular polarization of γ_n ($\tau_n = \pm 1$ for ($^{\text{right}}_{\text{left}}$) - circularly polarized gammas), and A is the asymmetry parameter given by eq. (1b). Eq. (1b) is formulated as a product of three

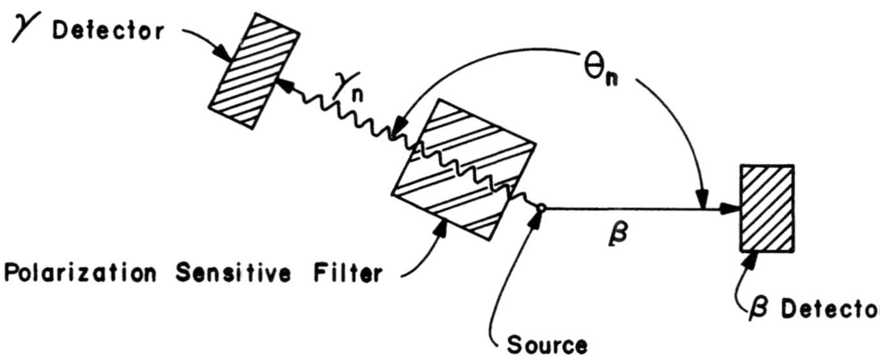

Fig. 1b. Idealized geometry pertaining to eq. (1). The polarization filter (ideally) permits only $\tau = +1$ or -1 γ - rays thru to the γ - detector. The β - detector and γ - detector are both energy - sensitive. In the former case this affects the measured polarization since the degree of polarization is proportional to the β - velocity (see eq. (1)).

factors, which separates into three stages the decay sequence being studied (see fig. 1). The first stage is the (allowed) β - decay, $J_0^\pi \rightarrow J_1^\pi$, where $J_1 = J_0 - 1, J_0$, or $J_0 + 1$ and π is the parity (initial parity and final parity are the same). The detection of the β - particle determines the population of the substates J_{1z} in the residual nucleus, which is expressed by the factor B $(J_0 J_1 y)$, closely related to the particle parameters defined by Schopper[3,4]). As already indicated this can be reliably calculated because we now have precise knowledge of the nature of the β - interaction, namely the well-substantiated V - A theory.

For allowed β - decay this theory[2,3]) yields the following familiar formula for B $(J_0 J_1 y)$,

$$B(J_0J_1y) = \pm \frac{0.2887}{(1+y^2)} \left[\frac{J_1(J_1+1) - J_0(J_0+1) + 2}{[J_1(J_1+1)]^{\frac{1}{2}}} \pm 4y \right]$$

$$+ \equiv \text{positrons}, \; - \equiv \text{electrons} \tag{2a}$$

where, neglecting second - forbidden corrections (see section 3), y is given as follows:

$$y \equiv C_V < 1 > / \, C_A < \underline{\sigma} > \tag{2b}$$

$\langle 1 \rangle$ is the Fermi or vector matrix element, $\langle \underline{\sigma} \rangle$ is the Gamow - Teller or axial - vector matrix element*, and C_V/C_A is the ratio of the coupling constants for these two modes of β - decay. Note that,

$$< 1 > = 0 \text{ unless } J_0 = J_1, \tag{2c}$$

the familiar selection rule pertaining to the allowed Fermi decay.

The second factor consists of iterated products of $U_p\ (J_pJ_{p+1}\delta_p)$, where p covers the range 1 to n - 1, i.e. the unobserved gamma rays (see fig. 1). Each U_p produces an alteration in the population of the magnetic substates $J_{p+1,z}$, appropriate to the γ_p transition $J_p \to J_{p+1}$. U_p is given by,

$$U_p(J_pJ_{p+1}\delta_p) \equiv \frac{1}{(1+\delta_p^2)} \left\{ V(L_pJ_pJ_{p+1}) + \delta_p^2\, V(L'_pJ_pJ_{p+1}) \right\} \tag{3a}$$

$$V(LJ_aJ_b) \equiv \frac{J_a(J_a+1) + J_b(J_b+1) - L(L+1)}{2\,[\,J_a(J_a+1)\,J_b(J_b+1)\,]^{\frac{1}{2}}} \tag{3b}$$

$$L' \equiv L + 1 \tag{3c}$$

$$\delta \equiv < X(L') > / < Y(L) > \tag{3d}$$

In eqs. (3) we have permitted the possibility of mixed multipolar electromagnetic radiation for all the γ's. This is expressed by the ratio of the electromagnetic matrix elements (or amplitudes) pertaining to γ_p, which we call δ_p. (Usually $\delta = \langle E2 \rangle / \langle M1 \rangle$, i.e. the ratio of electric quadrupole to magnetic dipole amplitudes). It is no surprise to find $|U_p|$ is always $\leqslant 1$, and thus the more intervening gamma rays there are the more the correlation is attenuated. It is also noteworthy the U_p can be < 0, which means an intervening gamma ray can change the sign of the correlation.

The last factor of eq. (1b) pertains to the observed γ_n. As is well known, this factor is given by,

* I have preferred to use the more descriptive symbols $\langle 1 \rangle$ and $\langle \underline{\sigma} \rangle$ rather than M_V and M_A for these matrix elements. Both sets of symbols are commonly used (see ref. 4).

$$F(J_nJ_{n+1}\delta_n) = (1+\delta_n^2)^{-1} \times F_1(L_nL_nJ_{n+1}J_n) + \delta_n^2 F_1(L'_nL'_nJ_{n+1}J_n) + 2\delta_n F_1(L_nL'_nJ_{n+1}J_n) \quad (4)$$

where δ_n is defined exactly as in eq. (3d) and the coefficients F_1 $(LL'J_aJ_b)$ are tabulated in several references[5]).

Eq. (1) has the notable virtue that its factored representation makes particularly transparent the kind of nuclear problems which can be studied using an experimentally determined value for A, the asymmetry parameter. (It is perhaps gratuitous to remark that a similar representation can be made for a β - transition of any degree of forbiddenness). First we note that for $J_1 = J_0 \pm 1$, i.e. for a pure Gamow - Teller transition, B has a uniquely defined value, since $y = 0$ in this case. This type of transition was, in fact, one of the earliest means of demonstrating the break - down of parity in the β - interaction[3]). Thus a measurement of A ought to be a very useful means of getting information about spins and/or multipole mixing ratios in the subsequent γ - decay cascade. In fact several experiments of this type have been done, but they hardly number ten in all. These experiments are discussed in the next section, where some reasons for this low number are mentioned. It is worthwhile noting that in many cases information regarding spin and multipole orders can also be deduced from mixed transitions, i.e. when $J_0 = J_1$ (see below).

In fact by far the most frequently investigated transitions is the mixed transition $J_1^{\pi} = J_0^{\pi}$. In this case $\langle 1 \rangle$ and therefore y may be non - zero, at least in principle. Thus from eq. (1) we see that, given all the requisite knowledge concerning spins, parities, and multipoles, a value of y can be deduced from a measurement of A. The number of measurements of this type is now of the order of a hundred or more (not all of different transitions). The reason for this is that a sufficiently accurate value for y combined with the ft value, which is inversely proportional to $[C_V^2 \,|\langle 1 \rangle|^2 + C_A^2 \,|\langle \underline{\sigma} \rangle|^2\,]$, can lead to a determination of $|\langle 1 \rangle|$ [6]), a most interesting but usually very elusive nuclear datum. In all cases to date $|\langle 1 \rangle|$ has been found to be $\ll 1$ (see section 3). This is now understood to be principally in consequence of the operation of another selection rule pertaining to the Fermi matrix element, namely that the isospins of the initial and final states connected by the allowed β - decay must be the same, i.e. T_0 muxt equal T_1. There is nothing trivial or obvious about this statement, since its proof depends upon a deeper consequence of the V - A theory than we have used so far. It is a result of the conserved vector current or CVC feature of the complete V - A interaction, where by 'complete' we mean extended to include all particles, including the strongly interacting particles, particularly mesons. The final issue is that the β - interaction by itself cannot introduce any breakdown of the $\Delta T = 0$ selection rule[7]) for $\langle 1 \rangle$. Thus a non - zero value for $\langle 1 \rangle$ means that of the two (nuclear) states characterized pre - eminently by $T_<$ and $T_>$, where $T_<$ denotes the lesser and $T_>$ the greater of the two isospin values T_0 and T_1 (see fig. 1), the $T_<$ state must contain some impurity of $T = T_>$. We can further say that $T_>$ must $= T_< + 1$, mainly because we expect (for example) that the nuclear transition $(Z, N) \to (Z+1, N-1)$ will be characterized (in the large) by an isospin transition of $(T_0 \equiv (N - Z/2)\,)$ to $(T_0 - 1)$. Therefore

Table 1. List of known cases in which β-γ(CP) correlation measurement yielded information concerning spin and/or multipole mixing ratios (δ_1). The references are to the earliest work on each particular case and do not necessarily represent the latest work on the subject, the main purpose here being to display the early history.

Initial Nucleus	Cascade Sequence			Decay ($\beta^{\pm}$)	γ-Energy	Extracted Info.	Ref.	Date
	J_0^π	J_1^π	$J_2^{\pi'}$					
$^{20}_{9}F_{11}$	2^+	2^+	0^+	β^-	1.63	$J_0^\pi = 2^+$	9	1961
$^{27}_{12}Mg_{25}$	$1/2^+$	$3/2^+$	$5/2^+$	β^-	1.01	$\delta_1 = -0.322$	15	1969
$^{28}_{13}Al_{15}$	3^+	2^+	0^+	β^-	1.77	$J_0^\pi = 3^+$	8	1965
$^{48}_{21}Sc_{27}$	6^+	6^+	4^+	β^-	1.04	$J_0^\pi = 6^+$	41	1958
$^{49}_{20}Ca_{29}$	$3/2^-$	$3/2^-$	$7/2^-$	β^-	3.11	$J_1^\pi = 3/2^-$	16	1970
$^{52}_{23}V_{29}$	2^+	2^+	0^+	β^-	1.43	$J_0^\pi = 2^+$	17	1967
$^{57}_{28}Ni_{29}$	$3/2^-$	$3/2^-$	$7/2^-$	β^+	1.38	$J_0^\pi = 3/2^-$	18	1967
$^{64}_{31}Ga_{33}$	0^+	1^+	0^+	β^+	3.40	$J_0^\pi = 0^+$	13	1966
$^{95}_{41}Nb_{54}$	$9/2^+$	$7/2^+$	$5/2^+$	β^-	0.76	$J_1^\pi = 7/2^+$	19	1961
						$\delta_1 \lesssim 0.2$		
$^{103}_{44}Ru_{59}$	$5/2^+$	J_1^+	$7/2^+$	β^-	0.50	$J_1 = 5/2, 7/2$	14	1970
						$\delta_1 > 2.0$		

a determination of ⟨1⟩ is a measurement of isospin purity, since it must 'travel' via the $T_> \rightarrow T_>$ route. Needless to say isospin purity is a topic of much past and current concern in nuclear physics. This is the reason, of course, for all the activity in this field. Also, the consistently observed smallness of ⟨1⟩ and hence y explains why, in some cases at least, a $J_0^\pi \rightarrow J_0^\pi$ β - transition may be usable for getting information on spins and multipole mixing. It is because of the fact that $y \approx 0$, practically speaking, leads to an (almost) unique value for B in eq. (1) (see above).

The above, in toto and in brief, summarizes the basics for the delineation of the two general types of nuclear physics which can be done by measuring A in allowed β - γ cascades. A review of each of these two areas is presented in the next two sections. The last section is devoted to general comments and a provisional prognosis for the future of this type of research.

2. Spins and Multipole Moments

In table 1 I have given a list of all the earliest, different cases I could discover in the literature wherein an explicit measurement of A in an allowed β - γ cascade yielded significantly new information on spins and/or multipole mixing ratios. Parity, of course, is never explicitly mentioned in eq. (1), and in fact the β - γ (CP) correlation (by itself) is parity independent. This statement is evidently true only for the 'γ' parts of eq. (1), but it is also true for B (eq. (2a)), the 'β' part, as was shown by Weidenmuller[8]) in the ξ - approximation. However ft - values plus basic knowledge of

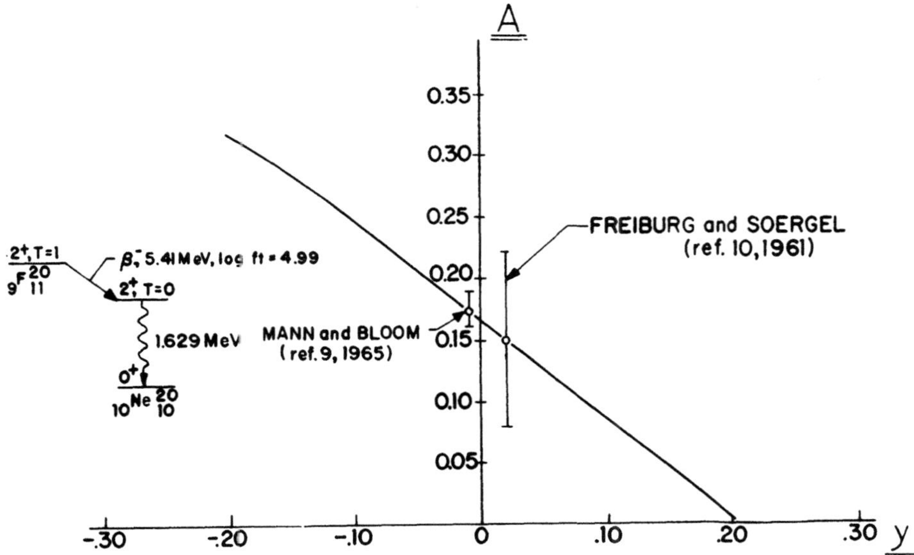

Fig. 2. Plot of A (correlation asymmetry parameter) vs y (see eq. (2b)) for the case of ^{20}F. Another measurement (ref. 11, 1962) gave essentially identical results to the 1961 work.

the systematics of nuclear structure usually gives the parity for low - lying states, where not otherwise known. For states of high excitation as in ^{64}Ga (see below), the parity assignment can be a problem. It should also be noted that almost all (allowed) β - γ (CP) correlation measurements, regardless of the value of $J_1 - J_0$, have some bearing on spin assignments and/or multipole determinations. Thus, for example, even the earliest measurements[3]) on ^{60}Co and ^{22}Na provided at least confirmatory evidence for the ground state spins of both these nuclei. (Also see the case of ^{49}Ca in the next section). However, in these cases, as in many others, other means were (and usually still are) used to make the final spin assignments.

One of the earliest and particularly convincing examples of the subject of this section is the case of ^{20}F. The original object in this case was the determination of the spin of the ground state of ^{20}F from the possible sequences $(3^+, 2^+) \xrightarrow{\beta} 2^+ \xrightarrow{\gamma} 0^+$. This is an ideal case from many points of view. To start with J_1^π and J_2^π were well - known and the assignment $J_0^\pi = 1^+$ was eliminated by the non - observation of any β-decay to the ground state ($J_2^\pi = 0^+$). Furthermore the ^{20}F decay was unencumbered by any other γ's besides the J_1^π to J_2^π transition in ^{20}Ne. Finally ^{20}F was spectacularly easy to manufacture, via the ^{19}F (d, p) ^{20}F reaction (Q = +4.4 MeV). A deuteron beam of energy $\approx$ 1.5 MeV and intensity $\approx$ 0.5 μA was more than adequate[9]). The first measurement was done in 1961 [10]). The results of it and the latest measurement (ref. 9) are shown in fig. 2. The $3^+ \rightarrow 2^+ \rightarrow 0^+$ sequence would have required A = - 1/3.

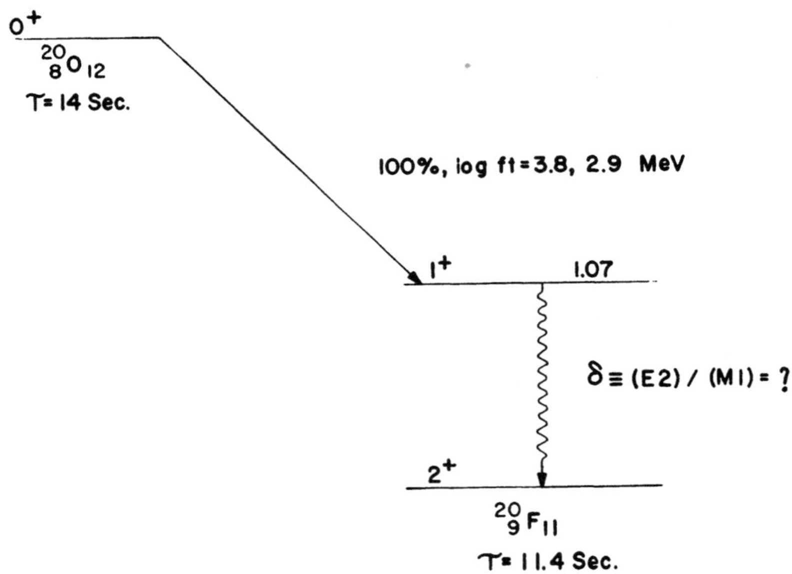

Fig. 3a. Present status of decay scheme of ^{20}O. The β - γ (CP) correlation measurement in this case would provide an accurate value for δ, the multipole mixing parameter. The super - allowed value of the log ft rules out any spin - parity assignment other than 1^+ for the 1.07 MeV level in ^{20}F.

The very first result[9]) therefore ($A = +.15 \pm .07$) was tantamount, in my opinion, to a spin measurement. The subsequent measurements, however, served more than a confirmatory exercise (see section 3).

Needless to say cases as ideal as ^{20}F are rare, but there were a few which were almost as good (see ^{52}V and ^{28}Al in table 1, for example). It now appears that most of the very easy cases have been done, though maybe not all (see below). However, let me continue to use the atomic weight $A = 20$ as a specific illustration of what can yet be done if just the problem of manufacturing the source could be solved. The next $A = 20$ case of interest in this context is, I believe, the $\beta \cdot \gamma$ decay of ^{20}O, shown in fig. 3a. It is evident from the super - allowed log ft value of 3.8 that J_1^π must be 1^+ in this case (see section 3), although the various nuclear level compilations consider the $J_1 = 1$ assignment as in question. From the point of view of nuclear structure the log ft value is almost surely saying that a major fraction of the transition must be $(1d\ 3/2)^4_{0^+} \longrightarrow (1d\ 3/2)^4_{1^+}$, a rather basic kind of a super - allowed transition. The measurement of A, the asymmetry parameter, would quickly yield a value for $\delta \equiv \langle E2 \rangle / \langle M1 \rangle$ for the γ - transition to the ground state, as is seen from fig. 3a. Considering the highly likely pre - eminence of $(1d\ 3/2)^4$ in the structure of the 1.06 MeV state and the degree of sophistication of present day calculations in the s - d shell[12]) it is clear that this could be a most useful piece of information. I hasten to add the comment that other techniques could be used, (e.g. the (d, pγ) correlation), but they are not necessarily easier (see section 4) and usually they are less accurate. It is the possibility of using the $\beta \cdot \gamma$ (CP) technique which I feel needs re - empha-

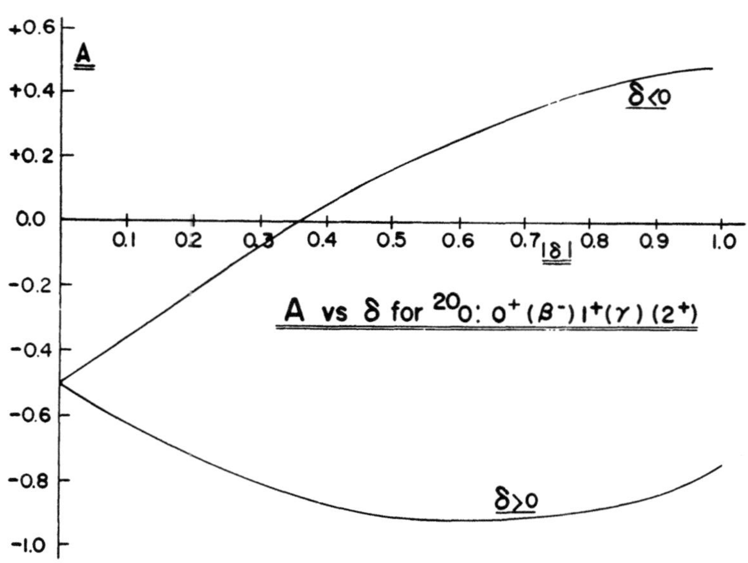

Fig. 3b. A vs $|\delta|$ curve for ^{20}O, $\delta \equiv \langle E2 \rangle / \langle M1 \rangle$. Note the two branches of the curve for $\delta >$ and $\delta < 0$.

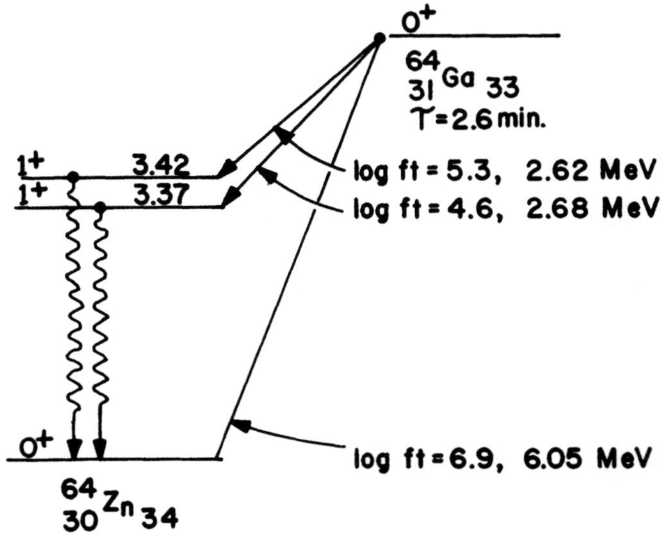

Fig. 4a. Much simplified decay scheme for ^{64}Ga. β - γ (CP) correlation measurement using doublet at $E_x \approx 3.4$ MeV helped solidify 0^+ assignment for ^{64}Ga, which established ground - state β - transition as a pure Fermi transition.

sis, and, of course, ^{20}O is only one example (see below).

^{64}Ga is a case which is as complex as ^{20}F and ^{20}O are simple. The question here was the ground state spin of ^{64}Ga. A much simplified decay scheme is shown in fig. 4a. The arguments are, like the decay scheme, complex, and can be found in detail elsewhere[13]). The main point I wish to make here is that in spite of such basic experimental difficulties as the impossibility of disentangling the two contributions to the asymmetry measurement (see fig. 4b), an essentially unambiguous 0^+ assignment was possible by combining the β - γ (CP) measurement, high - resolution Ge(Li) spectroscopy, and a thorough - going understanding of the systematics of the behaviour of the magnitude of the allowed Fermi matrix element ⟨1⟩. The importance of this case was in establishing the pure Fermi character of the ground state β - transition. (See the discussion in the next section).

As my last example[14]) for this section I have chosen the recently measured case of ^{103}Ru. A simplified decay scheme is shown in fig. 5. This measurement is notable for several reasons. For one it employs a relatively new technique which uses the magnet of the polarimeter as a crude but highly effective β - spectrometer[20,21]). In the case of ^{103}Ru, where there are five β - groups, this technique effectively filtered out all β - branches except the one of interest, which is shown in fig. 5. This particular case is further notable because it could involve mixed transitions in both the β and γ parts of the decay cascade. Nonetheless by utilizing the condition $y \approx 0$ and combining the measurement of the asymmetry parameter with measurements on the internal conversion coefficient Behrens and Zernial[14]) were able to come up with

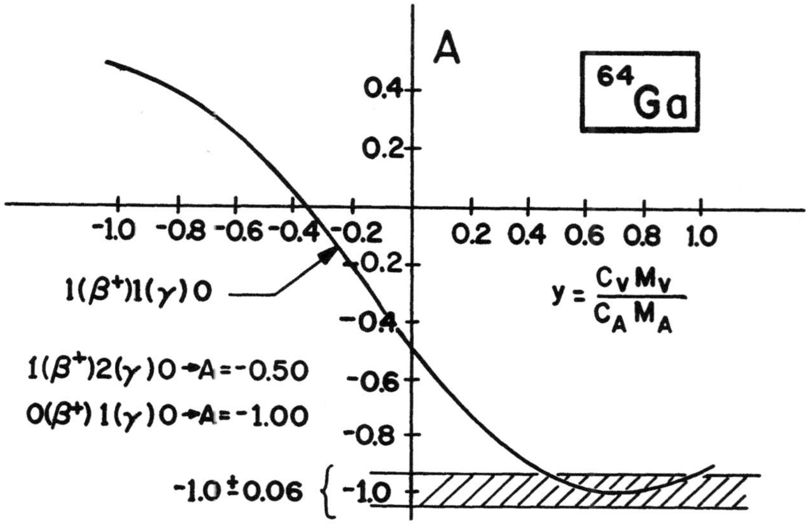

Fig. 4b. A vs y plot for ^{64}Ga case. The 1 → 2 → 0 and 0 → 1 → 0 decay sequences are characterized by unique values for A, as shown on the diagram. The experimental result shown on the diagram is from ref. 13.

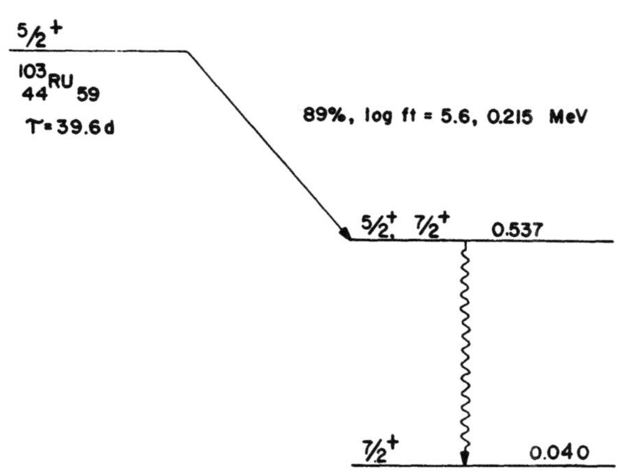

Fig. 5. Simplified decay scheme for ^{103}Ru (see ref. 14). The β · γ (CP) correlation measurement in combination with other experimental data lead to the spin assignments for the 537 keV level and the multipole mixing parameter |δ| for the 497 keV γ - transition. Notice the β - transition could be mixed. Nonetheless the condition y ≈0 (see eq. (2a)) permitted a successful analysis (see ref. 14).

Table 2. A list of some eligible candidates for a β-γ (CP) correlation measurement. All information in the table is self-explanatory except that under 'Intermediate State' and 'Final State' headings; the entries there refer to fig. 1 and describe the number of unobserved intermediate γ-rays. Source reactions are suggested possibilities only.

	Initial Nucleus	$\beta^{\mp}$	log ft	E_β MeV	$t_{1/2}$ mins.	Interm. State	E_γ MeV	δ	Final State	Questions to be answered
	$^{97}_{41}$Nb	β^-	5.4	1.3	53.0	1	0.7	?	2	J_0^π=?
J^π,T	$(9/2^+)$, 15/2					$(7/2)^+$, 13/2			$5/2^+$, 13/2	J_1=?
E_x,MeV	G.S.					0.665			G.S.	δ= ?
	Source reaction: ^{96}Zr(d,n), Q=+5.4 MeV									
	$^{54}_{27}$Co	β^+	5.2	4.3	1.5	3	1.4	0	4	J_0^π=?
J^π,T	(7^+),0					2^+,1			0^+,1	y=? or
E_x,MeV	0.220					1.409			G.S.	δ_1=?
	Source reaction: ^{54}Fe(p,n), Q=-11.4 MeV									J_1^π=?
	$^{98}_{45}$Rh	β^+	~5.0	2.5	8.7	1	0.7	0	2	J_0=?
J^π,T	$(?)^+$,4					2^+,5			0^+,1	and/or
E_x,MeV	G.S.					0.66			G.S.	y=?
	Source reaction: ^{98}Ru(p,n), Q=-5.0MeV									
	$^{37}_{16}$S	β^-	4.2	1.6	5.1	1	3.1	-0.18	2	
J^π,T	$7/2^-$,5/2					$7/2^-$,3/2			$3/2^-$,3/2	y=?
E_x,MeV	G.S.					3.09			G.S.	
	Source reaction: ^{36}S(d,p), Q=2.1 MeV									

the spin assignment $J_1 = 5/2$ or $7/2$ and the lower limit for the multipole mixing parameter of $|\delta| \gtrsim 0.8$ (see fig. 5), i.e. an almost pure E2 transition. From the measured life - time of the 0.537 MeV state[22]) it appears there is an enhancement of more than ten over the single - particle estimate of the E2 - strength, which betokens an important collective character for the low - lying states of ^{103}Ru, a useful datum of nuclear physics.

From the above sampling it is clear that most of the 'easy' cases have been done. However I will close this section with some examples which have yet to be studied and might still be relatively accessible. As will be seen they are mainly short - lived radioactivities ($\lesssim$ 5 mins.) and require that the polarimeter and accelerator (or reactor) generating the radioactivity be very closely associated. Such an operation has been achieved successfully many times (see refs. 15, 16, and 17, for example), but probably not yet in any optimal way (see section 4). I emphasize that many more cases than I give may still be investigated. I have chosen a few which are no doubt highly colored by a particular research background, and I am sure others would find many more eligible candidates. The suggested list is given in table 2. One of the entries (^{37}S) pertains to the next section but has been included for convenience. In all cases the major problem is the manufacturing of the source in adequate activities for the present purpose ($\gtrsim$ 100 μCi). Accordingly the source reactions given in the table are merely suggested possibilities, which in some cases may well not be the best way to go.

3. Fermi Matrix Elements and Isospin Impurity

As noted in the introduction, in mixed allowed transitions, wherein $J_1^{\pi} = J_0^{\pi}$ (see fig. 1), a measurement of A, the asymmetry parameter, can lead to a determination of $y \equiv C_V \langle 1 \rangle / C_A \langle \sigma \rangle$. If we neglect second - order corrections (see below) this can, in turn, lead to a determination of $|\langle 1 \rangle|$, the absolute value of the Fermi matrix element, in terms of y and the ft value. Using the latest available information[23]) on the absolute value of C_V the following formula is easily derived[6]),

$$|\langle 1 \rangle|^2 = \frac{(6173 \pm 16)\, y^2}{ft(1 + y^2)} \tag{5}$$

The interpretation of $\langle 1 \rangle$ as denoting an isospin impurity is illustrated in fig. 6 for the case of β^- decay. (For β^+ decay the roles of the initial and final states connected by the β - decay are reversed). As shown in the figure the Fermi decay proceeds by virtue of the mixing into the final state of some small portion, α_T^2, of (and only of) the **analog** of the initial state. The notation $|T, T_3\rangle$ is used to denote states in this figure, where T is, of course, the total isospin and $T_3 \equiv (N - Z) / 2$. If we lump into one term all the charge dependent elements of the nuclear Hamilton and call it V_{CD}, first order perturbation theory immediately yields[6,7])

$$\alpha_T(T-1) = \frac{\langle A |\, V_{CD} \,| L \rangle}{\Delta E} \tag{6}$$

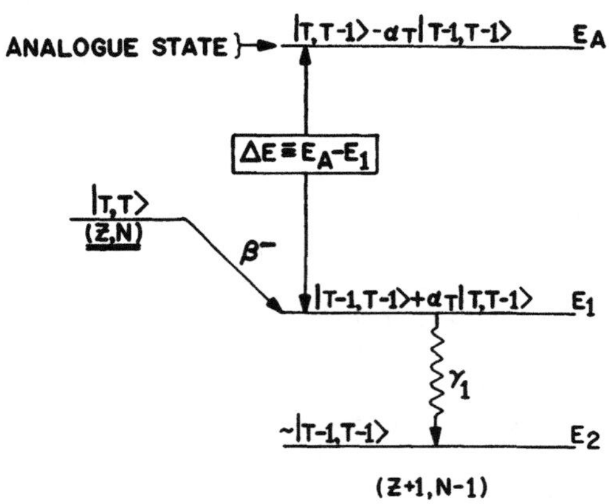

Fig. 6. Diagram of a β^- decay with a Fermi component, $<1>$, which proceeds strictly via the small admixture into state 1 of the analog of the initial state. (In the case of β^- decay state 1 is re - labelled the L - state in section 3). This small admixture, $\alpha_T\,|T, T-1>$ must also be characterized by all the other quantum numbers (spin, parity, etc.) of the initial state, except, of course, for T_3.

where $|A\rangle$ denotes the analog state and $|L\rangle$ the state of (mainly) lesser isospin in the β - decay. (This would be the final state in the β^- decay of fig. 1. Note that $T_< \equiv T-1$). Since $\langle 1\rangle$ is also related to α_T (T - 1) by the formula[6,7])

$$\alpha_T(T-1) = <1>/(2T)^{\frac{1}{2}} \tag{7}$$

we have finally,

$$<A|\,V_{CD}\,|\,L> = (\Delta E) \times <1>/(2T)^{\frac{1}{2}} \tag{8}$$

Thus knowledge of the Fermi matrix element is equivalent to knowledge of the isospin mixing matrix element. Since the Coulomb force is usually the main contributor to V_{CD} this is sometimes called the Coulomb matrix element. However, the experimental determination of this matrix element,which we shall abbreviate as $\langle V_{CD}\rangle$, is based only on eqs. (5) - (8), which say nothing about the form or origin of V_{CD} itself. It is therefore clear that additional arguments are needed to determine if a charge - dependent force of purely nuclear origin could contribute significantly to $\langle V_{CD}\rangle$. These arguments have been reviewed thoroughly by Blin - Stoyle[7]) very recently. The general tenor of his findings is that a nuclear charge - dependent force of as much as 1% or so of the charge - independent force is expected to exist, and such a force could give rise to contributions in $\langle V_{CD}\rangle$ from $\approx$ zero to some tens of keV. Since the charge - dependent force is, strictly speaking, unknown, it is not possible to make precise estimates as yet. For this reason I shall not take specific account of this effect in the discussion here. However, it should be emphasized that low values of $|\langle V_{CD}\rangle|$, say < 20 keV, might require consideration of the charge - dependent nuclear force and could indeed help in elucidating it.

At this point we will introduce an important assumption concerning V_{CD} which has been an essential ingredient of most of the recent theoretical studies stemming from the discovery of analog states and the states which can be, structurally speaking directly derived from them (see below) [24-27]. The assumption is that only the monopole part of V_{CD} need be considered, meaning that only single particle states of the same spin (j) and parity (π) can be connected by V_{CD}. For the Coulomb part of V_{CD},

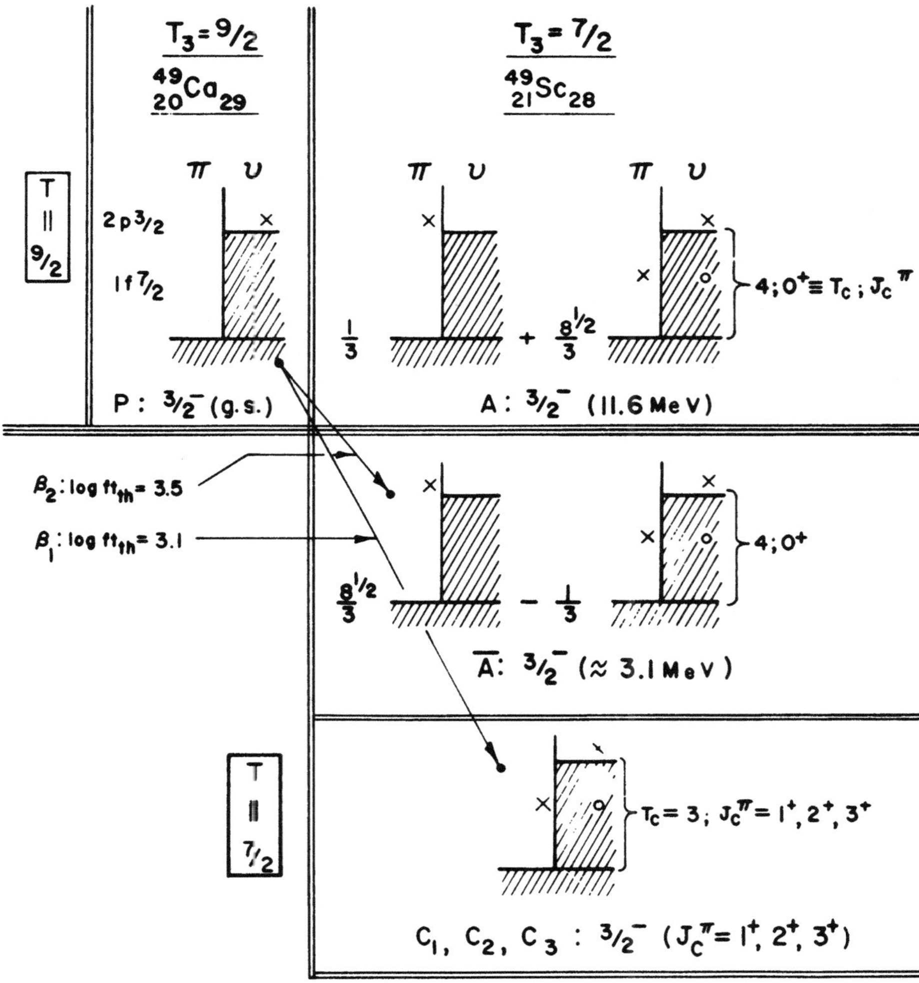

Fig. 7. Schematized representation of P (P ≡ parent), A, Ā, and C - states in ^{49}Ca and ^{49}Sc, as discussed in section 3 and defined in refs. 16 and 18. The T_3 nuclear states are separated vertically and the states of distinct T are separated horizontally. All shell structures are based on the simplest and energetically lowest configuration outside of a closed ^{40}Ca core, indicated by the base - line in each shell - picture. the structure outside this core is $(1f7/2)^8$ (sp3/2). Protons (π) and neutrons (ν) are relegated to the left (π) and right (ν) halves of each shell - picture.

which we call V_C, this is frequently expressed by the single particle approximation based on the charged uniform sphere model of the nucleus. (More accurate expressions can easily be used). This leads to,

$$V_C(r) \approx \frac{(Z-1)e^2}{R}\left\{\frac{3}{2} - \frac{1}{2}\frac{r^2}{R^2}\right\} \quad , r < R \qquad (9a)$$

$$V_C(r) \approx \frac{(Z-1)e^2}{r} \quad , r > R \qquad (9b)$$

where it is common to take the following definition for R,

$$R \equiv \{(5/3)\langle r^2\rangle\}^{\frac{1}{2}} \qquad (9c)$$

$\langle r^2 \rangle$ being usually taken from experiment. Since β - decay generally (but not always) involves only bound states the further approximation is sometimes made that the 'ex-

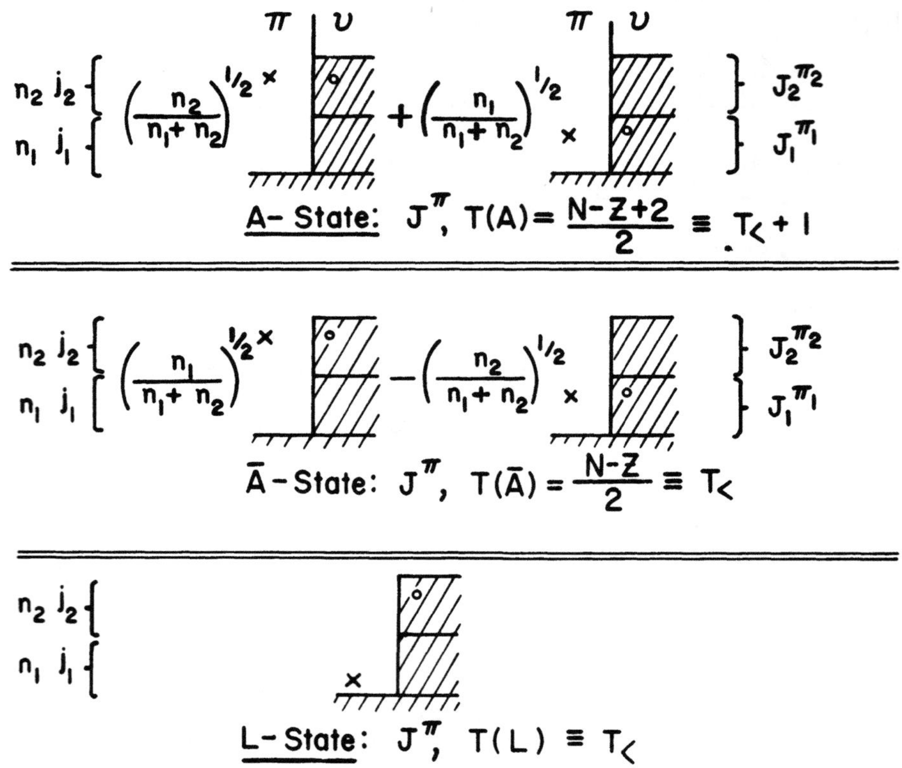

Fig. 8. Simplified diagram of unperturbed situation involving the $T_<$ state or L - state of the β-decay (see fig. 6), Ā - state, and A - state (see section 3 and fig. 7). The Coulomb force mixes A and Ā weakly and the nuclear force mixes the L and Ā states strongly. Many simplifications have been made in this figure, particularly in the structure of the L - state (see section 3). Also only two different shells have been used, j_2 and j_1, containing respectively n_2 and n_1 nucleons. The number of protons (π) or neutrons (ν) in each of the configurations is, of course, arbitrary. However, n_2 and n_1 denote only the neutron excess in each shell. The closed self - conjugate core is defined by the base - line of each shell - picture.

ternal' region, $r > R$, can be ignored. This means that (9a) fully defines V_C (r). This has the useful numerical consequence that a reasonable approximation for the single particle matrix element $\langle n\ell j \mid V_C \mid n'\ell j \rangle$ may be immediately written down,

$$< n\ell j \mid V_C \mid n'\ell j > \approx - \frac{Ze^2}{2R} < n\ell j \mid (r^2/R^2) - 3\delta_{nn'} \mid n'\ell j > \qquad (10)$$

It is obvious from the structure of eq. (10) that it has its maximum value when n = = n′, i.e. for the diagonal single particle matrix elements. For this reason the largest isospin mixing introduced by V_C is between the analog state and the so - called anti - analog states[24]). Fig. 7 illustrates the structure of analog, anti - analog, and the closely related core or configuration states[18,28]) in the particularly simple and important case[16]) of ^{49}Sc. I have abbreviated the designation of these states by A, $\bar{A}$, and C. From this figure it can be seen that the mixing of the A and $\bar{A}$ states occurs via the difference between single particle diagonal matrix elements of the form of eq. (10) (ref. 16). (See the discussion below regarding ^{49}Sc). In this particular case no other source of isospin mixing exists. However, the situation in ^{49}Ca is fairly typical and so in fig. 8 we have generalized the picture to show how the major mixing between the A - state and the L - state takes place. The mixing of the $\bar{A}$ - state and the L - state occurs through the action of the residual two - body nuclear force and should be a very strong mixing because it is not inhibited by any selection rule, in general. The isospin mixing then occurs through that portion of the (new) L - state which now has the structure of the $\bar{A}$ - state. The amplitude of the $\bar{A}$ - state (in the L - state) we call $\beta_L(\bar{A})$. β_L will in general be of the order of magnitude of 1, and could be positive or negative. The isospin mixing we have already designated as α_T, and this is always very small ($|\alpha_T|^2 < 10^{-5}$, see below).

Fig. 8 has been much simplified for the sake of this presentation, particularly in the representation of the L - state, which has been shown as a more - or - less pure ground - state type of configuration. Actually it may have a strong admixture of states of the C or configuration type as well as some $\bar{A}$ state. This is, of course, important in understanding the Gamow - Teller part of the decay, which accounts for ≈ 99% of the transition probability, but it is also important for the Fermi part, our main concern here. In fact we can summarize everything pertaining to the Fermi decay by a new expression for α_T, the impurity coefficient, as follows,

$$\alpha'_T = \beta_L(\bar{A}) \frac{< A \mid V_{CD} \mid \bar{A} >}{\Delta_A} \qquad (11)$$

where Δ_A is the energy of splitting between the A and unperturbed $\bar{A}$ state, often called the symmetry energy. Various empirical formulae for this energy have been given[16,29]) and a good average of these appears to be,

$$\Delta_A \approx (100 \pm 20) \frac{(N-Z+2)}{2A} \qquad (12)$$

where A is the atomic weight. Eq. (11) is the counter part of eq. (7). The counterpart of eq. (8) is,

$$<A| V_{CD} | \bar{A}> = \frac{\Delta_A}{\beta_L(\bar{A})} \quad \frac{<1>}{(2T)^{1/2}} \tag{13}$$

which is perhaps the most useful expression for our purpose. It differs significantly from eq. (8) in that a knowledge of the anti - analog amplitude β_L is required (in addition to a determination of ⟨1⟩) in order to get $\langle V_{CD}\rangle$. Δ_A is given adequately by eq. (12). Eq. (13) also makes clear that extraordinarily low values of ⟨1⟩ will coincide with low anti - analog strength and/or an anomalously low value of $\langle A \mid V_{CD} \mid \bar{A} \rangle$. **Both** of these situations can occur (see below). The neglections in eq. (13) are two in number: (a) Mixing between the A - state and states other than the $\bar{A}$ - state has been left out, and (b) The second - order corrections to ⟨1⟩ have not yet been made. In general (a) has been ignored hitherto because, in the main, if it is **not** ignorable it is somewhat doubtful as to whether much can be learned from $\langle V_{CD}\rangle$ regarding nuclear structure, since many other higher - order and competitive effects will then have to be included for a realistic calculation. The situation with regard to (b), the second - order corrections to ⟨1⟩, is considerably more accessible, as we now proceed to show.

Second - order corrections to allowed and first - forbidden β - decay have been studied in some detail in several publications[3,30,34]). For present purposes our interest is only in second - order corrections to the Fermi or vector part of allowed β - decay. If the Gamow - Teller part also requires second - order corrections the situation gets too complex for useful purposes[33]). Because the Fermi allowed decay can be mediated only by operators carrying zero spin only two matrix elements, $\langle i\underline{\alpha} \cdot \underline{r}\rangle$ and $\langle r^2 \rangle$, have to be added to ⟨1⟩ in the second order expression for the vector matrix element, which we will call ⟨1⟩′. This number is reduced to one due to another and highly fortunate consequence of the CVC feature of the V - A theory. This yields the following relation[31,32]) between $\langle i\underline{\alpha} \cdot \underline{r}\rangle$ and $\langle r^2 \rangle$,

$$<i\underline{\alpha} . \underline{r}> \approx - \frac{\alpha Z}{4} \left[3<\rho^2>_L - <\rho^4>_L \right] \tag{14a}$$

$$\rho \equiv r/R \tag{14b}$$

$$<\rho^p>_L \equiv <L| \sum_i t_3^{(i)} \rho_i^p | A> \tag{14c}$$

where α is the fine structure constant, R is defined by eq. (9c), and $t_3^{(i)}$ is the 'z' or '3' component of the isospin for a single - nucleon. The state vectors | L ⟩ and | A⟩ are defined above. Several approximations were made in arriving at (14) which can be largely eliminated by using a more realistic representation of V_C (r) than eq. (9a)[32]). This could be important under some circumstances, though we do no more than note this point here. Using the complete expression for ⟨1⟩′, which is given in several sources[3,30,33]) in combination with eqs. (14) and (9a), yields the following,

$$<1>' \approx - \frac{(Z\alpha)}{R\Delta_A} \left(\frac{T_>}{2}\right)^{1/2} \left\{ <\rho^2>_L + \frac{R\Delta_A(\alpha Z)}{4} <\rho^4>_L \right\} \tag{15}$$

where we have used the fundamental units $\hbar = c = m_e = 1$. Using the same monopole approximation described earlier in this section leads again to the conclusion that only the $|\bar{A}\rangle$ part of $|L\rangle$ described by the amplitude β_ℓ will contribute significantly to $\langle\rho^p\rangle_L$. This leads immediately to

$$<1>' \approx -\left(\frac{\alpha Z}{\Delta_A R}\right)\left(\frac{T_>}{2}\right)^{1/2} \beta_L(\bar{A}) \left\{<\rho^2>_{\bar{A}} + C_2 <\rho^4>_{\bar{A}}\right\} \qquad (16a)$$

$$C_2 \equiv (R\Delta_A(\alpha Z))/4 \qquad (16b)$$

where C_2 is the (dimensionless) correction factor for the 2nd forbidden matrix element $\langle\rho^4\rangle_{\bar{A}}$. It is plotted in fig. 9 for the stable - valley of nuclides. Although C_2 is

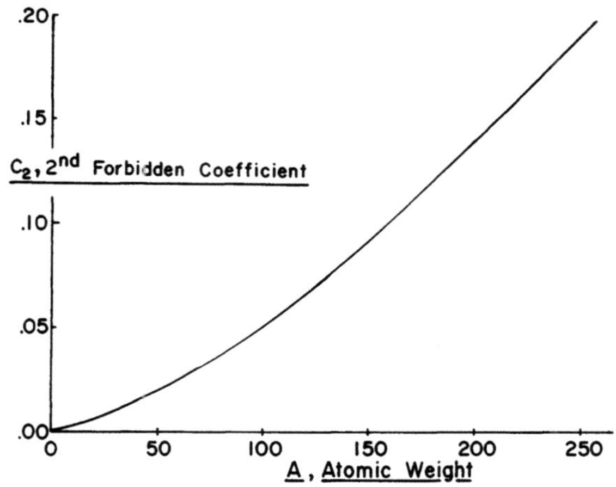

Fig. 9. Plot of C_2 vs A, atomic weight. C_2 is the correction coefficient multiplying the dimensionless 2nd forbidden matrix element $<\rho^4>_L$. See eq. (16) in section 3 and the concomitant discussion.

always < 0.2 it cannot be concluded that the 2nd forbidden correction is always less than 20%. The reason may be seen by giving the explicit formula for $\langle\rho^p\rangle_{\bar{A}}$ in the case of ^{49}Sc,

$$<\rho^p>_{\bar{A}} = \frac{\sqrt{8}}{9}\left(<1f7/2|\rho^p|1f7/2> - <2p3/2|\rho^p|2p3/2>\right) \qquad (17)$$

This is of course exactly the kind of difference between large single particle matrix elements encountered in the evaluation of $\langle A\,|V_C|\,\bar{A}\rangle$, as noted above. In fact it is clear from eq. (17) that a difference of two large numbers is always involved in the evaluation of $\langle\rho^p\rangle_{\bar{A}}$. Therefore a near cancellation in $\langle\rho^2\rangle_{\bar{A}}$ might lead to a considerable enhancement of the importance of $C_2\,\langle\rho^4\rangle_{\bar{A}}$. Thus it may be expected that 2nd forbidden corrections may well be non - ignorable for $A \gtrsim 100$ and $\langle V_{CD}\rangle \lesssim 20$ keV.

There is no way to tell without calculating the particular case of interest, but the calculations are not difficult and should be fairly reliable.

The preceding summarizes rather briefly the state of our present understanding concerning the origins of the Fermi matrix element ⟨1⟩ and the charge - dependent matrix element $\langle V_{CD} \rangle$. It is clear that the two are not exactly proportional to each other and furthermore the Coulomb part of $\langle V_{CD} \rangle$, which we have called $\langle V_C \rangle$, may not constitute the major part of $\langle V_C \rangle$. Therefore we rewrite eq. (8) as follows

$$< U_{CD} > \equiv < A | U_{CD} | L > \equiv \Delta E \times < 1 >' / (2T)^{1/2} \tag{18}$$

where ⟨1⟩′ is given by eq. (16) and, of course, should replace ⟨1⟩ in eq. (5). In other words it is ⟨1⟩′ that is given by experiment and the 'effective' charge - dependent matrix element $\langle U_{CD} \rangle$ is defined in terms of it by eq. (18). In fig. 10 I have plotted

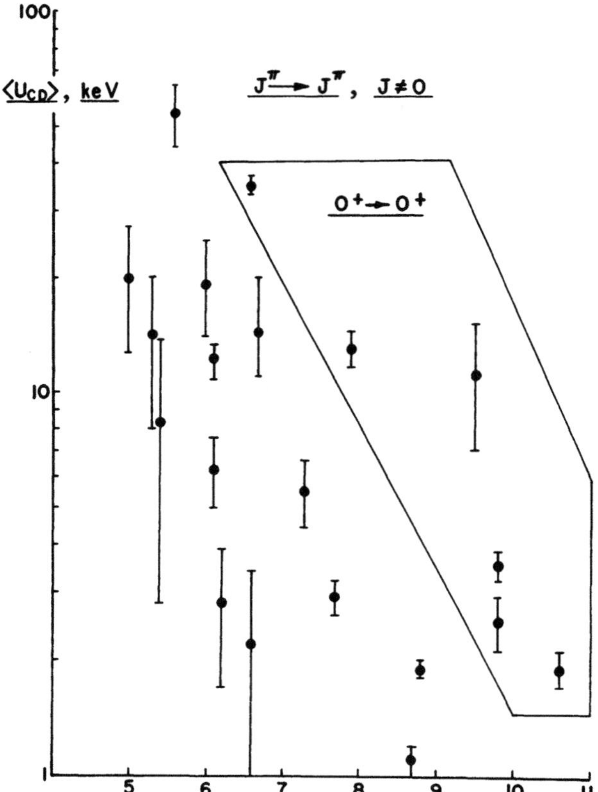

Fig. 10. Plot of $< U_{CD} >$ in keV vs log ft. Not all published data is shown. $< U_{CD} >$ is defined by eq. (18) in section 3.

$\langle U_{CD}\rangle$ vs log ft for most of the allowed - mixed β - transitions which have been measured to date[35]). I have included the $\langle U_{CD}\rangle$ values deduced from $0^+ \to 0^+$ transitions but these are blocked off from the $J \neq 0$ transitions (see below). I have specifically excluded results yielding only upper limits for $\langle U_{CD}\rangle$ since they do not alter the general trend of the figure and are also difficult to display.

It is clear from the figure that there is a rather strong general tendency for $\langle U_{CD}\rangle$ to decrease with log ft. In those cases with $J \neq 0$ the ft value is determined almost completely by the Gamow - Teller part and thus a large ft means a retarded β - decay. The meaning of the trend in the figure is obviously that a retarded β - decay implies a reduced value of $|\langle U_{CD}\rangle|$. Gross properties of this kind have been the subject of some theoretical study and this work is reviewed by Blin - Stoyle[7]). By and large it is reasonable to attribute this behaviour to a systematic fall - off with atomic weight of $\beta_L(\bar{A})$ in eq. (16). This is to be expected, since the mean excitation of the $\bar{A}$ - state increases with atomic weight, a consequence of the fact that the excitation of the analog state increases roughly as $A^{2/3}$, whereas the splitting Δ_A (see eq. (12)) increases by only $\approx 50\%$ in the range of atomic weight 40 to 240. Thus the $\bar{A}$ state becomes inaccessible to the lower - lying states at $A \approx 50$. Since these lower - lying states are exactly the ones populated, for example, in β^- - decay (the arguments for β^+ are similar), which we have called L - states (see above), they mix less and less with the $\bar{A}$ - state as the atomic weight increases. The $0^+ \to 0^+$ transitions require separate consideration, of course, since for them the entire ft value is determined by $\langle 1\rangle'$ (see eq. (16)). However even though they are displaced it is clear they show the same kind of trend as the other transitions, and very likely for the same reasons[37]). It should also be noted that a_T^2 , the strength of the isospin impurity, follows a similar trend, starting with values around 5×10^{-5} when log ft is 5 - 6 and going to values around 10^{-7} to 10^{-8} when log ft is 8 - 9.

There have been several theoretical calculations of $\langle V_C\rangle$ and some of $\langle V_{CD}\rangle$ also. These have been reviewed by Blin - Stoyle[7]) and it appears that the agreement achieved so far is encouraging. There appear to be no calculations of $\langle U_{CD}\rangle$ as yet, i.e. calculations using realistic nuclear wave functions which include charge - dependence of the nuclear force and 2nd order β - moment corrections. There is no question that such a complete calculation is really not very difficult, and it is to be hoped that at least for certain key cases, such as, for instance, ^{49}Sc, ^{57}Ni, and also the hitherto unmeasured case of ^{37}S (see table 2 and ref. 39),it will not be long before they are done. Also in the case of ^{49}Sc a more accurate β - γ (CP) correlation measurement is needed, since at present only an upper limit, $|\langle U_{CD}\rangle| < 31$ keV, is available[16]). This number, however, was sufficient to pin down the $3/2^-$ assignment for the 3.1 MeV level in ^{49}Sc (see section 2). Finally we mention the question of the sign of $\langle U_{CD}\rangle$. This can be inferred from y if the sign of Gamow - Teller matrix element, M_A, is given (see eq. (2b)). In some cases, e.g. the ubiquitous ^{49}Sc, this can be given with some reliability[37,38]). Needless to say an experimental determination of the sign as well as the magnitude of $\langle U_{CD}\rangle$ is of prime importance in confirming any theoretical prediction.

In summary, work to date yields good qualitative to semiqualitative insight into the phenomenon of isospin mixing in bound nuclear states. A relatively moderate extension of both theoretical and experimental work might yield a sharp quantitative understanding not only of isospin mixing but also of the charge - dependent nuclear force as well as some of the more esoteric consequences of the CVC theory of β - decay. These goals are well worth the effort of achievement.

4. Comments and Conclusions

It is evident from all the foregoing that the use of the β - γ (CP) correlation technique on allowed transitions has been a major factor in advancing our empirical knowledge of isospin in bound nuclear states. The technique has also been reasonably useful in the highly utilitarian business of spin assignments and the measurement of electromagnetic multipole mixing. At the present time the method has reached a kind of pause. However a very good possibility for a new and fertile field exists in the close linking of present day ion accelerators and γ - polarimeters in general. For example a γ - polarimeter of the Karlsruhe type[20]) working essentially at the point of beam activation could be used to do β - γ (CP) correlation measurements in the half - life range of seconds to a few minutes. More conventional γ - polarimeters with certain simple modifications[40]) could do the same thing, of course. The point in either case is that many more new radionuclides become available than before, especially in the lighter nuclei. From the experimental point of view this kind of work becomes even more convenient as the number of 5 - 30 MeV accelerators goes up and their reliability increases. Thus it might be expected that some of this kind of work will begin to be done in the not - to - distant future, even without the present exhortation.

Finally mention should be made of the possibility of a break - through in γ - polarimeter design. The limitation to the 7.9% fraction of polarized electrons in magnetized iron is a very severe one. No significant break - through has surfaced as yet, unfortunately, but perhaps a more vigilant and searching attitude towards the newer elements of our present technology could change the situation more rapidly than now suspected.

Note added in proof: ^{97}Nb has just been measured by H. Behrens and F. Brodt, Nucl. Phys. A150 (1970) 269. The measured A - value leads to J_1^π (0.758) = $7/2^+$ and $0 < \delta < 0.07$

† Work performed under the auspices of the U.S. Atomic Energy Commission.

References:

1. H. Schopper, Nucl. Instr. 3 (1958) 158.
2. R.M. Steffen and H. Frauenfelder, 'Alpha-, Beta-, and Gamma-Ray Spectroscopy', Vol. 2, p. 1453 (1965). Editor K. Siegbahn.
3. H. Schopper, 'Weak Interactions and Nuclear Beta - Decay', Amsterdam: North - Holland Publ. Co., 1966.
4. H. Behrens and J. Janecke, Landolt - Bornstein Neue Serie Bd. I/4, Springer - Verlag, Berlin, 1969. This work contains an extensive tabulation of particle parameters and β - decay matrix elements.
5. H. Appel, Landolt - Bornstein Neue Serie Bd. I/3, Springer - Verlag, Berlin, 1968.

6. See, e.g. S.D. Bloom, 'Isobaric Spin in Nuclear Physics', Tallahasse, 1966, p. 123 (Academic Press, 1966).

7. R.J. Blin - Stoyle, 'Isospin in Nuclear Physics', p. 115, ed. by D.H. Wilkinson, Amsterdam, North - Holland Publ. Co., 1966.

8. H. Weidenmuller, Rev. Mod. Phys. 118 (1960) 540.

9. L.G. Mann and S.D. Bloom, Phys. Rev. 139 (1965) B540.

10. E. Freiburg and V. Soergel, Z. Physik 162 (1961) 114.

11. V.M. Loboshov and V.A. Nazarenko, Zhur. Eksptl. i Teoret. Fiz. 42 (1962) 370; (Translation: Soviet Phys. JETP 15 (1962) 257).

12. See e.g. E.C. Halbert, J.B. McGrory, B.H. Wildenthal, and S.P. Pandya, 'Advances in Nuclear Physics, Vol. 4', to be published, Plenum Press, New York.

13. J.P. Deutsch and L.Grenacs, 'Isobaric Spin in Nuclear Physics', Tallahasse (1966), p. 881; L.G. Mann, K.G. Tirsell and S.D. Bloom, Nucl. Phys. A97 (1967) 425.

14. H. Behrens and W. Zernial, Z. Physik 233 (1970) 458.

15. R.A. Bigoni, S.D. Bloom, and K.G. Tirsell, Nucl. Phys. A134 (1969) 620.

16. L.G. Mann and S.D. Bloom, Nucl. Phys. A140 (1970) 598.

17. S.D. Bloom and L.G. Mann, Nucl. Phys. A93 (1967) 252.

18. J. Atkinson, L.G. Mann, K.G. Tirsell, and S.D. Bloom, Nucl. Phys. A114 (1968) 143.

19. L.G. Mann, S.D. Bloom, and R.J. Nagle, Nucl. Phys. 30 (1962) 636.

20. H. Schopper, H. Behrens, H. Muller, J. Gorres, W. Jungst, and H. Appel, Nucl. Instr. and Meth. 49 (1967) 277.

21. H. Behrens, Z.F. Phys. 201 (1967) 153.

22. M.T. Rama Rao, V.V. Ramamurty, and V. Lakshminarayana, Phys. Rev. 168 (1968) 1406.

23. J.M. Freeman, D.C. Robinson, and G.L. Wick, Physics Letters 30B (1969) 240.

24. A.F.R. de Toledo Piza, A.K. Kerman, S. Fallieros, and R.H. Venter, Nucl. Phys. 89 (1966) 369.

25. D. Robson, Phys. Rev. 137 (1965) B535.

26. 'Isobaric Spin in Nuclear Physics', Tallahasse (1966). Editors, J.D. Fox and D. Robson. Academic Press, 1966.

27. 'Nuclear Isospin', Asilomar (1969). Editors J.D. Anderson, S.D. Bloom, J. Cerny, and W.W. True. Academic Press, 1969.

28. A.M. Lane, Nucl. Phys. 37 (1962) 506.

29. S. Maripuu, Nucl. Phys. A123 (1967) 357.

30. L. van Neste, R. Coussement, and J.D. Deutsch, Nucl. Phys. A102 (1967) 369.

31. J. Damgaard, 'Proceedings of Symposium on Nuclear Beta Decay and Weak Interactions', Zagreb, 1967.

32. H. Behrens and R. Wischhusen, Zert. f. Physik, to be published.

33. H. Behrens and J. Janecke, Landolt - Bornstein, Neue Serie Bd. I/4, Springer - Verlag, Berlin, 1969.

34. J. Damgaard and A. Winther, Physics Letters 23 (1966) 345.

35. H. Behrens, contribution to this conference. See page 321.

36. L. van Neste, R. Coussement, and J.P. Deutsch, Nucl. Phys. A102 (1967) 369.

37. S. Maripuu, Phys. Letters 31B (1970) 181.

38. J.B. McGrory, S.A. Moszkowski, and S.D. Bloom, private communication.

39. A.K. Hyder, Jr., G.I. Harris, J.J. Porrizo, and F.R. Kandzioski, Phys. Rev. 169 (1968) 899.

40. R.E. Pechacek, L.G. Mann, S.D. Bloom, and R.M. Rodrigues, Rev. Sci, Instruments 35 (1964) 58.

41. B. van Nooijen, Thesis Delft University of Technology 1958 (see also Nuclear Data Sheets); B. van Nooijen, H. van Krugten, W. Lourens, and A.M. van der Kraan, Nucl. Phys. 84 (1966) 380.

Discussion

Coussement: In the formalism of second forbidden contributions you used the single body potential to approximate the Coulomb forces. This approximation is not always correct as is shown in our ^{66}Ga experiment.

Bloom: I agree. It depends on the signs of the matrix elements that you are working with.

Coussement: The second point is that you have also to consider other charge dependent forces.

Bloom: That I know, that I talked about at the beginning. If you have got to deal with very small matrix elements you have to put it in and you could possibly calculate it. I would say that if you are dealing with matrix elements of the order of say less than 10 kilovolts, the number of corrections might get to be so prohibitively high (perhaps 10 kilovolts is a little bit high and 5 kilovolts is a better lower limit), that I despair personally of extracting any reasonable value that you can believe. Even regarding isospin impurities or nuclear structure, unless the case is spectacularly simple; in that case you put it in and you hopefully believe that you can calculate the contributions of the charge dependent nuclear forces, exactly like you say, and the Coulomb force, exactly like you say. If you pick a case like ^{49}Ca, you will come out with what I was saying, namely you come out with information regarding the contribution of the charge dependent force to that matrix element and then you put it in and then it is an advantage, not a disadvantage. And it may be that we'll be able to do that.

Coussement: I would like to mention that we have done an experiment on ^{66}Ga, in which, if we take these Coulomb forces, we can make a relation between two matrix elements. We were able to measure these matrix elements by the measurement of the spectrum shape.

Bloom: And this is in allowed transitions?

Coussement: Yes, in allowed transitions: zero - zero isospin forbidden transitions. Well, the disagreement is so large that we have to be very careful.

Bloom: First of all you need to believe the shell model for the nucleus. You have to have wave functions. If you believe the wave functions then you have a tool. But if you have not got either the wave functions or a proper knowledge of all the matrix elements of all the operators involved, then you are nowhere and you will not be able to make a calculation and you should stop. That is what I believe, I think that is what you are saying.

Coussement: But in our experiment there was no calculation; the only approximation was the fact that we assumed that the charge dependent force was only a Coulomb force of the single body potential type.

Bloom: That may not be enough. Maybe that case is too complex.

Coussement: This disagreement means that we may not use this potential for the charge dependent forces.

Bloom: You must put in the nucleon - nucleon force, the possibility of there being a charge dependent nucleon - nucleon force. And also, you may not be able to use the monopole approximation for the Coulomb force. That may be important too. And finally, I hate to say it, but maybe you might have to include configuration mixing if you are ever going to compare theory with experiments. If it begins to get that complicated you may not be able to arrive at a believable result. So the case may be too complex to derive information concerning the existence of a charge dependent nucleon-nucleon force. But there may be other cases which the analysis would work on.

Coussement: Another question. In your last figure you show the $\langle U_{CD}\rangle$ as a function of the ft - value. Has this already been corrected for the second forbidden contribution?

Bloom: No. $\langle U_{CD}\rangle$ has everything in it. It is what comes out of the raw experiment.

Coussement: Yes, but did you use corrected Fermi matrix elements? I mean corrected for second order contributions?

Bloom: No, that is why I put that off in a separate square, you noticed it did not seem to be in the same place. If you correct it for the second forbidden matrix elements, then you should not plot, because it should just be a straight line, because a 0^+ to 0^+ transition is determined entirely by the Fermi matrix element. But all I plotted there was $\langle U_{CD}\rangle$, the thing that you get out of the experiment, uncorrected either for second forbidden contributions or for the charge dependent nucleon - nucleon force. And to see if there was a trend with the ft - value particularly for those cases where y is not equal to zero. The second forbidden corrections have not yet been made, but they could be made on a simple basis and then we would get a better basis, at least for comparing the 0^+ to 0^+ with the log ft - value. That is something I did not have time for to do.

Boehm: You concluded your lecture with a cryptic allusion to possible new technological developments for the efficiency of the polarization analyzer. Do you elaborate on that?

Bloom: There are two reasons I did that. One of them was that I have spent many hours, and I am sure you have, to think of some way to get more polarized electrons into a box, so we could do a better job. Even a factor of 2 would be a very significant achievement. And I think I mentioned to some of the people here that, and you know probably about it too, liquid oxygen has about 30% more polarized electrons in a magnetized state than iron. The only unfortunate thing is that it is so tenuous even in the liquid state that you cannot pack enough electrons into a cubic centimeter so as to get a polarimeter of a reasonable size. But at Livermore we have been thinking of a stuff like a plasma. It seems that the rate of technological advancement in fields of which I know nothing about is so rapid and so enormous that perhaps one of these days, maybe by using very low temperatures and by some kind of plasma techniques, this will happen. And I am just pointing out that everybody must be alert. It may happen. Those are the two reasons: I was thinking about the oxygen in the first place and I was thinking about what is happening in other fields. I do not have anything to offer, I wish I did.

Frauenfelder: I was wondering if it would not be worthwhile to note that in fact the charge - dependent elements in the nucleon - nucleon force must themselves be based on electromagnetic effects, since they arise from the mass differences between the charged and uncharged iso - triplet mesons that mediate the force between nucleons. Thus, ultimately, all the charge - dependent contributions to the nucleon Hamiltonian is of electromagnetic origin.

Bloom: That is absolutely correct, as far as I know, and is, in fact, what I did mean when I referred to the charge - dependent force worked out by Blin - Stoyle and used in his calculations.

Contributed Paper

BETA MATRIX ELEMENTS FOR SOME FIRST FORBIDDEN DECAYS FROM ANGULAR CORRELATIONS

J.C. Manthuruthil and C.P. Poirier

Aerospace Research Laboratories, Wright - Patterson AFB, Ohio, U.S.A.

From the theoretical standpoint, nuclear structure information is contained in the matrix elements, which dictate the β - decay properties of individual nuclei. A simultaneous fit of all β - decay observables (such as β - γ directional correlations) allows one to extract the relevant matrix elements from the experimental data in a consistent manner. We have performed such an analysis on the 3^- (β) 2^+ (γ) 0^+ decays of ^{124}Sb and ^{152}Eu, the 1^- (β) 2^+ (γ) 0^+ decays of ^{186}Re and ^{188}Re, and the 2^- (β) 2^+ (γ) 0^+ decays of ^{84}Rb and ^{86}Rb.

A detailed description of the techniques used to obtain the β - matrix elements

TABLE I. Matrix element parameters. Numbers in parenthesis are errors in last digit(s).

ELEMENT	x	u	$\xi' y$	w	$\xi' v$
Sb^{124}	0.05(2)	0.10(3)	2.6(4)		
	0.56(3)	0.05(3)	8.5(6)		
Eu^{152}	0.08(2)	0.18(3)	3.6(6)		
	0.58(3)	0.08(3)	9.0(7)		
Re^{186}	-0.03(1)	0.02(1)	0.27(2)		
Re^{188}	-0.06(1)	0.012(3)	0.35(3)		
Rb^{86}	-0.02(1)	-0.9(2)	-7(3)	0.7(5)	8(2)[a]
	-0.02(1)	-0.8(2)	-7(2)	-0.5(3)	3(2)
Rb^{84}	-0.01(1)	0.6(2)	-5(2)	-4(3)	-25(10)
	-0.3(1)	0.3(1)	0.0(1)	-5(3)	-31(11)
	-0.1(1)	0.3(1)	-0.5(2)	-0.7(4)	-4(1)
	-0.1(1)	0.3(1)	-0.3(1)	-2(1)	-14(6)

[a] Solutions for statistical and non-statistical shapes respectively.

has been given in a previous report[1]). The analysis is based on the theoretical expressions of Morita and Morita[2]) using exact radial wave - functions including nuclear finite size corrections. Care was taken to insure proper normalization by taking weighted energy averages when required. In the search procedure, a large number of starting points were used to insure that all solutions were found.

All results for the individual nuclei are given in table 1 and 2 where we have listed the matrix element parameters and matrix element values. For both ^{124}Sb and

TABLE II. Matrix element values.

ELEMENT	$\lvert\int B_{ij}\rvert/\rho$	$\lvert\int r\rvert/\rho$	$\lvert\int i\sigma\times r\rvert/\rho$	$\lvert\int i\alpha\rvert$	$\lvert\int \sigma\cdot r\rvert/\rho$	$\lvert\int i\gamma_5\rvert$
Sb^{124}	0.03(1)	0.002(1)	0.003(1)	0.0010(2)		
	0.02(1)	0.020(5)	0.001(1)	0.0010(2)		
Eu^{152}	0.01(1)	0.0010(2)	0.0020(3)	0.0007(2)		
	0.007(2)	0.005(1)	0.0006(1)	0.0010(3)		
Re^{186}	0.77(5)	0.03(1)	0.015(3)	0.005(1)		
Re^{188}	0.22(2)	0.016(3)	0.003(1)	0.0020(2)		
Rb^{86}	0.6(1)	0.010(3)	0.5(1)	0.07(3)	0.5(3)	0.07(3)
	0.6(1)	0.020(5)	0.5(1)	0.07(3)	0.3(1)	0.02(1)
Rb^{84}	1.5(3)	0.02(1)	0.9(2)	0.11(4)	6(3)	0.5(2)
	1.0(2)	0.3(1)	0.3(1)	0.0(1)	5(2)	0.4(1)
	2.2(4)	0.3(1)	0.6(1)	0.02(1)	2(1)	0.10(5)
	2.2(4)	0.4(1)	0.5(1)	0.01(1)	4(1)	0.4(1)

^{152}Eu, two distinct sets of matrix elements fit all available experimental data. However, for both transitions this ambiguity can be eliminated by a measurement of the β - circularly polarized γ - correlation as a function of energy [hereafter referred to as P_γ (E)]. In the analysis of ^{124}Sb, we have used the β - γ directional correlation (hereafter referred to as ϵ) of Manthuruthil et al.[3]), the shape factor measurement of Hsue et al.[4]), and the β - circularly polarized γ - correlation [hereafter referred to as P_γ (θ)] of Alexander and Steffen[5]). For ^{152}Eu, the analysis included the ϵ data reported in ref. 3, the shape measurements of Smith and Langer[6]) and the P_γ (θ) results of Alexander and Steffen[7]) and Berthier et al.[8]). For both ^{186}Re and ^{188}Re, a unique set of matrix elements fit all available experimental data. In the analysis of ^{186}Re, we have used the ϵ data reported in ref. 3, the shape measurements of Novey et al.[9]), the P_γ (θ) measurements of Delabaye[10]), and the measurements of the γ - ray angular distribution for oriented ^{186}Re nuclei by Greenebaum et al.[11]). A recent measurement[12]) of the ^{186}Re shape factor indicated a statistical shape. We attempted an analysis of the ^{186}Re data using a statistical shape but found no satisfactory solutions. In the case of ^{188}Re, the analysis included the ϵ data of Grenacs et al.[13]), the shape measurements of Van der Werf et al.[14]), and the P_γ (θ) measurements of Gygax and Hess[15]). For ^{86}Rb, we have obtained unique solutions for both statistical and non - statistical shapes. In the analysis we have included the non - statistical shape reported by Robinson and Langer[16]), the ϵ data of Fischbeck and Wilkinson[17]), the

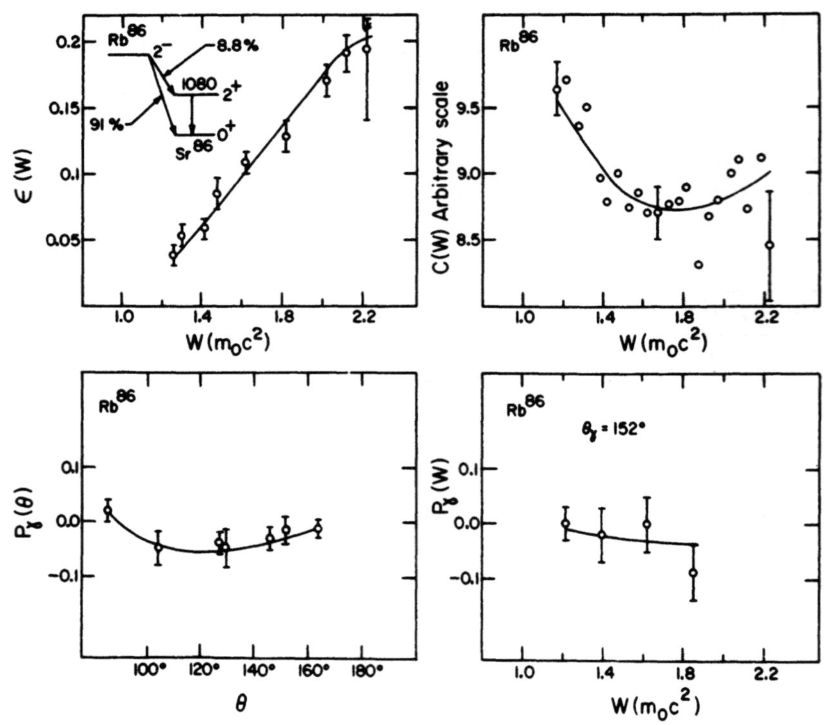

TABLE III. Comparison of the ratio $|\int i\alpha / \int r|$.

ELEMENT	V-A	CVC	EXPERIMENT
Sb^{124}	15	32	52^{+35}_{-18}
			15 ± 3
Eu^{152}	15	35	46^{+28}_{-17}
			16^{+2}_{-3}
Re^{186}	16	38	9^{+5}_{-3}
Re^{188}	16	38	6^{+2}_{-1}
Rb^{86}	11	26	350^{+650}_{-200}

P_γ (θ) measurements of Viano et al.[18]), and the P_γ (E) measurements of Kneissl[19]). For a report on a statistical shape measurement, see ref. 20. In the accompanying figure, we show our fit to all available experimental data for the transition in ^{86}Rb. Similar fits are obtained for all the studied nuclei. For ^{84}Rb, we obtained four distinct solution types which fit all available experimental data. We have included in the analysis the ϵ data reported in ref. 3, the shape measurements of Langer et al.[21]), and the P_γ (θ) measurements of Boehm and Rogers[22]). Table 3 contains a comparison of the ratio $\Lambda = |\int i\,a / \int r|$ obtained from theory and experiment. For ^{124}Sb and ^{152}Eu, the agreement with CVC theory is good. This is not the case for ^{186,188}Re and ^{86}Rb. For these nuclei, we have attempted a search in which the parameters are constrained to satisfy the CVC prediction, but were unable to obtain satisfactory solutions consistent with all the available experimental data. This seems to imply that either the estimate of the ratio Λ based on CVC theory is incorrect (see ref. 23 for details) or some of the experimental measurements are in error.

References

1. J.C. Manthuruthil and C.P. Poirier, Nucl. Phys. A118 (1968) 657.
2. M. Morita and R.S. Morita, Phys. Rev. 109 (1958) 2048.
3. J.C. Manthuruthil, C.P. Poirier, K.S.R. Sastry, R.F. Petry, B.K. Cantrell and R.G. Wilkinson, to be published.
4. S.T. Hsue, L.M. Langer, S.M. Tang, and D.A. Zollman, Nucl. Phys. 73 (1965) 375.
5. P. Alexander and R.M. Steffen, Phys. Rev. 124 (1961) 150.
6. L.M. Langer and D.R. Smith, Phys. Rev. 119 (1960) 1308.
7. P. Alexander and R.M. Steffen, Phys. Rev. 128 (1962) 1783.
8. J. Berthier, R. Lombard, and J.W. Sunier, Compt. Rend. 252 (1961) 257.
9. F.T. Porter, M.S. Freedman, T.B. Novey, and F. Wagner, Phys. Rev. 103 (1956) 921.
10. M. Delabaye, Thesis Louvain University (1965) (Unpublished).

11. G. Greenebaum, J. Hess, W. Weyhmann, and F.M. Pipkin, Bull. Am. Phys. Soc. 10 (1965) 120 and private communication.

12. M. Trudel, E.E. Habib, and H. Ogata, Phys. Rev. C1 (1970) 643.

13. L. Grenacs, R. Hess, and F.C. Rohmer, Helv. Phys. Act. 33 (1965) 374.

14. S.Y. Van der Werf, H. de Waard and H. Beekhuis, Nucl. Phys. A134 (1969) 215.

15. F. Gygax, and R. Hess, Ber. Schw. Physik Ges. 39 (1966) 209.

16. R.L. Robinson and L.M. Langer, Phys. Rev. 112 (1958) 481.

17. H.J. Fischbeck and R.G. Wilkinson, Phys. Rev. 120 (1960) 1762.

18. J.B. Viano, J.C. Renard, J. Menet, P. de Saintignon, A. Laverne, and P. Depommier, Jour. de Phys. 30 (1969) 763.

19. V. Kneissel, Z. Naturforschg. 20a (1965) 1364.

20. E.H. Spejewski, Nucl. Phys. 82 (1966) 481.

21. L.M. Langer, E.H. Spejewski, and D.E. Wortman, Phys. Rev. 133 (1964) B1145.

22. F. Boehm and J. Rogers, Nucl. Phys. 45 (1963) 392.

23. J. Damgaard and A. Winther, Phys. Lett. 23 (1966) 345.

Contributed Paper

$\beta\gamma$ ANGULAR CORRELATION MEASUREMENTS ON FIRST FORBIDDEN β - TRANSITIONS

R. Wischhusen and H. Behrens

Institut für Experimentelle Kernphysik der Universität und des Kernforschungszentrums Karlsruhe, Karlsruhe, Germany

In the past a great number of $\beta\gamma$ angular correlation measurements on first forbidden beta transitions has been carried out. In many cases the accuracy of these measurements is not sufficient to compare the results with theoretical calculations. Also the agreement between different measurements was not always satisfactory. We, therefore, developed an experimental setup which allows to reduce the statistical as well as the systematical errors considerably. In this arrangement a set of 4 β - counters in different positions and 2 periodically exchangeable γ - counters is used. The details can be taken from fig. 1.

To eliminate the $\beta\gamma$ coincidence efficiency it is necessary to interchange the two γ - counters with respect to their positions. The chamber with the β - counters can be turned by angles of 45° to both directions from the symmetric position. The advantage is that the influence of geometrical misadjustments and of γ scattering between β- and γ - counters can be diminished.

In fig. 2 a schematic diagram of the different possible positions of the β- and γ - counters is shown.

The data are then evaluated in the following way:

$$A_2 = \frac{10}{7} \cdot \frac{4 - 9\sqrt[4]{\frac{Q_{m,A}}{Q_{m,B}}} + 5\sqrt[4]{\frac{Q_{n,A}}{Q_{n,B}}}}{8 + 6\sqrt[4]{\frac{Q_{m,A}}{Q_{m,B}}} + \sqrt[4]{\frac{Q_{n,A}}{Q_{n,B}}}} \qquad A_4 = \frac{48}{7} \cdot \frac{-2 + \sqrt[4]{\frac{Q_{m,A}}{Q_{m,B}}} + \sqrt[4]{\frac{Q_{n,A}}{Q_{n,B}}}}{8 + 6\sqrt[4]{\frac{Q_{m,A}}{Q_{m,B}}} + \sqrt[4]{\frac{Q_{n,A}}{Q_{n,B}}}}$$

The first indices are in position I: m, n = 1,2
in position II: m, n = 3,4
in position III: m, n = 2,1
the second index A, B denotes the position of the γ - counters. The quantities Q are defined as:

$$Q_1 = \frac{K_{12} \cdot K_{21}}{K_{11} \cdot K_{22}}; \quad Q_2 = \frac{K_{32} \cdot K_{41}}{K_{31} \cdot K_{42}}; \quad Q_3 = \frac{K_{11} \cdot K_{42}}{K_{12} \cdot K_{41}}; \quad Q_4 = \frac{K_{22} \cdot K_{31}}{K_{21} \cdot K_{32}};$$

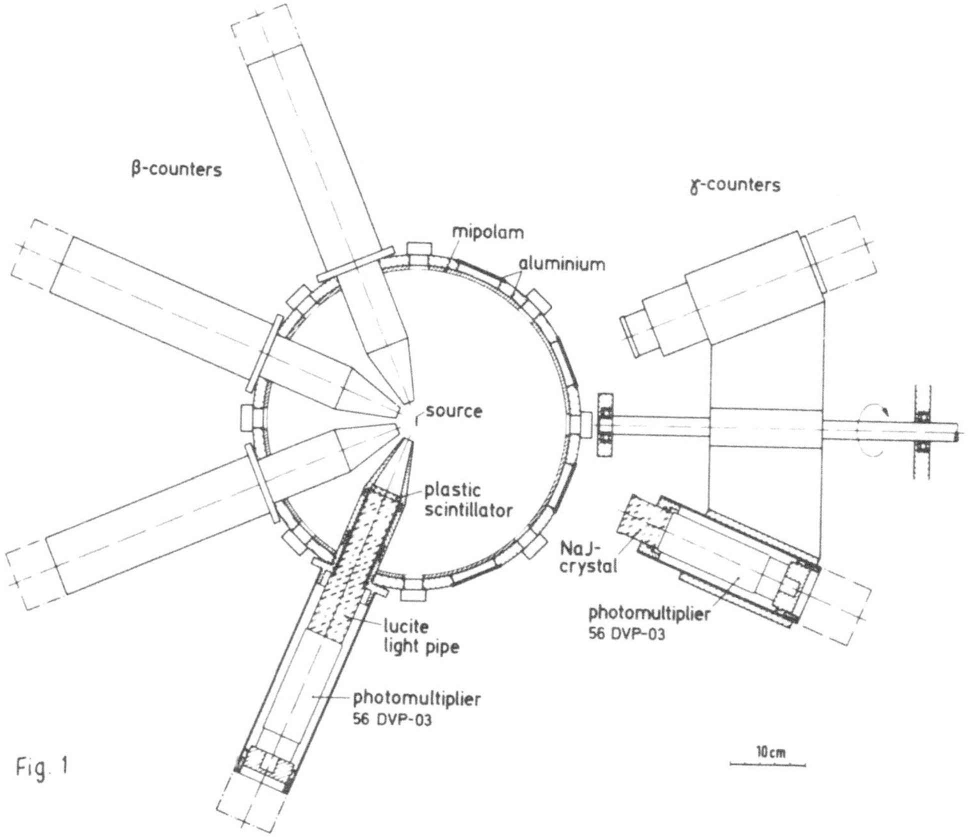

Fig. 1

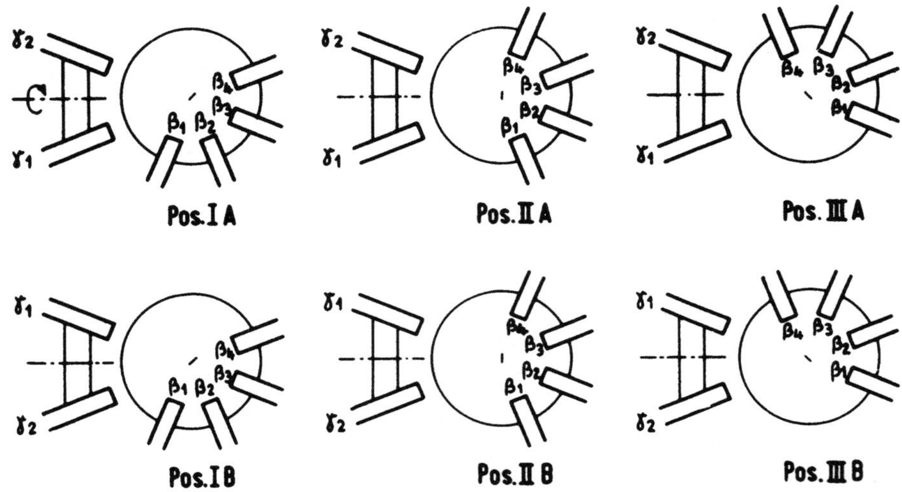

Fig.2

where K_{ij} is the true coincidence rate between the β - counter i and the γ - counter j.

It is evident that in comparison to a conventional set-up the advantages of this apparatus are the following:

A_2 and A_4 are measured simultaneously,

Higher statistical accuracy can be achieved by means of the 8 coincidence possibilities,

Systematic errors are reduced since the counter efficiencies cancel out and no corrections for the half life of the source are necessary.

For calibration purposes we measured the well known A_2- and A_4-coefficients of the $\gamma\gamma$ angular correlation of ^{60}Co. The results are

	exp.	theor.
A_2	0.1010 ± 0.0011	0.1020
A_4	0.0092 ± 0.0007	0.0091

Small corrections for finite solid angles were applied.

To test the experimental set-up we have also investigated the $\beta\gamma$ angular correlation of ^{60}Co. We obtained, as expected, $A_2 = 0.0 \pm 0.0003$, $A_4 = -0.0005 \pm 0.0003$. The statistical accuracy is typical for this arrangement.

First measurements of $\beta\gamma$ angular correlations have been carried out on the following first forbidden transitions:

$$^{86}Rb \quad 2^- \xrightarrow{\beta} 2^+ \xrightarrow{\gamma} 0^+$$

$$^{198}Au \qquad 2^- \xrightarrow{\beta} 2^+ \xrightarrow{\gamma} 0^+$$

Preliminary results are:

Nucleus	Energy (keV)	A_2	A_4
^{86}Rb	300 - 420	+0.104 ±0.002 (±0.005) 0.006	+0.009 ±0.001 (±0.001) 0.002
	360 - 480	+0.121 ±0.002 (±0.005) 0.006	+0.011 ±0.002 (±0.001) 0.003
	420 - 540	+0.147 ±0.004 (±0.005) 0.007	+0.016 ±0.003 (±0.001) 0.004
^{198}Au	375 - 475	+0.0172 ±0.0005 (±0.0008) 0.0010	+0.0017 ±0.0003 (±0.0002) 0.0004
	475 - 575	+0.0201 ±0.0006 (±0.0005) 0.0008	+0.0018 ±0.0004 (±0.0001) 0.0005
	575 - 675	+0.0238 ±0.0010 (±0.0006) 0.0012	+0.0028 ±0.0007 (±0.0001) 0.0007

The values for the A_2 - coefficients are in agreement with recent results. See, e.g., references 1, 2, 3 and 4. Other measurements of A_4 - coefficients have, up to now, not come to our knowledge.

References

1. J. Lachkar, Thèse Paris, Rapport CEA - R - 3659 (1969).
2. P.C. Simms, A. Namenson, T.H. Wei and C.S. Wu, Phys. Rev. 138 (1965), B777.
3. R.M. Steffen, Phys. Rev. 118 (1960) 763.
4. W. Deinet, H. Müller and H. Schopper, Z. Physik 188 (1965) 23.

Contributed Paper

INFLUENCE OF LINEAR POLARIZATION ON THE MEASUREMENT OF CIRCULAR POLARIZED γ - RAYS

H. Appel, K. Bürk and H. Behrens

Institut für Experimentelle Kernphysik der Universität und des Kernforschungszentrums Karlsruhe, Karlsruhe, Germany

In recent years numerous experiments have been carried out on circular polarized γ - rays. Emphasis has been laid on investigations after β - transitions and on oriented nuclei. While allowed β - transitions leave the nucleus in a pure polarized state, forbidden β - decays and all dynamical orientation methods include the possibility that the resulting state is also aligned. Consequently, after allowed β - transitions one observes purely circular polarized γ - rays. Radiation emitted in all the other cases shows also components of linear polarization. Alignment and linear polarized γ - radiation is characterized by an axis, polarized states and circular polarized γ - radiation involve a single preferred direction in space.

The Compton effect has been the primary tool for the detection of polarized γ - radiation. The differential scattering cross section is sensitive to both kinds of polarization:

$$\frac{d\sigma}{d\Omega} \sim \frac{d\sigma_o}{d\Omega} + P_\ell \frac{d\sigma_\ell}{d\Omega} + f P_c \frac{d\sigma_c}{d\Omega} \tag{1}$$

$\frac{d\sigma_0}{d\Omega}$ gives the ordinary Compton cross section, $\frac{d\sigma_\ell}{d\Omega}$ and $\frac{d\sigma_c}{d\Omega}$ represent the linear and circular polarization dependent parts, respectively, P_ℓ and P_c are the degrees of polarization, and f is the fraction of oriented electrons in the scatterer. Commonly magnetized iron is used as scattering material. It is the purpose of this note to draw attention to the linear polarization dependent part in circular polarization measurements.

The quantity P_ℓ is defined by $P_\ell = |P_\ell| (\sin^2 \psi - \cos^2 \psi)$, where ψ is the angle between the electric vector and the plane of scattering (see fig. 1). $\frac{d\sigma_\ell}{d\Omega}$ is proportional to $\sin^2 \theta$, with θ as the scattering angle. The dependence of P_c and $\frac{d\sigma_c}{d\Omega}$ on the energy and the relevant angles is thouroughly treated, e.g. in ref. 1.

There are in principle three methods to observe the γ - rays: after forward scattering, after back scattering and after transmission through magnetized iron. If a rotational symmetry is used the angle ψ is averaged out and the experimental setup is insensitive to linear polarization. In $\beta\gamma$ - polarization correlations involving forbidden β - decays it is in general requested to limit and define the angles more carefully as it is possible with a rotational symmetric arrangement. In such cases either plane scattering magnets have been used[2]) or sectors from a scattering cylinder[3]).

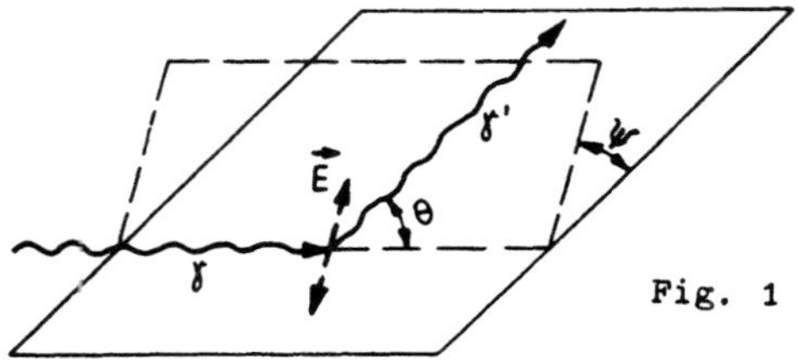

Fig. 1

Then the sensitivity of the system to detect linear polarization has to be taken into consideration.
If

$$W(\theta,W_e,\tau)=\frac{1}{S}\Big[\sum_{k=\text{even}} B_k(\beta)\,A_k(\gamma)\,P_k(\cos\theta)-\tau\sum_{k=\text{odd}} B_k(\beta)\,A_k(\gamma)\,P_k(\cos\theta)\Big] \tag{2}$$

is the $\beta\gamma$ polarization correlation (see e.g. ref. 4) then the circular polarization of the γ - radiation is given by

$$P_c(\theta,W_e)=\frac{W(\theta,\tau=+1)-W(\theta,\tau=-1)}{W(\theta,\tau=+1)+W(\theta,\tau=-1)} \tag{3}$$

$\tau=+1$ (-1) for right (left) hand circular polarized γ - radiation. The polarization P_c is related to the measurable quantity

$$E=\frac{d\sigma^- - d\sigma^+}{d\sigma^- + d\sigma^+}=\frac{f\,P_c < d\sigma_c/\,d\sigma_o >}{1+P_\ell < d\sigma_\ell/\,d\sigma_o >} \tag{4}$$

where $d\sigma^-$ and $d\sigma^+$ are the scattering cross sections for γ - rays in both magnetization directions, respectively.

$$P_\ell(\theta,W_e)=\frac{1}{S}\sum_{k=\text{even}} B_k(\beta)\,E_k(\gamma)\,P_k^{|2|}(\cos\theta).\,\cos 2\psi \tag{5}$$

with

$$E_k=\Big\{(-1)^{\sigma_{L_1}}\cdot F_k(L_1L_1j_2j_1)\,\frac{2k(k+1)L_1(L_1+1)}{k(k+1)-2L_1(L_1+1)}+2\delta(-1)^{\sigma_{L'_1}}\cdot F_k(L_1L'_1j_2j_1)$$

$$\times(L'_1-L_1)(L'_1+L_1+1)+\delta^2(-1)^{\sigma_{L'_1}}\cdot F_k(L'_1L'_1j_2j_1)\,\frac{2k(k+1)(L'_1)(L'_1+1)}{k(k+1)-2L'_1(L'_1+1)}\Big\}\frac{(k-2)!}{(k+2)!}$$

$$\begin{aligned}\sigma_{L_\gamma}&=+1 \text{ for electric } (2)^L \text{ - pole radiation}\\ \sigma_{L_\gamma}&=-\,1 \text{ for magnetic } (2)^L \text{ - pole radiation.}\end{aligned} \tag{6}$$

For γ - radiation from oriented nuclei a closely similar correlation formalism can be applied.

The size of the ratio $< d\sigma_\ell / d\sigma_0 >$ in eq. (4) can be estimated according to methods described in ref. 1.

The $2^- \xrightarrow{\beta} 2^+ \xrightarrow{\gamma} 0^+$ cascade in ^{86}Rb is one of the most interesting forbidden β-decays as far as $\beta\gamma$ angular correlations and polarization correlations are concerned. It has been investigated repeatedly[5,3,2]). This transition offers a remarkably large $B_3(\beta)\, A_3(\gamma)$ - term for the geometry used in ref. 2, namely $< d\sigma_1 / d\sigma_0 > = 0.45 \pm 0.05$. The coefficient $E_2 = B_2(\beta)\, E_2(\gamma) = -0.08$. The correction which has to be applied to circular polarization measurements due to linear polarization is in this case of the order of the present experimental errors. With improving experimental techniques it might increase in its importance.

References

1. H. Schopper, Nucl. Instr. and Methods 3 (1958) 158.
2. H. Appel, K. Bürk, and H. Behrens, to be published.
3. J.B. Viano, J.C. Renard, J. Menet, P. de Saintignon, A. Laverne and P. Depommier, Le Journal de Physique 30 (1969) 763.
4. H. Appel, Landolt - Börnstein New Series Vol. I/3, Springer - Verlag Berlin, Heidelberg, New York, 1969.
5. P.C. Simms, A. Namenson, T.H. Wei and C.S. Wu, Phys. Rev. 138 (1965) B777.

Contributed Paper

β - γ (CP) - ASYMMETRY COEFFICIENTS IN ALLOWED ISOSPIN FORBIDDEN TRANSITIONS

H. Behrens

Institut für Experimentelle Kernphysik der Universität und des Kernforschungszentrums Karlsruhe, Karlsruhe, Germany

During the past years a large amount of $\beta \cdot \gamma$ (CP) - asymmetry coefficients have been measured for isospin forbidden allowed mixed transitions in order to determine fermi matrix elements, isospin impurity coefficients and effective Coulomb matrix elements. Up to 1967 these results have been listed more or less completely several times by different authors[1-4]). A great number of new measurements have been carried out since with much smaller errors than in previous investigations. In the following table 1 a new compilation of all experimental results and their averages are given. In some cases it is possible to determine the fermi matrix elements also by measurements on oriented nuclei. Therefore the following notation is used to indicate this: The letter C means $\beta \cdot \gamma$ (CP) correlation and 0 oriented nuclei. The data ob-

Table 1

Nucleus	Asym. coeff.	Error	Method	CNO	Author and year		No.
^{20}F	0.15000	0.0700	C	0	Freiberg and Soergel	1961	1)
	0.14000	0.0700	C	0	Lobashov and Nazarenko	1962	2)
	0.17400	0.0140	C	0	Mann and Bloom	1965	3)
	0.17185	0.0135					
^{24}Na	-0.06800	0.0470	C	1	Appel and Schopper	1957	1)
	0.07000	0.0400	C	0	Boehm and Wapstra	1957	2)
	0.12000	0.0300	C	0	Mayer-Kuckuk and Nierhaus	1959	3)
	0.06000	0.0300	C	0	Steffen	1959	4)
	0.08500	0.0270	C	0	Bloom et al.	1960/62	5)
	0.10400	0.0260	C	0	Haase et al.	1963	6)
	0.09100	0.0170	C	0	Bloom et al.	1964	7)
	0.09300	0.0090	C	0	Schopper et al.	1965	8)
	0.10200	0.0060	C	0	Behrens	1967	9)
	0.09761	0.0045					
^{24}Al	-0.08900	0.0570	C	0	Haase et al.	1963	1)
	-0.08400	0.0540	C	0	Bloom et al.	1964	2)
	-0.08636	0.0392					
^{27}Mg	-0.22100	0.0190	C	0	Bigoni et al.	1969	1)
	-0.22100	0.0190					
^{41}A	0.33000	0.0700	C	1	Mayer-Kuckuk et al.	1960	1)
	0.06100	0.0700	C	0	Bloom et al.	1960/62	2)
	0.09000	0.0400	C	0	Chabre and Depommier	1962/63	3)
	0.06600	0.0250	C	0	Behrens	1967	4)
	0.07500	0.0075	C	0	Behrens and Koenig	1967	5)
	0.07461	0.0070					
^{44}Sc	-0.02000	0.0400	C	1	Boehm and Wapstra	1957	1)
	-0.15100	0.0300	C	0	Bloom et al.	1960/62	2)
	-0.12700	0.0140	C	0	Mann et al.	1965	3)
	-0.13129	0.0127					
^{46}Sc	0.33000	0.0400	C	1	Boehm and Wapstra	1957	1)
	0.29000	0.1100	C	1	Lundby et al.	1958	2)
	0.24000	0.0400	C	1	Juengst and Schopper	1958	3)
	0.24000	0.0200	C	1	Steffen	1959/60	4)
	0.07500	0.0180	C	0	Bloom et al.	1960/62	5)
	0.10000	0.0200	C	0	Daniel and Kuntze	1961	6)
	0.21500	0.0190	C	1	Boehm and Rogers	1962	7)
	0.11000	0.0200	C	0	Berthier	1962	8)
	0.07900	0.0130	C	0	Haase	1962	9)
	0.19000	0.0300	C	1	Singru and Steffen	1963	10)

Nucleus	Asym. coeff.	Error	Method	CNO	Author and year		No.
	0.11300	0.0080	C	0	Daniel et al.	1963	11)
	0.09700	0.0120	C	0	Miskel et al.	1963	12)
	0.11000	0.0100	C	0	Chabre	1963	13)
	0.07700	0.0100	C	0	Wei et al.	1964	14)
	0.09200	0.0070	C	0	Mann et al.	1965	15)
	0.13000	0.0200	C	0	Mitra and Padhi	1966	16)
	0.10300	0.0040	C	0	Behrens	1967	17)
	0.06000	0.0300	C	0	Meulenberg et al.	1967	18)
	0.08430	0.0030	C	0	Pingot	1969	19)
	0.09286	0.0020					
^{48}Sc	0.06000	0.0400	C	0	Van Nooijen	1958	1)
	0.06300	0.0130	C	0	Mann et al.	1965	2)
	0.11000	0.0600	C	0	Van Nooijen et al.	1965	3)
	0.05800	0.0030	C	0	Mann et al.	1968	4)
	0.05530	0.0055	C	0	Behrens and Wischhusen	1969	5)
	0.05771	0.0026					
^{48}V	0.06000	0.0500	C	1	Boehm and Wapstra	1957	1)
	0.0	0.0400	C	0	Daniel and Kuntze	1961	2)
	-0.06600	0.0350	C	0	Mann et al.	1962	3)
	-0.03600	0.0050	C	0	Bloom and Mann	1966	4)
	-0.05900	0.0100	C	0	Behrens	1967	5)
	-0.00200	0.0240	C	0	Van Nooijen et al.	1968	6)
	-0.03925	0.0043					
^{49}Ca	-0.13200	0.0170	C	0	Mann and Bloom	1970	1)
	-0.13200	0.0170					
^{52}Mn	-0.03300	0.0040	O	0	Postma et al.	1957	1)
	-0.16000	0.0500	C	1	Boehm	1958	2)
	-0.02100	0.0040	O	0	Ambler et al.	1958	3)
	-0.10000	0.0300	C	1	Boehm	1962	4)
	-0.09100	0.0280	C	1	Bloom et al.	1962	5)
	0.0	0.0200	C	0	Daniel et al.	1962	6)
	-0.06000	0.0080	C	1	Mann et al.	1965	7)
	-0.01900	0.0100	C	0	Behrens	1967	8)
	-0.02700	0.0090	C	0	Sawyer	1969	9)
	-0.02602	0.0026					
^{52m}Mn	-0.14100	0.0160	C	0	Bloom and Mann	1966	1)
	-0.14100	0.0160					
^{56}Co	-0.07400	0.0090	O	1	Ambler et al.	1957	1)
	-0.01000	0.0500	C	0	Daniel et al.	1961/64	2)
	0.0	0.0300	C	0	Mann et al.	1962	3)
	0.01400	0.0220	C	0	Bhattacherjee et al.	1967	4)
	0.00200	0.0100	C	0	Behrens	1967	5)
	0.00331	0.0086					
^{57}Ni	0.07100	0.0110	C	0	Atkinson et al.	1968	1)
	0.07100	0.0110					

Nucleus	Asym. coeff.	Error	Method	CNO	Author and year		No.
^{58}Co	0.11000	0.0500	0	1	Griffing and Wheatley	1956	1)
	-0.17300	0.0440	0	0	Postma et al.	1957	2)
	-0.17100	0.0240	0	0	Ambler et al.	1957	3)
	-0.14000	0.0700	C	0	Boehm and Wapstra	1957	4)
	-0.09300	0.0250	0	1	Dagley et al.	1958	5)
	-0.10300	0.0320	C	1	Collin et al.	1965	6)
	-0.18500	0.0110	C	0	Mann et al.	1965	7)
	-0.10600	0.0230	C	1	Behrens	1967	8)
	-0.16000	0.0400	C	0	Bhattacherjee et al.	1967	9)
	-0.18000	0.0400	C	0	Meulenberg et al.	1968	10)
	-0.18012	0.0091					
1 ^{59}Fe	-0.46000	0.0800	C	1	Forster and Sanders	1957	1)
	-0.06600	0.0370	C	1	Haase	1962	2)
	-0.13000	0.0400	C	0	Collin et al.	1964	3)
	-0.20500	0.0190	C	0	Mann et al.	1962/64	4)
	-0.25000	0.0700	C	0	Kneissl and Schneider	1965	5)
	-0.15800	0.0070	C	0	Behrens and Zernial	1970	6)
	-0.16348	0.0065					
2 ^{59}Fe	-0.04000	0.1100	C	0	Forster and Sanders	1960	1)
	-0.40000	0.2000	C	0	Haase	1962	2)
	-0.17000	0.1000	C	0	Collin et al.	1964	3)
	-0.04200	0.0430	C	1	Mann et al.	1962/64	4)
	-0.02000	0.1500	C	0	Kneissl and Schneider	1965	5)
	-0.15400	0.0230	C	0	Tirsell and Mann	1967	7)
	-0.15044	0.0216					
^{65}Ni	0.23000	0.0400	C	0	Atkinson et al.	1968	1)
	0.29200	0.0370	C	0	Behrens and Wischhusen	1969	2)
	0.26341	0.0272					
^{110m}Ag	0.07000	0.0200	C	0	Daniel et al.	1963	1)
	0.05800	0.0070	C	0	Behrens	1967	2)
	0.05931	0.0066					
^{124}Sb	0.22600	0.0160	C	1	Mann et al.	1965	1)
	0.35600	0.0300	C	0	Tirsell and Talbert	1965	2)
	0.28000	0.0100	C	0	Mann et al.	1965	3)
	0.30000	0.0300	C	0	Bhattacherjee et al.	1965	4)
	0.27800	0.0130	C	0	Behrens and Zernial	1970	5)
	0.28523	0.0074					
^{134}Cs	-0.07400	0.0220	C	0	Mann et al.	1962	1)
	-0.13000	0.0200	C	1	Daniel et al.	1963	2)
	-0.07700	0.0130	C	0	Tirsell and Talbert	1965	3)
	-0.07200	0.0180	C	0	Bhattacherjee et al.	1967	4)
	-0.06900	0.0070	C	0	Behrens	1967	5)
	-0.07113	0.0056					
^{152}Eu	0.10800	0.0320	C	0	Berthier et al.	1960	1)
	0.14000	0.0400	C	0	Collin et al.	1964	2)
	0.15000	0.0200	C	0	Bhattacherjee et al.	1965	3)
	0.10700	0.0120	C	0	Behrens and Wischhusen	1969	4)
	0.11869	0.0095					

tained with oriented nuclei are listed as if measured with $\beta \cdot \gamma$ (CP) correlation. Since, especially for the older measurements, there are discrepancies between different experimental results, which lie far outside the quoted errors, all the results could not be adopted for the weighted mean of the individual values. The number CNO indicates, if the value is adopted or not: O is adopted and 1 not adopted. The weights are the reciprocals of the squares of the errors σ_i. The error σ of the mean is then given by

$$\sigma = \sqrt{\frac{1}{\sum_i \frac{1}{\sigma_i^2}}}$$

The other notations in table 1 are self - evident. From the average asymmetry coefficients listed the Fermi matrix elements $|M_F|$, the isospin impurity coefficients $|a|$ and the effective Coulomb - matrix elements $|V_C|$ have been calculated and compiled in table 2. The notation in table 2 is:

$$\text{MF} \mathrel{\hat{=}} |M_F| \;;\; \text{AP} \mathrel{\hat{=}} |a| \;;\; \text{V (keV)} \mathrel{\hat{=}} |V_C| \text{ in keV}$$

TABLE 2

Nucleus	Y	Error	ME(*1000)	Error	AP(*1000)	Error	V(keV)	Error
^{20}F	-0.00636	0.0165	1.587	4.130	1.122	2.920	9.725	25.304
^{24}Na	-0.01921	0.0061	1.305	0.413	0.923	0.292	4.898	1.548
^{24}Al	0.00407	0.0528	0.203	2.622	0.143	1.854	0.747	9.662
^{27}Mg	0.00317	0.0494	1.118	17.371	0.645	10.029	3.843	59.733
^{41}A	0.02722	0.0093	6.789	2.307	3.036	1.032	21.659	7.361
^{44}Sc	-0.04303	0.0154	7.590	2.699	3.795	1.349	15.309	5.443
^{46}Sc	-0.01281	0.0027	0.803	0.167	0.402	0.084	2.873	0.598
^{48}Sc	-0.00299	0.0036	0.374	0.447	0.153	0.183	1.128	1.349
^{48}V	-0.05896	0.0058	4.134	0.404	2.067	0.202	11.051	1.080
^{49}Ca	0.00258	0.0329	0.723	9.210	0.241	3.070	2.014	25.665
^{52}Mn	-0.04096	0.0036	5.741	0.500	2.870	0.250	17.205	1.500
^{52m}Mn	-0.03127	0.0194	4.920	3.048	2.460	1.524	9.690	6.002
^{56}Co	-0.11629	0.0117	0.405	0.040	0.202	0.020	1.137	0.111
^{57}Ni	-0.11877	0.0209	14.653	2.509	8.460	1.449	55.204	9.452
^{58}Co	0.01654	0.0113	0.654	0.446	0.267	0.182	1.742	1.188
1 ^{59}Fe	-0.05949	0.0130	2.090	0.455	0.790	0.172	6.690	1.455
2 ^{59}Fe	-0.03345	0.0428	2.892	3.686	1.093	1.393	9.041	11.523
^{65}Ni	-0.00305	0.0178	0.191	1.114	0.064	0.371	0.602	3.508
^{110m}Ag	-0.00522	0.0092	0.033	0.058	0.008	0.014	0.106	0.187
^{124}Sb	-0.10542	0.0069	1.036	0.066	0.221	0.014	3.029	0.193
^{134}Cs	0.21150	0.0082	0.636	0.023	0.130	0.005	1.881	0.067
^{152}Eu	0.04137	0.0082	0.016	0.003	0.003	0.001	0.051	0.010

With exception of ^{65}Ni, where a mixing ratio δ (E2/M1) = -0.43 has been taken, the γ - transitions have been assumed as pure transitions with the lowest possible multipole order. The methods used for this calculation have been extensively described, for example, in ref. 3 and 5. The Coulomb energy differences between parent and analogue states, which are necessary for the calculation of the effective Coulomb matrix elements, have been calculated with the help of a formula given in ref. 6. The log ft - values and the maximum β - energies also needed for the calculations have been taken from standard compilations of decay schemes, for ft (^{14}O) the value 3116 sec has been used.

Taking table 2 one can establish that especially in the cases of ^{20}F, ^{24}Al, ^{27}Mg, ^{49}Ca and ^{59}Fe 2 the experimental accuracy should be improved. From table 1 one must conclude, that there are only two cases, where greater differences between different recent measurements exist: ^{48}V and ^{58}Co. In ^{59}Fe 2 the situation is much clearer after the measurement of Tirsell and Mann 1967, who used the transmission method, while the other groups used the forward scattering method, which is not suitable in this case.

References:

1. S.D. Bloom, Nuovo Cim 32 (1964) 1023.
2. H. Daniel and H. Schmitt, Nucl. Phys. 65 (1965) 481.
3. H. Schopper, Weak interactions and nuclear beta decay, Amsterdam North - Holland Publ. Co. 1966.
4. S.K. Bhattacherjee, S.K. Mitra and H.C. Padhi, Nucl. Phys. A96 (1967) 81.
5. R.J. Blin - Stoyle in: Isospin in nuclear physics, ed. by D.H. Wilkinson, Amsterdam, North-Holland Publ. Co. 1969.
6. J. Jänecke in: Isospin in nuclear physics, ed. by D.H. Wilkinson, North - Holland Publ. Co. 1969, p. 341.

REVIEW OF ALL QUADRUPOLE MOMENTS MEASURED BY COULOMB EXCITATION

B. Povh

Physikalisches Institut der Universität Heidelberg, Germany

1. Introduction

In the last few years a large number of static quadrupole moments of excited nuclear states have been determined using the reorientation effect[1,2]). The method itself has been suggested many years ago[1]), but the expansion of the experimental activity using this method has to wait for accelerators with suitable heavy ion beams and the development of a program for the evaluation of the multiple Coulomb excitation probabilities[3]). The quadrupole moments which have been measured predominantly by this method are those of the first excited 2^+ states of even nuclei. The interest in the reorientation method is obvious. It opens up a new possibility to measure static quadrupole moments of excited states in a large number of nuclei. In this way the shape of even nuclei becomes accessible to systematic investigation.

After summarizing some of the relevant formulae for the multiple Coulomb excitation I will briefly explain the different possibilities to measure quadrupole moments with the reorientation method. Then I will discuss some difficulties in deducing the values of quadrupole moments from the Coulomb excitation measurements. Finally I will try to review and discuss the results of the reorientation measurements up to now.

2. Method

The cross - section for the Coulomb excitation from the ground state i to the final state f is given by the product of the Rutherford cross - section and the excitation probability $P_{i \to f}$

$$\frac{d\sigma_c}{d\Omega} = \frac{d\sigma_R}{d\Omega} P_{i \to f} \tag{1}$$

$P_{i \to f}$ is usually calculated in the semiclassical approximation where the projectile moves along the classical trajectory. The semiclassical treatment will be good if the Sommerfeld parameter η is large compared to unity, η being the ratio of the distance of closest approach and the de Broglie wavelength of the projectile. In this frame the Coulomb excitation probability can be calculated exactly by solving the time - dependent coupled channel equations for the excitation amplitudes. This has been done in the computer code of Winther - de Boer[3]) commonly used in the analysis of multiple Coulomb excitation and reorientation experiments.

For the sake of plausibility I will discuss the reorientation effect in the perturbation expansion. In first order, which is a good approximation for Coulomb excitation with light projectiles (protons, alphas), the excitation probability for electric quadrupole excitation is directly proportional to the B (E2)

$$P^{(1)}_{i \to f} = B(E2)\ F(\theta,\xi) \tag{2}$$

Here θ is the scattering angle and ξ is the so called adiabaticity parameter taking into account the dynamics of the process.

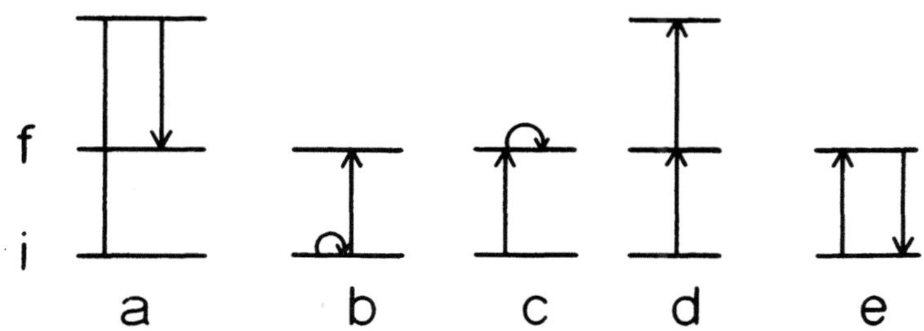

Fig. 1. Schematic representation of the second order excitation amplitudes.

In the bombardment with heavy ions second order effects become important. The second order corrections are schematically shown on fig. 1. Fig. 1b and 1c represent the reorientation effect, which is a second order contribution to the excitation probability and is proportional to the static quadrupole moment in the ground and excited state. In order to deduce an accurate value of the quadrupole moment the contribution of other second order effects should be kept small. This is most likely fulfilled in the even nuclei for the transition from the 0^+ ground state to the first excited 2^+ state. In this particular case the excitation probability of the target nucleus can be written as

$$P_{0 \to 2} = P^{(1)}_{0 \to 2} \left\{ 1 + 1.32 \frac{A_P}{Z_T} \frac{\Delta E}{1 + A_P/A_T} Q_2\ K(\theta,\xi) \right\} \tag{3}$$

In this expression A_P, A_T and Z_P, Z_T are the mass and charge of the projectile and target nucleus, respectively. The excitation energy of the 2^+ state is ΔE (MeV) and its static quadrupole moment is Q_2 (ebarn.). One arrives at the expression for projectile excitation by replacing Z_T by Z_P. The second term within the parentheses is given by the interference between the first and second order amplitudes corresponding to fig. 1c and depends linearly on the static quadrupole moment Q_2. In order to determine Q_2 one needs at least two independent measurements, one measurement where the reorientation effect is large and a second one where it is small. Since the function K (θ, ξ) does not depend strongly on ξ, i.e. on the projectile energy, we are left with two types of possible experiments: one experiment which uses the strong θ - dependence of the function K (θ, ξ), which has its maximum at $\theta = \pi$, and a second one which uses the A_P dependence of the interference term.

The reorientation effect is usually small. For a typical case of $Q \approx 1$ (ebarn) and for an oxygen ion beam, it amounts to about 10% of the excitation probability. Therefore, an accuracy of about 1% of $P_{0 \to 2}$ is required in order to extract Q_2 with reasonable error. This is in general not a trivial problem and many experimental techniques have been developed to meet the accuracy requirement for the measurements in different mass regions.

The straightforward technique is the direct measurement of inelastically scattered particles. Although this measurement is in principle very simple, the resolution of the detector system and the purity of the target material represent serious problems. Therefore many experiments are done by measuring the particle - gamma correlations. On the other hand these measurements are influenced by the deorientation effect as discussed in the next chapter.

3. Corrections (headache of the reorientists).

As was pointed out, the reorientation effect gives only a small contribution to the excitation probability. Therefore a careful investigation of possible sources of errors by evaluation of the quadrupole moments is necessary.

a) The deviation of the semiclassical approach from the quantum - mechanical treatment was studied carefully in the last years[4,5]). The quantum - mechanical correction to the reorientation effect is proportional to the quadrupole moment and to η^{-1}. In typical cases this correction is less than 10%. This correction is usually not taken into account because it is smaller than the experimental errors in most of the measurements.

b) The reorientation is only one of the second order effects in the Coulomb excitation. In the programs for multiple Coulomb excitation all the second order effects could be considered in principle. But to do this, all the relevant matrix elements should be known in magnitude as well as in sign. This is hard to achieve. Therefore the measurements done so far are restricted to nuclei with first excited states well separated from the higher excited states. In table 1 the influence of the virtual excitations to upper levels as well as other possible corrections to the excitation probability are shown for ^{114}Cd. The values are calculated[6]) for bombardment with oxygen ions of 40 MeV and 180° scattering. From table 1 it is seen that in the vibrational nuclei an important correction comes from the virtual excitation to the second 2^+ state. For the light nuclei the main correction comes from the virtual excitation of the giant resonance and can amount to a 10% correction of Q_2 in the case of ^{20}Ne when bombarded with 40 MeV sulphur. It seems that at present ignorance about the matrix elements relevant to the virtual excitation of the upper states brings about the main uncertainty in the determination of quadrupole moments with the reorientation method.

c) The probability for Coulomb excitation increases strongly with the bombarding energy. Therefore one tends to use the highest possible energy consistent with the requirement that the interaction is purely electromagnetic. Since the inelastic scattering cross - section is very sensitive to nuclear interaction, the bombarding energy

Effect	ΔP/P	ΔP/ΔP(Q_2)
Quadrupole Moment assuming $Q_2=Q_{ROT}=-0.68$ ebarn	-12%	
Higher 2^+ State	±1.2%	±10%
Giant Dipol Resonance	±1.0%	±8%
Hexadecapole Moment $Q_4=0.011$ ebarn^2	-0.9%	-7%
Semi-Classical Approximation	-0.5%	-4%
Atomic Screening	+0.7%	+6%
Vacuum Polarisation	-1.3%	-11%

Table 1. The contribution of various effects to the reorientation measurement for 44 MeV ^{16}O ions on ^{114}Cd and at $\theta = 180^\circ$. The values are taken from ref. 6.

should be chosen so low that the nuclear interference can be neglected as compared to the reorientation effect. Experimental investigations show[7,8]) that this requirement is fulfilled if the minimum distance of closest approach (scattering at 180°) is larger than [1.25 $(A_T^{1/3} + A_P^{1/3}) + 5.1$] fm. The question about a 'safe' bombarding energy seems to be settled now and should not be a source of error in measurements anymore.

d) One of the serious corrections to be taken into account in the gamma measurements is the deorientation effect. Nuclei recoiling out of the target into vacuum are highly ionized. The magnetic field produced by unpaired strongly bound electrons at the nucleus causes a fast attenuation of the nuclear alignment. In medium heavy nuclei the deorientation was investigated in connection with the measurements of the magnetic moments[9]). The magnitude of the deorientation cannot be calculated in general and should therefore be determined experimentally. This requires an additional measurement of the attenuation of particle - gamma correlations under the same conditions as the reorientation measurement. In light nuclei this correction can become very significant. At typical recoiling velocities of $\beta = 0.07$ in reorientation experiments, for example, 40% of the Ne - ions are in the 9^+ charge state. The magnetic field of the single 1s - electron at the nucleus is 170 MGauss. In this particular case[10]) the deorientation was studied and the attenuation was found to be due to the static magnetic interaction between the electrons and the nucleus. In general, however, the effective field in many electron systems is not known and the deorientation should be determined again by an independent measurement.

4. Summary of quadrupole moments measured by reorientation

So far only three reorientation measurements are known to me which do not concern the first excited 2^+ state of even nuclei. These are the measurements of the quadrupole moment of the first 3^- state in ^{208}Pb ($Q = -1.3 \pm 0.6$ ebarn) [11]), the moment of the second 2^+ state in ^{166}Er ($Q = 1.17 \pm 0.33$ ebarn) [12]); further, the Heidelberg group attempted to measure the ground state quadrupole moment of ^{7}Li.

It is interesting to compare the quadrupole moments of the 2^+ first excited states with the values calculated assuming the rotational model. The rotational value of the quadrupole moment Q_{ROT} of a 2^+ state is given by $Q_{ROT} = 0.911\,[B(E2,0\to2)]^{1/2}$. The results are obvious for the rotational nuclei. For a pure harmonic vibrator the quadrupole moment of the 2^+ state should vanish. Introducing an harmonicity into the model, the quadrupole moment will get a finite value[13]) but will hardly exceed Q_{ROT}. From this plausible argument one expects the quadrupole moments of the first 2^+ states to be limited by $0 < |Q_2| < |Q_{ROT}|$. Therefore in the summary (figs. 2, 3, 4 and 5) all Q_2 - values are given together with Q_{ROT}.

In fig. 2 the results of all reorientation measurements carried out at the s - d - shell nuclei are shown. These results have been obtained by three groups (Berkeley, Chalk River, Heidelberg), which used different experimental techniques. The agree-

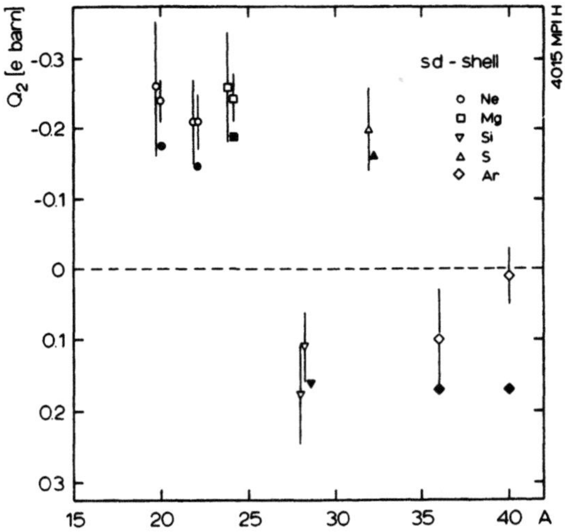

Fig. 2. Quadrupole moments of the first excited 2^+ states of the sd shell nuclei measured by reorientation (open symbols) and corresponding Q_{ROT} values (closed symbols). The sign of Q_{ROT} is arbitrary and in the figure its definition has been chosen for convenience.

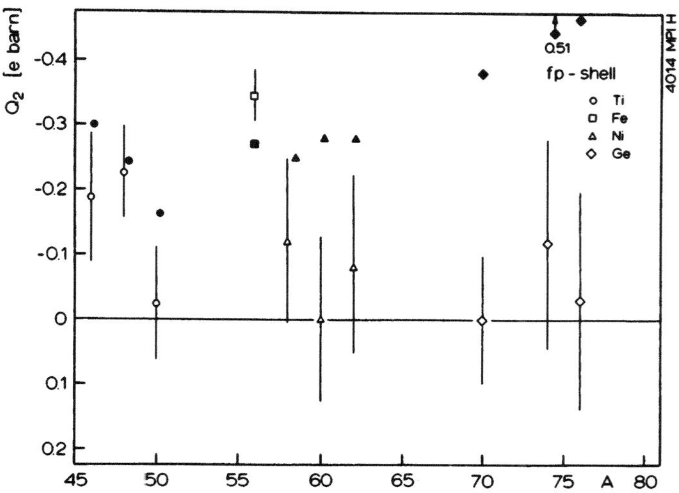

Fig. 3. Quadrupole moments of the first excited 2^+ states of the f - p - shell nuclei (for notation see fig. 2).

ment demonstrates the high reliability of the present experiments at least for the light nuclei. It is interesting to point out that the quadrupole moments for deformed light nuclei obtained by the reorientation method are about 30% larger than Q_{ROT}. A similar trend is observed in the Nd region (see fig. 5). At present it is hard to decide, whether this is a nuclear physics effect or some unknown correction omitted in the evaluation of the reorientation measurement. To clarify this situation, it would be desirable to use the reorientation method to determine a quadrupole moment which has also been measured by an independent, well - established method. The termination of the strong deformation for s - d - shell nuclei near A = 28. corresponds to the theoretical predictions.

The quadrupole moments Q_2 which have been measured up - to - now in heavier nuclei are summarized in figs. 3, 4 and 5. For this summary the compilation of Smilansky[14]) was used as well as information obtained from recent publications and preprints. All measured nuclei in the region $45 < A < 145$ belong to the generally called vibrational nuclei. The values of the quadrupole moments scatter uniformly between 0 and Q_{ROT}. This obviously requires a modification of the vibrational model. A simple introduction of anharmonicities in the model, however, does not give consistent results for the quadrupole moments and B (E2)'s connecting different phonon states[6]). It seems that there is no simple generalization of the vibrational excitation, and calculations have to be carried out for individual nuclei in order to understand the B (E2)'s and the observed quadrupole moments of the first excited 2^+ states. The transition to the deformed nuclei is well reproduced for Nd and Sm isotopes as well as the predicted transition from prolate to oblate shape around

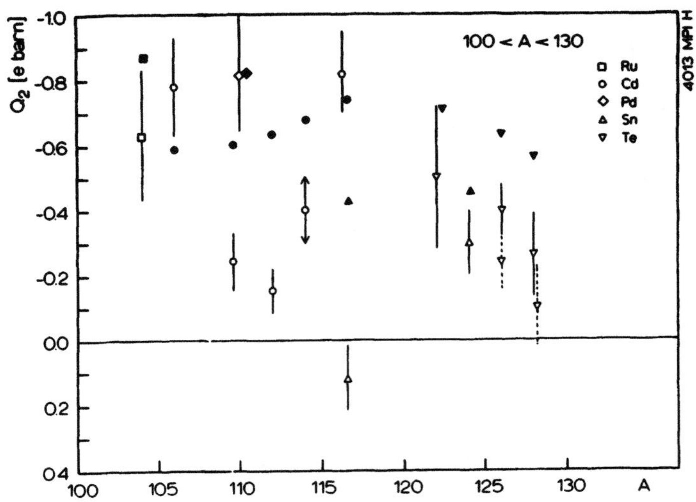

Fig. 4. Quadrupole moments of the first excited 2^+ states for the nuclei $100 < A < 130$ (for notation see fig. 2).

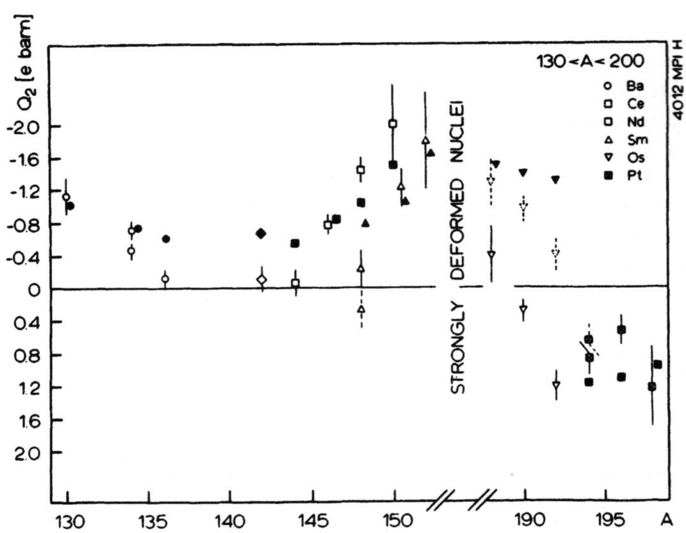

Fig. 5. Quadrupole moments of the first excited 2^+ states for nuclei $130 < A < 200$ (for notation see fig. 2).

A = 195.

In conclusion, I would like to say that there are many possibilities to improve the output of reorientation measurements; the method will certainly be brought to a high level of sophistication. But already in this stage we find the old rule confirmed: by introducing a new experimental method into physics we are rewarded by a new puzzling surprise.

References

1. G. Breit, R.L. Gluckstern and J.E. Russel, Phys. Rev. 103 (1956) 727.
2. J. deBoer and J. Eichler, in 'Advances in Nuclear Physics', vol. 1 (Plenum Publ. Co., New York, 1968).
3. A. Winther and J. deBoer, 'Coulomb Excitation', Perspectives in Physics (Academic Press, New York 1966).
4. U. Smilansky, Nuclear Physics A112 (1968) 185.
5. K. Alder and H. Pauli, Nuclear Physics A128 (1969) 193.
6. D. Cline, UR - NSRL - 22.
7. D. Cline, H.S. Gertzman, H.E. Gove, P.M.S. Lesser and J.J. Schwartz, Nuclear Physics A133 (1969) 445.
8. M. Samuel and U. Smilansky, Physics Letters 28B (1968) 318.
9. I. Ben - Zvi et al., Nuclear Physics A121 (1968) 592.
10. M. Faessler, Diplomarbeit 1970, Heidelberg.
11. A.R. Barnett and W.R. Phillips in 'Nuclear Reactions Induced by Heavy Ions' (North - Holland), p. 461.
12. R.O. Sayer, P.H. Stelson, F.K. McGowan, W.T. Milner and R.L. Robinson, Bull. Am. Phys. Soc. 13 (1968) 895.
13. T. Tamura and T. Udagawa, Phys. Rev. 150 (1966) 783.
14. U. Smilansky, in 'Nuclear Reactions Induced by Heavy Ions', (North - Holland) p. 392.

Discussion

Alder: If one tries to measure the quadrupole moment of the second excited state, one has to be very careful. The quantum - mechanical correction of the double excitation cross - section is of the order of 10%. Therefore, in the measurements of the quadrupole moments of the second 2^+ state or 4^+ state in even - even nuclei the quantum - mechanical correction must be included, especially if α - particles are used as projectiles.

Povh: In a well - adjusted experiment the reorientation effect can be as large as 50% of the first order. In other words, the effect can be macroscopic.

Thompson: I would like to ask if you looked ever at the cases where you have excitation of both the projectile and target nucleus or if you included the possibility of virtual excitation of the projectile during the collision?

Povh: Yes, the effects you are referring to constitute only a minor correction which is not going to influence the results by more than one per cent.

Thompson: If I may make a comment. I would guess the following. If you would use identical projectiles and excite them both to a 2^+ state then you probably only excite the even magnetic substates of your excited nucleus. Instead of having five substates, you should have three in the case of a 2^+ level.

Povh: You really have three substates in case of 2^+ states: 0, 2 and 4. I do not think you can perform any reorientation measurement with identical particles since you cannot distinguish between particles coming symmetrically with respect to 90 degrees. And what you want to see is the asymmetry between the scattering at zero degrees and 180 degrees.

Murnick: I have a comment and a question. First I think that one would like to get a confirmation of the reorientation effect results using a perturbed angular correlation technique. I believe that experiments with suitable single crystals as will be described by Alonso and myself tomorrow, may allow us to do that. And my second point is a question. Do you have any theoretical justification for plotting the rotational values with a sign. That is, you plotted positive and some negative values.

Povh: I must apologize. I forgot to point out that the sign of Q_{ROT} is arbitrary and it is the absolute value that has been plotted. The sign of the moment is clearly model dependent.

Contributed Paper

THE STUDY OF THE NEUTRON - PROTON CORRELATION IN THE DISINTEGRATION OF THE DEUTERON

P. Tomaš, M. Furić, V. Valković, D. Miljanić and B. Antolković
Institute 'Ruder Bošković', Zagreb, Yugoslavia

1. Introduction

The deuteron break - up induced by the neutron has been investigated for a long time[1,2]) to obtain information about the neutron - neutron interaction. However, two different mechanisms are present in that process: quasifree scattering (QF) and final state interaction (FSI). In single counter experiments on the deuteron break - up (kinematically incomplete) at low energies the contributions of these mechanisms could not be resolved. Therefore, nucleon - nucleon correlation experiments should be performed to identify each of these mechanisms separately. Coincidence spectra of a single mechanism can then be compared with theoretical predictions to extract some two - body data.

In this work, neutron - proton correlation spectra from the deuteron break - up will be presented. An experimental method using a counter telescope, a neutron detector and the associated alpha particle technique in combination with a three - parameter analyzing system was developed[3]) for the study of neutron induced (n, n charged particle) reactions.

At the angular setting $\theta_n = \theta_p = 30^\circ$ for the neutron and proton detectors, respectively, in the D (n, np) n reaction at 14.4 MeV, the pronounced maximum was observed at the minimum value of the spectator particle energy. The shape of the spectrum was fitted using the spectator model[4]).

When the direction of the proton telescope was changed to $\theta_p = 80^\circ$, the final state interaction between two neutrons was observed. The measured coincidence spectrum was compared with the simple FSI prediction[5]).

The technique used in the experiment is described in section 2. Section 3 is devoted to the QF mechanism. Section 4 contains the discussion of the FSI data.

2. Experimental Technique

A collimated neutron beam was obtained from the Cockcroft - Walton accelerator by the associated alpha particle method. The $(CD_2)_n$ target (10 mg cm^{-2} thick, 2 cm in diameter) was placed inside the beam. A NE 218 liquid scintillator was used for neutron detection. For identification and energy measurement of protons a standard counter telescope was used. For that purpose pulses proportional to the energy loss ΔE and the energy of the charged particle E were registered. The time difference T between proton and neutron pulses was measured to ensure a ~ 10 ns coincidence requirement between the two particles. Three pulses T, ΔE and E were led into a 100 X 100 X 100 analyzer.

The calibration of the experimental set - up was performed by measuring neutron - proton coincidences in n - p elastic scattering. In this way the position and the width of the peak in the T spectrum associated with true neutron - proton events was examined as a function of energies of the outgoing particles. Thus the allowed region in the T variable, where the events from the deuteron break - up should be present, was determined.

3. Quasifree mechanism of the deuteron break - up

In fig. 1 the shape of the experimental spectrum of protons at $\theta_n = \theta_p = 30^\circ$ in the D (n, np) n reaction at 14.4 MeV is shown. The solid line represents the spectator model predictions[4]) divided by a factor of 3. The disagreement between theory and experiment in the absolute value of the cross - sections cannot be interpreted as an off - shell effect. The numerical calculation was performed to estimate the influence of the off - shell effect. The potential model, which describes the n - p elastic scattering up to 10 MeV very well, was used. It has been shown that the off - shell effect contribution is a few per cent of the total effect[6]). However, it has been recently found by Paić et al.[7]) that theory and experiment can be brought together by introducing the cut - off radius in the Fourier transform of the deuteron Hulthén wave function.

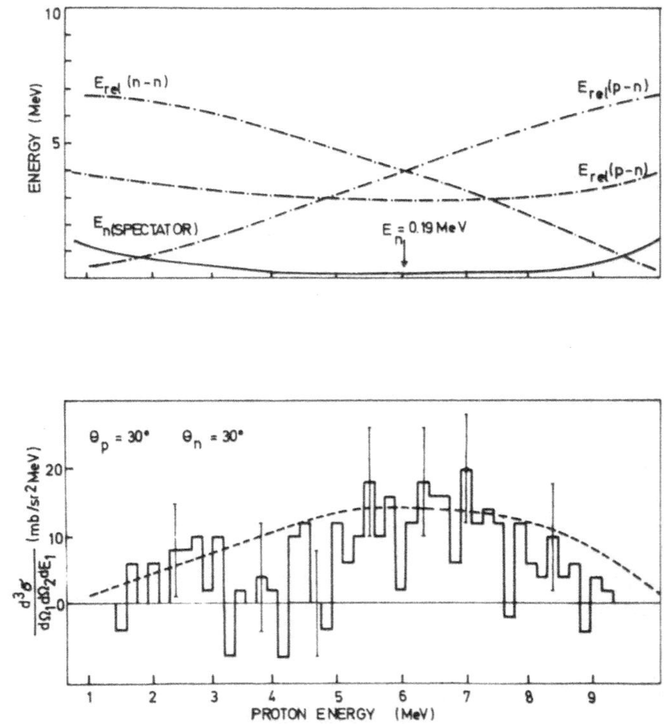

FIG. 1

4. Final state interaction mechanism of the deuteron break - up

Figure 2 shows the experimental spectrum of protons for $\theta_n = 30°$ $\theta_p = 80°$ in the D (n, np) n reaction at 14.4 MeV. In the upper part of the figure, relative energies of two - particle subsystems are shown as a function of the proton energy. The pronounced maximum in the proton spectrum does coincide with the minimum relative energy of the two neutrons. The theoretical predictions calculated by Perrin et al.[5]) using an inhomogeneous term and the first iteration of the Faddeev equations correspond to the values for the neutron - neutron scattering length $a_{nn} = -16$ fm (dashed line) and $a_{nn} = -23$ fm (solid line). Comparison of theoretical predictions and experimental results shows that higher values for the absolute value of a_{nn} are favoured. However, it should be mentioned that experimental statistics should be improved and the inclusion of higher order terms in the theory should be performed to extract more precise information on a_{nn}.

We may conclude that the proper way to extract neutron - neutron data from the deuteron break - up is to measure nucleon - nucleon correlations, since the two

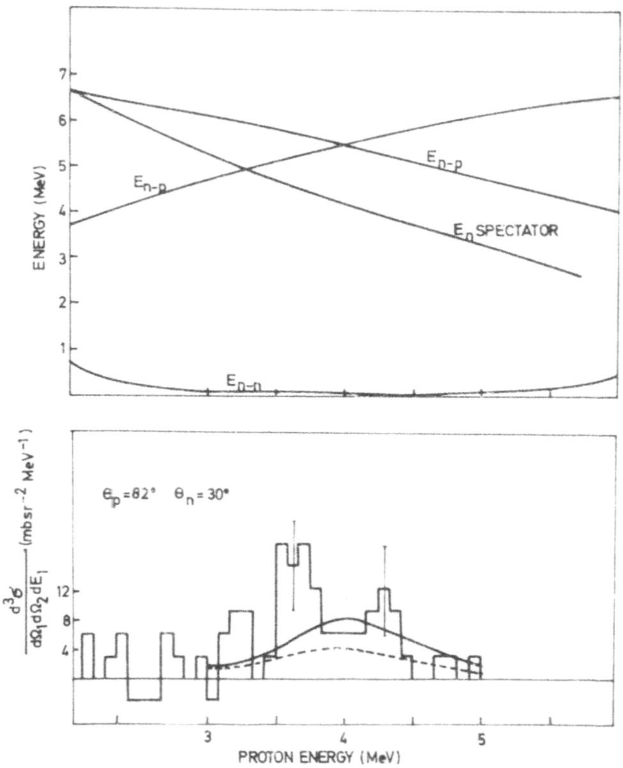

FIG. 2

competing mechanisms, QF and FSI, are of the same order of magnitude, and they cannot be resolved in single counter experiments at this energy.

References

1. K. Ilakovac, L.G. Kuo, M. Petravić and I. Šlaus, Phys. Rev. 124 (1961) 1923.
2. Few body problems, light nuclei and nuclear interactions, eds. G. Paić and I. Šlaus (Gordon and Breach 1968), pp. 83 - 153.
3. V. Valković, D. Miljanić, P. Tomaš, B. Antolković and M. Furić, Nucl. Instr. and Meth. 76 (1969) 29.
4. A.F. Kuches, R. Wilson and P.F. Cooper, Annals of Physics 15 (1961) 193.
5. C. Perrin, J.S. Gondraud, M. Durand, S. Desreumaux, R. Bouchez, Three body problem in nuclear and particle physics, eds. J.S.C. McKee and P.M. Rolph (North-Holland, 1970).
6. M. Furić, Ph. D. Thesis, University of Zagreb.
7. G. Paić, J.C. Young, D.J. Margaziotis, to be published in Phys. Lett.

A SURVEY OF CORRELATION EXPERIMENTS TESTING PARITY AND TIME REVERSAL AND A DISCUSSION OF RECENT PARITY TESTS†

F. Boehm

California Institute of Technology, Pasadena, California, U.S.A.

1. Introduction

In this lecture I shall discuss the correlation experiments aimed at testing parity and time reversal in nuclear interactions. Along with other experiments such as detailed balance tests or measurements of the static moments, the correlation experiments provide an important tool for the study of discrete symmetries.

In the first part of this lecture I shall give a survey on the principles of both P and T correlation studies. In the second part I shall confine myself to a discussion of the parity experiments only, leaving the time reversal experiments to the next speaker, W.D. Hamilton. Some considerations relative to nuclear beta decay are presented at the end in an appendix.

Interest in the investigations of the validity of invariance under space inversion dates back to 1957 when parity conservation was found to be violated in the weak interaction. The early experiments were designed to examine the validity of this symmetry in the strong and electromagnetic interactions. The upper limit for the ratio of the parity non - conserving and parity conserving interaction gradually decreased, passed the range corresponding to the electromagnetic interaction and settled at a value of $\sim 10^{-6}$. At this level parity non - conservation has been already expected through the intervention of the strangeness conserving, non - leptonic weak interaction as predicted by the current - current theory[1]). At present there is good evidence for the existence of parity - non - conservation at this level.

Simultaneously, numerous tests on time reversal have been performed during this period. The interest into these experiments was vitalized greatly when in 1964 it was discovered that CP is violated in the decay of the long - lived neutral K - meson. To the extent that the CPT theorem is valid, and we shall firmly assume its validity, this implied violation of time - reversal invariance. A great number of theories have been proposed to explain this phenomenon, the strength of the predicted CP violation ranging from 10^{-13} (with respect to the strong interaction) to 1. One of the reasons for the wide variety of possible theories is the existing, relatively poor limit for T. There is no evidence so far for T violation in nuclear processes. The upper limit is about 3×10^{-3}.

The principle of parity tests by means of correlation experiments is quite simple: Conservation of parity implies the following relation between the transition matrix elements

$$S_{fi} = S_{f'i'} \tag{1}$$

where the initial and final states i′ and f′ are the same as i and f, except that all momenta have been changed in sign. (| i′ > = P | i >, | f′ > = P | f >, where P is the parity operator). For the transition probabilities, ω, eq. (1) implies

$$\omega\,(\underline{p}, \underline{J}) = \omega\,(-\underline{p}, \underline{J}) \tag{2}$$

Thus the presence of a parity - non - conserving correlation ($\underline{p} \cdot \underline{J}$, for example) implies the violation of eqs. (2) and (1), and consequently the violation of parity conservation*.

In the case of time - reversal the situation is somewhat more complicated. The time - reversal operator T, unlike P which is unitary, is an anti - unitary operator, and time - reversal invariance implies therefore instead of eq. (1) the relation

$$S_{fi} = S_{i'f'} \tag{3}$$

where i′ and f′ denote the time - reversed states, having opposite signs of momenta and spins[2]). The analogue of eq. (2) would read

$$\omega\,(\underline{p}, \underline{J}) = \tilde{\omega}\,(-\underline{p}, -\underline{J}) \tag{4}$$

Because of the reversal of the order of initial and final state in (3), $\tilde{\omega}$ (which describes the reversed reaction) is not necessarily the same function of the kinematical variables as ω. An observation of a time - reversal non - invariant correlation ($\underline{J} \cdot \underline{p}_1 \times \underline{p}_2$, for example) does not, therefore, necessarily imply violation of time - reversal invariance. An exception is the case when the Hamiltonian of the system can be written as

$$H = H_0 + V,$$

where V is a hermitean potential, weak compared to H_0. To the extent that first - order perturbation theory is applicable we have

$$S_{i'f'} = S^{*}_{f'i'} \tag{5}$$

and consequently

$$|S_{fi}|^2 = |S_{f'i'}|^2$$

and also

$$\omega\,(\underline{p}, \underline{J}) = \omega\,(-\underline{p}, -\underline{J}) \tag{6}$$

meaning the absence of time - reversal non - invariant correlations. Thus eq. (6) can be violated not only by time - reversal non - invariant interactions, but also in time reversal invariant theory by higher order corrections, for which case eqs. (5) and (6)

* Take for example a simple correlation $\omega\,(\underline{p}, \underline{J}) = 1 + a\,\underline{p} \cdot \underline{J}$, and consequently $\omega\,(-\underline{p}, \underline{J}) = 1 + a\,(-\underline{p}) \cdot \underline{J}$. In view of (2) we must have a = 0.

no longer hold. These corrections limit the precision by which time reversal invariance can be tested in correlation experiments unless, of course, the corrections are known to be negligible at a given level of precision or can be calculated.

In the presence of P and T violating interactions the Hamiltonian of the nucleus will be

$$H_{\text{nucleus}} = H_{\text{kin}} + V_{P_{\text{even}}T_{\text{even}}} + V_{P_{\text{odd}}T_{\text{even}}} + V_{P_{\text{even}}T_{\text{odd}}}$$

where $V_{P_{\text{even}}T_{\text{even}}}$ is the regular part. If time reversal is violated in electromagnetic interactions[3]), in addition to the appearance of a $P_{\text{even}}T_{\text{odd}}$ nuclear force[4]) the electromagnetic transition operator will also be modified by a T - violating contribution.

2. Survey

Let us now consider the quantities we need to measure in a correlation experiment to obtain information on terms odd in P and T.

Consider the different multipole matrix elements for nuclear gamma transitions between two definite nuclear states, a and b,

$$\text{E1, M2, E3, ...,} \tag{7}$$

$$\text{M1, E2, M3, ...,} \tag{8}$$

If parity and time reversal invariance holds the lower or the upper line vanishes depending on the parity change, and the relative phases of all non - zero matrix elements are 0 or π. If P - odd and T - even forces are present, such as $\underline{\sigma} \cdot \underline{p}$ forces, the situation is changed and all matrix elements (7) and (8) may contribute. The phases between set (7) and (8) become fixed at 0 or π. Further in the case of P - even T - odd terms, the relative phases between matrix elements in (7) or (8) differ from 0 or π by an amount η which is of first order in $H_{P_{\text{even}}T_{\text{odd}}}$

For experiments exploring P - odd terms, one has to look for an interference between matrix elements (7) and (8) (such as E1, M1). The largest interference effect is expected if the regular transition, being dominating, is retarded (from nuclear structure effects, for example), while the irregular transition is unretarded or even enhanced. The strength of the parity admixing forces operating in the nuclear states is usually characterized by $F = V_{P_{\text{odd}}}/V_{P_{\text{even}}}$

Just as for $P_{\text{odd}}T_{\text{even}}$ forces, we can characterize the $P_{\text{even}}T_{\text{odd}}$ forces by a strength $F = V_{T_{\text{odd}}}/V_{T_{\text{even}}} \approx \frac{\eta}{\pi/2}$. In the case of $P_{\text{even}}T_{\text{odd}}$ we must look for a difference in phase, η, between regular transitions, such as (M1, E2). We must design an experiment which is sensitive to the sine of the phase difference η. Theoretical predictions for the size of η are not as advanced as for the parity force.

The quantities we can measure are **polar vectors** characterizing the photon momenta $\underline{k}_i$, **axial vectors** characterizing the nuclear spins $\underline{j}_i$ and the photon spin

Table 1. Gamma ray a → b, of mixed multipolarity whose circular polarization $\vec{s}$, or linear polarization $\vec{\epsilon}$ is measured. (Quantities measured to reveal breakdown of symmetries are listed in columns 2, 5, 6, and 7)

Odd under	Gamma ray a → b measured	Orientation Ω_a	Orientation Ω_b	Second gamma ray (analyzer) measured*	Preceding gamma ray (polarizer) and subsequent gamma ray (analyzer) measured	Allowed beta decay (polarizer) and subsequent gamma ray (analyzer) measured
	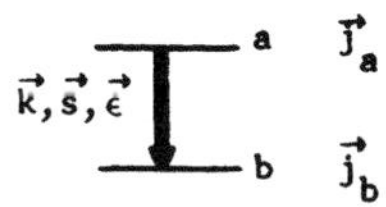			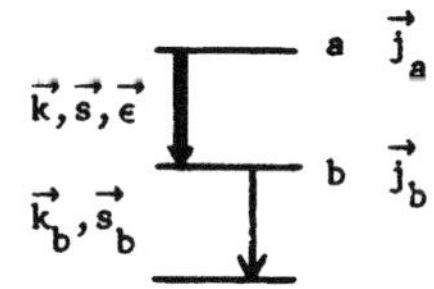	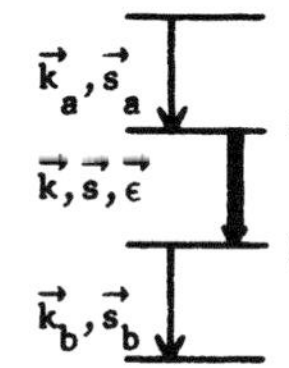	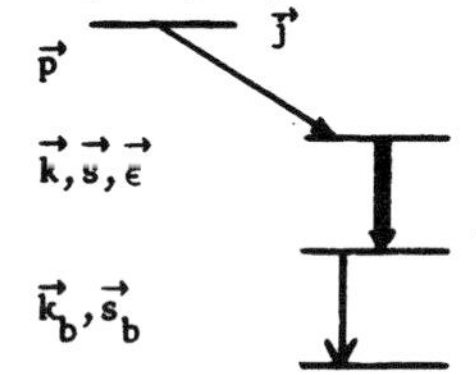
P	$\vec{k}\cdot\vec{s}$	0	0	-	-	-
	$\vec{k}\cdot\vec{j}_a$	1	0	-	$\vec{k}\cdot\vec{s}_a$	$\vec{k}\cdot\vec{p}$
	$\vec{k}\cdot\vec{j}_b$	0	1	$\vec{k}\cdot\vec{s}_b$	-	-
T	$(\vec{k}\cdot\vec{j}_a \times \vec{j}_b)(\vec{k}\cdot\vec{j}_a)$	2	1	$(\vec{k}\cdot\vec{j}_a \times \vec{s}_b)(\vec{k}\cdot\vec{j}_a)$	$(\vec{k}\cdot\vec{k}_a \times \vec{s}_b)(\vec{k}\cdot\vec{k}_a)$	-
	$(\vec{k}\cdot\vec{j}_b \times \vec{j}_a)(\vec{k}\cdot\vec{j}_b)$	1	2	$(\vec{k}\cdot\vec{k}_b \times \vec{j}_a)(\vec{k}\cdot\vec{k}_b)$	$(\vec{k}\cdot\vec{k}_b \times \vec{s}_a)(\vec{k}\cdot\vec{k}_b)$	$(\vec{k}\cdot\vec{k}_b \times \vec{p})(\vec{k}\cdot\vec{k}_b)$
	$(\vec{s}\cdot\vec{j}_a \times \vec{j}_b)$**	1	1	$(\vec{s}\cdot\vec{j}_a \times \vec{s}_b)$	$(\vec{s}\cdot\vec{s}_a \times \vec{s}_b)$	$(\vec{s}\cdot\vec{p} \times \vec{s}_b)$
	$(\vec{s}\cdot\vec{j}_a \times \vec{j}_b)(\vec{j}_a\cdot\vec{j}_b)$	2	2	$(\vec{s}\cdot\vec{j}_a \times \vec{k}_b)(\vec{j}_a\cdot\vec{k}_b)$	$(\vec{s}\cdot\vec{k}_a \times \vec{k}_b)(\vec{k}_a\cdot\vec{k}_b)$	-
	$(\vec{k}\cdot\vec{j}_a \times \vec{\epsilon})(\vec{j}_a\cdot\vec{\epsilon})(\vec{j}_a\cdot\vec{k})$	3	0	-	$(\vec{k}\cdot\vec{k}_a \times \vec{\epsilon})(\vec{k}_a\cdot\vec{\epsilon})(\vec{s}_a\cdot\vec{k})$	$(\vec{k}\cdot\vec{p} \times \vec{\epsilon})(\vec{p}\cdot\vec{\epsilon})(\vec{p}\cdot\vec{k})$†
	$(\vec{k}\cdot\vec{j}_a \times \vec{\epsilon})(\vec{j}_a\cdot\vec{\epsilon})(\vec{j}_b\cdot\vec{k})$	2	1	$(\vec{k}\cdot\vec{j}_a \times \vec{\epsilon})(\vec{j}_a\cdot\vec{\epsilon})(\vec{s}_b\cdot\vec{k})$	$(\vec{k}\cdot\vec{k}_a \times \vec{\epsilon})(\vec{k}_a\cdot\vec{\epsilon})(\vec{s}_b\cdot\vec{k})$	-
	$(\vec{k}\cdot\vec{j}_a \times \vec{\epsilon})(\vec{j}_a\cdot\vec{\epsilon})(\vec{j}_b\cdot\vec{k})$	1	2	$(\vec{k}\cdot\vec{k}_b \times \vec{\epsilon})(\vec{j}_a\cdot\vec{\epsilon})(\vec{k}_b\cdot\vec{k})$	$(\vec{k}\cdot\vec{k}_b \times \vec{\epsilon})(\vec{s}_a\cdot\vec{\epsilon})(\vec{k}_b\cdot\vec{k})$	$(\vec{k}\cdot\vec{k}_b \times \vec{\epsilon})(\vec{p}\cdot\vec{\epsilon})(\vec{k}_b\cdot\vec{k})$
	$(\vec{k}\cdot\vec{j}_b \times \vec{\epsilon})(\vec{j}_b\cdot\vec{\epsilon})(\vec{j}_b\cdot\vec{k})$	0	3	$(\vec{k}\cdot\vec{k}_b \times \vec{\epsilon})(\vec{k}_b\cdot\vec{\epsilon})(\vec{s}_b\cdot\vec{k})$	-	-

* Experimental situation: state "a" oriented by preceding nuclear reaction, polarized neutron capture, Coulomb excitation, Mössbauer effect, or by cryogenic means.

** This can also be written $(\vec{k}\cdot\vec{j}_a \times \vec{j}_b)(\vec{k}\cdot\vec{s})$.

† First-forbidden beta decay with $\Omega = 3$ required.

$\underline{s}_i$, and the plane of linear polarization $\underline{\epsilon}_i$ ($\underline{\epsilon}_i$ must occur an even number of times). Both polar and axial vectors mentioned above change sign under T; polar vectors, by definition, change sign under P. We conclude that to measure a **P - odd term we need an odd number of polar vectors,** and to measure **P - even T - odd we need an odd total number of (polar plus axial) vectors containing an even number of polar vectors.** In order to have a T - odd term we have to introduce at least one triple product $\underline{a} \cdot \underline{b} \times \underline{c}$.

A survey of the relevant vector combinations describing different symmetry tests is given in table 1. The transition from state a to b with momentum $\underline{k}$, photon spin $\underline{s}$, and linear polarization $\underline{\epsilon}$ serves as a probe of the symmetry tests. The vector combinations are shown to lowest order only. For those representing T violation we have profited from the work of Jacobsohn and Henley[5]). As usual, the quantities Ω_a, Ω_b are the ranks of the orientation tensors. The table serves as a guide of what quantity can be measured under what geometrical condition. Quantitative calculations of the sizes of the expected asymmetries have been carried out only in a few cases. If the reader is also interested in $P_{odd}T_{odd}$ terms we give reference to an earlier survey[6]).

3. Parity tests

As the average distance of nucleons is comparatively large, the parity - non - conserving potential, $V_{P_{odd}}$ can be well approximated by contributions from the exchange of light mesons, namely the low - lying pseudoscalar mesons ($\pi, \eta, ..$) and the vector mesons (ρ, ω, Φ). Selection rules permit only $\pi^{\pm}$, ρ, ω, Φ to be exchanged if CP is conserved. We can write, therefore

$$V_{P_{odd}} = V_\rho + V_\omega + V_\phi + V_{\pi^\pm} \tag{9}$$

The strangeness conserving non - leptonic Hamiltonian in the current - current theory* is

$$H = -(G/\sqrt{2})[\, \underset{\Delta T=0,2}{J_{\Delta S=0}\, J^{+}_{\Delta S=0} \cos^2\theta} + \underset{\Delta T=1}{J_{\Delta S=1}\, J^{+}_{\Delta S=1} \sin^2\theta}\,] \tag{10}$$

where G is the weak coupling constant, and θ is the Cabibo angle.

Historically the first calculations of V_ρ are due to Blin - Stoyle[9]) and Michel[10]). More recent attempts to calculate $V_{\rho,\omega,\phi}$ are based on the application of current field identities and field algebra[11]). V_π has been calculated first by McKellar[12]) using current algebra methods. Although the π exchange has a larger range, and consequently should be more effective in nuclear matter, it is suppressed with respect to the ρ exchange by a factor $\mathrm{tg}^2\,\theta \cong 1/15$, since, having $\Delta T = 1$, it originates only from

* Recent reviews on this subject are by Henley[7]) and by Blin - Stoyle[8]).

the last term in (9). McKellar[13]) concludes that V_π contributes about 25% of V_ρ in nuclear matter.

As V_π and $V_{\rho,\omega,\varphi}$ are evaluated employing different theoretical assumptions, it would be desirable to be able to study each term independently. In practice this is possible because V_π and $V_{\rho,\omega,\phi}$ obey different isospin selection rules ($\Delta T = 1$, $\Delta T = 0, 2$, respectively)[14]). Suggestions of specific experiments on the very light nuclei have been made. In the p + n capture process, for example, the circular polarization depends only on the $\Delta T = 0,2$ component, while the asymmetry in the angular distribution of the gamma ray with respect to the polarization axis established by the neutron capture process depends on the $\Delta T = 1$ component of V[15]). Experiments on the p + n reaction performed by Lobashov et al.[16]) are just starting to give us the first results on the fundamental question of the relative size of V_ρ and V_π.

As can be seen from table 1, two correlation experiments are particularly suited to detect a P - odd term. One is the distribution, A_γ, of gamma rays from a polarized nucleus, and another one is the $\beta\gamma$ - correlation, making use of the parity - non - conserving beta decay to polarize the nuclear state, a. In addition, there is the possibility of measuring the correlation $\underline{s} \cdot \underline{k}$ corresponding to a net circular polarization of a gamma ray from an unpolarized source.

A summary of the more recent experimental results on PNC in electromagnetic transitions is given in table 2. Column 3 shows the regular and irregular electromagnetic multipoles giving rise to the interference term. Columns 4 and 5 are the anisotropy coefficients (coefficients of cos θ) and circular polarizations of the gamma rays, respectively. The theoretical predictions are shown in column 6; they are quoted in the same units as shown on the heading of columns 4 and 5. The symbol V_ρ refers to the ρ meson exchange model by Michel[10]).

A word of explanation is in order for those experiments where A_γ is listed. The ^{3}H and ^{4}He experiments[17]) are angular correlation studies of the proton with respect to the incoming polarizing deuteron. The ^{19}F results[18]) were obtained from a study of the emission of the 110 keV ground state gamma ray with respect to the ^{19}F nucleus polarized by Coulomb excitation. The A_γ coefficients in ^{41}K, ^{133}Cs, ^{177}Lu and ^{203}Tl [20]) refer to the β - γ correlation in each case. In ^{114}Cd, A_γ describes the angular distribution[21,22]) of the $1^- - 0^+$ gamma ray from a ^{114}Cd nucleus polarized by capture of a polarized neutorn.

There is evidence for a parity effect from the experiments on ^{41}K, ^{114}Cd, ^{175}Lu, ^{180}Hf, and ^{181}Ta. Below let us look more closely into the ^{175}Lu, ^{181}Ta and ^{114}Cd experiments.

Recent Experiments designed to detect circular Polarization from unpolarized Sources.

The circular polarization of photons can be investigated, using Compton scattering off magnetized iron and a pulse - counting technique[27,38,39]). The most extensive pulse - counting work which approaches the limits of the digital counting scheme is by Boehm and Kankeleit[27]). Their results give upper limits for the parity admixture for several nuclear gamma transitions. These limits are consistent with the findings of all other quoted authors.

In all mentioned experiments the primary limitation has been that of obtaining adequate statistics. Although in favourable circumstances nuclear structure effects

Table 2. Summary of Results of Parity Experiments

Nucleus	E_γ keV	$L\ \tilde{L}$	Experiment: $A_\gamma \times 10^4$	Experiment: $P_\gamma \times 10^6$	Theory: A_γ or P_γ respectively	
H^2	~2000			- 1.5 ± 1.5[16]	± 0.1 to 1[15]	
He^4	$H^2(dp(\theta))$		0 ± 5† [17]			
Li^5	$He^3(dp(\theta))$		1 ± 5† [17]			
F^{19}	110	E1 $\tilde{M}$1	+4.3 ± 5.2[18]		± 4.3[23]	V_ρ
K^{41}	1291	M2 $\tilde{E}$2	24 ± 36[19]			
K^{41}	1291	M2 $\tilde{E}$2		19 ± 3[24]	6 ± 20[24]	
Fe^{57}	14	M1 $\tilde{E}$1		20 ± 60[25]	26[26]	V_ρ
Cd^{114}	9050	M1 $\tilde{E}$1	-3.6 ± 0.7[21] -0.6 ± 1.8[22]		± 2[13]	$V_\rho + V_\pi$
Cs^{133}	81	M1 $\tilde{E}$1	15 ± 15[19]			
Lu^{175}	343	M1 $\tilde{E}$1		20 ± 30[27]	-15 ± 10[13] 30 ± 20[26] -70[28]	$V_\rho + V_\pi$ V_ρ $V_{2\pi}$
Lu^{175}	396	E1 $\tilde{M}$1		40 ± 10[29] 63 ± 10[30]	±90 ± 60[10] ±45 ± 30[13]	V_ρ $V_\rho + V_\pi$
Lu^{177}	208	E1 $\tilde{M}$1	3 ± 13[19]			
Hf^{180m}	505	M2 $\tilde{E}$2		-2800 ±450[31] -2300 ±600 [a]		
Ta^{181}	481	M1 $\tilde{E}$1		-6 ± 1[32] -3.9 ± 1.2[30] -6 ± 2.5[33] -4 ± 8[34,b] -28 ± 6[39]	-70[28] -60 ± 30[26] -20 ± 10[13] -0.8[35]	$V_{2\pi}$ V_ρ $V_\rho + V_\pi$ V_π
Tl^{203}	273	M1 $\tilde{E}$1	0.6 ± 0.8[20]	20 ± 30[27] 2 ± 10[36] 0 ± 10[b]	-90 ± 30[37] -30 ± 10[13]	V_ρ $V_\rho + V_\pi$

† Coefficient of P_3 (cos θ). Recent References: a) E. Lipson, F. Boehm and J.C. Vanderleeden, to be published, b) E. Kankeleit and E. Kuphal, private communicat. (1970).

can enhance the quantities to be measured by factors from 10 to 100, the P_1 (cos θ) term in angular distribution, or the circular polarization of gamma rays, will typically be on the order of 10^{-5}. There is the additional task of differentiating the sought-after effect from various instrumental asymmetries with the help of careful control measurements.

It is only fairly recently that a new technique has made it possible to overcome the problem of inadequate statistics in measurements of the circular polarization of gamma rays from unpolarized sources. This method consists of replacing the pulse-counting process by a current-measuring process, and by increasing the strength of the radioactive sample by a factor of 100 to 1000, abandoning the requirement for energy resolution. Lobashov et al.[32,29,24]) using this technique have obtained evidence of a parity admixture in ^{181}Ta, ^{175}Lu, and ^{41}K.

Instead of reporting on Lobashov's experiment, which has been described extensively in literature and reviews, I am briefly describing some recent work with a similar technique from Caltech[30,40]). The transitions studied were the 482 keV transition in ^{181}Ta and the 396 keV transition in ^{175}Lu.

The circular polarization is related to the matrix elements M_L of the regular

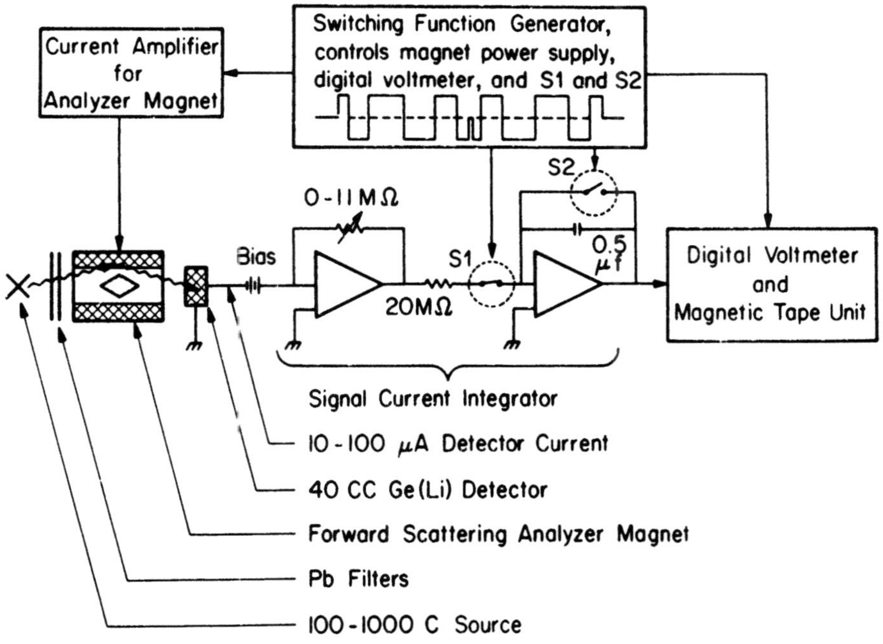

Fig. 1. Experimental set-up of parity experiments. Photons from 100 - 1000 Ci unpolarized sources are forward-scattered in an analyzer magnet and detected with the Ge (Li) detector. The detector current is integrated in a charge sensitive circuit and a voltage signal, proportional to the total charge, is read-out with the DVM. The magnet polarization is reversed in a sequence of unequal time intervals repeated continuously by the switching function generator.

transition of multipolarity L and the parity admixed transition $\widetilde{M}_L$ by[9])

$$P_\gamma = 2\,\frac{\widetilde{M}_L}{M_L}\,\frac{1}{1+\delta^2}$$

where δ is the mixing ratio of the multipole of order L and L + 1 of the regular transition. The circular polarization of the 482 keV transition in ^{181}Ta is due to the interference of the regular, strongly hindered M1 and the irregular - parity $\widetilde{E}1$ transitions. The 396 keV gamma ray in ^{175}Lu is a hindered E1 transition and the interference is caused by an irregular $\widetilde{M}1$. In both transitions the retardation of the regular component is expected to significantly enhance the circular polarization. A general description of the new method for measuring the circular polarization is given in form of the block diagram of fig. 1. Gamma rays from an unpolarized radioactive source of several hundred Curies strength are forward scattered on the inside surface of a hollow cylindrical iron magnet and are detected with a Ge(Li) counter. Upon reversal of the magnetization the spin dependent part of the Compton scattering cross section changes sign, resulting in a change in intensity of the scattered circularly polarized photons. Whereas in most previous circular polarization experiments the counting rate in the detector was low ($< 10^7$ counts / sec) and the individual detector pulses were resolved and counted in scalers, one uses in the experiment I am describing very high rates ($10^9 - 10^{10}$ counts / sec) and replaces the pulse - counting technique by a current - measuring technique. The detector current (10 - 100 μA) is integrated in a charge sensitive circuit, and at the end of preset time intervals a digitized voltage signal, proportional to the accumulated charge, is recorded on magnetic tape. A timer with associated logic controls the measuring process.

The magnet is energized with a current - stabilized power supply, and the polarization is reversed in a sequence of unequal time intervals. The sequence lasts 120.8 seconds, and is continuously repeated under control of the timer.

The switching function which was used in this experiment is shown in fig. 2. As illustrated in this figure the function, depicted in the last line of the figure, representing one half of the complete pattern (see also fig. 1) is designed to cancel drifts up to and including fourth order in time by a proper selection of the intervals t_1, t_2, t_3 [40,41]). Each interval is followed by 1.4 sec of dead - time during which data are read out, the capacitor in the charging circuit is discharged and the magnetization is (or is not) reversed.

A host of experimental problems must be carefully considered[40]), including the effects of magnetostriction, higher order terms in the polarization asymmetry, azimuthal asymmetries of the geometry,[42]), stray magnetic fields, electronic asymmetries, source impurities, bremsstrahlung, and source decay.

The results obtained in this experiment are collected in table 2. It can be seen that they agree well with those by Lobashov[32]) and by Bock and Jenschke[33]). Control experiments with a ^{103}Ru source that emits a fast, unhindered E2 gamma ray of comparable energy were carried out in the experiment described.

Comparing the present results on ^{181}Ta with theory it appears that the observed value is somewhat smaller than the calculations based on V_ρ, $V_\rho + V_\pi$, and $V_{2\pi}$, however it has the right sign. It is larger than the calculated value with V_π [35]).

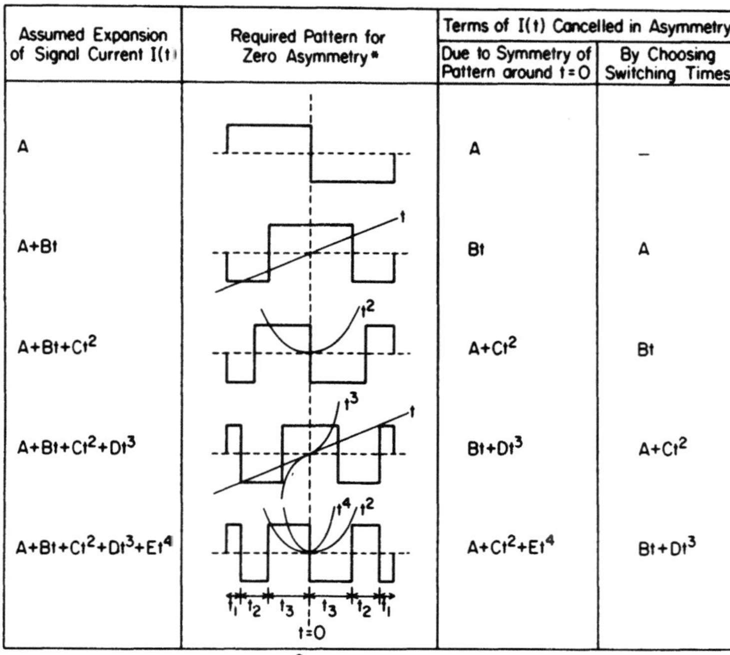

* Also shown are curves labelled t, t^2,..., representing terms in first column that are cancelled due to symmetry of pattern around t=0.

Fig. 2. Effects of slow drifts on the asymmetry are eliminated by the correct choice of the switching function. The five switching functions shown will cancel successively higher terms of the assumed polynomial expansion in t of the signal current.

Experiments on gamma Distribution from polarized Nuclei.

Experiments designed to study the distribution of gamma rays from polarized ^{114}Cd nuclei have been conducted since 1959. More recently there have been two experimental groups engaged in this work, one is the Moscow group of Abov et al.[21]) and the other is the group of Warming et al.[22]) (There is also a preliminary report by Forte et al.[43])). The quoted paper by Abov et al.[21]) summarized all the results of the Moscow group and contains also previous references. The Copenhagen results and previous references are summarized in ref. 22.

The principle of both experiments is the same. Slow neutrons from a reactor polarized by reflection on a cobalt mirror are captured by ^{113}Cd forming polarized ^{114}Cd + gamma. The distribution of the capture gamma ray from the 1^+ state at 9.05 MeV with respect to the polarization axis is measured. The neutron polarization axis can be turned into or out of the momentum direction by means of a guide field. The guide field and thus the polarization direction can be reversed in rapid succession to cancel instrumental instabilities. In control experiments the neutron beam was depolarized with a magnetic shim. While NaI counters were used in ref. 21 to detect the 9.05 MeV gamma ray, a Ge counter was employed in ref. 22. The advantage of Ge over NaI at these energies is however not very significant, and has to be weighed against the loss in detection efficiency.

Both experiments measure an anisotropy A_γ defined by

$$W(\theta) \approx 1 + A_\gamma \, \Omega \cos\theta$$

where Ω is a factor that includes the degree of polarization of the neutron as well as corrections from source and counter geometry.

The average value of the two Moscow experiments[21]) carried out in an interval of several years and with quite different techniques is quoted in table 2. The Copenhagen result[22]) also shown in table 2 has fairly large standard deviations and does not exclude zero asymmetry. It is, however, only 1.3 standard deviations apart from the error limit of the Moscow result. Thus we can conclude that the two experiments are in fair agreement; they also agree with the estimated strength of the weak parity - non - conserving potential (see table 2).

Appendix

Symmetry tests in nuclear beta Decay

It is well known that parity is not conserved in nuclear beta decay and the pertinent experimental evidence is summarized in review articles[44]). The question of time - reversal, invariance, however, is still an open one. The best limit on η at present is about 1.5 degrees[45]).

The distribution in electron and neutrino directions and electron energy for allowed transitions, displaying only those terms which change sign under T, is given (ref. 46) by

$$\omega \, dE_e \, d\Omega_e \, d\Omega_\nu = \frac{F(\pm Z, E_e)}{(2\pi)^5} \, p_e \, E_e \, (E^0 - E_e)^2$$

$$\times \; \xi \left\{ 1 + .. + D \frac{\underline{J}}{J} . \frac{\underline{p}_e \times \underline{p}_\nu}{E_e E_\nu} + L \underline{s} . \frac{\underline{p}_e \times \underline{p}_\nu}{E_e E_\nu} + 2R \underline{s} . \frac{\underline{J}}{J} \times \frac{\underline{p}_e}{E_e} \right.$$

$$+ M \left[\frac{\underline{s} . \underline{p}_e \times \underline{p}_\nu}{3E_e E_\nu} - \frac{(\underline{j} . \underline{s}) \, (\underline{j} . \underline{p}_e \times \underline{p}_\nu)}{E_e E_\nu} \right] \left[\frac{J(J+1) - 3 \langle (\underline{J} . \underline{j})^2 \rangle}{J(2J-1)} \right]$$

$$\left. + V \underline{s} . \frac{\underline{J}}{J} \times \frac{\underline{p}_\nu}{E_\nu} + W \frac{\underline{s} \cdot \underline{p}_e}{E_e + m} \, \frac{\underline{J}}{J} . \frac{\underline{p}_e \times \underline{p}_\nu}{E_e E_\nu} \right\} dE_e \, d\Omega_e \, d\Omega_\nu \qquad (11)$$

Here $\underline{p}_e$ and $\underline{p}_\nu$ are the electron and neutrino momenta, $\underline{J}$ is the polarization of the initial nucleus, $\underline{j}$ is a unit vector in the direction of $\underline{J}$, $\underline{s}$ is the electron spin, and

$$\xi = | M_F |^2 \, |g_V|^2 + | M_{GT} |^2 \, |g_A|^2$$

Table 3 : Asymmetry coefficients D,L,R,M,V, and W appearing in Eq. (11) are shown in lines 1 to 6, respectively.

Correlation \ Asymmetry Coefficient	Contribution of the T-violating interactions	Contribution of the final state electromagnetic interactions
$\vec{J}\cdot\vec{p}_e \times \vec{p}_\nu$	$-4\delta_{JJ'}\left(\frac{J}{J+1}\right)^{1/2} \mathrm{Im}(g_V g_A^* M_F M_{GT}^*)$	0
$\vec{s}\cdot\vec{p}_e \times \vec{p}_\nu$	0	$\pm\eta\left(\lvert g_V M_F\rvert^2 - \frac{1}{3}\lvert g_A M_{GT}\rvert^2\right)$
$\vec{s}\cdot\vec{J} \times \vec{p}_e$	0	$2\eta\left[\left(\lambda_{JJ'}\lvert g_A M_{GT}\rvert^2\right) \pm 2\delta_{J'J} \mathrm{Re}(g_V g_A^* M_F M_{GT}^*)\right]$
$\vec{s}\cdot\vec{p}_e\times\vec{p}_\nu - 3(\vec{j}\cdot\vec{s})(\vec{j}\cdot\vec{p}_e\times\vec{p}_\nu)$	0	$\pm 2\eta\, \Lambda_{J'J}\lvert g_A M_{GT}\rvert^2$
$\vec{s}\cdot\vec{J} \times \vec{p}_\nu$	$\pm 4\,\frac{m}{E_e}\,\delta_{JJ'}\left(\frac{J}{J+1}\right)^{1/2} \mathrm{Im}(g_V g_A^* M_F M_{GT}^*)$	0
$(\vec{s}\cdot\vec{p}_e)(\vec{J}\cdot\vec{p}_e \times \vec{p}_\nu)$	$\pm 4\delta_{JJ'}\left(\frac{J}{J+1}\right)^{1/2} \mathrm{Im}(g_V g_A^* M_F M_{GT}^*)$	0

The coefficients D, L, R, M, V, and W are given in table 3. In column 2 the contributions from the time reversal non - invariant interactions are listed, and in column 3 the contributions from final - state electromagnetic interactions, neglecting recoil effects[48]), are given[46,47]). The quantities $\lambda_{J'J}$ and $\Lambda_{J'J}$ are defined in ref. 46, $\eta = \frac{Z\alpha m}{p_e}$, as usual.

Clearly, the most appropriate experiments seem to be those listed on lines 1, 5, and 6 of table 3. They all involve a measurement of the neutrino momentum, i.e. in an experiment the nuclear recoil must be measured.

As the precision of experiments testing time reversal invariance in beta decay increases, recoil corrections to the electromagnetic final state interactions will have also to be considered[48]).

† This work was performed under the auspices of the U.S. Atomic Energy Commission. Prepared under Contract AT (04 - 3) - 63 for the San Francisco Operations Office, U.S. Atomic Energy Commission.

Acknowledgment

The generous help of Dr. P. Herczeg and Dr. J.C. Vanderleeden is gratefully acknowledged.

References

1. R.P. Feynman and M. Gell - Mann, Phys. Rev. 109 (1958) 193.
2. See for example P. Roman, Advanced Quantum Mechanics, Addison - Wesley (1965).
3. J. Bernstein and G. Feinberg and T.D. Lee, Phys. Rev. 139 (1965) B1650.
4. A.H. Huffman, Phys. Rev., to be published.
5. B.A. Jacobsohn and E.M. Henley, Phys. Rev. 113 (1959) 234.
6. F. Boehm, Hyperfine Structure and Nuclear Radiation, p. 279, North - Holland (1968).
7. E. Henley, Ann. Rev. Nucl. Sci. 19 (1969) 367.
8. R.J. Blin - Stoyle, Topical Conference on Weak Interactions, CERN (1969), p. 495.
9. R.J. Blin - Stoyle, Phys. Rev. 118 (1960) 1605; 120 (1960) 181.
10. F.C. Michel, Phys. Rev. 133 (1964) B329.
11. P. Olesen and J.S. Rao, Phys. Lett. 29B (1969) 233;
M. Feuer, Thesis, Harvard University (1969);
E. Fishbach, D. Tadic and K. Trabert, to be published (1969);
B.H.J. McKellar, preprint.
12. B.H.J. McKellar, Phys. Lett. 26B (1967) 107.
13. B.H.J. McKellar, Phys. Rev. Lett. 20 (1968) 1542.
14. R.F. Dashen, G. Frautschi, M. Gell - Mann, and Y. Hara, The Eightfold Way, Benjamin New York (1964).
15. G.S. Danilov, Phys. Lett. 18 (1965) 40.
16. V.M. Lobashov et al., JETP Lett. 11 (1970) 76.
17. B.P. Ad'Yasevich and V.G. Antonenko, Sov. J. Nucl. Phys. 10 (1970) 278.
18. A. Moline, J. Morris, P. Dyer and C.A. Barnes, to be published (1970).

19. F. Boehm and U. Hauser, Nucl. Phys. 14 (1960) 615.
20. K.D. Baker and W.D. Hamilton, Phys. Lett. 31B (1970) 557.
21. Yu.C. Abov et al., Sov. J. Nucl. Phys. 10 (1970) 320.
22. E. Warming, Phys. Lett. 29B (1969) 564.
23. E. Maqueda, Phys. Lett. 23 (1966) 571.
24. V.M. Lobashov et al., Phys. Lett. 30B (1969) 39.
25. E. Kankeleit, Conf. Intern. Phys. Nucl., Paris 1964, 2, 1206 (CNRS, Paris, 1964).
26. S. Wahlborn, Phys. Rev. 138 (1965) B530.
27. F. Boehm and E. Kankeleit, Nucl. Phys. A109 (1968) 587.
28. E. Maqueda and R.J. Blin - Stoyle, Nucl. Phys. A91 (1967) 460.
29. V.M. Lobashov et al., Proc. of Int. Conf. on Nuclear Structure, Tokyo Phys. Soc. of Japan, (1968) 443.
30. J.C. Vanderleeden and F. Boehm, Phys. Lett. 30B (1969) 467.
31. B. Jenschke and P. Bock, Phys. Lett. 31B (1970) 65.
32. V.M. Lobashov et al., JETP Lett. 5 (1967) 59.
33. P. Bock and B. Jenschke, Nucl. Phys., to be published (1970).
34. H. Diehl, G. Hopfersitz, E. Kankeleit and E. Kuphal, Proc. of the Third Int. Conf. on High Energy Physics and Nuclear Structure, New York (1969), Plenum, New York.
35. N. Vinh Mau and A.M. Bruneau, Phys. Lett. 29B (1969) 408.
36. J.C. Vanderleeden and F. Boehm, unpublished.
37. Z. Szymanski, Nucl. Phys. 76 (1966) 539.
38. P. Bock and H. Schopper, Phys. Lett. 16 (1965) 284;
J.J. van Rooijen, P. Pronk, S.U. Ottevangers and J. Blok, Physica 37 (1967) 32;
D.W. Cruse and W.D. Hamilton, Nucl. Phys. 125 (1969) 241.
39. E. Bodenstedt, L. Ley, H.O. Schlenz, U. Wehman, Phys. Lett. 29B (1969) 165.
40. J.C. Vanderleeden and F. Boehm, Phys. Rev., to be published (1970).
41. J.D. Bowman and J.C. Vanderleeden, Nucl. Instr. and Meth., to be published (1970).
42. P. Bock, Phys. Lett. 30B (1969) 628.
43. M. Forte and O. Saavedra, EUR 3053 (1966).
44. See for example H. Schopper, Weak Interaction and Nuclear Beta Decay, North - Holland (1966).
45. B.G. Erzolinsky et al., Phys. Lett. 27B (1968) 557;
F.P. Calaprice, Phys. Lett. 18 (1967) 918.
46. J.D. Jackson, S.B. Treiman, and H.W. Wyld, Phys. Rev. 106 (1957) 517;
M.E. Ebel and G. Feldman, Nucl. Phys. 4 (1957) 213.
47. J.D. Jackson, S.B. Treiman, and H.W. Wyld, Nucl. Phys. 4 (1957) 206.
48. C.G. Callan and S.B. Treiman, Phys. Rev. 162 (1967) 1494.

Discussion

Bloom: I have two questions. The first one has to do with the time reversal method, not the time reversal principle, that was utilized in this technique recently by Lobashov. I wondered what the period, total period of the time reversal cycle was for the magnet.

Boehm: The total period was 120 seconds. They were subdivided into the intervals I have shown you in which each one corresponds to a subperiod between 1 and 5 seconds.

Bloom: My second question has to do with the utilization of another idea, that is old now at least comparing to the time reversal idea, namely to measure in periods of 1/40 of a second. We have checked electronic drifts and we have found them to be extremely small. Of course we were not explicitly looking for these effects but I think we found something like one part into 10^4 or less, conceivably. I was wondering what your opinion would be about utilizing fast reversal, real fast reversal to obviate electronic drifts as we have done. Would that be perhaps a useful principle also in this method?

Boehm: We have indeed looked into that. We have for the last two years performed experiments with a magnet. It is a laminated magnet switching 10 times per second. It is a little slower than what you say. And we have used a Princeton Applied Research phase interlock amplifier to detect the signal which is directly in phase with the switching of the magnet. And this sounds like a marvelous and very ingenious way of doing the experiment but unfortunately we have found snags which we do not understand. We still find some other asymmetries. Possibly this laminated magnet which consists of 8 pieces arranged in an octagon fashion is not symmetric enough and one has small azimuthal asymmetries such as described by the Karlsruhe group. There might be other causes. We do not understand the problem yet. I agree with you it is a very intriguing and very appealing approach.

McKellar: First of all a comment. If the potential is of one or other of the isospin types, in other words if it is not significantly mixed, then it is not necessary to measure light nuclei and find out which type it is. It would suffice to make a measurement in an odd - neutron nucleus for in this case the two types of force would give predictions of opposite sign. However, I understand that it would be as hard to measure an odd - neutron nucleus as it would be to measure a light nucleus. A couple of other comments: concerning the current state of the theoretical art here. Unfortunately, I cannot report any more recent calculations but I thought you might like to know what is interesting to theoreticians in this field. First of all there is a question about the rho exchange potential. This is not precisely tied down in the same sense as the pion exchange potential is tied down and there is some controversy about the magnitude of the vertex here. There has been a little bit of progress in that one can discover models in which the vertex or rather the diversions of the vertex is given by the Michel expression. Unfortunately, this is not the same thing as saying that the force is given by the Michel expression. A rather related question which is currently being investigated is the question of how much of the two pion exchange force, e.g. the two pion exchange force that Blin - Stoyle calculated originally, can be accounted for by rho exchange. And the various handwaving arguments which suggest that in Cabibbo's theory at least the corrections to rho exchange are small. We are about half - way through a calculation to try and confirm whether this is true or not. And finally of course one would like to see the nuclear physics calculations done a whole lot better and I know there are groups at Sydney, Stonybrook and also in Leningrad which are working on this problem.

Kaufmann: Could you elaborate on experiments involving coincidence techniques and whether they are still valuable for this kind of investigation. Because of the high currents involved in the singles rates, it occurred to me that one might use the ideas of R.I. Lewis to do coincidence experiments using the correlated currents. Has that been looked into?

Boehm: The circular polarization experiments are single counting experiments and the experiments from polarized nuclei are also single counting experiments. The coincidence experiments are used for time reversal experiments and those will be discussed in the next lecture.

Kaufmann: As I recall Lewis also proposed the possibility of magnetic and electric fields applied to a nucleus like ^{181}Ta and doing a perturbed angular correlation, where some parity violating term would be apparent but that would require coincidences.

Boehm: I did not mention those experiments, because they are rather esoteric and probably quite a bit more difficult than what has been done so far considering the size of the effect. But I should probably have mentioned these experiments for completeness.

Contributed Paper

THE WEAK PARITY - VIOLATING NUCLEON - NUCLEON POTENTIAL*

B.H.J. McKellar
University of Sydney, Sydney, Australia

The weak parity - violating nucleon - nucleon potential may be regarded as a meson exchange potential arising from the reactions

$$N\bar{N} \xrightarrow[\text{p.v.}]{\text{weak}} \text{mesons} \quad (\pi,\rho,2\pi,\text{etc.}) \quad \xrightarrow{\text{strong}} N\bar{N}$$

The main problem in calculating the single meson exchange terms is in the calculation of the weak meson - nucleon vertex. For 2π exchange the strong amplitude is no longer trivial, introducing additional complications. I now wish to discuss the π, ρ, and 2π potentials in turn.

The π exchange potential. In this case the p.v. πNN vertex can be related, using SU3 and current algebra, to the p.v. hyperon decays and can be determined to about 15%

(ref. 1).

The ρ exchange potential presents a more difficult problem. The Feynman Diagrams for the p.v. NNρ vertex can be classified as in fig. 1. Since only the first diagram is finite it has been customary to follow Michel[2]) and ignore the others. Using the re-

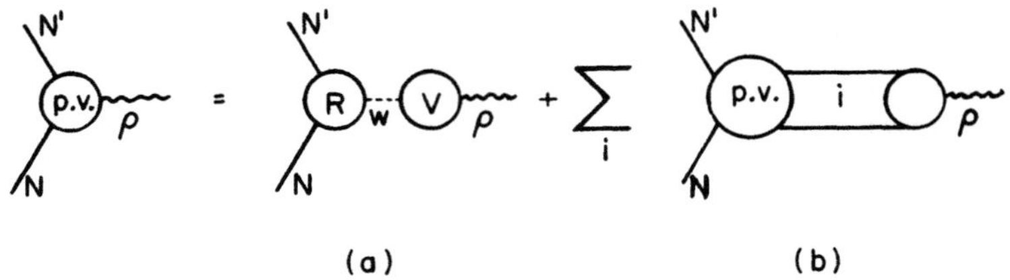

Fig. 1. Feynman diagrams for the p.v. ρNN vertex. Diagram (a) was evaluated by Michel.

duction technique and the field - current identities we can write the vertex function M_λ in a form which formally exhibits this separation. In the Cabibbo theory one can then demonstrate that $k^\lambda M_\lambda$ is dominated by the Michel term in the limit $k \to 0$ [3]). Unfortunately this does not tell us that the ρ exchange force is dominated by Michel's expression. To see this write M_λ out in terms of spinor invariants,

$$M_\lambda = \bar{u}(\gamma_\lambda \, f(k^2) + k_\lambda g(k^2) + \sigma_{\lambda\mu} k h(k^2)) \, \gamma_5 u.$$

Then our result tells us only about f (0). The conservative nature of the strong ρNN vertex prohibits a contribution from g to the NN potential, but h can contribute and we know nothing about it.

The 2π exchange potential. If we forget that the theory of the ρ force is in an unsatisfactory state and pretend that we know all about it, it is reasonable to ask the question 'How much of the 2π exchange force is already counted in ρ exchange? '

As a step towards the answer McKellar and Pick[4]) have calculated the weak πN scattering amplitude. Our result is that in the soft pion limit the amplitude is given by the single pole term, and that the major hard pion correction comes from ρ exchange in the t channel. These results are summarized in fig. 2.

We are at present combining these results with the strong πN amplitude to obtain the 2π exchange force. We have no results yet, but I will tell you what I expect.

Clearly the ρ term in fig. 2 corresponds to ρ exchange. The pole term has to be joined on to the strong πN amplitude, but in the soft pion limit the Adler consistency condition says that the symmetric part of the strong amplitude is just given by the single nucleon intermediate state. In this limit, therefore, the 2π force is the ρ force together with the diagram of fig. 3. But fig. 3 is included when we take the matrix elements of the π force between correlated nuclear wavefunctions. Of course, in real nuclei we are not at the soft pion limit, but we are not too far away so I conjecture that most of the 2π force is already taken into account when we calculate with the ρ and π forces.

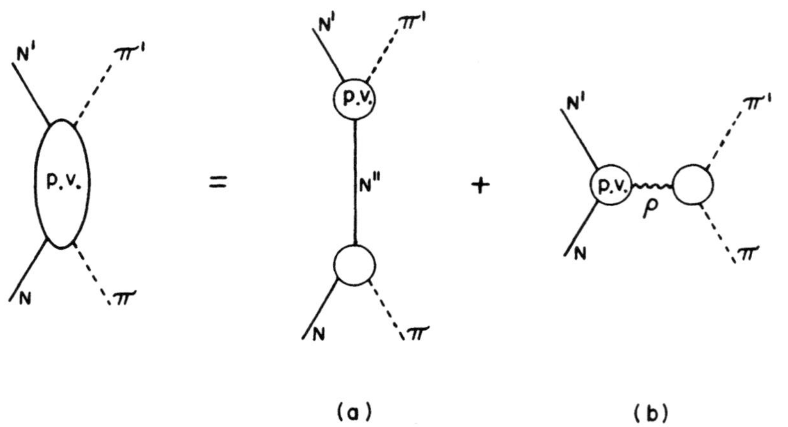

Fig. 2. The p.v. πN scattering amplitude. Diagram (a) is the soft pion limit, and diagram (b) is the major hard pion correction.

It is instructive to contrast this situation with that of the strong NN force, when the 2π force gives most of the nuclear binding energy. The difference is that the important part of the strong 2π force behaves like a σ meson, and the p.v. σNN vertex must vanish.

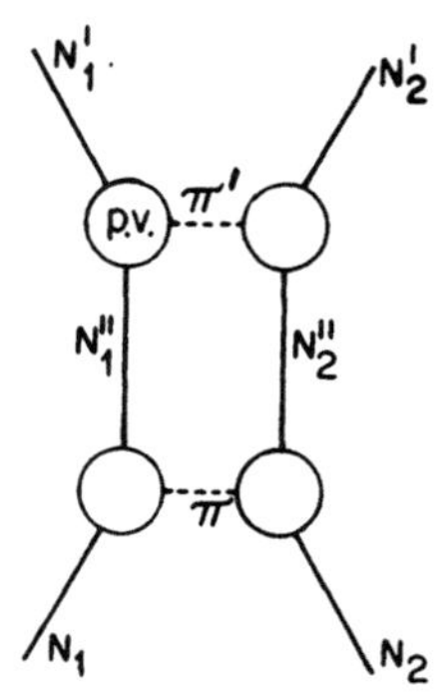

Fig. 3. The p.v. 2π exchange diagram in the soft pion limit.

* Work supported in part by the Science Foundation for Physics within the University of Sydney.

References

1. B.H.J. McKellar, Physics Letters 26B (1967) 107.
2. F.C. Michel, Phys. Rev. 133 (1964) B329.
3. B.H.J. McKellar, to be published. See also
   P. Olesen and J.S. Rao, Physics Letters 29B (1969) 233.
   E. Fishbach, D. Tadic and K. Trabert, Phys. Rev. 186(1969) 1688.
4. B.H.J. McKellar and P. Pick, Nuclear Physics B, in press.

ANGULAR CORRELATION TESTS OF TIME REVERSAL INVARIANCE

D. Hamilton†

University of Sussex, School of Mathematical and Physical Sciences, Brighton BN1 9QH, England

1. Introduction

The theoretical basis for angular correlation tests of time reversal invariance was first presented by Henley and Jacobsohn[1]). Subsequent work by Lobov[2]) and Szymanski[3]) has analyzed in some detail the main features of the method. There have been several review articles[4,5]) that have covered aspects of these tests and have extended the original proposals, and in addition the general experimental situation has been recently presented[6,7]).

Tests of time reversal invariance are approaching the accuracy which will be required to distinguish between the various predictions about the source of a possible T - violating interaction. So far there is no good theoretical reason for uniquely identifying this interaction, but on the basis of CP - violation, as demonstrated by the decay of the long - lived neutral K meson[8]), $K_2^0 \rightarrow \pi^+ + \pi^-$, the combined CPT - invariant operation implies that either C or T is violated. Present experiments cannot distinguish between the various proposals, which may be briefly summarized as follows:

a) T - Violation may occur in strong processes by a new interaction which has an amplitude $\sim 10^3$ times that of the weak interaction[9]). It can be regarded as a small C- or T-violating part of the strong interaction, and under the most favourable circumstances ($\Delta S = 0$ transitions), could introduce a T - violating amplitude of 10^{-2} to 10^{-3}.

b) The electromagnetic interaction between strongly interacting particles may contain a large violation of C- and T-invariance[10]). Higher order effects will then allow the strong interaction to show T - violation with an amplitude in the range 10^{-2} to 10^{-3}.

c) CP - violation may be large within the V - A theory of weak interactions[11]). T - violation by the lepton currents is restricted by the two component theory, but strongly interacting particles may interact via a weak T - violating current. Except for β - decay processes, in which effects as large as $\sim 10^{-3}$ in amplitude may occur, the T - violating contribution in nuclear states will be small.

As Professor Boehm[12]) has discussed, the precision of P - violation tests has increased by several orders of magnitude during the past few years. However, there has been only a small improvement in the accuracy of experiments sensitive to T - violation, and in the case of angular correlation tests, there has been little change. The developments in experimental techniques and data handling imply that substantial changes are possible, and there are indications that angular correlation tests may be one of the best methods for establishing the existence of a T_{odd} term in the Hamiltonian of the nucleon system*. Recent detailed balance[15]) and e - pol. p scattering[16]) experiments suggest that substantial improvements in accuracy will be difficult in these types of measurements. It should also be added that tests in the strong interaction such as p - p scattering[17]) are limited in precision to a few per cent.

Since T - violation is insensitive to selection rules, and since both axial vectors, e.g. angular momentum, and polar vectors, e.g. linear momentum, change sign under T - operation, it is necessary to construct experiments that measure the scalar triple product of three such quantities. This may be invariant under rotation, but will be non - invariant with respect to T, i.e. quantities similar to

$$\underline{J} \cdot \underline{p}\,(1) \times \underline{p}\,(2)$$

are required, and by their very nature these are more difficult to measure than a P - violating term such as $\underline{J} \cdot \underline{p}$. An asymmetry, associated with the latter, may be determined with a precision of one part in 10^6, while the present limit on T - reversal tests is at most one part in 10^3. This large difference, in some respects, is due to the different emphasis that has been placed on these tests, but it is also due to the greater complexity of the T - violation measurement.

2. Weak interaction tests for T - invariance

Since it is the weak decay of K_2^0 that first showed CP - violation, it is perhaps natural to examine firstly the angular correlation tests involving the weak interaction, Cabibbo[11]) suggested that CP - violation may be large within the V - A theory of the weak interaction. Since parity is violated by the leptons, and we also know that charge conjugation is violated, then the amplitude of a T - violating process involving lepton currents is restricted. It may be assumed, however, that the weak currents of strongly interacting particles are not invariant under T. A consequence of this will be that the Fermi and Gamow - Teller coupling constants will contain imaginary parts, and show a relative phase difference; it is commonly assumed that there is no phase difference between the matrix elements.

The different types of experiment that may be used to detect an asymmetry associated with this phase difference were discussed by Jackson, Treiman and Wylde (ref. 18). They should be independent of any Coulomb distortion which can produce asymmetries. Four quantities are available for measurements: (i) the polarization $\langle \underline{J} \rangle$ of the decaying nucleus; (ii) the electron polarization direction $\underline{\sigma}$; (iii) the momenta of the electron $\underline{p}_e$; (iv) the momenta of the neutrino $\underline{p}_\nu$ (in practice the nuclear recoil momentum). The quantities $\underline{\sigma} \cdot (\underline{p}_e \times \underline{p}_\nu)$ and $\underline{\sigma} \cdot (\underline{J} \times \underline{p}_e)$ have not been looked for so far, possibly because it is so difficult to obtain reliable measurements of elec-

* The search for the neutron magnetic dipole moment tests the occurence of T_{odd} - P_{odd} interactions, and the present limit[13]) is approaching the accuracy necessary to distinguish between some theoretical proposals[14]). Angular correlation tests of a T_{odd} - P_{odd} term appear well beyond the range of available techniques, since the amplitude of such an effect would be the product of the P- and T-violating processes, and could thus be as small as $\sim 10^{-9}$.

tron polarization. It should also be remarked that the second quantity, containing two axial vectors, violates both T and P. However, several tests have measured the electron recoil nucleus correlation from polarized nuclei, i.e. they have searched for an asymmetry due to the quantity $\underline{J} \cdot (\underline{p}_e \times \underline{p}_R)$.

The asymmetry coefficient D has the form

$$D = 2\,|\rho|\,\sin\phi \,/\, \sqrt{3}(1 + |\rho|^2),$$

where $\rho = C_A \langle\sigma\rangle / C_V \langle 1\rangle$. The first attempt to measure this coefficient[19]) searched for an asymmetry in the proton - electron correlation following the β^- decay of polarized neutrons, the neutron spin being reversible. If T is non - invariant then on reversing the spin one goes from the situation (a) to (b) in fig. 1.

Erozalimsky et al.[20]) repeated the experiment with a greater precision, and measured

$$D = +0.01 \pm 0.01.$$

Using the value $t_{1/2} = 10.80 \pm 0.16$ min for the neutron half - life[21]) from which is obtained $\rho = 2.13$ then

$$\phi = 178.7 \pm 1.3^\circ.$$

A similar experiment by Calaprice et al.[22]) measured the positron - neutrino asymmetry in the β^+ decay of polarized ^{19}Ne. This was in effect done by measuring the positron - recoil $^{19}F^-$ ion correlation with respect to the ^{19}Ne polarization. A Stern - Gerlach method separated the two polarization states of the spin 1/2 Ne, and a collimator system could select those that were polarized parallel or antiparallel with respect to their momentum. The degree of polarization was approximately 90%. Positron - ion pairs were recorded by a multidetector system that had shared elements,

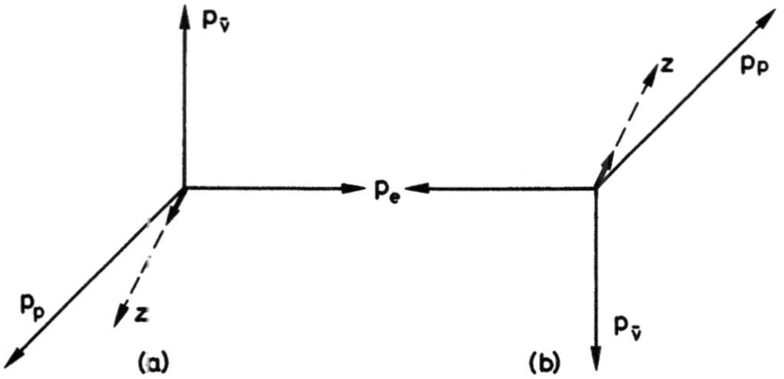

Fig. 1. The normal correlation is shown in (a) when the n polarization is out of the plane, containing the electron and neutrino momenta ($p_\nu \cong p_e$). When the n polarization is reversed and T is invariant the situation is given by (b), cf. ref. 19.

so that efficiency differences and experimental asymmetries largely cancelled. They obtained the value

$$D = +0.002 \pm 0.014$$

and substituting for $\rho = 1.603 \pm 0.006$ from the ft value of ^{19}Ne this gives

$$\phi = 180.2 \pm 1.6^\circ$$

The corrections[23]) that it may be necessary to apply to these results for non - symmetric interactions are the Coulomb correction, which for ^{19}Ne gives an asymmetry not greater than 0.0017, and weak magnetism, which is negligible. The total electromagnetic correction for the neutron data is also negligible.

A significant improvement, in the accuracy of these or similar experiments, will be difficult. A more important aspect is that it is unlikely that these measurements are sensitive to a T - violating weak interaction. In the first place, Cabibbo[11]) considers that for a β - decay with a low Q - value, such as neutron decay, the T - violating amplitude would be expected to be small, since the interaction occurs through a first - forbidden matrix element. Furthermore, as pointed out by Marchioro et al. (ref. 24), null results would be found for these measurements if T - reversal stemmed from second class currents. Kim and Primakoff[25]) have developed the argument further, and show that only when the accuracy approaches 1 in 10^4 will a $\underline{J} \cdot (\underline{p}_e \times \underline{p}_\nu)$ measurement be useful for isodoublet β - decays such as n or ^{19}Ne decay. On the other hand, they argue that a much more favourable case exists, if one were to look for a similar asymmetry in an isotriplet decay. They chose as an example the β - decay

$$^{32}P \to {}^{32}S + e^- + \bar{\nu}_e$$

In such a non - mirror decay, where the spin difference is unity, there is an anomalously small G - T matrix element, and the asymmetry coefficient becomes large. Under favourable circumstances they obtain $D = 0.3$ for this decay.

The required accuracy is not too great, but the difficulty rests in the experimental method.

3. Electromagnetic tests for T - invariance

If T is invariant, then the matrix elements, describing electromagnetic transitions, are real[26]), and the phase difference between the reduced matrix elements, describing a mixed transition, will be 0 or π. A non - vanishing phase difference, i.e. $\Delta\eta \neq 0$, implied T - non - invariance. This is not necessarily in the e.m. interaction, but may arise from a small imaginary part in the nuclear wave functions. We may write the mixing ratio as

$$\delta = \frac{\langle I_f || L' || I_i \rangle}{\langle I_f || L || I_i \rangle} = |\delta| \, e^{i\Delta\eta} = |\delta| \sin \eta$$

and low energy tests to measure this quantity are, for the most part, easier to perform than detailed balance experiments, aimed at detecting the presence of non - real contributions in the matrix elements. The latter also require a careful analysis, since, if the reaction is well described by the Born approximation, the transition matrix elements will be Hermitian, and detailed balance will occur irrespective of T - invariance or not. Uncertainties in nuclear reaction theory may also limit the method. Jacobsohn and Henley have set out the conditions for low energy tests, and these have been further discussed by Szymanski[3]). Boehm[4]) has tabulated the possible test in a convenient form. The angular correlation tests have been confined to the searching for an asymmetry associated with the quantity

$$\underline{J} \cdot [\underline{k}(1) \times \underline{k}(2)],$$

which conserves P but violates T. Other possible tests include the measurement of either the circular or linear polarization of a photon. Because of the poor efficiency of a circular polarimeter (typically a few per cent) and the requirement of two detectors to form a conventional linear polarimeter, only the angular correlation test offers the best chance of success. The Mössbauer effect does allow measurements involving linear polarization to be made.

Following Jacobsohn and Henley[1]) and the development by Lobov[2]) the angular correlation between successive γ - rays from a polarized nucleus may be written

$$W(\theta,\eta) = \sum_{k_{\text{even}}} A_k(1)\, A_k(2)\, P_k(\cos\theta) - 6P\left(\frac{j_o}{j_o + 1}\right)^{1/2} \sin\eta_1\, |\delta_1|\, \underline{J}\cdot[\,\underline{k}(1) \times \underline{k}(2)]$$

$$\sum_{k=2,4} \left\{\frac{2k+1}{k(k+1)}\right\}^{1/2} F^k_{1k}(1)\, A_k(2)\, P'_k(\cos\theta)\,,$$

where the A_k coefficients are written

$$A_k(1) = F_k(LLj_0j_1) + 2\cos\eta\; \delta(1) F_k(LL'j_0j_1) + |\delta(1)|^2 F_k(L'L'j_0j_1)$$

$$A_k(2) = F_k(LLj_2j_1) - 2\cos\eta\; \delta(2) F_k(LL'j_2j_1) + |\delta(2)|^2 F_k(L'L'j_2j_1)$$

The sign change before the second term for radiation (2) is consistent with the Biedenharn definition of δ [27]), when the first member of the γ - γ cascade is represented by an absorption matrix element and the second member by an emission matrix element. (This should be compared with the Rose and Brink definition[28]); the two definitions are consistent only when the first member of the cascade is mixed). In practice, it is usual that the second member is a pure multipole transition, and since η is expected to be small, then $\cos\eta$ may be taken as unity. The A_k coefficients are then those normally encountered in γ - γ correlations.

The factor F^k_{1k} is the triple angular correlation factor[29]), P the degree of polarization, and P' (cos θ) is dP_k (cos θ) / d (cos θ), i.e. $P'_2(\cos\theta) = 3\cos\theta$.

It should be noted that both F_k and F^k_{1k} vanish, if the spin of the intermediate state in the γ - cascade is 0 or 1/2, and if the first transition is pure, then F^k_{1k} is also zero.

The angular correlation function may be conveniently written

$$W(\theta,\eta) = W(\theta) + V(\theta,\eta)\ ,$$

where W (θ) is the usual angular correlation function, and considering only the first term in the second function

$$V(\theta,\eta) = -9\sqrt{\tfrac{5}{6}}\ P\left(\frac{j_0}{j_0+1}\right)^{1/2} F^2_{12}(1)A_2(2)\,|\delta(1)|\sin\eta\,\sin 2\theta$$

Apart from the values of the various coefficients which are determined by the spin sequence, the maximum anisotropy will occur for $|\delta| \cong 1$ and $\theta = 45°$. There is not yet any indication what nuclear features, if any, of the mixed γ - ray will give maximal phase difference. This is an important aspect of the problem, and it would be useful to have some guide when selecting correlations that appear to satisfy the other requirements.

An interesting feature of this type of measurement is that there are three distinct methods available for produdng polarized nuclei (cf. fig. 2):

a) A measurement of a β - decay electron emission direction may be used to indicate the polarization of the state j_0;
b) The state j_0 may be formed polarized by the capture of polarized neutrons;
c) Direct polarization may be obtained by applying a sufficiently large magnetic field at low temperatures.

It is useful to consider the relative merits of these different methods, in so far as a particular method may provide the largest asymmetry. It is convenient to define the asymmetry as

$$A = \frac{W(45,\eta) - W(135,\eta)}{W(45,\eta) + W(135,\eta)} = E\sin\eta\ , \tag{1}$$

and assume that the first γ - ray is an E2 + M1 transition with $|\delta| = 1$.

In the case of neutron capture the degree of polarization is given by the orientation parameter B_1 (j_0) (cf. Blin - Stoyle and Grace[30])), and for a polarized nucleus

$$B_1(j_0) = \frac{\sqrt{3}}{j_0(j+1)}\sum_m mp(m)$$

where the summation is over the magnetic hyperfine states with population p (m). If

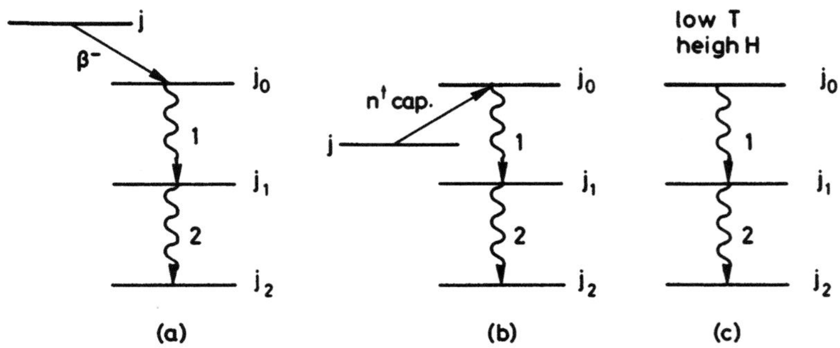

Fig. 2. The state j_0 may be polarized by:
a) Allowed β - decay;
b) Polarized n capture;
c) A high magnetic field at a low temperature.

it is assumed that the neutron polarization is 100%, then

$$A_n = 9\sqrt{\tfrac{5}{6}} \left(\frac{j_0}{j_0+1}\right)^{1/2} B_1(j_0) \left\{ \frac{F_{12}^2(12j_0j_1)F_2(L_2L_2j_2j_1)|\delta| \sin\eta}{W(45)} \right\}$$

When the polarization is produced by the interaction of a magnetic field with the magnetic moment of the nucleus at a low temperature, the asymmetry expression differs from the above equation only in respect of the orientation parameter which is replaced by the factor f_1.

$$A_p = 9\sqrt{\tfrac{5}{6}} \left(\frac{j_0}{j_0+1}\right)^{1/2} f_1 \left\{ \frac{F_{12}^2(12j_0j_1)\ F_2(L_2L_2j_2j_1)\ |\delta| \sin\eta}{W(45)} \right\}$$

This factor, f_1, depends on the temperature and magnetic field that the nucleus experiences. In the limit f_1, tends to unity, but typically for a source implanted in a ferromagnet, such as iron, and at a temperature of ~ 30 m°K, then $f_1 \cong 0.5$.

The expression for the β - γ - γ angular correlation is discussed in some detail by Krupchitsky and Lobov[6]), and it is shown that the anisotropy may be written for an allowed β - transition as:

$$A_\beta = 3\sqrt{\tfrac{5}{6}}\ \rho \left\{ \frac{F_{12}^2(12j_0j_1)F_2(L_2L_2j_2j_1)|\delta| \sin\eta}{W(45)} \right\}$$

where

$$\rho = (2 \operatorname{Re} x + |x|^2 \Lambda_{jj_0})/(1 + |x|^2)$$

and

$$\Lambda_{jj_0} = \frac{1}{2} \left[j_0(j_0 + 1) - j(j + 1) + 2 \right] / \left[j_0(j_0 + 1) \right]^{1/2}$$

$$x = C_A \int \sigma / C_V \int 1$$

The value of ρ is optimum for an allowed G - T transition when $\rho = \Lambda_{jj0}$, and from the expression for Λ_{jj0} then further optimisation occurs when $j + 1 = j_0$. The Coulomb interaction between the nucleus and electron and the finite size of the nucleus are negligible[6]). In order to compare the relative effectiveness of these methods we may, by way of illustration, apply them to j_0 (M1 + E2) j_1 (E2) j_2 γ - γ cascades, and to have as fair a comparison as possible optimum conditions are chosen, i.e. the β - transition is an allowed G - T; $j = j_0 - 1$ and the spin of the n capture state is $j = = j_0 - 1/2$ (100% polarization of the n - beam is also assumed and so is the condition $f_1 = 1$). Fig. 2 illustrates the processes, and in table 1 the polarization coefficients and the terms common to the γ - γ part of the correlation are given.

Table 1. The polarization coefficients for γ - γ angular correlation tests of T - invariance in the electromagnetic interaction. They are evaluated for several specific cases.

Polarizing method	β-γ-γ correl.	Polarized n capture	Direct nuclear polarization	Common term $3\sqrt{\frac{5}{6}}\{\ \}$
Polarization coefficient	Λ_{jj_0}	$3\left(\frac{j_0}{j_0 + 1}\right)^{\frac{1}{2}} B_1(j_0)$	$3\left(\frac{j_0}{j_0 + 1}\right)^{\frac{1}{2}} f_1$	
γ-γ spin sequence 2(M1 + E2)2(E2)0	1.224	1.732	2.449	0.063
3(M1 + E2)2(E2)0	1.154	1.732	2.598	0.067

It is apparent that there is no great difference between the methods, since in practice f_1 will seldom be greater than 0.5 and a 100% polarized neutron beam is difficult to achieve. The choice of a particular method, assuming that the selected γ - γ cascade can be reached by any of these processes, is thus determined by which ever is capable of giving the greatest accuracy. Both the n capture and the direct polarization methods are basically a two - detector coincidence experiment that can be extended to multidetector systems with great advantage, although shielding considerations may limit space for detectors in the n capture experiment. Both the direct polarization technique and the $\beta \cdot \gamma \cdot \gamma$ correlation measurement use radioactive sources, and if these have convenient half - life, may be simpler to control than monitoring the polarization and flux of a neutron source. However, the $\beta \cdot \gamma \cdot \gamma$ experiment has the disadvantage of three solid angle terms contributing to its coincidence yield, each of which is typically 10%. The direct polarization method has some attractions, but it does require that the low temperature conditions can be maintained for long periods. It might also be added that once high field superconducting magnets become readily available, then in principle any nucleus may be polarized.

3.1 *Polarization following β - decay*

The first suggestion to test T - non - invariance was made by Lee and Yang[31]), who proposed a $\beta \cdot \gamma \cdot \gamma$ correlation measurement, and this was also the first method that was tried. Fuschini et al.[32]) used a three - detector system to measure the $7/2^-(\beta)\,5/2^-(E2+M1)\,3/2^-(E2)\,7/2^-$ correlation following the decay of ^{47}Ca. They were limited by statistics to a precision of 5 in 10^4, which corresponds to one standard deviation, and it is doubtful if their method, in which they moved one of the γ - ray detectors, was capable of a greater accuracy. They also measured the 1 (β^-) 2 (E2+M1) 2 (E2) 0 correlation from ^{106}Rh with a similar precision. This correlation was subsequently remeasured by Perkins and Ritter[33]), who used a β - detector, above and below the plane of the γ - detectors which remained fixed (fig. 3). A single coincidence unit was used for all triple coincidence events, and their method of data analysis removed most unknown geometrical effects. As a check for experimental asymmetries a $\beta \cdot \gamma \cdot \gamma$ correlation, containing only pure E2 multipoles and originating in the same decay, was also measured simultaneously. Their main measurement and other check correlations were all of approximately equal precision of about 2 parts in 10^4.

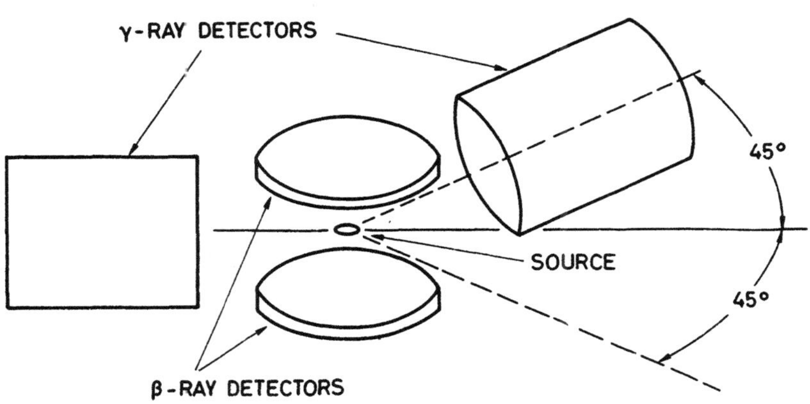

Fig. 3. The counter geometry used in a multicoincidence $\beta \cdot \gamma \cdot \gamma$ correlation measurement is illustrated.

Table 2

Angular correlation tests of T-invariance in electromagnetic interaction processes

Experiment	Source	Asymmetry × 10^{-4}	$\eta \times 10^{-2}$	Reference
β-γ(E2 + M1)-γ(E2) correlation	^{47}Ca	−0.4 ± 5.7	−2 ± 29	Fuschini et al. [32]
"	^{106}Rh	+5 ± 7	+3 ± 4	"
"	^{106}Rh	+0.7 ± 3.3	+0.4 ± 1.8	Perkins and Ritter [33]
"	^{56}Mn	−4.3 ± 2.5 −3.8 ± 2.0	+2.6 ± 1.4(δ = −0.28) −4.5 ± 2.7(δ = +0.18)	Garrell et al. [34]
Pol. n capture γ(E2 + M1)-γ	^{49}Ti	−0.29 ± 0.41	−17 ± 25	Kajfosz et al. [37]
	^{36}Cl	5 ± 15	+8 ± 23	Eichler [38]

The latest reported measurement in this class by Garrell et al.[34]) examined two cascades following the decay of ^{54}Mn. The nice feature is that the two cascades are identical in respect of spin sequence 3 (β^-) 2 (E2 + M1) 2 (E2) 0, but their mixing ratios have opposite sign, so that the asymmetries should be opposite. Thus a difference in the measured asymmetries will indicate T - reversal. Again a four - detector system of two β- and two γ - detectors was used, and coincidences were combined algebrically, so that instrumental effects were largely eliminated. In addition, the accidental coincidence contribution was measured simultaneously. Both correlations, as may be seen from the data in table 2, gave similar asymmetries and differed from zero by 2σ. However, the difference between the measurements was consistent with a zero effect and it was through that the measured asymmetries were instrumental.

In this group of measurements the coincidence counting rates were $\sim 10\,\text{sec}^{-1}$, and the quoted errors correspond to a measuring time of approximately 1 year. It is not acceptable to increase the measuring period, and it is difficult to increase the triple coincidence rate and maintain acceptable single and random rates, and so it is perhaps fair to say that the method does not offer much chance of testing T - non - invariance to a lower limit.

3.2 *Polarization following polarized neutron capture*

The production of polarized neutrons is now an established technique. It offers a high degree of polarization and has the ability to reverse neutron spin direction quickly and easily without disturbing the experimental system, and introduces usually negligible asymmetries into the measuring procedure. The method has been exploited fully testing parity violation[35,36]).

The experiments measure the quantity $[\underline{k}(1)\cdot\underline{k}(2) \times \underline{j}_a]\,[\underline{k}(1)\cdot\underline{k}(2)]$. The first one by Kajfosz et al. attempted to detect an asymmetry in the $1/2^-(341)3/2^-(1.378)7/2^-$ γ - γ cascade in ^{49}Ti, which is formed following the capture of polarized thermal neutrons. The system is illustrated in fig. 4. The advantage of a simpler experiment,

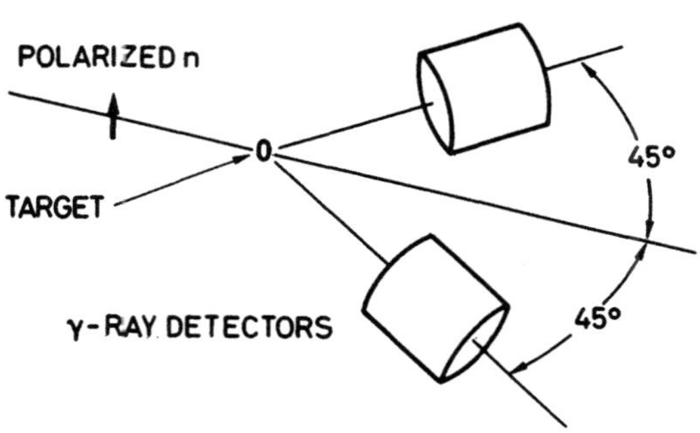

Fig. 4. A two counter system used in a polarized n capture experiment is illustrated. The extensive shielding is omitted.

a two fold coincidence, gave a much greater coincidence rate (50 sec^{-1}) than in β- γ-γ measurements, but this increase was more than off - set by a much shorter measuring period, and the final precision was no greater. Kajfosz obtained an asymmetry of

$$A = (-2.9 \pm 4.1) \times 10^{-4} .$$

A similar experiment by Eichler[37]) examined the cascade 2^+ (7.79) 3^+ (0.79) 2^+ in ^{36}Cl in which both members are mixed, and it gave

$$A = (5 \pm 15) \times 10^{-4} .$$

This large error reflects the low coincidence rate of 10 sec^{-1}. The evaluation of η from these measurements may be made using eq. (1), and the results are given in table 2.

The large difference in the values of η, which is given by these results, is due to the spin - dependent term which contributes to the asymmetry. For Kajfosz's measurement this was small, E = 0.0165, while for the cascade in ^{36}Cl, $E = 0.75^{+0.57}_{-0.40}$ where the large error arises from the uncertainty of the mixing ratio of the second analyzing transition. Since E can have the value unity[37]), and the second transition be pure, the accuracy of η will be just that of the asymmetry measurement. However, finding suitable cascades has not been possible, and the full sensitivity of the method has not yet been realized, nor have multiple detector systems been used.

There have been no reports so far of successful experiments using nuclei polarized at low temperatures, although such work is being prepared.

It should also be mentioned, that the presence of the cos η term in the A_k coefficient will lead to asymmetries in a normal γ - γ correlation involving a mixed transition. However, the accuracy of such measurements is too poor to draw any conclusions, and no experiments have been specifically made to search for such a term.

3.3 Mössbauer tests of T - invariance

The final group of experiments do not involve measuring an angular correlation, but since these experiments constitute one of the most accurate methods used so far, and since they also involve a basic nuclear spectroscopy technique, it is appropriate that they are briefly discussed.

In the classification by Henley and Jacobsohn it was shown that distribution measurement of linearly polarized radiation from a polarized nucleus could have an asymmetry indicative of T - violation. The quantity being measured is

$$[\,\underline{k}(1)\cdot \underline{j}_a \times \underline{\epsilon}\,]\,[\,\underline{k}(1)\cdot \underline{j}_a\,]\,[\,\underline{\epsilon}\cdot \underline{j}_a\,]$$

where $\underline{\epsilon}$ is the linear polarization. A direct method for observing an asymmetry is to compare the intensity of two components, $\Delta m \pm 1$, of the hyperfine spectrum for alternate nuclear polarizations. The h.f. spectrum is produced using a Mössbauer source and a magnetized foil, and a further magnetized foil acts as analyzer. The great sensitivity of the method is the comparison of intensities upon changing the field direction in the analyzer, while all other conditions remain unaltered. A full discussion of the concept and conditions for these sophisticated experiments has been given by Kistner[39]).

In fig. 5 the experimental arrangement used by Kistner is shown. The absorbing foil is positioned with its magnetization axis H orientated with respect to the field H′ in the first foil in order to achieve maximum effect when the field H is reversed. (In practice both fields are reversed in order to reduce magnetic interference between them; the field direction in the first foil does not affect the h.f.s. produced by it).

The orientation of the analyzing foil introduces serious complications into the experiment, since the radiation reemitted by the first layer of the foil will have a different polarization from the incident radiation, and its subsequent absorption will be different also. The emitted and incident radiation are coherent, and the effect produces a change in the polarization direction; i.e. the polarization is rotated (Faraday rotation). Blume and Kistner[40]) discuss fully the implications of the effect in such symmetry measurements.

The Faraday rotation affects both h.f. components in the same manner upon changing the field direction, and its contribution to the asymmetry thus cancels.

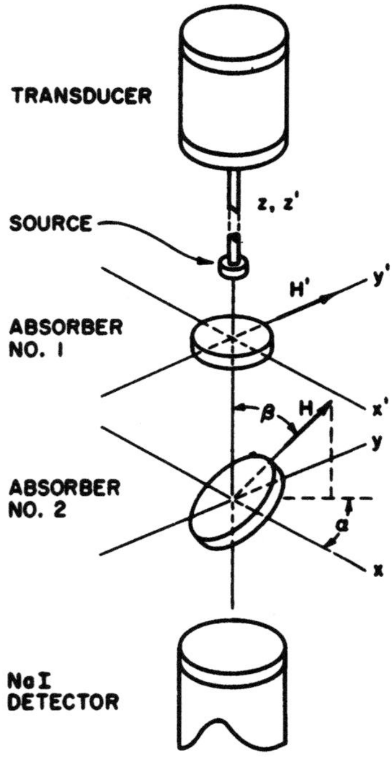

Fig. 5. A schematic of the experimental arrangement used by Kistner and taken from ref. 39. Absorber I produces the hyperfine spectrum from the single line source. The intensity of the $\Delta m \pm 1$ components are compared for both magnetic field directions in absorber 2.

However, it is a large effect, and, under optimum conditions, is approximately an order of magnitude greater than the one expected from T - non - invariance. It is thus necessary to be able to assess its possible contribution, and to be certain that experimental conditions cannot lead to an asymmetry in it.

Kistner examined the $\Delta m = \pm 1$ components of the $3/2^+ \rightarrow 5/2^+$ 90 keV transition in ^{99}Ru. His results are given in table 3. A similar experiment by Zech et al.[41]) used the 73 keV transition in ^{193}Ir.

Table 3. Mössbauer tests of T - invariance. The E2 : M1 phase difference is denoted by η.

Source	Transition	$\sin \eta \times 10^{-3}$	Corrected result	Reference
^{99}Ru	$^3 2^+ \rightarrow {}^5 2^+$; 90 keV	1.0 ± 1.7	0 ± 1.7	Kistner
^{193}Ir	$^1 2^+ \rightarrow {}^3 2^+$; 73 keV	< 5	-	Zech et al.
"	"	1.1 ± 3.8	0.2 ± 3.8	Atac et al.

Atac et al.[42]) have also studied this transition by an emission experiment, and removed the Faraday effect by having the analyzing field perpendicular to the photon direction. These results are also given in table 3.

These Mössbauer transitions, because of their low energy, are internally converted to a significant extent and Hannon and Trammell[43]) have considered an asymmetry that might arise because of the conversion process. The nuclear M1 and E2 currents excite currents in the inner atomic electron shells, and these induced E2 and M1 conversion currents have a different relative phase, ξ, which is not associated with T - non - invariance. This phase difference will combine with η , and, in the case of an emission experiment, the measured effect will be $\eta' = \xi_E + \eta_E - (\xi_M + \eta_M)$ where $\xi_E - \xi_M = \xi$. Their calculations give a correction to the results of Atac et al., which is comparable to the measured effect.

For an absorption experiment, such at that by Kistner, it is difficult to separate the Faraday rotation and the ξ phase difference, although in principle this may be done, since the T - effect is proportional to the absorber thickness, while the Faraday effect depends on the square of the thickness. Hannon and Trammell conclude that the effect observed by Kistner is consistent with the one caused by internal conversion currents.

In the other classes of experiments where one is dealing with much higher energy transitions, so that the conversion coefficient is perhaps 10^{-2} to 10^{-3}, this 'atomic' effect should be negligible. The nuclear effects were estimated by Henley and Jacobsohn[44]). The electromagnetic experiments assume that the transition matrices are Hermitian, and any lack of Hermiticity may introduce asymmetries. They conclude for typical transitions used in such tests, e.g. E2 + M1, that the test will become insensitive, only if the T - odd amplitude is less than 1 part in 10^6.

Once firm experimental evidence indicating T - non - invariance is obtained from one of these tests, a much more detailed examination will have to be made for the existence of sources of asymmetry which can mimmic T - non - invariance. The experimentalist will have to be very certain that his angular correlation technique cannot introduce asymmetries[4 5]), and also that the method he uses has an accuracy consistent with the precision that may be achieved.

† At present, Research Associate, CERN, Geneva.

Acknowledgement

I am grateful to Mr. M.J. Homes for checking and evaluating some of the numerical data.

References

1. E.M. Henley and B.A. Jacobsohn, Phys. Rev. 113 (1959) 225; B.A. Jacobsohn and E.M. Henley, Phys. Rev. 113 (1959) 234.
2. G.A. Lobov, JETP Letters 1 (1965) 157.
3. Z. Szymanski, Nuclear Physics A113 (1968) 385.
4. F. Boehm, in Hyperfine Structure and Nuclear Radiations (Ed. Matthias and Shirley, North - Holland Publ. Co., Amsterdam, 1968), p. 279.
5. E.M. Henley, in Annual Review of Nuclear Science 19 (1968) 367.
6. P.A. Krupchitsky and G.A. Lobov, in Atomic Energy Review, Vol. 7, no. 3 (1969) 91.
7. D. Hamilton, in High Energy Physics and Nuclear Structure (Ed. Devons, Plenum Press, New York, 1970) p. 696.
8. J.H. Christenson, J.W. Cronin, V.L. Fitch and R. Turlay, Phys. Rev. Letters 13 (1964) 138.
9. T.D. Lee and L. Wolfenstein, Phys. Rev. 138B (1965) 1490.
10. J. Bernstein, G. Feinberg and T.D. Lee, Phys. Rev. 139B (1965) 1650.
11. N. Cabibbo, Phys. Letters 12 (1964) 137.
12. F. Boehm, paper submitted to this conference.
13. J.K. Baird, P.D. Miller, W.B. Dress and N.F. Ramsey, Phys. Rev. 179 (1969) 1285.
14. G. Barton and E.D. White, Phys. Rev. 184 (1969) 1660; D.J. Broadhurst, Nuclear Phys. B20 (1970) 603.
15. B.L. Schrock, J.-F. Detoeuf, R.P. Haddock, J.A. Helland, M.J. Longo, K.K. Young, S.S. Wilson, D. Cheng, J. Sperinde and V. Perez - Mendez, in High Energy Physics and Nuclear Structure (Ed. Devons, Plenum Press, New - York, 1970), p. 727.
16. S. Rock, M. Borghini, O. Chamberlain, R.Z. Fuzsey, C.C. Morehouse, T. Powell, G. Shapiro, H. Weisberg, R.L.A. Cottrell, J. Litt, L.W. Mo and R.E. Taylor, Phys. Rev. Letters 24 (1970) 748.
17. L. Dobrzynski, Nguyen - Huu Xuong, L. Montanet, M. Tomas, J. Duboc and R.A. Doanld, Phys. Letters 22 (1966) 105.
18. D. Jackson, S. Treiman and H. Wylde, Phys. Rev. 106 (1957) 517.
19. M.J. Burgy, V.E. Krohn, T.B. Novey, G.R. Ringo and V.L. Telegdi, Phys. Rev. Letters 1 (1958) 324.

20. B.G. Erozalimsky, L.N. Bondarenko, Yu.A. Mostovoy, B.A. Obinyakov, V.P. Zacharova and V.A. Titov, Phys. Letters 27B (1968) 557.
21. C.J. Christensen, A. Nielsen, A. Bahnsen, W.K. Brown and B.M. Rustad, Phys. Letters 26B (1967) 11.
22. G.L. Wick, Ph.D. Thesis UCRL - 17708 (1967); F.P. Calaprice, E.D. Commins, H.M. Gibbs, G.L. Wick and D.A. Dobson, Phys. Rev. 184 (1969) 1117.
23. D. Jackson, S. Treiman and H. Wylde, Nuclear Phys. 4 (1957) 213; C.G. Callan and S.B. Treiman, Phys. Rev. 162 (1967) 1494.
24. C. Marchioro, A. Prosperetti and A. Pugliese, Nuovo Cimento Letters 1 (1969) 473.
25. C.W. Kim and H. Primakoff, Phys. Rev. 108 (1969) 1502.
26. S.P. Lloyd, Phys. Rev. 81 (1951) 161.
27. L.C. Biedenharn, in Nuclear Spectroscopy (Ed. F. Ajzenberg - Selove, Academic Press, New York and London, 1960), part B, p. 732.
28. H.J. Rose and D.M. Brink, Rev. Mod. Phys. 39 (1967) 306.
29. A.J. Ferguson, Angular Correlation Methods in Gamma - Ray Spectroscopy (North - Holland Publ. Co., Amsterdam, 1965).
30. R.J. Blin - Stoyle and M.A. Grace, Encyclopedia of Physics 42 (1957) 555.
31. T.D. Lee and C.N. Yang, Report BNL - 443 (1957).
32. E. Fuschini, V. Gadjokov, C. Maroni and P. Veronesi, Nuovo Cimento 33 (1964) 1309.
33. R.B. Perkins and E.T. Ritter, Phys. Rev. 174 (1968) 706.
34. M. Garrell, H. Frauenfelder, D. Ganek and D.C. Sutton, Phys. Rev. 187 (1969) 1410.
35. Y.G. Abov, P.A. Krupchitsky, M.I. Bulgakov, O.N. Yermakov and I.L. Karpikhin, Phys. Letters 27B (1968) 16.
36. E. Warming, Phys. Letters 29B (1969) 564.
37. J. Kajfosz, J. Kopecky and J. Honzatko, Phys. Letters 20 (1966) 284; Nucl. Phys. A120 (1968) 255.
38. J. Eichler, Nucl. Phys. A120 (1968) 535.
39. O.C. Kistner, Phys. Rev. Letters 19 (1967) 872. O.C. Kistner, in Hyperfine Structure and Nuclear Radiation (Ed. Matthias and Shirley, North - Holland Publ. Co., Amsterdam, 1968), p. 295.
40. M. Blume and O.C. Kistner, Phys. Rev. 171 (1968) 417.
41. E. Zech, F. Wagner, H.J. Körner and P. Kienle, in Hyperfine Structure and Nuclear Radiation (Ed. Matthias and Shirley, North - Holland Publ. Co., Amsterdam, 1968), p. 314.
42. M. Atac, B. Chrisman, P. Debrunner and H. Frauenfelder, Phys. Rev. Letters 20 (1968) 691.
43. J.P. Hannon and G.T. Trammell, Phys. Rev. Letters 21 (1968) 726.
44. E.M. Henley and B.A. Jacobsohn, Phys. Rev. Letters 16 (1966) 706.
45. E. Breitenberger, Phil. Mag. 45 (1954) 497.

Discussion

Medicus: You pointed out that presently the only evidence for T - violation is in the decay of the K_L^0. An experiment is now being performed at the Rutherford High Energy Laboratory to test if there is indeed a difference in the energy distributions of the positive and negative pions in the three - pion decay of the η meson. Would an outcome of this experiment establishing a small asymmetry have serious consequences for the interpretation of the K_L^o - decay?

W.D. Hamilton: I am not sure of this, but I do not see how it does.

Boehm: On the danger of repeating myself I should like to point out again that in addition to the coincidence experiments to check on T you mentioned, there is a single counting experiment quoted in my table. It only involves the measurement of the linear polarization $\underline{\epsilon}$ in one counter which also defines the momentum $\underline{k}$ of the gamma ray, and it requires a polarized source with $\Omega = 3$.

W.D. Hamilton: That's right. And the main problem really is to have a big enough slice of germanium which is going to be an efficient polarimeter. And that might have a comparable difficulty with the two - counter polarimeter of good efficiency.

Coussement: I would like to ask what the present situation is in regard to the search for electric dipole moments. Do you know something about this?

W.D. Hamilton: Well, the last results were published roughly a year ago by Ramsey and there has been no improvement in those. There have been several theoretical papers examining or anticipating the setting of a lower limit by an order of a magnitude. I can give you references to those.

Coussement: Have some new experiments been proposed?

W.D. Hamilton: I know that Dr. Pendlebury at Sussex is planning to carry out an experiment but it is still in the testing stage, although, like many experiments in the testing stage, it offers great hopes and a high degree of accuracy. But I think it would be unfair to say that this experiment would necessarily be able to improve Ramsey's latest value by an order of magnitude.

Frauenfelder: The situation with the correction for the Hannon - Trammell effect in the Mössbauer experiment is not as serious as you said because the dependence on the thickness of the absorber is different and also the angular dependence. What it means is essentially a doubling of the measuring time to take the two effects apart. But the correction can be made. And the difficulty with the Mössbauer experiment is simply solid state: the production of absorber and source with the proper magnetic field. We are trying to improve by about a factor of 5 our present results. All the nuclear problems seem to be quite simple compared to the solid state problems. For the last two years we have been stuck with solid state problems. So the difficulties always come from an unexpected side.

INTERFERENCE PHENOMENA IN PERTURBED ANGULAR CORRELATIONS (THEORY OF TIME - INDEPENDENT INTERACTIONS) †

H.J. Leisi
Laboratory for High Energy Physics, Swiss Federal Institute of Technology, Zurich, Switzerland

Abstract

The theory of perturbed angular correlations of time - independent interactions is formulated in such a way as to display explicitly the interference properties of the individual energy levels of the intermediate state. In time - differential angular correlation experiments, interfering energy eigenstates give rise to oscillations whose period is a direct measure of the energy difference of the corresponding levels. Two energy levels that cross as a function of an external magnetic field may produce a resonance in a time - integrated angular correlation experiment, because of interference between the two states. From general symmetry properties of the extranuclear interactions, a number of selection rules for interference is derived. Based on these rules and the level structure of the intermediate system in the extranuclear field, general properties of the angular correlation function can be derived without calculation. The method is illustrated by discussing hitherto unobserved effects in angular correlations due to the hyperfine interaction in free atoms and in paramagnetic crystals. Characteristic effects produced by the mixing of states in the vicinity of repulsive energy levels are also discussed.

1. Introduction

The aim of a reformulation of the theory of perturbed angular correlations is to gain a simple physical picture of the correlation process which allows one to predict important features of an angular correlation experiment without having to perform detailed calculations. The method is particularly useful in discussing strongly perturbed angular correlations, whereby it is possible to fully predict the qualitative behaviour of the correlation function in many cases. The difference with the usual formulation of the theory[2]) consists in considering, as a first step, the dependence of the correlation function on energy (and time, in the case of the time - differential correlation experiment), without yet introducing the complications that arise from the multipole expansion of the planewave states of the photons. This formulation is given in section 2. We write the angular correlation function as a sum of a direct part

† This work is in most parts an abbreviated version of ref. 1.

and an interference part, the latter arising from the indefiniteness of the intermediate states. The dependence on energy and time is fully contained in the interference terms; they are responsible for the appearance of oscillations and resonances in time - differential and time - integrated angular correlation experiments, respectively. In general, each pair of energy eigenstates contributes an interference term, unless the corresponding interference amplitude vanishes. The existence of the various interference amplitudes for a specific problem depends on the symmetry properties of the extranuclear interaction and can conveniently be described by selection rules. In order to derive such rules it is necessary to express the interference amplitude explicitly in terms of the physical properties of the two states. Appropriate formulae are given in section 3, and a number of selection rules is derived. The method is illustrated in section 4 by discussing some as yet unobserved interference effects. These involve angular correlations perturbed by the hyperfine interactions in free atoms or ions and in paramagnetic single crystals, in the presence of an external magnetic field. The properties of level crossings with regard to resonances are more closely examined in section 5. A small term in the interaction Hamiltonian may cause a mutual repulsion of the energy levels. In the vicinity of closest approach, the corresponding states are strongly mixed. Under appropriate conditions, characteristic effects in angular correlations may be produced.

2. Interference between individual energy states

Although the interference phenomena to be discussed do not depend on the type of nuclear radiation involved,we consider here explicitly only directional correlations of γ - rays, since, in general, the most precise results are obtainable from such experiments.

The natural description of the nuclear system in the initial-, intermediate- and final state is in terms of energy eigenstates $|\,a>$, $|\,b>$ and $|\,c>$ respectively. Depending on the problem, these states may be chosen to be the eigenstates of the nuclear system in an external electromagnetic field, (e.g. a magnetic field or the crystalline field in a solid), or the eigenstates of the whole atom either in free space or in an external electromagnetic field, or even the eigenstates of a whole crystal. In our treatment we assume that the interaction between the nucleus and its environment is stationary. This means that the system in the intermediate state is not affected by dynamical interactions other than the dynamical coupling of the nucleus to the electromagnetic field.* Furthermore, the initial substates $|\,a>$ are assumed to be equally populated. This condition is fulfilled, e.g. for systems in thermal equilibrium, down to temperatures of liquid helium for any conceivable extranuclear coupling.

We first proceed to calculate the angular correlation function describing a transition from a specific initial substate $|\,a>$ through the intermediate state to some specific final state $|\,c>$, whereby γ - rays of fixed circular polarizations σ_1 and σ_2 and directions $\underline{\Omega}_1$ and $\underline{\Omega}_2$ are emitted. The corresponding transition amplitude may be written as[3])

* This implies, in particular, the absence of relaxation phenomena on a time scale of the order of the lifetime τ of the intermediate nuclear state.

$$A_{ac}^{(\sigma_1\sigma_2)}(t) = \sum_b < c\underline{\Omega}_2\sigma_2|H'|b_t>< b\underline{\Omega}_1\sigma_1|H'|a> \tag{1}$$

where H ′ is the interaction between the nucleus and the radiation field and $|b_t>$ stands for the evolution in time of the stationary state $|b>$ that was formed at the time of emission of the first γ - ray (t = 0), i.e.

$$|b_t> = e^{-\frac{iE_b}{\hbar}t}|b> e^{-\frac{t}{2\tau}} \tag{2}$$

Here, E_b is the energy of the state $|b>$. The exponential factor in eq. (2) describes the decay of the intermediate state, τ being the lifetime. The transition probability for the specific transition from $|a>$ to $|c>$ is then proportional to the square of the absolute value of $A_{ac}^{(\sigma_1\sigma_2)}$ (t), eq. (1). The total angular correlation function W (t) is obtained by averaging incoherently over initial states† and summing over all final states (including γ - ray polarizations):

$$W(t) = \sum_{a,c,\sigma_1,\sigma_2} |A_{ac}^{(\sigma_1\sigma_2)}(t)|^2 \tag{3}$$

We introduce the abbreviation

$$\mathscr{F}_{bb'} = \sum_{a,c,\sigma_1,\sigma_2} <c\underline{\Omega}_2\sigma_2|H'|b>^*<b\underline{\Omega}_1\sigma_1|H'|a>^*$$

$$\times <c\underline{\Omega}_2\sigma_2|H'|b'><b'\underline{\Omega}_1\sigma_1|H'|a> \tag{4}$$

and write the time - differential angular correlation function (3) with the aid of eqs. (1) and (2), as

$$W(t) = e^{-t/\tau}\sum_{b,b'}\mathscr{F}_{bb'}\, e^{i\frac{E_b - E_{b'}}{\hbar}t} \tag{5}$$

It is appropriate to break up expression (5) into a direct part, containing all terms with b′ = b, and an interference part.

† This follows from taking the statistical average over the ensemble of the nuclei in the source since the latter are independent of each other and hence their wave functions have random phase distributions.

$$W(t) = e^{-t/\tau} \{ \sum_b \mathscr{F}_{bb} + 2 \sum_{(b,b')} |\mathscr{F}_{bb'}| \cos (\phi_{bb'} + \frac{E_b - E_{b'}}{\hbar} t) \} \tag{6}$$

Here, $\phi_{bb'}$ is the phase of the complex quantity $\mathscr{F}_{bb'}$ and the second sum is over all pairs of states (b, b'). We have used the property $\mathscr{F}_{b'b} = \mathscr{F}_{bb'}^*$. Formula (6) shows that each pair of intermediate states gives rise to an oscillatory interference term, the period of the oscillation being directly related to the energy difference of the two states. The quantity $\mathscr{F}_{bb'}$, that governs the amplitude of the oscillation, may be called interference amplitude and depends, of course, on the emission directions $\underline{\Omega}_1$ and $\underline{\Omega}_2$ of the γ - rays. The existence of the interference part has its origin in the coherent sum of the amplitudes in eq. (1) and reflects the fact that the intermediate state in a cascade transition is, in general, not determined. From the structure of formula (6) it is apparent that the energy splittings, $E_b - E_{b'}$, of the intermediate state may be determined by a time - differential angular correlation experiment. In principle, a knowledge of the properties of the nuclear states and the radiations is not required, nor is it necessary to know the nature of the interaction with the extra-nuclear field. In practice, however, such knowledge is usually useful. The range of energy splittings accessible to observation is restricted, on the one hand by the lifetime τ, and on the other hand by the experimental time resolution Δt:

$$\Delta t < \frac{h}{|E_b - E_{b'}|} \lesssim \tau \tag{7}$$

Integration of eq. (6) over the time variable leads directly to the time - integrated angular correlation function

$$W = \sum_b \mathscr{F}_{bb} + 2 \sum_{(b,b')} |\mathscr{F}_{bb'}| \frac{\cos \phi_{bb'} - \frac{E_b - E_{b'}}{\Gamma} \sin \phi_{bb'}}{1 + (\frac{E_b - E_{b'}}{\Gamma})^2} , \tag{8}$$

where we have introduced the natural width of the intermediate state, $\Gamma = \hbar/\tau$. The correlation function (8) is again the sum of a direct part and an interference part. In the limit

$$|E_b - E_{b'}| \gg \Gamma \tag{9}$$

the corresponding interference term vanishes. If there is no degeneracy among the levels of the intermediate state and condition (9) is satisfied for all pairs of states, then the interference part of eq. (8) vanishes completely. In this case the emission of the second γ - ray can be considered to occur from a well - defined state $|b\rangle$ with 'sharp' energy. Indeed, a precise measurement of the energy of the second γ - ray (apart from its direction) would enable one to localize the energy E_b (and hence the state

$|b\rangle$) to within the natural width Γ. The disappearance of interference terms may be utilized to observe sharp resonances in time - integrated correlation experiments. Let us assume that, for some value of an external parameter (such as a magnetic field), the energy levels E_b and $E_{b'}$ cross, and all other energy splittings are much larger than Γ. At exact crossing the interference part in eq. (8) reduces to $2\,|\mathscr{F}_{bb'}|\cos\phi_{bb'}$. Upon changing the magnetic field the interference term decreases sharply, due to the resonance denominator, until the term vanishes completely for $|E_b - E_{b'}| \gg \Gamma$. Thus, a measurement of the coincidence rate, at some fixed position of the detectors, as a function of the magnetic field exhibits a resonance[4]), centered around the value of the field for which $E_b = E_{b'}$.

3. Selection rules for interference

From eq. (4) it is apparent that two states $|b\rangle$ and $|b'\rangle$ interfere whenever there exists at least one pair of states $|a\rangle$ and $|c\rangle$, such that all four matrix elements in eq. (4) are non - zero. This means that the two states $|b\rangle$ and $|b'\rangle$ must be able to share the same two photons (characterized by $\underline{\Omega}_1, \sigma_1$ and $\underline{\Omega}_2, \sigma_2$, respectively) in a transition from some initial state $|a\rangle$ to some final state $|c\rangle$. In order to investigate the interference properties more closely, and also for the purpose of calculating the angular correlation functions explicitly, we have to evaluate the interference amplitude $\mathscr{F}_{bb'}$ in terms of the properties of the states $|b\rangle$ and $|b'\rangle$. We express the states $|b\rangle$ as

$$|b\rangle = \sum_{m,n} |Im\rangle |n\rangle \langle mn|b\rangle \quad , \tag{10}$$

where $|I\,m\rangle$ are angular momentum states of the nucleus (in the intermediate state) and $|n\rangle$ is a complete set of states describing the nuclear environment. The coefficients $\langle m n|b\rangle$ in eq. (10) may be obtained by solving the eigenvalue problem for the Hamiltonian K which describes the interaction between the nucleus and the environment. A number of steps are necessary in order to show that the interference amplitude, eq. (4), can be written as[1]) *

$$\mathscr{F}_{bb'} = 4\pi \sum_{k_1,k_2,\mu_1,\mu_2} A_{k_1}(1)\, A_{k_2}(2)\, \mathscr{S}_{bb'}(k_1,\mu_1) \mathscr{S}^{*}_{bb'}(k_2,\mu_2) \times$$

$$\times\, Y_{k_1\mu_1}(\vartheta_1,\phi_1)\, Y^{*}_{k_2\mu_2}(\vartheta_2,\phi_2) \quad , \tag{11}$$

* The steps leading from eq. (4) to eq. (11) may be mentioned briefly. An expansion of the states $|a\rangle$ and $|\underline{c}\rangle$ similar to eq. (10) and a multipole expansion of the plane - wave photon states permits the matrix elements of eq. (4) to be transformed into multipole matrix elements between nuclear angular momentum states. The Wigner - Eckhart theorem can be applied and summation over magnetic quantum numbers leads to expressions equal to A_{k_i} (i).

where A_{k_i} (i) are the usual coefficients in the Legendre polynominal expansion of the unperturbed angular correlation, $Y_{k_i\mu_i}(\vartheta_i, \phi_i)$ are spherical harmonics, (ϑ_i, ϕ_i) are the polar angles describing the emission directions of the two γ - rays (i = 1, 2) and the quantities $\mathscr{S}_{bb'}(k,\mu)$ are given by †

$$\mathscr{S}_{bb'}(k,\mu) = \frac{1}{\sqrt{2J+1}} \sum_{m,m',n} (-1)^{I+m} <mn|b><m'n|b'>^* \times \begin{pmatrix} I & I & k \\ -m & m' & \mu \end{pmatrix} \quad . \tag{12}$$

Here, (2J + 1) is the multiplicity of the states |n>. The amplitudes $\mathscr{F}_{bb'}$ in eq. (11) are normalized such that the correlation function (8) transforms into

$$W(\theta) = \sum_k A_k(1)\, A_k(2)\, P_k(\cos\theta) \tag{13}$$

for vanishing perturbation. The quantities $\mathscr{S}_{bb'}$ display the dependence of the interference amplitude on the states |b> and |b'>. For many applications it is appropriate to express $\mathscr{S}_{bb'}$ in a different representation. The states |n> may be chosen to be eigenstates of some (in general effective) angular momentum operator $\underline{J}$. In some problems $\underline{J}$ corresponds to the true angular momentum of the nuclear environment, e.g. in the case of a free atom. In general, it is an effective angular momentum in the sense of the spin - Hamiltonian formalism (see e.g. ref. 7). We then define a total angular momentum operator $\underline{F} = \underline{I} + \underline{J}$ and expand the states |b> in terms of simultaneous eigenstates of F_z (eigenvalue M) and $\underline{F}^2$ (eigenvalue F (F + 1)):

$$|b> = \sum_{F,M} |FM><FM|b> \quad . \tag{14}$$

The old coefficients <mn |b> are related to the new ones by

$$<mn|b> = \sum_{F,M} \sqrt{2F+1}\,(-1)^{J-I-M} \begin{pmatrix} I & J & F \\ m & n & -M \end{pmatrix} <FM|b> \quad . \tag{15}$$

Eq. (15), inserted into eq. (12) and summed over m, m' and n yields

† The quantity $\mathscr{S}_{bb'}(k,\mu)$ is related to $S_{bb'}(k,\mu)$ of refs. 5 and 6 by

$$\mathscr{S}_{bb'}(k,\mu) = \frac{1}{\sqrt{(2I+1)(2J+1)}} S_{bb'}(k,\mu).$$

$$\mathscr{S}_{bb'}(k,\mu) = \sum_{F,F',M,M'} (-1)^{I+J-M} \sqrt{\frac{(2F+1)(2F'+1)}{2J+1}}$$

$$\times \begin{pmatrix} F & F' & k \\ -M & M' & \mu \end{pmatrix} \begin{Bmatrix} F & F' & k \\ I & I & J \end{Bmatrix} <FM|b><F'M'|b'>^* \quad . \qquad (16)$$

In what follows we will always describe the nuclear environment $|n>$ by eigenstates of $\underline{J}$, where n is then interpreted as the eigenvalue of $\underline{J}_z$, but we shall still occasionally use eq. (12) instead of eq. (16).

A. *Nuclear motion decoupled from environment*

Inspection of eq. (12) shows that two components $<mn|b>$ and $<m'n'|b'>$ of states $|b>$ and $|b'>$, respectively, with $n \neq n'$ do not interfere. This property reflects the assumption, built into the theory, that the nuclear environment remains unchanged after the emission of a γ - ray. In particular, if the Hamiltonian K commutes with J_z, i.e. each state $|b>$ is characterized by a single quantum number n, then two states $|b(n)>$, $|b'(n')>$ with $n \neq n'$ cannot interfere. Thus, we can formulate **selection rule (I)**: if $[K, J_z] = 0$, two eigenstates of K with different n - values can never interfere. Such a decoupling of the nuclear motion from the environment (expressed by $[K, J_z] = 0$) exists e.g. in the Paschen - Back regime of the hyperfine interaction.

B. *Time reversal invariance*

We consider here that group of angular correlation problems which does not involve an external magnetic field. The interaction Hamiltonian K is then invariant under time reversal. This implies that each eigenstate $|b>$ of K has associated with it a time - reversed state $|\tilde{b}> = T|b>$ that is also an eigenstate of K, both states having the same energy. According to Kramer's theorem[8]) the states $|b>$ and $|\tilde{b}>$ are linearly independent whenever the system consists of an **odd** total number of (spin 1/2) particles, which means that the eigenstates of K are always degenerate. We shall prove a theorem of similar generality concerning the interference property of time - reversed states in an angular correlation process. The states $|b>$ may be expressed by eq. (14). If the phase convention of the states $|FM>$ is chosen suitable[9]) we have

$$T|FM> = (-1)^{F+M} |F,-M> \quad . \qquad (17)$$

Time reversal applied to eq. (14) then gives

$$T|b> = \sum_{F,M} (-1)^{F+M} |F,-M><FM|b>^* = |b'> =$$

$$= \sum_{F',M'} |F'M'\rangle\langle F'M'|b'\rangle \quad . \tag{18}$$

Hence one has the relation

$$\langle F'M'|b'\rangle = (-1)^{F'-M'} \langle F', -M'|b\rangle^* \quad . \tag{19}$$

Inserting expression (19) into eq. (16) gives

$$\mathscr{S}_{bb'}(k,\mu) = \sum_{F,F',M,M'} (-1)^{I+J-M+F'+M'} \sqrt{\frac{(2F+1)(2F'+1)}{2J+1}}$$

$$\times \begin{pmatrix} F & F' & k \\ -M & -M' & \mu \end{pmatrix} \begin{Bmatrix} F & F' & k \\ I & I & J \end{Bmatrix} \langle FM|b\rangle\langle F'M'|b\rangle \quad , \tag{20}$$

where we have changed the summation index M' to $-M'$. We insert the symmetry relation

$$\begin{pmatrix} F & F' & k \\ -M & -M' & \mu \end{pmatrix} = (-1)^{F+F'} \begin{pmatrix} F' & F & k \\ -M' & -M & \mu \end{pmatrix} \tag{21}$$

into eq. (20) and interchange primed and unprimed summation indices. By comparing the expression thus obtained with eq. (20), one obtains the relation

$$\mathscr{S}_{bb'} = (-1)^{2F} \mathscr{S}_{bb'} \quad . \tag{22}$$

Eq. (22) implies that $\mathscr{S}_{bb'}$ is identically zero whenever F is half - integral. Since F is integral or half - integral depending on whether the total number of particles in the system is even or odd, we can formulate a **selection rule (II)** : time - reversed states never interfere for a system consisting of an odd total number of particles.

In the case of the hyperfine interaction, e.g. in a free atom, the total number of particles is equal to the number of nucleons plus the number of electrons. Selection rule (II) is also valid for paramagnetic ions in a crystal, in which case the states $|n\rangle$ correspond for instance to the lowest ionic level in the crystalline field. Indeed, the proof of relation (22) can be generalized by starting from eq. (12) and expanding the states $|n\rangle$ in terms of angular momentum states $|J m_J\rangle$ of the free paramagnetic ion. The time reversal properties of both, the hyperfine states $|b\rangle$ and the ionic states $|n\rangle$, have to be used. The relevant number of electrons is then that of the free ion. In problems involving the coupling of the nucleus to an external (classical) field, e.g. by the quadrupole interaction in non - magnetic crystals, the nuclear spin I plays the role of F and the total number of particles is equal to the number of nucleons.

C. *Axial symmetry*

There are a number of problems for which the Hamiltonian K is axially symmetric[5,6]). This property is expressed by $[K, F_z] = 0$, i.e. each state $|b(M)\rangle$ is characterized by a quantum number M. According to eqs. (16) and (11) the interference amplitude is zero unless $\mu_1 = \mu_2 = M - M'$. Since μ_i cannot exceed the maximum value of k_i, we can state **selection rule (III)**: two states with quantum numbers M and M′ cannot interfere if

$$|M - M'| > \text{minimum}\ (k_{1\,max}, k_{2\,max}) \quad . \tag{23}$$

This selection rule can be related to the conservation of angular momentum along the z - axis. In cases of a coupling of the nucleus to a (classical) external field, axial symmetry means $[K, I_z] = 0$, and from eqs. (12) and (11) we have $\mu_1 = \mu_2 = m - m'$. We therefore obtain for these situations **selection rule (III′)**: two states with quantum numbers m and m′ cannot interfere if

$$|m - m'| > \text{minimum}\ (k_{1\,max}, k_{2\,max}) \tag{24}$$

The selection rules considered so far are independent of the emission directions of the γ - rays. It is also useful to consider the frequently occurring case of γ - rays observed perpendicularly to the symmetry axis ($\vartheta_1 = \vartheta_2 = \pi/2$). Since

$$Y_{k\mu}(\pi/2, \phi) = 0 \text{ for } k + \mu \text{ odd} \tag{25}$$

and $\mu_i = M - M'$ for axial symmetry we infer from eq. (11) **selection rule (IV)**: two states with quantum numbers M and M′ cannot interfere if

$$|M - M'| = \text{odd integer} \tag{26}$$

for observation perpendicular to the symmetry axis. A similar rule is obtained in the case of coupling to a classical field; M and M′ in selection rule (IV) have to be replaced by m and m′, respectively.

4. Examples

With the tools presented in the two preceding sections we can calculate the angular correlation function for any time - independent interaction in a highly efficient way. With the aid of the selection rules, a number of interference terms in the correlation function can be discarded by careful inspection, thus reducing the computational work considerably. Instead of performing explicit calculations here, we prefer to demonstrate the method by discussing two situations for which only few experimental observations have yet been reported.

A. *Hyperfine interaction in a free atom*

The magnetic hyperfine interaction of an atom or ion in the presence of an external magnetic field H is described by the Hamiltonian

$$K = a\,\underline{I} \cdot \underline{J} + g_J \mu_B\, H\, J_z \quad , \tag{27}$$

where a is the magnetic hfs constant, g_J is the atomic splitting factor and μ_B is the Bohr magneton. In the Zeeman region (weak magnetic field) the energy levels are

$$E_{FM} = \frac{a}{2}\,[F(F+1) - J(J+1) - I(I+1)] + g_F\, \mu_B\, HM \quad , \tag{28}$$

where

$$g_F = g_J\, \frac{F(F+1) + J(J+1) - I(I+1)}{2F(F+1)} \quad . \tag{29}$$

In the Paschen - Back region (strong magnetic field) we have the levels

$$E_{mn} = g_J\, \mu_B\, H\, n + amn. \tag{30}$$

The energy levels for an arbitrary magnetic field are shown schematically in fig. 1, for the specific case $I = 3/2$, $J = 1$. We assume that the angular correlation is strongly perturbed, i.e. $a \gg \Gamma$, and observe the γ - rays perpendicularly to the field axis. According to selection rules (III) and (IV), only states with $|M - M'| = 2$ can interfere; the corresponding pairs are marked by arrows in fig. 1. Except in field regions where the energy levels of interfering states are close, the interference part of the correlation function is vanishingly small. Hence the second γ - ray is always emitted from a well - defined eigenstate of K. Since these states are also eigenstates of F_z, the second γ - ray carries a well - defined amount of angular momentum with respect to the symmetry axis, and hence its emission probability is isotropic. Therefore, the anisotropy of the angular correlation in these field regions is zero (see fig. 1). There are a number of level crossings at zero magnetic field that give rise to a resonance. This effect has been observed experimentally[10,4]). At non - zero magnetic field we expect one more resonance, due to the crossing of levels with $M = -1/2$ and $M' = -5/2$ (see fig. 1). The observation of such resonances would probably allow to determine precisely the hfs constants of excited nuclear states from magnetic field measurements alone. Another as yet unobserved interference effect occurs at large fields. In the Paschen - Back region, all states with $n = 0$ are degenerate, provided the direct interaction between the magnetic field and the nucleus can be neglected. The corresponding interference terms produce a non - zero anisotropy of the angular correlation which gradually decreases to zero in the field region where the energy splitting of the interfering states is of order Γ (see fig. 1). The magnetic field value around which this

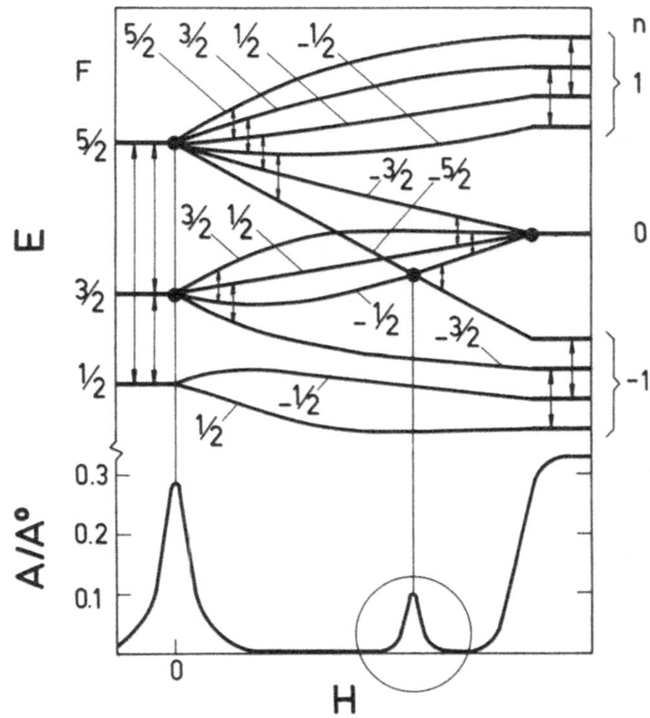

Fig. 1. Upper part: energy levels of the hyperfine interaction in a free atom with J = 1 and I = 3/2 as a function of an external magnetic field (schematic). The levels are labelled by the quantum number M. Lower part: anisotropy $A = [W(\pi) - W(\pi/2)] / W(\pi/2)$ divided by the anisotropy A^0 of the unperturbed angular correlation as a function of an external magnetic field. The height of the resonance at non - zero field is not to scale.

drop is centered contains information on the hfs constant. The limiting value of the anisotropy at large fields can be readily calculated, and one finds[1])

$$A(\infty) = \frac{A^0}{2J+1} \quad , \tag{31}$$

where A^0 is the anisotropy of the unperturbed correlation. The effect is obviously not present for J half - integral.

In a time - differential angular correlation experiment one would observe, at zero magnetic field, three oscillations corresponding to all the energy differences between three levels. In the Paschen - Back region the nuclear spin is decoupled from the atomic shell and selection rule (I) is applicable. One therefore expects to observe a single frequency, corresponding to an energy difference of 2a. It should be interesting to perform a time - differential correlation experiment in the vicinity of a level crossing in order to determine precisely the crossing point.

B. Hyperfine interaction in a paramagnetic single crystal

The axially symmetric hyperfine coupling of a Kramers doublet state without quadrupole interaction in an external magnetic field can be written as

$$K = A\, I_z J_z + B(I_x J_x + I_y J_y) + g_{//}\mu_B H_z J_z + g_{\perp}\mu_B (H_x J_x + H_y J_y); \qquad J = 1/2 \ , \tag{32}$$

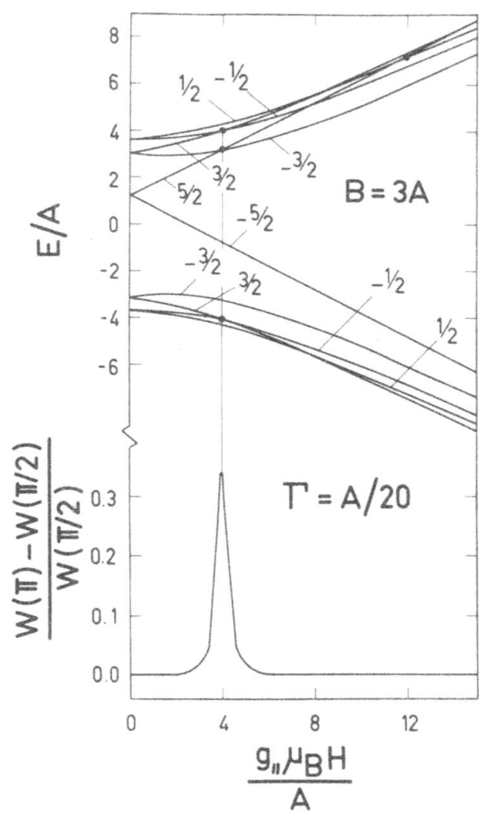

Fig. 2. Upper part: energy levels of the hyperfine interaction in a paramagnetic single crystal as a function of an external magnetic field parallel to the symmetry axis, for I = 2. Lower part: anisotropy of angular correlation with spin sequence 0 - 2 - 0 as a function of an external magnetic field. The resonance at these particular angles is due to the two crossings with $\Delta M = 2$.

where A and B are magnetic hfs constants, and $g_{//}$ and $g_{\perp}$ are electronic splitting factors. The energy levels of the Hamiltonian (32) are shown in fig. 2 for an orientation of the crystalline axis parallel to the magnetic field[5]) ($H_x = H_y = 0$) and a nuclear spin I = 2. In this geometry, each eigenstate is again characterized by a quantum number M. We consider a time - integrated angular correlation experiment in the strong - perturbation limit, i.e. $\Gamma \ll A$. As in the preceding example the angular correlation is isotropic for observation of the γ - rays perpendicular to the magnetic field, unless interfering states are close in energy. As a consequence of time - reversal invariance (selection rule (II)) the crossings at zero magnetic field do not produce a resonance, regardless of the emission directions of the γ - rays. The only crossings of interfering states compatible with selection rules (III) and (IV) are those marked in fig. 2 ($\Delta M = 2,4$). An explicit calculation shows that the interference of the states $M = 5/2, M' = 1/2$, vanishes in the case of a cascade with spin sequence 0 - 2 - 0. The remaining resonance at a critical magnetic field

$$H_c = \frac{A}{2g_{//}\mu_B} [(B/A)^2 - 1] \tag{33}$$

is shown in the lower part of fig. 2.

It is worth noting that, in a time - differential correlation experiment, the hfs constants A and B may be measured directly. A strong magnetic field along the symmetry axis of the crystal produces a level scheme similar to the one in fig. 1. The splitting of the equidistant levels in this case is given by A. In a strong magnetic field perpendicular to the symmetry axis, the same level structure results, but this time the level splitting is given by B.

5. Mutually repulsive energy levels

In this section, properties of level crossings are further investigated in view of their decisive rôle in resonance experiments. The behaviour of crossing levels may be changed drastically by a very small component of the interaction Hamiltonian K. The Hamiltonian (32) may, e.g., contain a small term

$$K' = \epsilon I_x J_x - \epsilon I_y J_y \tag{34}$$

which violates the axial symmetry of the total Hamiltonian. Generally, we expect to find a small effect of K′ on the eigenstates and energies, provided $\epsilon \ll B$. This, however, is not necessarily the case in the vicinity of a level crossing. We consider two states, ϕ_1 and ϕ_2, which are eigenstates of the dominant part K^0 of the Hamiltonian

$$K = K^0 + K' \ . \tag{35}$$

The energies E_i^0, defined by

$$K^0 \phi_i = E_i^0 \phi_i \; ; \quad i = 1, 2, \tag{36}$$

are assumed to cross as a function of an external parameter. The true eigenfunctions and energies are given by

$$K \psi_i = E_i \psi_i \tag{37}$$

The eigenstates ψ_i may be expressed in terms of the eigenstates of K^0. Since K' is small we expect to be able to describe ψ_1 and ψ_2, to a good approximation, as a superposition of ϕ_1 and ϕ_2 alone. Within this approximation the eigenstates can be calculated by diagonalizing a 2 X 2 matrix, and one finds

$$\begin{aligned} \psi_1 &= \frac{1}{\sqrt{1+|R|^2}} \; (R\phi_1 + \phi_2) \\ \psi_2 &= \frac{1}{\sqrt{1+|R|^2}} \; (\phi_1 - R^*\phi_2) \quad , \end{aligned} \tag{38}$$

where

$$R = \frac{K_{12}}{|K_{12}|^2} \left\{ \sqrt{\left(\frac{K_{22}-K_{11}}{2}\right)^2 + |K_{12}|^2} - \frac{K_{22}-K_{11}}{2} \right\} . \tag{39}$$

Here, we have used the abbreviation

$$K_{ik} = \langle \phi_i \,|\, K \,|\, \phi_k \rangle . \tag{40}$$

The corresponding eigenvalues are

$$E_{1,2} = \frac{K_{11}+K_{22}}{2} \pm \sqrt{\left(\frac{K_{11}-K_{22}}{2}\right)^2 + |K_{12}|^2} \quad , \tag{41}$$

where the plus sign refers to E_1 and the minus sign to E_2. If the perturbation K' is such that

$$K_{12} = K'_{12} = 0, \tag{42}$$

the eigenfunctions of K are identical with ϕ_1 and ϕ_2 and the energies E_1 and E_2 are at most slightly different from E_1^0 and E_2^0. Thus, the angular correlation function is practically unaffected by K'. On the other hand, if

$$K'_{12} \neq 0 \tag{43}$$

the states ψ_1 and ψ_2 are strong mixtures of ϕ_1 and ϕ_2 in the vicinity of the crossing and the corresponding energy levels, eq. (41), repel each other. The effects of level repulsion are best discussed by considering our specific example, eq. (34). We may write K′ as

$$K' = \frac{\epsilon}{2}\,(I_+ J_+ + I_- J_-), \tag{44}$$

where we have introduced

$$I_\pm = I_x \pm i\, I_y$$

and

$$J_\pm = J_x \pm i\, J_y. \tag{45}$$

Hamiltonian (44) mixes states with $\Delta M = 2$ and is thus expected to affect the resonance shown in fig. 2. Since the states $\phi_{1,2}$ in our example are eigenstates of F_z, they are of the form

$$\phi = a\,|M - 1/2, +1/2> + \beta\,|M + 1/2, -1/2>, \tag{46}$$

where $|m\ n>$ are eigenstates of I_z and J_z. From eqs. (46) and (44) we obtain

$$K'_{11} = K'_{22} = 0. \tag{47}$$

Therefore, $K_{11} = E_1^0$ and $K_{22} = E_2^0$, and the problem is governed by a single (real) parameter

$$\Delta = K'_{12} = K'_{21} \tag{48}$$

With eqs. (47) and (48) the energies, eq. (41), and the mixing parameter, eq. (39), can be written as

$$E_{1,2} = \frac{E_1^\circ + E_2^\circ}{2} \pm \sqrt{\left(\frac{E_1^\circ - E_2^\circ}{2}\right)^2 + \Delta^2} \tag{49}$$

$$R = \frac{1}{\Delta}\left\{\sqrt{\left(\frac{E_2^\circ - E_1^\circ}{2}\right)^2 + \Delta^2} - \frac{E_2^\circ - E_1^\circ}{2}\right\} \tag{50}$$

At the crossing point, $E_1^0 = E_2^0$, the energy levels are separated by the amount 2Δ, and the states are fully mixed ($R = 1$ in eq. (38)). Far away from the crossing point ($|E_1^0 - E_2^0| \gg \Delta$) the energies and the eigenstates merge into $E_{1,2}^0$ and $\phi_{1,2}$, respectively.

We now have to investigate the effect of the level repulsion on the angular correlation. Since we are interested in the resonance behaviour, we consider a time-integrated angular correlation and assume that in the magnetic field region of interest all energy differences of interfering states are much larger than Γ, except for the two states that cross for $K' = 0$. Under these conditions the only interference term appearing in the correlation function is the one between the states ψ_1 and ψ_2 and any deviations from the K^0 problem must be fully contained in that part of the correlation function which depends on ψ_1 and ψ_2, i.e.

$$W_{12} = \mathscr{F}_{11} + \mathscr{F}_{22} + 2|\mathscr{F}_{12}| \frac{\cos\phi_{12} - \frac{E_1 - E_2}{\Gamma}\sin\phi_{12}}{1 + (\frac{E_1 - E_2}{\Gamma})^2} \quad . \tag{51}$$

We consider first the limiting case

$$|E_1 - E_2| \ll \Gamma \quad . \tag{52}$$

Eq. (51) then takes the form

$$W_{12} = \mathscr{F}_{11} + \mathscr{F}_{22} + 2|\mathscr{F}_{12}|\cos\phi_{12} \tag{53}$$

which can be written as

$$W_{12} = \sum_{a,c,\sigma_1,\sigma_2} |F_1 + F_2|^2 \quad , \tag{54}$$

where F_1 and F_2 are defined by

$$\begin{aligned} F_1 &= \langle c\underline{\Omega}_2\sigma_2|H'|\psi_1\rangle\langle\psi_1\underline{\Omega}_1\sigma_1|H'|a\rangle \\ F_2 &= \langle c\underline{\Omega}_2\sigma_2|H'|\psi_2\rangle\langle\psi_2\underline{\Omega}_1\sigma_1|H'|a\rangle \quad , \end{aligned} \tag{55}$$

and eq. (4) has been used. By inserting eqs. (38) into expressions (55) we find

$$\begin{aligned} F_1 + F_2 = &\langle c\underline{\Omega}_2\sigma_2|H'|\phi_1\rangle\langle\phi_1\underline{\Omega}_1\sigma_1|H'|a\rangle + \\ &+ \langle c\underline{\Omega}_2\sigma_2|H'|\phi_2\rangle\langle\phi_2\underline{\Omega}_1\sigma_1|H'|a\rangle \end{aligned} \tag{56}$$

From this result we conclude that the angular correlation is not affected by the K′ part of the Hamiltonion. In fact, condition (52) implies $|E_1^0 - E_2^0| \ll \Gamma$, and the step leading from eq. (51) to eq. (54) can also be performed for the angular correlation perturbed by the Hamiltonian K^0 alone. Eq. (56) shows that both expressions for W_{12} are identical. We now consider the other limiting case

$$|E_1 - E_2| \gg \Gamma. \tag{57}$$

The interference term in expression (51) vanishes and we have

$$W_{12} = \mathscr{F}_{11} + \mathscr{F}_{22}. \tag{58}$$

The resonance in fig. 2, being the result of the interference term, no longer exists if

$$\Delta \gg \Gamma. \tag{59}$$

There is, however, another mechanism which may produce a resonance in the region of closest approach of mutually repulsive levels. Far away from this region, the direct terms in eq. (58) are not affected by K′ since the states $\psi_{1,2}$ are pure ϕ_1 and ϕ_2 states. As one moves into the region, the states become mixtures of ϕ_1 and ϕ_2. This may cause a resonance - like variation in the direct terms of eq. (58) as a function of the magnetic field. An example of a resonance of this type has been discussed in ref. 11. If we have on the other hand

$$\Delta \ll \Gamma \quad , \tag{60}$$

condition (52) applies and the correlation is not affected by K′ in the vicinity of closest approach. Some distance away, when condition (52) no longer holds, the states are already pure ϕ_1 and ϕ_2 states, and for this reason no effect of K′ is felt. For $\Delta \ll \Gamma$ we therefore have a pure resonance due to 'crossing levels'. In the intermediate regime, when

$$\Delta \sim \Gamma \tag{61}$$

both mechanisms are effective and one may expect a modification of the line shape of the resonance in fig. 2.

Appendix: Interference amplitudes and perturbation factors

In previous theoretical treatments of perturbed angular correlations the correlation function has been written in the form[2,3] †

$$W(t) = e^{-t/\tau} \sum_{k_1,k_2,\mu_1,\mu_2} \frac{1}{\sqrt{(2k_1+1)(2k_2+1)}} A_{k_1}(1) A_{k_2}(2) \times$$

† The normalization of ref. 2 is used.

$$\times\, G_{k_1k_2}^{\mu_1\mu_2}(t)\, Y_{k_1\mu_1}^{*}(\vartheta_1,\phi_1)\, Y_{k_2\mu_2}(\vartheta_2,\phi_2) \quad , \tag{62}$$

where the effect of the extranuclear interaction is described by the perturbation factors $G_{k_1k_2}^{\mu_1\mu_2}$. The connection between the perturbation factors and the interference amplitudes introduced in the present work is apparent by comparing eq. (62) to eq. (5). Inserting expression (11) for $\mathscr{F}_{bb'}$ into eq. (5) leads to the following relation between the perturbation factors and the quantities $\mathscr{S}_{bb'}$ that describe the dependence of the interference amplitudes on the properties of the intermediate states:

$$\frac{1}{\sqrt{(2k_1+1)(2k_2+1)}}\, G_{k_1k_2}^{\mu_1\mu_2\,*}(t) = \sum_{b,b'} \mathscr{S}_{bb'}(k_1,\mu_1)\, \mathscr{S}_{bb'}^{*}(k_2,\mu_2)\, e^{i\frac{E_b-E_{b'}}{\hbar}t} \tag{63}$$

References

1. H.J. Leisi, Habilitationsschrift, Swiss Federal Institute of Technology, Zurich, 1967 (unpublished).
2. H. Frauenfelder and R.M. Steffen, in Alpha-, Beta- and Gamma-Ray Spectroscopy, edited by K. Siegbahn (North - Holland, 1965) Chapt. 14.
3. A. Abragam and R.V. Pound, Phys. Rev. 92 (1953) 943.
4. H.J. Leisi, Th. v. Ledebur and F.N. Gygax, this conference.
5. H.J. Leisi and R.T. Deck, Phys. Rev. 129 (1963) 2117; 131 (1963) 2840.
6. H.J. Leisi, in Hyperfine Structure and Nuclear Radiations, edited by E. Matthias and D.A. Shirley (North - Holland, Amsterdam, 1968), p. 384.
7. K.W.H. Stevens, in Magnetism, Vol. I, edited by G.T. Rado and H. Suhl (Academic Press, New York and London, 1963), Chapt. I.
8. H.A. Kramers, Quantum Mechanics (North Holland, Amsterdam, 1957), p. 384.
9. A.R. Edmonds, Angular Momentum in Quantum Mechanics (Princeton University Press, 1957).
10. H.J. Leisi, Phys. Rev. 1A (1970) 1654.
11. H.J. Leisi, Hyperfine Structure and Nuclear Radiations, edited by E. Matthias and D.A. Shirley (North - Holland, Amsterdam, 1968), p. 397.

Discussion

Thieberger: I would like to mention that we have been thinking along somewhat similar lines. And we just concluded an experiment which we designed in order to illustrate this new point of view. What we did was to look at the time - differential perturbed angular correlation of the 122 - 14.4 keV cascade in ^{57}Fe with a source which was magnetized perpendicular to the plane of the detectors. And then by introducing two resonant Mössbauer absorbers, we selectively attenuated pairs of hyperfine components of the 14.4 keV γ - ray and observed how the amplitude of oscillations on the decay curve was effected by this filtering. Now, of course, if we interpret these oscillations as interference phenomena, then it is clear that it is sufficient to eliminate one of the two components which contribute to this interference in order to reduce or eliminate the oscillations and that is precisely what we observe. Then again, if you eliminate by means of absorption other components which do not contribute to the interference, then the relative amplitude of the oscillations is expected to increase and within the experimental errors we have agreement with these very simple considerations.

Bodenstedt: I would like to mention that we have performed in our group two years ago an experiment on ^{181}Hf (R. Lieder et al., to be published in Zeitschrift für Physik). We have observed the resonance - like behaviour of the anisotropy of 133 - 482 keV angular correlation in ^{181}Ta in a single crystal of metallic hafnium in an external magnetic field. The c - axis was perpendicular to the detector plane and the resonance happens at the level crossing of the $m = +3/2$ and $m = -1/2$ Zeeman components. We wondered at that time about the selection rules which you mentioned in your talk and which we did not understand at that time. I would like to point out that everything can now be understood nicely on basis of your lecture.

C.T. Alonso: I wish to point out that an example of pure quadrupole interaction was reported by me in the Washington meeting of the A.P.S. last spring, and that a theoretical explanation was presented along with experimental confirmation at that time, although the theory was not as beautifully presented as yours was.

ANGULAR CORRELATIONS PERTURBED BY DYNAMIC INTERACTIONS WITH THE ENVIRONMENT

H. Gabriel and J. Bosse
I. Physikalisches Institut der Freien Universität Berlin, Germany

1. Introduction

The present survey will cover the main theoretical contributions to the problem of angular correlations perturbed by non - static extranuclear interactions, which have been published since the classic paper by Abragam and Pound[1]). Although, of course, static interactions cannot be excluded from the discussion, emphasis is placed on the dynamic part of nucleus - environment interactions leading to the well - known damping effects seen in perturbed angular correlation (PAC) experiments. Since the Abragam - Pound treatment of the problem one usually speaks of interactions of the excited nuclei with time - dependent fields. We should, however, keep in mind that in many cases 'interaction with time - dependent field' is nothing but a convenient way of expressing the complex dynamic effect the surroundings have on the nuclear states. The present talk is based on a treatment where the influence of the environment on the nuclei is taken into account by reducing the density operator for the entire system (nucleus and surroundings mutually interacting) to a density operator for the nucleus alone. Use is made of the Liouville representation of quantum mechanical problems described in detail in the appendix, but presented in a more illustrative way in the first parts of this contribution. It will be shown that radiative detection of electronic relaxation effects by PAC or by nuclear orientation methods is a natural and direct way of studying multipole relaxation phenomena.

The outline of the talk is as follows: a brief introduction to the mathematical framework is given in section 2. Fano's state multipole representation is introduced in section 3. The angular correlation function will be defined in terms of state multipoles and certain matrix elements of the evolution operator for the intermediate state. In section 4 we give a sketch of the model leading to the equation of motion for the particular state multipoles of interest (the ones characterizing the intermediate state in a double cascade decay). The last sections are devoted to examples of practical importance.

2. Liouville representation of quantum mechanical problems

We immediately consider the general situation where we do not have maximal information about the physical system gained by a complete measurement. That is to say, we deal with a system which cannot be characterized by a single state vector. In such cases we use instead a density operator which is often defined as an incoherent superposition of pure states:

$$\rho = \sum_i |i\rangle w^{(i)} \langle i| \; ; \; \sum_i w^{(i)} = 1 \tag{1}$$

($w^{(i)}$ are statistical weight factors for the pure state $|i\rangle$), but which might be introduced in a completely different way. The **main point** is that knowledge of the density operator of a system enables us to evaluate the quantum mechanical average of any observable A with respect to the state characterized by the density operator $\rho = \rho^+$:

$$\langle A \rangle = \mathrm{Tr}\,[\rho A] \tag{2}$$

Expectation values with respect to pure states follow from (2) with the special density operator given by $w^{(j)} = 1$, $w^{(i)} = 0$ for $i \neq j$

$$\rho_{\text{pure}} = |j\rangle w^{(j)} \langle j| = |j\rangle\langle j| \quad .$$

Let us introduce a representation for operators and their matrix elements with respect to certain Hilbert space (or unitary vector space) base vectors which is the complete analog to what we are used to in the Dirac representation of quantum mechanics. In the latter case we have complete sets of base vectors (e.g. eigenvectors of a particular hermitian operator) and a general state vector $|\psi\rangle$ can be expanded as

$$|\psi\rangle = \sum_i |i\rangle\langle i|\psi\rangle , \quad 1 = \sum_i |i\rangle\langle i| \quad (= \text{unit operator}).$$

Assume a unitary vector space $\mathscr{U}$ of finite dimension N and consider the set of linear operators A, B, C, . . . which transform state vectors into different state vectors of the same space: $|\phi\rangle = A\,|\psi\rangle$. Matrix elements of A with respect to $|i\rangle$ are usually arranged in a quadratic array. There exists, however, a different way of looking at quantum mechanical operators or their matrix elements. Instead of using a matrix scheme we rewrite the matrix elements as a column or row vector, e.g.

$$[\langle 1|A|1\rangle, \langle 1|A|2\rangle, \ \ldots \quad \langle 2|A|1\rangle, \langle 2|A|2\rangle, \ldots] .$$

We interpret this (in slightly modefied notation) as the decomposition of a vector $|A)$ in a new Hilbert space of higher dimension with respect to an orthonormal set of base vectors $|ij)$ related to the old state vectors by

$$|ij) \stackrel{\text{def}}{=} |\,|i\rangle\langle j|\,) . \tag{3}$$

We see that the basis in our new space, which has been called Liouville space ($\mathscr{L}$) by Fano[2]) is formed from all operators $|i\rangle\langle j|$ which one can combine from Dirac bras and kets. The correspondence between the two notations

$$(ij\,|\,A) = \langle i|A|j\rangle \tag{4}$$

is given in a unique way by defining a metric in our Liouville space by

$$(A \mid B) = \mathrm{Tr}\,[A^{+}B]. \tag{5}$$

Eq. (2) for the expectation value of a quantum mechanical operator can be rewritten as $\langle A \rangle = (\rho \mid A)$. Now we are ready to define **superoperators** in the same pictorial way as we did for operators in the ordinary unitary vector space, namely by $\mid Y) = \mathscr{R} \mid X)$, i.e. by the action of the superoperator $\mathscr{R}$. Also, we are now able to understand the role of a special set of super base vectors in $\mathscr{L}$, which merely means introduction of a special coordinate system in Liouville space. If we have a complete set of basis superkets the closure relation just defines the unit superoperator

$$1 = \sum_{ij} \mid ij)\,(ij \mid . \tag{6}$$

All relations known for quantum mechanics in Dirac's notation are literally valid also in Liouville space. A special type of superoperator is the so - called Liouville operator defined by

$$\frac{d\rho}{dt} = -i\,[H,\rho] \overset{\mathrm{def}}{=} -iL\rho$$

$$\mid \rho\,(t)) = e^{-iLt} \mid \rho\,(t=0)). \tag{7}$$

Eq. (7) represents a unitary transformation (rotation) of $\mid \rho\,(t=0))$ in $\mathscr{L}$.

3. Expansion in multipole operators

For our purpose it is sufficient to restrict ourselves to the special case where the matrix elements of the density operator ρ with respect to angular momentum eigenstates and additional quantum numbers, indicated by a, are diagonal in (I, a). The state of orientation is either represented by the matrix elements $\langle aIm \mid \rho \mid aIm' \rangle \equiv$ $\equiv \langle Im \mid \rho \mid Im' \rangle$ or, according to Fano and Racah[3]), by a linear substitution with new quantities $\rho q^{(k)}$ that transform under rotation like the complex conjugates of spherical harmonics Y_{kq}^{*}. In Liouville representation this is equivalent to expanding $\mid \rho)$ into the complete orthonormal set of irreducible tensor operators introduced in appendix A:

$$(U_q^{(k)} \mid U_{q'}^{(k')}) = \delta_{kk'}\,\delta_{qq'} \qquad k \text{ integral with } k_{max} = 2I,\ -k \leqslant q \leqslant k \tag{8a}$$

$$\sum_{kq} \mid U_q^{(k)})\,(U_q^{(k)} \mid = 1\,. \tag{8b}$$

Including the normalized unit operator $U_0^{(0)} = (2I + 1)^{-1/2}\, 1$, we have $N^2 = (2I + 1)^2$ different operators $U_q^{(k)}$ in $\mathcal{U}$. The expansion for $|\rho)$ reads

$$|\rho) = \sum_{kq} |U_q^{(k)})(U_q^{(k)}|\rho) = \sum_{kq} |U_q^{(k)})\rho_q^{(k)} \tag{9}$$

where $\rho_q^{(k)}$ are Fano's **statistical tensors** or **state** multipoles ($= R_q^{(k)}$ in the notation of ref. 4). In terms of $<Im|\rho|Im'> = (mm'|\rho)$ the statistical tensors are given by

$$\rho_q^{(k)} = \sum_{m,m'=-I}^{+I} (U_q^{(k)}|m\,m')(m\,m'|\rho) =$$

$$= \sum_{m,m'} (-1)^{I-m} (2k+1)^{1/2} \begin{pmatrix} I & k & I \\ -m & q & m' \end{pmatrix} <Im|\rho|Im'>. \tag{10}$$

Each $\rho_q^{(k)}$ is proportional to the complex conjugate of the corresponding expectation value of the nuclear 2^k - pole moment component $<M_q^{(k)}>^*$. This follows from

$$|M_q^{(k)}) = M_k\, |U_q^{(k)}) = \frac{<I||M^{(k)}||I>}{\sqrt{2k+1}}\, |U_q^{(k)}) \tag{11}$$

(which is valid for any irreducible tensor operator) with the help of the Wigner - Eckart theorem. The coefficient M_k entering

$$<M_q^{(k)}> = M_k\, (\rho_q^{(k)})^* \tag{12}$$

is itself proportional to the multipole moment in the usual definition.

The Hamiltonian of the nucleus is written as sum of terms representing multipole interactions of various orders:

$$H = \sum_{kq} (-1)^q M_q^{(k)} A_{-q}^{(k)} \tag{13}$$

with spherical 'field' operators $A_q^{(k)}$ constructed with space derivatives of order (k-1) of the electric or magnetic field. The 2^k - pole moment operator $M_q^{(k)}$ has already been introduced above.

The physical significance of the statistical tensors $\rho_q^{(k)}$ is clear from (12), whereby the state of orientation of the nucleus is directly related to the mean value of the various multipole moment operators.

4. Description of relaxation effects

Most theories describe relaxation effects in terms of time - dependent fluctuating magnetic fields, electric field gradients or their quantum analogs. The system is then characterized by a Hamiltonian which is the sum of H_0 and H_1 (t), describing, respectively, static and stochastic interactions of the system of interest with perturbative 'fields' or 'field operators'.

A different approach, reported here, is based on a model which guarantees from the outset that results known from magnetic relaxation theories (used e.g. in NMR) can safely be compared with results found from Mössbauer- and PAC-work. The model is based on the decomposition of the entire system into a small subsystem (the system of interest, e.g. nucleus or atom) and surroundings, mutually interacting.

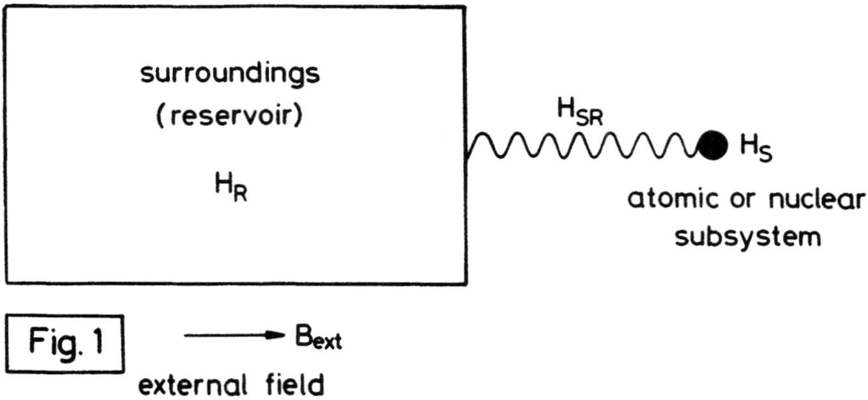

Fig. 1

The Hamiltonian is the sum of three terms

$$H = H_S + H_R + H_{SR} \tag{14}$$

where H_S describes the atomic or nuclear subsystem in an external magnetic field and includes static quadrupole interactions if there are any. H_R is a complicated Hamiltonian which would specify all the properties of the surroundings. The most important part is the interaction Hamiltonian H_{SR} that needs no specification here except that we assume it to be of the form

$$H_{SR} = \sum_{kq} (-1)^q \, T_q^{(k)} (S) \, V_{-q}^{(k)} (R) \, . \tag{15}$$

An important point for a well - functioning theory lies in the separation of the subsystem from the surroundings. In some cases it is reasonable to consider the nucleus, characterized by the sequence of levels and states of interest, as the subsystem and attribute the electronic shells to the surroundings. This approximation does not, however, imply that the influence of the environment occurs via classical fluctuating fields, although the latter case is contained in our treatment. Dynamic interaction effects are expressed, in general, in terms of correlation functions for electronic effective spins, total shell angular momenta, etc. In more general situations a correct description of the static and dynamic interactions can only be obtained by treating the combined electron - nucleus system quantum mechanically. But, except for the difference in the 'size' of the subsystem S the underlying ideas of handling the relaxation phenomena are exactly the same. We assume that the isotopes are dilutely dissolved in the host, so that their direct interaction can be ignored. The surroundings are supposed to act upon the atoms as a heat bath.

We start out with a density operator W (S, R, t) for the entire system and try to get rid of the unobserved reservoir variables. The elimination procedure is described in detail in ref. 5. The generalization for time - dependent Hamiltonians is of interest in itself, and therefore, given in appendix B. The density operator for the system of interest ρ (S, t) (either the nucleus or the atom, symbolically written as $S \hat{=} n$ or a) follows from the non - Markoffian integrodifferential equation

$$\frac{d}{dt} |\rho(t)) = -iL' |\rho(t)) - \int_0^t d\tau\, \mathcal{M}(\tau) |\rho(t-\tau)) \tag{16}$$

which has been derived under the assumption that at t = 0 the combined S - R system was uncorrelated, i.e. W (S, R; 0) = ρ_T (R) ρ (S). The first factor ρ_T (R) is the equilibrium density operator for the reservoir. The first term on the right - hand side gives the static interactions where

$$L' = L_S + \langle L_{SR} \rangle_R \; ; \; \langle L_{SR} \rangle_R = \mathrm{Tr}_R \, [L_{SR}\, \rho_T(R)] \, . \tag{17}$$

The ensemble average of L_{SR}, the Liouville operator belonging to the interaction Hamiltonian H_{SR}, is added to the Liouville operator L_S for the subsystem. The influence of the heat bath on the subsystem is condensed in the relaxation (super) operator

$$\mathcal{M}(t) = \mathrm{Tr}_R \, \{\Delta L_{RS}\, e^{-it(L'+L_R)}\, \Delta L_{RS}\, \rho_T(R)\}; \; \Delta L_{SR} = L_{SR} - \langle L_{SR} \rangle_R \, . \tag{18}$$

The dynamic effects of the interactions between the system of interest and the environment are, to lowest order in L_{SR}, completely given by (18) which is still a superoperator with respect to S - variables since the trace is over the reservoir variables, only.

For times $t \gg \tau_c$ (this requires $\tau_n \gg \tau_c$ for the excited nuclei), the generalized master equation (16) can be approximated by a Markoffian equation of motion

$$\frac{d}{dt} |\rho(t)) = -(iL' + \mathcal{R}) |\rho(t)) \tag{19}$$

where the now time - independent averaged relaxation operator is defined by

$$\int_0^\infty \mathscr{M}(t)\, e^{itL'}\, dt = \mathscr{R} + i\tilde{\mathscr{R}} \; . \tag{20}$$

The second - order energy renormalization is neglected in (19). The nuclear lifetime of the excited (intermediate) state and the largest correlation time of the dissipative system are τ_n and τ_c, respectively. For times comparable to τ_c, i.e. $\tau_n \approx \tau_c$, the exact equation of motion (16) has to be used. We return to this case later and deal with the Markoffian theory first.

To be specific let us consider the n - model which, in the case of PAC, means that the subsystem is the nucleus in the intermediate state characterized by spin I. Interpreting the density operator ρ as Liouville ket $|\rho)$ we immediately end up with the equation of motion for the statistical tensors by 'multiplying by $(U_q^{(k)}|$ from the left' and by using (8b)

$$\frac{d(U_q^{(k)}|\rho)}{dt} = -\sum_{k'q'} (U_q^{(k)}|iL' + \mathscr{R}\,|U_{q'}^{(k')})\,(U_{q'}^{(k')}|\rho)$$

$$\frac{d\rho_q^{(k)}(t)}{dt} = -\sum_{k'q'} (U_q^{(k)}|iL' + \mathscr{R}\,|U_{q'}^{(k')})\,\rho_{q'}^{(k')} \; . \tag{21}$$

Formal integration yields (for $t > 0$)

$$\rho_q^{(k)}(t) = \sum_{k'q'} (U_q^{(k)}|e^{-(iL'+\mathscr{R})t}|U_{q'}^{(k')})\rho_{q'}^{(k')}(0) = \sum_{k'q'} G_{kk'}^{qq'}(t)\,\rho_{q'}^{(k')}(0) \; . \tag{22}$$

It has been shown in ref. 5 that the Liouville matrix elements in multipole representation are nothing but the usual attenuation factors $G_{kk'}^{qq'}(t)$ well known from PAC theory. The intimate relation between multipole relaxation and angular correlation perturbed by dynamic interaction with the environment is now obvious.

The general expression for the directional angular correlation function of a double cascade decay can be written as

$$W(\underline{k}_1\underline{k}_2;t) = \mathrm{Tr}\{\rho(\underline{k}_2)\rho(\underline{k}_1;t)\} = (\rho(\underline{k}_2)|\Omega(t)|\rho(\underline{k}_1)) =$$

$$= \sum_{kqk'q'} (\rho(\underline{k}_2)|U_q^{(k)})\,(U_q^{(k)}|\Omega(t)|U_{q'}^{(k')})\,(U_{q'}^{(k')}|\rho(\underline{k}_1)) =$$

$$= \sum_{kqk'q'} \rho_q^{(k)*}(\underline{k}_2)\, G_{kk'}^{qq'}(t)\, \rho_{q'}^{(k')}(\underline{k}_1) \tag{23}$$

where $\rho\,(\underline{k}_2)$ and $\rho\,(\underline{k}_1)$ describe e.g. the nuclear γ - decays (wave vectors $\underline{k}_1$ and $\underline{k}_2$) without any perturbation other than interaction with the radiation field causing the radiative transition. In Markoffian approximation the evolution operator $\Omega\,(t)$ is seen from comparison with (22) to be

$$\Omega\,(t) = e^{-(iL' + \mathscr{R})t} \qquad \text{or}$$

$$(U_q^{(k)}|\Omega(t)|U_{q'}^{(k')}) = (U_q^{(k)}|e^{-(iL'+\mathscr{R})t}|U_{q'}^{(k')}) = G_{kk'}^{qq'}(t)\ . \tag{24}$$

The reality condition

$$G_{kk'}^{qq'}(t)^* = (-1)^{q+q'}\, G_{kk'}^{-q-q'}(t) \tag{25}$$

implies that the following relations for the real and imangiary parts hold:

$$\mathrm{Re}\, G_{kk'}^{qq'}(t) = (-1)^{q+q'}\ \mathrm{Re}\, G_{kk'}^{-q-q'}(t) \tag{26}$$

$$\mathrm{Im}\, G_{kk'}^{qq'}(t) = (-1)^{q+q'+1}\ \mathrm{Im}\, G_{kk'}^{-q-q'}(t)\ . \tag{27}$$

The behaviour of $G_{kk'}^{qq'}$ depends, of course, on the properties of Ω and, consequently, on those of $(iL' + \mathscr{R})$. Define the unitary superoperator $\mathscr{D}X = DXD^+$ (X an arbitrary operator of $\mathscr{U}$) associated with the rotation D in the vector space $\mathscr{U}$. In addition to the transformation law of an operator A, which can be rewritten as $A' = DAD^+ = \mathscr{D}A$, we get for the transformed superoperator after the rotation

$$\mathscr{R}' = \mathscr{D}\, \mathscr{R}\, \mathscr{D}^+\ . \tag{28}$$

It follows from the requirement $(A'|\mathscr{R}'|B') = (A|\mathscr{R}|B)$ that this is in fact the general transformation law of a superoperator under an arbitrary unitary transformation $\mathscr{O}$ ($\mathscr{O}\mathscr{O}^+ = \mathscr{O}^+\mathscr{O} = 1$); unitary with respect to the spur metric (5) of $\mathscr{L}$). The matrix elements of $\mathscr{D}$ with respect to the multipole representation $|U_q^{(k)})$ are just the rotation matrices:

$$(U_q^{(k)}|\mathscr{D}|U_{q'}^{(k')}) = (U_q^{(k)}|DU_{q'}^{(k')}D^+) = \sum_{\bar{q}} (U_q^{(k)}|U_{\bar{q}}^{(k')})D_{\bar{q}q'}^{(k')} =$$

$$= D_{qq'}^{(k)} \cdot \delta_{kk'}\ . \tag{29}$$

It is obvious from (29) and (24) that $G_{kk'}^{qq'}$ transforms under rotation (α, β, γ) as

$$(U_q^{(k)}|\Omega'|U_{q'}^{(k')}) = \sum_{q_1 q_2} D_{qq_1}^{(k)}(\alpha,\beta,\gamma)\, D_{q'q_2}^{(k')*}(\alpha,\beta,\gamma)\, (U_{q_1}^{(k)}|\Omega|U_{q_2}^{(k')}). \tag{30}$$

In the important case in which the physical system is **invariant** under a certain transformation $\mathcal{O}$, eq. (28) is modified to $\mathcal{R}' = \mathcal{O}\mathcal{R}\mathcal{O}^{+} = \mathcal{R}$. One of the simplest examples is that of axial symmetry C_∞. The invariance condition for rotation about the symmetry axis applied to (30) leads to the well - known restriction

$$G_{kk'}^{qq'} = \delta_{qq'}\, G_{kk'}^{qq} \quad (\text{for } C_\infty)\ . \tag{31}$$

It follows from the unitarity of $\mathcal{O}$ and series expansion of (24) that in general

$$\mathcal{O}\Omega\mathcal{O}^{+} = \exp\,[(-i\mathcal{O}L'\mathcal{O}^{+} - \mathcal{O}\mathcal{R}\mathcal{O}^{+})\, t] \tag{32}$$

holds. Therefore the symmetry of $G_{kk'}^{qq'}$ is determined by the lowest symmetry of the operators in the exponent.

If the ensemble average over the various orientations of the microcrystals of a polycrystalline source can be expressed as an average over angles (which is almost always tacitly assumed), we find for a polycrystal without a preferential axis, by averaging (30) over all Eulerian angles,

$$\overline{G_{kk'}^{qq'}}^{\,\alpha\beta\gamma} = \delta_{qq'}\delta_{k'k}\, G_{kk} = \delta_{q'q}\,\delta_{kk'}\,(2k+1)^{-1} \sum_{q_1} G_{kk}^{q_1 q_1} \tag{33}$$

which has been proven for several special cases (see ref. 4 FS 231 and FS 234). In polycrystalline sources with a preferential direction (as in the presence of an external magnetic field) the only general restriction is again (31), as in single crystals.

There are other relations for the perturbation factors which follow directly from the multipole expansion in (24). The relation

$$G_{ok}^{oq} = 0 \quad (k \neq 0) \tag{34}$$

is generally true and not restricted to static extranuclear interactions[6]). Eq. (34) expresses the trivial fact that the operator $U_0^{(0)}$ commutes with every operator, so that $(U_0^{(0)}|(iL' + \mathcal{R}) = 0$. However, the relation

$$G_{k0}^{q0} = 0 \quad (k \neq 0) \tag{35}$$

is **not** generally valid, if relaxation processes must be considered.

5. Dynamic effects on PAC in the Markoffian limit, examples

It has been shown in ref. 5 how typical results of the randomly fluctuating field models can be found within the present approach. The case of static magnetic fields and magnetically - caused relaxation processes in the high - temperature limit $\hbar\omega_L \ll k_B T$ is of considerable importance; therefore the main results will be collected here. It follows that the evolution operator is then diagonal in the multipole representation which immediately leads to an explicit expression for the perturbation factors

$$G_{kk}^{qq}(t) = e^{-itq\omega_L - \lambda_{kq} t} \qquad (t > 0) \tag{36}$$

where ω_L is the Larmor frequency in the static magnetic field and

$$\lambda_{kq} = (U_q^{(k)} | \mathscr{R} | U_q^{(k)}) = k(k+1)/2T_1 + q^2 (1/T_2 - 1/T_1) \, . \tag{37}$$

The longitudinal and transverse spin - lattice relaxation times T_1 and T_2, respectively, have been defined in exactly the same way as is usually done in magnetic relaxation theory. In the simplest case of an isotropic hyperfine interaction

$$H_{hf} = -\gamma A \, \underline{I} \cdot \underline{J} \tag{38}$$

($\underline{J}$ the electronic angular momentum operator) the relaxation times are expressed in terms of electronic correlation functions with $\Delta\underline{J} = \underline{J} - \langle \underline{J} \rangle_R$ as

$$\frac{1}{T_1} = \gamma^2 A^2 \frac{1}{2} \int_{-\infty}^{+\infty} dt e^{-i\omega t} \langle \tfrac{1}{2} [\Delta J_+(t), \Delta J_-(0)]_+ \rangle_R \tag{39a}$$

$$\frac{1}{T_2} - \frac{1}{2T_1} = \gamma^2 A^2 \frac{1}{2} \int_{-\infty}^{+\infty} dt e^{i\omega t} \langle \tfrac{1}{2} [\Delta J_z(t), \Delta J_z(0)]_+ \rangle_R \, . \tag{39b}$$

The main result is that the damping constants are factored into a geometrical part, depending only on the multipole order k and the multipole orientation q, and the usual relaxation times. For the particular relaxation mechanism under consideration, we thus have a definite answer to the question: how do the damping constants measured in a PAC or NO experiment compare to relaxation measurements via NMR or spin echo? The result (36), (37) differs from the Abragam - Pound solution whenever the term proportional to q^2 is different from zero (i.e. if $T_1 \neq T_2$). The Abragam - Pound method has recently been extended (see ref. 7) yielding, of course, the same result as ours[5]).

For the case of axial symmetry (and $k_{max} = 2$) we explicitly give some angular correlation functions of practical importance.

1) magnetic field perpendicular to the counter plane; θ = angle between the two counters

$$W_\perp(\pi;t) = e^{-t/\tau_n}\{1+\tfrac{1}{4}A_{22}e^{-3t/T_1}[1+3e^{-4t(1/T_2-1/T_1)}\times\cos 2\omega_L t]\} \quad (40a)$$

$$W_\perp(\pi/2;t) = e^{-t/\tau_n}\{1+\tfrac{1}{4}A_{22}e^{-3t/T_1}[1-3e^{-4t(1/T_2-1/T_1)}\times\cos 2\omega_L t]\} \quad (40b)$$

$$W_\perp(3\pi/2;t) = e^{-t/\tau_n}\{1+\tfrac{1}{4}A_{22}e^{-3t/T_1}[1-3e^{-4t(1/T_2-1/T_1)}\times\sin 2\omega_L t]\} \quad (40c)$$

2) magnetic field parallel to one of the counters; the latter at 90° with respect to the second counter

$$W_{//}(\pi/2;t) = e^{-t/\tau_n}\{1-\tfrac{1}{2}A_{22}e^{-3t/T_1}\}\ . \quad (41)$$

A combination of (40) and (41) should under favourable conditions lead to a simultaneous determination of T_1 and T_2 from PAC.

We have also looked into the low - temperature problem in which case the evolution operator Ω is no longer diagonal with respect to the $|U_q^{(k)})$ basis. Furthermore, the result for $G_{kk}^{q\dot{q}}$ is no longer independent of the nuclear spin. The final formulae are lengthy analytic expressions, except for spin I = 1, and it is more reasonable to resort to numerical analysis. (This has been done for the NO case in cooperation with J. Barclay[8])). We collect the result of ref. 5 for I = 1 and use it in the expression for the angular distribution (axial symmetry assumed):

$$W(\theta;t) = \sqrt{2I+1}\sum_{k,k'} U_k F_k P_k(\cos\theta)\, G_{kk'}^{00}(t)\, \Delta\rho_0^{(k')}(t=0), \quad (42)$$

in standard notation. The coefficients $U_k F_k$ depend on the decay scheme of the particular nucleus and P_k (cos θ) are the Legendre polynomials. The statistical tensors $\Delta\rho_0^{(k)}$ (t = 0) characterize the deviation of the initial state of orientation from the one corresponding to thermal equilibrium. Under the same condition as in the beginning of the present section the attenuation coefficients read (k_{max} = 2):

$$G_{11}^{00}(t) = \tfrac{1}{2}\{1-\cosh(\beta\omega_L/2)\}e^{-a_+t} + \tfrac{1}{2}\{1+\cosh(\beta\omega_L/2)\}e^{-a_-t} \quad (43a)$$

$$G_{12}^{00}(t) = (1/2\sqrt{3})\sinh(\beta\omega_L/2)\{e^{-a_-t} - e^{-a_+t}\} \quad (43b)$$

$$G_{21}^{00}(t) = (\sqrt{3}/2)\sinh(\beta\omega_L/2)\{e^{-a_+t} - e^{-a_-t}\} \quad (43c)$$

$$G_{22}^{00}(t) = \frac{1}{2}\{1+\cosh(\beta\omega_L/2)\}e^{-a_+t} + \frac{1}{2}\{1-\cosh(\beta\omega_L/2)\}e^{-a_-t} . \quad (43d)$$

The damping constants contain an extra temperature dependence (as an outgrowth of the fluctuation - dissipation theorem)

$$a_\pm = \frac{1}{T_1}\{2 \pm 1/\cosh(\beta\omega_L/2)\}; (\beta = k_B T; \hbar = 1) . \quad (44)$$

Notice that the temperature dependent cofactors either disappear (G_{12}^{00} and G_{21}^{00}) or approach unity (G_{11}^{00} and G_{22}^{00}) in the high - temperature limit. At extremely low temperatures $a_+ \approx a_-$ and the solution then becomes

$$G_{11}^{00}(t) = (1 + \frac{t}{T_1})\, e^{-2t/T_1} \quad (45a)$$

$$G_{12}^{00}(t) = \frac{1}{\sqrt{3}} \cdot \frac{t}{T_1}\, e^{-2t/T_1} \quad (45b)$$

$$G_{21}^{00}(t) = -\sqrt{3}\frac{t}{T_1}\, e^{-2t/T_1} \quad (45c)$$

$$G_{22}^{00}(t) = (1 - \frac{t}{T_1})\, e^{-2t/T_1} . \quad (45d)$$

Axially Symmetric Static Interaction and Isotropic Relaxation Operator

Recently Andrade et al.[9]) have applied the Redfield theory of nuclear relaxation to PAC. Generalizations of their results are easily obtained within our formalism: The evaluation of the perturbation factors (24) is considerably simplified if we approximate the relaxation operator by:

$$\mathcal{R}' = \sum_{m,\bar{m}} |mm)(mm|\mathcal{R}|\bar{m}\bar{m})(\bar{m}\bar{m}| + $$
$$+ \sum_m \sum_{m' \neq m} |mm')(mm'|\mathcal{R}|mm')(mm'| \approx \mathcal{R}$$

where the $[\,|mm')\,]$ are eigenvectors of the static perturbation L': $L'|mm') = \omega_{mm'}|mm')$. In many practical applications $\mathcal{R}'$ is a good approximation for $\mathcal{R}$ (see ref. 18, 17 and 9). $\mathcal{R}'$ commutes with L', and we may write:

$$G_{kk'}^{qq'}(t) = (U_q^{(k)}|e^{-itL'}e^{-\mathcal{R}'t}|U_{q'}^{(k')}) = \sum_{mm'} (U_q^{(k)}|mm')\, e^{-it\omega_{mm'}}(mm'|e^{-\mathcal{R}'t}|U_{q'}^{(k')}) .$$

If the static perturbation is axially symmetric (e.g. an axially symmetric field gradient or a constant magnetic field $\underline{B}$), we may interpret the eigenvectors of L' as angular momentum eigenvectors: $||Im><Im'|)$. The coefficients $(mm'|U_q^{(k)})$ are then given by (A39a) and we have:

$$\omega_{mm'} = \begin{cases} -g\mu_N B(m-m') & \text{for magn. field } \underline{B} \text{ (axis of quant. along } \underline{B}) \\ 3\omega_Q (m^2 - m'^2) & \text{for axially symm. fieldgradient (axis of quant. along symm. axis; } \omega_Q \text{ quadrupole frequency).} \end{cases}$$

If, moreover, the relaxation operator $\mathscr{R}'$ is assumed isotropic, i.e. $\mathscr{R}' = \mathscr{D}\mathscr{R}'\mathscr{D}^+$ the perturbation factors are easily evaluated, since isotropic operators are diagonal in the multipole basis:

$$(U_q^{(k)}|\mathscr{R}'|U_{q'}^{(k')}) = \delta_{qq'}\delta_{kk'}\frac{1}{2k+1}\sum_{\bar{q}}(U_{\bar{q}}^{(k)}|\mathscr{R}'|U_{\bar{q}}^{(k)}) \equiv \delta_{qq'}\delta_{kk'}\lambda_k \quad .$$

Hence, for axially symmetric static perturbations and isotropic relaxation we have:

$$G_{kk'}^{qq'}(t) = \sum_{mm'}(U_q^{(k)}|mm')\,e^{-it\omega_{mm'}}(mm'|U_{q'}^{(k')})\,e^{-\lambda_{k'}t}, \text{ inserting (A39a):}$$

$$\text{(I)} \quad G_{kk'}^{qq'}(t) = \delta_{qq'}\sqrt{(2k+1)(2k'+1)}\,e^{-\lambda_{k'}t}\sum_{mm'}\begin{pmatrix} I & k & I \\ -m & q & m' \end{pmatrix}\begin{pmatrix} I & k' & I \\ -m & q & m' \end{pmatrix}e^{-it\omega_{mm'}}$$

and for the attenuation factors in the case of polycrystal sources:

$$\text{(II)} \quad G_{kk}(t) = \frac{1}{2k+1}\sum_q G_{kk}^{qq}(t) = e^{-\lambda_k t}\sum_q\sum_{mm'}\begin{pmatrix} I & k & I \\ -m & q & m' \end{pmatrix}^2 e^{-it\omega_{mm'}} \quad .$$

If $\mathscr{R}'$ is anisotropic, we must diagonalize $\mathscr{R}'$ in general. However, note the following properties of $\mathscr{R}'$:

$$(mm|e^{-\mathscr{R}'t} = \sum_{\bar{m}}(mm|e^{-\mathscr{R}'t}|\bar{m}\bar{m})(\bar{m}\bar{m}| \quad ,$$

and for $m \neq m'$:

$$(mm'|e^{-\mathcal{R}'t} = e^{-\mathcal{R}_{mm'}t}(mm'| \quad \text{with } \mathcal{R}_{mm'} = (mm'|\mathcal{R}|mm') \quad .$$

With the help of these relations we get:

$$\text{(III)} \quad G^{qq'}_{kk'}(t) = \delta_{qq'}\{\delta_{q,0}(U_0^{(k)}|e^{-\mathcal{R}'t}|U_0^{(k')}) +$$

$$+ (1-\delta_{q,0})\sqrt{(2k+1)(2k'+1)} \sum_{mm'} \begin{pmatrix} I & k & I \\ -m & q & m' \end{pmatrix} \begin{pmatrix} I & k' & I \\ -m & q & m' \end{pmatrix} e^{-it\omega_{mm'} - \mathcal{R}_{mm'}t}\}$$

which is the general form of the perturbation factors for anisotropic (as well as isotropic, of course) relaxation $\mathcal{R}'$ and axially symmetric L'. (The sums over m' are redundant in (I), (II), (III). The results of [9] are contained in (I), (II) and (III).
Let us mention two special cases only:
(a) no static interaction ($L' = 0$) and isotropic relaxation ($\mathcal{R}' = \mathcal{D}\,\mathcal{R}'\,\mathcal{D}^+$); from (I) (II) we have:

$$G^{qq'}_{kk'}(t) = \delta_{qq'}\delta_{kk'}e^{-\lambda_k t}; \quad G_{kk}(t) = e^{-\lambda_k t}$$

(b) static axially symmetric field gradient and isotropic relaxation, from (II) we have

$$G_{kk}(t) = e^{-\lambda_k t} \sum_{mm'q} \begin{pmatrix} I & k & I \\ -m & q & m' \end{pmatrix}^2 e^{-3i\omega_Q(m^2-m'^2)t} =$$

$$= \frac{e^{-\lambda_k t}}{2k+1} + \sum_{m \neq m'} \begin{pmatrix} I & k & I \\ -m & q & m' \end{pmatrix}^2 e^{-3i\omega_Q(m^2-m'^2)t - \mathcal{R}_{mm'}t}; \quad (q=m-m') \quad .$$

The latter expressions are identical to eqs. (3.3), (3.4) of ref. 9, however, in our derivation the assumptions are less restrictive.

6. Time - integral measurements in the non - Markoffian case ($\tau_n \approx \tau_c$)

The existence of a natural time window of the order of the nuclear lifetime in PAC experiments leads, under certain conditions, to results with respect to dynamic extranuclear effects which cannot reasonably be compared with those gained from ground state experiments, like NMR. Whereas, in the latter case, a long - time limit $t \gg \tau_c$ can always be defined, this is not true for excited state experiments if $\tau_n \approx \tau_c$. Stochastic theories of line shape (of the Markoffian type) applied to PAC or Mössbauer relaxation[10]) do not suffer from restrictions of the ratio of nuclear lifetime and reciprocal jump frequency. The approach to relaxation phenomena is based on a model completely different from the present one. Blume[11]) has given a generalized version of Markoffian modulation theory using a very general form for the Hamiltonian of a system whose surroundings fluctuate in time between a finite number of values. Within our approach we have to resort to the non - Markoffian equation of motion (16), whenever $\tau_n \approx \tau_c$. The memory term appearing there means essentially that the nuclear density operator is influenced not only by an instantaneously acting time averaged relaxation operator as in (19) but by the entire continuous sequence of dynamic interactions within a given time interval. (We also have to account for the time evolution due to static interaction, if present). We would like to report on some recent work wherein the non - Markoffian equation of motion has been applied to time - integral PAC. In order to get an explicit expression for the relaxation operator (t) the following assumptions have been made: (1) negligible quadrupole interaction, (2) magnetic interactions with cylindrical symmetry, (3) high - temperature approximation, $\hbar\omega_L \ll k_B T$, (4) exponentially damped correlation functions. The further assumption that spin - lattice relaxation is due to isotropic magnetic interactions has been used both in ref. 12 and ref. 13, but could be removed easily. The theoretical work contained in ref. 13 was motivated by experiments on ^{169}Tm in aligned ferromagnetic iron down to 4° K. Details of the analysis will be reported elsewhere. The experimental quantities (for the axially symmetric case)

$$A_\perp \equiv \frac{W_\perp(\pi) - W_\perp(\pi/2)}{W_\perp(\pi/2)} \tag{46}$$

(a better choice being $B_\perp \equiv \dfrac{W_\perp(\pi) - W_\perp(\pi/2)}{(1/2)[W_\perp(\pi/2) + W_\perp(\pi)]}$

$$R \equiv \frac{W_\perp(B) - W_\perp(-B)}{(1/2)\,[W_\perp(B) + W_\perp(-B)]} \tag{47}$$

can be expressed in terms of the real and imaginary part of

$$\overline{G_{kk}^{qq}} = (1/\tau_n) \int_0^\infty e^{-t/\tau_n}\, G_{kk}^{qq}(t)\,dt = \mathrm{Re}\,\overline{G_{kk}^{qq}} + i\,\mathrm{Im}\,\overline{G_{kk}^{qq}} \tag{48}$$

and for $k_{max} = 2$ by

$$A_\perp = \frac{3}{2} A_{22} \frac{\mathrm{Re}\,\overline{G_{22}^{22}}}{1+(1/4)A_{22}\{\overline{G_{22}^{00}} - 3\,\mathrm{Re}\,\overline{G_{22}^{22}}\}} \tag{49}$$

$$\left.\begin{aligned} B_\perp &= \frac{3}{2} A_{22} \frac{\mathrm{Re}\,\overline{G_{22}^{22}}}{1+(1/4)A_{22}\,\overline{G_{22}^{00}}} \qquad (50) \\ R &= \frac{3}{2} A_{22} \frac{\mathrm{Im}\,\overline{G_{22}^{22}}}{1+(1/4)A_{22}\,\overline{G_{22}^{00}}} \qquad (51) \end{aligned}\right\} \quad \frac{B_\perp}{R} = \frac{\mathrm{Re}\,\overline{G_{22}^{22}}}{\mathrm{Im}\,\overline{G_{22}^{22}}}$$

A_{22} is the anisotropy parameter.

The behaviour of G_{kk}^{qq} depends on the properties of the extranuclear interactions. Without going into the details here, we merely present the theoretical results based on assumptions (1) to (4) and assuming isotropic hyperfine interaction, e.g. $H_{hf} = -\gamma \underline{I} \cdot \underline{B}_{hf}$, where $\underline{B}_{hf}$ is the effective hyperfine field operator. The final results for the real and imaginary parts of G_{kk}^{qq} are given in terms of τ_c/τ_n, the ratio of electronic correlation time to the nuclear lifetime, and the parameter $P = \omega_L \tau_n$ (ω_L = Larmor frequency) by

$$\mathrm{Re}\,\overline{G_{kk}^{qq}} = \frac{1+v_q}{(1+v_q)^2 + P^2 q^2 (1-w_q)^2} \qquad (52)$$

$$\mathrm{Im}\,\overline{G_{kk}^{qq}} = -P \frac{q(1-w_q)}{(1+v_q)^2 + P^2 q^2 (1-w_q)^2} \qquad (53)$$

$$v_q = K_0 \frac{q^2 x(1+x)}{(1+x)^2 + q^2 P^2 x^2} \qquad (54)$$

$$w_q = K_0 \frac{q^2 x^2}{(1+x)^2 + q^2 P^2 x^2} \qquad (55)$$

$$K_0 = \gamma^2 \{ < B_{hf,z}^2 > - < B_{hf,z} >^2 \} \tau_n^2 = < (\Delta\omega_L)^2 > \tau_n^2 \,. \qquad (56)$$

Without any effects due to fluctuating interaction ($v_q = w_q = 0$) eqs. (52) and (53) are, respectively, Lorentzians and the derivative of Lorentzians as a function of $P = \omega_L \tau_n$. The presence of relaxation effects causes deformation of the curves and shifts of the maximum for Im $\overline{G_{kk}^{qq}}$. This is shown in fig. 2, where the real and imaginary part of $\overline{G_{22}^{22}}$ is plotted as a function of P for a fixed value of K_0 and

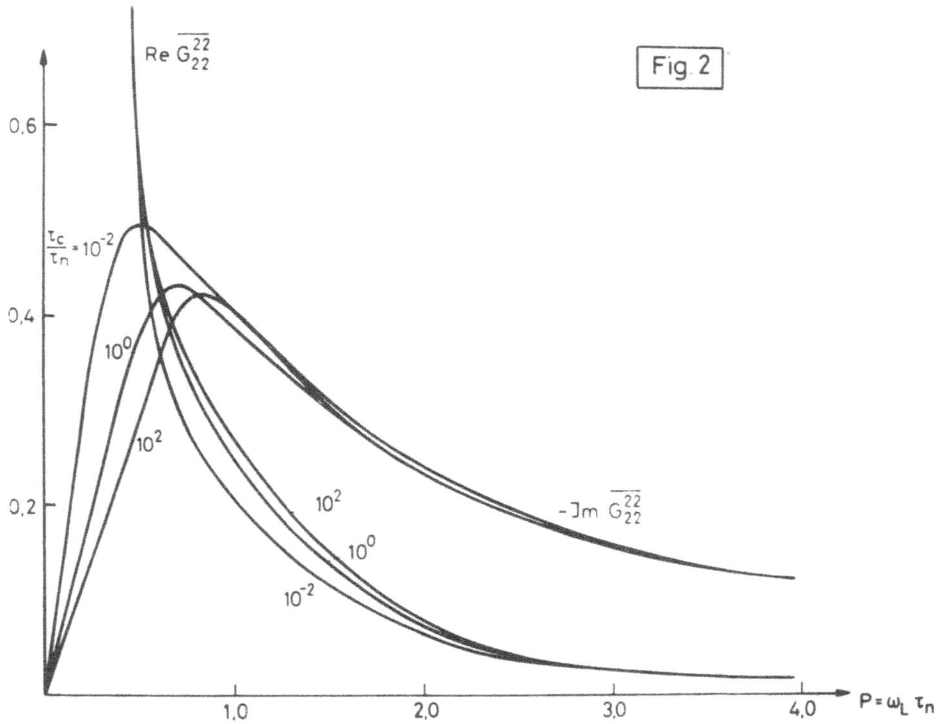

various values of $x = \tau_c/\tau_n$. (The curves have been chosen rather arbitrarily from a set of computer runs made in connection with the Fe - Tm - analysis. Plots of $A_\perp$ and R look similar to the ones given in fig. 2 for the 5/2 – (118 keV) state of ^{169}Tm. For details see ref. 13. In fig. 3 we have plotted $\overline{G}_2 = (1/5) \sum_q \overline{G_{22}^{qq}}$ as a function of τ_c/τ_n for various parameters $\omega_L\tau_n$ and $(1/3) < (\Delta\omega_L)^2 > \tau_n^2$. Fig. 4 displays the field - free case, in which situation G_2 is calculated from

$$\overline{G}_k = [\,1 + (1/3)\,k\,(k+1) < (\Delta\omega_L^2) > \tau_n^2\ x/(1+x)]^{-1}\,. \tag{57}$$

Let us discuss several limiting cases of (52) and (53). It is obvious that the case of purely static magnetic interaction is contained in the latter equations. On the other hand, if there is no static field ($P = \omega_L\tau_n = 0$) we find $\mathrm{Im}\ \overline{G_{kk}^{qq}} = 0$ and

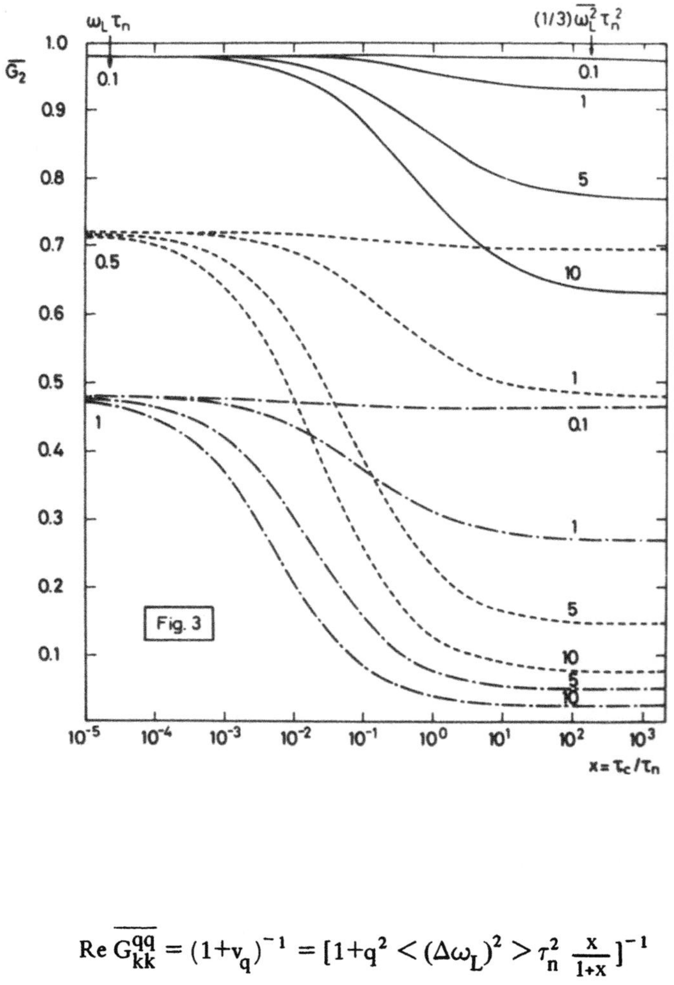

$$\mathrm{Re}\,\overline{G_{kk}^{qq}} = (1+v_q)^{-1} = [1+q^2 < (\Delta\omega_L)^2 > \tau_n^2 \frac{x}{1+x}]^{-1} \tag{58}$$

which has exactly the same mathematical structure as Blume's eq. 23 in ref. 10. Micha's result[14]) for a multidomain ferromagnet with both an average static magnetic field in each domain and a randomly time - fluctuating component has already been deduced in the Markoffian approximation[5]), but follows also from (52) for $\omega_L \tau_c \ll 1$, $x \ll 1$, in which case the changes due to the non - Markoffian character of (52) disappear. It is obvious from (52) to (53) that we get, indeed, the largest effects on the G's for $x = \tau_c/\tau_n \approx 1$.

The new type of analysis briefly discussed for integral PAC here, can easily be extended to the time - differential case, at least for simple relaxation models. The main field of application will be for all systems with strong variation of the correlation time as a function of certain external parameters (temperature, viscosity, pressure,

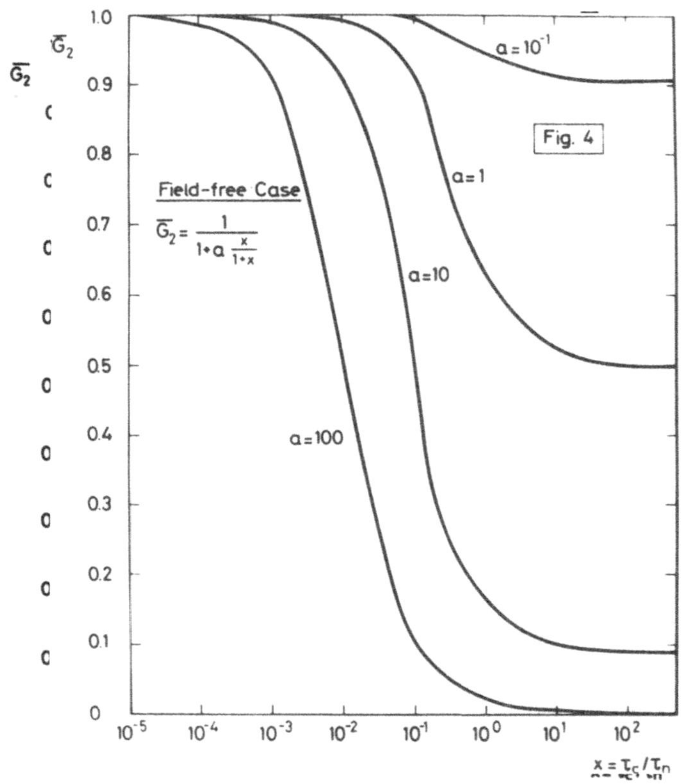

. . .) thereby crossing the region $\tau_c/\tau_n \approx 1$. Possible examples will be found from PAC measurements on molecules in liquid solutions, recoil - into - gas experiments and, in special cases, also from measurements of nuclear spin - lattice relaxation phenomena by the PAC technique.

Appendix A

1. Introduction to the superoperator formalism:

Let [|n>] be a complete orthonormal set of basis vectors which span the unitary space $\mathcal{U}$ of dimension N:

$$\langle m|n\rangle = \delta_{n,m} \,, \quad 1 = \sum_{n=1}^{N} |n\rangle\langle n| \,. \tag{A1}$$

The linear operators A, B, C, ... of the space $\mathcal{U}$ may be considered as elements of a linear vector space $\mathcal{L}$ (Liouville - space) of dimension N^2, which is unitary if we define the scalar product of two vectors $|A) \in \mathscr{L}$ and $|B) \in \mathscr{L}$ as:

$$(A\,|\,B) = \mathrm{Tr}\,[A^{+}B] \quad \text{(spur metric), see e.g. ref. 15.} \tag{A2}$$

It is easily verified that this scalar product has all the properties of a positive definite hermitian metric:

$$(A\,|\,B)^{*} = (B\,|\,A) \tag{A3}$$

$$(A\,|\,\beta B + \gamma C) = \beta\,(A|B) + \gamma(A\,|C); \quad (\alpha A + \beta B\,|C) =$$
$$= \alpha^{*}\,(A\,|C) + \beta^{*}\,(B\,|C) \tag{A4}$$

$$(A\,|\,A) \geqslant 0\;[\;(A\,|\,A) = 0 \text{ implies } A = 0]\;. \tag{A5}$$

Moreover, we have the important relation:

$$(A\,|B)^{*} = (A^{+}\,|B^{+}) \tag{A6}$$

which follows from (A2) and the invariance property of the trace under cyclic permutation of the operators. Using (A1) and (A2) we may write down the following expansion for any vector $|A) \in \mathscr{L}$:

$$|A) = \sum_{m,n} \,||m\rangle\langle m\,|A\,|n\rangle\langle n\,|) = \sum_{m,n} \,||m\rangle\langle n|)\,(|m\rangle\langle n||A)\,. \tag{A7}$$

It is easily verified by (A2) that the N^2 Liouville - vectors, $||m\rangle\langle n\,|) = |\,mn)$ (m, n = 1, 2, . . . , N), form a complete orthonormal basis in $\mathscr{L}$ (referred to as the trivial basis in the following):

$$(mn\,|\,m'n') = \delta_{mm'}\delta_{nn'}\,; \quad 1 = \sum_{m,n} |mn)\,(mn| \quad \text{(Identity operator in } \mathscr{L}) \tag{A8}$$

and that the coordinates $(mn\,|A)$ of the vector $|A)$ with respect to this basis are just the ordinary matrix elements $\langle m\,|A\,|n\rangle$ of A with respect to the basis $[\,|n\rangle]$. Now any linear transformation of the coordinates $(mn\,|A)$ of the form:

$$(mn\,|B) = \sum_{kl} (mn\,|\mathscr{R}|kl)\,(kl\,|A) \quad (m, n = 1, 2, \ldots N) \tag{A9}$$

where the coefficients of the transformation are complex numbers denoted by $(mn\,|\mathscr{R}|\,kl)$), defines a linear superoperator $\mathscr{R}$, which acting on a vector $|A) \in \mathscr{L}$ generates the vector $|B) \in \mathscr{L}$. We write symbolically:

$$|B) = \mathscr{R}\,|\,A) \tag{A10}$$

using either script letters ($\mathcal{R}$, $\mathcal{S}$, . . .) for a general linear operator in $\mathcal{L}$ or capital letters with a caret or bar ($\hat{A}$, $\underline{B}$, $\bar{C}$, . . .) for special superoperators to be defined later. The coefficients $(mn|\mathcal{R}|kl)$ in (A9) are the matrix elements of $\mathcal{R}$ with respect to the trivial basis [|nm)].

The adjoint of an operator $\mathcal{R}$ is defined in the usual way by:

$$(A|\mathcal{R}|B)^* = (B|\mathcal{R}^+|A) . \tag{A11}$$

Due to the property (A6) of the scalar product, we may define another conjugation relation in Liouville space, called 'Liouville - conjugation'. $\mathcal{R}^{\mathcal{L}}$ is by definition the Liouville - conjugate operator of $\mathcal{R}$, if

$$(A|\mathcal{R}|B)^* = (A^+|\mathcal{R}^{\mathcal{L}}|B^+) \tag{A12}$$

The Liouville conjugate of an operator can be expressed explicitly if we define the unitary antilinear superoperator $C_{\mathcal{L}}$ by:

$$|A^+) = C_{\mathcal{L}}|A) \quad \text{with } C^+_{\mathcal{L}} = C^{-1}_{\mathcal{L}} = C_{\mathcal{L}} . \tag{A13}$$

Then it follows:

$$\mathcal{R}^{\mathcal{L}} = (C_{\mathcal{L}} \mathcal{R}\, C^+_{\mathcal{L}}) . \tag{A14}$$

Note that:

$$(\mathcal{R}_1\mathcal{R}_2)^+ = \mathcal{R}^+_2\mathcal{R}^+_1 \qquad \text{but: } (\mathcal{R}_1\mathcal{R}_2)^{\mathcal{L}} = \mathcal{R}^{\mathcal{L}}_1\mathcal{R}^{\mathcal{L}}_2 \tag{A15}$$

$$(|A)(B|)^+ = |B)(A| \qquad \text{but: } (|A)(B|)^{\mathcal{L}} = |A^+)(B^+| .$$

Liouville conjugation in $\mathcal{L}$ is analogous to time - reversal in $\mathcal{U}$ as pointed out in ref. 16.

2. Special superoperators

We will now give some simple examples of superoperators; let A, B be two operators of $\mathcal{U}$; then the superoperator $\underline{A}$ is defined by:

$$\underline{A}\,|B) = |AB) \tag{A16}$$

and its matrix elements with respect to the trivial basis are given by:

$$(mn|\underline{A}|kl) = (mn|A|k\rangle\langle l|) = \langle m|(A|k\rangle\langle l|)|n\rangle =$$

$$= \langle m|A|k\rangle \cdot \delta_{ln} \quad . \tag{A17}$$

Another example is the superoperator $\overline{A}$ defined by:

$$\overline{A}|B) = |BA) \tag{A18}$$

with matrix elements:

$$(mn|\overline{A}|kl) = (mn||k\rangle\langle l|A) = \langle m|(|k\rangle\langle l|A)|n\rangle = \delta_{mk} \langle l|A|n\rangle \, . \tag{A19}$$

As a third example let us mention the so called 'Liouville - operator' $\hat{A}$ which is defined by:

$$\hat{A} = \underline{A} - \overline{A} \quad \text{or} \quad \hat{A}|B) = |[A, B]) \tag{A20}$$

with matrix elements:

$$(mn|A|kl) = \langle m|A|k\rangle \delta_{ln} - \delta_{mk} \langle l|A|n\rangle \, . \tag{A21}$$

If H is the Hamiltonian of a physical system, the corresponding Liouville operator $\hat{H}$ is usually denoted by $L \equiv \hat{H}$. The trivial basis constructed from the eigenbasis of H constitutes an eigenbasis of $L \equiv \hat{H}$ where the eigenvalues are given by corresponding energy differences:

$$(mn|L|kl) = (E_k - E_l)\, \delta_{mk} \delta_{ln}, \quad \text{if } H|n\rangle = E_n|n\rangle \; . \tag{A22}$$

Note that the identity operator in $\mathscr{L}$ can be written $\mathbf{1} = \underline{1}\,\overline{1}$. The following relations are easily verified:

$$(\underline{A})^{\mathscr{L}} = \overline{(A^+)} \; ; \qquad (\underline{A})^+ = \underline{(A^+)} \tag{A23}$$

$$(\hat{A})^{\mathscr{L}} = -\widehat{(A^+)} \; ; \quad (\hat{A})^+ = \widehat{A^+} \tag{A24}$$

$$(e^{a\hat{A}})^{\mathscr{L}} = e^{-a^*\widehat{A^+}} ; \quad (e^{a\hat{A}})^+ = e^{a^*\widehat{A^+}} \tag{A25}$$

$$\underline{f(A)} = f(\underline{A}) \; ; \tag{A26}$$

$$\overline{f(A)} = f(\overline{A}) \tag{A27}$$

$$\underline{AB} = \underline{A}\,\underline{B}\ ; \quad \overline{AB} = \overline{B}\,\overline{A}\ . \tag{A28}$$

From the definitions (A16), (A18) it follows that the two superoperators $\underline{A}$ and $\overline{B}$ commute:

$$[\underline{A}, \overline{B}] = 0\ . \tag{A29}$$

(Note that this is true no matter what [A, B] is). Relation (A29) can be of great help when deriving the following formulae (see ref. 17)

$$|e^{aA} B e^{-aA}) = e^{a\hat{A}}|B) = |B + a[A, B] + \frac{a^2}{2!}[A, [A, B]] + \ldots) \tag{A30}$$

$$(\hat{A})^n |B) = |[A, [A, \ldots [A, B] \ldots]\,]) = |\sum_{\nu=0}^{n} (-)^\nu \binom{n}{\nu} A^{n-\nu} B A^\nu)\ . \tag{A31}$$

The proof of (A30), (A31) becomes very simple if we note that $[\underline{A}, \overline{A}] = 0$ and therefore

$$e^{a\hat{A}} = e^{a\underline{A} - a\overline{A}} = e^{a\underline{A}} e^{-a\overline{A}} = \underline{(e^{aA})}\ \overline{(e^{-aA})}$$

$$\text{and } (\hat{A})^n = (\underline{A} - \overline{A})^n = \sum_{\nu=0}^{n} (-)^\nu \binom{n}{\nu} \underline{A}^{n-\nu} \overline{A}^\nu\ .$$

3. Unitary transformations in $\mathscr{L}$; Basis of multipole operators

A unitary transformation S of the kets $|\psi\rangle$ of space $\mathscr{U}$ effects a unitary transformation

$$\mathscr{S} = \underline{S}\,\overline{S^+}\ ; \qquad \mathscr{S}^+\mathscr{S} = \mathscr{S}\mathscr{S}^+ = 1 \qquad (S^+ = S^{-1}) \tag{A32}$$

of the Liouville - kets |A). For example the density - operator at time t (Liouville state vector of the system) is generated from the Liouville state vector at time t = 0 for conservative systems by the unitary transformation:

$$|\rho\,(t)) = e^{-itL}|\rho\,(0)) \quad \text{with} \quad \mathscr{S} = \underline{e^{-itH}}\ \overline{e^{+itH}} = e^{-itL}\ . \tag{A33}$$

Now let us look at a physical system of angular momentum $\underline{J}$ described by a set of standard basis vectors in $\mathscr{U}$:

$$|JM\rangle \text{ with } \{J = J_{min}, J_{min} + 1 \ldots, J_{max}; -J \leqslant M \leqslant +J\} . \quad (A34)$$

(For simplicity we assume that the quantum numbers J, M describe the state completely). $\underline{J}$ may be thought of e.g., as the sum of two vectors j_1, j_2 so that $J_{min} = |j_1 - j_2|, J_{max} = j_1 + j_2$ and $|JM\rangle$ represents the coupled basis. If, as another example, we take $j_2 = 0$, we have $J_{min} = J_{max} = J$ for a system which is fully described by one angular momentum $\underline{J}$ of magnitude J (J + 1). The dimension of $\mathscr{U}$ is given by

$$\dim \mathscr{U} = N = \sum_{J=J_{min}}^{J_{max}} (2J + 1) . \quad (A35)$$

The corresponding Liouville - space $\mathscr{L}$ spanned by the linear operators of $\mathscr{U}$ has the dimension $\dim \mathscr{L} = N^2$ and a (trivial) complete orthonormal basis is made up of the vectors:

$$||JM\rangle\langle J'M'|) \equiv |JM; J'M') \text{ with } \begin{cases} J, J' = J_{min}, \ldots, J_{max} \\ -J \leqslant M \leqslant J; -J' \leqslant M' \leqslant J' \end{cases} \quad (A36)$$

$$(J_1M_1; J_2M_2|J_3M_3; J_4M_4) = \delta_{J_1J_3}\delta_{M_1M_3}\delta_{J_2J_4}\delta_{M_2M_4} \quad (A37a)$$

$$1 = \sum_{J_1J_2} \sum_{M_1M_2} |J_1M_1; J_2M_2)(J_1M_1; J_2M_2| \quad (A37b)$$

Now let us define a set of N^2 vectors in $\mathscr{L}$

$$|U_q^{(k)}(J_1J_2)) \text{ with } \begin{cases} J_1, J_2 = J_{min}, J_{min} + 1, \ldots, J_{max} \\ k = |J_1 - J_2|, |J_1 - J_2| + 1, \ldots, J_1 + J_2 \\ -k \leqslant q \leqslant +k \end{cases} \quad (A38)$$

by giving the coordinates of each vector with respect to the trivial basis (A37):

$$(JM; J'M'|U_q^{(k)}(J_1J_2)) = \delta_{JJ_1}\delta_{J'J_2}(-)^{J-M}\sqrt{2k+1}\begin{pmatrix} J & k & J' \\ -M & q & M' \end{pmatrix} \quad (A39)$$

Due to the properties of the 3j - symbols the transformation defined in (A39) is a unitary transformation in $\mathscr{L}$ and the N^2 vectors (A38) form a complete orthogonal basis in $\mathscr{L}$. We have:

$$(U_q^{(k)}(J_1J_2)|U_{q'}^{(k')}(J_1'J_2')) = \delta_{kk'}\delta_{qq'}\delta_{J_1J_1'}\delta_{J_2J_2'} \quad (A40)$$

$$1 = \sum_{J_1,J_2=J_{min}}^{J_{max}} \sum_{k=|J_1-J_2|}^{J_1+J_2} \sum_{q=-k}^{+k} |U_q^{(k)}(J_1J_2))\,(U_q^{(k)}(J_1J_2)| = \qquad \text{(A41)}$$

$$= \sum_{J_1,J_2=J_{min}}^{J_{max}} 1_{J_1J_2} \,.$$

(Note the decomposition of the space $\mathscr{L}$ into mutually orthogonal subspaces $\mathscr{L}(J_1J_2)$). If the system is fully described by one value of angular momentum $J = I$, $J_{min} = J_{max} = I$, we specialize (A38) - (A41) to:

$$|U_q^{(k)}) \quad \text{with} \quad \begin{cases} k = 0, 1, 2, \ldots, 2I \\ -k \leqslant q \leqslant +k \end{cases} \qquad \text{(A38a)}$$

dropping the (I, I) in the vectors for shortness.

$$(Im; Im'|U_q^{(k)}(I, I)) \equiv (m; m'|U_q^{(k)}) = (-)^{I-m}\sqrt{2k+1}\begin{pmatrix} I & k & I \\ -m & q & m' \end{pmatrix} \qquad \text{(A39a)}$$

$$(U_q^{(k)}|U_{q'}^{(k')}) = \delta_{kk'}\,\delta_{qq'} \qquad \text{(A40a)}$$

$$1 = \sum_{k=0}^{2I} \sum_{q=-k}^{+k} |U_q^{(k)})\,(U_q^{(k)}| \,. \qquad \text{(A41a)}$$

It is easily shown that the operators $U_q^{(k)}(J_1J_2)$ corresponding to the Liouville - kets (A38) are the (2k + 1) standard components of an irreducible tensor operator $U^{(k)}(J_1J_2)$ of rank k. For this we give the matrix elements of the Liouville rotation operator $\mathscr{D} = \underline{D}\,\overline{D}^{+}$ (see (A32)) which are especially simple with respect to the basis (A38):

$$(U_{q'}^{(k')}(J_1'J_2')|\mathscr{D}|U_q^{(k)}(J_1J_2)) = \delta_{J_1'J_1}\,\delta_{J_2'J_2}\,\delta_{kk'}\,D_{q'q}^{(k)} \qquad \text{(A42)}$$

From (A42) it follows:

$$\mathscr{D}|U_q^{(k)}(J_1J_2)) = \sum_{q'=-k}^{+k} |U_{q'}^{(k)}(J_1J_2))\,D_{q'q}^{(k)} \,. \qquad \text{(A43)}$$

which is by definition the transformation property required for the standard components of an irreducible tensor operator. The adjoint of the multipole operators $U_q^{(k)}(J_1J_2)$ can be deduced by taking the complex conjugate of (A39), using relation (A6) and noting that the coefficients (A39) are real; we get:

$$(U_q^{(k)}(J_1J_2))^{+} = (-)^{J_1-J_2+q}\, U_{-q}^{(k)}(J_2J_1)\,. \tag{A44}$$

The expansion of an arbitrary vector $|A) \in \mathscr{L}$ with respect to the multipole basis (A38) is:

$$|A) = \sum_{JJ'} \sum_{k=|J-J'|}^{J+J'} \sum_{q=-k}^{+k} |U_q^{(k)}(JJ'))\, a_q^{(k)}(JJ') \tag{A45}$$

where the coordinates

$$a_q^{(k)}(JJ') \equiv (U_q^{(k)}(JJ')|A) \tag{A46}$$

form a set of (2k + 1) contrastandard components of an irreducible spherical tensor of rank k, for each triple (k, J, J') [3]). This is shown by the behaviour of the $a_q^{(k)}(JJ')$ under rotation:

$$(a_q^{(k)}(JJ'))_{rot} = (U_q^{(k)}(JJ')|\mathscr{D}^{+}|A) = \sum_{q'} a_{q'}^{(k)}(JJ')\, D_{qq'}^{+(k)}\,. \tag{A47}$$

If A is the standard component of an irreducible tensor operator $T^{(k)}$, we may apply the Wigner - Eckart - Theorem to get for the coordinates of $|T_q^{(k)})^q$ with respect to the basis (A39):

$$(U_{q'}^{(k')}(JJ')|T_q^{(k)}) = \delta_{kk'}\delta_{qq'} \frac{1}{\sqrt{2k+1}} <J||T^{(k)}||J'> \delta(J\,k\,J')\,. \tag{A48}$$

(The symbol δ (JkJ') equals 1 if $|J - J'| \leq k \leq J + J'$ and vanishes in all other cases). The expansion (A45) for $|T_q^{(k)})$ reads:

$$|T_q^{(k)}) = \sum_{JJ'} |U_q^{(k)}(JJ'))\, \frac{\delta(JkJ')}{\sqrt{2k+1}}\, <J||T^{(k)}||J'>\,. \tag{A49}$$

Note that the identity operator in $\mathscr{U}$ - space is given by:

$$1 = \sum_{J=J_{min}}^{J_{max}} \sqrt{2J+1}\; U_0^{(0)}(JJ) \tag{A50}$$

which is easily verified by the definition (A39). (A50) provides us with a useful tool for expressing the trace of an operator A, because in Liouville space the trace of A is just the projection (scalar product) of |A) on the vector |1):

$$\mathrm{Tr}\,\{A\} = (1|A) = \sum_{J=J_{min}}^{J_{max}} \sqrt{2J+1}\; a_0^{(0)}(JJ) \tag{A51}$$

and with (A48) the trace of a component $T_q^{(k)}$ of an irreducible tensor operator vanishes unless k = 0:

$$\mathrm{Tr}\,\{T_q^{(k)}\} = \delta_{k0} \sum_J \sqrt{2J+1}\; \langle J||T^{(0)}||J\rangle \;. \tag{A52}$$

One can also write the expectation value of an operator A in terms of its multipole components:

$$\langle A\rangle \equiv (\rho|A) = \sum_{J,J'} \sum_{k=|J-J'|}^{J+J'} \sum_{q=-k}^{+k} \rho_q^{(k)*}(JJ')\, a_q^{(k)}(JJ') \tag{A53}$$

where ρ is the density operator and the $\rho_q^{(k)*}(JJ')$ are the statistical tensors (Rose) or state multipoles (Fano) which appear here just as the complex conjugates of the Liouville state - vector coordinates with respect to the multipole basis (A38):

$$\rho_q^{(k)*}(JJ') = (\rho|U_q^{(k)}(JJ')) \;. \tag{A54}$$

Appendix B

1. Reduction of the density operator

The Liouville - von Neumann equation for the density operator ρ of a system with Hamiltonian H reads in Liouville formalism:

$$\frac{d}{dt}\,|\rho(t)) = -iL(t)|\rho(t)) \qquad (\hbar = 1;\; L \equiv \hat{H}) \;. \tag{B1}$$

We refer to $|\rho(t))$ as the (Liouville) state vector of the system in the following and may call (B1) its Schrödinger equation. In general the operator L may be time - dependent (e.g., due to an external rf - field acting on the system). The formal solution of (B1) is given by

$$|\rho(t)) = \Omega_0^t(-iL)|\rho(t{=}0)) \tag{B2}$$

where the evolution (super) operator $\Omega_0^t(-iL)$ is defined by the well - known series:

$$\Omega_0^t(-iL) = 1 - i\int_0^t d\tau L(\tau) + i^2 \int_0^t d\tau \int_0^\tau d\tau' L(\tau) L(\tau') \mp \ldots \quad \text{(B3)}$$

which is often written as a time - ordered exponential and which reduces to the exponential function

$$\Omega_0^t(-iL) = e^{-itL} \quad \text{(B4)}$$

for time - independent Hamiltonians. Note the following properties of time - ordered exponentials (analogous to the ordinary exponential function):

$$\frac{d}{dt}\Omega_0^t(-iL) = -iL(t)\,\Omega_0^t(-iL)\,; \quad \text{(B5)}$$

if $L = L' + L''$, we have:

$$\Omega_0^t(-iL) = \Omega_0^t(-iL') - i\int_0^t d\tau\, \Omega_\tau^t(-iL')\, L''(\tau)\, \Omega_0^\tau(-iL) \quad \text{(B6)}$$

$$(\Omega_0^t(-iL))^{-1} = \Omega_t^0(-iL)\,. \quad \text{(B7)}$$

Now in general we need not calculate the whole state vector $|\rho(t))$ from (B1) or (B2), - i.e. all its matrix elements with respect to some orthogonal basis in $\mathscr{L}$; we need only some special matrix elements for the following reason: in quantum mechanics all statements or predictions we make on a physical system can be put into the form of an expectation value of some Hermitian operator (observable) $H = H^+$. The expectation value of H is the projection of $|H)$ onto the state vector $|\rho)$, i.e.

$$\langle H \rangle = (\rho|H) = (H|\rho)\,. \quad \text{(B8)}$$

(The last equality is due to the hermiticity of H and ρ). Now let $|k\alpha\rangle$ denote an eigenvector of H in $\mathscr{U}$ (there always exists an orthonormal set of eigenstates for an hermitian operator) with:

$$H|k\alpha\rangle = H_k|k\alpha\rangle \qquad (\alpha \text{ denotes possible degeneracy}). \quad \text{(B9)}$$

Thus we may write (B8) as:

$$\langle H \rangle = \sum_k H_k \left(\sum_\alpha |k\alpha\rangle\langle k\alpha|\,|\rho\right) \quad \text{(B10)}$$

from which we conclude that only expectation values of the form

$$\langle P \rangle = (P|\rho) \quad \text{with } P^2 = P;\, P = P^+ \quad \text{(B11)}$$

have to be calculated, where P is a projector onto a certain subspace of $\mathscr{U}$:

$$P = \sum_a |ka\rangle\langle ka| \; . \tag{B12}$$

As an example we ask for the probability to find the system in a state $|f\rangle$ (e.g. an eigenstate of the unperturbed Hamiltonian) at time t. This is expressed as the expectation value of the projector $P = |f\rangle\langle f|$ onto the state $|f\rangle$:

$$W_f(t) = (|f\rangle\langle f| \, |\rho(t)) \; . \tag{B13}$$

Now let us mention a situation, though formally included in (B11) already, which deserves special attention. If the system consists of two parts (S) and (R) with hamiltonians H_S and H_R respectively, the space $\mathscr{U}$ is the direct product space $\mathscr{U} = \mathscr{U}_S \otimes \mathscr{U}_R$ and the Liouville - space $\mathscr{L} = \mathscr{L}_S \otimes \mathscr{L}_R$ is a direct product of $\mathscr{L}_S$ and $\mathscr{L}_R$ respectively. We often have to calculate the expectation value of an operator $0^{(S)}$ of $\mathscr{U}_S$ alone, i.e. of the operator $0 = 0^{(S)} 1_R$ of $\mathscr{U}$:

$$\langle 0\rangle_t = \langle 0^{(S)} 1_R \rangle_t = (0^{(S)} 1_R | \rho(S, R, t)) \equiv (0^{(S)} | \rho(S; t)) \tag{B14}$$

where following Fano we call

$$|\rho(S;t)) = (1_R|\rho(S; R; t)) = |\mathrm{Tr}_R\, [\rho(S, R, t)]) \tag{B15}$$

the reduced state vector of the S - system, which is a vector in $\mathscr{L}_S$ only. In many applications the R - system is identified with a heat - bath surrounding the system of interest S, in which case the reduced state vector (B15) may be calculated by making some reasonable assumptions about the R - system and the interaction H_{SR}. From (B11) and (B15) we see that in any case we need to calculate only

$$\mathscr{P}|\rho(t)) \qquad \text{with} \qquad \mathscr{P}^2 = \mathscr{P} \tag{B16}$$

where $\mathscr{P}$ is an appropriately chosen idempotent superoperator. In (B11) we would take

$$\mathscr{P} = \underline{P}\,\overline{P}\,; \quad \mathscr{P}^2 = \underline{P}\,\overline{P}\,\underline{P}\,\overline{P} = \underline{P}^2\,\overline{P}^2 = \mathscr{P} \; ; [(P|\mathscr{P}|\rho) = (P|\rho)] \tag{B17}$$

and in (B15), for reasons that will become clear very soon, set:

$$\mathscr{P} = 1_S |\rho(R, 0))\,(1_R| \, ; [(0^{(S)} 1_R | \mathscr{P} | \rho(R, S, t)) =$$
$$= (0^{(S)} 1_R | \rho(R, S, t)] \tag{B18}$$

here $\mathscr{P}$ acts in $\mathscr{L}_R$ only (it is the identity operator in $\mathscr{L}_S$) and its idempotent character is due to the property of density operators $\mathrm{Tr}_R\,[\rho(R)] = (1_R|\rho(R)) = 1$.

To get the vector (B16) we must calculate the operator $\mathscr{P}\Omega_0^t$ (-iL) according to (B2). Let the complement of $\mathscr{P}$ be defined by

$$\mathscr{P} + Q = \mathbf{1}\,; \quad \mathscr{P}Q = 0\,; \quad \mathscr{P}^2 = \mathscr{P} \tag{B19}$$

and separate L according to:

$$L = L' + L'' \tag{B20}$$

with

$$L' = \mathscr{P}L\mathscr{P} + QLQ\,; \quad L'' = \mathscr{P}LQ + QL\mathscr{P}\,. \tag{B21}$$

From (B19) we easily verify:

$$\mathscr{P}L' = \mathscr{P}L\mathscr{P} = L'\mathscr{P}\,; \quad QL' = QLQ = L'Q \tag{B22}$$

$$\mathscr{P}L'' = \mathscr{P}LQ = L''Q\,; \quad QL'' = QL\mathscr{P} = L''\mathscr{P}\,. \tag{B23}$$

Using (B6) and (B22), (B23) we get:

$$\mathscr{P}\Omega_0^t(\text{-iL}) = \{\Omega_0^t(\text{-i}\mathscr{P}L\mathscr{P})\mathscr{P} - i\int_0^t d\tau\,\Omega_\tau^t(\text{-i}\mathscr{P}L\mathscr{P})\mathscr{P}L(\tau)Q\Omega_0^\tau(\text{-iL})\}\,(\mathscr{P}+Q)\,. \tag{B24}$$

The identity $\mathbf{1} = \mathscr{P} + Q$ has been inserted to split (B24) into two terms:

$$\mathscr{P}\Omega_0^t(\text{-iL})\mathscr{P} = \Omega_0^t(\text{-i}\mathscr{P}L\mathscr{P}) - \int_0^t d\tau \int_0^\tau d\tau'\,\Omega_\tau^t(\text{-i}\mathscr{P}L\mathscr{P})\mathscr{P}\mathscr{M}(\tau,\tau')\,\mathscr{P}\Omega_0^{\tau'}(\text{-iL})\mathscr{P} \tag{B25}$$

and

$$\mathscr{P}\Omega_0^t(\text{-iL})Q = -i\int_0^t d\tau\,\Omega_\tau^t(\text{-i}\mathscr{P}L\mathscr{P})\mathscr{P}L(\tau)Q\Omega_0^\tau(\text{-iL})Q \tag{B26}$$

with

$$\mathscr{M}(\tau,\tau') = \mathscr{P}L(\tau)\,Q\Omega_{\tau'}^{\tau}(\text{-iQLQ})\,QL(\tau')\mathscr{P}\,. \tag{B27}$$

(To get (B25), one applies (B6) again to Ω_0^τ (-iL) in (B24)). Now in many cases of practical importance the operator $\mathscr{P}$ can be chosen so that:

$$\mathscr{P}\,|\rho\,(t=0)\,) = |\rho\,(t=0)\,) \tag{B28}$$

which means $Q|\rho\ (t=0)\,)=0$, and all we need to know is (B25). Two examples are:
(a) calculating the probability for a system to be found in a state $|\,i>$ at time t, when we know it was in $|\,i>$ at $t=0$. (More generally $|\,i>$ may be substituted by a group of initial states). Then

$$\mathscr{P} = \underline{P_i}\,\overline{P}_i\ ;\quad P_i = \sum_a |ia><ia| \tag{B29}$$

where P_i is the projector onto the initial states.
(b) Calculating the reduced state vector (B15) with the assumption that at $t=0$ the density operator $\rho\ (S, R, t)$ factors into $\rho_T\ (R)\ \rho\ (S;0)$; i.e., the S- and R-systems are uncoupled at $t=0$. Then we choose:

$$\mathscr{P}_R = |\rho_T\ (R))\ (1_R|1_S \tag{B30}$$

which explains the choice in (B18). It is useful to define an 'interaction representation' form of the reduced evolution operator (B25) by:

$$\widetilde{\Omega}_0^t\,(-iL) \equiv [\Omega_0^t\,(-i\mathscr{P}L\mathscr{P})]^{-1}\,\mathscr{P}\,\Omega_0^t\,(-iL)\ . \tag{B31}$$

Differentiating (B25) and inserting (B31) we get:

$$\frac{d}{dt}\,\widetilde{\Omega}_0^t\,(-iL) = -\int_0^t d\tau\,\mathscr{R}(t,\tau)\,\widetilde{\Omega}_0^\tau\,(-iL) \tag{B32}$$

with

$$\mathscr{R}\,(t,\tau) = [\Omega_0^t\,(-i\mathscr{P}L\mathscr{P})]^{-1}\,\mathscr{P}\mathscr{M}(t,\tau)\mathscr{P}\Omega_0^\tau\,(-i\mathscr{P}L\mathscr{P})\ . \tag{B33}$$

To be specific consider case (b) above and the Liouville - operator

$$L = L_S + L_R + L_{SR}\quad \text{(time - independent)} \tag{B34}$$

describing the total system. Denote the average over bath variables of an operator $\mathscr{S}$ by:

$$<\mathscr{S}>_R \equiv (1_R|\,\mathscr{S}(S,R)\,|\,\rho_T\ (R)) \equiv \mathscr{S}'(S) \tag{B35}$$

and note that:

$$a)\ [\mathscr{P}_R, L_S] = 0\ ;\quad b)\ \ L_R\,\mathscr{P}_R = 0 = \mathscr{P}_R L_R \tag{B36}$$

$$(\text{since: } L_R|\rho_T\ (R)) = 0 = L_R|1_R)\,)\ .$$

From (B32) we then get for the reduced Liouville state vector in the interaction representation:

$$\frac{d}{dt}\,|\tilde{\rho}(S;t)) = -\int_0^t d\tau < \mathcal{R}(t,\tau)>_R |\tilde{\rho}(S,\tau)) \tag{B37}$$

with the abbreviations:

$$\text{a)}\quad |\tilde{\rho}(S;t)) = e^{i\,tL'_S}|\rho(S;t))\,; \qquad \text{b)}\quad L'_S = L_S + <L_{SR}>_R\,;$$

$$\text{c)}\quad \Delta L_{SR} = L_{SR} - <L_{SR}>_R \tag{B38}$$

$$<\mathcal{R}(t,\tau)>_R = e^{i\,tL'_S} <\Delta L_{SR}\, e^{-i(t-\tau)(L'_S + L_R + Q_R \Delta L_{SR} Q_R)}\, \Delta L_{SR}>_R\, e^{-i\tau L'_S}\,. \tag{B39}$$

References

1. A. Abragam and R. Pound, Phys. Rev. 92 (1953) 943.
2. U. Fano, Rev. Mod. Phys. 29 (1957) 74.
3. U. Fano and G. Racah, Irreducible Tensorial Sets (Academic Press Inc., New York (1959)).
4. H. Frauenfelder and R.M. Steffen, in Alpha-, Beta-, and Gamma-Ray Spectroscopy, edited by K. Siegbahn (North - Holland Publishing Co., Amsterdam, 1966).
   We refer to the latter review as FS and denote equations therein by e.g. (FS 231).
5. H. Gabriel, Phys. Rev. 181 (1969) 506.
6. K. Alder, E. Matthias, W. Schneider, R.M. Steffen, Phys. Rev. 129 (1963) 1199.
7. D. Spanjaard and F. Hartmann - Boutron, J. Physique 30 (1969) 975.
8. J. Barclay and H. Gabriel, (unpublished).
9. P. da R. Andrade, J.D. Rogers and A. Vasquez, Phys. Rev. 188 (1969) 571.
10. M. Blume, in Hyperfine Structure and Nuclear Radiations, edited by E. Matthias and D.A. Shirley, page 911.
11. M. Blume, Phys. Rev. 174 (1968) 351.
12. H. Gabriel, Physics Letters 32A (1970) 202.
13. H. Bernas and H. Gabriel, (to be published).
14. D.A. Micha, Phys. Rev. 156 (1966) 627.
15. H. Primas, Helv. Phys. Acta 34 (1961) 331.
16. A. Ben - Reuven, Phys. Rev. 141 (1966) 38.
17. J. Bosse, Diplomarbeit Braunschweig, 1969 (unpublished).
18. A.G. Redfield, IBM J. Res. Develop 1 (1957) 19.

Discussion

Leisi: I would like to ask two questions. Under which conditions is the multipole expansion that you use for describing the interaction between the subsystem and the environment valid? More specifically, does it apply to the two systems we have studied experimentally, namely gases at various pressures and paramagnetic single crystal sources measured at various temperatures? And second, how does one geneıally determine the size of the subsystem?

Gabriel: The main point is to make up your mind about the subsystem. Whenever you can be sure that you have to start out with the combined quantum mechanical system consisting of the nucleus and the electronic shell belonging to it, as e.g. in a rare - earth ion under certain conditions, you have to introduce spherical tensor operators which take care of this. These spherical operators are no longer simply defined in terms of the nuclear spin I, but you have to introduce at least the I and J, or the coupling of these two. This would influence only the details of your calculations. Nothing is changed with respect to the way you handle the problem: the reduction of your entire system to the subsystem. You have to be careful to use the correct vector space before you can go to the Liouville space which has a dimension of n^2. I had a quick look at the Xenon data, you reported at Asilomar, and I tried to apply the non - Markoffian equation of motion to this case. I have not finished it in a quantitative way, but I think it explains just the minimum which you observe and which you could not explain at that time.

Leisi: The interaction term is very general; is it not so that you have to have a closed system?

Gabriel: Yes, but the only thing I have really used here is the fact that you can write the interaction Hamiltonian as a scalar product of tensor operators; one of them acts only on the subsystem and the other on the surroundings. It is just a matter of making up your mind about which part you attribute to the subsystem and which to the surroundings.

Contributed Paper

THEORETICAL ASPECTS OF QUADRUPOLE PERTURBATIONS ON TIME - INTEGRATED ANGULAR CORRELATIONS*

C.T. Alonso

Heavy Ion Accelerator, Yale University, New Haven, Connecticut, U.S.A.

Recent advances in the dynamic reorientation effect[1]) have stimulated this examination of the possibility of performing static measurements of nuclear quadrupole moments. Comparisons of results from the two methods could lead to new insights concerning the static or dynamic nature of nuclear deformations during Coulomb excitation. The electric field gradients required for such static experiments are readily supplied by the lattice field gradients found in commercially available single crystals.

Although the interpretation of time - differential experiments is much less ambiguous than in the time - integrated case, the majority of nuclear states must be studied by the latter technique because of their short lifetimes. Therefore we concentrate here on the problems and possibilities involved in performing time - integrated experiments.

The familiar techniques of implantation using accelerator or isotope separator beams[2]) now make it possible to embed a radioactive nucleus into almost any single crystal. However radiation damage may remain as a not infrequent attendant problem in such implantations. The internal electric field gradient acting upon such an impurity nucleus is the derivative of the resultant electric field produced by the charge points in the lattice and the electron cloud surrounding the nucleus. The various coupling interactions usually ensure that the directional symmetries of the lattice are retained in the total field gradient acting upon the nucleus. In this paper it is assumed that this field gradient has axial symmetry about the crystalline c - axis.

In the standard notation a general axially symmetric perturbed angular correlation is given by[3])

$$w(\theta,t) = \sum_{k_1 k_2}^{\infty} \frac{4\pi A_{k_1 k_2}}{\sqrt{(2k_1+1)(2k_2+1)}} \sum_q Y_{k_1}^{q*}(\theta_1\phi_1)\, G_{k_1k_2}^{qq}\, Y_{k_2}^{q}(\theta_2\phi_2) \tag{1}$$

where θ_i and ϕ_i are the spherical angles of radiation (i) relative to the symmetry axis of the field gradient. The perturbation parameters are wholly contained in the attenuation coefficient

$$G_{k_1k_2}^{qq} = \sqrt{(2k_1+1)(2k_2+1)} \sum_{m_1 m_2} \begin{pmatrix} I & I & k_1 \\ m_1 & -m_2 & q \end{pmatrix} \begin{pmatrix} I & I & k_2 \\ m_1 & -m_2 & q \end{pmatrix} e^{-i(E_2-E_1)t/\hbar} \tag{2}$$

Some interesting properties of this coefficient can be derived by writing out

$$w(\theta) = w^*(\theta) \tag{3}$$

In the magnetic dipole case, for example, application of the orthogonality of the 3 - j symbols yields

$$G_{k_1k_2}^{qq} = \delta_{k_1k_2}\, e^{-iq\omega_B t} \tag{4}$$

where ω_B is the usual Larmor precession frequency. This interaction is diagonal, and thus only the familiar coefficients A_{22} and A_{44} appear in the magnetic dipole equations. The diagonality also allows one to use the addition theorem to write the spherical harmonics as ordinary Legendre polynomials.

In the case of an electric quadrupole interaction the reality of w (θ) yields the following hermiticity relation for the G coefficients:

$$G^{qq}_{k_1k_2} = (-)^{k_1+k_2} G^{qq*}_{k_1k_2} \tag{5}$$

This implies that G is real when $(k_1 + k_2)$ is even and G is imaginary when $(k_1 + k_2)$ is odd. Thus if parity is conserved then k_1 and k_2 are both even and G must be real. Since an imaginary part of a G coefficient is necessary to produce a precession of an angular correlation, we conclude that under parity conservation a quadrupole perturbation produces only an attenuation of the angular correlation. This sheds some light on the popular remark that quadrupole interactions do not precess an angular correlation because the interaction energy goes as m^2. If parity is not conserved, then under certain conditions a precession should be observable.

The G coefficient for a quadrupole perturbation is usually written[4])

$$G^{qq}_{k_1k_2} = \sum_n S^{k_1k_2}_{nq} e^{-in\omega_0 t} \tag{6}$$

where ω_0 is the usual quadrupole frequency and the coefficients S are real sums of 3 - j coefficients. Under parity conservation we have shown that G is real, so it can be written

$$G^{qq}_{k_1k_2} = \sum_n S^{k_1k_2}_{nq} \cos n\omega_0 t \tag{7}$$

If a magnetic field is applied parallel to the electric field gradient then

$$G^{qq}_{k_1k_2} = e^{-iq\omega_B t} \sum_n S^{k_1k_2}_{nq} \cos n\omega_0 t \tag{8}$$

Taking the limit $\omega_0 = 0$ and applying eq. (4) yields a useful sum rule:

$$\sum_n S^{k_1k_2}_{nq} = \delta_{k_1k_2} \tag{9}$$

Assuming parity conservation, the time - integrated perturbed angular correlation becomes

$$w(\theta,\infty) = \sum_{k_1k_2}^{\infty} \frac{4\pi A_{k_1k_2}}{\sqrt{(2k_1+1)(2k_2+1)}} \sum_q Y^{q*}_{k_1}(\theta_1\phi_1) Y^q_{k_2}(\theta_2\phi_2) \sum_n \frac{S^{k_1k_2}_{nq}}{1+(n\omega_0\tau)^2} \tag{10}$$

This equation is not diagonal and cannot be written in terms of ordinary Legendre polynomials. When written out, eq. (10) contains twelve terms, each composed of a factor containing an inverse square function of the perturbation parameter $\omega_0\tau$, which is proportional to three important quantities:

$$|\omega_0\tau| \sim |Q\phi\tau| \tag{11}$$

where Q is the nuclear quadrupole moment, ϕ is the magnitude of the electric field gradient, and τ is the lifetime of the nuclear state. The magnitude $|\omega_0\tau|$ is the sole perturbation number that can be pulled out of these experiments.

In order to deduce one of the three quantities Q, ϕ, or τ, the other two must be known. Usually the lifetime is known. It is common practice to deduce the field gradient by substituting the B (E2) quadrupole moment for Q. Note however that there can be a substantial difference between the B (E2) value and the static quadrupole moment. The B (E2) value is a transition amplitude,

$$B\,(E2) \sim |\langle f|Q_{op}|i\rangle|^2 \tag{12}$$

whereas the quadrupole moment which we measure is

$$|Q|^2 \sim |\langle i|Q_{op}|i\rangle|^2 \tag{13}$$

These two quantities are equal only if the initial and final wave functions overlap sufficiently; that is, for a purely rotational nucleus.

Note finally in eq. (10) that there is a hard core value at $\omega\tau = \infty$ given by the n = 0 terms.

Eq. (10) has been studied extensively by computer with some encouraging and

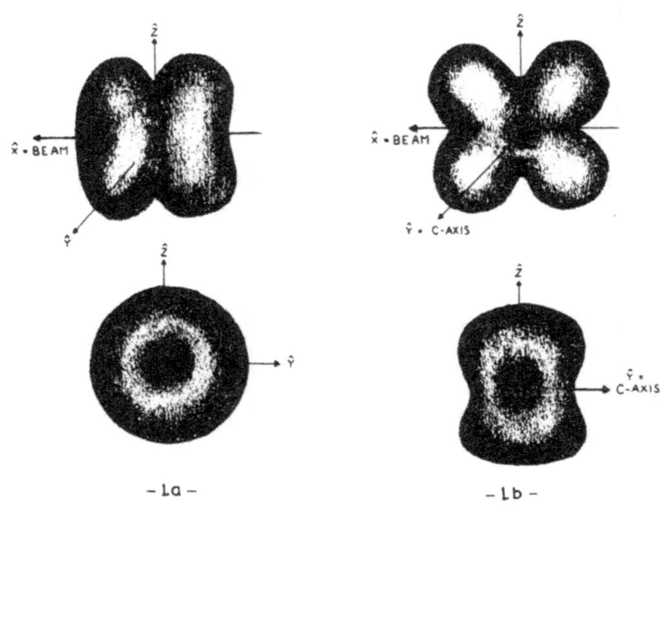

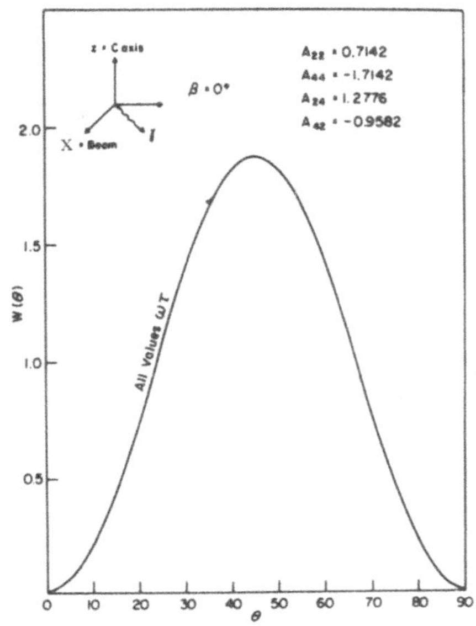

– 2.a –

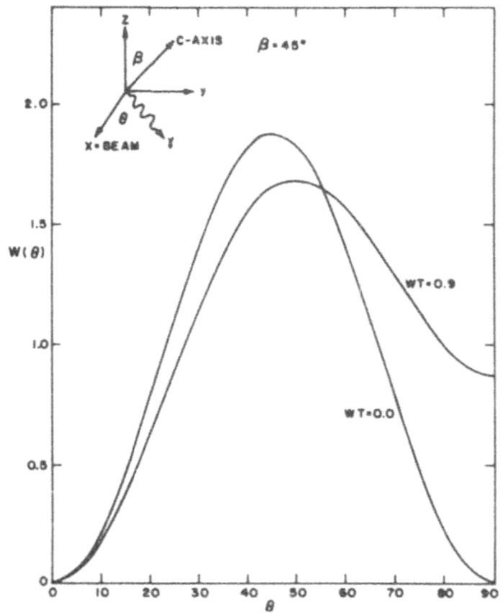

– 2.b –

unexpected results[5]). Two programs will soon be available from the author; one (FASOLT) predicts quadrupole perturbed angular correlations for any geometry and the other (FAFNER) performs a non - linear least squares fit of data for such perturbations. Various radioactivity gamma - gamma cascades have been studied with these programs, but the emphasis has been on the particle - gamma 0 - 2 - 0 cascade because of the widespread applicability of Coulomb excitation recoil implantation.

Fig. 1a and 1b show three - dimensional views of the 0 - 2 - 0 correlation, the upper view looking along the beam and the lower view looking into the beam. Fig. 1a shows the unperturbed 0 - 2 - 0 correlation, which is a double torus about the beam, and fig. 1b shows this same torus distorted into four lobes by the quadrupole interaction.

The figures labelled 2a, 2b, and 2c show perturbed 0 - 2 - 0 angular correlations for the geometrically simple case in which the c - axis is rotated on angle β in the plane perpendicular to the beam (see insets). When $\beta = 0^\circ$ (fig. 2a) the quadrupole field has no effect on the angular correlation. This useful result will be discussed later. At $\beta = 45^\circ$ (fig. 2b) the angular correlation shows some attenuation, and this attenuation is maximized at $\beta = 90^\circ$ (fig. 2c). In these figures the angle θ is the angle between the beam and the gamma detector, and the curves are for increasing values of $\omega_0\tau$. At $\beta = 90^\circ$, when the c - axis is in the detector plane and perpendicular to the beam, the angular correlation attenuates steadily downward as $\omega_0\tau$ increases. The hard - core values for $\omega_0\tau = \infty$ lie not too far below the curve shown for $\omega_0\tau = 0.9$. The non - linear character of the interaction means that the lower limit

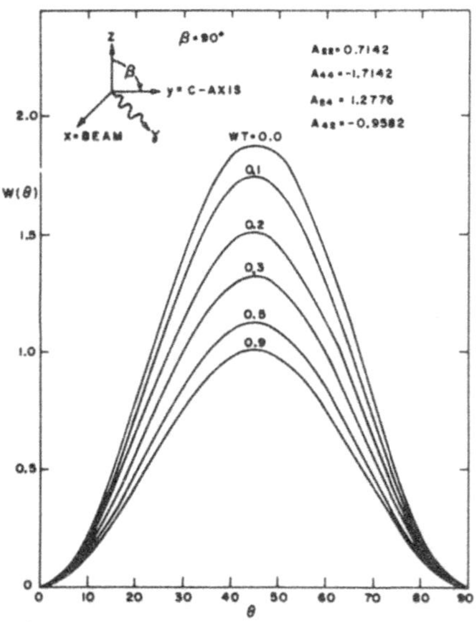

— 2.c —

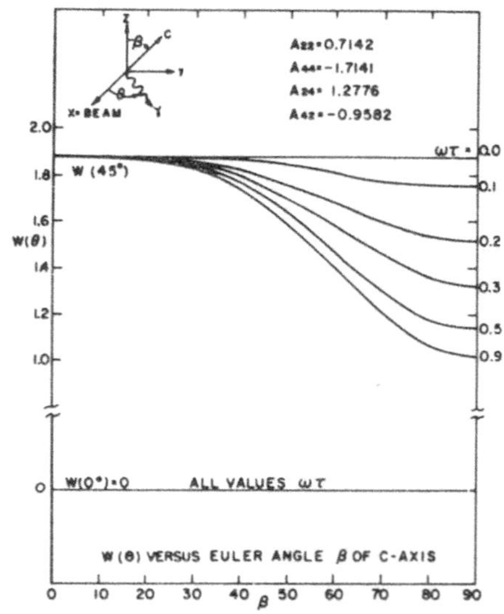

— 3 —

of $\omega_0\tau$ will always be known to greater accuracy than the upper limit. Note that only values of $\omega_0\tau$ between about one and one tenth are easily measured, but that for these values there is a very measurable attenuation. For quadrupole moments of a few barns and electric field gradients of about 10^{17} volts/cm^2 these values correspond to a lifetime of nanoseconds or less. Therefore many interesting nuclear states can be studied by this technique.

Figures 2c and 3 show two good experiments to perform. Fig. 3 has the advantage that gamma absorption in the target does not enter because the gamma detectors are held fixed at 45° and 0° with respect to the beam. The target is rotated in the plane perpendicular to the beam. The data at 45° should move from an unperturbed value at $\beta = 0°$ to a maximum attenuation at $\beta = 90°$. The data at 0°, that is along the beam, should not be affected and can be used to monitor the experiment.

Figures 9a through 9d show just such an experiment performed in collaboration with Drs. Daniel Murnick and Elton Kaufmann and Professor Lee Grodzins at the Rutgers - Bell Laboratories tandem Van de Graaff. The geometry for these experiments is indicated in fig. 8. We have performed these experiments for various isotope and crystal combinations, but our best results came from tungsten isotopes on Gd single crystals[6]). The data shown in figures 9a and 9b follow the theoretical predictions of fig. 3 very well. Fig. 9c shows data from four orientations of the c - axis which follow the predictions of figures 2a through 2c. In each case the data behave according to the predictions. The dotted lines are least square fits using only one set of seven parameters: the four coefficients $A_{k_1k_2}$, the perturbation $\omega_0\tau$, a scale factor, and a

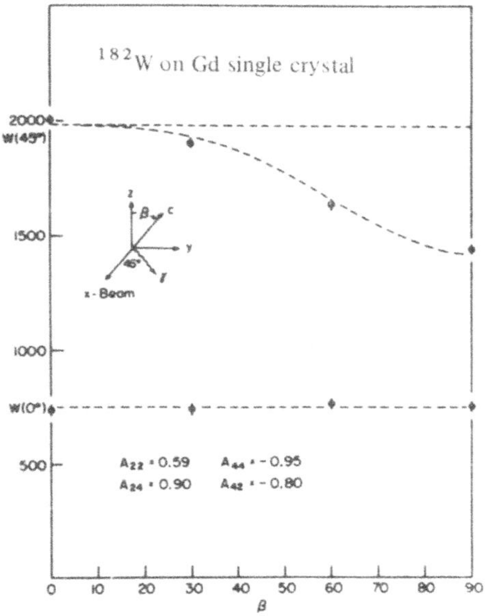

^{182}W on Gd single crystal
W(45°)
W(0°)
$A_{22} = 0.59$ $A_{44} = -0.95$
$A_{24} = 0.90$ $A_{42} = -0.80$
x - Beam
β

-9.a-

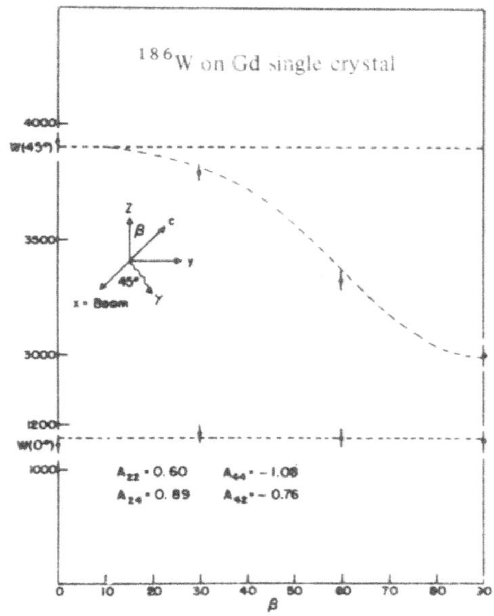

^{186}W on Gd single crystal
W(45°)
W(0°)
$A_{22} = 0.60$ $A_{44} = -1.08$
$A_{24} = 0.89$ $A_{42} = -0.76$
x - Beam
β

-9.b-

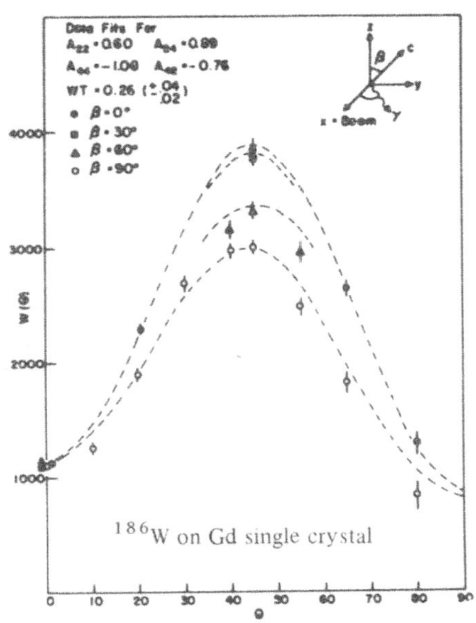

Data Fits For
$A_{22} = 0.60$ $A_{24} = 0.89$
$A_{44} = -1.08$ $A_{42} = -0.76$
$\omega T = 0.26$ ($^{+.04}_{-.02}$)
β = 0°
β = 30°
β = 60°
β = 90°
x - Beam
W(θ)
^{186}W on Gd single crystal
θ

-9.c-

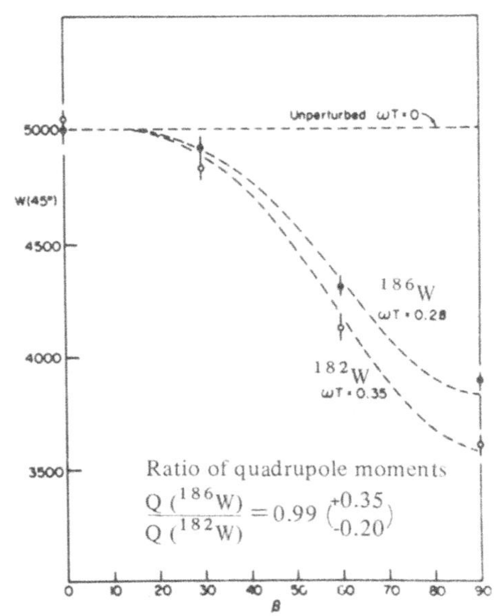

Unperturbed ωT = 0
W(45°)
^{186}W
ωT = 0.28
^{182}W
ωT = 0.35
Ratio of quadrupole moments
$\frac{Q(^{186}W)}{Q(^{182}W)} = 0.99 \binom{+0.35}{-0.20}$
β

-9.d-

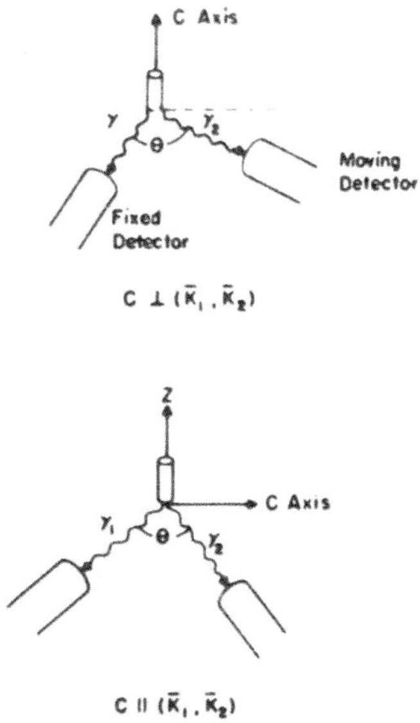
C Axis
Moving Detector
Fixed Detector
C ⊥ (K̄₁ , K̄₂)
Z
C Axis
C ∥ (K̄₁ , K̄₂)

– 7 –

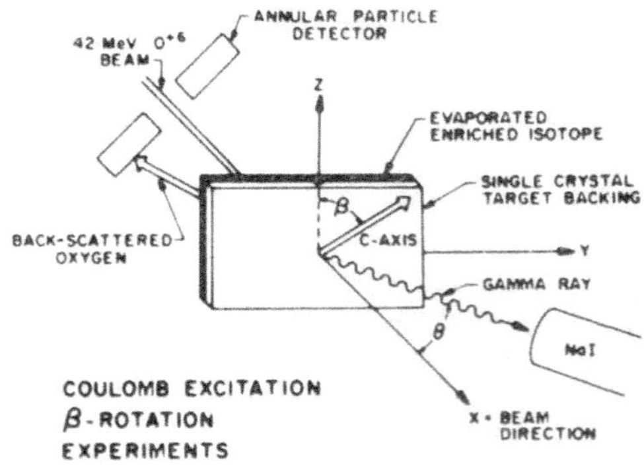
ANNULAR PARTICLE DETECTOR
42 MeV O+6 BEAM
EVAPORATED ENRICHED ISOTOPE
SINGLE CRYSTAL TARGET BACKING
BACK-SCATTERED OXYGEN
C-AXIS
GAMMA RAY
NaI
X = BEAM DIRECTION
COULOMB EXCITATION
β-ROTATION
EXPERIMENTS

– 8 –

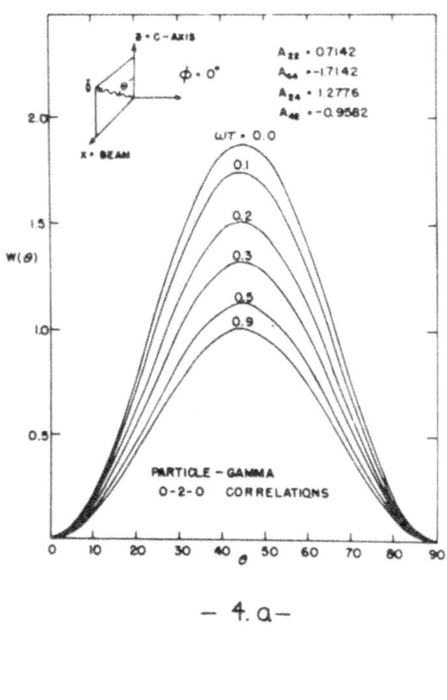
z = C-AXIS
φ = 0°
A22 = 0.7142
A44 = -1.7142
A24 = 1.2776
A42 = -0.9582
X = BEAM
ωT = 0.0
0.1
0.2
0.3
0.5
0.9
W(θ)
PARTICLE – GAMMA
0-2-0 CORRELATIONS
θ

– 4.a –

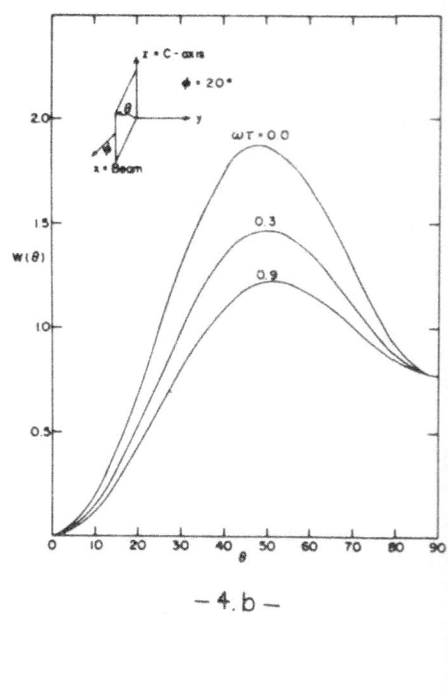
z = C-axis
φ = 20°
y
x = Beam
ωT = 0.0
0.3
0.9
W(θ)
θ

– 4.b –

gamma absorption coefficient.

Since electric field gradients are formed by the outer electron cloud and the lattice, they are independent of nuclear properties to first order. Therefore the field gradients acting on all isotopes of the same element should be equal. Thus the ratios of quadrupole moments of various isotopes can be measured directly. Fig. 9d shows some preliminary results which measure the ratios of the quadrupole moments of ^{182}W and ^{186}W. We have also measured the corresponding ratio for ^{184}W.

Predictions for out - of - beam geometries are also interesting because the attenuation can be negative as well as positive. That is, for certain geometries the angular correlation is enhanced by the quadrupole interaction. Figures 4a through 4e trace the evolution of such an out - of - beam experiment. By out - of - beam we mean that the gamma detector plane does not include the beam. Fig. 4a is basically the in - beam experiment discussed before (fig. 2c). As the detector plane rotates an angle ϕ about the c - axis, various things happen. As ϕ changes from 0° (fig. 4a) to 45° (fig. 4c) the attenuation of the angular correlation steadily decreases. At 45° is one of those cases which remain unperturbed no matter how strong the quadrupole field. Beyond 45° the unperturbed angular correlation ($\omega_0\tau = 0$ curves) diminishes its anisotropy while the perturbed correlation enhances its anisotropy, until at 90° (fig. 4e) the unperturbed correlation is isotropic but the perturbed correlation has a maximum enhancement. This particular 90° geometry is equivalent to holding the gamma detector perpendicular to the beam and rotating the c - axis between beam and detector, so it need not be an out-of-beam experiment. A combination of the maximum enhancement of fig. 4e with the maximum attenuation of fig. 4a results in the sensitive experiment indicated in fig. 5, where small values of $\omega\tau$ are specially measurable.

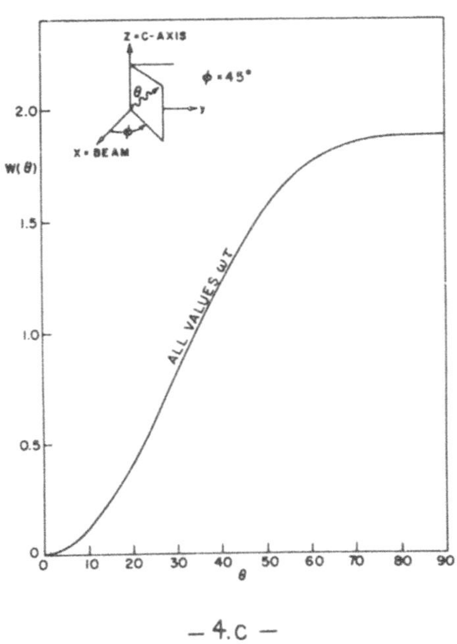

— 4.c —

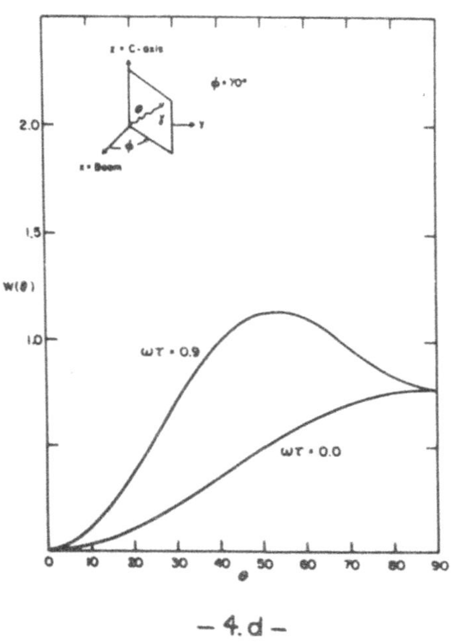

— 4.d —

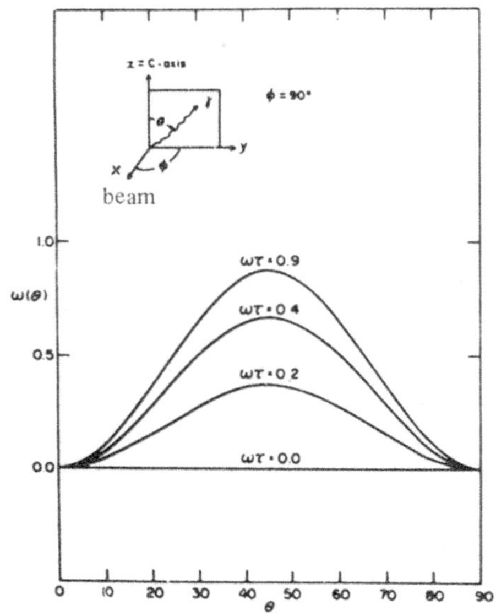
z = C-axis
φ = 90°
beam
ωτ = 0.9
ωτ = 0.4
ωτ = 0.2
ωτ = 0.0
ω(θ)
θ

– 4. e –

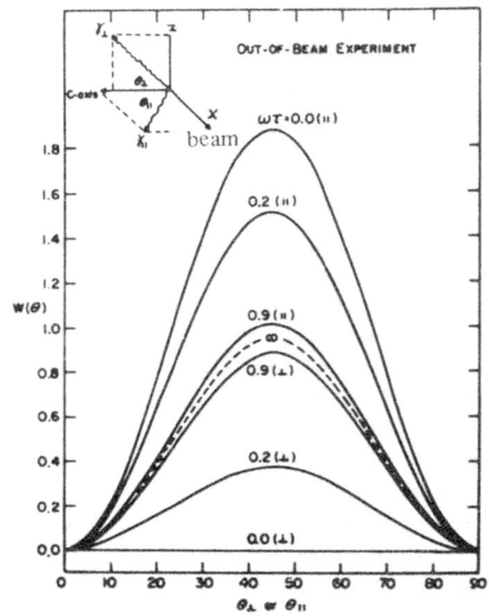
OUT-OF-BEAM EXPERIMENT
C-axis
beam
ωτ = 0.0 (∥)
0.2 (∥)
0.9 (∥)
∞
0.9 (⊥)
0.2 (⊥)
0.0 (⊥)
W(θ)
θ⊥ or θ∥

– 5 –

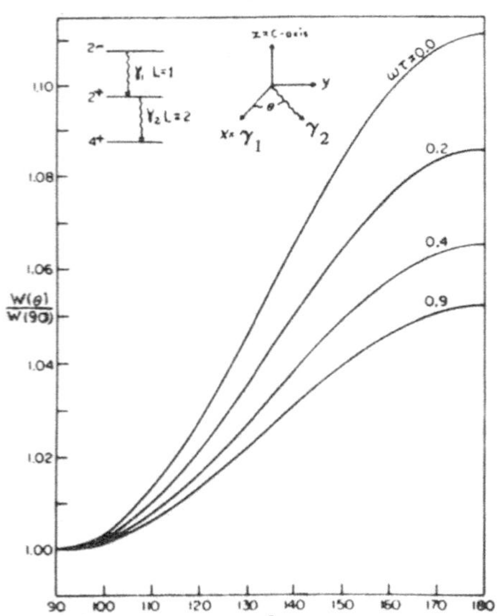
z = C-axis
2−
γ1 L=1
2+
γ2 L=2
4+
x = γ1
γ2
ωτ = 0.0
0.2
0.4
0.9
W(θ)/W(90)
θ

– 6. b –

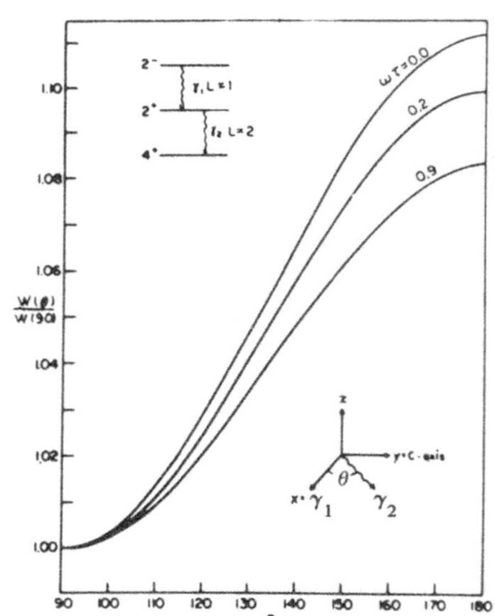
2−
γ1 L=1
2+
γ2 L=2
4+
ωτ = 0.0
0.2
0.9
W(θ)/W(90)
z
y = C-axis
x = γ1
γ2
θ

– 6. a –

Predictions for radioactive gamma - gamma cascades show the same encouraging trends. Figures 6a through 6d show predictions for two cascades in the two geometries indicated in fig. 7. Figures 6a and 6b show a $2^- - 2^- - 4^+$ cascade, and figures 6c and 6d a $4^+ - 2^+ - 0^+$ cascade. In all these predictions except fig. 6c the effect of the quadrupole field is very measurable.

Fig. 6c, along with fig. 2a, is an example of a rather peculiar property of the quadrupole perturbation[7]). Stating it in physical terms, if a gamma ray is emitted to or from an intermediate state with equally populated even m - states and equally populated odd m - states, then the time - integrated angular correlation remains unperturbed over all space if the interaction is pure quadrupole. Fig. 10a illustrates some cases for which this happens. If, for example, a 0 - 2 gamma ray is perpendicular to the c - axis then the angular distribution of the other gamma is completely independent of any perturbing quadrupole field.

This is contrary to our experience in magnetic perturbations, which exhibit a maximum precession when the perturbing field is perpendicular to the detector plane. Our intuition expects that the torques will be maximized with this geometry, whereas in fact no attenuation occurs at all. A simple physical description is still lacking, but a theoretical explanation is given below.

The matrix elements for a quadrupole perturbation, written in terms of the irreducible tensorial components of the electric field gradient, are given by

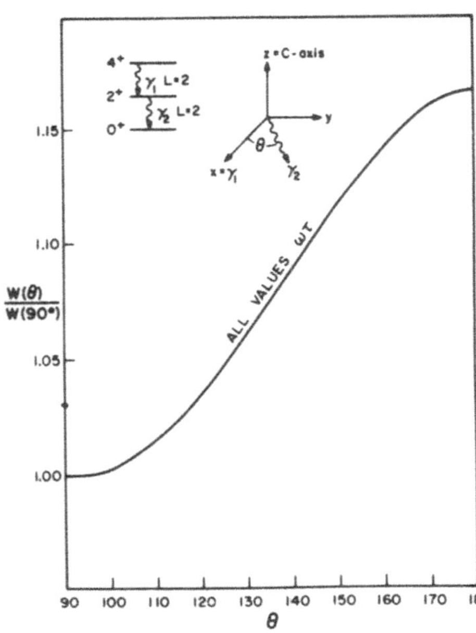

–6.c–

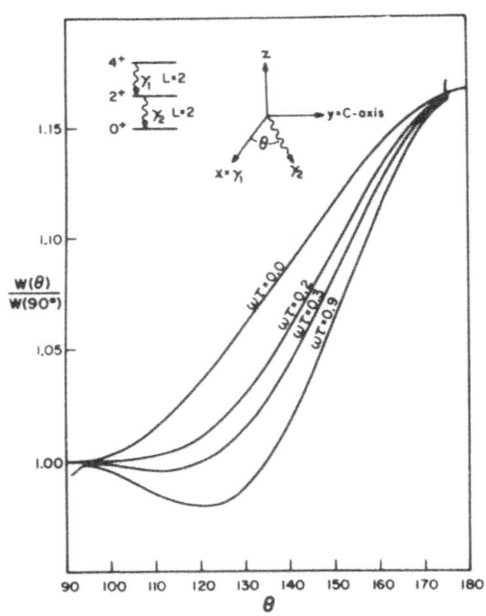

–6.d–

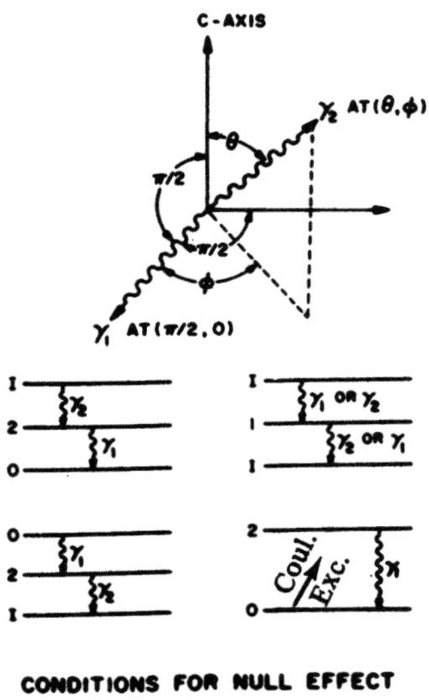

-10.a-

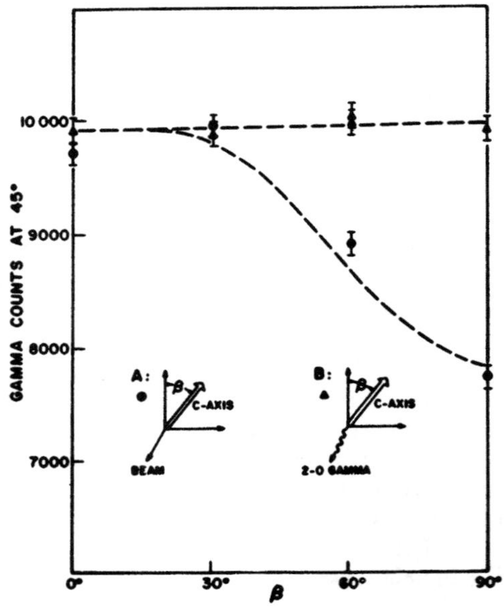

-10.b-

$$(m|Q{\cdot}\nabla E|m) \quad = a\,(\nabla E)_0\,[3m^2 - I(I+1)] \tag{14}$$

$$(m|Q{\cdot}\nabla E|m\pm1) = \pm a\frac{\sqrt{6}}{2}\,(\nabla E)_{\pm1}\,(2m\pm1)\sqrt{(I\pm m+1)(I\mp m)}$$

$$(m|Q{\cdot}\nabla E|m\pm2) = a\frac{\sqrt{6}}{2}(\nabla E)_{\pm2}\sqrt{(I\mp m)(I\mp m-1)(I\pm m+1)(I\pm m+2)}$$

where $a = eQ/2I\,(2I - 1)$. The field gradient components are written in terms of the spherical angles θ and ϕ of the field symmetry axis:

$$\begin{aligned}(\nabla E)_0 &= 1/4\; e\,q\,(3\cos^2\theta-1)\\ (\nabla E)_{\pm1} &= \pm\frac{\sqrt{6}}{4}\,e\,q\,\sin\theta\cos\theta\,e^{\pm i\phi}\\ (\nabla E)_{\pm2} &= \frac{\sqrt{6}}{8}\,e\,q\,\sin^2\theta\,e^{\pm2i\phi}\end{aligned} \tag{15}$$

where $eq = \sum_f e_j\,(3\cos^2\theta_j-1)/r_j^3$

Note that, when $\theta = \pi/2$, that is when the c - axis is perpendicular to the z - axis then the ± 1 components of the field gradient vanish. This means that $\Delta m = \pm 1$ is forbidden, so we have a new selection rule, namely

$$\Delta m = \pm 2, 0 \qquad (\underline{c} \perp \underline{z}) \qquad (16)$$

If all the even m - states have the same population a, and all the odd m - states have the same population b, then under $\Delta m = \pm 2$ nothing happens because evens go into evens and odds go into odds. There have been m - state transitions, but no total change in the m - state populations, so the angular correlation remains unperturbed.

The most important cascade which results in the even - odd population equality necessary for this effect is a 0 - 2 gamma ray from or to an intermediate state of spin two. A choice of z - axis along the 0 - 2 gamma ray shows this clearly. For, since the 0 - 2 gamma is along the z - axis, it can only radiate by $\Delta m = \pm 1$. Since the initial state had $m = 0$, the intermediate state is left in $m = \pm 1$ equally. Thus the odd m - states are equally populated and the evens ($\pm 2, 0$) are equally unpopulated. Therefore we predict that when a 0 - 2 gamma ray is emitted perpendicular to the c - axis, no matter where the other cascade gamma goes, no matter what its multipolarity, and no matter what spin it goes to, its angular distribution is totally independent of any static axially symmetric electric field gradients present at the nucleus. This effect works for the important particle - gamma 0 - 2 - 0 correlations discussed above, and it works for any pair of gammas through intermediate spin one.

Moreover, because this effect depends on a delicate balance of m - state populations which can be upset by other interactions such as magnetic fields, its presence would seem to indicate a pure quadrupole interaction. Exceptions are the very special cases described by Dr. Leisi at this conference[8]). Note that if another field changes the m - state populations then the quadrupole interaction itself can contribute to an attenuation of the angular correlation. It appears that this effect will be a useful test for determining whether the attenuation one observes is due to a pure quadrupole interaction. The sensitivity of the effect to the directional properties of the crystal also make it a good probe for radiation damage studies.

This phenomenon has been experimentally verified by Drs. Murnick and Kaufmann at Bell Laboratories. The results are shown in fig. 10b. In experiment A the c - axis was rotated in a plane perpendicular to the beam, where it should show a normal quadrupole attenuation. The data (circles) showed a large attenuation. In experiment B the c - axis was rotated in a plane perpendicular to the 2 - 0 gamma ray. Here the data should remain unperturbed by the crystalline field because of the effect discussed above. The actual data (triangles) remained at the unperturbed value for all rotations of the c - axis, confirming the predictions.

In conclusion, we have shown that it is possible to perform time - integrated perturbed angular correlation measurements of the static quadrupole parameter $|\omega_0 \tau|$ by using implantation into single crystals. We hope that similar measurements may be possible for a wide range of nuclei, though it remains to be seen how much radiation damage will hinder such experiments. Since our data consistently followed the predictions for all c - axis orientations and for the null effect, we believe that we have seen a pure quadrupole effect. We have shown that states of vibrational or transitional character as well as deformed nuclei can be studied by this technique. It can also be used as a probe for solid state field gradients, although the time - differential method should be used for such studies whenever possible.

* Work supported by the United States Atomic Energy Commission.

Acknowledgement

The author wishes to acknowledge many fine discussions with Professor Lee Grodzins.

References

1. Paper presented by B. Povh, This Proceedings, p. 327.
2. Paper presented by D.E. Murnick, This Proceedings, p. 455.
3. A. Abragam and R.V. Pound, Phys. Rev. 92 (1953) 943.
4. R.M. Steffen and H. Frauenfelder, Perturbed Angular Correlations, E. Karlsson, E. Matthias and K. Siegbahn, eds., Amsterdam, North - Holland, 20 (1964).
5. C.T. Alonso, Bull. Am. Phys. Soc. 14 (1969) 1202.
6. C.T. Alonso, L. Grodzins, E. Kaufmann, H. Lie and D.E. Murnick, Bull. Am. Phys. Soc. 14 (1969) 1202.
7. C.T. Alonso, J.R. Alonso, and L. Grodzins, Bull. Am. Phys. Soc. 15 (1970) 628.
8. Paper presented by H.J. Leisi, This Proceedings, p. 375.

Discussion

Alder: It has become customary here to refer to old papers written 20 years ago. I like to confess a sin: there is an error in the tabulation of the angular dependence of the crystal - axis for the quadrupole interaction in the paper by Albers - Schönberg, Alder, Heer and Novey. Did you use the corrected formula? Since such experiments have now become feasible, I like to urge the experimental physicists to use only the corrected formula.

C.T. Alonso: I don't understand which formula you are referring to.

Alder: The tabulation in this old paper about the quadrupole interaction.

C.T. Alonso: No, I did not use that paper. I have used the initial equation with the spherical harmonics themselves. I did not use any of these coefficients.

Alder: But anyhow, if somebody would want to use this paper I would like to warn him a little bit.

C.T. Alonso: Yes, we did notice the error and that is why we did not use the paper you are referring to.

Boehm: In line with the tradition established at this meeting to remind the speaker of earlier results, I like to mention that the ratios of the static, as well as the dynamic quadrupole - moments in tungsten isotopes, have been determined years ago to high accuracy by Persson et al. with the Mössbauer technique.

C.T. Alonso: Yes, that is why we did them; we wanted to test this technique against so well - known ratios.

Boehm: Your errors are about 30%, I remember.

C.T. Alonso: Yes, we have since improved them. I just wanted to show here that one could do such an experiment.

Leisi: Do you conclude from your experiment that the implanted particles are actually at lattice sites?

C.T. Alonso: The one conclusion you can make is that the atoms landed interstitionally rather than substitutionally. That the electric field gradient on them has somehow preserved the symmetry of the crystal. We don't know enough to say whether that is likely or not. Intuitively it seems unlikely. We think that because all these experiments indicate that the c - axis of the crystal is also the symmetry axis of the electric field gradient, that it is a fairly good argument for saying that they must have landed substitutionally. That is all we can say at this time.

Contributed Paper

A GEOMETRY FOR OBTAINING UNPERTURBED ANGULAR CORRELATIONS IN THE PRESENCE OF ELECTRIC FIELD GRADIENTS*

C.T. Alonso and J. Alonso

Heavy Ion Accelerator, Yale University, New Haven, Connecticut,U.S.A.

L. Grodzins

Laboratory for Nuclear Science, Massachusetts, Institute of Technology, Cambridge, Massachusetts, U.S.A.

An experimental geometry has been found which makes possible the measurement of unperturbed angular correlations in the presence of axially symmetric static electric field gradients. If an angular distribution through an I = 2 intermediate state involves a 0 - 2 quadrupole gamma radiation, then the angular distribution will be unperturbed by the quadrupole interaction whenever the momentum of the 0 - 2 radiation is perpendicular to the field symmetry axis, independent of the multipole nature of the second radiation. Similarly all correlations through an intermediate spin I = 1 are unperturbed whenever the momentum of one of the radiations is perpendicular to the field symmetry axis.

The emphasis of study in the field of nuclear moments has recently been shifting from magnetic moments to electric quadrupole moments. In view of this we have made a detailed exploration of the predictions of the theory of quadrupole perturbations of angular correlations and have found good experimental geometries for measuring quadrupole interaction frequencies[1]). Such experiments have been successfully carried out using the electric field gradients in single metallic crystals; these results will be reported in another paper[2]).

This letter calls attention to a rather peculiar property of the theoretical angular correlation which has hitherto gone unnoticed and which should be of great use to experimentors in the field of perturbed angular correlations in general. We have discovered a geometry for which there cannot be a static quadrupole perturbation of an angular correlation, no matter where it is measured in space. The prerequisite geometrical conditions are shown in fig. 1. The experiment must be set up in such a way that the symmetry axis (c - axis) of the field gradient remains perpendicular to a certain gamma ray. This gamma ray must be either from a spin one intermediate state or it must be a gamma ray emitted from intermediate spin two to spin zero. The latter case is of some importance in Coulomb excitation experiments. If these conditions are fulfilled, the theoretical angular correlation is not perturbed by a static electric field gradient, regardless of its magnitude.

Fig. 1.

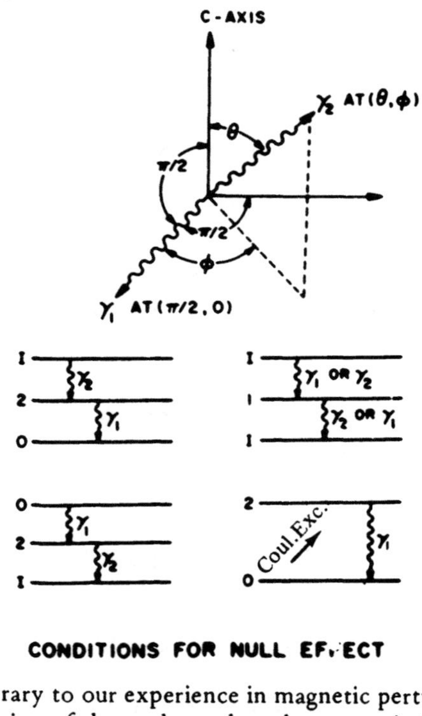

CONDITIONS FOR NULL EF. ECT

This effect is contrary to our experience in magnetic perturbations, which exhibit a maximum precession of the nucleus when the magnetic field is perpendicular to the detector plane. Our intuition expects that the torques involved will be maximized with this geometry, whereas our prediction indicates that no net precession has occurred. A simple physical description is still lacking, but we outline below a quantum mechanical explanation of this null behaviour.

The matrix elements for the m - state transitions under a quadrupole perturbation are given by[3])

$$(m|\vec{Q}\cdot\vec{\nabla}E|m) = a\,(\nabla E)_0\,[3m^2 - I\,(I+1)]$$

$$(m|\vec{Q}\cdot\vec{\nabla}E|m\pm1) = \pm a\frac{\sqrt{6}}{2}(\nabla E)_{\pm1}\,(2m\pm1)\sqrt{(I\pm m+1)(I\mp m)}$$

$$(m|\vec{Q}\cdot\vec{\nabla}E|m\pm2) = a\frac{\sqrt{6}}{2}(\nabla E)_{\pm2}\sqrt{(I\mp m)(I\mp m-1)(I\pm m+1)(I\pm m+2)}$$

where $a = \dfrac{eQ}{2I\,(2I-1)}$.

The five irreducible tensorial components of the electric field gradient are written in terms of the spherical angles θ and ϕ of the field symmetry axis, with the z axis along the emitted gamma ray:

$$(\nabla E)_0 = 1/4\ eq\,(3\cos^2\theta - 1)$$

$$(\nabla E)_{\pm1} = \pm\frac{\sqrt{6}}{4}\,eq\,\sin\theta\cos\theta\,e^{\pm i\phi}$$

$$(\nabla E)_{\pm2} = \frac{\sqrt{6}}{8}\,eq\,\sin^2\theta\,e^{\pm 2i\phi}$$

where $eq = \sum_j e_j\,(3\cos^2\theta_j - 1)/r_j^3$.

When $\theta = \frac{\pi}{2}$ the conditions for the null effect are fulfilled and the matrix elements reduce to the following:

$$(m|\vec{Q}\cdot\vec{\nabla}E|m) = -1/4B\,[3m^2 - I(I+1)]$$

$$(m|\vec{Q}\cdot\vec{\nabla}E|m\pm1) = 0$$

$$(m|\vec{Q}\cdot\vec{\nabla}E|m\pm2) = 3/8B\sqrt{(I\mp m)(I\mp m-1)(I\pm m+1)(I\pm m+2)}\;e^{\pm 2i\phi}$$

where $B = \dfrac{e^2Qq}{2I(2I-1)}$.

We deduce that $\Delta m = \pm 1$ is forbidden, and the perturbation only induces $\Delta m = \pm 2$ transitions under the geometrical conditions of the null effect.

Consider now the two cases which display the null effect. In the case of intermediate spin one, the intermediate state is intially populated in the $m_i = \pm 1, 0$ sublevels. Under the $\Delta m = \pm 2$ selection rule these substates go to $m_f = \mp 1, 0$ respectively. Assuming that the initial $m_i = \pm 1$ sublevels were equally populated, this means that no net change in m - state population has taken place, since the probabilities for $m_i = \pm 1 \rightarrow m_f = \mp 1$ are equal. We conclude that there has been no observable perturbation of the angular correlation.

The second case of the 0 - 2 (or 2 - 0) gamma ray follows a similar argument. We have defined the z axis to be along this gamma ray, so it can only radiate by $\Delta m = \pm 1$. Hence the 0 - 2 gamma ray initially populates only the $m_i = \pm 1$ sublevels of the intermediate state with spin I = 2. Our $\Delta m = \pm 2$ selection rule then implies that these $m_i = \pm 1$ substates transform equally to $m_f = \mp 1$ substates. Again there is no net m - state population change. For the 2 - 0 gamma ray the same argument follows in reverse: since it goes to spin zero, it can only come from $m_f = \pm 1$ sublevels in the intermediate I = 2 state. Our $\Delta m = \pm 2$ selection rule then implies that these $m_f = \pm 1$ substates were transformed under the interaction from $m_i = \mp 1$ substates. Thus there is no perturbation of an angular correlation through intermediate spin I = 2 when a 0 - 2 (or 2 - 0) gamma ray is perpendicular to the field symmetry axis.

This phenomenon has been well confirmed in the course of our experiments: the results will be presented in the paper mentioned earlier[2]). We have found several uses for this effect which we point out in conclusion. (1) It allows a systematic elimination of the static axially symmetric quadrupole effect, thus pointing out the presence of other perturbations such as relaxation or non - axial field gradients. In general other perturbing fields or asymmetries remove the null effect, since the m - state populations are then not locked into the ± 1 sublevels, and the quadrupole field can act on m - states populated by the other perturbations. (2) It makes possible a study of progressive in - beam radiation damage, since it depends sensitively on the preservation of the directional order of the lattice field gradient, as well as on the absence of other perturbing fields. (3) Finally, it allows a direct measurement of the perturbed and the unperturbed angular correlations in a single crystal quadrupole moment experiment, without having to direct the c - axis along one of the radiations. This is important since the c - axis may be the symmetry axis of other static fields which are contributing attenuations to the angular correlation; these fields will remain undetected when the c - axis points along one of the radiations, but can be detected through their removal of our null effect.

* Supported in part by U.S. Atomic Energy Commission.

References

1. C.T. Alonso, Ph.D. Thesis, L.N.S. - M.I.T., 1970, to be published.
2. C.T. Alonso, L. Grodzins, E. Kaufmann and D. Murnick, paper in preparation.
3. R.V. Pound, Phys. Rev. 79 (1950) 685.

Contributed Paper

RESULTS FROM OBSERVED RESONANCES IN PERTURBED ANGULAR CORRELATIONS DUE TO CROSSING HYPERFINE LEVELS

H.J. Leisi, Th. v. Ledebur and F.N. Gygax
Laboratory for High Energy Physics, Swiss Federal Institute of Technology, Zürich, Switzerland

Resonances in perturbed angular correlations due to crossing hyperfine levels have, so far, been observed on two cascade transitions, the 172 - 203 keV cascade[1]) in ^{127}I and the 55 - 188 keV cascade[2]) in ^{125}I. According to the theory[3]) the shape of the resonance signal in the usual geometry (magnetic field perpendicular to the detector plane) depends strongly on the angle Φ between the two γ rays. In particular at angles $\Phi = 5\pi/4$, $3\pi/4$ a pure dispersive shape is expected. We have measured both γ - γ angular correlations at these angles and obtained the results shown in figs. 1 and

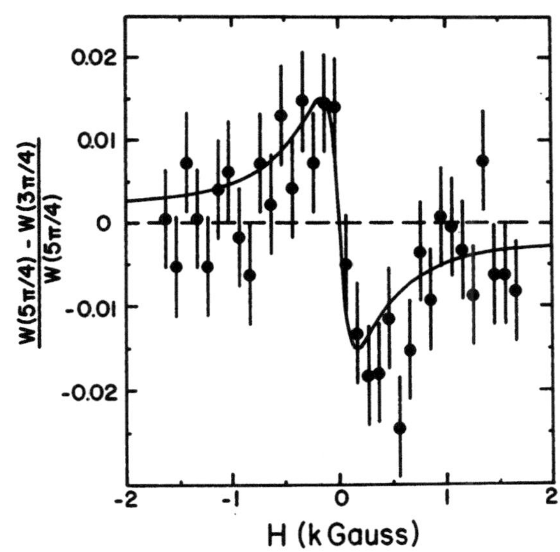

Fig. 1. 172 - 203 keV cascade measured as a function of a magnetic field perpendicular to the detector plane.

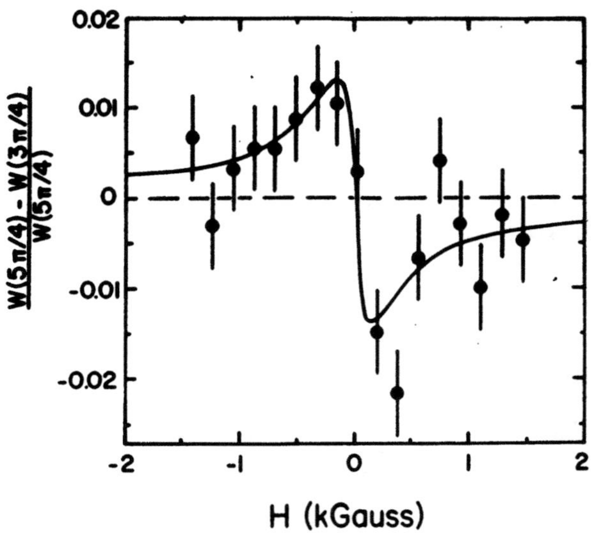

Fig. 2. 55 - 188 keV cascade measured as a function of a magnetic field perpendicular to the detector plane.

2. With the same ^{127}Xe source we have remeasured the resonance at angles $\Psi = \pi$ and $\pi/2$ (fig. 3). The results from these and from all earlier experiments are listed in table 1. All experiments have been analyzed with a single - width curve, and the resonance parameters in table 1 are defined by the expressions

$$\frac{W(\pi) - W(\pi/2)}{W(\pi/2)} = \frac{A_0}{1 + (H/\Delta H_1)^2} + A(\infty) \tag{1}$$

$$\frac{W(5\pi/4) - W(3\pi/4)}{W(5\pi/4)} = -B_0 \frac{H/\Delta H_2}{1 + (H/\Delta H_2)^2} + B(\infty) \tag{2}$$

It has been observed that ΔH_1 and ΔH_2 are equal within experimental errors and that the amplitude B_0 is only slightly smaller than A_0. The difference between $A(\infty)$ and $B(\infty)$ is explained by a fraction of decays which does not show a resonance[4]). These decays do not contribute to expression (2) and hence $B(\infty)$ is very small.

If the hyperfine interaction is strictly stationary during the lifetime of the intermediate state, the resonance width can be expressed as

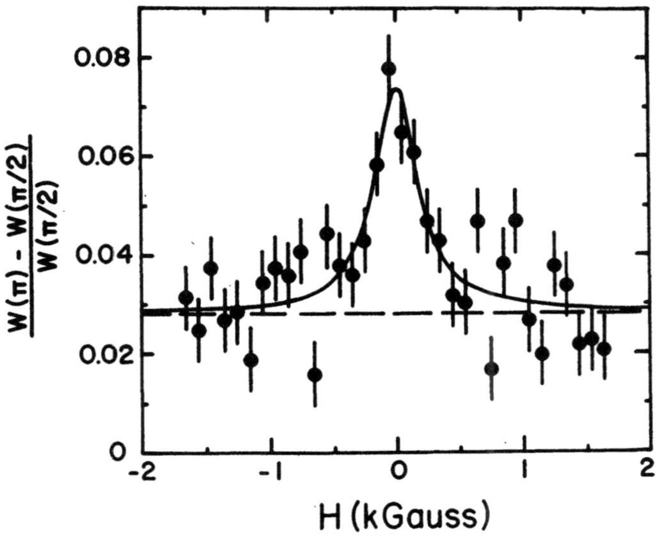

Fig. 3. 172 - 203 keV cascade measured as a function of a magnetic field perpendicular to the detector plane.

$$\Delta H = \frac{\Gamma}{2\mu_B < |\gamma_F g_J| >} \tag{3}$$

where Γ is the natural width of the nuclear state, g_J is the atomic splitting factor, γ_F is the ratio of the splitting factor g_F and g_J, and μ_B is the Bohr magneton. If an experiment is performed on a single atomic state, g_J is known and the average in eq. (3) can be evaluated. The measured width ΔH then determines directly the lifetime of the nuclear state. In the experiments performed so far two complications arise. First, the systems consist of an ensemble of many atomic states which are produced by the preceding K - capture decays. Second, the hyperfine interaction is not fully stationary, as was shown[4]) from an analysis of A_0 and $A(\infty)$. Time dependence in principle increases the resonance width above the value given by eq. (3). We conjecture that this effect is small since very special conditions on atomic lifetimes would have to be realized in order to change ΔH appreciably. We therefore suspect that the average value

$$< |\gamma_F g_J| > = 0.045 \pm 0.05 \tag{4}$$

deduced from the ΔH values of table 1 by eq. (3) is representative for other K -

Table 1: Resonance parameters as defined by eqs. (1) and (2) (see text)

Cascade (Energies in keV)	Gas pressure (Torr)	ΔH_1 (Gauss)	A_o	$A(\infty)$	Reference
172 - 203	< 1	237 ± 46	0.053 ± 0.006		present work
	2 ± 1	195 ± 41	0.045 ± 0.004	0.054 ± 0.009	ref. (1)
		214 ± 31	0.048 ± 0.003		weighted average
55 - 188	2 ± 1	400 ± 160	0.040 ± 0.007	0.049 ± 0.009	ref. (2) [a)]

Cascade (Energies in keV)	Gas pressure (Torr)	ΔH_2 (Gauss)	B_o	$B(\infty)$	Reference
172 - 203	< 1	204 ± 52	0.034 ± 0.006	-0.003 ± 0.001	present work
55 - 188	3.6 ± 1.4	234 ± 69	0.029 ± 0.005	0.003 ± 0.001	present work

a) Values slightly different from ref. (2) as a result of a more thorough evaluation of the data.

capture decays. Since the hyperfine interaction after K - capture is extremely large[4]) the resonance condition (hf splitting large compared to Γ) may still be realized for very short nuclear lifetimes, particularly for heavy atoms. Thus, resonance experiments may possibly be used to measure short nuclear lifetimes.

References

1. H.J. Leisi, Phys. Rev. 1A (1970) 1654.
2. Th. v. Ledebur, F.N. Gygax and H.J. Leisi, Helv. Phys. Acta 42 (1969) 581.
3. H.J. Leisi, in Hyperfine Structure and Nuclear Radiations, edited by E. Matthias and D.A. Shirley (North - Holland, Amsterdam, 1968) p. 384.
4. H.J. Leisi, Phys. Rev. 1A (1970) 1662.

Contributed Paper

PERTURBED ANGULAR CORRELATIONS IN PARAMAGNETIC SINGLE CRYSTALS OF RARE - EARTH ETHYLSULFATES

H. Dirren and H.J. Leisi
Laboratory for High Energy Physics, Swiss Federal Institute of Technology, Zürich, Switzerland

We have studied the γ - γ directional correlation of the 1380 - 81 keV cascade in the decay of ^{166}Ho which was magnetically diluted in single crystals of La - ethylsulfate (LaES) and Y - ethylsulfate (YES). The angular correlation function was measured at room temperature and at 6° K for different orientations of the single crystals. Holmium oxide was irradiated and the crystals were grown afterwards from aqueous solutions, according to different chemical procedures (see table 1). The measurements were evaluated in terms of attenuation factors G_k which are listed in table 1. The main features of the results may be summarized as follows:

(i) All experiments show a large temperature effect[1]) on G_2 and G_4.

(ii) The effect of crystal orientation is small in all cases.

(iii) No decisive differences for different chemical preparations are observed.

Samples obtained from chemical procedures I, II and III have, in addition, been investigated with EPR, using the Stark modulation technique[2]). Signals from Ho^{3+} were detected and interpreted as arising from a site with inversion symmetry[2]) which does not correspond to the La or Y lattice site.

The strong temperature effect observed is clearly due to the temperature dependence of the spin - lattice relaxation time. If time - dependent interactions due to

Table 1

Attenuation factors at zero magnetic field for different orientations of the cristalline axis.

Chemical preparation a)	Crystal orientation b)	Temperature [°K]	G_2	G_4
I	⇕	7 ± 1 300	0.185 ± 0.056 0.471 ± 0.084	0.078 ± 0.030 0.703 ± 0.047
II	⇔	5.5 ± 0.5 300	0.218 ± 0.053 0.646 ± 0.057	0.146 ± 0.030 0.530 ± 0.031
III	⇔	5.5 ± 0.5 300	0.342 ± 0.045 0.591 ± 0.076	0.176 ± 0.023 0.592 ± 0.041
IV	⇔	5.5 ± 0.5 300	0.280 ± 0.064 0.577 ± 0.056	0.118 ± 0.028 0.572 ± 0.030
	⇕	5.5 ± 0.5 300	0.126 ± 0.048 0.541 ± 0.050	0.120 ± 0.025 0.564 ± 0.027

a) I: Ho oxide → Ho chloride + LaES
II: Ho oxide → Ho chloride → HoES + LaES
III: Ho oxide → Ho chloride → HoES + YES
IV: Ho oxide → Ho sulfate → HoES + YES

b) ⇕ : c axis perpendicular to the detector plane
⇔ : c axis parallel to one detector axis

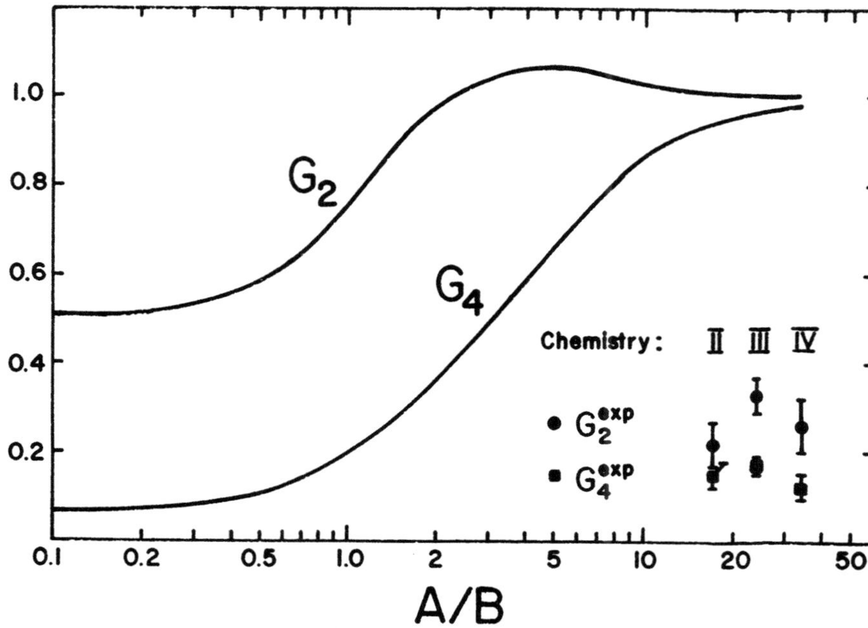

Fig. 1. Attenuation factors G_2 and G_4 for single crystal orientation parallel to one detector axis as a function of the anisotropy parameter A/B for spin sequence 0 - 2 - 0. The curves represent lower limits (strong hyperfine interaction). The experimental values are obtained from various samples prepared by chemical procedures II, III and IV, respectively.

relaxation no longer occur at helium temperature and the hyperfine interaction is axially symmetric, then the attenuation factors depend only on the magnetic anisotropy parameter A/B in the strong perturbation limit[3]). Values of G_k as a function of A/B are shown in fig. 1, corresponding to a geometry with the c - axis oriented parallel to one detector axis. None of the experimental G_2 values are compatible with any value of the anisotropy parameter. We therefore conclude that the active ion does not occupy a site with axially symmetric hyperfine interaction, provided relaxation effects are absent at 6° K. Further investigations are in progress.

References

1. H. Dirren, Th. v. Ledebur and H.J. Leisi, Helv. Phys. Acta 41 (1968) 1287.
2. P. Wysling, Ph. D. Thesis, University of Zurich, 1969 (unpublished).
3. H.J. Leisi, and R.T. Deck, Phys. Rev. 129 (1963) 2117; 131 (1963) 2840.

Contributed Paper

EFFECTS OF PAC ON HALF - LIFE MEASUREMENTS

U. Bäverstam and M. Höjeberg
University of Stockholm, Institute of Physics, Stockholm, Sweden

1. Introduction

The delayed coincidence technique is now a standard method in measuring half - lives of excited nuclear levels. The quantity measured is the time delay between a feeding transition and a de - exciting transition. There are mainly three different methods used to analyze the recorded time spectrum:

1) the slope method, where the decay constant is calculated from the slope of the time spectrum. This can be done when the measured half - life is longer than the time resolution.

2) the (first) moment shift method, where the centroid shift between the prompt and delayed time spectra is used. This can be done for half - lives down to a few per cent of the resolution.

3) the convolution method[1]), where the numerical convolution of the prompt and delayed curves is fitted to the measured spectrum.

These methods have been extensively discussed by several authors (see refs. 2 and 3). The under - lying assumption in all these methods is that the decay curve is a simple exponential. However, if there is a PAC connected to the cascade in question, this is certainly not the case. It will be shown in this paper that, in fact, such perturbations can give errors in the calculated half - lives as large as 20 - 30% in unfavourable cases.

2. Static and time - dependent perturbations

Generally the perturbations are caused by electric or magnetic fields from the atomic surroundings. The mechanism of these perturbations have been carefully examined [4]) , though it is in general not feasible to give analytical expressions. But in many cases the perturbations can be approximately described by parameters such as the precession frequency ω or the relaxation time t_r. The coincidence counting rate can then be written:

$$f(\theta, t) = e^{-\lambda \cdot t} \sum_k G_k A_k P_k (\cos \theta) \qquad k = 0, 2, 4 \ldots \tag{1}$$

where G_k is a function of ω and/or t_r and t. When no perturbations are present G_k =

= 1 and the decay is purely exponential, otherwise this is not valid. The detailed structure of G_k depends on the spin sequence and the character of the perturbation. For an axially symmetric electrostatic field gradient it can be written[5]) as

$$G_k = \sum_n s_{kn} \cos(n\,\omega\,t) \tag{2}$$

When the direction and/or magnitude of the perturbing field vary G_k cannot be given analytically or has to be given by complicated functions, but in the simplest case where the field strength is constant, while the direction fluctuates randomly, the attenuation factor can be approximated by

$$G_k = e^{-t/t_r} \tag{3}$$

Even in cases where G_k could not be given analytically, (3) might be a rather good approximation. The effect of these perturbations in actual life - time measurements will be investigated below.

3. Calculations

Usually $A_2 \gg A_4, A_6$, etc., so for simplicity only the first two terms of eq. (1) are considered. Computer simulated time spectra were analyzed with a computer program[1]). The time spectra were obtained by numerical convolution of a prompt spectrum and a delayed curve of the following form:

$$f(\theta, t) = e^{-t}\,[1 + e^{-t/t_r} . \cos(\omega t) \,.\, A_2 P_2(\cos\theta)] \tag{4}$$

where $A_2 = 25\%$. Parts of our results are summarized in table 1. Errors greater than 20% occur although the parameters used are by no means unreasonable.

Table 1

θ	t_r	ω	λ_1	λ_2
180°	2.0	0	1.02	1.05
90°	2.0	0	0.99	1.00
180°	∞	0.5	1.24	1.21
90°	∞	0.5	0.90	0.93

Table 2

θ	λ	$\omega \cdot t$
	1.00	0
180°	1.10	0.33
95°	0.91	0.33

(λ_1 calculated from the slope, λ_2 from the convolution fit). Decay curves were also calculated from (1) using G_2 for I = 5/2 from (2) and table 5 (ref. 5), $A_2 = 0.80$ which is large, but observed in for example ^{75}As, $5/2^+ - 5/2^- - 3/2^-$ cascade and $\theta = 90$ and 180°. A least square fit to the decay, assuming an exponential shape, was performed, and the results are given in table 2.

4. Experiments

Experimental evidence for the effect discussed has also been obtained. We have measured the 165L - 134 gamma transition in ^{197m}Hg, where the intermediate level has a half - life of 7.0 ns, the A_2 is .21 and A_4 vanishes (see Bäverstam and Othaz, results presented at this conference). Moreover, the intermediate level is strongly perturbed in some cases[6]). The coincidences were recorded at three angles (180°, 135° and 95°) with a conventional fast - slow coincidence set - up. The time - resolution was approximately 3 ns, and the electronic drift was negligible (moreover, the convolution method is very slightly dependent on drifts). The maximum number of coincidences per channel was about 10^5. The decay constants calculated from the convolution fit for the different angles is shown in table 3. (These results are not given in proper scale). The errors given are the statistical standard deviations. If other possible errors are taken into account, the errors may become twice the quoted values, but the 7% difference between the results can definitely not be explained by poor accuracy in the measurements, but are certainly a consequence of the perturbation in the source.

Table 3

θ	λ
180°	5.71 (6)
135°	5.48 (6)
95°	5.35 (4)

References

1. L. Boström, B. Olsen, W. Schneider, and E. Matthias, Nucl. Instr. and Meth. 44 (1966) 61.
2. B. Olsen and L. Boström, Nucl. Instr. and Meth. 44 (1966) 65.
3. P. Sparrman and F. Falk, Arkiv Fysik 32 (1966) 447.
4. H. Frauenfelder, and R.M. Steffen, Alpha-, Beta- and Gamma-Ray Spectroscopy, Ch. XIX A (ed. K. Siegbahn, North - Holland Publ. Co., Amsterdam, 1965).
5. op. cit., p. 1118.
6. B.G. Pettersson, J.E. Thun, and T.R. Gerholm, Nucl. Phys. 24 (1961) 223; U. Bäverstam, A. Johansson and T.R. Gerholm, Arkiv Fysik 35 (1967) 451.

Discussion

Steffen: I don't think the assertion of the speaker that the angular correlation people have failed to point out the influence of perturbations on life - time measurements is correct. In a paper published 15 years ago by P.C. Simms and myself we discussed not only this effect, but also showed experimentally that this effect is important, by measuring the life - time of the 247 keV state of ^{111}Cd at 90° and 180° in viscous liquid sources. A pronounced difference of the apparent life - time was found in these measurements.

Bäverstam: Let me add that this mercury level which I mentioned and where we found a difference of 7% was measured in Uppsala 10 years ago by Lindskog who tried to find this effect but he found nothing. What I want to say is that although the effect was pointed out and measured 15 years ago, still people measure half - lives without taking any care of these things.

Van Middelkoop: Also in plunger - type lifetime measurements one should be aware of possible systematic errors due to a perturbation of the angular distribution of emitted γ - rays from the stopped ions.

J. Alonso: I think it is a very obvious point, but for many experiments the simplest method of eliminating angular correlation effects is to increase the solid angle of the detectors by bringing them closer to the source. If each detector has a solid angle of 2π, there will be no angular correlations at all.

Bäverstam: That is true. But if you make use, for example, of two magnetic electron spectrometers to measure electron - electron correlations, you have a fixed geometry and you cannot increase the solid angles.

IMPLANTATION PERTURBED ANGULAR CORRELATION TECHNIQUES IN FERROMAGNETIC AND NON - FERROMAGNETIC MATERIALS

D.E. Murnick

Bell Telephone Laboratories, Incorporated, Murray Hill, New Jersey,U.S.A.

1. Introduction

The field of ion implantation has experienced a major growth of interest in the last several years. To a large extent this growth has been stimulated by semiconductor research with the realization that ion implantation provides a means for precisely controlling the concentration and depth distribution of impurities in semiconductor crystals. As a result of this research, a body of information and understanding of heavy ion ranges, stopping power and radiation damage has accumulated. In addition, diagnostic methods for evaluating implanted samples have been developed. Technological advances in ion sources and accelerators have also made it possible for experimenters to obtain beams of virtually any isotope.

This paper is concerned with ion implantation techniques relevant to perturbed angular correlation (PAC) studies. Much of what will be covered applies equally well to Mössbauer effect and nuclear orientation studies, and examples from each of these fields will be used. The primary impetus to the use of ion implantation in these fields has been the study of hyperfine interactions (hfi) on dilute impurities in various materials. The hyperfine magnetic field on iron nuclei in iron metal is -339 kOe[1]), where the minus sign reflects the orientation with respect to the external polarizing field. Samoilov and others discovered that the magnetic field on gold nuclei in iron is -1280 kOe[2]). Subsequently, many workers measured other large fields on impurity nuclei in ferromagnetic materials. Fig. 1 shows our present knowledge of hyperfine fields on impurities in iron metal.

It is clear that these large and easily obtainable magnetic fields allow a study of nuclear magnetic moments of states having much shorter lifetimes than were previously possible. It is equally clear that known nuclear magnetic moments can be used to study these hyperfine fields, and indeed, this has been the case. For many elements, Pd, Au, Cd, Ru, for example, radioactive alloys for angular correlation, Mössbauer effect or nuclear orientation studies can be prepared by alloying techniques or by diffusion. For other very interesting systems, specifically the rare earths in iron, no such techniques are available. Ion implantation either using a beam of radioactive ions from a particle accelerator or recoil implantation from a nuclear reaction or alpha decay are the only preparation methods available to date.

Similar statements can be made about electric quadrupole interaction studies in crystals. Information about quadrupole fields as well as nuclear quadrupole moments is obtainable from such experiments. For many important cases, growing

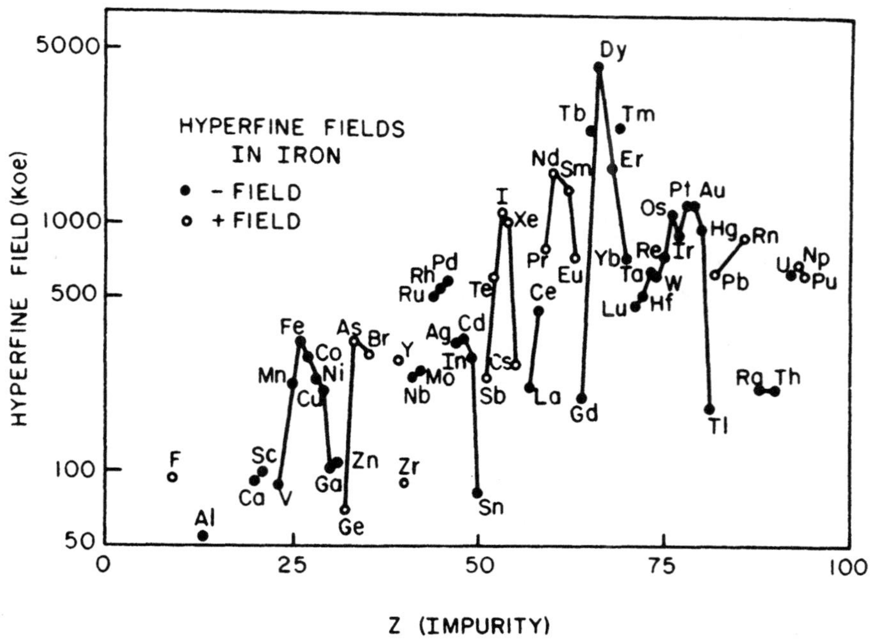

Fig. 1. Hyperfine fields on dilute impurities in iron at room temperature. Data available to July 1970 are used.

single crystals with low radioactive impurity concentrations may be impossible. And, the multitude of interesting systems which could be studied might require many varied techniques. The universality of ion implantation for any impurity - host combination is quite apparent.

The primary advantage of ion implantation for hyperfine interaction studies is the just mentioned free choice of host material and impurity atoms. Virtually any atom can be impelled with sufficient energy into any host lattice. A secondary advantage is that extremely low dilutions are possible with implantation techniques. In fact, for reasons discussed below, implantation concentrations greater than a few parts per thousand are to be avoided.

Several aspects of ion implantation hyperfine interaction studies will be discussed in this paper. Of primary importance is the site of the implanted impurity and the uniqueness of that site. In addition, what damage, if any, has been caused so that the impurity is not in normal surroundings? And finally, for recoil experiments what happens during the particle flight to change the nuclear alignment? These questions with extensive (though not completely inclusive) examples from the literature are considered in turn. Section 2 deals with methods of implantation emphasizing similarities common to all. Depth of penetration, radiation damage and lattice location are considered in section 3. The string effect is discussed and applied to the study of impurities in iron. Section 4 deals with effects unique to recoil implantation, in particular the transient field effect on fast ions in ferromagnetic media. Section 5 briefly reviews recent single crystal implantation hfi experiments.

2. Implantation methods

Isotope Separator

Recent advances in ion source and accelerator technology have made it possible to obtain beams of virtually any element. Reviews of isotope separator techniques for ion implantation can be found in ref. 3 and 4. As discussed by Freeman[3]), ion implantation with radioactive beams present some special requirements and problems. Separator efficiencies are only of the order of 5% so radioactive contamination presents a serious problem. For this reason many installations designed for stable isotope work are unprepared to implant radioactive species, and others will only work with relatively short lived nuclides. Nevertheless, because of the great amount of interesting physics to be done, particularly in the PAC field, better techniques are being developed and the next few years will undoubtedly see an increase in this area of research.

Depth of penetration and impurity concentration are sometimes a problem for separator produced sources. Except for surface studies, it is important that all atoms in a hfi study be in an environment as equivalent as possible to that of the bulk material. The range of 30 keV Yb in iron is approximately 80 Å with a range straggling of about 25 Å. Even carefully prepared iron samples have oxide layers of about 50 Å and magnetization of surface domains is sometimes difficult. Reid et al.[5]) (table 1) find a large variation in hyperfine fields for Au and Sb implanted in iron at different energies. Ravn et al.[6]), on the other hand, found no variation of PAC effects for Yb implanted into iron at energies down to 30 keV (fig. 2). Other experiments, however, have confirmed a variation in effective fields as a function of implantation energy[5,7]). As many existant ion accelerators are only in the 30 - 60 keV range, an improvement may be obtained by post - acceleration by biasing the target at 30 to 100 keV.

Radioactivity hfi experiments demand low ($\leqslant 1\%$) impurity concentrations. As the amount of radioactive material to begin with is usually small, this places very stringent requirements on the isotope separator. There is usually a significant amount of stable isotope present at a neighbouring mass to the desired radioactive mass, and in addition there are possible hydride ions and other contaminents. Poor separation together with relatively low energy implants may lead to much higher actual impurity concentrations than the amount of radioactive material present indicates[3,8]).

Alpha Recoil

After emission of an alpha particle, the typical recoil energy imparted to the daughter nucleus is about 100 keV. Effectively, then, the daughter is implanted at about 100 keV into its surroundings. Two types of hfi experiments have recently been reported in the literature. In one, a gamma - gamma PAC experiment is done on a nuclide which was formed in an alpha decay chain. Pramila et al.[9]), and inde-

Table 1. Implantation h.f. interaction data of Reid et al.[5]

Source	Fe - foil	Surface	Energy (keV)	Dose per cm^2	Annealing	$R=\overline{H}/H$ (diffused)
^{198}Au I	Polycrystal	Etched	30	4×10^{13}	None	0.1
^{198}Au II	Polycrystal	Etched	100	4×10^{13}	None	1.00 ± 0.05
					100 min, 800°C	1.00 ± 0.05
^{198}Au III	Oriented polycrystal	Etched	100	4×10^{13}	None	0.75 ± 0.05
					100 min, 800°C	1.00 ± 0.05
^{198}Au IV	Oriented polycrystal	Slightly oxidized	100	4×10^{13}	None	0.28 ± 0.04
					13 h, 300°C	0.28 ± 0.04
^{122}Sb I	Polycrystal	Etched	40	1×10^{12}	None	0.45 ± 0.05
					100 min, 800°C	0.85 ± 0.05
^{122}Sb II	Polycrystal	Polished, weak etch	80	4×10^{12}	None	0.80 ± 0.05
					100 min, 800°C	1.00 ± 0.05
^{122}Sb III	Single crystal <110> along beam	Polished, weak etch	80	2×10^{11}	None	1.00 ± 0.05

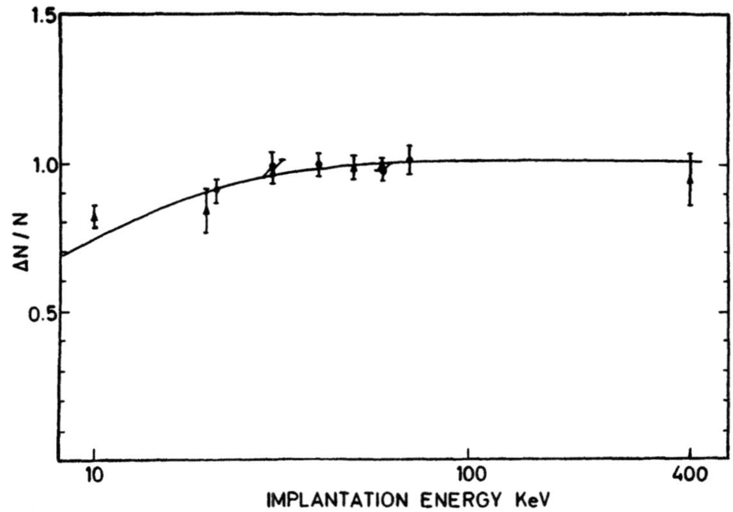

Fig. 2. The PAC effect relative to the value at 60 keV implantation energy for ^{169}Tm (circles) and ^{175}Lu (triangles) as a function of implant energy (from ref. 6). In both cases radioactive Yb was the implant.

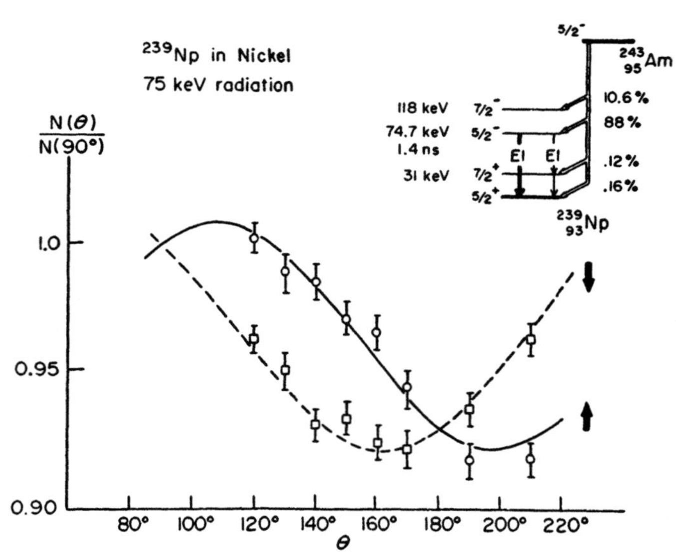

Fig. 4. Example of $\alpha \cdot \gamma$ recoil PAC experiments of Ansaldo et al. (ref. 12).

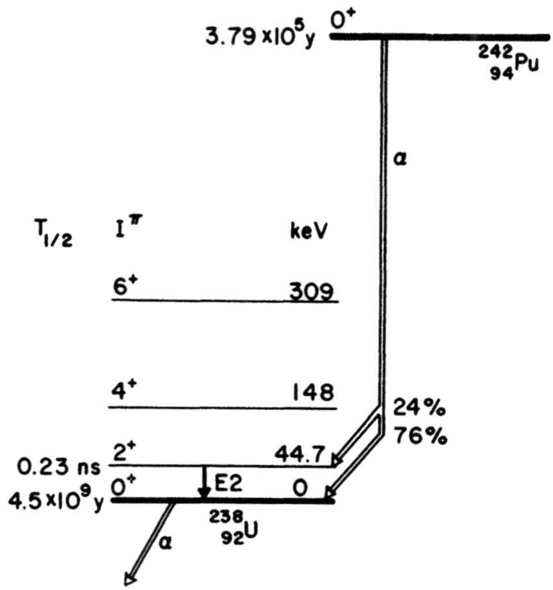

Fig. 3. ^{238}Pu decay and ^{238}U level scheme. Mössbauer studies on the ^{238}U 2^+ state are after alpha recoil (ref. 11).

pendently, Bowman and Zawislak[10]), studied states in ^{208}Pb in iron in this way. In both experiments, the effective hyperfine fields on lead differed from that on carefully prepared annealed diffused samples.

The second type of experiment which is similar to reaction recoil experiments discussed below, is α - γ or α - e^- PAC experiments. Similar from an ion - implantation standpoint, are Mössbauer effect studies where the resonance level is formed via alpha decay[11]) (fig. 3). Ansaldo et al.[12]) (fig. 4) have been studying alpha gamma correlations on impurities in iron. In addition to pure aligned magnetic field effects, they have observed other attenuating perturbations. Similar to accelerator produced sources, surface effects in this case might be a problem. In addition, of course, the 'implanted' ion loses energy in traversing the source thickness itself. For this reason thin sources with high specific activities must be used.

Nuclear Recoil Implantation

Nuclear recoil implantation perturbed angular correlation (IMPACT) studies have been reviewed in several publications in the last few years[13,14]). Fig. 5 typifies the experimental arrangement used in several laboratories. A beam of high energy particles excites the nuclear state of interest and implants it by recoil into a suitable

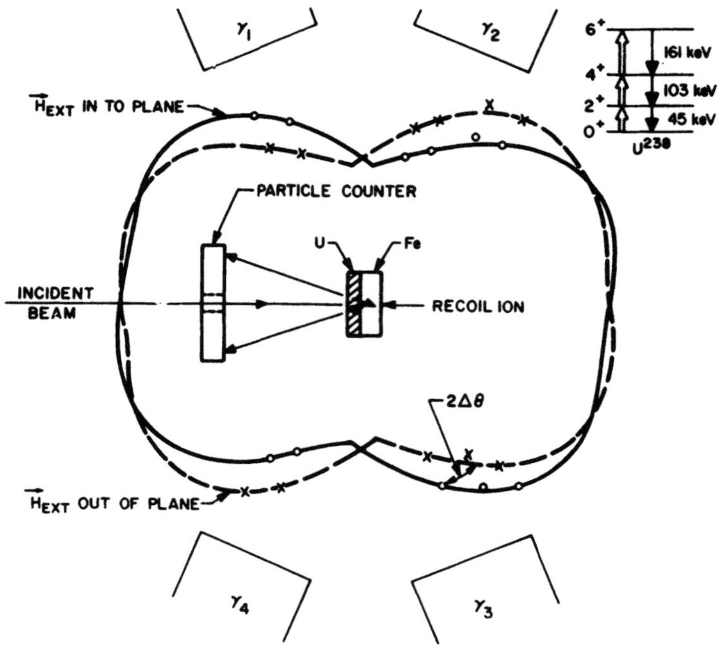

Fig. 5. Schematic experimental arrangement for recoil implantation perturbed angular correlation (IMPACT) experiment. The case illustrated is the $4^+ \rightarrow 2^+$ gamma correlation for ^{238}U implanted into iron.

host lattice. In the experiment illustrated rotational levels of ^{238}U are Coulomb excited by 80 MeV ^{32}S ions and implanted into iron metal. The gamma ray angular correlation is measured in coincidence with backscattered sulfur ions. The coincidence condition ensures implantation energies of the order of several MeV as well as an enhanced anisotropic correlation. The first such experiments by Scharenberg et al.[15]) were on tungsten implanted into tungsten with an external magnetic field applied. In that case unperturbed correlations were measured. Subsequently, Goldring et al.[16]) determined that Cu was a suitable perturbation - free environment for many excited nuclei. Grodzins, Borchers and Hagemann[17]), and Boehm, Winther and Hagemann[18]) were first to utilize the large hyperfine fields available by implanting rare earths into iron.

Coulomb excitation is by no means the only reaction which can be used for these experiments. Christiansen et al. studied the Mössbauer effect on iron implanted into iron via the (d, p) reaction[19]). Yamazaki and Matthias[20]), implanted Po into lead Experiments have also been reported using (p, p′) reactions[22]). In all of these cases with the exception of (p, p′), implantation energies are in the

MeV range so that surface effects can be ignored. Furthermore, the impurity concentrations involved are smaller than in any other type of experiment, less than .01%.

3. Range straggling, radiation damage and lattice location of implanted ions

The total dose of implanted ions can usually be stated with fair accuracy. The volume concentration, however, is much less certain. One usually assumes a gaussian distribution about some mean range with a mean square straggling determined by theoretical considerations. Lindhard, Scharff and Schiott[23]) discuss straggling formulae for high energy heavy ions. For implants of energies less than 100 keV, $\frac{\Delta R}{R} \approx 0.3$ (see ref. 24). In any given situation, however, the actual impurity distribution is a function of implant energy, sample purity and, especially temperature history. Many implanted sources have been annealed with differing results. The hyperfine field on implanted antimony and gold in iron increased upon annealing[5]), whereas that on Tm and Cs in iron tended to decrease[6,8]). No simple rule can be stated, the annealing behaviour of each element in each host must at present be considered independently.

Implanted ions slow down and stop in their host medium losing energy as they do so to the lattice. In addition, in recoil implantations, the primary accelerated particle beam, or alpha particles, also pass through the host lattice. The dissipated energy in addition to heating the host lattice produces radiation damage in the form of vacancies and interstitials and clusters of these from thousands of atomic collisions per ion. Nevertheless, Mössbauer effect and PAC studies on many implanted impurities in metals, particularly iron and copper, indicate unperturbed environments for the decaying nucleus.

The most important reason for this effect is the truly microscopic probing of the hyperfine interaction. The implanted atom is only affected by its nearest and perhaps second nearest neighbours. This is not quite true for quadrupole interactions, but these should be small in any case in a cubic lattice. Thus, though there may be significant damage in its wake, it is only the environment near where the implant is stoppped that is significant for the hfi.

Fig. 6 shows the stopping power in dimensionless units for heavy ions in dense media[23]). For recoil implantation energies most slowing down is electronic in nature, causing minimal lattice damage. As the recoiling atom slows down, nuclear stopping finally becomes important and host atoms are displaced from their lattice sites by the collisions. Some lattice atoms recoil with large energies (> 100 eV) and undergo further collisions that contribute to a collision cascade. The total number of host atom displacements is approximately $n_d = E/2E_d$, where E_d is the displacement energy, about 50 eV for iron in iron, and E is the energy expended in nuclear collisions.

At very low energies (< 1 keV) the collision mean free path is determined largely by the interatomic spacing. In close packed lattices in this energy region, focused collision sequences are likely to occur – i.e., each atom having a replacement collision with its neighbour. Thus the primary implant may be in a lattice site with a vacancy and interstitial far from its location. In heavy closed packed materials such as copper and iron, this type of process is thought to lead to defect clusters which

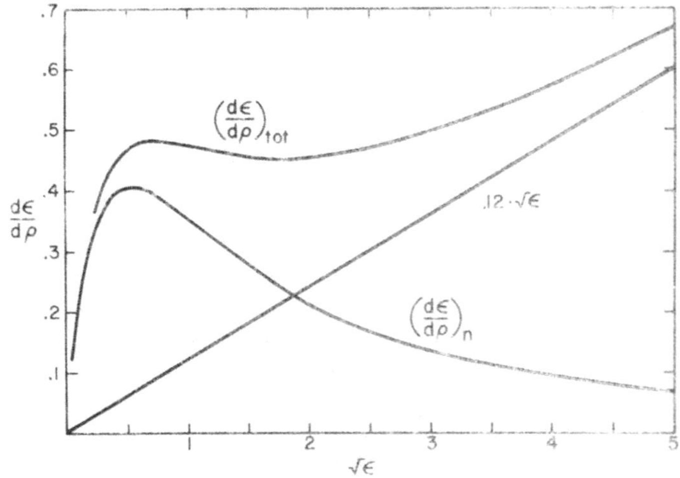

Fig. 6. Electronic, nuclear, and total stopping power as a function of reduced energy. The dimensionless variables are defined in reference 23.

are difficult to anneal out[25]).

Erginsoy and others have made detailed dynamical calculations for the case of iron recoiling in iron at energies up to 1.5 keV [26,27]). Their conclusions corroborate the arguments above and indicate that the iron atom implant always ends up sub-

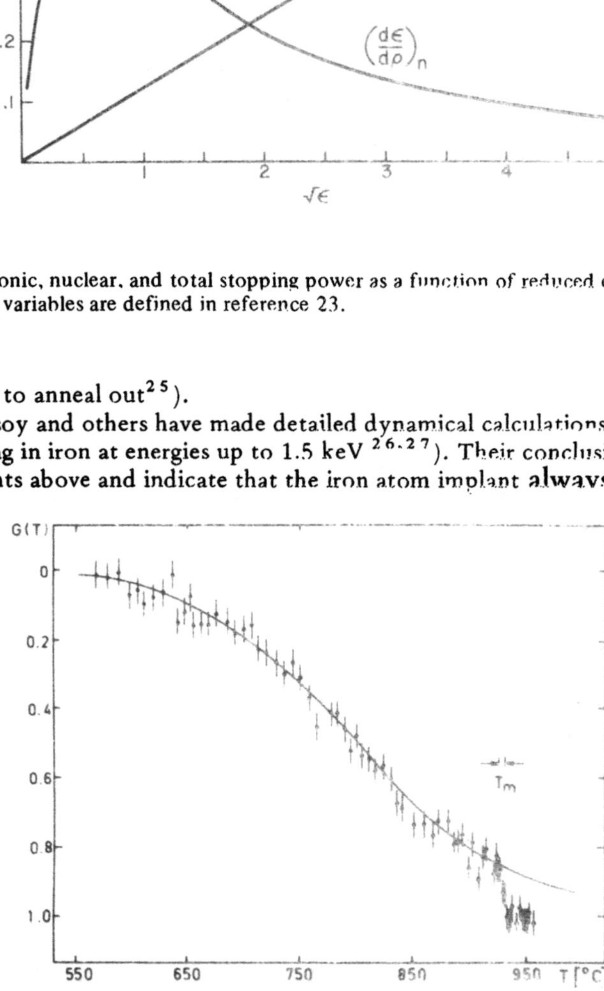

Fig. 7. Attenuation factor of the 426 keV gamma PAC from ^{72}Ge (d, n) ^{73}As recoil implantation. The quadrupole attenuation is associated with vacancy diffusion (ref. 29).

stitutionally, at worst, with one or two nearest neighbour vacancies. Implantation Mössbauer effect results corroborate this conclusion. One cannot, however, extend this conclusion to other implants without further evidence.

Using the stroboscopic resonance technique[28]), Bertschat et al.[29]) have studied radiation induced vacancies near ^{73}As implanted into germanium. Their results, shown in fig. 7 indicate vacancies associated with the implanted ion site induce quadrupole interactions which vary as a function of temperature. These results are not directly applicable to implants in copper and iron, though the technique may be useful in studying effects in these metals. As this experiment demonstrates indirect evidence on the lattice site and surrounding radiation damage of implanted ions can be obtained from analysis of the PAC experiments themselves.

To a large extent, heavy ions implanted into non - magnetic cubic metallic lattices have shown unperturbed correlations. Notable exceptions to this are the rare earths in copper which Waddington and others have shown to exhibit paramagnetic relaxation effects[30]) – an effect probably not due to any special effects of the

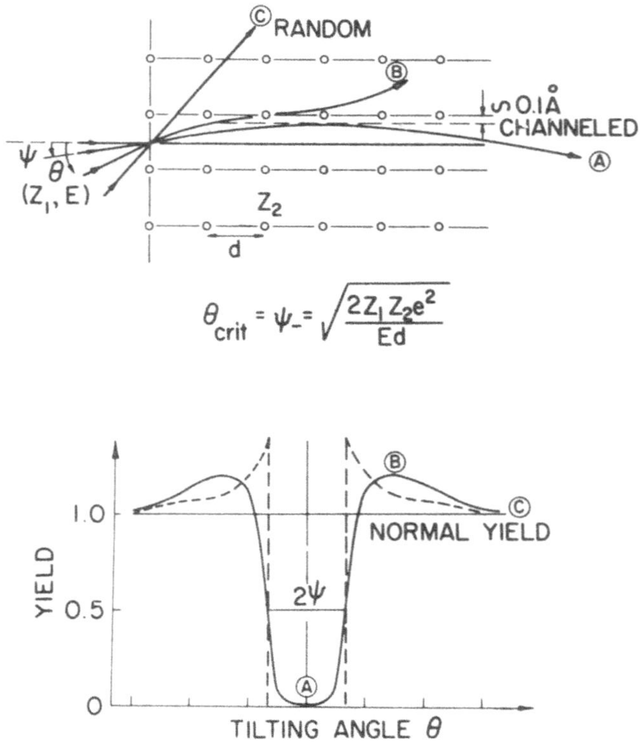

Fig. 8. Schematic representation of the channeling effect showing particle trajectories and yield of a low impact parameter interaction as a function of particle incidence angle.

implantation process. Most impurities implanted into iron and nickel have shown hyperfine interactions consistent with a unique magnetic field. Exceptions, of course, exist. Fluorine implanted into Ni was shown to experience a non - unique field[31]) as was xenon in iron[8]). Light impurities like fluorine, are expected to be interstitial and the rare gases are undoubtedly special cases.

A direct determination of the lattice location of implanted ions, as well as their depth distribution and the degree of radiation damage in the host lattice, can be made using the channeling effect in single crystals[32]). Fig. 8 illustrates the effect which has been the subject of much research during the past several years. A beam of charged particles incident parallel to a 'string' of lattice atoms is effectively channeled along that axis by the continuum potential of the string of atoms. The critical angle for channeling is given by

$$\psi_1^2 = \frac{2 Z_1 Z_2 e^2}{E\, d}$$

where E is the incident ion energy and d is the interatomic distance. The channeled beam does not 'see' either host lattice atoms or impurity atoms substitutionally imbedded into the lattice. It does see interstitial impurity atoms and lattice atoms substantially displaced from their lattice sites, as for example by radiation damage. A beam incident at large angles to the crystal can interact with all host and impurity atoms. Comparison of some reaction yield, such as backscattering in channeled and nonchanneled directions, thus gives information on impurity atom location as well as lattice damage.

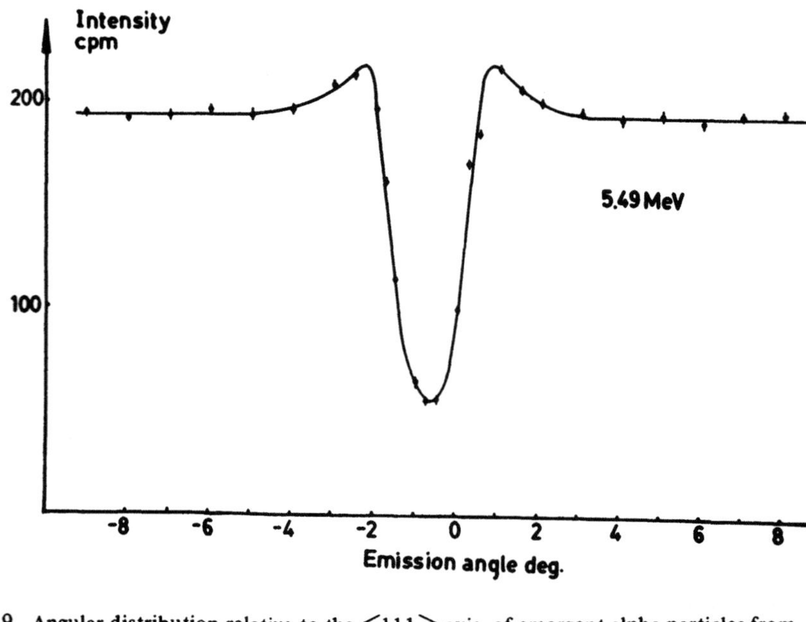

Fig. 9. Angular distribution relative to the <111> axis, of emergent alpha particles from ^{219}Rn implanted into a tungsten crystal (from ref. 33).

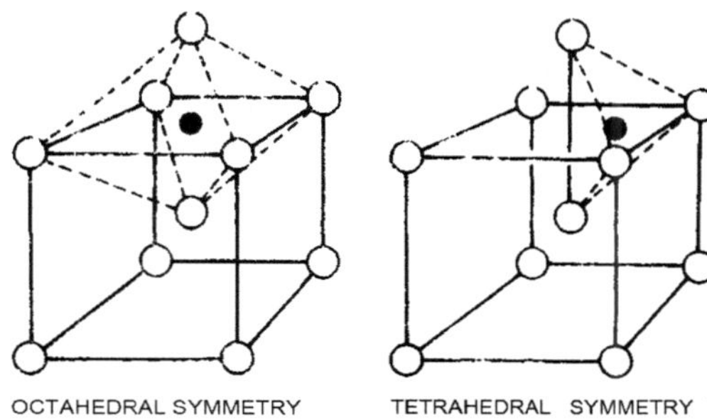

Fig. 10 bcc iron lattice illustrating posible stable interstitial positions

Domeij and Bjorquist[33]) were first to utilize the channeling, or more correctly biocking in this case, effect for the location of impurity ions implanted into a single crystal. Radon was implanted into tungsten and the emergent alpha particle distribution observed. The resuits (fig. 9) indicated that 80% of the radon atoms were at lattice sites.

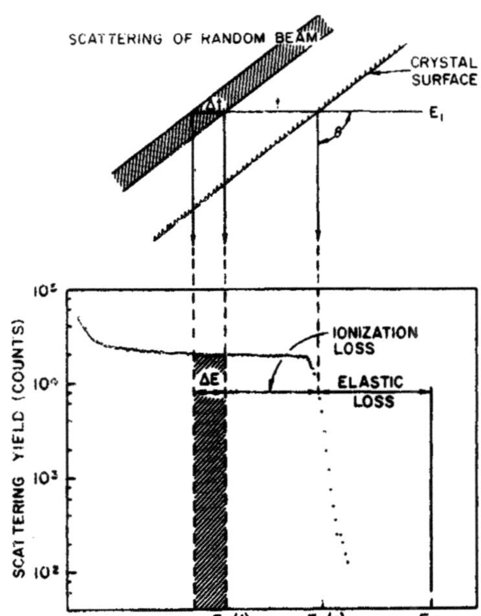

Fig.11. Schematic represantation of elastic scattering illustrating depth resolution determined from the scattered particle's energy.

Iron, like tungsten is body centered cubic, with lattice spacing a = 2.86 Å (fig. 10). The face center site, which is the center of an octahedron might be a stable interstitial site. Less obvious is the (120) site of tetrahedral symmetry, which for heavy ions might be more favourable. Channeling experiments are capable of probing these sites and determining the percentage of impurities substitutionally on them. In addition, careful channeling experiments give information on the amount of implanted ions, the depth distribution and the amount and type of radiation damage in the crystal.

For a heavy impurity, Rutherford backscattering of heavy ions is the usual probing reaction. As illustrated in fig. 11, the energy at which a particle is backscattered can be correlated with the mass of the scattering center and the depth of scattering. Because of the larger specific energy loss and momentum transfer heavy particles

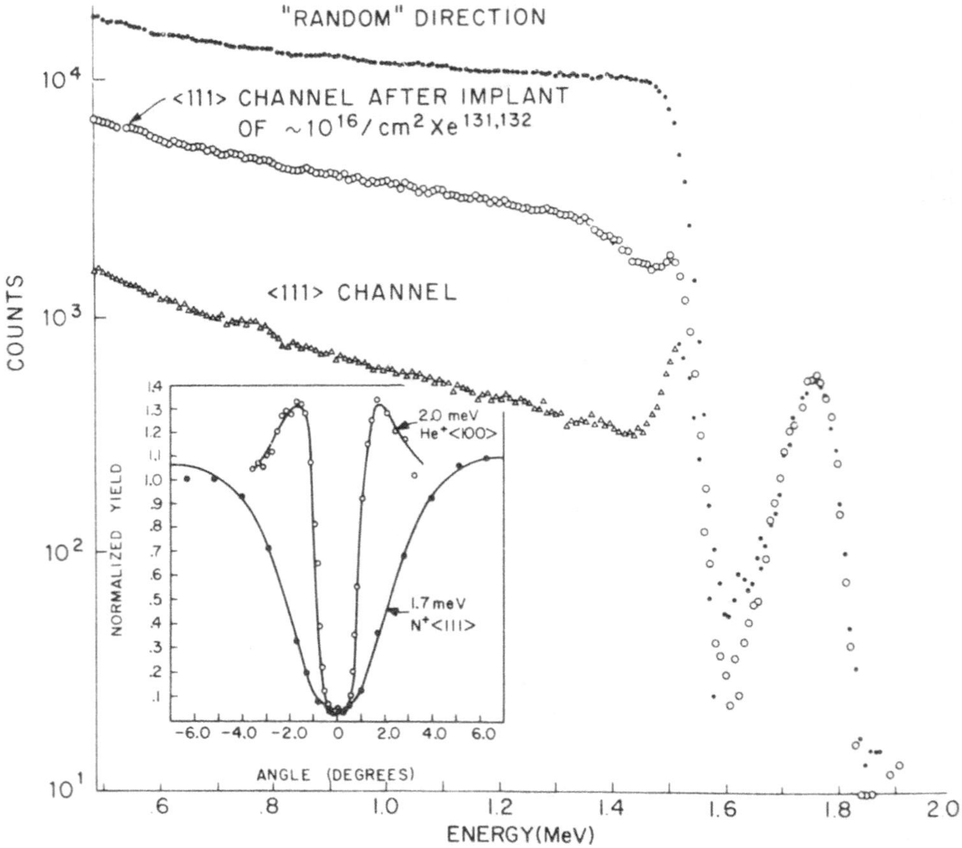

Fig. 12. Backscattered particle spectra pre- and post-implant of 2 x 10^{16} Xe atoms/cm^2 into iron crystal. Inset: angular distributions for He and N beams.

are best for mass separation and depth resolution.

Lattice location studies on implanted iron single crystals have been reported by Shmorak and Bøgh[34]), by Bøgh[35]) and by Feldman and coworkers[36]). Fig. 12 shows channeling spectra and angular scans for carefully prepared iron crystals. The critical angles for He and N ions are consistent with theoretical predictions. The best minimum yields, ~ 2%, are somewhat higher than predicted but consistent with results on other good crystals. In the channeling spectra, oxygen peaks are evident, and comparison with the surface peak width allows an estimate of surface oxide thickness and stochiometry. Bøgh, using the double alignment technique, has confirmed a surface layer of about 50 Å Fe_2O_3.

The channeling spectra of fig. 12 are shown pre- and post-implant of 2×10^{16} atoms / cm^2 xenon into iron. The extraordinarily high dose is chosen here for illustrative purposes. Several conclusions can be drawn from the figure. **First,** one sees a large increase in minimum yield post - implant. The increase can be attributed to scattering from interstitial iron atoms displaced during the implantation. The number of interstitials displaced during the implantation is calculated in this case to be about 6×10^{21} / cm^2. For similar implants of about 10^{14} atoms / cm^2, the increase in minimum yield was ≈ 2%. **Second,** one sees evidence of damage corresponding to the maximum in nuclear stopping as the implanted ions have slowed down. The depth of the damage maximum is less than the range of the implanted xenon. Finally, the xenon peak area in this case is hardly affected by crystal orientation. Careful analysis indicates that only about 15% of the xenon atoms are on lattice sites. Experiments at lower doses indicate that about 50% of implanted xenon goes substitutional and that this fraction decreases with high temperature annealing. Comparisons with Mössbauer effect results on ^{133}Cs in iron[8]), from ^{133}Xe implanted into iron, offer some insight into the correlation of lattice site with hyperfine field. De Waard et al.[8]) found high and low field components in the Mössbauer hfi spectra at xenon doses equivalent to those used in our lattice location measurement. Reasonable assumptions for ratios of recoil free fractions led to an estimate of about 60% of the implanted xenon at a high field site. Furthermore, this fraction decreased with increasing dose and after a 450° C anneal. The high field site thus seems correlated with the lattice site.

In another case, Feldman et al.[37]) have found, via channeling, that implanted bismuth and thallium are predominantly at iron lattice sites, and that the fraction of implanted bismuth at a lattice site decreases with increasing temperature. Hfi experiments at Cal Tech[10]) indicate that the average field on lead in iron is less after alpha recoil ($^{212}Po \xrightarrow{\alpha} {}^{208}Tl \xrightarrow{\beta} {}^{208}Pb$) implantation than after bismuth alloying and annealing ($^{207}Bi \xrightarrow{\epsilon_k} {}^{207}Pb$). These results seem to indicate that the lattice site is a lower field site in this case.

Shmorak and Bøgh[34]) found via channeling, implanted Lu at lattice sites and a unique hyperfine field. Similarly Nielson found that Sn implanted into iron was ~ 85% substitutional[35]). In the latter study an interesting surface effect was noted (fig. 13). A high dose implant resulted in sputtering away of more iron than the mean range of the implant, thus the impurity concentration was approximately uniform to the surface of the crystal, yet no Sn surface peak was observed. The explanation is that the tin content in the oxide, formed after implantation, is low because of low solubility or preferential sputtering of tin. A similar explanation might be given for

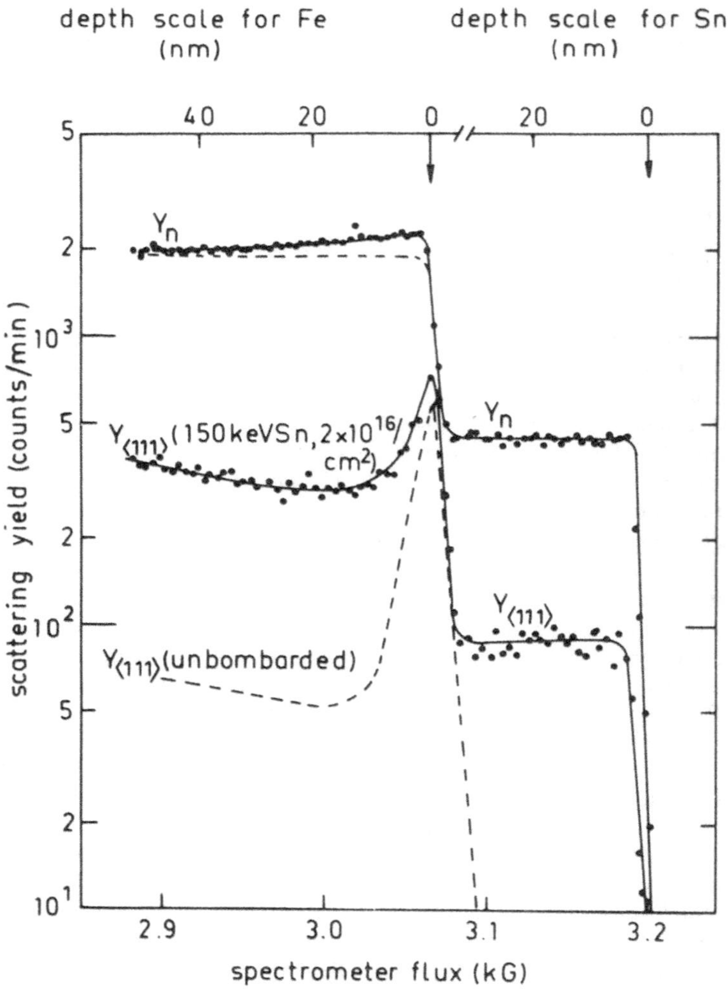

Fig. 13. 90° scattered particle spectra for Sn implanted iron sample (from ref. 35).

the insensitivity of the Yb in iron results to implant energy[6]) whereas Au or Hf in iron have reduced hyperfine fields at low implant energy.

Channeling studies of impurities in crystals of hfi interest have not yet been carried out on enough systems for general trends to have become apparent. Nevertheless it is evident that this technique yields important insights as to the physical state of ion implanted samples. Further studies and comparisons with relative atomic size and valency as discussed in ref. 38 are essential to the analysis of experimental hyperfine interaction results.

4. Reaction recoil implantation

In addition to its similarities to other types of implantation experiments, nuclear recoil implantation has several unique and interesting aspects. As described previously, and illustrated in fig. 5 an excited nucleus is implanted via recoil, and the angular correlation of the de - exciting radiation is subsequently measured. In many cases the gamma radiation alone is anisotropic by virtue of the nuclear alignment and/or polarization induced by the reaction. In other cases, significant anisotropies are only obtained in coincidence with a scattered particle. Most Coulomb excitation studies have been done with a backscatter geometry to enhance the angular correlation as well as to maximize the recoil energy. If double layer targets are used in non - coincidence experiments, it is important to consider the distribution in momentum of the recoil nuclei in order to ensure that all are indeed implanted into the host material.

The excited nucleus is initially moving with energies of a million electron volts or greater*; slows down in a time of the order of 10^{-12} seconds or less, and during its flight is exchanging electrons with the host crystal approximately every 10^{-16} seconds. For cadmium in iron, Coulomb excited by 35 MeV oxygen for example, the initial recoil energy is 14 MeV the equilibrium charge decreases from 12, and the slowing down time is 0.8×10^{-12} seconds. Nevertheless, for many systems the initial alignment is retained, and angular correlations consistent with minimal perturbations or static interactions only are observed.

The extreme rapidity of electron pickup and loss for fast ions in dense media is sufficient to explain the constancy of alignment, or m - state population, during the slowing down process. Though the field due to any one unpaired electron is enormous, the time averaged value is zero, or more pictorially the nucleus does not have time to interact in one field before the field changes in direction as well as in magnitude. The experiments of Goldring and collaborators[39]) implanting into vacuum and gases have experimentally illustrated this effect. As gas pressure is increased, the mean free path decreases, and the perturbation factor increases to unity. The effect is relatively insensitive to the atomic number of the gas and is clearly a density effect.

Perturbations on the stopped atom are more complex, and must be considered separately for various host impurity combinations. There is evidence that cubic metallic hosts provide a relatively perturbation free environment for most implants. The early experiments of Scharenberg et al. indicated that Coulomb excited tungsten nuclei implanted into tungsten retained their original alignment. The work of Goldring et al.[16]) and Murnick et al.[40]) indicated that copper was a suitable perturbation free host for a wide variety of implanted species. Sugimoto et al.[41]) showed that polarization could be retained for ten seconds or more on ^{17}F implanted into CaF_2. Berlijn et al.[42])

* For (p, p′) studies the recoil energy is generally much lower and these experiments are more closely related to alpha recoil implantation studies. The (p, p′) reaction for implantation PAC work is reviewed in ref. 22.

have done a particularly complete study of polarization retention in ^{12}B implanted via the ^{11}B (d, p) ^{12}B reaction into various materials. Their not unexpected conclusion was that fcc metals were best.

In a cubic lattice, the major perturbations are due to electronic relaxation effects which will be different for each impurity - host combination. The major experimental work to date has been on paramagnetic relaxation on the rare earths in copper[30]). This work has already been reviewed here by Heestand and Deutch[43]). The most recent time dependent perturbed angular correlation studies[44]) have indicated the presence of some electric quadrupole relaxation in addition to the paramagnetic effect.

Transient Magnetic Fields

Though in - flight perturbations in solids were expected to average to zero, Borchers et al.[45]) discovered the existence of an enhanced aligned magnetic field on fast ions moving through polarized iron metal. This transient field effect which is present on all nuclei moving through any ferromagnetic lattice was first deduced from apparent systematic effects in experimental studies of 2^+ state magnetic moments of even - even nuclei in the vibrational region[46]).

The static hyperfine field on tellurium in iron is +625 kOe. Using this value, Borchers et al.[47]) found that the 2^+ state g factors of even - even Te isotopes were apparently inversely proportional to time. On the other hand, data for elements with

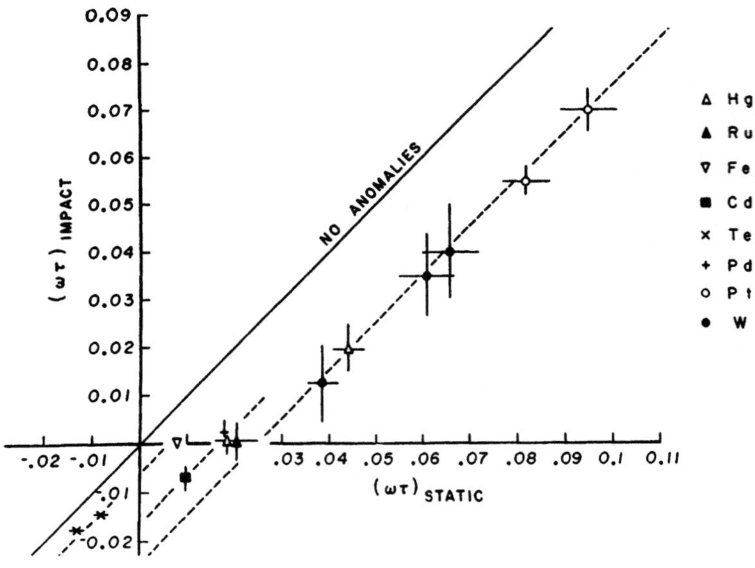

Fig. 14. Implantation recoil (IMPACT) obtained precession angles for short lived 2^+ states versus radioactivity results for the same levels.

negative static fields in iron showed reduced effective hyperfine interactions. An alternate interpretation of the data was that an additional aligned positive hyperfine field was acting on the implanted nuclei for a time short compared to the nuclear lifetimes involved. The data for even - even isotopes of Mo, Ru, Pd, Cd and Te was most consistent with an effective additional interaction, $H_1 t_1 \approx 10^{-5}$ gauss - sec, that is, an average field of ten million gauss acting for about one picosecond.

Striking evidence of the existence of the transient field effect is obtained by comparing Larmour precession angles for the same nuclear state in iron obtained from gamma - gamma correlations with Coulomb excitation recoil - gamma correlations, (fig. 14). The precession angles are always more negative for the implantation experiment. The transient nature of the additional positive hyperfine field is evident from the fact that it is independent of nuclear lifetime for the states investigated, all having $\tau \gtrsim 2 \times 10^{-12}$ sec. A particularly interesting case is Ge in iron, cobalt and nickel (table 2). The static fields are small, yet the correlation precession angle is relatively large and constant for each host over an order of magnitude change in lifetime. Clearly this precession angle is not simply defined by $\Delta\theta = - g\mu_n H\tau/h$, where H is the static effective hyperfine field and τ is the nuclear lifetime. For constant g - factors, the best fit is for a field which disappears after about 1 ps.

Table 2. Summary of IMPACT precession angles for first 2^+ states of Ge-isotopes in iron, cobalt and nickel illustrating the transient field effect

Isotope	E 2^+(keV)	τ(p sec.)	Host	Angular Shift (miliradians)
^{70}Ge	1040	1.92 ± 0.20	Fe	-11.2 ± 1.5
			Co	-8.9 ± 2.3
			Ni	-4.3 ± 1.7
^{72}Ge	835	4.54 ± 0.50	Fe	-9.9 ± 1.7
			Co	-8.9 ± 1.4
^{74}Ge	596	17.2 ± 1.4	Fe	-11.3 ± 1.3
			Co	-8.7 ± 1.3
			Ni	-7.4 ± 1.6
^{76}Ge	563	25.2 ± 2.6	Fe	-9.9 ± 1.5
			Co	-7.2 ± 2.0
			Ni	-5.3 ± 2.6

The slowing down time of the recoil nuclei studied is approximately one pico-second, hence, the transient effect is an in - flight or possibly an end of flight effect. Triple layer target experiments have proven the in - flight nature of the effect. Fig. 15 shows results of Cd - Cu - Fe experiments. The transient field effect is reduced when the recoil cadmium nuclei spend most of their flight time in non - magnetic copper. The Fe - Fe - Cu experiments of Herskind et al.[48]) attempt to isolate the transient field by having recoils slow down in polarized iron, but stop in copper. Preliminary results are consistent with an inflight transient field effect. One more significant experimental property of the transient field is that it seems to scale with the atomic moment of the host material for impurities in iron, cobalt and nickel.

Lindhard and Winther have been developing a theoretical formulation of the transient field effect based on the scattering of polarized electrons in the unscreened

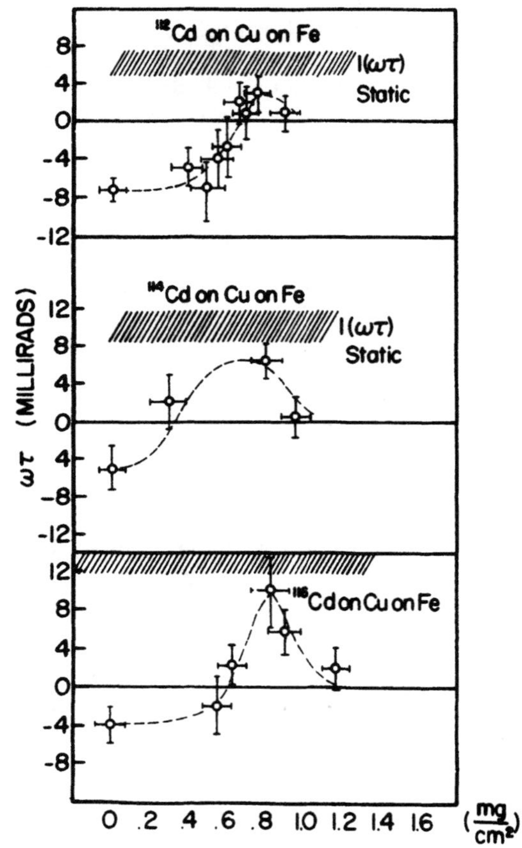

Fig. 15. Precession angles from triple layer target (Cd - Cu - Fe) IMPACT experiments as a function of moderator (Cu) thickness.

Coulomb field of the moving ion[49]). Electron scattering leads to an enhanced s electron density at the nucleus, and since there exists a net polarization in ferromagnetic material, the enhanced Fermi contact term gives rise to the large effective magnetic field ($|\psi s^2(0)\uparrow| \neq |\psi s^2(0)\downarrow|$) at the scattering center.

For an ion moving with a velocity v, in a polarized solid with ζ polarized electrons per atom, the magnetic field at the nucleus is given by

$$B(v) = \frac{16\pi^2}{3}\,\mu_0\,\zeta\,N\cdot C_r\cdot C_a\cdot Z_1 \begin{cases} v_p/v, & v > v_p \\ 1, & v < v_p \end{cases} \tag{1}$$

μ_0 is the Bohr magneton, N is the atomic density and v_p is the polarized electron velocity in the host lattice. Z_1 is the atomic number of the fast ion. C_r is a relativistic correction factor currently estimated to be

$$C_r \approx 1 + \left(\frac{Z_1}{84}\right)^{5/2}$$

and C_a is an atomic binding correction factor. In an experiment, one measures a precession angle due to the integral effect of B (v) throughout the slowing down. The total transient field precession angle can be expressed

$$\phi(v_1) = -g\,\frac{\mu_n}{\hbar}\,M_1 \int_0^{v_1} \frac{B(v)\,dv}{N\,S(v)} \tag{2}$$

where g is the nuclear g factor, M_1 is the recoil mass and S (v) is the stopping cross section per atom.

Several aspects of the Lindhard Winther model have been studied experimentally. The velocity dependence in particular has been investigated. Eqs. (1) and (2) predict that most of the effect occurs at low velocities consistent with the fact that equal effects have been found for various target thicknesses. Heestand et al.[50]) have carefully studied $\phi(v_1)$ for Pt in Fe. Fig. 16 shows their results. The trend is consistent with theory, yet magnitudes are greater than predicted by the equations above. Other aspects of the transient field under experimental study are the Z_1 dependence (fig. 17, see page 476) and the polarized host dependence. The fields scale as predicted in iron, cobalt and nickel. Recent experiments of Heestand et al.[50]) indicate that the transient field in a gadolinium host is also consistent with the Lindhard Winther model.

From an operational standpoint, it is sufficient for analysis of implantation PAC experiments to include the transient field effect as an additional precession angle due to a field H_1 acting for a time t_1:

$$\Delta\theta = -g\,\frac{\mu_n}{\hbar}\,(H_0\tau + H_1 t_1) \tag{3}$$

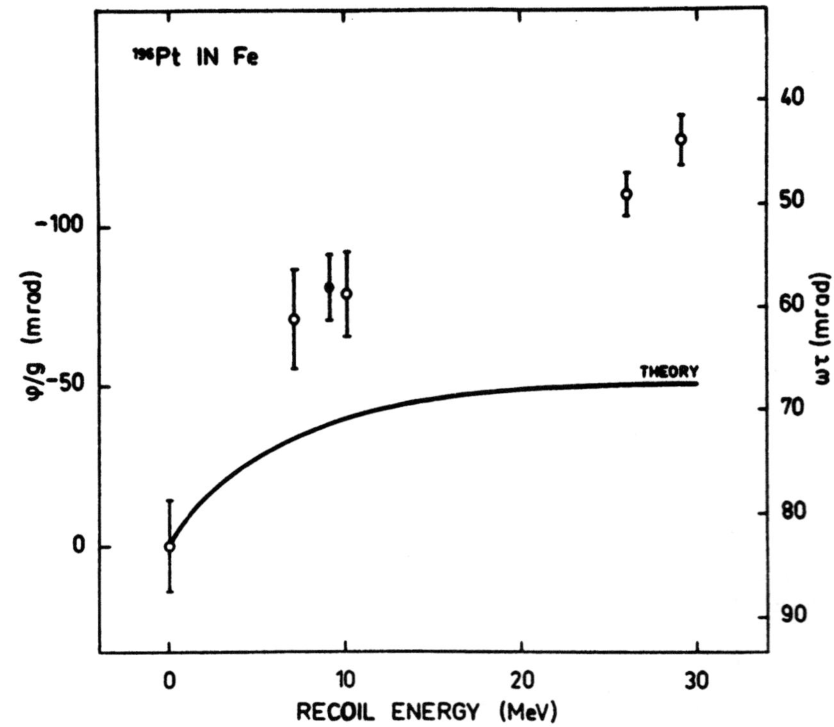

Fig. 16. Velocity dependence of the transient field for Pt implanted into iron. The solid curve is calculated from equation (2) (from ref. 50).

The value of $H_1 t_1$ can be determined experimentally for a given element if H_0 is known as well as g for some state. Or it can be obtained by demanding internal consistency for many states of the same element. By using these 'bootstrap' techniques, g factors of more than 25 short lived 2^+ states were obtained for isotopes of Ge, Se, Mo, Ru, Pd, Cd and Te [51]). Fig. 18 (see page 477) shows these results which show g to be rather insensitive to Z and A in contradistinction to microscopic calculations from nuclear models.

The transient field is always positive yet static fields are usually negative. Eq. (3) predicts a zero precession angle, independent of g for the case $|H_0 \tau| =$
$= |H_1 \tau_1|$.
This was approximately the case for the first excited state of ^{56}Fe, where g ≈ .5 yet $\Delta\theta$ was approximately zero, τ for this state is 11 ps., coincidently the same as for the 367 keV $3/2^-$ state of ^{57}Fe. A recoil implantation experiment in iron is thus incapable of determining g for this state. Sprouse et al.[52]) recently quoted an upper limit for g from such an experiment neglecting the transient field effect.

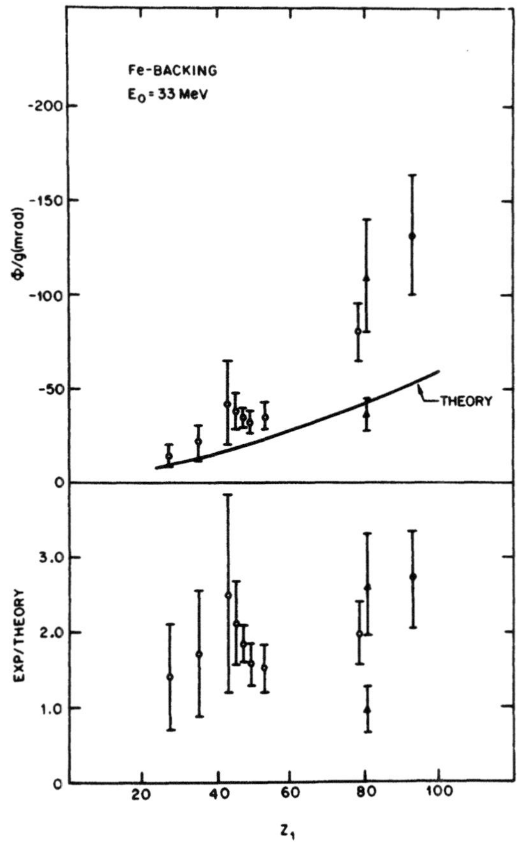

Fig. 17. Z dependence of the transient field in iron using equations (1) and (2) (top). Experimental divided by theoretical values (bottom) (from ref. 50).

5. Single crystal quadrupole moment studies

Electric quadrupole interactions studied via Coulomb excitation and perturbed angular correlation techniques are the subject of several contributions to these proceedings[53]). In this section, certain unique advantages of ion implantation techniques for these studies are presented, and some recent experimental data obtained at the Bell - Rutgers tandem accelerator are reviewed.

The major uncertainty associated with electric quadrupole PAC studies of short lived excited states lies in the difficulty of unambiguously proving the presence of quadrupole interactions only. The experimentally measured parameter is an

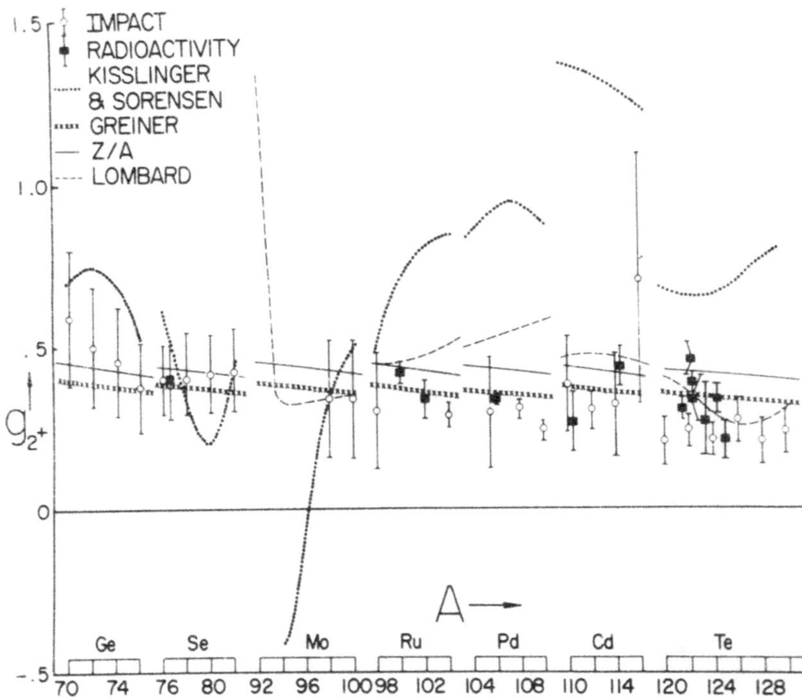

Fig. 18. 2^+ state g factors for even - even spherical nuclei and various theoretical predictions (from ref. 51).

attenuation factor which may be due in part to other effects such as various electron relaxation perturbations. Some of the ambiguity is removed if single crystals are used. The analysis and interpretation of data becomes more straight - forward if the quadrupole induced attenuation factor has the expected functional dependence on crystal orientation. The difficulties of preparing suitable single crystal samples, however, has heretofore limited such experiments to a very few cases. Furthermore even when an interaction frequency,

$$\omega_Q = -e\, q_{zz}\, Q/4\hbar\, I\,(2I-1)$$

(q_{zz} is the field gradient component, Q is the nuclear quadrupole moment and I is the nuclear spin) is measured it consists of a product of unknown parameters. The calculation of field gradients is uncertain at best, so for useful nuclear parameter information it is necessary to do at least two experiments on different excited states of the same element in the same host. In this way ratios of quadrupole moments can accurately be determined. Similarly ratios of field gradients in different crystals or

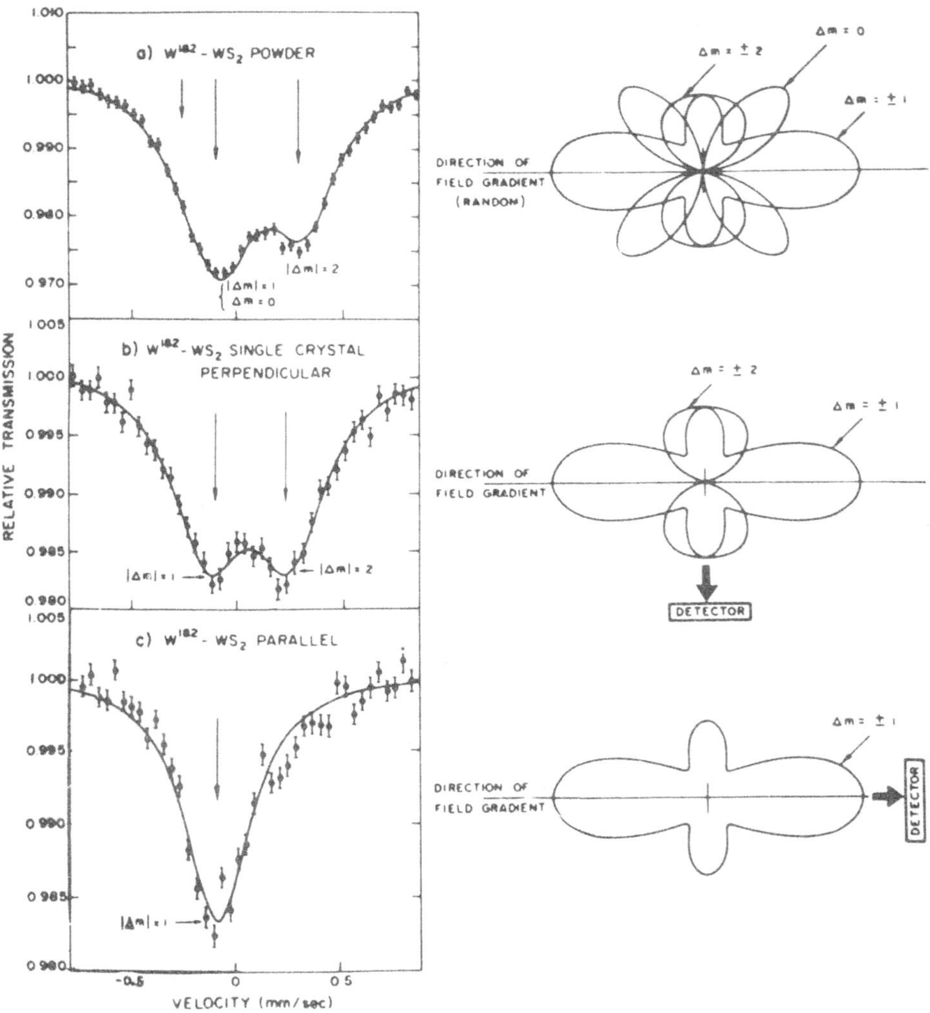

Fig. 19. Mössbauer effect studies of ^{182}W using single crystal WS_2 absorbers (from ref. 54).

the same crystal at different temperatures can be precisely determined.

Analogous experiments using the Mössbauer effect can be done with somewhat less sample preparation difficulty. A single non - radioactive crystal can be used as an absorber with unsplit sources. Chow et al.[54]) have used this technique to measure quadrupole moment ratios of ^{182}W: ^{184}W: ^{186}W as shown in fig. 19. Alignment of the absorber eliminates certain transitions so that the quadrupole splitting can be more accurately and unambiguously determined. Unfortunately, useful Mössbauer levels are limited to a small fraction of all interesting nuclear states.

Ion implantation here offers the possibility of obtaining any interesting nucleus in any desired crystalline host. If single crystals are used, the 'goodness' of implant site can be determined from the quadrupole interaction's behaviour with crystal orientation as well as by supplementary channeling studies. Furthermore, ratios of

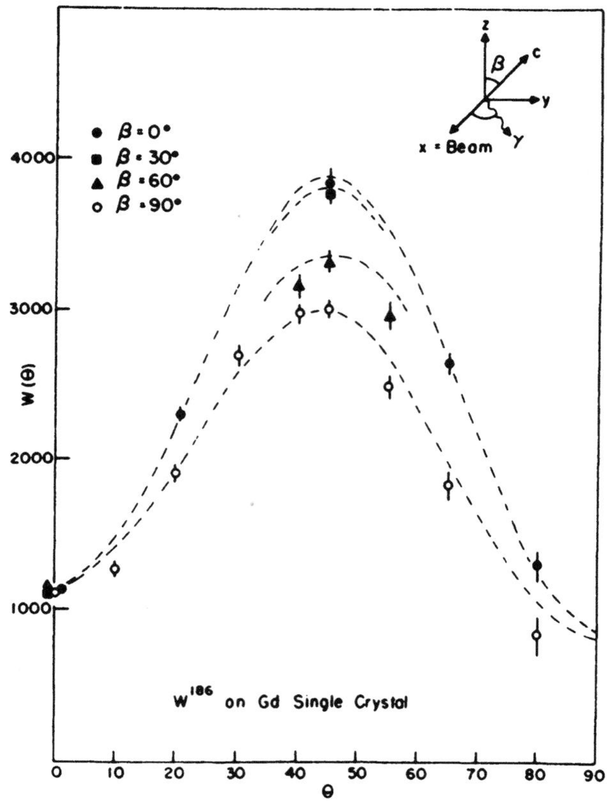

Fig. 20. $0 \to 2 \to 0$ angular correlations for ^{186}W implanted into gadolinium single crystals for various orientations of the gadolinium crystal.

field gradients in many different materials can be determined.

In collaboration with Alonso and Grodzins[55]), we have begun Coulomb excitation recoil implantation studies of 2^+ states of even - even nuclei. To verify the efficacy of the technique, first experiments have been performed on the even - even tungsten isotopes. Exactly as in the magnetic moment studies, previously described, two layer targets are used. The thin front layer consists of the isotope of interest, and the backing is a carefully prepared non - cubic single crystal. Angular correlations in coincidence with backscattered particles are measured as a function of crystal orientation. If the implant is at a lattice site and/or the effective field gradient is parallel to the c - axis, the interaction frequency can be determined from attenuation versus orientation data. The relevant equations for the backscatter geometry used can be found in

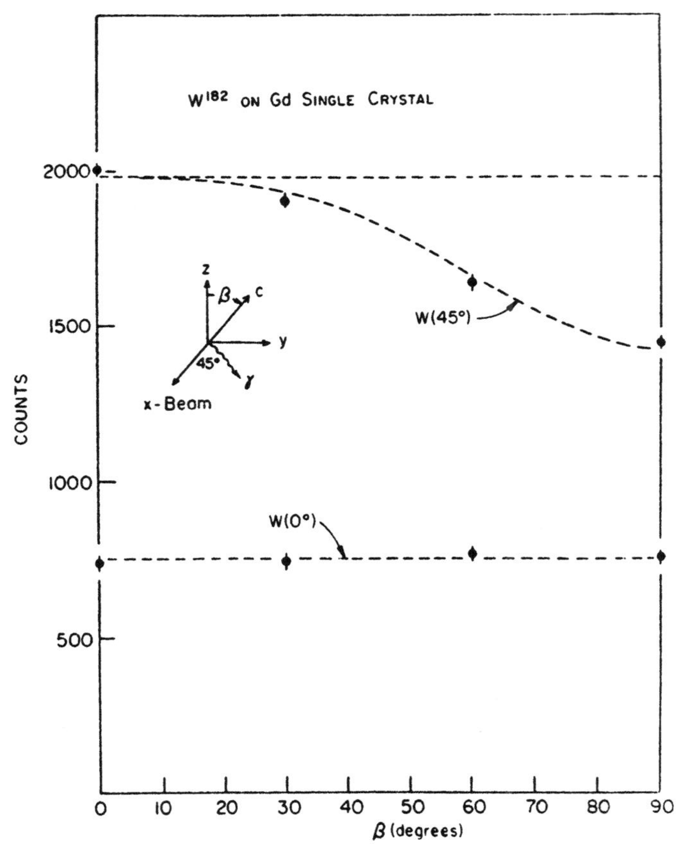

Fig. 21. Normalized coincidence count rates at 45° and 0° as a function of crystal orientation from data like that of figure 20 for ^{182}W. The solid line is the best theoretical fit for $\omega_Q\tau =$ $=0.36$.

ref. 56 and 57. The PAC is now a function of (θ, ϕ), the c - axis orientation as well as Ω_1, Ω_2, the radiation directions. The sensitivity of electric quadrupole interactions can be maximized for various spin sequences, however, certain geometries are more sensitive to systematic errors than others[57]).

Various crystals were tried as backings in an attempt to maximize the expected effect. Tellurium was first tried as it has the most anisotropic elemental crystal structure. The results, however, could not be explained by a simple axially symmetric interaction. There was an orientation dependent quadrupole attenuation, but it did not disappear for the c - axis parallel to the beam or perpendicular to the counting plane. A possible explanation is that the field gradient symmetry axis is not parallel to the c - axis for tellurium. Alternatively the implanted tungsten may not all be at equivalent sites.

Günther et al.[57]) have used the same technique described here for ^{178,180}Hf implanted into hafnium single crystals. Similar to the W in tellurium case, the attenuation did not vanish for the c - axis parallel to the beam direction. In addition, the best fit $\omega_Q \tau$ value was smaller than that obtained via Mössbauer effect experiments on the same system without ion implantation.

Fig. 20 and 21 show results obtained at the Bell Laboratories - Rutgers tandem for W implanted into gadolinium crystal backings. The correlation is unattenuated for $\beta = 90°$ as predicted, and $\omega_Q \tau$ can be extracted from the variation of attenuation with β. The quadrupole moment ratio results are consistent with the Mössbauer effect values. Work is continuing in this area to improve precision in the final results. Measurements on states having lifetimes $\gtrsim$ .1 ns should be possible. Particularly interesting will be comparisons of this type of PAC experiment with reorientation results for the same state. Should good agreement be obtained, it would be possible to measure many more absolute quadrupole moments as well as absolute values of electric field gradients.

6. Summary

Various aspects of ion implantation techniques of relevance to perturbed angular correlation studies have been presented. Of particular interest are the correlation of hyperfine interactions with details of lattice location and damage, and the transient effects on implanted ions. The high transient magnetic fields on fast ions in ferromagnetic material provides the only way for nuclear magnetic moment studies of states having lifetimes less than one picosecond.

Similarly, ion implantation studies in suitable single crystals will greatly extend PAC studies of electric quadrupole moments. Experiments such as those described here, and reorientation effect results may soon provide as much information on nuclear electric quadrupole moments as was known a few years ago about magnetic moments. This should provide a real challenge to theorists.

Acknowledgements

The author is indebted to his collaborators and colleagues for many helpful discussions, and particularly to L.C. Feldman, W.L. Brown, and L. Grodzins for comments on this paper.

References

1. S.S. Hanna, J. Heberle, C. Littlejohn, G.J. Perlow, R.S. Preston, and D.H. Vincent, Phys. Rev. Letters 4 (1960) 177.
2. B.N. Samoilov, V.V. Sklyarevskii and E.P. Stepanov, JETP 9 (1959) 448.
3. J.H. Freeman, Proc. Roy. Soc. A311 (1969) 123.
4. K.O. Nielsen, Int. Conf. Appl. of Ion Beams to Semiconductor Techn., ed. P. Glotin, Grenoble (1967).
5. P.G.E. Reid, M. Sott, N.J. Stone, D. Spanjaard, and H. Bernas, Phys. Letters 25A (1967) 396.
6. H. Ravn, F. Abildskov, K.H. Johansen, and B.I. Deutch, Proc. Int. Conf. on Radioactivity in Nuclear Spectroscopy, Nashville (1969).
7. D.E. Murnick, S. Hufner and B. Herskind, Z. Physik 226 (1969) 175.
8. H. de Waard and S.A. Drentje, Proc. Roy. Soc. A311 (1969) 139.
9. G.C. Pramila, S.G. Cohen, and L. Grodzins, Phys. Letters 24A (1967) 7.
10. J.D. Bowman and F.C. Zawislak, Nucl. Phys. A138 (1969) 90.
11. S.L. Ruby, G.M. Kalvius, B.D. Dunlap, G.K. Shenoy, D. Cohen, M.B. Brodsky and D.J. Lam, Phys. Rev. 184 (1969) 374.
12. E.J. Ansaldo, L. Grodzins and R. Kalish, Phys. Letters 30B (1969) 538.
13. D.E. Murnick in Hyperfine Interactions, ed. A.J. Freeman and R. Frankel, Academic Press, New York (1967).
14. L. Grodzins in Nuclear Reactions Induced by Heavy Ions, ed. R. Bock and W. Hering, North Holland, Amsterdam (1970).
15. G. Goldring and R.P. Scharenberg, Phys. Rev. 110 (1958) 701.
16. I. Ben Zvi, P. Gilad, G. Goldring, R. Herber, and R. Kalish, Nucl. Phys. A96 (1967) 138.
17. L. Grodzins, R. Borchers, and G. Hagemann, Phys. Letters 21 (1966) 214.
18. F. Boehm, G. Hagemannn, and A. Winther, Phys. Letters 21 (1966) 217.
19. J. Christiansen, E. Recknagel and G. Weyer, Phys. Letters 20 (1966) 46.
20. T. Yamazaki and E. Matthias, Phys. Rev. 175 (1968) 1476.
21. D.B. Fossan, and A.R. Poletti in Hyperfine Structure and Nuclear Radiations, ed. E. Matthias and D. Shirley, North Holland, Amsterdam (1968).
22. D.E. Murnick, J.R. MacDonald, R.R. Borchers, G. Heestand and B. Herskind, Proc. Roy. Soc. A311 (1969) 111.
23. J. Lindhard, M. Scharff and H.E. Schiott, Mat. Fys. Medd. Dan. Vid. Selsk 33 (1963) 14.
24. H.E. Schiott, Radiation Effects, 3(1970) 1.
25. R.S. Nelson, Proc. Roy. Soc. A311 (1969) 53.
26. C. Erginsoy, G. Vineyard and A. Englert, Phys. Rev. 133A (1964) 595.
27. C. Erginsoy, G. Vineyard and A. Shimizu, Phys. Rev. 139 (1965) A118.
28. J. Christiansen, V.E. Krohn, and G.R. Ringo, Phys. Rev. C1 (1970) 1693.
29. H. Bertschat, J. Christiansen, H.E. Mahnke, E. Recknagel, G. Schatz, R. Sielemann and W. Witthuhn, Phys. Rev. Letters 25 (1970) 102.
30. J.C. Waddington, K.A. Hagemann, S. Ogaza, D. Kiss and B. Herskind in Nuclear Reactions Induced by Heavy Ions, ed. R. Bock and W. Hering, North Holland, Amsterdam (1970).
31. R. Stokstad, R. Moline, C. Barnes, F. Boehm, and A. Winther, in Hyperfine Structure and Nuclear Radiations, ed. E. Matthias and D. Shirley, North Holland, Amsterdam (1968).

32. J. Lindhard, Mat. Fys. Medd. Dan. Vid. Selsk. 34 (1965) 14.

33. B. Domeij and K. Björkquist, Phys. Letters 14 (1965) 127.

34. M. Schmorak and E. Bøgh in Hyperfine Structure and Nuclear Radiations, eds. E. Matthias and D. Shirley, North Holland, Amsterdam (1968).

35. E. Bøgh, Proc. Roy. Soc. A311 (1969) 35.

36. L.C. Feldman and D.E. Murnick, Bull. Am. Phys. Soc. 14 (1969) 82 and P.R., to be published.

37. L.C. Feldman, W.M. Augustyniak, and E.N. Kaufmann, Proc. Int. Conf. on Hyperfine Interactions detected by Nuclear Radiation, Rehovot (1970).

38. H. de Waard, P. Schurer, P. Inia, L. Niesen, and Y.K. Agarwal, op. cit.

39. G. Goldring in Hyperfine Structure and Nuclear Radiations, eds. E. Matthias and D. Shirley, North Holland, Amsterdam (1968).

40. D.E. Murnick, L. Grodzins, J. Bronson, B. Herskind, and R. Borchers, Phys. Rev. 163 (1967) 254.

41. K. Sugimoto, A. Mizobuchi, and K. Nakai, Phys. Rev. 134 (1964) B539.

42. J. Berlijn, P. Keaton, Jr., L. Madansky, G. Owen, L. Pfeiffer, and N. Roberson, Phys. Rev. 153 (1967) 1152.

43. G. Heestand and B. Deutch, these proceedings.

44. P. Ryge, H. Kugel and R. Borchers, Proc. Int. Conf. on Hyperfine Interactions detected by Nuclear Radiations, Rehovot (1970).

45. R. Borchers, B. Herskind, J. Bronson, L. Grodzins, R. Kalish and D. Murnick, Phys. Rev. Letters 20 (1968) 424.

46. B. Herskind, R. Borchers, J. Bronson, D.E. Murnick, L. Grodzins and R. Kalish in Hyperfine Structure and Nuclear Radiations, ed. E. Matthias and D. Shirley, North Holland, Amsterdam (1968).

47. R. Borchers, J. Bronson, D.E. Murnick, and L. Grodzins, Phys. Rev. Letters 17 (1966) 1099.

48. B. Herskind, K. Hagemann, P. Hvelplund, and D.E. Murnick, to be published.

49. J. Lindhard, Proc. Roy. Soc. A311 (1969) 11;
J. Lindhard and A. Winther, Nucl. Phys. A166 (1971) 413.

50. G. Heestand, P. Hvelplund, B. Skaali, and B. Herskind, Phys. Rev. B2 (1970) 3698.

51. G. Heestand, R. Borchers, B. Herskind, L. Grodzins, R. Kalish and D.E. Murnick, Nuclear Physics A133 (1969) 310.

52. G.D. Sprouse and S.S. Hanna, Nuclear Physics A137 (1969) 658.

53. C. Alonso, L. Grodzins, E. Kaufmann, H. Lie and D.E. Murnick, Bull. Am. Phys. Soc. 14 (1969) 1202.

54. Y. Chow, E. Greenbaum, R. Howes, F. Hsu, P. Swerdlow and C.S. Wu, Physics Letters 30B (1969) 171.

55. E. Kaufmann, D.E. Murnick, C. Alonso, and L. Grodzins, Proc. Int. Conf. on Hyperfine Interactions detected by Nuclear Radiations, Rehovot (1970).

56. C.T. Alonso, Ph.D. Thesis, Massachusetts Inst. of Techn. (1970), unpublished.

57. C. Günther, B. Skaali, R. Bauer and B. Herskind, Nuclear Physics A164 (1971) 321.

Discussion

Deutch: There are several ambiguities with respect to the errors involved in the location of atoms via channeling. First, in assessing the damage, the beam usually does not have the depth resolution for one to tell the damage at the implanted region and thus it is difficult to correct for these damages in densely implanted sources ($\geqslant 10^{15}$ ions/cm^2). Second, there is an ambiguity with respect to crystal symmetry in that, if its atom is not shadowed, it may lie on any other axis through the substitutional site, hence also on its shadowed axis being measured, thus increasing the possible errors in the determination of the position. Third, the determination of the random direction will be complicated by channeling effects through the larger Miller indices, which introduces errors of at least 10%.

Murnick: Some of these points are mentioned in the text of my paper and are considered in greater detail in several review articles on lattice location via channeling; primarily considering a silicon lattice which has been most studied by people in this field. I would refer interested people to ref. 35 and 36 and the references therein. It is possible though to get around many of these effects by looking at more than one axis as can be inferred from fig. 10. That is looking at a ⟨110⟩ - axis as well as a ⟨111⟩ - axis, eliminates certain ambiguities due to the symmetry of the crystal. Another very precise method, though also quite difficult, is the double alignment technique, where one looks at beam channelled both incident and emergent from the crystal.

De Waard: I have one question and two comments. The question is: can you tell where the non - substitutional fraction of Xe - atoms in an iron crystal sits?

Murnick: Unfortunately, we cannot say where the non-substitutional Xe-atoms do sit.

De Waard: I see. Well, my first comment is that a more recent analysis of our Mössbauer spectra of Cs in Fe indicate that there are not only a high and a low field component but also an intermediate field component. This we could find out by making Mössbauer spectra as a function of source temperature. And we see now that the three components have three different Debye temperatures meaning that these three kinds of atoms have different space in the lattice. When the Debye temperature is low, they have more space. What we suggest is that the xenon atoms are associated with zero, one or 2 vacancies, or maybe with other numbers of vacancies. But at any rate, the data are rather unambiguous. And of course, when there is a vacancy next to a xenon atom, it will shift from its substitutional position. My second comment is that in general there is a relation between, what I would call, implantibility (by which I mean regular substitutional situation in the crystal) and solubility. And for cases like Cs or Xe where the xenon is not at all soluble in the metal, but also for the rare - earth elements, the solubility is very bad and therefore also the implantibility. Whereas in the cases where there is some solubility, I think the vacancies, I just talked about, anneal out, because in this way a more stable situation is reached, as you have in an alloy. This is just a hypothesis. But it fits so far, also for the cases you have shown, where both implantation, IMPACT, and other measurements have been done. It always turns out that if you have the same result for the diffused activity, let us say an implanted activity, the diffused activity is soluble. So it looks like something that is completely insoluble, that does not fit in the lattice and likes to sit there associated with vacancies. I suggest that this might explain why the fraction, you talked about, is not substitutional. But it is not a complete proof I must admit.

Murnick: The solubility is an equilibrium process and an ion implantation of such low doses of impurity materials is very far from equilibruim and it is not clear how one can extrapolate over orders of magnitude in concentration.

De Waard: What I want to say is that if something is unsoluble, it does not want to sit there. That means that a lower energy is reached by some distortion and that even at room temperature vacancies can anneal out quite well in iron.

Rasera: Both you and the previous speakers use the term 'pure quadrupole interaction', that you were looking for. Would you mind enlarging briefly what you mean by the term and why they are so hard to find?

Murnick: What we mean is that the angular correlation data can be fit unambiguously with a static quadrupole interaction. And with single crystals that means that the angular correlation is unattenuated under the geometrical conditions previously described and it follows the dependence on β which came out of the equation described. I would guess that the quadrupole interaction is hard to find because there are other effects going on in a crystal and it is the quadrupole interaction which depends strongly on second and third nearest neighbours in the lattice. Since the radiation damage is worse a little further from the site of the final implanted atom this would show up more in a quadrupole interaction than in a magnetic interaction. More data need to be obtained. I would say one has to do channeling experiments on the tungsten in gadolinium case for example which has not yet been done.

Stone: I would like to say something which is in nature of a comment or perhaps a warning to people here, and that is to say that even without the risks of implantation there is a danger in using the hyperfine interaction fields indiscriminately. The effective field is not necessarily the same for all levels or isotopes of a given element. This is because the effective field is due primarily to a contact interaction with the s - electrons and the s - electron wave functions will vary over the nuclear volume. The density of nuclear magnetisation also varies over the nuclear volume and H_{eff} is really an integral of the magnetic hyperfine structure over the nucleus. In different nuclei the variation in magnetic distribution can lead to what is called a hyperfine anomaly. But for two nuclei, the ratio of the magnetic hyperfine structure is not equal to the ratio of their magnetic moments. Another way of saying this is that H_{eff} varies for different isotopes. While this effect is normally of the order of 1% or less, recent work has shown that in certain cases it can be as large as 8%, particularly for nuclei in which the magnetic moment is the small resultant of opposed spin and orbital contributions. Particularly, for example, in the case of iridium and gold, one must be careful about using H_{eff}, measured for ground states, in measurements on rotational levels. As an example for the first excited state in gold 197 there should be a 13% anomaly in the field with respect to the ground state. That is the H_{eff} would be in fact 13% greater in the excited state than in the ground state.

Frauenfelder: I would like to make one remark on this. Rather than shifting the blame on H_{eff}, it would probably be better to introduce form factors and place the blame where it belongs.

Stone: I would agree with that, except that essentially one has a measurement on one isotope and you cannot correct for the hyperfine anomaly. If it were possible to correct for the hyperfine anomaly, with any accuracy, then it might be worth doing it. But in practice, correction for hyperfine anomaly is something you can scarcely do, because the nuclear wave functions are not well enough known. If we could explain perfectly nuclear magnetism and the distribution of magnetization within the nucleus, then we could do as you say.

Frauenfelder: At least we should know where the problem lies. From electron scattering data the distribution of magnetization is now becoming known, at least for ground states.

Zawislak: The hyperfine fields on ^{207}Pb and on ^{208}Pb in a Ni - matrix agree in value, in contradiction to the fields in a Fe - matrix. It seems that it would also be interesting to perform channeling experiments on Bi and Tl in Ni, and compare the results with those obtained for Fe.

Murnick: Nickel is an fcc crystal whereas iron is an pcc crystal and cobalt is hexagonal. So each case is probably different.The experiments you suggest are being carried out at Bell Labs.

PERTURBED ANGULAR CORRELATIONS WITH IMPLANTED SOURCES

B.I. Deutch and G.M. Heestand
University of Aarhus, Institute of Physics, Aarhus, Denmark

1. Introduction

There are several rather complete recent articles, proceedings, and reviews on the subject of perturbed angular correlations (PAC), hyperfine interactions (hf), and g factor measurements[1-7]). This article will therefore limit itself to some of the latest information concerning PAC measurements with atoms implanted either by isotope separators or via Coulomb excitation in various substrates, both ferromagnetic and non - ferromagnetic, that have not been emphasized in the previous surveys. Among the topics concerned are the limitations of the method, relaxation processes, and experiments with emitted particles.

The subject of PAC combines the nuclear physics techniques of the angular correlation with the atomic physics topic of hyperfine interactions. Each subject has its own history, somewhat, but not quite independent of each other. The discovery of the megagauss hf of gold in iron by Samoilov et al.[8]) (1959), the determination via the Mössbauer effect of the hf in iron metal of H = −333 kG at room temperature[9]) led to the first g - factor measurement of the 2^+ state in ^{56}Fe using this internal field of iron, Metzger[10]) (1961). The implantation of ^{241}Am, ^{243}Am, and ^{243}Cf nuclei in iron using the recoil momentum trom an α - γ cascade (1962), by Flamm and Asaro[11]), the Coulomb excitation experiments with external magnetic fields on copper targets by Goldring et al.[12]) (1966) were among the developments that lead to the recent hf investigations with ferromagnetic foils by use of implantation techniques. In the recent studies, embedding was achieved by means of nuclear recoil following Coulomb excitation of rare earths into ferromagnetic backing (1966) by Grodzins et al.[13]) and Boehm et al.[14]). Later, isotope separators were used to implant radioactivity for Mössbauer studies of tellurium in iron by De Waard and Drentje[15]) (1966) and for PAC studies of ^{175}Lu in iron by Deutch et al.[16]) (1966).

Regarding the PAC technique, itself, in 1950, Brady and Deutsch[17]) pointed out that under the influence of a magnetic field, the precession of the magnetic moment of an intermediate short - lived state of a γ - γ cascade would result in the rotation of the γ - γ angular distribution. Frauenfelder and co - workers[18]) (1952) applied this PAC technique to the measurement of the magnetic moment of the 247 keV, 84 ns state in ^{111}Cd by determining the attenuation of the angular correlation from an external magnetic field. The history of the subject for the next 10 years followed with the determination of about 50 magnetic moments of excited nuclear states mainly with γ - γ cascades. For this reason the later emphasis in PAC research using hyperfine interactions has dealt predominantly with nuclear problems. It was natural first to utilize the megagauss effective magnetic fields to measure the g - factors of the shorter lived states. The knowledge of the solid state properties of the embedded substrate has lagged behind. In 1962, Caspari et al.[19]) published their rare - earth garnet work of ^{154}Gd and ^{152}Sm in which they observed time - dependent magnetic hyperfine interactions attenuating the measured angular correlations.

The present period witnesses an increased merging of the disciplines. Among the recent contributions has been the experimental discovery by Herskind et al.[20]) of a transient field following Coulomb excitation in the 10 megagauss range which lasts less than one picosecond. The theoretical explanation of these transient fields by Lindhard and Winther[21]) and by Lindhard[21]) (1969) indicates that the hf results from the scattering of polarized electrons by the screened Coulomb field of the slowing down ions; this process amplifies the average electron spin magnetic moment density at the site of the moving particle by a factor of about 300.

Magnetic time - dependent interactions have also been observed from embedded atoms in metallic foils. Such interactions can explain the temperature changes of the PAC involving the $5/2^+$ (63 ps) and $7/2^+$ (0.32 ns) intermediate states of ^{169}Tm in iron in the work of Deutch et al.[22]) (1968) and the temperature changes of the attenuations of the angular distribution from the 2^+ (2 ns) and 4^+ (0.1 ns) rotational states in a number of rare - earth isotopes in copper as indicated by Waddington et al.[23]) (1969). The theoretical treatment by Abragam and Pound[24]) (1953) and Blume[25]) (1968) of electron relaxation in PAC describes this effect of the instantaneous fluctuating hf at the nucleus due to flipping electron spins. However, details of fast relaxations in PAC have just begun to be explored.

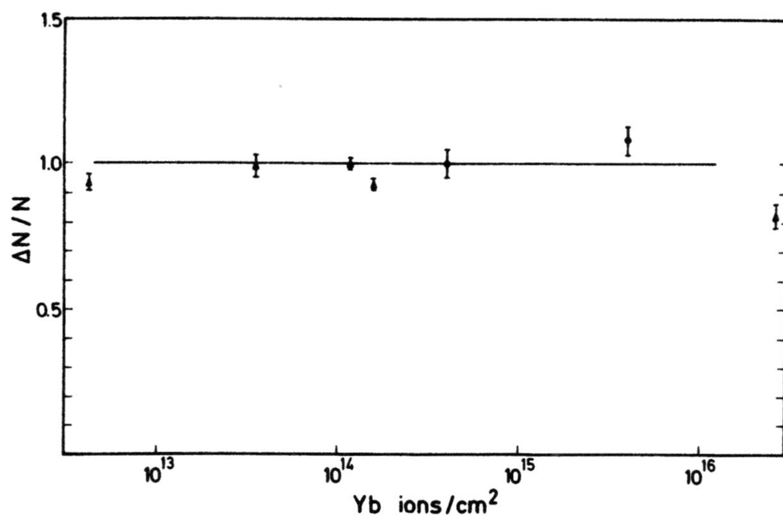

Fig. 1. The PAC effect[33]) relative to the value at a density of 10^{14} atoms/cm^2 of ^{169}Tm (circles) and ^{175}Lu (triangles) in iron as a function of implantation dose (radioactive and stable atoms).

There has been more stress, lately, on the nature of the implantation process itself, which may give rise to special problems. Precipitation may occur. Impurities which are normally substitutionally embedded if prepared by diffusion may end up in interstitial sites when implanted. There may be radiation damage after bombardment. Increased effort has been made to locate the foreign atom sites in various substrates.

Lindhard and co - workers[26]) have estimated the average implantation depth of the final implantation site. However, few publications exist with experimental information related to foreign atom location in ferromagnetic material. Bøgh[27]) indicated that after a dose of 2×10^{16} Sn ions/cm^2, iron single crystals are only moderately damaged. About 85% of these ions are on lattice sites. In a study with implanted lutecium, Schmorak and Bøgh[28]) report that the lutecium atoms are 38% substitutional when implanted in iron crystal. Nevertheless, much hyperfine work does not treat the interactions in the embedded substrates relative to the specific implantation conditions. Few follow up studies exist which relate implantation location to the element embedded, dose, implantation temperature, the treatment of the source after implantation etc. However, studies on iron substrate are in progress[29]). Investigations with implanted atoms in other materials such as silicon[30]) indicate a wide range in the final embedded sites of the foreign atoms. The percentage of atoms located on these sites are a function of the parameters mentioned. Work has begun which correlates the location of the site with the effective magnetic fields and gradients of the electric fields from crystal lattices for Xe in iron[29]) and hafnium in hafnium[31]).

Among the technical advances of the last years have been the adaptation of the

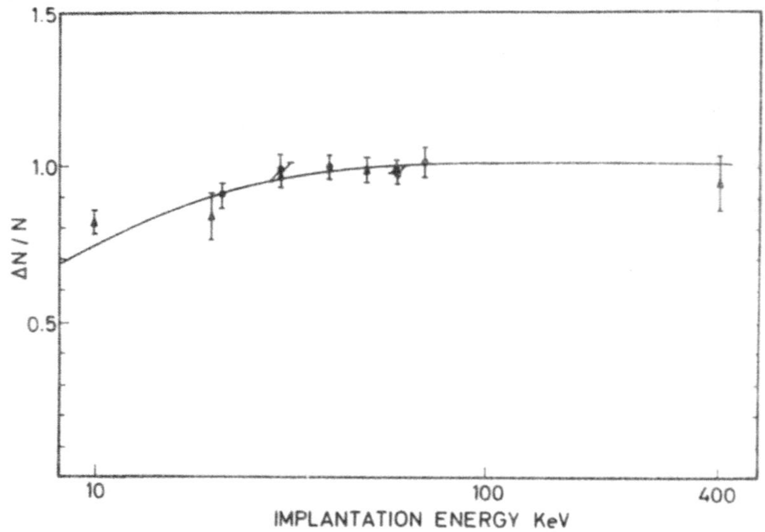

Fig. 2. The PAC effect[33]) relative to the value at 60 keV of ^{169}Tm (circles) and ^{175}Lu (triangles) in iron as a function of implantation energy.

PAC technique to electron cascades by Nielsen and Deutch[32,33]) (1967), the reintroduction of the α - γ cascade for hf studies with the decay of ^{228}Th in iron by Ansaldo et al.[34]) and increased effort to measure the hf in the actinide elements by several groups.

2. Implantation limitations

Some early PAC measurements with isotope implanted sources did not yield the values of the hf's which were determined by other methods. The main difficulties may have been: incomplete magnetization of the ferromagnetic foils, relocation of the hf site of the foreign atom through the processing of the source such as annealing, improper implantation energies and dosages, and, or incorrect analysis due to electron relaxation effects or the effects of other interactions.

This does not mean that the conditions for measurement are more stringent than with other methods. In fact as shown in figs. 1 and 2 the PAC effect[33]) is invariant in a range of densities $10^{12} - 10^{16}$ atoms/cm^2 and above energies of 30 keV.

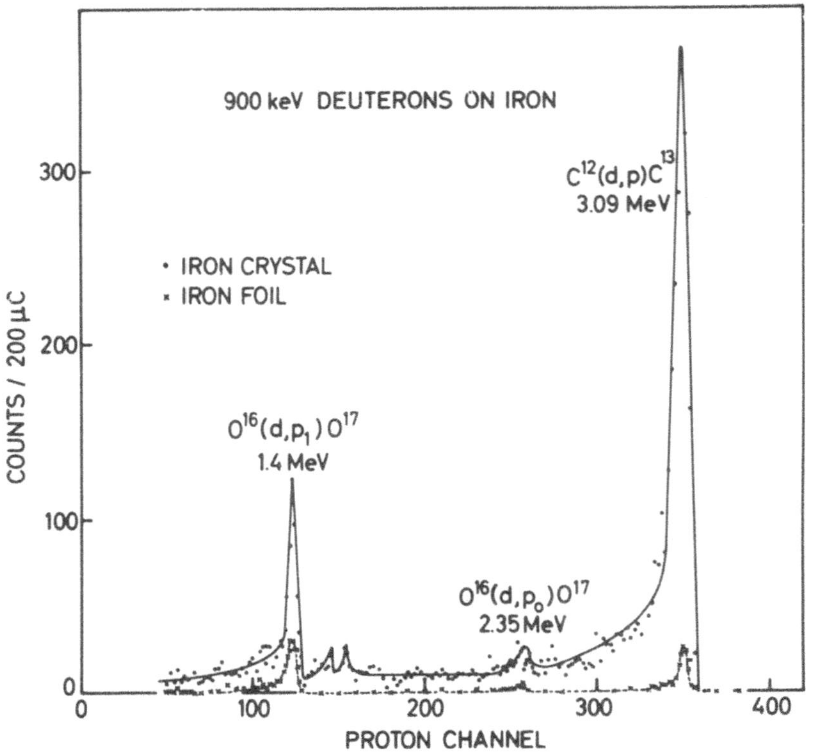

Fig. 3. The proton spectrum[36]) at 135° from 900 keV deuterons on iron crystal or iron foil. The peaks are the nuclear reactions indicated where p_1 is the proton from the excited state and p_0 the proton from the ground state of ^{17}O.

The Coulomb recoil techniques use energies of about 10^4 keV.

The shallow average implantation depth[26]) of the final embedded site creates special problems. Surface oxidation and surface treatment are dangers. Ferromagnetic material, in particular iron foil, may have an oxide layer that can seriously effect location and hyperfine determinations. The ~ 50 Å oxide layer competes as a substrate for the embedded impurity (range ~ 150 Å) with the ferromagnetic material itself. Surfaces have been studied by a comparison of the known oxidation layer produced in W from a given anodizing voltage[35]) with oxidation layers in various ferromagnetic materials. By comparing the number of protons produced via the ^{16}O (d, p) ^{17}O reaction to the excited ^{17}O state using a 900 keV D beam (see fig. 3), the oxide thicknesses listed in table 1 were determined[36]). Note that although the iron foil was annealed by a standard laboratory technique, the oxide layer nevertheless considerably increased as a result of the procedure. This can seriously affect attempts to anneal out the damage from the implantation into Fe. The range distribution of a 50 keV implanted Yb in iron is shown in fig. 4 indicating that $\leqslant 1\%$ of the atoms are in the oxide layer of an iron foil; about 15% would be in this layer after annealing. For a comparison, fig. 5 illustrates that there is a 20% loss of the PAC effect from cascades in Tm and Lu as a function of annealing time. When the surface oxide is removed away, the PAC effect is still slightly below the unannealed value indicating that some, but not all, of the loss results from the thick surface oxide. Since the average energy of a nucleus after α recoil is about 100 keV, α - γ experiments will be effected by these same considerations (see fig. 6 which illustrates the average

Material	Oxide Depth (Å)
Ni foil	10
" " (etched)	6
Ni crystal (epitaxially grown)	50
Fe foil	37
" " (etched)	37
" " (annealed 1/2 hr, 200°C)	92
Fe crystal (A)	52
(B)	75
(C)	137

Table 1. The oxide depth in various materials[36]). The depth was determined from a comparison of the ^{16}O (d, p) ^{17}O reaction yield from a 900 keV deuteron beam on these materials with the yield from a 200 Å tungsten oxide layer.

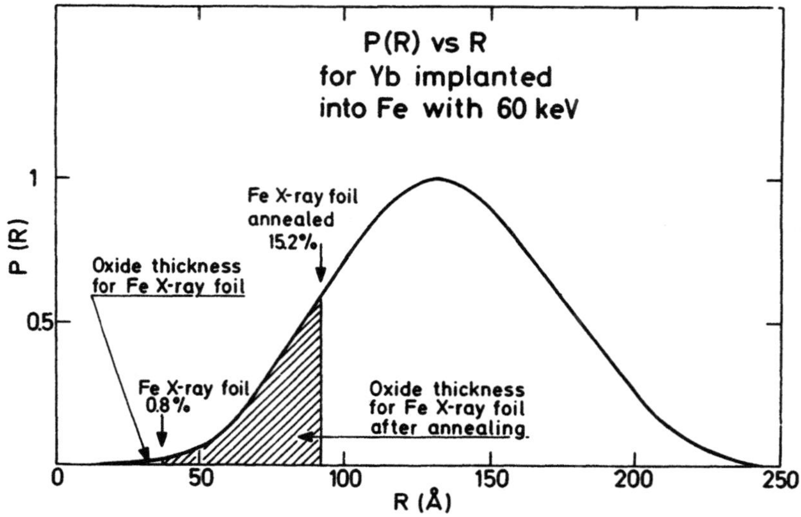

Fig. 4. The range profile of ytterbium in iron implanted at 50 keV showing the normal oxide depth and the oxide depth after an annealing of the foil in argon at 200 °C for 1/2 hr. The data have been calculated from the curves of ref. 26.

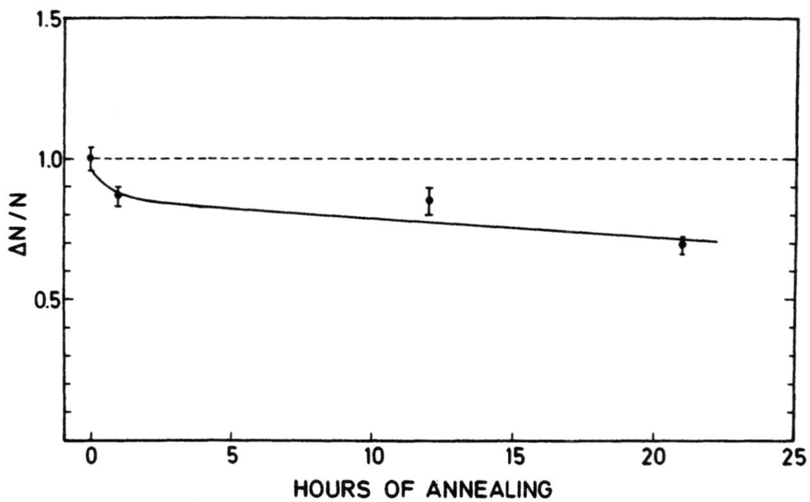

Fig. 5. The PAC effect relative to the unannealed value of ^{169}Tm in iron as a function of annealing time at 200 °C after implantation. The data is taken from ref. 33.

range as a function of implantation energy).

An essential prerequisite of PAC determinations is the complete alignment of the iron domains in the direction of the applied magnetic field. The difficulty in magnetizing even bulk iron samples has been reported by Johansson et al.[37]). They demonstrated that the PAC effect of palladium in iron increased up to external magnetizing fields of 10,000 gauss. Other laboratories have experienced these same difficulties. Fig. 7 shows the magnetization curve of our own foil calibrated with a known field. In general, the magnetization saturation problem can be solved by use of larger magnets. For the special case of the electron - gamma cascade, the requirement of a minimization of the stray field to prevent large deflections of the emitted electron by the stray field has been solved via a magnetization procedure which retains a residual field in the iron foil at zero coil current (see the hysteresis curves in ref. 33).

In early investigations of the rare - earth region, foil saturation apparently was not reached. Measurements[38]) in ^{175}Lu, ^{166}Er, ^{168}Er and ^{170}Er with a more powerful magnet resulted in measured hf's as much as 80% higher than yielded in the original determinations. As the precision of g - factor measurements increases care must be exercised that the measurements are undertaken substantially above the slowly rising plateau of the magnetization curve.

In general, the hf in ferromagnetic material follow the host magnetic moment ratio rule. Table 2 shows how the internal magnetic fields of Lu, a closed 4f - electron shell rare - earth, embedded in Fe, Co, and Ni are proportional to the magnetic moment of Fe, Co, and Ni, respectively. Sect. 3 discusses that Tm, an unclosed 4f - electron shell rare-earth, has a local moment and thus deviates from the above rule.

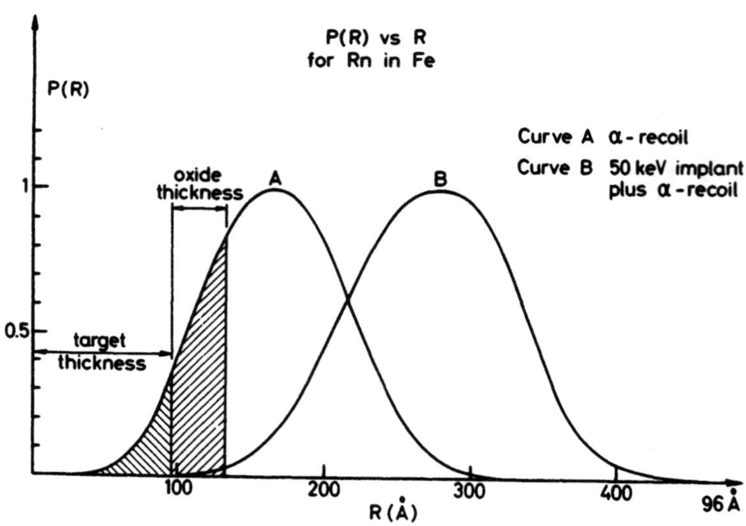

Fig. 6. The range profile of ^{220}Rn recoil from α - decay in iron (curve A) and ^{220}Rn recoil after a 50 keV implant of ^{224}Ra (curve B) through an 8 μg/cm^2 source and a 37 A iron oxide layer.

As previously mentioned, heating an iron foil after implantation lowers the PAC effect. Surface and internal oxidation as well as precipitation of the rare - earth may compete in these annealing difficulties. Also, the location of the atomic sites may change after heat treatment. Radioactive sources which have been produced from neutron irradiation after the isotope implantation procedure consequently may have entirely different properties due to irreversible heating effects than those sources being discussed in this paper. One limitation imposed by the annealing difficulties, so far, has been to accept radiation damaged implants in PAC investigation. In the case of iron, the damage is not great at even high dosages[39]). As stated in the introduction after a dose of 2×10^{16} Sn ions/cm^2, Bøgh[27]) found that iron single crystals are only moderately damaged. However, on other substrates most location studies are done at dosages $\leqslant 10^{15}$ ions/cm^2 because of the heavy damage at the higher dosages. For example, J.A. Davies et al.[30]) find that the individual damage regions around Sb - ion tracks produce a completely disordered region at 0.5×10^{14} Sb ions/cm^2 in Si - crystal. Channeling* experiments with Cu - crystal are often not possible above doses of 10^{15} ions/cm^2 without annealing or hot implantation. Table 3 lists a pre-

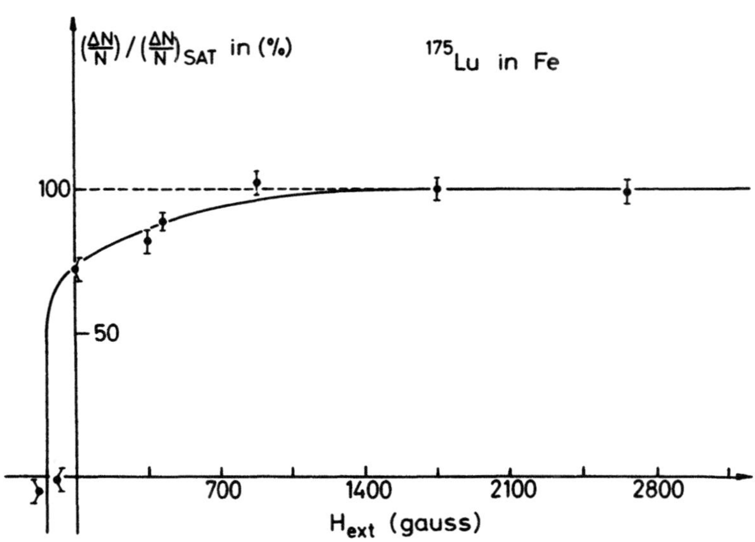

Fig. 7. Variation of the perturbed angular correlation in a thin iron foil with magnetization. The differences between coincidences measured with polarizing external magnetic field 'up' minus 'down' (at 135°) for the 283γ - 114γ cascade in ^{175}Lu. The curve was calibrated with equivalent measurements in a known magnetic field. The calibration points are not shown.

* For review of channeling see: J.W. Mayer, L. Eriksson and J.A. Davies: Ion Implantation in Semiconductors, Ch. 3 and 4 (Academic Press, N.Y.) 1970 or Location of Implanted Atoms via Channeling, B.I. Deutch, Proc. of the Intern. Conf. on hf Detected by Nucl. Rad., Israel 1970 (to be published) or, D. Murnick, these proceedings.

^{175}Lu T = 300°K

Material	H (kG)	H/H_{Fe}	μ/μ_{Fe}
Fe	- 483	1	1
Co	- 392	.81 ± .05	.77
Ni	- 122	.25 ± .02	.27

Table 2 A comparison of the effective magnetic field of ^{175}Lu in Fe, Co, and Ni with the magnetic moments of the same materials.

liminary result[40]) indicating the decrease of 2 MeV He Rutherford backscatter from a Dy - implant in a Cu single crystal along three axial directions at two different implantation dosages. At a density of 10^{15} Dy - ions/cm^2, crystal damage results in the appearance of the axial scattering as pure random. At the lower dosage, 50% of the foreign Dy - atoms appear shadowed when observed in the < 111 > direction. (20% can lie substitutional, 30% in the space along the < 111 > row). However, there is still sufficient crystal damage to complicate this deduction. More atoms may lie on lattice sites than the string dip indicates. In the silicon studies, an anneal at 735° C for 10 minutes was sufficient to remove 96% of the original crystal damage and change the location of a 5×10^{14} ion/cm^2 Sb - implant from completely 'interstitial' lattice sites to 90% substitutional sites. It is clear that in the future, techniques for

Dy in copper

Density (ions/cm²)	3×10^{14}	10^{15}
⟨111⟩/random	.53	.97
⟨110⟩/ "	.77	.91
⟨100⟩/ "	.81	.94

Table 3. The decrease in the aligned relative to the random backscatter of a 2 MeV He - beam from Dy implanted in copper at different dosages. (60 keV implant at 300 °K).

annealing away damage must be adapted for high dosage PAC studies.

The interpretation of hf measurements on implanted sources is critically dependent on the location of the implanted ions[15,16]). Low temperature nuclear orientation[41]) of Ce in Fe indicate the possible existence of more than one hyperfine field. Mössbauer spectra[42] of ^{161}Dy in Fe can be decomposed into 3 spectra, one (55%) with a well defined 4.5 MG field, a second corresponding to a smeared out relaxation shape, and a third which is a single line. Our own example[43]) of the time - dependent PAC of Dy in Cu (see table 3, and discussion in sect. 3) can be interpreted on the basis of 2 sites relaxing at two different frequencies. A preliminary result[40]) in table 4 shows the decrease in the 2 MeV He Rutherford backscattering from Yb in Fe single

Yb in Iron	10^{15} ions/cm^2
< 100 >/random	.50
< 110 >/random	.59
< 111 >/random	.54

Table 4. The decrease in the aligned relative to the random backscatter of a 2 MeV He - beam from Yb implanted into iron. (10^{15} Yb - ions/cm^2 implanted at 60 keV at 300 °K).

crystal along the < 100 >, < 110 > and < 111 > directions relative to a random direction. The data indicates that at least 50% of the Yb may lie in substitutional sites (it is difficult to assess the maximum expected error due to crystal damage). This data is consistent with the Mössbauer studies of Dy in Fe (the Yb - atom has the same atomic radius as the Dy - atom and will be implanted into the same lattice sites). However, within experimental uncertainties, there is an ambiguity whether the high field component (H = 4.5 MG, 55% abundant) lies on the substitutional site, or whether the remaining 45% of the atoms are substitutional. Many of the above examples will be better understood when the research programmes correlating PAC with location information become fully matured.

3. Time - dependent angular correlations with hyperfine fields

It will be assumed that the reader is familiar with the general equations related to perturbed angular correlations (see refs. 1 - 7). They will therefore not be repeated except for descriptive purposes.

When a magnetic field H is applied to the nucleus with g - factor g, it will spin with a Larmor precession

$$\omega = -\frac{g\mu_n H}{\hbar} \tag{1}$$

(μ_n being the nuclear magneton).

If analogous to the case of an applied external magnetic field, the hyperfine fields were ordered (complete alignment of the atomic spins) perpendicularly to the plane of two gamma rays emitted in cascade, the second gamma ray would be emitted at a slightly rotated angle $\pm\,\Delta\theta$ due to this nuclear precession. For small rotations the angular shift would be $\Delta\theta = \omega\tau_n = -\,4.7 \times 10^3\ gH\tau_n$ radians. Shifts as small as 0.001 radians have been determined, hence with $g = 1/3$, $H = 3$ MG, g - factors of states with lifetimes as short as 0.2×10^{-12} s can be conceivably measured. With a known g and τ_n, the integral PAC technique can be utilized to study the hyperfine fields.

In general with ordered interactions, the integral rotation is accompanied by an appreciable attenuation due to the large magnitude of the hyperfine fields. The integral angular correlation becomes

$$W(\theta) = 1 + \sum_{N=2}^{k_{max}} b_N \cos N\Delta\theta \cos N\,(\theta\text{-}\Delta\theta) \tag{2}$$

where $\tan N\Delta\theta = N \cdot \omega\tau_n = \frac{-Ng\mu_n H_{int}\tau_n}{\hbar}$ and the b_N are given functions[6]). In the absence of further interactions the observed attenuation $\cos N\Delta\theta$ and the rotation $N\Delta\theta$ must yield consistent values of $\omega\tau_n$.

In recoil implantation studies, the precession angle $\Delta\theta$ is composed of rotations from transient and 'static' magnetic hyperfine fields and is written as follows:

$$\Delta\theta = \phi_{transient} - gH\tau_n \frac{\mu_n}{\hbar} \tag{3}$$

The $\phi_{transient}$ is being calculated for relativistic electron energies by Lindhard and Winther[21]) and is analyzed by Murnick in these proceedings. The special case of a time varying 'static' field will be discussed in the following pages.

Expression (2) describes the situation in magnetic material whose domains are aligned in an external field. Below the Curie temperature the atoms are almost completely aligned via the exchange interaction. The correlation (relaxation) time τ_c is slow compared to τ_n. However for impurities, for example, rare - earths embedded in iron, the alignment may not be complete even far below the Curie temperature of iron. Likewise the correlation time may be fast. Abragam and Pound[24]) have derived an expression appropriate for the case of an external magnetic field applied to a liquid source in which the nuclear quadrupole moment interacts with fluctuating electric field gradients. The formula (taken from ref. 7) is

$$W(\theta) = 1 + \frac{1}{4} A_{22}G_{22} \{ 1 + \frac{3 \cos 2(\theta - \Delta\theta_{22})}{[1+(2\omega\tau_n G_{22})^2]^{1/2}} \} +$$

$$+ \frac{1}{64} A_{44}G_{44} \{9 + \frac{20 \cos 2(\theta - \Delta\theta_{24})}{[1+(2\omega\tau_n G_{44})^2]^{1/2}} + \frac{35 \cos 4(\theta - \Delta\theta_{44})}{[1+(4\omega\tau_n G_{44})^2]^{1/2}} \} \quad (4)$$

where $\tan \Delta\theta_{Nk} = N\omega\tau_n G_{kk} = \frac{N\omega\tau_n}{1+\lambda_k\tau_n}$. As discussed by Caspari et al.[19]) the application of this formula can be extended to describe the case of a partially ordered system showing fast relaxation provided that $< H^z_{int} > \ll < H^2_{int} >^{1/2}$; the brackets denote the sample average of the internal field. The relation between these average fields and the frequencies ω and ω_m are $\omega = - g\mu_n < H^z_{int} > / \hbar$ which as stated previously describes the rotation and $\omega_m = - g\mu_n < H^2_{int} >^{1/2}$ which accounts for the attenuation of the angular correlation. The conditions for the validity of (4) to be fulfilled are, $\omega_m\tau_c \ll 1$ and $\tau_c \ll \tau_n$. To first order in the exchange field, the explicit formula for λ_k in the magnetic case is the same as given by Abragam and Pound:

$$\lambda_k = \frac{1}{3} k (k + 1)\omega_m^2 \tau_c \quad (5)$$

For $\lambda_k\tau_n \ll 1$, (4) takes the form of (2). If $\tau_c \gg \tau_n$ and the alignment is not complete, one has a static distribution of fields and (2) and (4) lose their validity. The measurement of the rotation and attenuation of the angular correlation would not yield consistent values of $\omega\tau_n$ when evaluated from (2). The break - down of (2) and (4) could be due also to an additional effect of static quadrupole interactions, say, from a non-cubic crystalline environment of the embedded ions or from a fraction of the embedded ions in damage sites. With fast relaxation, the individual atoms pass through their possible electronic levels many times during the decay of the intermediate nuclear state. All atoms in the sample appear identical with respect to the hf interaction. Therefore both for rapid relaxation and its antithesis, complete alignment, (2) applies. Even in the case of rapid relaxation, it is possible to have more than one hyperfine field. This depends upon the sites into which the impurities are implanted.

For paramagnetic ions in non - magnetic materials an external magnetic field plays the same role as the exchange interaction. It partially lines up the atoms resulting in an enhanced field at the nucleus. When the external field is removed the $< H^z_{int} >$ is zero and no rotation is observed. However attenuations due to the fluctuating field are still present. For that case and the case of a fluctuating electric field gradient the angular correlation is (dropping the double indices of G_{kk})

$$W(\theta) = \sum_{k=0}^{k_{max}} A_k G_k P_k (\cos\theta) \quad (6)$$

where

$$G_k = \frac{1}{1+\lambda_k \tau_n} \tag{7}$$

For the magnetic interaction λ_k is given by (5) while the electric quadrupole interaction yields

$$\lambda_k = \frac{3}{5} k\,(k+1)\,[4I\,(I+1) - k\,(k+1) - 1]\,\omega_e^2\,\tau_c \tag{8}$$

where ω_e is the quadrupole frequency and I is the nuclear spin.

In the derivation of (7) the nucleus is described as experiencing classical fields which continually reorient themselves. The validity is limited to cases where $\omega_{m,e}\tau_c \ll 1$ and $\tau_c \ll \tau_n$. In a later treatment by Blume[25]) these conditions are dropped and the hf hamiltonian expresses explicitly the effect of stochastic fluctuations among a number of atomic states. In the special case of fluctuations between two states corresponding to an electron with an effective spin of 1/2 the magnetic dipole interaction results in the attenuation

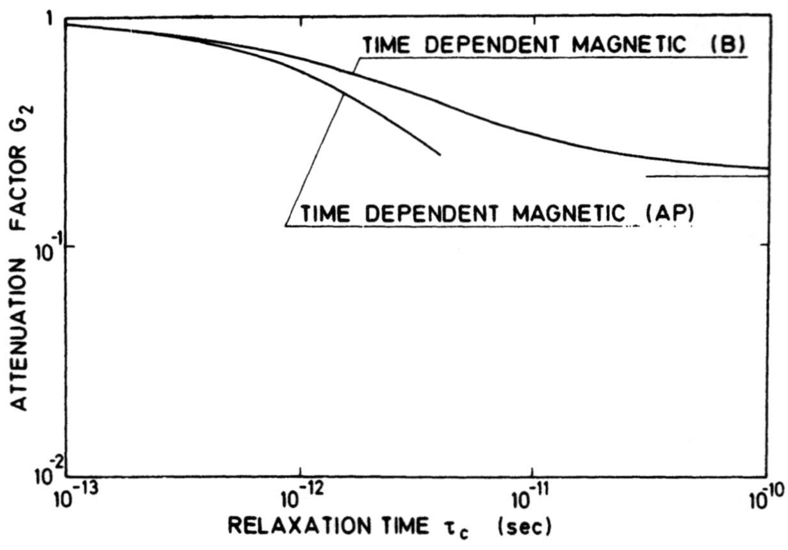

Fig. 8. The integral magnetic time dependent attenuation factor G_2 vs. correlation (relaxation) time τ_c. The curve marked AP is for Abragam and Pound's theory (eq. (7)) and that marked B is for Blume's calculations (eq. (9)) from implanted ^{166}Er in iron where $\omega_m = 1.2 \times 10^{10}$ cps and $\tau_n = 2.6 \times 10^{-9}$ s. The plot of G_4 vs. τ_c is similar.

$$G_k = \frac{1}{2k+1} \sum_{N=-k}^{k} \frac{1}{1+\dfrac{\omega_m^2 \tau_n^2 N^2}{1+\dfrac{\tau_n}{\tau_c}}} \qquad (9)$$

A similar result is obtained in the case of a flipping electric quadrupole interaction. With the appropriate limits the Abragam and Pound result is yielded in both cases. Recent work by Gabriel[44]) generalizes the Bloom theory to spins greater than 1/2 and to non-Markovian fluctuations. Fig. 8 displays G_2 as a function of the correlation time τ_c and illustrates the utility of the PAC technique to investigate relaxation when τ_c is small. Note the hard core value (N = 0) in the Blume expression.

The analysis becomes complicated when there are simultaneous static quadru-

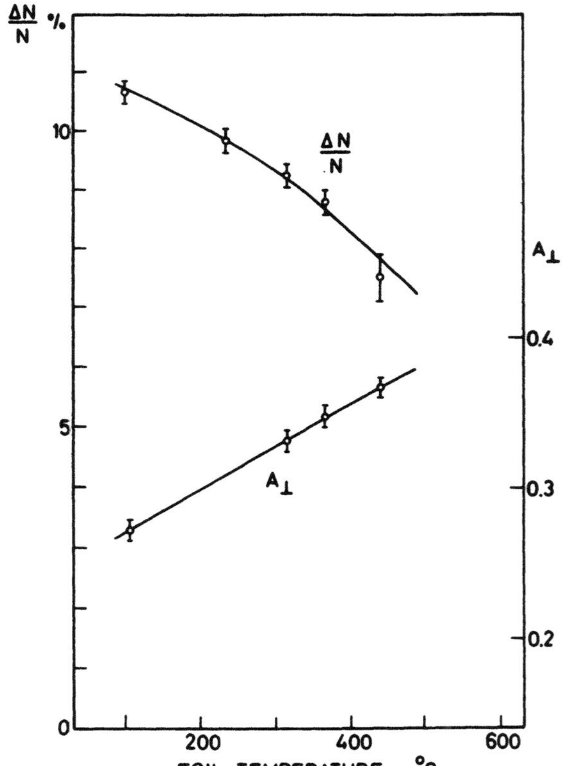

Fig. 9. The PAC ΔN/N and anisotropy $A_\perp$ of the 198γ - 110γ keV cascade of ^{169}Tm embedded in annealed magnetized iron foil as a function of temperature[22]).

pole and magnetic interactions present. In general the matrix elements have to be evaluated numerically using computer methods. For a pure static quadrupole randomly oriented interaction the simplicity is regained and the correlation is attenuated by the factor

$$G_k = \sum_n s_{kn} \frac{1}{1 + (n\omega_0 \tau_n)^2} \tag{10}$$

where the coefficients s_{kn} are given in ref. 7 ($\omega_0 = 3\omega_e$ for I even and $\omega_0 = 6\omega_e$ for I odd). The crystal field acting on the nuclear quadrupole moment when the nucleus is embedded in a non - cubic environment results in an attenuation of the form (10). In general, exchange polarization and distortion of closed shells by the crystalline fields or by aspherical outer electrons will cause quadrupole effects.

Static interactions may be distinguished from the time dependent interactions by observing the temperature dependence of the perturbed angular correlation. The relaxation phenomena has characteristic temperature variations. The λ_k appears to vary as ~ 1/T where T in degrees Kelvin is above liquid nitrogen temperature. This has been observed in PAC work for YbIG[19]) and for Yb implanted in iron[22]). Fig. 9 taken from ref. 22 displays the variation of ΔN/N and $A_\perp$ (defined below) as a function of temperature. In this case the third term in (4) is negligible and one finds corresponding to a correlation with just A_2 terms

$$\frac{\Delta N}{N} = \frac{W(\uparrow) - W(\downarrow)}{\frac{1}{2}(W(\uparrow) + W(\downarrow))} = \frac{3A_2}{4a + A_2} \sin 4\Delta\theta' \tag{11}$$

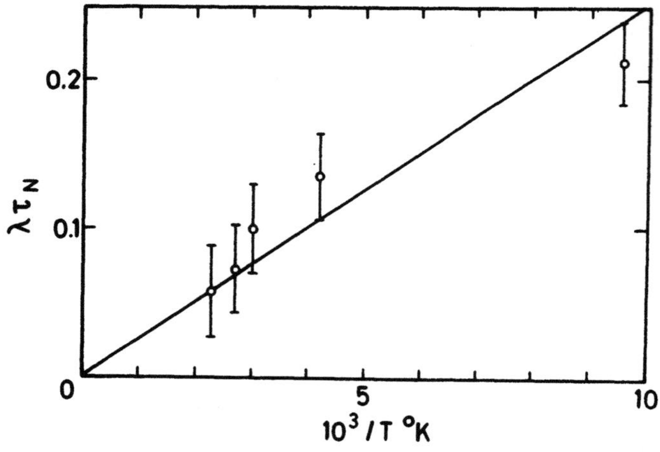

Fig. 10. The parameter $\lambda_2\tau_m$ from the 5/2⁺ state of ^{169}Tm embedded in iron as a function of 1/T.

and

$$A_\perp = \frac{W(\uparrow\downarrow,\pi) - W(\uparrow\downarrow,\frac{\pi}{2})}{W(\uparrow\downarrow,\frac{\pi}{2})} = 2\,(\frac{3A_2}{4a+A_2})\cos^2 2\Delta\theta'/(1-(\frac{3A_2}{4a+A_2})\cos^2 2\Delta\theta') \tag{12}$$

where

$$a = 1 + \lambda_2\tau_n \text{ and } \Delta\theta' = \frac{1}{2}\text{ arc tan}\frac{2\omega\tau_n}{1+\lambda_2\tau_n}$$

with $\tau_n = 0.09$ ns one deduces $\lambda_2\tau_n = 0.08$ and a relaxation time of Tm in iron $\tau_c = 2 \times 10^{-12}$ s at room temperature. The $\lambda_2\tau_n$ varies as 1/T (see fig. 10). Further temperature experiments[45]) with the longer (0.32 ns) $7/2^+$ state in ^{169}Tm confirm the consistency of the data with the Abragam and Pound time - dependent magnetic relaxation model. Although the data does not rule out time dependent quadrupole effects and possible small influences of static quadrupole interactions[46]), relaxation (mainly magnetic) seems to occur and is observable in the time range of the PAC technique.

The interpretation of a dominant magnetic relaxation behaviour of Tm in Fe has been recently confirmed by Bernas and Gabriel[47]) in an extension of the measurements to liquid He temperatures. However in the temperature range below 100° C the Abragam and Pound assumptions no longer hold and a more general model[45]) is applicable. For the Tm case, the Gabriel calculation is equivalent to that of Blume.

In another study[23]) of electronic relaxation effects, copper was used as a target backing. The integral angular correlations of the de - exciting gamma rays following Coulomb excitation were measured in copper from the rotational 2^+ states of ^{150}Nd, ^{150,152,154}Sm, ^{158}Gd, ^{160}Dy, ^{166}Er, and ^{174}Yb and from the 4^+ states of ^{152}Sm, ^{160}Dy and ^{166}Er at different temperatures. The correlations were all attenuated, in several cases by a large factor (see fig. 11). It is possible to compare the experimental values with the various theoretical predictions calculated from the interactions represented in eqs. 4 - 9 by the ratio of the G_4/G_2 attenuations. The conclusion that can be drawn from this data (fig. 8) is that relaxation is the main cause of the attenuation. An analysis using (6) and (7) does not display the 1/T temperature dependence of λ_k as was found for Tm in iron. Due to the large hf of the unfilled 4f shell which yields $\omega_m\tau_c \geqslant 0.3$, the Abragam and Pound assumptions are hardly fulfilled. Application of Blume's general formula (for which a special case (9) is shown in the figure) may give a more consistent picture, although a small admixture of an electric quadrupole interaction would bring the Abragam and Pound model in agreement with the data.

Recent PAC experiments by Skaali et al.[48]) and Heestand et al.[43]) with radioactive sources of ^{166}Dy implanted in Cu have attempted to untangle the ambiguities. In the former work, the paramagnetic correction factor β was measured as a function of temperature to see if crystalline field effects are important. Their conclusion was that β closely followed the free ion value; this implies that crystalline field effects are

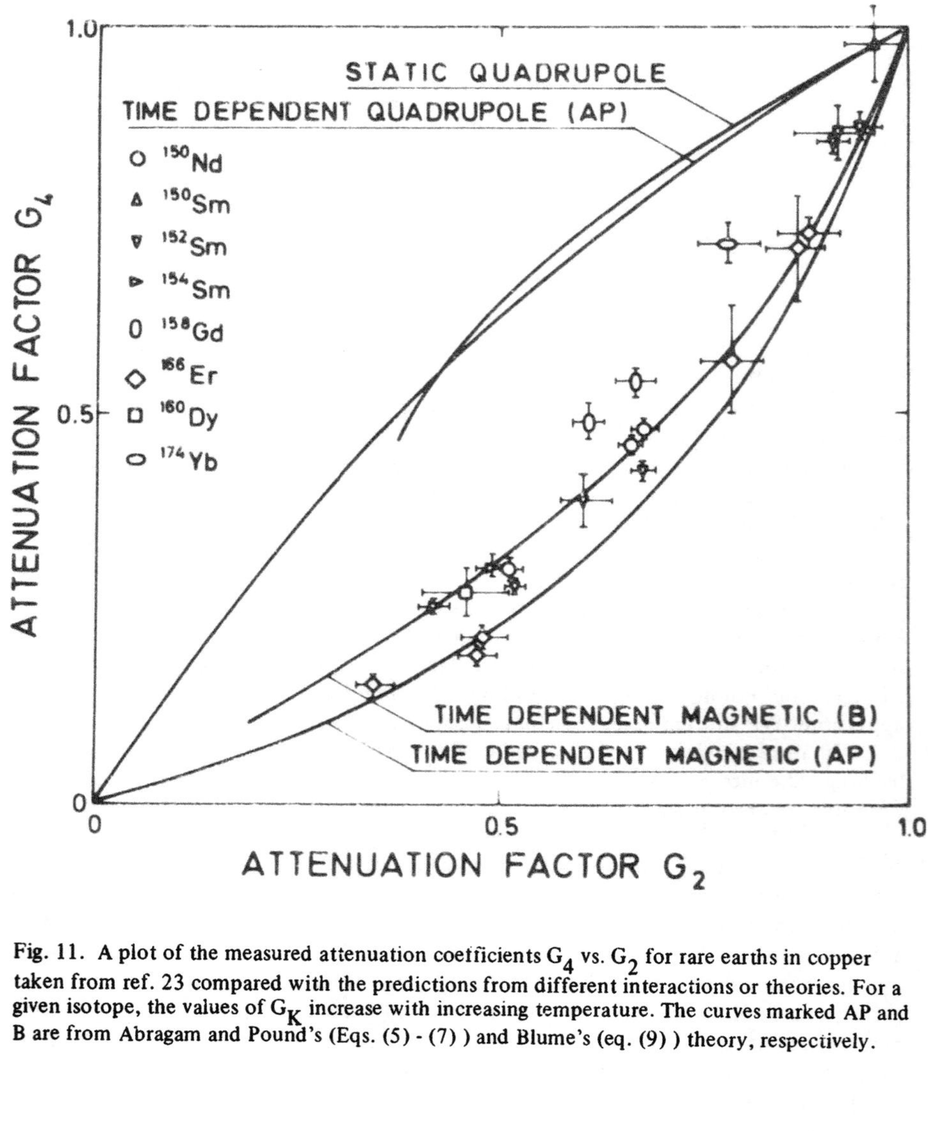

Fig. 11. A plot of the measured attenuation coefficients G_4 vs. G_2 for rare earths in copper taken from ref. 23 compared with the predictions from different interactions or theories. For a given isotope, the values of G_K increase with increasing temperature. The curves marked AP and B are from Abragam and Pound's (Eqs. (5) - (7)) and Blume's (eq. (9)) theory, respectively.

small. This conclusion is supported also in the measure of β as a function of external field for the alloy $HoAl_2$ at 80° K by the Uppsala group[49]). In ref. 43, the time - differential PAC of ^{166}Er in Cu was investigated to distinguish the purely exponential decay of the differential attenuation coefficient $G_k(t)$ in the Abragam and Pound model (see eq. (13)) from the residual non - time dependent term of the Blume theory (see eq. (14)).

$$G_2\,(t) = e^{-\lambda_2 t} \tag{13}$$

(Magnetic time dependence)

$$G_2\,(t) = 1/5 + 2/5 \sum_{N=1,2} \left(\cosh\ xWt + \frac{1}{x}\ \sinh\ xWt\right) e^{-Wt} \tag{14}$$

where $\lambda_2 = 2\omega_m^2\tau_c$, $x = (1 - 4N^2\omega_m^2\tau_c^2)^{1/2}$, $W = 1/2\tau_c$, ω_m is the Larmor frequen-

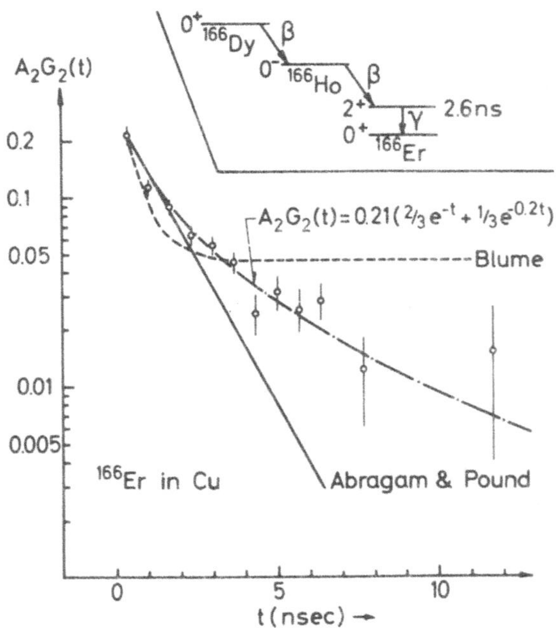

Fig. 12. The differential angular correlation A_2G_2 (t) for the 1760 keV (β) – 81 keV (γ) angular correlation[43]) of ^{166}Er implanted in Cu. Implant conditions: $\leqslant 1 \times 10^{14}$ Dy ions/cm^2 at 300 °K, 50 keV. The solid curve represents the predictions of the Abragam and Pound theory while the dashed curve represents the Blume theory for a one site implant (see ref. 43). The experimental points are consistent with a 2 site Abragam and Pound calculation with the Yb - atoms in the 2 sites in the ratio of 2 to 1 and having 2 different relaxation constants λ_1 = 0.2 and λ_2 = 1,respectively. The insert shows the relevant levels involved in the decay from the mother ^{166}Dy to the granddaughter ^{166}Er.

cy due to the magnetic field produced by the unfilled 4f - electron shell, i.e. $\omega_m^2 = \frac{g^2}{\hbar^2} \mu_N^2 \langle H_{4f}^2 \rangle$, and τ_c is the relaxation time. The Blume formula given above in eq. (14) was derived for the case of j = 1/2 electronic states and in principle should only be applied to those cases.

The results show in fig. 12 that although the angular correlation is, indeed, time - dependent, neither calculation agrees with the data. An admixture of electric quadrupole and magnetic dipole interactions cannot be ruled out. Curves based on the Abragam and Pound model can be fit through the experimental points if one assumes two different sites for the implanted ions with populations in the ratio 2 to 1 and relaxation rates in the ratio 5 to 1. Incidentally, the Dillenburg - Maris[50]) picture also describes relaxation with two different decay constants.

An assumption of 2 sites is consistent with the location data of Dy in Cu contained in table 3 (see discussion in sect. 2), which, however, is preliminary and subject to uncertainties due to crystal damage. A recent Mössbauer experiment with implanted Dy in Fe by De Waard et al.[42]) (see discussion in sect. 2) which indicates that 55% of the foreign atoms lie in high field sites (there are also 2 other 'sites') is consistent with the preliminary location data of Yb in Fe contained in table 4. The latter data indicates that at least 50% of the rare - earths are substitutional.

To speculate, it seems proper to consider the 3 possible sites of implanted foreign atoms: the substitutional, the interstitial, and a small surface oxide region when analyzing nuclear orientation, Mössbauer and PAC data. Of course, the relaxation calculations themselves are in need of generalization but this is no substitute to the physical reality of the actual foreign atomic positions in the substrate.

4. Charged particle cascades

The implantation techniques offer unique possibilities to utilize charged particles from radioactive decay in PAC studies of hyperfine fields or of g - factors of excited states. The small path lengths of the emitted charged particles in the effective stray magnetic fields result in a vanishingly small stray deflection $\Delta\theta_s$ of the particle. The angular deflection of a charged particle in an average magnetic field B over a path length d with radius of curvature ρ is approximately

$$\Delta\theta_s \cong \frac{B \cdot d}{B\rho} \text{ radians} \tag{15}$$

In experiments with conventional magnets, α particles, β particles, or conversion electrons would be deflected from there rectiliniar path to the counter by the stray field. With the miniature magnets used in implantation studies, the product B·d thus $\Delta\theta_s$ is often 10^6 times smaller than with a conventional 50,000 gauss magnet. Rotations as small as 10^{-2} radian are then possible to measure via implantation techniques. A 20 keV conversion electron has a $B\rho = 500$. For such an electron, in a foil field and stray field of 2×10^4, and 10^2 gauss, with a 2×10^{-6}, and 5×10^{-2} cm path length, the apparent rotation $\Delta\theta_s$ would only be 10^{-4}, and 10^{-2} radians, respectively. This is a negligible correction to most determinations of hf rotations. Higher energy electrons would have even smaller corrections.

In the actinides, the 5f - atomic shell elements in the second region of nuclear deformity lying above lead, can be investigated via cascades involving a charged particle transition. The first 2^+ excited states of the even A nuclei in this region are highly converted ($a_T \sim 10^3$) which suppresses the usual gamma counting rate but is observable with conversion electrons. Many levels of nuclei in this region are also reached by a decay. (The rotational model predicts that the lifetime of the 2^+ and 4^+ states of the even - even nuclei in the region are comparable being about 0.2 ns. Thus precessions of these states would be about the same). Measurements in the actinides are under progress.

The first PAC electron measurements[32]) indicated that the rotations from the $1855\beta - 123\gamma$ and the $1281\gamma - 123\gamma$ cascade in ^{154}Gd in iron determined equivalent precessions. In an extension[33]) of this technique to low energy conversion electrons (20 keV) in the measure of the hf of ^{238}Pu in Fe, a magnetization procedure which minimizes the stray fields is reported.

The most recent electron experiment utilizes the cascade

$$2^- \xrightarrow{971\gamma} 2^+ \xrightarrow[L_{II},\, L_{III},\, M,\, N,]{47.6\ \text{(electron)}} 0^+ \ \text{in}\ ^{232}\text{U}$$

Ravn et al.[51]) find the precession of ^{232}Pa implanted in iron foil to be $\omega\tau =$ $= 0.34 \pm 0.037$. Under the assumptions that the gyromagnetic ratio $g = 0.3$, and the mean life of the 2^+ state $\tau_n = 0.37$ ns, the hf of uranium in iron at room temperature, H, was found to be negative and is $- 0.66 \pm 0.07$ Mg.

Alpha activity is evaporated or sputtered onto the ferromagnetic foil. The first a recoil experiments, in fact, were radioactive measurements[11]) with ^{241}Am, ^{243}Am, and ^{243}Cf. For most cases, the highly anisotropic $0 \rightarrow 2 \rightarrow 0$ correlation can be utilized. However, there is danger from errors due to surface oxidation if a thick chemical source

Hf in iron (M Gauss)

Method	Element		
	Thorium	Uranium	Plutonium
Coulomb Exc.	- .54 ± .10	- 0.66 ± .15	
Radioactivity		- 0.66 ± .07	+ 0.62 ± .12
Atomic shell (Seaborg)	$6d^2\ 7s^2$	$5f^3\ 6d\ 7s^2$	$5f^6\ 7s^2$

Table 5. Measured hyperfine fields in the actinide region[33,51,52]). A g - factor equal to 0.3 was assumed in the calculation of the fields from the measured precessions.

is used. We have found the precession of an evaporated ^{224}Ra source on Fe to be – 0.25 ± 0.03 radians whereas ^{224}Ra implanted into Fe at 50 keV via an electromagnetic isotope separator yields a precession of – 0.34 ± 0.05 radians. In fig. 6 the range distribution for both the implanted and evaporated Ra - source is shown. A source thickness of about 8 μgms/cm^2 plus a 37 Å oxide layer is sufficient to reduce the precession by 0.73, the ratio of the measured evaporated to implanted precessions. The measured hf's near the actinide region from the Danish laboratories[33,52,51]) are listed in table 5.

References

1. B.I. Deutch and K. Bonde Nielsen, Proc. of the Int. Conf. on Radioactivity in Nuclear Spectroscopy (Nashville 1969).
2. Discussion on ion implantation and hyperfine interactions. Proc. Roy. Soc. (London) A311 (1969) pps. 1 - 209.
3. Hyperfine Structure and Nuclear Radiations, Ed. by E. Matthias and D. Shirley (North Holland Publ. Co., Amsterdam, 1968).
4. L. Grodzins, Ann. Rev. Nucl. Sci. 18 (1968) 291.
5. Hyperfine Interactions. Ed. by A. Freeman and R. Frankel (Academic Press, New York, 1967).
6. Perturbed Angular Correlations. Ed. by E. Karlsson, E. Matthias, and K. Siegbahn (North Holland Publ. Co., Amsterdam, 1964).
7. Alpha-, Beta-, and Gamma-Ray Spectroscopy. Ed. by K. Siegbahn (North Holland Publ. Co., 1965) Chap. XIX.
8. B.N. Samoilov, V.V. Sklyarevskii, and E.P. Stepanov, Soviet Phys., JETP 36 (1959) 448.
9. S.S. Hanna, J. Heberle, C. Littlejohn, G.J. Perlow, R.S. Preston, and D.H. Vincent, Phys. Rev. Letters 4 (1960) 177.
   S.S. Hanna, J. Heberle, G.J. Perlow, R.S. Preston, and D.H. Vincent, Phys. Rev. Letters 4 (1960) 513.
10. F.R. Metzger, Nucl. Phys. 27 (1961) 612.
11. E. Flamm and F. Asaro, Phys. Rev. 129 (1962) 290.
12. G. Goldring, R. Kalish, and H. Spehl, Nucl. Phys. 80 (1966) 33.
13. L. Grodzins, R.R. Borchers, and G.B. Hagemann, Phys. Letters 21 (1966) 214.
14. F. Boehm, G.B. Hagemann, and A. Winther, Phys. Letters 21 (1966) 217.
15. H. de Waard, and S.A. Drentje, Phys. Letters 20 (1966) 38.
16. B.I. Deutch, P. Hornshøj, K. Bonde Nielsen, and F. Boehm, Phys. Letters 21 (1966) 659.
17. E.L. Brady and M. Deutsch, Phys. Rev. 78 (1950) 558.
18. H. Aeppli, H. Albers - Schönberg, A.S. Bishop, H. Frauenfelder, and E. Heer, Helv. Phys. Acta 25 (1952) 339.
19. M. Caspari, S. Frankel, and G.T. Wood, Phys. Rev. 127 (1962) 1519.
20. B. Herskind, R.R. Borchers, J.D. Bronson, D.E. Murnick, L. Grodzins, and R. Kalish, ref. 3, 735.
    R.R. Borchers, B. Herskind, J.D. Bronson, L. Grodzins, R. Kalish, and D.E. Murnick, Phys. Rev. Letters 20 (1968) 424
    R. Kalish, L. Grodzins, R.R. Borchers, J.D. Bronson, and B. Herskind, Phys. Rev. 161 (1967) 1196.
21. J. Lindhard and A. Winther, Nucl. Phys. A166(1971) 413.

J. Lindhard, Proc. Roy. Soc. (London) 311 (1969) 11.

22. B.I. Deutch, K. Bonde Nielsen, and H. Bernas, Phys. Letters 27B (1968) 209.

23. J.C. Waddington, K.A. Hagemann, S. Ogaza, D. Kiss, B. Herskind, and B.I. Deutch, Nuc. Reactions Induced by Heavy Ions, North Holland (1970) 438.

24. A. Abragam and R.V. Pound, Phys. Rev. 92 (1953) 943.

25. M. Blume, ref. 3, 911;
M. Blume, Phys. Rev. 174 (1968) 351.

26. J. Lindhard, M. Scharff, and H.E. Schiøtt, Mat. Fys. Medd. Dan. Vid. Selsk. 33 no. 14 (1963).

27. E. Bøgh, ref. 2, 35.

28. M. Schmorak and E. Bøgh, ref. 3, 712.

29. L.C. Feldman, and D.E. Murnick, Proc. of the Int. Conf. on Hyperfine Int. Detected by Nucl. Rad., Israel (1970) (to be published).
L.C. Feldman, W.M. Augustyniak, and E.N. Kaufmann, ibid.
R.B. Alexander, N.J. Stone, D.V. Morgan, and J.M. Poate, ibid.

30. J.A. Davies, J. Denhartog, L. Eriksson, and J.W. Mayer, Can. Journ. of Phys. 45 (1967) 4053.

31. B. Herskind, B. Skaali, and C. Günther, Proc. of the Int. Conf. on Hyperfine Int. Det. by Nucl. Rad., Israel (1970) (to be published).

32. K. Bonde Nielsen, and B.I. Deutch, Phys. Letters 25B (1967) 208.

33. H. Ravn, F. Abildskov, H.K. Johansen, and B.I. Deutch, Proc. of the Int. Conf. on Rad. in Nuc. Spect. Nashville (1969) (to be published).

34. E. Ansaldo, L. Grodzins, and R. Kalish, Phys. Lett. 30B (1969) 538.

35. M. McCargo, J.A. Davies, and F. Brown, Can. J. Phys. 41 (1968) 1231.

36. J. Bøttiger, J.A. Davies, and B.I. Deutch (to be published).

37. K. Johansson, E. Karlsson, L.O. Norlin, P.N. Tandon, and H.C. Jain, ref. 3, 471.

38. B.I. Deutch, G.B. Hagemann, K.A. Hagemann, S. Ogaza, ref. 3, 731.

39. L.C. Feldman, W.M. Augustyniak, and E.N. Kaufmann, Proc. Int. Conf. on hf Detected by Nucl. Rad., Israel 1970 (to be published).

40. B.I. Deutch, H. Ravn, and G.M. Heestand (to be published).

41. D. Spanjaard, Proc. Int. Conf. on hf Detected by Nucl. Rad. Israel (1970) (to be published).

42. H. De Waard, P. Schurer, P. Inia, L. Niesen, and Y.J. Agarwal (to be published).

43. G.M. Heestand, K. Bonde Nielsen, H. Ravn, F. Abildskov, and B.I. Deutch, these proceedings.

44. H. Gabriel, Phys. Rev. 181 (1969) 509.

45. B.I. Deutch, ref. 2, 151.

46. A. Becker and R. Steffen, BAPS 13 (1968) 1470.

47. H. Bernas and H. Gabriel, Proc. Int. Conf. on hf Int. Detected by Nucl. Rad., Israel (1970) (to be published).

48. B. Skaali, C. Günther, and B. Herskind, these proceedings.

49. R. Wäppling, E. Karlsson, G. Carlsson, and M.M. Bajaj (to be published).

50. D. Dillenburg and T.A.J. Maris, Nucl. Phys. 33 (1962) 208.

51. H. Ravn, K.H. Johansen, F. Abildskov, and B.I. Deutch (to be published).

52. C. Münchow, J.C. Waddington, B. Herskind, H. Ravn, and B.I. Deutch, Nuclear Reactions Induced by Heavy Ions, North Holland Publ. Co., Amsterdam (1970, 435 and to be published).

Discussion

Berkes: I would like to comment on the crystalline site in alloys of rare - earths metals. We prepared dilute Yb - Ag alloys by implantation and by classical fusion, and observed hyperfine interactions by DPAC and IPAC. We found that in the implanted source the electronic relaxation time was about an order of magnitude lower than in the melted one. In the latter an electric field gradient was also seen and the hf interaction did not change after annealing at 660° C under vacuum. Comparing our results with those reported by Deutsch and by De Waard, we conclude that after implantation we have a solid solution which is metastable. The stable configuration is something which corresponds to the normal crystalline structure and in the case of ytterbium in silver there is practical no solid solution at room temperature. Therefore, the melting process leads in fact to an Yb - Ag_5 - compound which is stable.

De Waard: You said that you can explain the relation between G_2 and G_4 as shown in one of your slides by the relaxation picture. Now, how is this affected if the atoms are not in one site but in two different sites that may have different interactions? Can you still do something about it then?

Deutch: Of course this picture will change; the particular slide was shown to try to explain just a small feature. Because the G_2 to G_4 is basically a dominant magnetic time - dependent interaction. So it is very easy to put 30% in another site, change the relaxation slightly and find some empirical lifting of the curve.

De Waard: I am not quite sure that the difference would be that small. I mean that the atoms occupy sites with quite different crystal fields.

Deutch: Well, O.K. Let me say something stronger. What we have shown is speculation. To seriously examine this two sites hypothesis that has just been raised, I think one has to do real proper redundant experiments.

Murnick: The curve of G_2 versus G_4 showed two points which were obviously off the theoretical curves. The points were for Gd which is an S - state ion, hence least likely to exhibit magnetic relaxation. I wonder whether the addition of some quadrupole relaxation might give a better fit to these data as well as to the erbium time - dependent data.

Deutch: That is correct, the two sites hypothesis is only one reasonable possibility. Quadrupole relaxation components might do equally as well.

Contributed Paper

INTERDEPENDENCE OF MöSSBAUER EFFECT AND γ - γ ANGULAR CORRELATION IN ^{119m}Sn*

H.L. Berger† and H.A. Medicus

Department of Physics and Astronomy, Rensselaer Polytechnic Institute, Troy, New York U.S.A.

It is well known that the degree to which extranuclear perturbations cause modifications of the γ - γ angular correlation function W (θ) in a crystalline solid depends on the nature of the nuclear environment. Recoil - less γ - ray emission is likewise known to be sensitive to the environment. Various nuclear properties are derivable in principle from either angular correlation or Mössbauer effect experiments (ref. 1). ^{119}Sn is one of the few isotopes in which the Mössbauer effect and angular correlation between γ - rays of a cascade can both be studied experimentally.

Regions in randomly oriented polycrystalline solids in which there is a high degree of crystal regularity should experience a smaller perturbation arising from defects and boundaries than regions where the lattice structure is poorly formed. In these regular regions, recoil - less emissions from the intermediate state should be more probable than from nuclei in less uniform regions of the crystal since the magnitude of the fraction f of γ - rays emitted recoil - lessly depends to a large extent on the degree of regularity of the lattice structure.

We have examined this relationship between recoil - less emission in the 23.8 keV transition and angular correlation of the 65 - 23.8 keV γ - γ cascade in a randomly oriented polycrystalline source of metallic ^{119m}Sn by measuring the difference in the angular correlation function W (θ) for cascades involving Mössbauer effect type transitions and those containing no recoil - less transitions. Because the Mössbauer effect can isolate the hyperfine components, there will in general be an observable anisotropic distribution of radiation for each of the hyperfine transitions. This effect is eliminated in metallic ^{119}Sn which exhibits only little splitting by using a thick metallic Sn absorber which has a broad resonance absorption width. We thus can examine the total directional correlation of the sum of the hyperfine components of the transitions.

Metallic ^{119m}Sn in the β - phase has a large Mössbauer effect at moderate temperatures and a relatively simple lattice structure. The decay scheme contains only the 11/2 (M4) 3/2 (M1) 1/2 cascade of the 65 keV and the partly recoil - less 23.8 keV transition. The first transition is highly converted ($\alpha = 5000$). The half - life of the intermediate state is 18.4 ns. The spin of I = 3/2 in the intermediate state limits W (θ) to an a_{22} term (in theory $a_{22} = - 0.1545$) so that measurements at 90° and 180° are sufficient. Because of the tetragonal lattice structure, an electric quadrupole interaction must be expected (ref. 2), but the internal magnetic field in tin is very weak and can be neglected.

We measured the recoil - less emission from ^{119m}Sn in a transmission arrangement, using a 90% enriched ^{119}Sn metal absorber of 0.005 cm thickness while simultaneously finding the anisotropy through coincidence measurements on the 65 - 23.8 keV cascade. The source had a strength of 100 μCi and was in the form of a foil of 0.2 cm diameter and 0.005 cm thickness. Two NaI - detectors were used, with the resonance absorber, which was attached to a loudspeaker coil, between source and the detector for the 23.8 keV quanta. Source and absorber were cooled to 154° K. Standard 'fast - slow' delayed coincidence techniques were used, the lower limit of the resolution time $2\tau_F$ being 16 ns.

Measurements were taken at 90° and 180° with the resonance absorber at rest and in motion. With the absorber at rest the resonance absorption is maximal and the coincidence rate and anisotropy A^{NM} will be primarily that due to 65 keV and non-recoil - less 23.8 keV γ - rays. By moving the absorber relative to the source at a sufficiently high velocity the recoil - free γ - rays as well as the other 23.8 keV γ - rays are detected. We define a quantity A^{ALL} as the anisotropy with a moving absorber. Assuming that appropriate geometric corrections have been made, the velocity dependent anisotropies can be deduced from the appropriate coincidence rates N (v, θ) and are given by

$$A^{NM} = \frac{W_{NM}(180^\circ) - W_{NM}(90^\circ)}{W_{NM}(90^\circ)} = \frac{N(0, 180^\circ) - N(0, 90^\circ)}{N(0, 90^\circ)} \quad (1)$$

$$A^{ALL} = \frac{N(\infty, 180^\circ) - N(\infty, 90^\circ)}{N(\infty, 90^\circ)} \quad (2)$$

Finally, for cascades with recoil - less 23.8 keV quanta only we have

$$A^{M} = \frac{W_{M}(180^\circ) - W_{M}(90^\circ)}{W_{M}(90^\circ)} = \frac{N(\infty, 180^\circ) - N(0, 180^\circ) - N(\infty, 90^\circ) + N(0, 90^\circ)}{N(\infty, 90^\circ) - N(0, 90^\circ)} \quad (3a)$$

In terms of the fraction of 23.8 keV γ - rays resonantly absorbed

$$A^{M} = A^{ALL}/\mathcal{A}_m - A^{NM}(1 - \mathcal{A}_m)/\mathcal{A}_m \quad (3b)$$

where $\mathcal{A}_m$ is the Mössbauer - effect absorption fraction. The coincidence rate due to Mössbauer - effect transitions in eq. (3a) is the difference N (∞, θ) − N (0, θ) in each case.

Careful measurements (ref. 3) of the Mössbauer - effect absorption fraction in metallic tin show $\mathcal{A}_m = 0.20$ at 154° K. We observed a smaller Mössbauer - effect absorption fraction of 0.13. One possibility for the discrepancy is that regions of our source have a poorly formed crystalline lattice resulting in a lowered probability for recoil - less emission there. In line with our earlier discussion, the angular correlation function for transitions in these regions should also show the effects of a major change in the local perturbing field. Our data show that the reduction in $\mathcal{A}_m$ is indeed consistent with the anisotropy difference between Mössbauer and non - Mössbauer - effect radiation A^M and A^{NM} respectively. With a sample of about 3000 coincidences at each of the four N (v, θ) values, we found (using eqs. (1), (2) and (3a)) $A^{ALL} = -0.122 \pm 0.027$, $A^{NM} = -0.051 \pm 0.030$ and $A^M = -0.72 \pm 0.28$.

Because the practical lower limit on resolution time is about the same as the intermediate - state half - life, no useful information could be obtained about the dependence of the anisotropy on the length of time between the two successive γ - ray emissions. The resulting anisotropies just given are therefore time averages over this interval. The time - averaged value of the angular correlation damping factor G_{kk} (t) denoted by G_{kk} (∞) in general has a non - zero value whose magnitude depends on the strength of the quadrupole interaction (ref. 4). Thus the time - averaged anisotropy, as measured in our work, should not be wiped out after long time intervals even by strong perturbing fields. As an example, the value of G_{22} (∞) for an interaction strength of 3×10^{-8} eV, a reasonable value for ^{119}Sn in the beta lattice (ref. 2), is 0.28.

It should be noted that A^{NM} is much smaller than the theoretical value for the unperturbed cascade, A = − 0.215, indicating a large perturbation acting upon these γ - rays. The anisotropy A^{ALL} contains the effects of all 23.8 keV γ - rays. That it is larger than A^{NM} suggests that the anisotropy involving recoil - less 23.8 keV γ - rays, A^M, is larger than A^{NM}. The statistics in the case of A^M are very poor since it is a quantity based on the difference of four numbers of the same magnitude. Qualitatively, however, it is apparent that A^M is substantially greater than A^{NM}, as predicted by our hypothesis. Taking into consideration the statistical uncertainty, it is likely that A^M is close to the unperturbed anisotropy of − 0.215 calculated on the basis of a pure 11/2 (4) 3/2 (1) 1/2 cascade.

It was established that the resonant absorber does not distort the values of the anisotropies, but acts only as a γ - ray filter, by measuring the anisotropy without a resonance absorber in the path of the 23.8 keV γ - rays. Using the experimental results for A^{NM}, A^{ALL}, and $\mathcal{A}_m$ we can calculate A^M by eq. (3b) obtaining $A^M =$ $= -0.58$. This calculation arrives at an independent estimate of A^M because here we employ the value of the Mössbauer effect fraction $\mathcal{A}_m$, obtained from the counting rate of the 23.8 keV γ - rays.

It would appear from the results of our experiments that a correlation exists between recoil - less emission and the lack of perturbation in the intermediate state. The large experimental value for A^M can be attributed to a somewhat lower than expected value for A^{NM} which, however, is still within the statistical error. The magnitude of the electric quadrupole interaction required to account for the damping of the non - recoil - less anisotropy A^{NM} can be estimated if one assumes that an anomalous electric field gradient is present in some regions of the ^{119}Sn polycrystal which is entirely responsible for the angular correlation attenuation. On that basis the electric field gradient acting there is calculated to be $V_{zz} = 3.5 \times 10^{18}$ V/cm^2. A more realistic calculation could be made if the exact nature of the perturbation were fully understood. We cannot yet distinguish between effects of boundary layer defects, point defects, and crystal - face effects. If some of the tin atoms are not sitting at their normal lattice sites we have the special circumstances of point defects. Defects produced by an atom leaving its lattice site due to recoil associated with the emission of radiation can be excluded in ^{119m}Sn because the only possible contributor is the 65 keV transition whose recoil energy is only 0.0195 eV, at least two orders of magnitude less than typical lattice binding energies. Natural point defects arising from the construction of the sample are possible and will contribute to the general perturbation level. Since the defect nuclei are not at regular lattice sites the field gradient seen by them will change the magnitude of the perturbation.

This combination of angular correlation and Mössbauer - effect measurements may provide a valuable and promising technique for investigations of the nature of perturbing interactions between crystal lattices and nuclei.

* Supported in part by the National Science Foundation.

† National Defense and Education Act Fellow; now at Espey Manufacturing and Electronics Corp., Saratoga Springs, New York.

References

1. H. Frauenfelder, in Perturbed Angular Correlations, E. Karlsson, E.M. Matthias, and K. Siegbahn, Eds., North Holland Publish. Co. (1964).
2. A.J.F. Boyle, D.St.P. Bunbury, and C. Edwards, Proc. Phys. Soc. (London) 79 (1962) 416.
3. C. Hohenemser, Phys. Rev., 139 (1965) 185.
4. A. Abragam and R.V. Pound, Phys. Rev. 92 (1953) 943.

Contributed Paper

TIME DIFFERENTIAL STUDIES OF ^{166}Er IN COPPER

G.M. Heestand, K. Bonde Nielsen, H. Ravn, F. Abildskov, and B.I. Deutch
Aarhus University, Institute of Physics, Aarhus, Denmark

Previous measurements indicate that the integral perturbed angular correlation (PAC) of rare earths implanted in Cu[1]) and Fe[2]) are highly attenuated. These attenuations have been attributed to magnetic time dependent interactions. In the case of the 5/2 (0.32 ns) and 7/2 (0.063 ns) states of ^{169}Tm in Fe[2]), these attenuations are well described by the Abragam and Pound[3]) model whereas for the slower 2^+ rare earths states ($\tau \sim 2$ ns) the stochastic model of Blume[4]) fits the data better. The Abragam and Pound theory agrees if a small amount of time dependent electric quadrupole interaction is added to the predominantly time dependent magnetic interaction.

More information concerning the nature of the relaxation processes can be obtained by using the highly sensitive differential PAC technique. Specifically, one may be able to distinguish between the purely exponential decay of the differential attenuation coefficient G_k (t) in the Abragam and Pound model (see eq. (1)) from the residual non - time dependent term of the Blume theory (see eq. (2)).

$$G_2(t) = e^{-\lambda_2 t} \qquad (1)$$

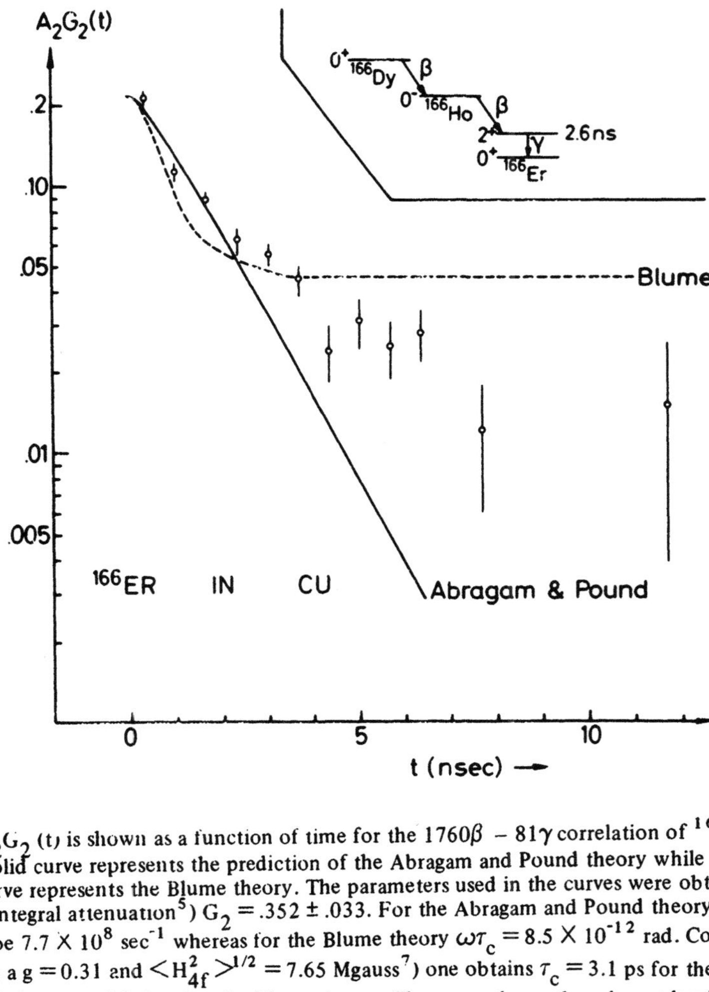

Fig. 1. A_2G_2 (t) is shown as a function of time for the 1760β – 81γ correlation of ^{166}Er in Cu. The solid curve represents the prediction of the Abragam and Pound theory while the dashed curve represents the Blume theory. The parameters used in the curves were obtained from the integral attenuation[5]) $G_2 = .352 \pm .033$. For the Abragam and Pound theory λ_2 was found to be 7.7×10^8 sec^{-1} whereas for the Blume theory $\omega\tau_c = 8.5 \times 10^{-12}$ rad. Combining these with a g = 0.31 and $<H^2_{4f}>^{1/2} = 7.65$ Mgauss[7]) one obtains $\tau_c = 3.1$ ps for the Abragam and Pound theory and 7.5 ps for the Blume theory. The insert shows the relevant levels involved in the decay from the mother ^{166}Dy to the granddaughter ^{166}Er.

(Magnetic time dependence)

$$G_2(t) = 1/5 + 3/5 \sum_{N=1,2} (\cosh xWt + \frac{1}{x}\sinh xWt)\, e^{-Wt} \qquad (2)$$

where $\lambda_2 = 2\omega^2\tau_c$, $x = (1 - 4N^2\omega^2\tau_c^2)^{1/2}$, $W = 1/2\tau_c$, ω is the Larmor frequency due to the magnetic field produced by the unfilled 4f - electron shell, i.e. $\omega^2 = \frac{g^2}{\hbar^2}\mu_N^2 \langle H_{4f}^2\rangle$, and τ_c is the relaxation time. The Blume formula given above in eq. (2) was derived for the case of j = 1/2 electronic states and in principle should only be applied to those cases.

In order to study the time dependent nature of G_2 (t), the angular correlation between the 1760 keV β - ray going from the 0^- ground state of ^{166}Ho to the first excited 2^+ state of ^{166}Er and the $2^+ \rightarrow 0^+$ 81 keV γ - ray (see insert in fig. 1) was measured time differentially. The source was prepared by implanting ^{166}Dy into Cu by means of a 50 kv electromagnetic isotope separator. The detection equipment was standard, including a thin plastic scintillator for β - rays and a 1″ X 1/8″ NaI crystal for the γ - rays. The time resolution was about 1.8 ns.

Fig. 1 displays the 1760 β – 81 γ time differential correlation for ^{166}Er in Cu. The curves in fig. 1 are the predictions of the Abragam and Pound model and the Blume model with the parameters in eqs. (1) and (2) adjusted to reproduce the integral attenuation[5]) G_2 = 0.352 ± 0.033. This number is consistent with the 1760 β – 81 γ correlation measured integrally in 30 degree steps. The experimental data have not been corrected for finite time resolution nor has the zero time been accurately determined. A time resolution of 1.8 ns has been folded into the theoretical predictions.

The rapid decay of G_2 (t) within the first few nanoseconds is consistent with the work of Ryge et al.[6]) who studied ^{154}Sm and ^{156}Gd implanted into Cu using the Coulomb excitation implantation technique. Although static quadrupole effects are similar, a pure static quadrupole interaction could not explain the integral G_2/G_4 ratio nor the temperature dependence previously observed[1]). The results shown in fig. 1 clearly show that neither the Blume theory nor the Abragam and Pound theory alone are sufficient. Perhaps what is being observed are quadrupole effects superimposed upon a predominantly time dependent magnetic interaction. Further investigations of the problem are clearly warranted.

Acknowledgement

The authors wish to thank B. Herskind and H.L. Nielsen for useful discussions.

References

1. J.C. Waddington, K.A. Hagemann, S. Ogaza, D. Kiss, B. Herskind, and B.I. Deutch, Nuclear Reactions Induced by Heavy Ions, North Holland (1970) 438.
2. B.I. Deutch, K. Bonde Nielsen, and H. Bernas, Phys. Letters 27B (1968) 209.
3. A. Abragam and R.V. Pound, Phys. Rev. 92 (1953) 943.

4. M. Blume, Hyperfine Structure and Nuclear Radiation, ed. by E. Matthias and D. Shirley, (North Holland Publ. Co., Amsterdam, 1968) 911.
5. C. Günther, B. Skaali, and B. Herskind (to be published this conference).
6. P. Ryge, D.W. Gebbie, and W.L. Randolf, Bull. Am. Phys. Soc. Jan. 1970, 101.
7. C. Günther, private communication.

Contributed Paper

PARAMAGNETIC RELAXATION FOR ERBIUM IN COPPER

B. Skaali, C. Günther, and B. Herskind
University of Copenhagen, The Niels Bohr Institute, Denmark
and
H. Ravn
University of Aarhus, Institute of Physics, Aarhus, Denmark

The perturbations acting on rare - earth ions in solids are not well understood. Waddington et al.[1]) have studied the perturbed angular correlations on rare - earth ions implanted into copper lattices after Coulomb excitation (IMPAC). They conclude that a pure magnetic interaction as calculated by Blume[2]) seems to explain most of the features of the measured perturbations. However, the results can also be explained in the framework of the theory of Abragam and Pound[3]), if two relaxation times are assumed to result from different lattice locations of the implanted ions, or if both magnetic and electric time - dependent interactions are present. The latter could result from unknown crystalline effects. Heestand et al.[4]) have recently performed a time - differential measurement on ^{166}Er in copper, and they have not observed the shape as predicted by the theory of Blume.

The presence of crystalline effects would also influence the effective magnetic field at the nucleus in an applied external field. Wäppling et al.[5]) have calculated the temperature dependence of the paramagnetic correction factor $\beta = B_{eff}/B_{ext}$ for Er^{3+} in a cubic lattice. They show that in a strong crystalline field β is reduced considerably.

In this paper we report a measurement of the perturbed angular correlation for Er in copper at 295° K and 77° K. A source was prepared by implanting radioactive ^{166}Dy ions at 50 keV energy into a copper backing with the Aarhus isotope separator. The integral angular correlation of the 1380 keV – 81 keV γ - γ cascade in ^{166}Er – 0^+ was studied. The spin sequence of this cascade is 0^+ - 2^+ - 0^+ where the intermediate state has a mean life - time $\tau_N = 2.63 \cdot 10^{-9}$ s and a g - factor[6]) of 0.31. The 81 keV γ - rays were detected with a thin - window planar Ge (Li) detector, whereas the angular distribution of the coincident 1380 keV radiation was measured simultaneously by four 3″ X 3″ NaI (Tl) crystals.

Without an external magnetic field, the attenuation factors G_k were determined to be $G_2 = 0.352 \pm 0.033$, $G_4 = 0.245 \pm 0.012$ for T = 295° K and $G_2 = 0.224 \pm 0.045$,

$G_4 = 0.148 \pm 0.020$ for $T = 77°$ K. The values of G_k obtained in this experiment are plotted in fig. 1 together with attenuation factors deduced from ref. 7 for the same cascade in ^{166}Er in liquid sources at $T = 295°$ K. Also plotted in fig. 1 are values of G_k as predicted by the theory of Abragam and Pound[3]) and Blume[2]) (see also ref. 1). Assuming the validity of the theory of Blume, relaxation times of τ_c $(T = 295° \text{ K}) \cong$ $\cong 6 \cdot 10^{-12}$ s and τ_c $(T = 77° \text{ K}) \geqslant 1.5 \cdot 10^{-11}$ s are obtained. These values can be compared with τ_c $(T = 295° \text{ K}) = 0.35 \cdot 10^{-12}$ s as determined for Er^{3+} ions in hydrochloric and perchloric solutions[7]). For these liquid sources an almost pure magnetic interaction as described by Abragam and Pound is observed.

The paramagnetic correction factors were determined by measuring the rotation of the angular correlation in an external magnetic field. The magnitude of the field was 8.5 kG and 6.7 kG at 295° K and 77° K, respectively. The attenuation factors G_k, measured at 77° K with applied field, are in agreement with the values obtained without field. A comparison with results from other measurements and theoretical

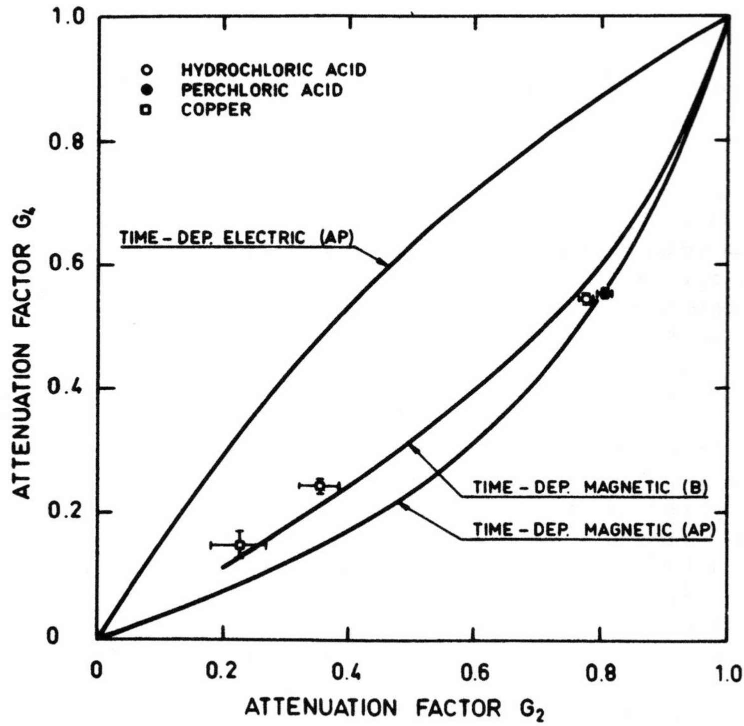

Fig. 1. Measured attenuation factors G_4 versus G_2 for Er^{3+} in copper (this work) and Er^{3+} in hydrochloric and perchloric acid (ref. 7). The curves represent theoretical predictions by Abragam and Pound for pure magnetic and pure electric time - dependent interactions, and by Blume for pure magnetic interaction.

predictions[8]) is presented in table 1. Although the errors are quite large, which is mainly due to the very strong perturbation observed, it can be concluded that the measured values of β for Er in copper are in agreement with the free ion predictions.

Table 1. Experimental and Theoretical Values of β for Er^{3+} Ions in Different Surroundings

Temp. ^{o}K	Liquid Source Ref.[9]	$Ho(Er)/Al_2$ Ref.[5]	Copper this exp.	Theory Ref.[8]
77		20.9±1.6	21.4±5.9	23.7
295	6.61±0.30[a]	4.4±0.6	8.8±1.8	6.9

a) At 300°K

The high effective magnetic field measured at low temperatures suggests application in IMPAC experiments free from transient field effects. Furthermore, the use of heavier backings with a cubic crystal structure, such as gold, would allow higher bombarding energies in Coulomb excitation experiments. An interesting aspect of this is e.g. the possibility of testing the constancy of the g - factor within a ground state rotational band.

References

1. J.C. Waddington, K.A. Hagemann, S. Ogaza, D. Kiss, B. Herskind, and B.I. Deutch, in Nuclear Reactions Induced by Heavy Ions (North - Holland Publishing Co., Amsterdam, 1970) p. 438.
2. M. Blume, in Hyperfine Structure and Nuclear Radiations, eds. E. Matthias, and D.A. Shirley (North - Holland Publishing Co., Amsterdam, 1968) p. 911.
3. A. Abragam and R.V. Pound, Phys. Rev. 92 (1953) 943.
4. G. Heestand, K. Bonde Nielsen, H. Ravn, F. Abildskov, and B.I. Deutch, to be presented at this conference.
5. R. Wäppling, E. Karlsson, G. Carlsson and M.M. Bajaj, private communication.
6. V.S. Shirley, in Hyperfine Structure and Nuclear Radiations, eds. E. Matthias and D.A. Shirley (North - Holland Publ. Co., Amsterdam, 1968) p. 985.
7. M. Forker, H.F. Wagner, and E. Bodenstedt, Nuclear Physics A138 (1969) 644.
8. C. Günther, and I. Lindgren, in Perturbed Angular Correlations, eds. E. Karlsson, E. Matthias and K. Siegbahn (North - Holland Publ. Co., Amsterdam, 1964) p. 357.
9. [illegible] Lange and H.F. Wagner, Z. Physik 234 (1970) 465.

Contributed Paper

HYPERFINE INTERACTIONS OF ^{161}Dy NUCLEI IMPLANTED IN IRON

P. Inia and H. de Waard

University of Groningen, Natuurkundig Laboratorium, The Netherlands

In recent years the magnetic field experienced by rare - earth nuclei implanted in ferromagnetic host lattices has been determined for many cases. Most of the fields have been measured with the aid of the IMPACT method in which the recoil energy after Coulomb excitation is used to implant excited nuclei into the host lattice and where the hyperfine field at the nucleus is derived from the perturbed angular distribution of gamma rays emitted by the implanted nuclei.

In this investigation we report on the hyperfine interactions derived from Mössbauer spectra of the 26 keV transition of ^{161}Dy implanted in iron. Sources were made by implantation of 130 keV ^{161}Tb ions into an iron foil by means of the Groningen isotope separator. A 35 mg/cm^2 DyF_3 absorber was used at liquid nitrogen temperature (LNT). Combined with a source of ^{161}Dy in CeO_3 at room temperature (RT) this absorber gives a single line of about 45 times the natural line width. Spectra were measured with the ^{161}DyFe source at RT and at LNT (fig. 1 and 2).

From the spectrum at RT we may conclude that there are at least two sites for the Dy nuclei. First, a fraction of about 55% experiences an effective magnetic hyperfine field $|H| = 4.50 \pm 0.05$ MOe and an effective quadrupole interaction with

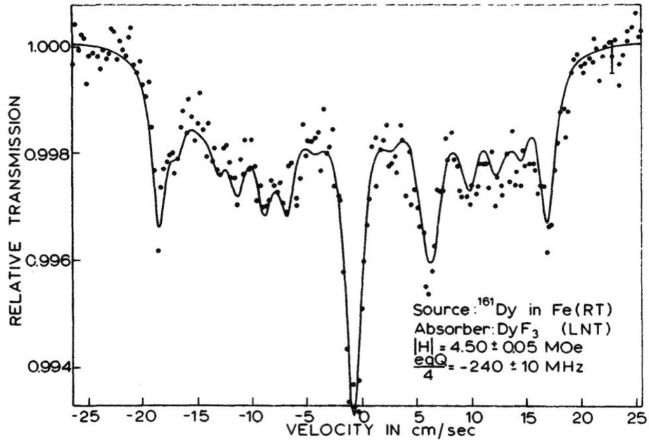

Fig. 1. Mössbauer spectrum of ^{161}DyFe at room temperature.

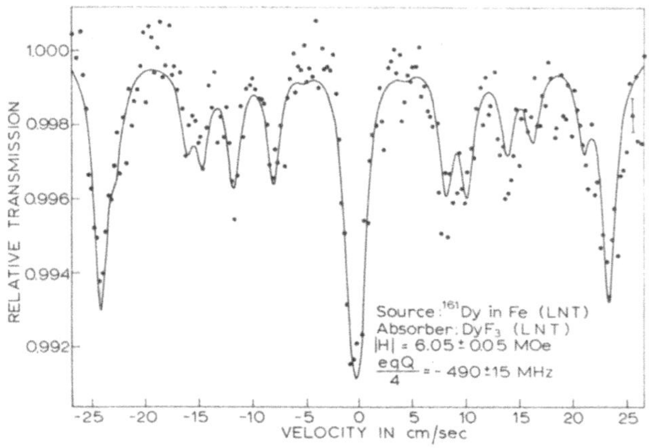

Fig. 2. Mössbauer spectrum of ^{161}DyFe at liquid nitrogen temperature.

a coupling constant eqQ/4 = 240 ± 10 MHz. A second fraction of about 40% exhibits a smeared out relaxation type of spectrum underlying the spectrum of the first fraction. Finally, there may be an extra central peak.

From the spectrum with the source at LNT we find a magnetic hyperfine field that has increased to |H| = 6.05 ± 0.05 MOe for about 90% of the Dy atoms, while the electric quadrupole coupling constant is now eqQ/4 = 490 ± 15 MHz. There is still an extra central peak, but the fraction with a relaxation type spectrum is no longer observed.

In most cases where rare - earth ions in solids are involved the internal fields arise mainly from the electrons in the partly filled 4f - shell. The free ion contribution to the magnetic field is given by[1])

$$H_{4f} = 2\beta < 1/r^3 > < J \| N \| J > \underline{J}$$

and to the electric field gradient by

$$q_{4f} = - e < J \| \alpha \| J > [3J_z^2 - J(J+1)]$$

in the nomenclature of Elliott and Stevens[2]).

In cases where crystal field effects can be neglected, the behaviour of rare - earth ions in magnetic compounds is determined by the exchange interactions with the surrounding magnetic ions, If the spin correlation times corresponding to transitions between the various J_z states are much shorter than the reciprocal of the nuclear Larmor precession frequency, the effective field H_{eff} will be equal to the Boltzmann average over all J_z states.

If we further assume that the exchange field is temperature independent, the temperature dependence of the field at the nucleus is given by a Brillouin function

$$H_{eff}(T) = H_{4f} \frac{\sum_{J_z} J_z e^{-J_z x}}{\sum_{J_z} e^{-J_z x}}$$

where $x = -\frac{g\beta H}{kT}$, the exchange interaction parameter.

Similarly the temperature dependence of the electric field gradient is given by

$$q_{eff}(T) = q_{4f}(1 - R) \frac{\sum_{J_z} [3J_z^2 - J(J+1)] e^{-J_z x}}{\sum_{J_z} e^{-J_z x}}$$

R is the Sternheimer shielding factor.

The temperature dependence given by these formulas is shown in fig. 3. As is shown in that figure, our results for the fraction of the dysprosium nuclei that experiences a well defined hyperfine interaction can be fitted very well to these curves. The best fit is obtained for an exchange interaction strength $g\beta H/k = 100°$ K, somewhat lower than the value of 135° K found by Bowden et al.[3]) for the intermetallic compound $DyFe_2$. An extrapolation of the magnetic and electric interaction to T = 0° K yields $H_{eff}(0) = 7.0 \pm 0.2$ MOe and $eq_{eff}(0)\, Q/4 = 670 \pm 40$ MHz, in reasonable agreement with the values found for $DyFe_2$. The magnetic field is about 20% higher than the free ion value calculated by Bleaney[4]). This is probably due to supplementary contributions by core and conduction electrons, polarized by the d

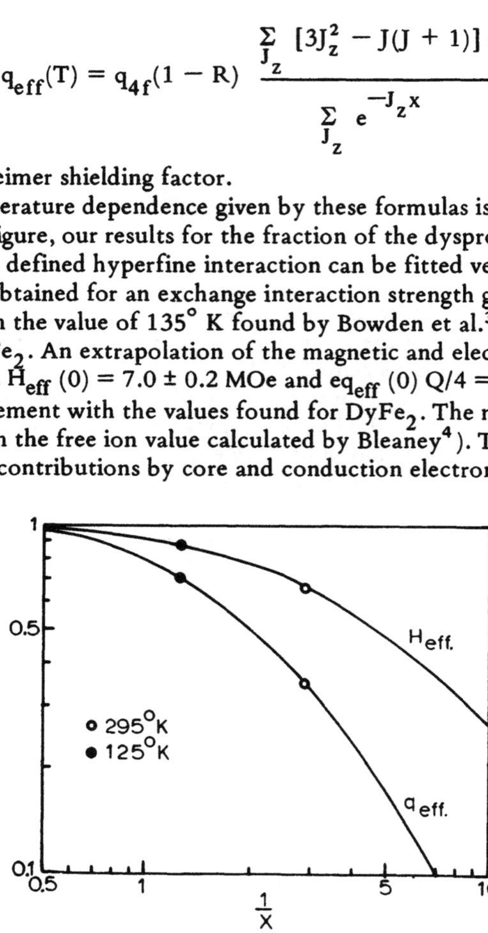

Fig. 3. Temperature dependence of the effective magnetic and quadrupole interactions on ^{161}Dy as a function of the exchange parameter $x = \frac{g\beta H}{kT}$.

electrons of the iron atoms.

Apparently, about one half of the implanted nuclei are in sites where crystal field effects can be neglected relative to the axial exchange field. Presumably, these are cubic sites, where each dysprosium atom is regularly surrounded by eight iron atoms. It should be noted, that the channeling work reported by Schmorak and Bøgh[5]) on ^{169}Tm in Fe indicates that about one half of the implanted nuclei are in substitutional sites.

The other half of the implanted atoms is probably associated with vacancies in the iron lattice, For such atoms, the crystal field splitting of the 4f - electron states is expected to be much larger; moreover, the exchange field may be different. These differences may well lead to longer electron spin correlation times, explaining why a relaxation type of Mössbauer spectrum is observed for this fraction at room temperature. This interpretation is supported by recent results of Heestand et al.[6]) on the angular correlation of gamma rays from ^{166}Er implanted in copper. The attenuation of the angular correlation pattern observed in that case may be explained by assuming a static quadrupole interaction superimposed on a time dependent magnetic interaction.

Our value of the magnetic hyperfine field at room temperature, $|H| = 4.50 \pm 0.05$ MOe, differs by a factor of about two from the value $H = -2.0 \pm 0.3$ MOe derived by Grodzins et al.[7]) from an IMPACT measurement on Coulomb excited dysprosium nuclei recoiled into magnetized iron. This discrepancy can possibly be explained by assuming that about half of the nuclei precesses around axes with directions that are randomly distributed in space. Such a situation is not unlikely for the vacancy associated fraction of the implanted atoms. For this fraction, where the crystal field interaction is expected to be rather large, the main axes of the field gradient should indeed by randomly oriented in space.

References

1. S. Ofer, I. Nowik, and S.G. Cohen in Chemical Applications of Mössbauer - spectroscopy, edited by V.I. Goldanskii and R.H. Herber (Academic Press, New York 1968), Chapter 8.
2. R.J. Elliott and K.W.H. Stevens. Proc. Roy. Soc. A218 (1953) 553.
3. G.J. Bowden, D.St.P. Bunbury, A.P. Guimaraes and R.E. Snyder, J. Phys. C 1 (1968) 1376.
4. B. Bleaney, Journ. Appl. Phys. 35 (1963) 1024.
5. M. Schmorak and E. Bøgh in Hyperfine Structure and Nuclear Radiations, edited by E. Matthias and D.A. Shirley (North - Holland Publ. Co., Amsterdam 1968), p. 712.
6. G.M. Heestand, K. Bonde Nielsen, H. Ravn, F. Abildskov and B.I. Deutch, this conference.
7. L. Grodzins, R. Borchers, and G.B. Hagemann, Phys. Lett. 21 (1966) 214.

ON THE THEORY OF MAGNETIC MOMENTS: ROTATIONAL NUCLEI

J. Krumlinde

Lund Institute of Technology, Department Mathematical Physics, Lund, Sweden

The regions of deformed nuclei are characterized in particular by the occurrence of low - lying rotational bands which very nearly obey the following rule

$$E_I = \frac{\hbar^2}{2\mathcal{J}} I(I+1) \tag{1}$$

The magnetic moment and M1 matrix elements of a rotating nucleus arise partly from the intrinsic nucleonic motion and partly from the rotational motion. The rotational part of the M1 matrix elements within a band may have components with $\Delta K = 0$ and $\Delta K = \pm 2K$. The $\Delta K = 0$ part can be expressed[1])

$$(\mathcal{M}(M1,\mu))_{\Delta K=0} = \sqrt{\frac{3}{4\pi}}\frac{e\hbar}{2Mc}(g_R I_\mu + (g_K - g_R) I_3 \mathcal{D}^1_{\mu 0}) \tag{2}$$

In the particle plus rotor model the magnetic moment is given by

$$\underline{\mu} = g_R \underline{R} + g_l \underline{l} + g_s \underline{s} = g_R \underline{J} + \{(g_l - g_R)\underline{l} + (g_s - g_R)\underline{s}\} \tag{3}$$

where $\underline{R}$ is the angular momentum of the rotor and $\underline{J} = \underline{R} + \underline{l} + \underline{s}$ is the total angular momentum of the nucleus. The expectation value of the last term for a rotational wave function consists only of contributions from the products $l_3 J_3$ and $s_3 J_3$ (except for $K = 1/2$).

One usually defines an intrinsic g - factor by

$$g_K K = \langle K|g_l l_3 + g_s s_3|K\rangle = g_l K + (g_s - g_l)\langle K|s_3|K\rangle \tag{4}$$

and thus we obtain as a special case of eq. (2):

$$\mu = g_R I + (g_K - g_R)\frac{K^2}{I+1} \tag{5}$$

for $K > 1/2$.

The coefficients g_K and g_R are most easily evaluated in the cranking model.

Suppose $|0>$ is the unperturbed nuclear ground state of an even - even nucleus and $|\Omega>$ is the 'cranked' state, representing the slow rotation of the nucleus with an angular velocity Ω, i.e.

$$H|0> = \mathscr{E}_0 |0>$$

and $|\Omega>$ is an eigenstate of the Hamiltonian in the rotating frame:

$$H'|\Omega> = (H - \Omega J_x)|\Omega> = \mathscr{E}'(\Omega)|\Omega>$$

The energy $\mathscr{E}(\Omega)$ is given by

$$\mathscr{E}(\Omega) = <\Omega|H|\Omega> = \mathscr{E}'(\Omega) + <\Omega|J_x|\Omega> \tag{6}$$

The expectation value of the angular momentum J_x is constrained by the semiclassical relation

$$<\Omega|J_x|\Omega> = \sqrt{I(I+1)} \tag{7}$$

This matrix element as well as the energy can then be evaluated by first and second order perturbation theory respectively:

$$<\Omega|J_x|\Omega> = \mathscr{I}\,\Omega \tag{8}$$

and

$$\mathscr{E}(\Omega) = \mathscr{E}_0 + \frac{1}{2}\mathscr{I}\Omega^2 = \mathscr{E}_0 + \frac{I(I+1)}{2\mathscr{I}} \tag{9}$$

where

$$\mathscr{I} = 2\sum_i \frac{<i|J_x|0>^2}{\mathscr{E}_i - \mathscr{E}_0} \tag{10}$$

is the moment of inertia.
For the magnetic moment of a even - even nucleus one has

$$\mu_x = g_R J_x$$

Then

$$<\Omega|\mu_x|\Omega> = 2\Omega \sum_i \frac{<0|\mu_x|i><i|J_x|0>}{\mathcal{E}_i - \mathcal{E}_0} \tag{11}$$

and using

$$\Omega = \frac{1}{\mathcal{J}} \sqrt{I(I+1)}$$

we get

$$g_R = \frac{2}{\mathcal{J}} \sum_i \frac{<0|\mu_x|i><i|J_x|0>}{\mathcal{E}_i - \mathcal{E}_0} \tag{12}$$

Our Hamiltonian is supposed to contain quadrupole deformation ($r^2 Y_{20}$) as well as pairing interactions. The excited states are then two - quasi - particle states,labelled $|\nu_1 \bar{\nu}_2>$. Using the BCS - approximation one arrives at[2])

$$\mathcal{J} = 2 \sum_{\nu_1 \nu_2} \frac{<\nu_1|j_x|\nu_2>^2}{E_1 + E_2} (U_1 V_2 - V_1 U_2)^2 \tag{13}$$

$$g_R = \frac{2}{\mathcal{J}} \sum_{\nu_1 \nu_2} \frac{<\nu_1|j_x|\nu_2><\nu_2|\mu_x|\nu_1>}{E_1 + E_2} (U_1 V_2 - V_1 U_2)^2 \tag{14}$$

Using

$$\mu_x = \begin{cases} l_x + g_s^{(p)} s_x & \text{(protons)} \\ g_s^{(n)} s_x & \text{(neutrons)} \end{cases}$$

we arrive at the following formula

$$g_R = \frac{\mathcal{J}_p}{\mathcal{J}} + (g_s^{(p)} - 1) \frac{W_p}{\mathcal{J}} + g_s^{(n)} \frac{W_n}{\mathcal{J}} \tag{15}$$

where

$$W = 2 \sum_{\nu_1 \nu_2} \frac{<\nu_1|j_x|\nu_2><\nu_2|s_x|\nu_1>}{E_1 + E_2} (U_1 V_2 - V_1 U_2)^2$$

The ratio g_R is nearly equal to $\mathcal{J}_p / \mathcal{J}$ since W is much smaller than $\mathcal{J}$ and the terms in W largely cancel each other owing to $g_s^{(p)} - 1 \approx -g_s^{(n)}$.

Still, within the adiabatic approximation, one can include corrections to rotational and magnetic quantities due to changes and fluctuations in the static nuclear field.

The nuclear Hamiltonian used here is characterized by 3 parameters: the quadrupole deformation β and the proton and neutron pairing parameters Δ_p and Δ_n

$$H_0(\beta,\underline{\Delta}) = H_0^0 - \beta\Sigma\, r_i^2 Y_{20}(\theta_i) - \Delta_p \Sigma_p (a_p^* a_{\bar{p}}^* + a_{\bar{p}} a_p) -$$

$$- \Delta_n \Sigma_n (a_n^* a_{\bar{n}}^* + a_{\bar{n}} a_n) \tag{16}$$

This is a linearized Hamiltonian or 'generator' Hamiltonian used to generate wave functions. The energy is given by the expectation value in the ground state of the full two - body Hamiltonian.

Let a denote any of the parameters used to characterize the nuclear field (β, Δ etc.) and let κ be the coupling constant (strength) [1,3])

$$H(a) = H_0 - \kappa a F$$

Here F is a nuclear field and a is called a deformation parameter. The deformation of the density, a_ρ is defined by

$$a_\rho(a) = \langle a|F|a\rangle \tag{17}$$

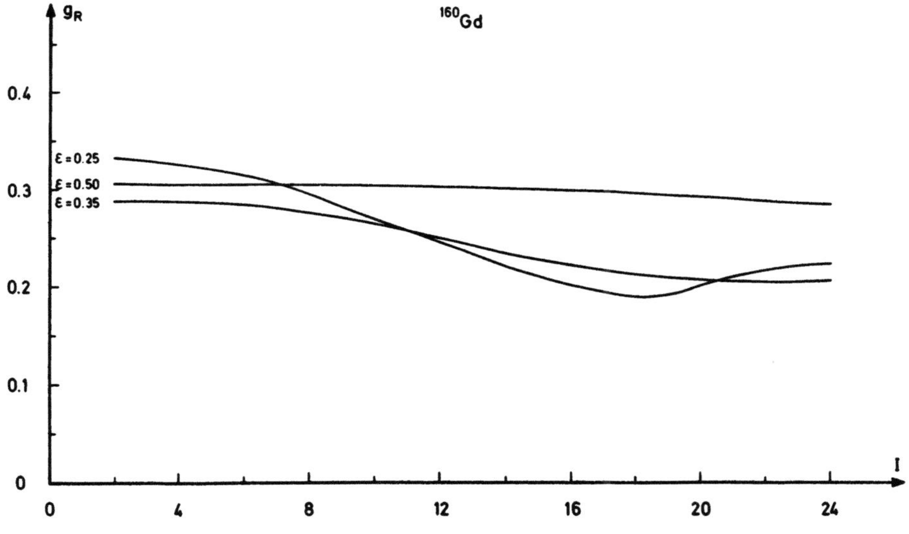

Fig. 1. Collective gyromagnetic ratio as a function of I for different deformations in the nucleide ^{160}Gd.

where $|a\rangle$ is an eigenstate of H (a) (the ground state). For self - consistency is required that $a_\rho\,(a) = a$. The energy is given by

$$\mathscr{E}(a) = \langle a\,|\mathrm{H}|\,a\,\rangle = \langle a\,|\mathrm{H}_0 + \tfrac{1}{2}\kappa\,\mathrm{F}^*\mathrm{F}|a\rangle \approx$$

$$\approx \langle a\,|\mathrm{H}\,(a)\,|a\rangle - \kappa\, a a_\rho + \tfrac{1}{2}\kappa\, a_\rho^2 \qquad (18)$$

It can be shown that the energy function $\mathscr{E}(a)$ is stationary at those values of a where self - consistency is fulfilled ($a_\rho = a$). When including rotational motion, we must use a generator Hamiltonian

$$\mathrm{H}\,(a, \Omega) = \mathrm{H}\,(a) - \Omega \mathrm{J}_x$$

generating states $|a, \Omega\rangle$ given by the cranking model.

The deformation of the density is then

$$a_\rho\,(a, \Omega) = \langle a, \Omega|\mathrm{F}\,|a, \Omega\rangle \qquad (19)$$

The energy is

$$\mathscr{E}(a, \mathrm{I}) = \mathscr{E}(a) + \frac{\mathrm{I}\,(\mathrm{I}+1)}{2\,\mathscr{I}(a)} + \kappa\,(a_\rho(a) - a)(a_\rho\,(a, \mathrm{I}) - a_\rho\,(a)) \qquad (20)$$

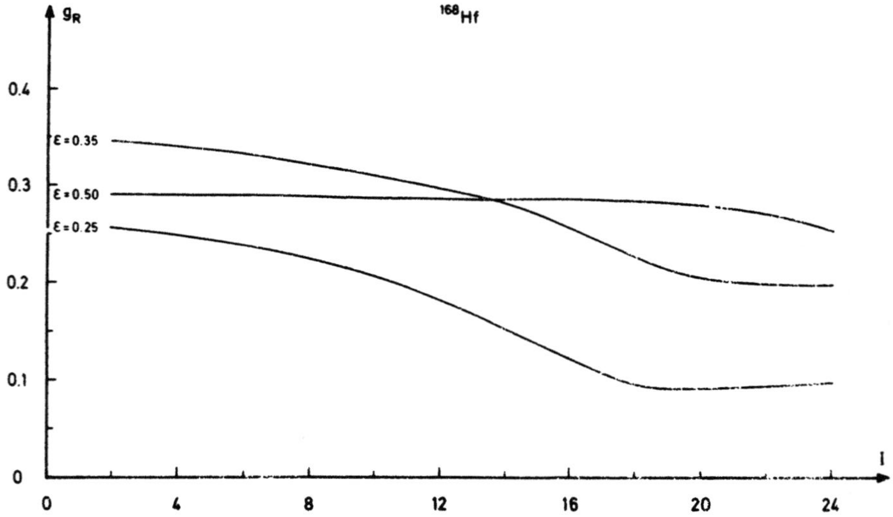

Fig. 2. Same as fig. 1 for ^{168}Hf.

where $\mathscr{I}(a)\,\Omega = \sqrt{I\,(I+1)}$ and $\mathscr{E}(a)$ is the energy for $I = 0, \Omega = 0$. The last term arises because the states $|a\rangle$ and $|a, \Omega\rangle$ are eigenstates of a generator Hamiltonian which is different from the energy operator.

For self - consistency at each I: $a_\rho\,(a, I) = a$, but $a \neq a_\rho\,(a)$ when $I > 0$.

From the fact that $\mathscr{E}(a, I)$ is stationary with respect to changes in the wave function, one can easily derive the following relation

$$a_\rho\,(a, I) = a_\rho\,(a) + \frac{1}{\kappa}\,\frac{I(I+1)}{2\mathscr{I}(a)^2}\,\frac{\partial \mathscr{I}(a)}{\partial a} \tag{21}$$

For the pairing case is $\kappa\; a\,G$ and $a \propto \Delta$ and we get a dispersion relation $(a_\rho \propto \sum_\rho U_\nu V_\nu)$

$$\frac{2}{G} = \sum_\nu \frac{1}{E_\nu} + \frac{I(I+1)}{\mathscr{I}^2}\frac{\partial \mathscr{I}}{\partial \Delta^2} \tag{22}$$

This means that when $I > 0$, we have a smaller Λ for a fixed G. Since pairing reduces the moment of inertia, we get an effective moment of inertia that is increasing as a function of I. It is a general feature both in the rare - earth and the actinide region that the equilibrium pairing parameter of neutrons $\Delta_n\,(I)$ decreases faster than that of the protons $\Delta_p\,(I)$.

Consequently, $\mathscr{I}_n$ increases faster than $\mathscr{I}_p$ as a function of angular momentum. The result for the collective gyromagnetic ratio is a decrease for $I = 0 - 12$ since $g_R \approx \frac{\mathscr{I}_p}{\mathscr{I}} = \frac{1}{1 + \frac{\mathscr{I}_n}{\mathscr{I}_p}}$. The behaviour for larger values of I is more complicated when $\Delta_n \to 0$ and Δ_p starts decreasing faster. Typical sketches of $g_R\,(I)$ are shown in fig. 1

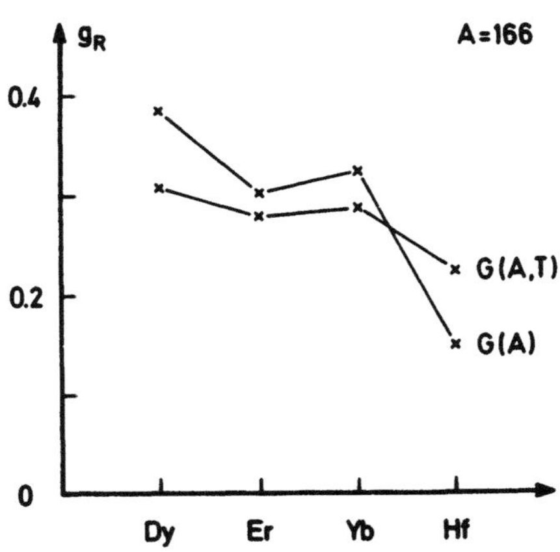

Fig. 3. The collective gyromagnetic ratio for isobars of A = 166 using a pairing force with and without isospin dependence respectively.

and 2. Here we assume an isospin dependent pairing force; the coupling constant G is

$$G_{p,n} = g_0\ (A) \pm g_1\ (A)\ T$$

At larger deformations the decoupling of the pair field is much weaker thus making g_R a much more slowly varying function of I, which can be seen in these figures. Values of g_R - factors have been calculated by the above mentioned isospin dependent pairing force and with a pairing force with pure A - dependence, normalized to give the same G for the stable nuclides. Discrepancies between the resulting g_R - factors are then mainly seen in neutron deficient nuclides since here the addition of a T - dependent term results in a smaller Δ_p and a larger Δ_n thus causing g_R to increase. This makes g_R a more slowly varying function of isospin. (See fig. 3 and 4). Experimental values of g_R in this region ($^{166-174}Hf$, $^{172-178}W$) would yield a good test of the isospin dependence of the pairing force.

The g_R - factors for the 4^+ rotational state have been measured in some nuclides yielding values slightly smaller than those of the 2^+ state in agreement with the theory. However, the large statistical errors in the measurements do not permit a definite conclusion on this point.

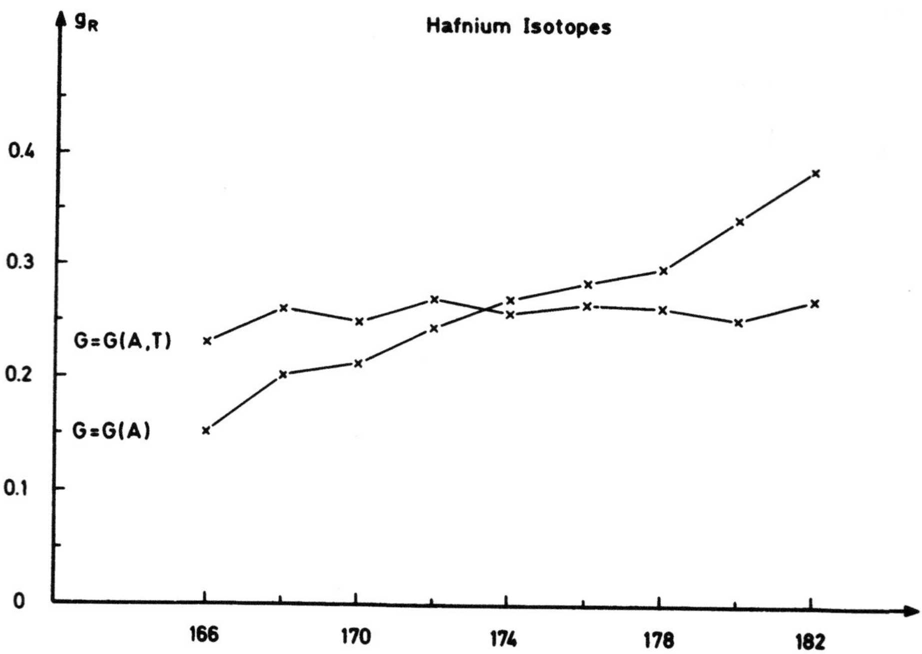

Fig. 4. Same as fig. 3 but for hafnium isotopes.

References

1. A. Bohr and B. Mottelson, Nuclear Structure and Energy Spectra, Vol. II, to be published; O. Nathan and S.G. Nilsson, 'Collective Nuclear Motion and the Unified Model', reprinted from 'Alpha, Beta and Gamma Ray Spectroscopy', ed. K. Siegbahn.
2. S.G. Nilsson and O. Prior, Mat. Fys. Medd. Dan. Vid. Selsk. 32, No 16 (1961); J.J. Griffin and M. Rich, Phys. Rev. 118 (1960) 850; O. Prior, F. Boehm, and S.G. Nilsson, Nucl. Phys. A110 (1968) 257.
3. J. Krumlinde, Nucl. Phys. A121 (1968) 306.

Discussion

Berkes: The expression $\mu = g_R I + (g_K - g_R) \frac{K^2}{I+1}$, given in your text is limited to the ground state rotational band and the β - vibrational band. However, for the γ - vibrational band the more complete expression, given by Nilsson in his original paper (1955), $\mu = g_R I + (g_K - g_R) \frac{\Omega K}{I+1}$ should be used.

Krumlinde: I am very sorry I did not mention this fact. However, I used the expression only in connection with the ground state band.

ON THE THEORY OF MAGNETIC DIPOLE MOMENTS OF SPHERICAL AND TRANSITIONAL NUCLEI

S. Ogaza
Institute of Nuclear Physics, Cracow, Poland

R.A. Sorensen, in his invited paper on nuclear moments at the Asilomar Conference, stated that 'in the continuing contest of experimentalist versus theorist, the experimentalist seems to be in the lead'. After three years this statement seems to be even more justified as the number of experimental data on nuclear magnetic moments and their accuracy have increased, while the theory of magnetic moments is still not capable of predicting moments better than a few tenths of the nuclear magneton. Not one of the main problems in the theory of nuclear magnetic moments has reached a decisive solution, although a constant development in their understanding is observed.

The magnetic dipole moment is defined as

$$\mu = \langle I, M = I | \mu_z | I, M = I \rangle \tag{1}$$

where μ_z is the z - component of the vector operator $\underline{\mu}$. Most theories of magnetic moments start from the single particle model, where

$$\underline{\mu} = g_l \underline{l} + g_s \underline{s} = g_l \underline{j} + (g_s - g_l) \underline{s} \tag{2}$$

with $g_l = 1$ (0) and $g_s = 5.58$ (-3.82) for proton (neutron), represents the dipole moment of a point nucleon moving in a static binding field. For a nucleon in a shell model orbit, the expectation value (eq. (1)) may be evaluated in the vector model and the well - known Schmidt - values are obtained from

$$\mu_{sp} = j \; \{ g_l \pm (g_s - g_l) \frac{1}{2l+1} \}, \; j = l \pm \frac{1}{2} \tag{3}$$

This expression is valid for single - particle and for single - hole configurations, as the operators $\underline{l}$ and $\underline{s}$ are even under particle - hole conjugation.

The expressions (1) – (3) are expected, however, to be modified, owing to nucleonic interactions which influence the intrinsic structure of nucleons in nuclei, and their velocity dependence gives rise to additional terms in expression (2). The smallness of the kinetic energies of nucleons in nuclei compared with the excitation energy of nucleons may suggest, however, that the interaction terms in the magnetic moment may be treated as corrections to the predominant one - particle moment (see ref. 1).

The nucleonic interaction effects were studied more or less tentatively by

TABLE 1. Comparison of experimental and theoretical data on magnetic moments for the ground states of odd-mass nuclei with jj-closed shells ± one nucleon

Nucleus	Configuration	μ_{sp} n.m.	$\delta\mu_{int}$ [3)] n.m.	$\delta\mu_{conf}$ [6)] n.m.	μ_{theor} [a)] n.m.	μ_{exp} n.m.
${}^{3}_{1}H_{2}$	$p(1s_{1/2})^{1}$	2.79	0.14		2.93	2.98
${}^{3}_{2}He_{1}$	$n(1s_{1/2})^{1}$	-1.91	-0.17		-2.08	-2.13
${}^{15}_{7}N_{8}$	$p(1p_{1/2})^{1}$	-0.26	0.06	-0.10	-0.30	-0.28
${}^{15}_{8}O_{7}$	$n(1p_{1/2})^{1}$	0.64	0.09 [b)]	0.14	0.87	0.72
${}^{17}_{8}O_{9}$	$n(1d_{5/2})^{1}$	-1.91	-0.07	0.28	-1.70	-1.89
${}^{17}_{9}F_{8}$	$p(1d_{5/2})^{1}$	4.79	0.26 [b)]	-0.30	4.75	4.72
${}^{27}_{13}Al_{14}$	$p(1d_{5/2})^{-1}$	4.79	0.10	-0.93	3.96	3.64
${}^{29}_{14}Si_{15}$	$n(2s_{1/2})^{1}$	-1.91	-0.04	0.73	-1.22	-0.56
${}^{31}_{15}P_{16}$	$p(2s_{1/2})^{1}$	2.79	0.04	-0.99	1.84	1.13
${}^{33}_{16}S_{17}$	$n(1d_{3/2})^{1}$	1.15	-0.05	-0.53	0.57	0.64
${}^{37}_{17}Cl_{20}$	$p(1d_{3/2})^{1}$	0.12	0.21	0.31	0.64	0.68
${}^{39}_{19}K_{20}$	$p(1d_{3/2})^{-1}$	0.12	0.21	-0.26	0.07	0.39
${}^{41}_{20}Ca_{21}$	$n(1f_{7/2})^{1}$	-1.91	-0.15	0.28	-1.78	-1.59
${}^{59}_{27}Co_{32}$	$p(1f_{7/2})^{-1}$	5.79	0.19	-1.03	4.95	4.62
${}^{71}_{31}Ga_{40}$	$p(2p_{3/2})^{-1}$	3.79	-0.01	-1.53	2.25	2.56
${}^{87}_{37}Rb_{50}$	$p(2p_{3/2})^{-1}$	3.79	0.11	-1.10	2.80	2.75
${}^{87}_{38}Sr_{49}$	$n(1g_{9/2})^{-1}$	-1.91	-0.19	0.66	-1.44	-1.09
${}^{89}_{39}Y_{50}$	$p(2p_{1/2})^{1}$	-0.26	0.02	0.02	-0.22	-0.14

TABLE 1 (cont.)

Nucleus	Configuration	μ_{sp} n.m.	$\delta\mu_{int}$ [3)] n.m.	$\delta\mu_{conf}$ [6)] n.m.	μ_{theor} [a)] n.m.	μ_{exp} n.m.
${}^{91}_{40}Zr_{51}$	$n(2d_{5/2})^1$	-1.91	-0.02	0.41	-1.52	-1.30
${}^{115}_{49}In_{66}$	$p(1g_{9/2})^{-1}$	6.79	0.27	-0.89	6.17	5.54
${}^{115}_{50}Sn_{65}$	$n(3s_{1/2})^1$	-1.91	0.05	0.81	-1.85	-0.92
${}^{199}_{79}An_{120}$	$p(2d_{3/2})^{-1}$	0.12	0.17	0.20	0.49	0.27
${}^{205}_{81}Tl_{124}$	$p(3s_{1/2})^1$	2.79	0.00	-0.61	2.18	1.63
${}^{207}_{82}Pb_{125}$	$n(3p_{1/2})^1$	0.64	-0.07	-0.01	0.56	0.59
${}^{209}_{83}Bi_{126}$	$p(1h_{9/2})^1$	2.62	0.50	0.53	3.65	4.08

a) $\mu_{theor} = \mu_{sp} + \delta\mu_{int} + \delta\mu_{conf}$ b) ref. 2), E.M. Nyman

several authors in different aspects and using different approaches[2]). Recently, Chemtob[3]), using the meson - theoretical appraoch in the one - meson exchange approximation, performed numerical calculations of the interaction moments for odd - mass nuclei with closed shells in the jj - coupling sense ± one nucleon. A comparison of experimental magnetic moments for this group of nuclei with theoretical prediction[3]) is shown in table 1 and fig. 1. It is seen immediately that the correction due to the interaction moments is quite insufficient to explain the experimental data. This correction causes essentially a small shift of both Schmidt - lines, upwards for odd - proton and downwards for odd - neutron nuclei. The satisfying points are the explanation of the excess moments of ^3H and ^3He and the fact that the very small correction for ^{207}Pb (– 0.07 n.m.) and the much larger (although not sufficient) one for ^{209}Bi (0.50 n.m.), are both improving the results with respect to the experimental values.

Although for the lightest nuclei ($A \leqslant 17$) the agreement between experiment and theory is fairly good (the agreement for the ^{16}O region may be fortuitous, as significant configuration mixings are suggested there), the large deviations for heavier nuclei must be explained by other effects. The observed trend of the deviations finds a simple explanation in terms of a polarization effect of the closed shells, due to forces exerted by the valence nucleon on the rest of the system. The polarization effect on magnetic moments depends decisively on the configuration of the closed

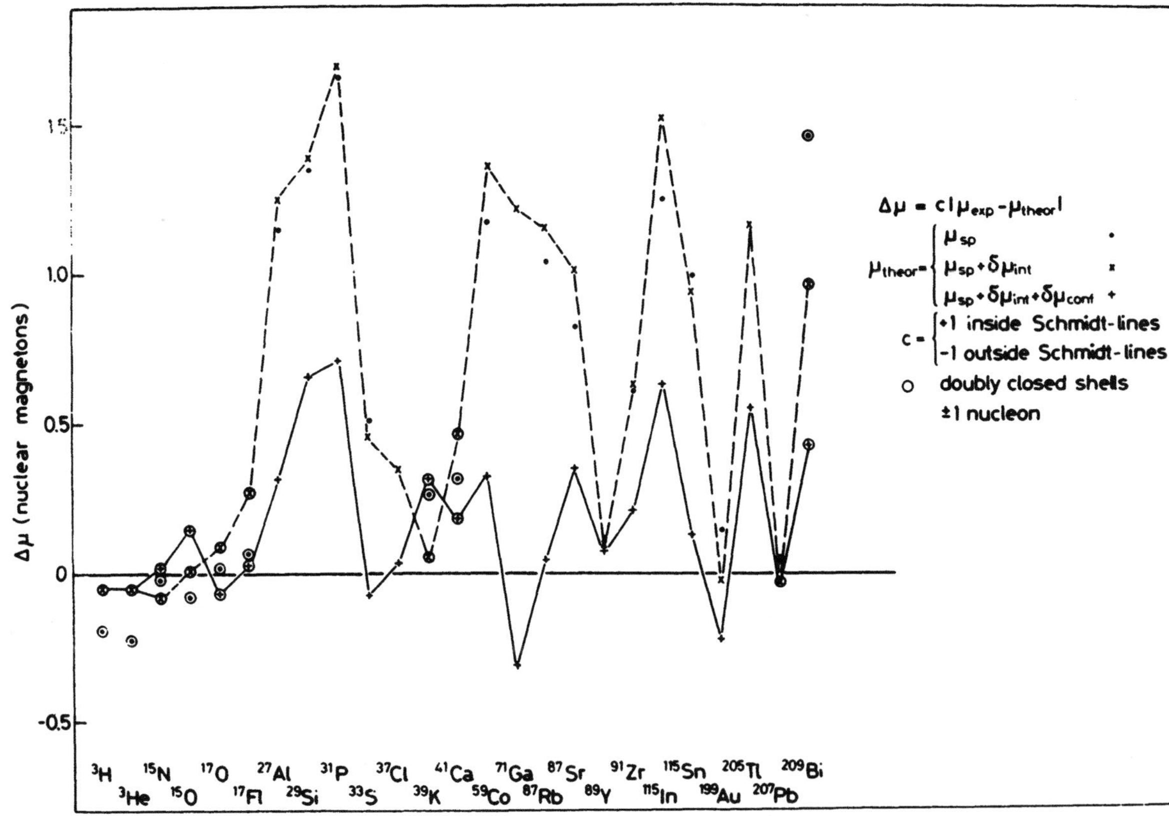

Fig. 1. Derivations of theoretical predictions on magnetic moments of odd - mass nuclei with jj - closed shells ± one nucleon from experimental values. The sign of the deviation is chosen to be positive if experimental magnetic moments are between the Schmidt - lines, negative for excess moments.

shells, since the main effect is associated with the presence of unsaturated spins in the closed shells, which can be partially aligned by the interaction with the spin of the valence nucleon.

In the treatment of the spin polarization, different equivalent methods can be used. One can use an average spin - dependent field in which the particles move in the closed shells (a method similar to that applied in the treatment of the quadrupole polarizability of nuclei). One can calculate the spin polarization effect in terms of a coupling of the valence particle to the collective spin - mode characterized by the quantum numbers $\lambda = 1$, $\sigma = 1$, $\tau = 1$, $\pi = +$, which might appear at about 5 - 7 MeV; this method is an extension of the treatment of electric dipole polarization. Finally, one can use the more conventional language of configuration mixing, in which we speak of the admixture of configurations with particle - hole excitations coupled with the valence - nucleon mode

$$|j\rangle = |(\text{closed shell})\, j\rangle + \sum_{ik} c_{ik} |[(j_i^{-1}\, j_k)\, J, j]\, I = j\rangle \tag{4}$$

The important restriction of the quantum numbers is that j_i and j_k are spin orbit partners, $j_i = l + 1/2$, $j_k = l - 1/2$.

The effect of the spin polarization on magnetic moments depends on the form of the residual interaction. The main effect is expected to come from a spin - exchange force of the type $V_{\sigma\tau}(r_{12})\,(\underline{\sigma}_1 \cdot \underline{\sigma}_2)\,(\underline{\tau}_1 \cdot \underline{\tau}_2)$, in which case the magnetic moment operator takes the simple form

$$\underline{\mu} = g_l\, \underline{l} + g_s\, (1 + a)\, \underline{s} = g_l\, \underline{l} + g_s^{\text{eff}}\, \underline{s} \tag{5}$$

where a is the spin polarizability ($a < 0$). Even rough estimates[1,4]) show that the first order perturbation theory, yielding $|a| \lesssim 1$, is not adequate, and higher order terms must be considered.

The total polarization effect cannot be described simply in terms of a renormalized g_s - factor, and the complete spin dependence of the actual interaction has to be considered in order to make quantitative estimates. Several attempts in this direction have been made with the use of realistic nucleon - nucleon interactions[5,6]). Results of calculations with the Kallio - Kolltveit and the Hamada - Johnston interactions[6]) are given in table 1 and fig. 1. With the exception of nuclei with ^{16}O and ^{40}Ca cores (the only ones closed in the L - S sense), where only higher order effects can contribute, the calculations are made in the first order perturbation theory. Second order contributions for ^{29}Si and ^{209}Bi worsen the agreement between theory and experiment. This is not disturbing for ^{29}Si, as it is a deformed nucleus and should be treated rather within the Hartree - Fock framework with a deformed basis, but ^{209}Bi remains an unsolved problem. Although an overall improvement in the comparison of experimental and theoretical results on magnetic moments is noted, an enormous amount of work has still to be done in order to test the ideas and reach a quantitative agreement.

Going away from closed shells, the problem of calculating magnetic moments of bound nuclear states becomes more complex and difficult, as we enter all the open problems of describing the state of a many - particle system with many degrees of freedom. Although the proper wave function is the crucial point in calculating nuclear magnetic moments, the broad scope of the problem does not allow me to go into details. Instead of trying to make a very incomplete review of the open questions in nuclear structure theory and of the recent attempts towards better understanding of nuclear structure and forces, let me draw some conclusions about the agreement of experimental data with theoretical predictions on nuclear moments.

We do not have 'good' wave functions for few - particle configurations, and the various attempts to describe collective states (even the first excited 2^+ states) in spherical and transitional nuclei have met with moderate success in reproducing the experimental magnetic moments. The harshest conclusion is that there is as yet no theory capable of predicting magnetic moments to better than a few tenths of the nuclear magneton. The 'phenomenological' as well as the 'microscopic' theories are only approximation schemes and are based on a number of parameters taken from observation. The different approaches result, of course, in a different ability to reproduce the magnetic properties of nuclei, but generally, the more internal parameters are taken from observation, the better the agreement with experiment. Nevertheless, the magnetic moments still remain an important and sensitive tool for testing the nuclear coupling scheme and for pinning down the main components of the nuclear wave function.

References

1. A. Bohr and B.R. Mottelson, Nuclear Structure, I, W.A. Benjamin Inc. (1969) 3c.
2. S.D. Drell and J.D. Walecka, Phys. Rev. 120 (1960) 1069;
   E.M. Nyman, Nucl. Phys. B1 (1967) 535;
   M. Ross, Phys. Rev. 88 (1952) 935;
   A. Rusek and L. Spruch, Phys. Rev. 87 (1957) 1111;
   H. Miyazawa, Prog. Theor. Phys. 5 (1951) 801;
   Y. Hara and E. Kuroboshi, Progr. Theor. Phys. 21 (1959) 768;
   M. Ichimura and K. Yazaki, Nucl. Phys. 63 (1965) 401.
3. M. Chemtob, Nucl. Phys. A123 (1969) 449.
4. B.R. Mottelson, Nordita Publ. No 288 (1969).
5. J. Blomqvist, N. Freed and H.O. Zetterström, Phys. Lett. 18 (1965) 47.
6. H.A. Mavromatis, L. Zamick and G.E. Brown, Nucl. Phys. 80 (1966) 545;
   H.A. Mavromatis and L. Zamick, Phys. Lett. 20 (1966) 171; Nucl. Phys. A104 (1967) 17.

Discussion

Alder: The calculation of magnetic moments with an exchange meson potential reminds me very much of the very old work by Villars on the magnetic moment of ^3He and ^3H. Is it correct, to say that the mentioned calculation is an extension of this old work?

Ogaza: Yes, it is.

Contributed Paper

SHELL MODEL CALCULATIONS OF MAGNETIC DIPOLE AND ELECTRIC QUADRUPOLE MOMENTS AND γ-RAY MIXING RATIOS

P.W.M. Glaudemans
Utrecht University, Utrecht, The Netherlands
and
B.H. Wildenthal
Michigan State University, U.S.A.

During recent years more sophisticated shell model calculations have been performed. This is mainly due to the fact that large computers are available and also that methods have been developed and programs written to handle wave functions in rather large shell model spaces[1]).

In collaboration with E.C. Halbert and J.B. McGrory (Oak Ridge National Laboratory) the sd - shell nuclei with mass A = 17 - 40 are investigated[2]). This is a particularly nice mass region, since the shell model seems to work well for most nuclei while, moreover, the experimental information is copious and easily accessible[3]). The sd - shell is also interesting, since there are many sets of isobaric multiplets (such as mirror nuclei) available. When enough experimental transition rates and mixing ratios are measured in these multiplets, one can test certain relations that are expected to exist between these data.

In this paper some results about electromagnetic transition rates and multipole moments will be discussed. The effective two - body matrix elements used in these calculations are obtained from a simple surface delta interaction[4]). The values of the strength parameters of this interaction as well as the values of the single - particle energies are obtained from a least - squares fit to many experimentally determined binding energies and excitation energies.

A very useful test of the wave functions obtained from these calculations consists of a comparison of calculated electro - magnetic transition rates and multipole moments with experiment. The strength of a transition expressed in Weisskopf units can be written in terms of matrix elements reduced in space and iso - space as:

$$|M_L|^2_{W.u.} = C_L\,[\delta_{T_iT_f} \langle\psi_f|||S_L|||\psi_i\rangle + \langle T_iT_{zi}10|T_fT_{zf}\rangle \langle\psi_f|||V_L|||\psi_i\rangle]^2 \qquad (1)$$

where L specifies the multipolarity of the transition and C_L is a normalization factor. The indices i and f refer to the initial and final state, respectively. The symbols S_L

and V_L denote the isoscalar and iso - vector operator, respectively, while $\langle T_iT_{zi}\ 10|T_fT_{zf}\rangle$ is a Clebsch - Gordan coefficient. Similarly a multipole moment can be written in terms of reduced matrix elements.

For $T_i = T_f$ the Clebsch - Gordan coefficient given in eq. (1) can be written as $T_z\,[T\,(T+1)]^{-1/2}$. One thus finds from eq. (1) the following expressions for M1 and E2 strengths of $\Delta T = 0$ transitions:

$$|M\,(M1)|^2_{W.u.} = [M_s + T_zM_v]^2 \qquad (2)$$

and

$$|M\,(E2)|^2_{W.u.} = [E_s + T_zE_v]^2 \quad , \qquad (3)$$

where the indices s and v denote the iso - scalar and iso - vector contributions, respectively. From detailed calculations in the A = 30 - 34 mass region in a large shell model space[5]) it follows that $M_s \approx 0.1\ M_v$ and $E_s \approx 10\ E_v$ for almost all calculated transit-

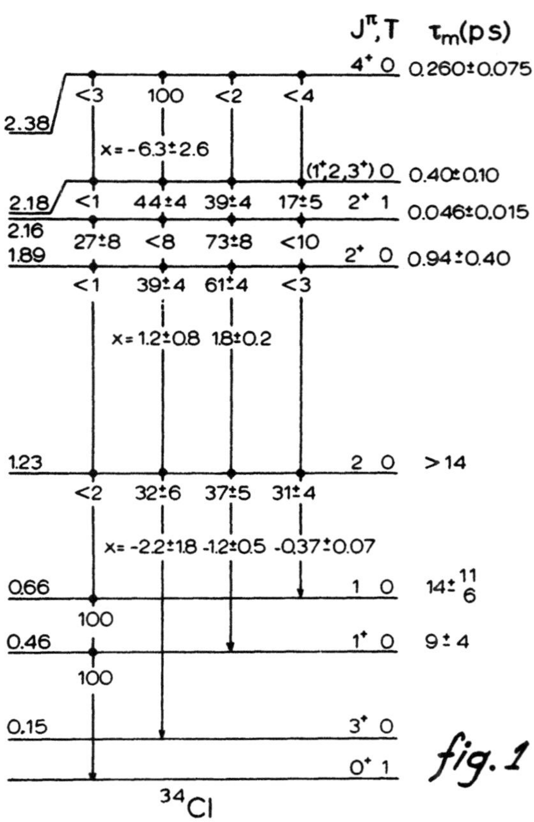

fig. 1

ions. This holds especially for the stronger transition probabilities.

E2/M1 mixing ratios for $\Delta T = 0$ transitions in mirror nuclei

An E2/M1 mixing ratio can be written in terms of the transition strengths given in eqs. (2) and (3) as:

$$\delta\,(E2/M1) \propto M(E2)/M(M1) = (E_s + T_z E_v)/(M_s + T_z M_v) \qquad (4)$$

The sign of the mixing ratio follows from the relative signs of the E2 and M1 amplitudes.

Table 1

Comparison between E2/M1 mixing ratios in mirror nuclei

A	transition	δ_{exp}	
		$T_z = +1/2$	$T_z = -1/2$
29	(1) → (0)	+0.22 ± 0.03	−0.13 ± 0.02
31	(1) → (0)	−0.28 ± 0.02	+0.37 ± 0.02
33	(2) → (0)	−0.6 ± 0.2	+0.36 ± 0.15

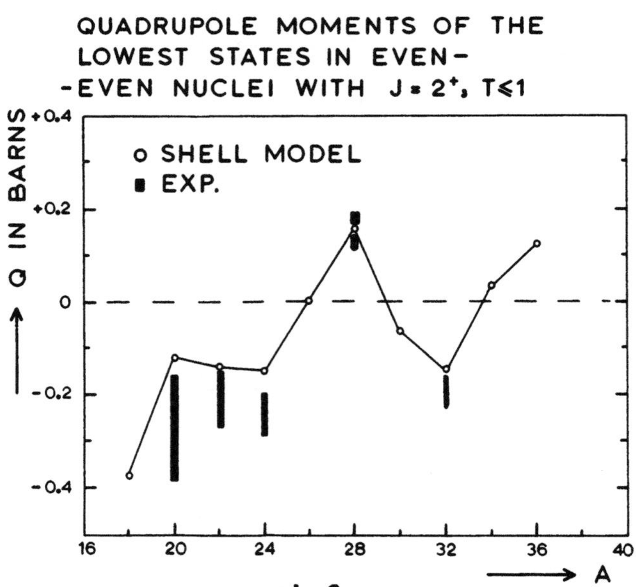

Fig. 2

When wave functions of corresponding states in mirror nuclei are assumed to be identical, only the value of T_z in eq. (4) changes sign. One thus finds that the matrix element for E2 does not change sign (since $E_s \gg E_v$). The M1 matrix element, however, will get the opposite sign (since $M_s \ll M_v$). Thus the value of the E2/M1 mixing ratio is expected to change sign, while the absolute value should not change very much. Some examples are given in table 1.

E2/M1 mixing ratios for $\Delta T = 0$ transitions in $T_z = 0$ nuclei

Another check of the ratio between iso - scalar and iso - vector contributions follows with eq. (4) from a study of E2/M1 mixing ratios in $T_z = 0$ nuclei. In this case only the iso - scalar part contributes. A reasonable estimate of an M1 transition strength is about 10^{-2} W.u. This strength is thus expected to be about 100 times smaller in a $T_z = 0$ nucleus (see eq. (2)). An E2 transition is generally of the order of one W.u. This value should also be valid for a $T_z = 0$ nucleus (see eq. (3)). A mixing ratio is defined as $|\delta\ (E2/M1)| = \{\Gamma\ (E2)/\Gamma\ (M1)\}^{1/2}$, where Γ specifies the width of

TABLE 2. Comparison between experimental and theoretical E2/M1 mixing ratios between π = + states

Nucleus	$E_i \rightarrow E_f$ (MeV)	$J_i \rightarrow J_f$	E2/M1 mixing ratio δ experiment	theory
^{30}P	2.94 → 1.45	2 → 2	(+)0.03 ± 0.03	+0.02
^{30}Si	3.51 → 2.23	2 → 2	+0.18 ± 0.05(I)	+0.11
	3.77 → 2.23	1 → 2	-0.08 ± 0.03(I)	-0.05
^{31}P	1.27 → 0	3/2 → 1/2	-0.28 ± 0.02(II)	-0.36
^{31}S	1.25 → 0	3/2 → 1/2	+0.35 ± 0.02(III)	+0.47
^{31}Si	1.69 → 0	5/2 → 3/2	+4.7 ± 1.0(IV)	+1.7
	2.79 → 0	5/2 → 3/2	+0.20 ± 0.05(IV)	+0.15
^{32}S	4.29 → 2.24	2 → 2	$\lvert\delta\rvert$ > 6	+16
^{33}S	0.84 → 0	1/2 → 3/2	$\lvert\delta\rvert$ = 0.18	-0.17
	1.97 → 0	5/2 → 3/2	-0.6 ± 0.2(I)	-0.69
^{33}Cl	2.85 → 0	5/2 → 3/2	-0.10 ± 0.03(V)	-0.13
^{33}P	1.43 → 0	3/2 → 1/2	-0.60 ± 0.09(III)	-0.46
	2.54 → 0	3/2 → 1/2	{+0.16 ± 0.04(III) / -2.6 ± 0.3 (III)}	+0.30
^{34}Cl	1.23 → 0.15	2 → 3	+2.2 ± 1.8 (II)	+0.16
	1.89 → 0.15	2 → 3	-1.2 ± 0.8 (II)	-1.4
	2.38 → 0.15	4 → 3	+6 ± 3 (III)	+4.0

the transition. One thus finds inserting the Weisskopf estimates with the assumptions about expected strengths given above, that $|\delta\ (E2/M1)| \approx 0.15\ A^{2/3}\ E_\gamma$, where E_γ is the transition energy expressed in MeV. For sd - shell nuclei with $A \approx 30$ one finds $|\delta\ (E2/M1)| \approx E_\gamma$ (expressed in MeV). An example is given in fig. 1 for ^{34}Cl, where this simple rule seems to work very well.

Signs of E2/M1 mixing ratios

From the study of M1 and E2 transition rates in A = 30 – 34 nuclei[5]) it was found that the signs of E2/M1 mixing ratios can be very well predicted from the shell model. Examples are given in table 2, showing that all calculated signs agree with experiment.

Multipole moments

A very nice test of shell model wave functions is the comparison of calculated and experimentally determined multipole moments. Such a comparison has the advantage that it tests directly the wave function of a single state instead of those for a combination of initial and final states. In fig. 2 the calculated quadrupole moments of the first excited state in even - even nuclei are compared with the experimental data. It is seen that the sign change at A = 26 and 30 is nicely reproduced by the shell model. In these calculations effective proton and neutron charges of about 1.6 e and 0.6 e are used. These effective charges can partly compensate for the truncation of the shell model space.

Conclusion

It is interesting to investigate the effect of configuration mixing and hole - state contributions on the calculated electro - magnetic properties. An illustration is given in tables 3 and 4, where some multipole moments and transition rates in ^{33}S are calculated with different assumptions about the shell model space that is taken into

Table 3
Magnetic dipole moments and transition rates of low-lying states in ^{33}S.

		theory[a)]			experiment
		I	II	III	
μ(in n.m.)	^{33}S(0)	+1.15	+1.32	+1.04	+0.6434 ± 0.0001
$\lvert M(M1)\rvert^2_{W.u.} \times 100$	^{33}S(1→0)	0	0.012	4.4	3.2 ± 0.4
"	^{33}S(2→0)	0	5.4	6.6	2.1 ± 0.5

a) The values are calculated with bare nucleon g-factors and the following assumptions:
I : pure one particle or two particle-one hole configurations;
II : configuration mixing and a ^{28}Si core;
III: up to two holes in the $d_{5/2}$ shell.

Table 4
Electric quadrupole moments and transition rates of low-lying states in ^{33}S

		theory[a)]						experiment
		I		II		III		
		Δ=0	Δ=1.3e	Δ=0	Δ=0.7 e	Δ=0	Δ=0.5 e	
Q(in e x barn)	$^{33}S(0)$	0	−0.057	−0.015	−0.057	−0.034	−0.066	−0.055 ± 0.00[illegible]
$\|M(E2)\|^2_{W.u.}$	$^{33}S(1\to0)$	0.2	5.6	1.6	7.0	1.4	4.7	6 ± 3
"	$^{33}S(2\to0)$	0.3	7.8	0.9	6.8	1.8	7.2	6 ± 4

account. It follows that the calculated values are sometimes sensitive to the model space, but also that the sign of multipole moments are already correctly reproduced by a rather simple model. The most complete model of which the results are shown in column III of tables 3 and 4 overestimates magnetic dipole moments and transition probabilities, whereas it underestimates electric quadrupole moments and transition rates. Excellent agreement for the latter can be obtained when an additional charge Δ is added to the proton and neutron charge with Δ = 1.3, 0.7 and 0.5 e for the calculations I, II and III, respectively (see table 4). This indicates that, as expected, the most complete wave functions need the smallest effective charge. A similar improvement can be obtained for magnetic dipole transitions and moments, when effective g - factors or single - particle matrix elements are used[5]).

It would be interesting to investigate shell model descriptions in the manner described above also in many other nuclei. Therefore, however, more multipole moments and mixing ratios should be measured.

References

1. J.B. French, E.C. Halbert, J.B. McGrory and S.S.M. Wong, Advances in Nuclear Physics, Vol. 3 (Plenum Publ. Co., New York), eds. M. Baranger and E. Vogt.
2. E.C. Halbert, in 'The Structure of low - medium mass nuclei', ed. by J.P. Davidson, The University Press of Kansas, Lawrence/London, 1968.
3. P.M. Endt, and C. van der Leun, Nucl. Phys. A105 (1967) 1 and private communication.
4. P.W.M. Glaudemans, P.J. Brussaard, and B.H. Wildenthal, Nucl. Phys. A102 (1967) 593.
5. P.W.M. Glaudemans, P.M. Endt, and A.E.L. Dieperink, Annals of Physics, to be published.

Contributed Paper

MAGNETIC MOMENTS FOR THE 265 keV AND 280 keV LEVELS IN ^{75}As*

A.J. Becker and F.C. Zawislak**

California Institute of Technology, Pasadena, Calif., U.S.A.

1. Introduction

The directional - correlation measurement of magnetic moments of excited nuclear levels with lifetimes much less than a nanosecond requires a strong interaction to produce a perceptable perturbation in such a short period. The relatively high internal magnetic field experienced by ^{75}As nuclei following electron capture by ^{75}Se impurities in a polarized Fe host lattice have been used in these experiments for this purpose. The presence of a longer - lived state in the same nucleus permits a more straightforward measurement of its magnetic moment and the determination of the internal field.

There have been results presented[1,2]) which indicate that the ^{75}As nucleus exhibits collective properties somewhat unexpected for a member of this region of the periodic table. The determination of these magnetic moments is of definite assistance in testing the extent to which this collective behaviour exists.

2. Experimental Procedure

The measurements were made using the time - integral perturbed directional - correlation technique[3]), with the sources in a magnetic field perpendicular to the detection plane. For those runs in which the field direction was held fixed and one detector moved, the true - coincidence rates were used to obtain the parameters b_{22} and $\omega\tau$ in the correlation function

$$W(\theta, H) = 1 + \frac{b_{22}}{\sqrt{1 + (2\omega\tau)^2}} \cos\left[2(\theta - \omega\tau)\right]$$

where the b_{44} term is negligibly small. The other runs involved keeping the detector angle fixed at $3\pi/4$ or $5\pi/4$ and reversing the sign of the magnetic field. The combination of both types of measurements is highly efficient when $\omega\tau \ll 1$.

The decay scheme for ^{75}As is shown in fig. 1. The detectors were a planar Ge (Li) and a 3″ × 3″ NaI (Tl), the former isolating either the 121 keV or the 136 keV gamma ray while the latter responded to the combination of the 265 keV and the 280 keV gammas. The electronics were the standard high - resolution, fast - slow coincidence type, with the TAC spectra accumulated in an MCA. A computer program was used to interpolate and subtract the accidentals background from the coincidence

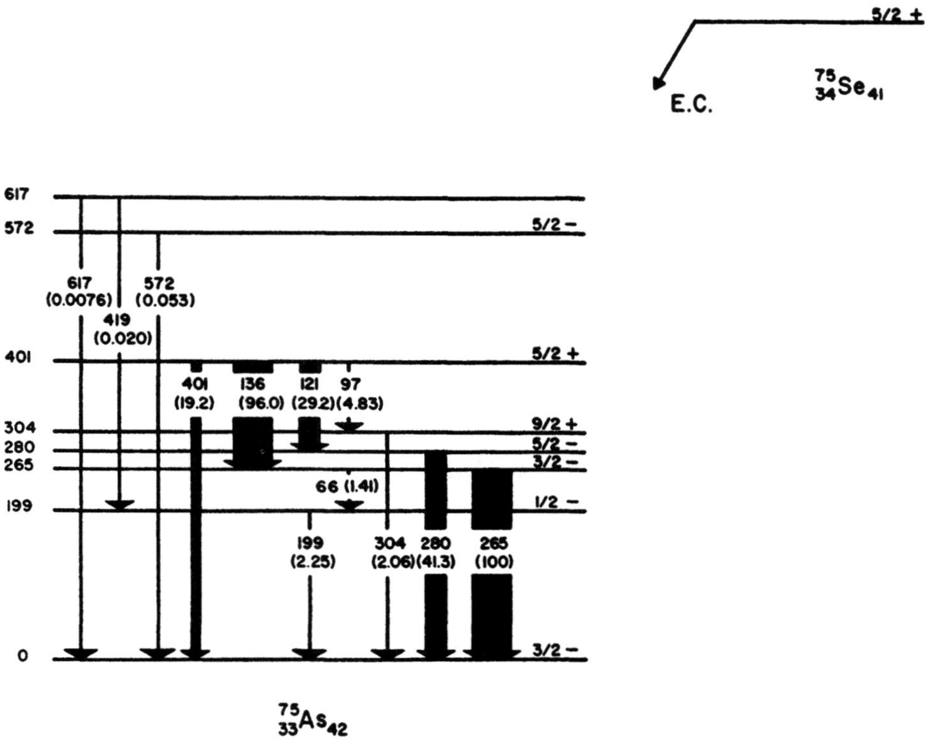

Fig. 1. Partial decay scheme of ^{75}As. From ref. 14.

peak and obtain the parameters in the correlation function.

The liquid sources were of the form H_2SeO_3 in a dilute HCl solution. Three such sources of widely different normalities were used and yielded consistent results. The source with Se in an Fe host was made by melting and then annealing a pellet of Se metal and Fe. An extensive study[4]) has been made of the effects of the mode of preparation and annealing on the perturbations experienced by the As impurities in Fe, Co, and Ni hosts. To ensure the reproducibility of the results for the internal field, the single As - in - Fe source used was not treated in any way during the measurements. There was no noticeable systematic change in the results during the course of the experiment.

The liquid sources were polarized using a large electromagnet capable of producing 20 kOe in a 1/4″ gap. The polarization of the sources in Fe was through a small, closed - loop 'mini - magnet' [5]) which reduces the stray field to a minimum.

3. Results and Discussion

The magnetic moment for the 280 keV ($5/2^-$)state was determined from the 121 - 280 keV correlation, using the liquid sources and the large electromagnet. The result of the experiment is $\omega\tau/H = (6.98 \pm 0.75) \times 10^{-4}$ kOe^{-1}. The current

Table 1. Lifetime data for the 280 keV level in As^{75}

Lifetime (nsec)	Method	Reference
0.78 ± 0.07	Delayed Coincidence	11
0.402 ± 0.030	Resonance Fluorescence	6
0.40 ± 0.03	Delayed Coincidence	12
0.81 ± 0.07	Delayed Coincidence	13

measurements of the lifetimes of this state are indicated in table 1. There are two distinct groups of results which are not at all in agreement. Using the value of Langhoff and Schumacher[6]) the g - factor is

$$g_{5/2} = 0.363 \pm 0.047$$

This number is consistent with that of Manning and Rogers[7]) of 0.42 ± 0.13.

For the source in Fe, using the mini - magnet, the result for this level is $\omega\tau =$ $= 0.1047 \pm 0.0035$. Combining this number with the $\omega\tau/H$ for the liquid sources yields the total field experienced by the As nuclei in this source: $H_{total} =$ $= (150 \pm 17)$ kOe. The internal field of (145 ± 18) kOe is consistent with that expected from the systematics of hyperfine fields in Fe as a function of the atomic number[8]).

The magnetic moment for the 265 keV ($3/2^-$) level was determined from the 136 - 265 keV correlation, using the source in Fe. Since the life - time[6]) of this state is very small: (17.1 ± 1.0) psec., the perturbation observed was correspondingly tiny, and large amounts of data had to be accumulated. Approximately 75 million true coincidences were registered, and yield $\omega\tau = 0.0090 \pm 0.0024$, which, using the internal field found from the study of the $5/2^-$ state and the lifetime quoted above, gives for the g - factor,

$$g_{3/2} = 0.74 \pm 0.22$$

A series of measurements have been performed to ensure that Compton - background contributions, magnetic - field influences on the detectors, and possible source mis-centering have not invalidated this result.

The use of first - order shell - model theory with the assumption that these two levels are single - particle states for the odd proton is inadequate to explain the values. An alternative is to analyze the levels in ^{75}As using the deformed - nucleus model and to determine the consistency of the results with calculations based on Nilsson wave - functions.

Considering the levels at 199, 265, and 280 keV as members of a K = 1/2 band and using as input data the lifetime[6]) of the 265 keV level, the intensity and conversion coefficient[9]) of the 66 keV transition, and the result given above for the magnetic moment of the 280 keV level, one solution for $g_{3/2}$ is + 0.77, a value completely consistent with the experimental result of this work. The corresponding g_K is 2.3, where $g_R = Z/A = 0.44$ has been used.

The calculations using the Nilsson wave - functions[10]) give a deformation $\eta =$ $= +3$ for this g_K, where $g_{s,eff} = g_s/2$. Other calculations based on the energy separation of these three levels and on the ground - state quadrupole moment yield the same value for η. Thus the results reported here for the two magnetic moments are in good agreement with a deformed - nucleus description and a deformation parameter of + 3. It must be added, however, that the agreement no longer is significant if one of the higher values for the lifetime of the 280 keV level, as given in the table, and its correspondingly adjusted g - factor of 0.18 ± 0.03 are used.

* Work performed under the auspices of the U.S. AEC Prepared under Contract AT (04 – 3) – 63.

** Present address: Instituto de Fisica, UFRGS, Porto Alegre, Brasil

References

1. N. Imanishi, M. Sakisaka and F. Fukuzawa, Nucl. Phys. A125 (1969) 626.
2. Th. Paradellis and S. Hontzeas, to be published.
3. R.M. Steffen and H. Frauenfelder, in Perturbed Angular Correlations (North - Holland, 1964), 3.
4. F.C. Zawislak and A.J. Becker, to be published.
5. J.D. Bowman, H.E. Henrikson and F. Zawislak, to be published.
6. H. Langhoff and H. Schumacher, Phys. Rev. 155 (1962) 1246.
7. G. Manning and J.D. Rogers, Nucl. Phys. 15 (1960) 166.
8. D.A. Shirley, in Hyperfine Interactions and Nuclear Radiations, ed. by E. Matthias and D.A. Shirley (North - Holland, 1968), Appendix B.
9. W.B. Ewbank et al., Nucl. Data B1, No. 6 (1966).
10. S.G. Nilsson, Mat. Fys. Medd. Dan. Vid. Selsk. 29 No. 16 (1955).
11. J. Varma and M.A. Eswaran, Phys. Rev. 125 (1962) 656.
12. M. Höjeberg and G.S. Malmskog, Nucl. Phys. A133 (1969) 691.

13. S.Sh. Azimov et al., Yadern. Fiz. 10 (1969) 3;
Translation: Sov. J. of Nucl. Phys. 10 (1970) 1.

14. D.E. Raeside et al., Nucl. Phys. A130 (1969) 677.

Contributed Paper

MAGNETIC MOMENT OF THE LOWEST 2^+ VIBRATIONAL STATE IN ^{102}Ru

M.R. Ahmed, K. Johansson, E. Karlsson and L.-O. Norlin
University of Uppsala, Institute of Physics, Uppsala, Sweden

1. Introduction

In the ion implantation perturbed angular correlation technique (IMPACT) g-factors from measurements on radioactive sources are used as calibration when the magnetic transient fields are calculated from the IMPACT - data. The existing g - factor[1]) of the 475 keV 2^+ level in ^{102}Ru has a relative error of 23% and therefore it was decided to remeasure it.

2. Source and Experiments

The ^{102}Rh activity was prepared by irradiating natural ruthenium metal powder with 20 MeV deuterons during 20 hours. Chemical separation was done 8 months later. After the separation the activity dissolved in HCl (pH = 1.5) was deposited onto an iron foil by the liquid drop method. The foil was melted down to a small sphere with an electron gun. The sphere was annealed in an evacuated quartz capsule at 1000° C for 50 hours, and then used in the g - factor measurement. The source consisted of two ^{102}Rh half - lives, 206 d and 1057 d, with different decay schemes[2]). Fig. 1 shows a small portion of the γ - singles spectrum taken with a Ge (Li) - detector. This part of the spectrum was used to calculate the relative number of 206 d and 1057 d decays and it was found that at the time of the g-factor measurement 84% of the activity consisted of ^{102}Rh nuclei with $T_{1/2}$ = 1057 d.

From the decay schemes of Konijn et al.[2]), fig. 1 and measured γ - ray intensities, the relative numbers of coincidences of the different γ - γ cascades were calculated. It was found that 8 cascades contributed to the total number of coincidences and that 92% of these were due to γ - γ cascades passing the 475 keV level and that the main contribution came from the 1047 – 475 keV (57%) and 1113 – (631) – 475 keV (33%) cascades.

The unperturbed angular correlation of all the γ - γ cascades present in the gate - settings used in the g - factor experiment was measured with an automatic two - channel setup and the result was $A_2 = -0.2364 \pm 0.0015$ and $A_4 = 0.0362 \pm 0.0037$.

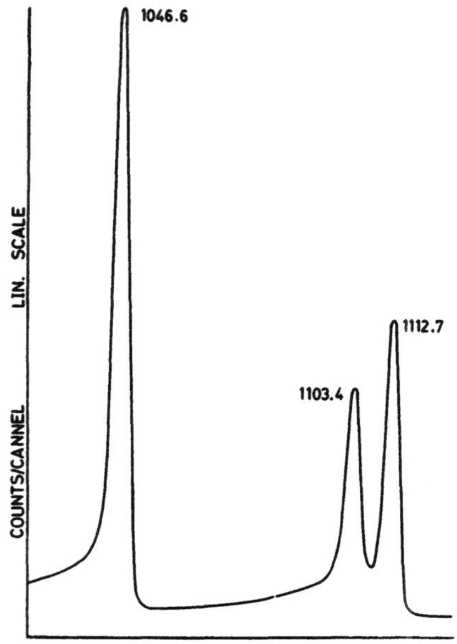

Fig. 1. A part of the gamma - ray spectrum of ^{102}Rh measured with a Ge (Li) detector.

The g - factor of the 475 keV level was determined by measuring the rotation of the angular correlation with the IPAC - method. The Ru - nuclei were situated in a ferromagnetic iron matrix and the sample was polarized by the application of an external magnetic field of 22 kG perpendicular to the plane of four detectors.

The temperature dependence of the magnetic hyperfine field deviates in some cases from the temperature dependence of the bulk magnetization of the host material[3]) and therefore the measurement was done at liquid helium temperature.

The result of the experiment was an angular rotation $\omega\tau = -(0.0217 \pm 0.0012)$ rad. The effective field at the ^{102}Ru nucleus can be expressed as $B_{eff} = B_{hf} + (\frac{4\pi}{3} - D)\,M + B_0$ where B_0 is the polarizing field, M the magnetization and D the demagnetization factor. Using the values $B_{hf} = -(503 \pm 9)$ kG (ref. 4), $B_0 = +22$ kG and $D = \frac{4\pi}{3}$ (a sphere) together with $\tau = (23.1 \pm 1.4)$ ps (ref. 4) and the $\omega\tau$ observed here, the results are $g = +0.408 \pm 0.034$ and $\mu = +0.82 \pm 0.07$.

3. Discussion

The g - factor of the 2^+ first excited state in ^{102}Ru agrees with the estimate Z/A = 0.43. Greiner[5]) has deduced a formula for the g - factors of vibrational nuclei based on a collective model with the assumption that the proton distribution is less deformed than the neutron distribution. This is due to a larger pairing force between the protons than between the neutrons. The formula proposed is

$$g = \frac{Z}{A} \cdot (1 - 2f) \cdot (1 + 2f/3) \approx \frac{Z}{A} (1 - \frac{4f}{3})$$

$$f \approx \frac{N}{A} (\sqrt{\frac{G_p}{G_n}} - 1)$$

The factors G_p and G_n are the strength of the pairing force for neutrons and protons respectively. The experimental value[6]) $G_p/G_n = 1.4$ gives $g = 0.37$ which is in agreement with the experimental result.

References

1. K. Auerbach, K. Siepe, J. Wittkemper and H.J. Körner, Phys. Letters 23 (1966) 367.
2. J. Konijn, E.W.A. Lingeman, F. Diederix, B.J. Meijer, P. Koldewijn and A.C. Klaasse, Nucl. Phys. A138 (1969) 514.
3. D.A. Shirley, S.S. Rosenblum and E. Matthias, Phys. Rev. 170 (1968) 363.
4. Hyperfine Structure and Nuclear Radiations, ed. by E. Matthias and D.A. Shirley (North - Holland Publ. Co., Amsterdam, 1968) p. 980 and p. 1061.
5. W. Greiner, Nucl. Phys. 80 (1966) 417.
6. S.G. Nilsson and O. Prior, Mat. Fys. Medd. Dan. Vid. Selsk, 32 (1961) No. 16.

Contributed Paper

MAGNETIC MOMENT AND HALF - LIFE OF THE FIRST EXCITED STATE IN ^{134}Cs

H. Drost, W. Weiss and G. Weyer
I. Physikalisches Institut der Freien Universität, Berlin, W. Germany

The spin of the ground state of ^{134}Cs, $I^\pi = 4^+$, is considered to arise from the coupling of a $g_{7/2}$ - proton and a $d_{3/2}$ - neutron, on the basis of the j - j - coupling model for odd - odd nuclei, with a possible admixture of a $d_{5/2}$ - configuration for the proton[1,2]). From the possible values for the coupling of the angular momenta of the proton and the neutron predicted by 'Nordheims weak rule' [3]) for ground states of odd - odd nuclei with $j_p = l_p - \frac{1}{2}$ and $j_n = l_n - \frac{1}{2}$, this is not the highest possible value. The ground state of the neighbouring odd nucleus ^{136}Cs has $I^\pi = 5^+$, and the first excited state of ^{134}Cs is assumed to have $I^\pi = 5^+$ [4]). It would therefore be of interest to compare the magnetic moment of the 5^+ state in ^{134}Cs with the known

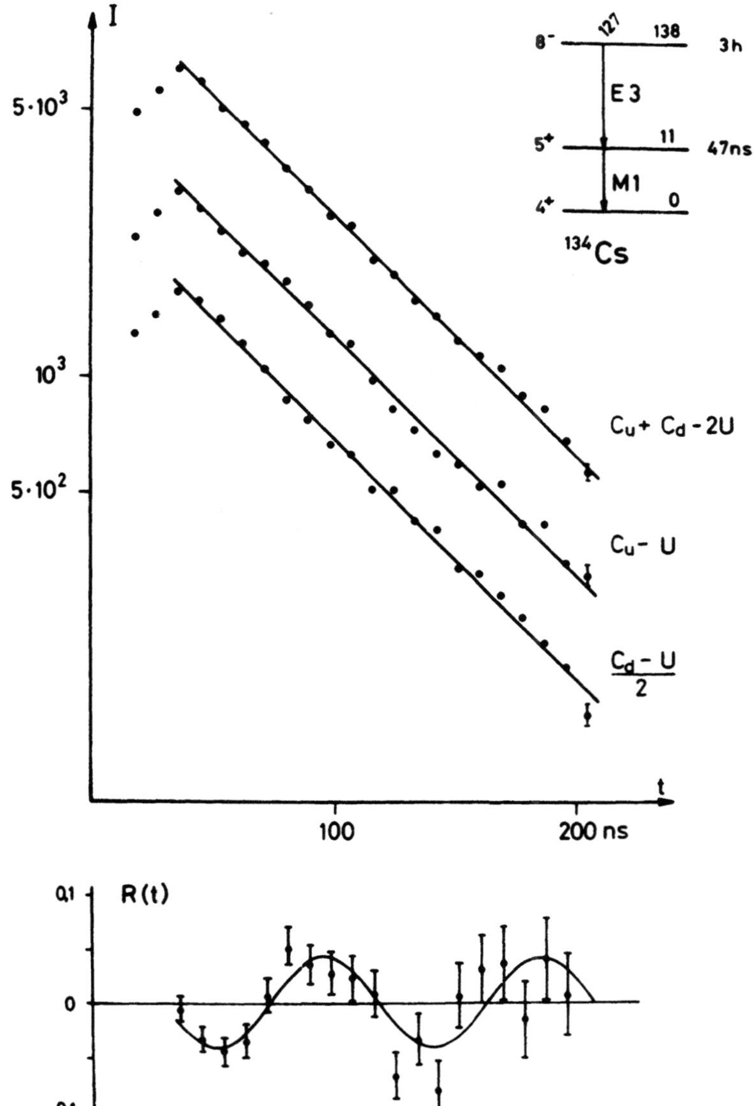

Fig. 1. Delayed coincidence spectra with spin rotation pattern in an external magnetic field up (C_u) and down (C_d). The lower graph shows the ratio R (t) with a fit (solid line) as defined in the text.

magnetic moments of the ground states of ^{134}Cs and ^{136}Cs.

We have measured the g - factor of the 5^+ state using the time differential perturbed angular correlation method. Sources of ^{134m}Cs were prepared by neutron - irradiation of CsCl, which then was dissolved in water. The level scheme of ^{134}Cs is shown in fig. 1. A Ge (Li) - detector (1 cm^2 X 0.5 cm) was used to detect the 11 keV γ - rays in a plane with three NaI (Tl) - detectors (1 1/2" X 2") at angles of 135, 180 and 225 degrees for the 127 keV γ - rays. Each detector was connected to a slow - fast coincidence circuit. The time resolution was about 25 ns FWHM. Delayed coincidence spectra of the three possible detector combinations were measured in external magnetic fields of 9.9 kG and 11.0 kG applied perpendicularly to the detector plane. Typical time spectra for the 225° - detector combination, with a magnetic field of 11 kG, are shown in fig. 1 together with the ratio $R(t) = [C_u(t) - C_d(t)] / [C_u(t) + C_d(t) - 2U]$, where $C_u(t)$ and $C_d(t)$ are the delayed coincidence counting rates for field up and down, respectively; U is the time independent background. Neglecting the A_4 coefficient and attenuation, this ratio is given by $R(t) = [3A_2/(4 + A_2)] \cdot \cos 2\omega_L t$, where A_2 is the correlation coefficient and ω_L is the Larmor frequency of the magnetic moment in the external field H: $\omega_L = -g\mu_n H/\hbar$. This frequency was obtained from a weighted least squares fit of the function R (t) to the experimental data.

The energy of the first excited state in ^{134}Cs was determined to be (11.2 ± 0.1) keV. The result for the half - life is $T_{1/2} = (47.0 \pm 1)$ ns in agreement with a previous γ - conversion electron delayed coincidence measurement using a parallel - plate - avalanche - counter for electron detection. The Weisskopf estimate for an M1 - transition, assuming $\alpha \sim 200$, is $T_{1/2} = 0.1$ ns. The weighted average value for the g - factor is g = + (0.665 ± 0.013). No time - dependent attenuation of the anisotropy was found within 180 ns. The anisotropy was found to have the theoretically expected value with an error of about 10% due to uncertain corrections for the finite time resolution of our equipment.

If we assume a $[(g\frac{7}{2}), (d\frac{3}{2})]\ 5^+$ configuration for the first excited state of ^{134}Cs, we can calculate the g - factor by using the experimental g - factors of the $\frac{7}{2}^+$ ground state of ^{133}Cs for the odd proton and of the $\frac{3}{2}^+$ ground state of ^{135}Ba for the odd neutron. The result is g = + 0.683. A similar calculation for the 5^+ ground state of ^{136}Cs using the adjacent nuclear states gives g = + 0.733, to be compared with the experimental value[6]) of g = + 0.736(8). The agreement is excellent in both cases. For the 4^+ ground state of ^{134}Cs the theoretical and experimental g - factors[7]) are g = + 0.690 and g = + 0.748, respectively. In this case a possible admixture of a $d_{5/2}$ - configuration for the proton state (about 9%), which is expected from the shell model, can account for the difference.

References

1. J.P. Elliot and A.M. Lane, Handbuch der Physik, edited by S. Flügge (Springer-Verlag, Berlin 1957) Vol. 39.

2. M.H. Brennan and A.M. Bernstein, Phys. Rev. 120 (1960) 927.

3. L.W. Nordheim, Phys. Rev. 23 (1951) 322.

4. A.W. Sunyar, J.W. Mihelich and M. Goldhaber, Phys. Rev. 96 (1954) 570.
5. J. Christiansen, E. Recknagel, D. Riegel and G. Weyer, unpublished.
6. O.B. Daboussi, M.H. Prior and H.A. Shugart, Bull. Am. Phys. Soc. 11 (1966) 775.
7. H.H. Stroke, V. Jaccarino, D.S. Edmonds, Jr. and R. Weiss, Phys. Rev. 105 (1957) 590.

Contributed Paper

ANOMALOUS ORBITAL MAGNETISM OF THE PROTON DEDUCED FROM THE MAGNETIC MOMENT OF THE 11^- STATE OF ^{210}Po

T. Yamazaki and T. Nomura
Department of Physics, University of Tokyo, Bunkyoku, Tokyo, Japan, and the Institute of Physical and Chemical Research, Yamatomachi, Saitamaken, Japan

S. Nagamiya* and T. Katou
The Institute of Physical and Chemical Research, Yamatomachi, Saitamaken, Japan

It is well known that the magnetic moment of the $h_{9/2}$ state in ^{209}Bi deviates largely from the Schmidt value. Theoretical calculations based on the magnetic core polarization mechanism[1,2]) could account for only half of its deviation and the remainder has been left as a question.

A measurement of the g - factor of the $[h^2_{9/2}]8^+$ state of ^{210}Po was recently reported by Yamazaki et al.[3]); the result turned out to be equal to that of the $h_{9/2}$ state of ^{209}Bi within an accuracy of one percent. From this agreement they demonstrated that the $[h^1_{11/2}\ h_{9/2}]\ 1^+$ type spin polarization is only a partial contributor to the large deviation of the ^{209}Bi magnetic moment from the Schmidt value. Here our interest is to search for the possible origins of this anomalous deviation.

One question is as to whether the proton g_l - factor is 1 or not. In order to investigate this point we tried to measure the g - factor of the $[h_{9/2}i_{13/2}]\ 11^-$ state of ^{210}Po. In this case the spin contributions are nearly cancelled out so that the g - factor should be nearly equal to g_l.

A metallic ^{208}Pb target of 30 mg/cm^2 thickness was bombarded with 30 MeV α - particles from the IPCR Cyclotron to populate the 24 nsec 11^- state of ^{210}Po [4]) in the reaction ^{208}Pb $(\alpha, 2n)$ ^{210}Po. A relevant level scheme is shown in fig. 1. A 30 cc Ge (Li) detector was placed at 135° to the beam direction to measure the time - differential angular distribution of delayed γ - rays between the natural beam bursts

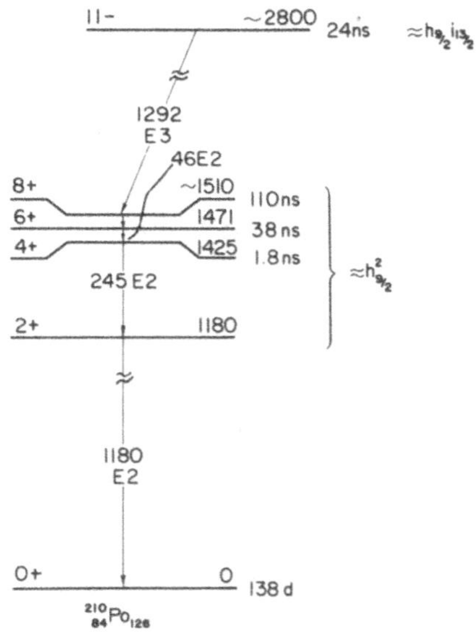

Fig. 1

of 120.0 nsec interval. An external magnetic field of 19.05 kG was applied up and down perpendicularly to the beam - detector plane. Energy and time spectra were taken by a 4096 - channel two - parameter PHA (256 time channels X 16 energy channels). The energy channels covered both the 1180 keV ($2^+ \to 0^+$) and the 1292 keV ($11^- \to 8^+$) γ - rays. Time - differential patterns $R(t) = \frac{(N^\uparrow - N^\downarrow)}{(N^\uparrow + N^\downarrow)}$ are shown in fig. 2. (Counts $N^\uparrow$ and $N^\downarrow$ correspond to the external magnetic field up and down, respectively). From the figure we obtain

$$g(11^-) = 1.090 \pm 0.016 \tag{1}$$

$$g(8^+) = 0.891 \pm 0.014 \tag{2}$$

after the Knight shift correction of 1.47 percent. The value (2) is determined predominantly by the Larmor frequency at the 8^+ state with a small correction due to partial feeding from the 11^- state. This value is compatible with the previous measurement[3]) within the experimental error.

The wave - function of the 11^- state of ^{210}Po is expressed as

$$|11^-) = \sqrt{1-a^2}\, |h_{9/2}\, i_{13/2};\ 11^-) + a|[h^2_{9/2}]8^+, (\hbar\omega_3)\, 3^-; 11^-) \tag{3}$$

The squared amplitude a^2 was estimated to be 0.05 from the experimental

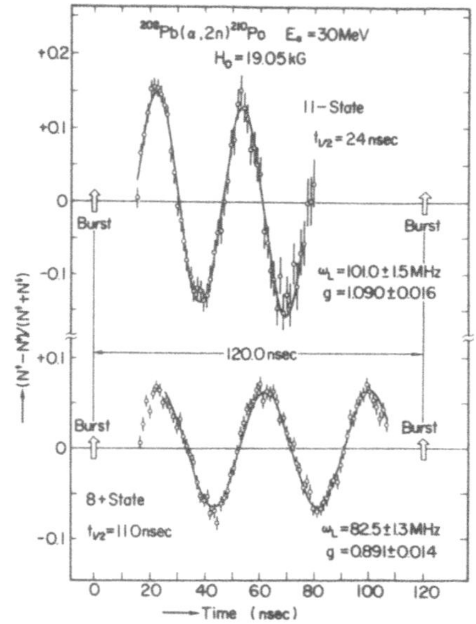

Fig. 2

B (E3; $11^- \to 8^+$) (ref. 4), thus the magnetic moment due to the second term can be estimated with the use of the g - factors of the 8^+ state of ^{210}Po (ref. 3) and of the 3^- state of ^{208}Pb (ref. 5). Hence we have for the first term

$$g\,([h_{9/2}\ i_{13/2}]\ 11^-) = 1.104 \pm 0.017 \tag{4}$$

From this expression we can deduce the g - factor of the $i_{13/2}$ state by using the known value of $g\,(h_{9/2}) = 0.9067$ as

$$g\,(i_{13/2}) = 1.241 \pm 0.029 \tag{5}$$

We show in fig. 3 the two lines corresponding to the experimental values $g\,(h_{9/2})$ and $g\,(i_{13/2})$ under the assumption that the single - particle moment is renormalized as

$$g\,(jl) = g_l^{eff} \pm \frac{1}{2l+1}\,(g_s^{eff} - g_l^{eff}), \quad \text{for } j = l \pm 1/2 \tag{6}$$

If we assume that g_s^{eff} as well as g_l^{eff} is the same for the two orbitals $h_{9/2}$ and $i_{13/2}$, then the cross point in fig. 3 determines a unique set of $g^{eff} = 1.09 \pm 0.02$ and $g_s^{eff} = 3.1 \pm 0.2$. Under this assumption the g - factor of the $[h_{9/2}\ i_{13/2}]\ 11^-$ state be-

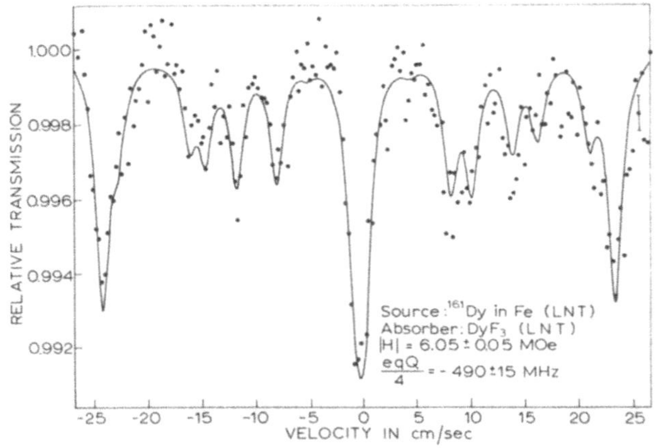

Fig. 3. A: g_s^{eff} $(i_{13/2})$ by Arima and Horie. C: g_s^{eff} $(h_{9/2})$ by Blomqvist et al.
B: g_s^{eff} $(h_{9/2})$

comes nearly equal to g_l^{eff}, since

$$g\,(j_1 = l_1 + 1/2, j_2 = l_2 - 1/2; J = j_1 + j_2) = g_l^{eff} + \frac{1}{J(2l_2+1)}(g_s^{eff} - g_l^{eff}) \cong g_l^{eff} \quad (7)$$

The orbital independence of g_s^{eff} is theoretically supported by the following facts:

(1) As far as the spin polarization of the core is concerned, the two - body interaction of the δ - function type leads to g_s^{eff}'s nearly independent of orbitals as shown by Arima and Horie[1]). Calculated values within the framework of Arima and Horie become to be g_s^{eff} $(h_{9/2}) = 3.62$ and g_s^{eff} $(i_{13/2}) = 3.49$. With the realistic two - body interaction we may expect that g_s^{eff}'s are almost independent of orbitals, since a refined calculation of Blomqvist et al.[2]) gave almost the same result for the $h_{9/2}$ orbital as that of Arima and Horie.

(2) Corrections caused by other effects such as two - body spin - orbit interaction or the mesonic effect are nearly independent of orbitals.

An experimental support to this conjecture is that the g - factor of the $s_{1/2}^{-1}$ state of ^{205}Tl, which should be g_s^{eff} itself, is very close to the cross point as shown in fig. 3.

The present experiment has thus shown clearly an anomalous value, 1.09 ± 0.02 for g_l^{eff}. The shell model calculation of the core polarization accounted for about half the deviation of the ^{209}Bi magnetic moment from the Schmidt value. Based on the present analysis the remainder is ascribed to the anomalous g_l^{eff} - factor.

The mesonic effect is believed to be the main source of such an anomaly. A recent calculation of Chemtob[6]) gave $\delta g_l = 0.10$ for odd protons in heavy nuclei. Miyazawa[7]) also calculated this quantity and found $\delta g_l \cong 0.10$ from the Fermi gas model. Bhaduri and Van Leuven[8]) derived $\delta g_l = 0.114$ by assuming a momentum - dependent internucleon force. These values are in good agreement with the present observation.

A characteristic feature of such effects is that δg_l is an iso - vector so that one expects $g_l^{eff} \cong -0.1$ for neutrons. Because of lack of more experimental data of good precision in the closed shell region this point is open to further investigation.

* On leave from Department of Physics, Osaka University.

References

1. A. Arima and H. Horie, Prog. Theor. Phys. 12 (1954) 623.
2. J. Blomqvist, N. Freed and H.-O. Zetterström, Phys. Letters 18 (1965) 47.
3. T. Yamazaki, T. Nomura, T. Katou, T. Inamura, A. Hashizume and Y. Tendou, Phys. Rev. Letters 24 (1970) 317.
4. T. Yamazaki, Phys. Rev. C 1 (1970) 290.
5. J.D. Bowman, F.C. Zawislak and E.N. Kaufmann, Phys. Letters 29B (1969) 226.
6. M. Chemtob, Nucl. Phys. A123 (1969) 449.
7. H. Miyazawa, Prog. Theor. Phys. 6 (1951) 801.
8. R.K. Bhaduri and P. van Leuven, Phys. Letters 20 (1966) 182.

Contributed Paper

STROBOSCOPIC DETERMINATION OF THE MAGNETIC MOMENT OF THE 8^+ ISOMERIC STATE IN ^{208}Po

S. Nagamiya*

The Institute of Physical and Chemical Research, Yamatomachi, Saitamaken, Japan

T. Nomura and T. Yamazaki

University of Tokyo, Department of Physics, Bunkyoku, Tokyo, Japan, and the Institute of Physical and Chemical Research, Yamatomachi, Saitamaken, Japan

To investigate the effects of neutron particle - holes on the magnetic dipole spin polarization in the Pb - region, we are measuring the g - factors of the $[h^2_{9/2}\ (p)]\ 8^+$ isomeric states[1]) in neutron - deficient even - A Polonium isotopes with mass numbers A = 202 - 208. Here we report on a determination of the g - factor of the 380 ns 8^+ state in ^{208}Po and discuss the result as compared with the g - factor[2]) of the 8^+ state in ^{210}Po.

The method we employed in measuring the g - factor was the stroboscopic resonance method. This method has recently been developed[3,4]) in the field of pulsed beam γ - ray spectroscopy, and has been proven to be a very convenient method to accurately determine the g - factor of an excited state if the lifetime of that state is longer than the period of excitations with a pulsed beam. The principle of this method is simple. Since an anisotropy of γ - rays from excited aligned nuclei in a magnetic field is smeared out unless the period of a pulsed beam coincides with an integral multiple of half the Larmor period of the excited nuclei, search for a maximum of the anisotropy as a function of the strength of the applied magnetic field should determine the Larmor period.

The stroboscopic resonance occurs when

$$n \frac{\pi}{\omega_L} = T_B \qquad (1)$$

where ω_L is the Larmor frequency, T_B the period of a pulsed beam and n an arbitrary integer. If we define the half width ΔH to be the interval between the two field strengths at which the anisotropy of γ - rays is reduced to half its maximum value then

$$\frac{\Delta H}{H_0} = \frac{T_B}{n\pi\tau} \tag{2}$$

where τ is the mean life of excited nuclei and H_0 the resonant magnetic field given by $g\mu_N H_0 = n\hbar(\pi/T_B)$. The accuracy of the determination of the g - factor can be estimated from this expression.

A 50 mg/cm^2 thick metallic foil of ^{206}Pb was bombarded with 29 MeV α - particles from the IPCR Cyclotron. The time interval of the natural beam bursts of the cyclotron at this energy was 123.43 ns, which is suitably shorter than the mean life ($\cong$ 550 ns) of the isomeric state of interest. The γ - rays following the ^{206}Pb (α, 2n) ^{208}Po reaction were detected with two cylindrical 4.5 cm X 4.5 cm NaI (Tl) counters placed at 60° and 150° with respect to the beam direction. Energy spectra of delayed γ - rays exhibited two remarkable peaks corresponding to 176 keV and unresolved 660 and 685 keV transitions. A relevant level scheme is shown in fig. 1. We used these two peaks to analyze the perturbing effect of the external magnetic field on the 8$^+$ isomeric state. In order to observe the anisotropy of γ - rays as a function of the strength of the applied magnetic field, time spectra of energetically selected γ - rays and the field strength were taken with a 4096 channel two - parameter PHA (64 time channels X 64 field channels). As a measure of the anisotropy of γ - rays we used a quantity like (A_1/B_1) / (A_2/B_2) which is schematically illustrated in fig. 2. (At the resonant field H_0, counts A_1 and B_2 have maximum values and A_2 and B_1 have minimum values). By sweeping the external field we searched for the resonant magnetic field at which the Larmor period of the 8$^+$ state

Fig. 1

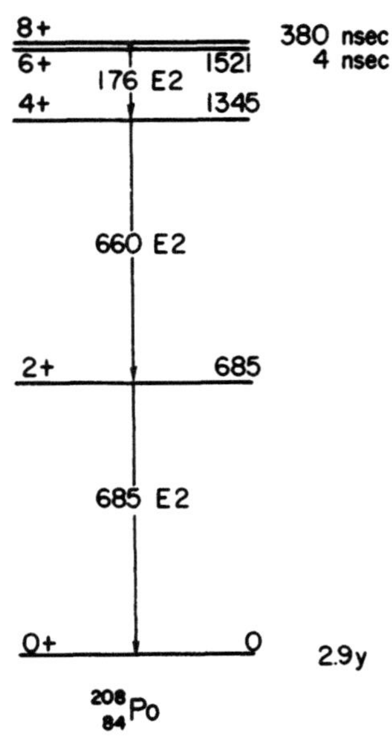

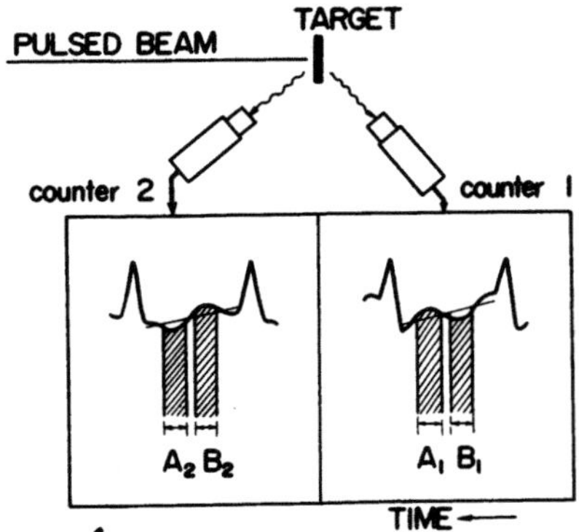

fig. 2

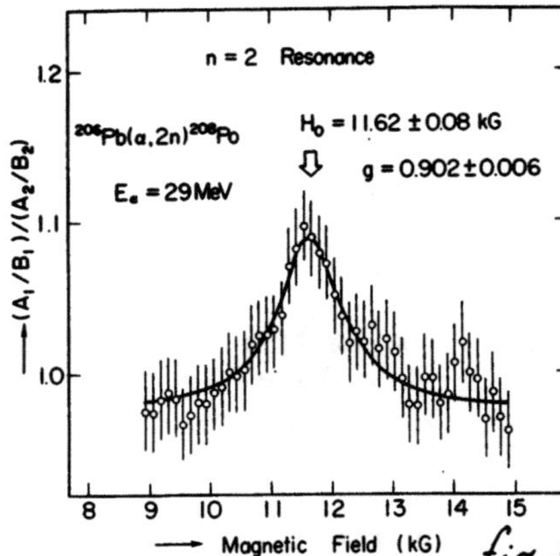

fig. 3

coincides with the period of the beam from the cyclotron.

Fig. 3 shows the n = 2 stroboscopic resonance from which we obtain

$$H_0 = 11.62 \pm 0.08 \text{ kG} \tag{3}$$

Since the period of a pulsed beam was 123.43 ns, the resonant Larmor frequency becomes from eq. (1) to be ω_L = 50.905 MHz. Thus we have

$$g\,(8^+) = 0.902 \pm 0.006 \tag{4}$$

after the Knight shift correction of 1.47 per cent[5]).

It was pointed by Treytl, Hyde and Yamazaki[6]) that the 8^+ state of ^{208}Po is described as a two - particle state weakly coupled to a surface vibration of the ^{206}Pb core. The correction for the g - factor due to this collective motion can be evaluated (ref. 7) by the use of the coupling constant $k\sqrt{\frac{\hbar\omega}{2C}} \sim 0.6$ MeV as adopted by Yamazaki[1]) and of the collective g - factor $g_R \sim Z/A \sim 0.4$. Hence we have

$$g\,([h^2_{9/2}\,(p)\,p^{-2}_{1/2}\,(n)]\,8^+) = 0.905 \pm 0.006 \tag{5}$$

The g - factors of protons in the $h_{9/2}$ orbitals are listed in table 1. The following points are to be noted.

Table 1

nucleus	^{209}Bi	^{210}Po	^{208}Po
configuration	$h_{9/2}$ (p)	$h^2_{9/2}$ (p)	$h^2_{9/2}$ (p) $p^{-2}_{1/2}$ (n)
g - factor	0.9067	0.911 ± 0.010	0.905 ± 0.006

(1) The shell model additivity for the magnetic moments, that is $g(h^2_{9/2}(p)) = g(h_{9/2}(p))$, holds well between ^{208}Po and ^{209}Bi as well as between ^{210}Po and ^{209}Bi.

(2) Concerning the M1 - spin polarization due to neutron particle - holes, the fact (1) also implies that the contribution of the $[p^{-1}_{3/2}(n)\, p_{1/2}(n)]\, 1^+$ type polarization to the magnetic moment is negligibly small. This is supported theoretically, since the deviation δg due to the $[p^{-1}_{3/2}\, p_{1/2}]\, 1^+$ type spin polarization is estimated to be

$$\delta g_{calc} \sim 0.01$$

under the assumption of the δ - function type interaction applied by Arima and Horie[8]).

* On leave from Department of Physics, Osaka University.

References

1. T. Yamazaki, Phys. Rev. C 1 (1970) 290. For ^{202}Po we have recently found the existence of such a $[h^2_{9/2}]\, 8^+$ isomeric state in the reaction Pt (^{12}C, 4n).
2. T. Yamazaki, T. Nomura, T. Katou, T. Inamura, A. Hashizume and Y. Tendou, Phys. Rev. Letters 24 (1970) 317.
3. J. Christiansen, H.-E. Mahnke, E. Recknagel, D. Riegel, G. Weyer and W. Witthuhn, Phys. Rev. Letters 21 (1968) 554;
Phys. Rev. C 1 (1970) 613.
4. S. Nagamiya and K. Sugimoto, Osaka University Laboratory of Nuclear Studies Report OULNS 69 - 3 (1969).
5. L.E. Drain, Met. Rev. 12 (1967) 195.
6. W.J. Treytl, E.K. Hyde and T. Yamazaki, Nucl. Phys. A117 (1968) 481.
7. A. Bohr and B.R. Mottelson, Met. Fys. Medd. Dan. Vid. Selsk. 27 (1953) No. 16.
8. A. Arima and II. Horie, Prog. Theor. Phys. 12 (1954) 623.

Contributed Paper

MAGNETIC MOMENT OF THE $(g_{9/2}{}^2)$ 8^+ STATE IN ^{90}Zr

T. Katou and S. Nagamiya*

The Institute of Physical and Chemical Research, Yamato - machi, Saitamaken, Japan

T. Nomura and T. Yamazaki

Department of Physics, University of Tokyo, Bunkyo - ku, Tokyo, Japan and The Institute of Physical and Chemical Research

Near the doubly closed ^{88}Sr nucleus, a number of high spin isomeric states, with lifetimes ranging from microseconds to nanoseconds such as the $(g_{9/2}\,(p)^2)$ 8^+ states in ^{90}Zr [1,4]), ^{92}Mo [1,2]) and ^{94}Ru [2]) or the $(g_{9/2}\,(n)^{-2})$ 8^+ states[3]) in ^{86}Sr, ^{88}Zr and ^{90}Mo have been identified. These states are suitable to investigate the magnetic core polarization in this region, since they can be explained by almost pure shell model considerations and since their lifetimes are suited for the conditions of the pulsed - beam spin - precession experiment in the natural beam bursts of a cyclo-

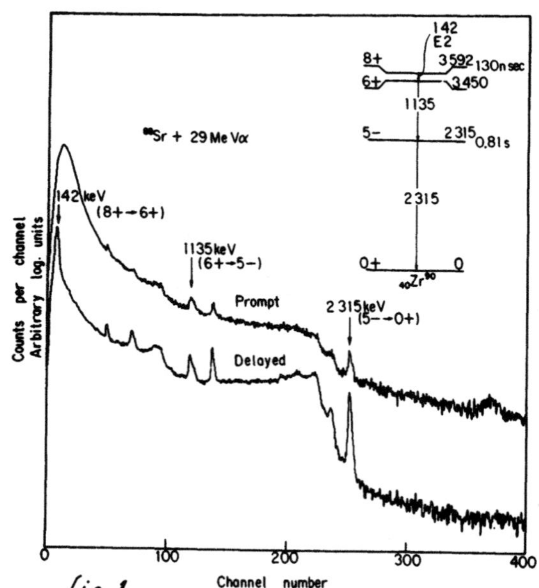

fig. 1.

tron.

Here, we report on a g - factor measurement of the $(g_{9/2}\ (p)^2)\ 8^+$ state (half-life 130 ns) in ^{90}Zr and discuss the magnetic core polarization.

A metallic ^{88}Sr target of 260 mg/cm^2 thickness was bombarded by 29 MeV α - particles from the IPCR cyclotron to populate the 8^+ state in ^{90}Zr by the $^{88}Sr\ (\alpha, 2n)\ ^{90}Zr$ reaction. Prompt and delayed gamma - ray spectra taken with a Ge (Li) - detector are shown in fig. 1. Since no peak was observed near the 142 keV peak, the 142 keV gamma - rays were detected with a 4.5 cm ∅ X 4.5 cm NaI (Tl) counter placed at 150 degrees with respect to the beam direction. The time distribution of the gamma - rays was measured between the natural beam bursts of 123.4 nsec interval. An external 11.96 kG (± 0.2%) magnetic field was applied perpendicularly to the beam - detector plane. The experimental procedures were similar to those described in ref. 5 and 6.

To check the linearity of the TAC (Time - to - Amplitude Converter), we took the time spectra of 3 prompt gamma - ray peaks. The non - linearity of the TAC was less than 1%.

The time differential pattern $R(t) = (N^{\uparrow}(t) - N^{\downarrow}(t)) / (N^{\uparrow}(t) + N^{\downarrow}(t))$ is shown in fig. 2 (the arrows indicate field 'up' or 'down'). The time differential pattern R (t) was fitted by

$$R(t) = 2B_2 \sin 2\theta \sin 2(\omega_L t + \theta') + 2B_4 \sin 4\theta \sin 4(\omega_L t + \theta')$$

where θ is the observation angle, θ' is the beam deflection angle due to the applied magnetic field and ω_L is the Larmor frequency. From the least - square fit, the Larmor frequency $\omega_L = 78.7 \pm 0.8$ MHz was obtained.

Since no data of Knight shift of ^{90}Zr in the Sr metal are available, we assume

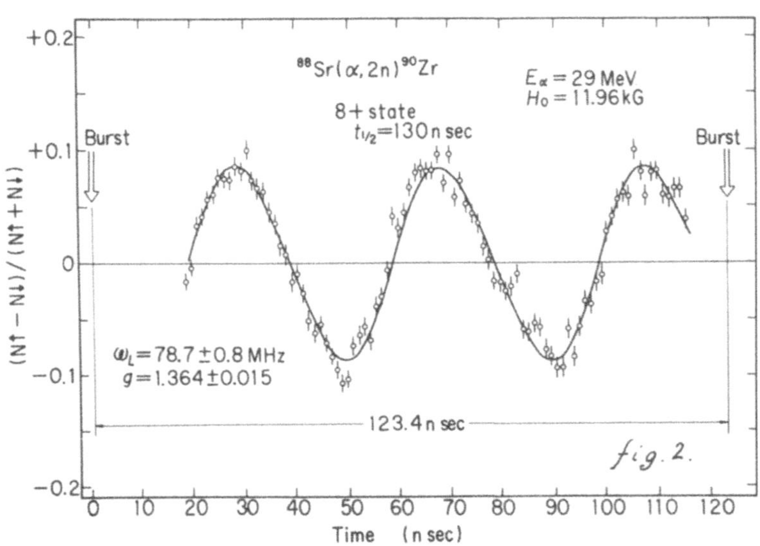

fig. 2.

the Knight shift to be 0.7% (deducted from the data of ^{93}Nb (0.875%), ^{95}Mo (0.582%) and ^{97}Rb (0.653%)). After the correction for the Knight shift, we obtain $g(8^+) = 1.364 \pm 0.015$. Yamazaki et al.[5]) reported that shell model additivity for one - and two - particle states, i.e., $g(|j^2, J\rangle) = g(|j\rangle)$, holds for $j = h_{9/2}$. If such an additivity holds for $j = g_{9/2}$, the g - factor of the 8^+ state in ^{90}Zr becomes

$$g(|g_{9/2}{}^2, 8^+\rangle) = g(|g_{9/2}\rangle)$$

The single particle estimate, g_{sp}, with the free proton g_s - factor of 5.5856, becomes 1.510. Then $\Delta g_{exp} = g_{exp} - g_{sp} = -0.150 \pm 0.015$. A theoretical interpretation of the deviation from the Schmidt value was given by Arima and Hórie[8]), who took into account the particle ($j = l - 1/2$) – hole ($j = l + 1/2$) excitation. According to their calculation, the deviation Δg_{th} due to the proton $(2P_{3/2}{}^{-1}, 2P_{1/2})\ 1^+$ excitation and the neutron $(g_{9/2}{}^{-1}, g_{7/2})\ 1^+$ excitation becomes $\Delta g_{th} = \Delta g_{proton} + \Delta g_{neutron} = -0.140 - 0.058 = -0.198$. Comparison of Δg_{exp} and Δg_{th} suggests that the deviation from the Schmidt value can be attributed to the core polarization. The small deviation of Δg_{th} is possibly caused by the overestimation of the mixing coefficient of core polarization. The coefficient is estimated by employing a δ - function type interaction.

Further investigation of g - factors in ^{92}Mo and ^{94}Ru will give us information about the effects of proton and neutron core polarization.

We would like to thank Profs. H. Kumagai and K. Matsuda for their interest and encouragement. We are indebted to Prof. A. Arima for valuable discussions. The help of the cyclotron crew of the Institute of Physical and Chemical Research is gratefully acknowledged.

* On leave of absence from Department of Physics, Osaka University.

References

1. K.E.G. Löbner, Nucl. Phys. 58 (1964) 49.
2. J.M. Jaklevic, C.M. Lederer and J.M. Hollander, Phys. Lett. 29B (1969) 179.
3. M. Ishihara, private communication.
4. N. Auerbach and I. Talmi, Nucl. Phys. 64 (1965) 458.
5. T. Yamazaki, T. Nomura, T. Katou, T. Inamura, A. Hashizume and Y. Tendou, Phys. Rev. Letters 24 (1970) 317.
6. T. Yamazaki and E. Matthias, Phys. Rev. 175 (1968) 1476.
7. See the compilation by L.E. Drain, Met. Rev. 12 (1967) 195.
8. A. Arima and H. Horie, Progr. Theoret. Phys. (Kyoto) 12 (1954) 623.

Contributed Paper

MAGNETIC MOMENTS AND LIFETIMES OF A (3^+) - 273 keV LEVEL IN ^{74}As AND THE 1^+ - 45 keV LEVEL IN ^{76}As

M. Becker, H. Bertschat, H.-E. Mahnke, E. Recknagel, D. Riegel,

R. Sielemann, B. Spellmeyer and T. Wichert

Hahn - Meitner - Institut für Kernforschung Berlin, Sektor Kernphysik, 1 Berlin 39, W. Germany

J. Christiansen and W. Witthuhn

Universität Erlangen, Physikalisches Institut, W. Germany

As a part of a systematic search for isomeric states in arsenic isotopes, populated by (p, n) or (d, n) reactions from stable germanium isotopes, a new level at E = 273 keV with a half - life of 27 ns in ^{74}As and the first excited state at E = = 45 keV with $T_{1/2}$ = 2.6 μs in ^{76}As were investigated and the magnetic moments of both levels were determined.

The experiments were carried out with the pulsed particle beam of the 5.5 MeV Van - de - Graaff accelerator of the Hahn - Meitner - Institut, Berlin, having a pulse width of ΔT = 10 ns and a repetition time T = 1 μs. These are excellent conditions for application of the method of time - differential observation of the nuclear Larmor precession (PBDDCA)[1]) on nuclear levels with lifetimes in the nanosecond - region. The time - dependent intensity distribution as a function of an external magnetic field is given by

$$I(t, B, \theta) \sim e^{-t/\tau} \{ 1 + \sum_{\substack{k \\ \text{even}}} A_K(1)\, A_K(2)\, P_K(\cos(\theta - \omega_L t)) \} \qquad (1)$$

with $\omega_L = - g\frac{N}{h} B$. In contrast to normal angular correlation measurements, the coefficients $A_K(1)$ describe the alignment of the nuclei by the nuclear reaction and are usually unknown. For isomeric states with lifetimes $\tau > 1$ μs, the application of the stroboscopic observation of the nuclear Larmor precession[2,3]) is advantageous. This method is in principle a resonant method in which one compares the Larmor frequency with the pulse frequency. With $k_{max} = 2$ and a proper choice of the observation time - window t_0, the intensity distribution is a Lorentzian:

$$I(t_0,\tau,T_0,B,\theta) \sim 1 + \frac{3/4A_2(1)A_2(2)}{1+1/4A_2(1)A_2(2)} \cdot \frac{(1/\tau)^2}{(1/\tau)^2 - 4(\omega-\omega_{Lo})^2} \qquad (2)$$

The resonance frequency is given by $\omega_{Lo} = n\pi/T_0$. The line - width of the curve is the natural one, $\Delta\omega = 1/\tau$.

The 273 keV - (3^+) - level in ^{74}As

The level diagram of ^{74}As was almost unknown, and detailed investigations following the reaction ^{74}Ge (p, n) ^{74}As were necessary to establish the fact that the delayed γ - radiations of 77 keV and 183 keV belong to the spectrum of ^{74}As.

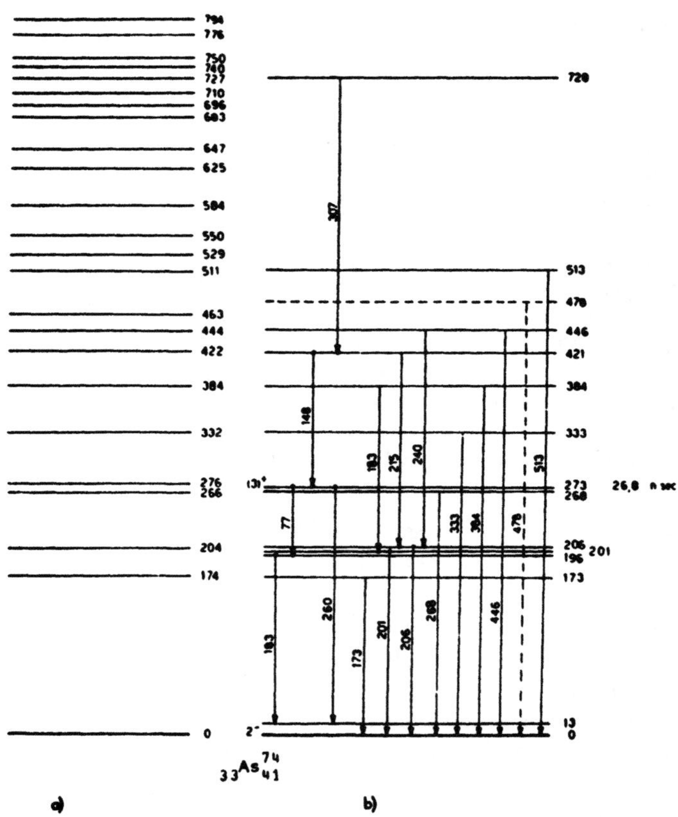

Fig. 1. Level scheme of ^{74}As.
a) Neutron time - of - flight measurements (Finckh et al.)
b) γ - transitions following the reaction ^{74}Ge (p, n) ^{74}As.

Yield and coincidence measurements result in the level scheme shown in fig. 1b. The 273 keV isomeric state with a half - life of $T_{1/2} = 26.8 \pm 0.6$ ns decays by a γ - cascade of 77 keV - 183 keV - (13 keV) to the ground state. Many states found by these measurements could be confirmed by neutron time - of - flight measurements performed by Finckh et al. (fig. 1a) [4]). In fig. 2a and 2b the normalized PBDDCA - pattern of the two delayed transitions are shown; a target of metallic Germanium was used. The Larmor frequencies of both transitions agree within the experimental errors. Using the known magnetic moment of the 198 keV - $5/2^-$ -

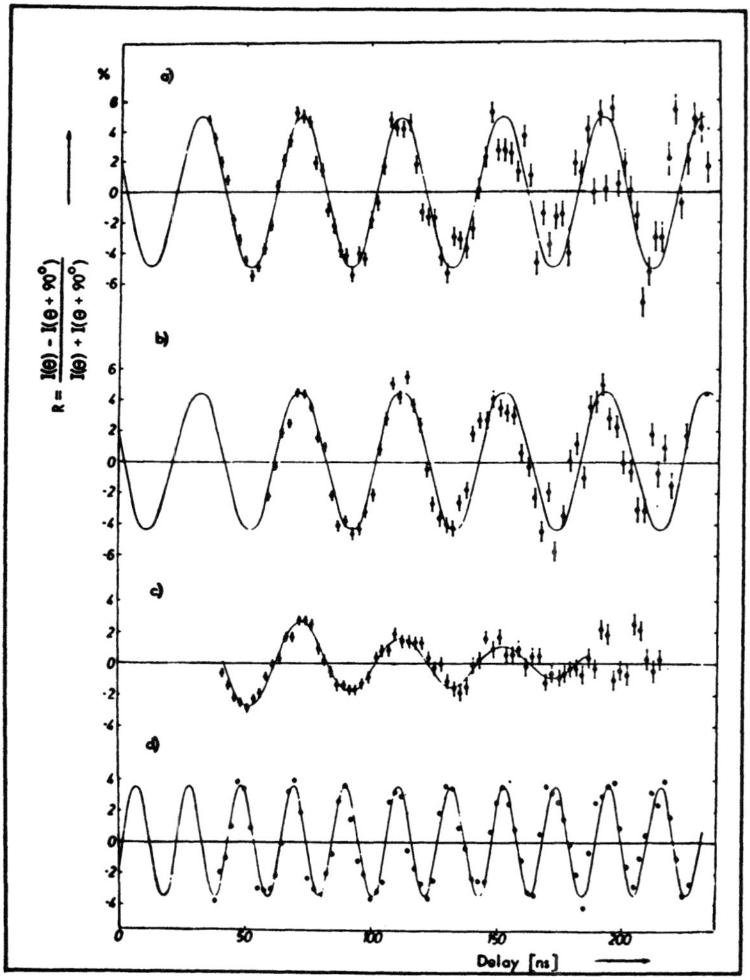

Fig. 2. ^{74}As: Spin precession pattern.
a) 183 keV and b) 77 keV γ - transition, natural Ge - target,
c) $^{74}Ge\,O_2$ - target, 77 keV transition,
d) Magnetic field calibration; $g = +1.445$; ^{19}F; $E = 198$ keV.

level of ^{19}F to calibrate the external magnetic field, the g - factor of the 273 keV level could be determined as g = + 0.807 ± 0.010.

A comparison of this g - factor with the theoretical expected values from shell - model calculations suggests a nucleon configuration for the 273 keV level of the (0, 0) nucleus ^{74}As of $[\pi\,(f_{5/2})^1; \nu\,(p_{1/2})^1\,(g_{9/2})^2\,]_{3^+}$ or $[\pi\,(p_{3/2})^{-1}; \nu\,(f_{5/2})^{-1}\,]_{3^+,\,4^+}$. The fact, that the level could be observed in the neutron time - of - flight spectrum, favours the lower spin assignment, but neighbouring spin values could not be definitely excluded.

The spin precession measurement in the non - cubic $^{74}GeO_2$ shows a damped amplitude (fig. 2c). Because of the short lifetime of the 273 keV level, no statement on an expected quadrupole interaction could be made.

The 45 keV - 1^+ - level in ^{76}As

The first excited state of ^{76}As was populated by the reaction ^{76}Ge (p, n) ^{76}As. A metallic target of enriched ^{76}Ge was used. From the delayed time spectrum the half - life of the 45 keV level was found to be in agreement with measurements of Tubbs[5]) as $T_{1/2}$ = 2.60 ± 0.05 μs.

The method of the stroboscopic observation of the nuclear Larmor pecression was applied to determine the g - factor. In fig. 3 a resonance curve is shown. From the resonance magnetic field B_0 = 1175 ± 2 Gauss, the g - factor is deduced as g = = + 0.558 ± 0.001.

The solid curve is a least squares fit using eq. (2). The line width of the curve is slightly broader than the expected natural line width, indicating an additional relaxation. According to estimations by Johnson and Kenneth[6]) the 45 keV transition is

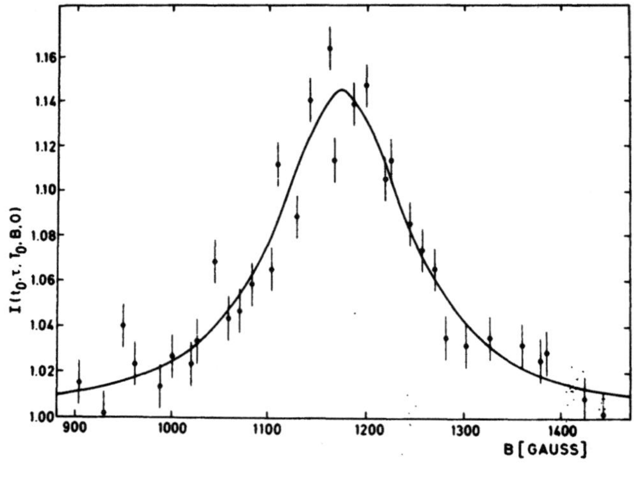

Fig. 3. Stroboscopic resonance of the perturbed angular correlation of the 45 keV γ - rays. B_0 = 1175 ± 2 Gauss. The solid line is a least - square fit.

E1, leading to I^{π} values of 1^+, 2^+, 3^+. A decision in favour of one of these values from a comparison of the measured g - factor with shell - model calculations, using the possible proton - neutron configurations $[\pi (f_{5/2})^1 ; \nu (p_{3/2})^{-1}]$ or $[\pi (p_{3/2})^{-1} ; \nu (p_{3/2})^{-1}]$, remains ambiguous because all calculated g - values are compatible with the measured one. But investigations of the neutron decay of analog states in ^{76}Se by Finckh et al.[7]) allow only 0^+ or 1^+. Therefore, 1^+ is the most probable spin - parity assignment for this state.

References

1. R.M. Freeman, Nucl. Phys. 26 (1961) 446.
2. J. Christiansen, H.-E. Mahnke, E. Recknagel, D. Riegel, G. Weyer and W. Witthuhn, Phys. Rev. Lett. 21 (1968) 554.
3. J. Christiansen, H.-E. Mahnke, E. Recknagel, D. Riegel, G. Schatz, G. Weyer and W. Witthuhn, Phys. Rev. C 1 (1970) 613.
4. E. Finckh, U. Jahnke, B. Schreiber and A. Weidinger, Nucl. Phys. A144 (1970) 67.
5. N.A. Tubbs (Ph. D. Thesis, Oxford Univ. 1966), Nuclear Data Sheets B 1 - 6 - 119 (1966).
6. L.V. Johnson and T.J. Kenneth, Nucl. Phys. A113 (1968) 104.
7. E. Finckh, U. Jahnke and B. Schreiber, Conference on Nuclear Isospin, Asilomar/Cal., Academic Press 1969.

Discussion

Günther: You mentioned, that you saw a time - dependent internal perturbation in your ^{76}As measurement. Do you have an explanation for the nature of this perturbation?

Mahnke: The time - dependent perturbation is due to some sort of radiation damage produced by the recoiling excited nuclei. We assume that near the As - nucleus after it has come to rest, there are one or two vacancies which produce fluctuating electric quadrupole interactions by diffusion. The diffusion mechanism could be confirmed by measuring the temperature dependence of the relaxation time using the 6 μs state in ^{73}As. The measured activation energy for the diffusion agrees quite well with the migration energy of a vacancy in solid germanium (Phys. Rev. Lett. 25 (1970) 102).

Günther: Can you give an estimate of the size of the field gradients resulting from the lattice damage and the relaxation time?

Mahnke: I do not know the values of the field gradients, but the value necessary to interpret the temperature dependent relaxation was quite reasonable. Within the measured temperature range from 500° C up to the melting point of germanium at 936° C, the relaxation time varies from about 100 ns up to 10 μs and up to about 100 μs for the liquid phase.

Contributed Paper

THE g - FACTOR OF THE 7/2⁻ LEVEL AT 1.611 MeV IN ^{37}Ar

D.W. Haag, W. Leitz, R. Michaelsen, W. Ribbe
Hahn - Meitner - Institut für Kernforschung Berlin, Sektor Kernphysik, Berlin - West, Germany

The mirror theorem of Sachs[1]) states that most of the corrections to the single particle operator $\underline{\mu} = g_l \underline{l} + g_s \underline{s}$ cancel out in the mean value $\overline{g} = \frac{1}{2}(g_n + g_p)$ of mirror states with $T_3 = \pm T$. Thus the experimental values for $\overline{g}$ should fit well to the respective single particle values. This simple relationship is remarkably well satisfied by the hitherto known pairs of magnetic moments[2,3]). Recent measurements were performed on the first excited pair of states[3]). The $7/2^-$ states in ^{37}Ar and ^{37}K can possibly provide for the first time a check of the abovementioned theorem for an excited pair of $f_{7/2}$ states.

The measurements of the g - factor were performed by the method of time - differential observation of the rotation of the γ - ray pattern in a magnetic field. The $7/2^-$ state in ^{37}Ar was populated through the reaction ^{37}Cl (p, n) in a solid KCl - target (enriched ^{37}Cl) by pulsed protons of 5.3 MeV (pulse width 6 ns, repetition rate 1 MHz). The γ - rays were detected in two 1.5″ X 1.5″ Ø NaI (Tl) crystals. They were analyzed with respect to energy and time delay. A magnetic field of approximately 40 kG was applied perpendicularly to the beam - detector plane. The field strength was calibrated by observing the spin precession of the $5/2^+$ state[4]) in ^{19}F. The half-life of the $7/2^-$ state in ^{37}Ar was found to be 4.6 ± 0.3 ns, which is slightly smaller than the values reported elsewhere[5]). The modulation of the time spectra (fig. 1a) was fitted by the function $R_t = A \cos 2 (\omega_L t - \phi)$. From six independent runs the mean value g (^{37}Ar, $7/2^-$, 1.611 MeV) = - 0.365 ± 0.020 is obtained.

For the measurement on the $7/2^-$ - state in ^{37}K (E_x = 1.380 MeV, $T_{1/2}$ = = 9.6 ± 1.4 ns (ref. 5)) a gaseous target of enriched ^{36}Ar was bombarded with deuterons of 5.9 MeV. The time spectrum of the 1380 keV γ - intensity was strongly perturbed by the 1611 keV γ - rays from the reaction ^{36}Ar (d, p) ^{37}Ar (fig. 1b). Although the spectra were corrected for the ^{37}Ar contribution there still may be systematic errors. A preliminary result from two independent runs is g (^{37}K, $7/2^-$, 1.380 MeV) = + 1.45 ± 0.12. For definite conclusions refined measurements are necessary.

The measured g - factors differ considerably from the single particle values $g_n = -0.547$ ($g_p = +1.655$) for a $f_{7/2}$ - neutron (proton). Their isoscalar part $\overline{g}_{exp} = \frac{1}{2}(-0.365 + 1.45) = 0.55 \pm 0.07$ however, agrees well with the predicted value for $(f_{7/2})^n$ configurations, $\overline{g}_{s.p.} = \frac{1}{2}(-0.547 + 1.665) = 0.554$. Thus for the $7/2^-$

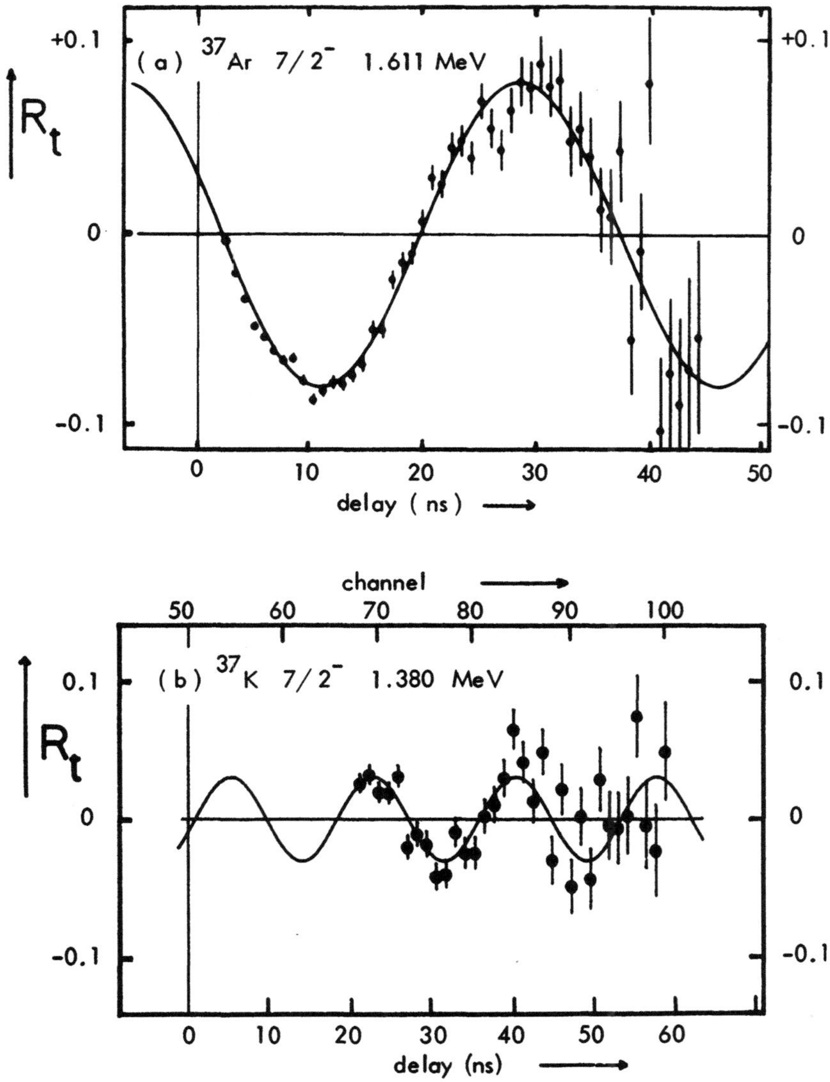

Fig. 1. Modulations of the γ-ray time spectra caused by the Larmor precession of the excited (a) ^{37}Ar and (b) ^{37}K nuclei. The solid lines are least-squares fits to the experimental points.

states in ^{37}Ar and ^{37}K the prediction of the mirror theorem proves to be correct within the present experimental accuracy.

The deviation form the Schmidt values indicate that there are admixtures of excited core configurations to the simple Schmidt configuration $^{36}Ar_{0^+} \times f_{7/2}$. This is supported by the spectroscopic factors of the transfer reactions ^{36}Ar (d, p) and ^{36}Ar (d, n) to the $7/2^-$ states, which lie between 0.6 and 0.8 (ref. 6). In a rough overall - treatment the influence of the core on the magnetic moment usually is taken into account by an effective g - factor for the single nucleon in a definite shell, which is fitted to the experimental data. This has been done by McCullen et al.[7]) for a series of states tentatively described as $^{40}Ca_{0^+} \times (f_{7/2})^n$. The results g_n^{eff} $(f_{7/2}) = -0.39$, g_p^{eff} $(f_{7/2}) = +1.50$ are close to our measured values. This leads to the conclusion that in the presence of $f_{7/2}$ nucleons both the ^{36}Ar and the doubly magic ^{40}Ca cores give similar contributions to the magnetic moment of the nucleus.

References

1. R.G. Sachs, Phys. Rev. 69 (1946) 611.
2. R. Leonardi and M. Rosa - Clot, Phys. Rev. Lett. 21 (1968) 377.
3. J.Bleck et al., Nucl. Phys. A123 (1969) 65.
4. J. Bleck et al., Nucl. Inst. and Meth. 67 (1969) 169.
5. P.M. Endt and C. Van der Leun, Nucl. Phys. A105 (1967) 1.
6. M. Hagen, U. Janetzki, K.H. Maier and H. Fuchs, Nucl. Phys., to be published; D.R. Goosman and R.W. Kavanagh, Phys. Rev. 161 (1967) 1156; B. Rosner and E.J. Schneid, Phys. Rev. 139 (1965) B66.
7. J.D. McCullen et al., Phys. Rev. 134 (1964) B515.

Contributed Paper

THE PERTURBED ANGULAR CORRELATIONS FOR THE (596 - 42) keV CASCADE IN ^{62}Cu

E. Bożek, J. Golczewski, A.Z. Hrynkiewicz, G. Polok, M. Rybicka, B. Styczeń, J. Styczeń

Institute of Nuclear Physics, Cracow, Poland

The ^{62}Cu nucleus has one proton and five neutrons outside the closed shells of Z, N = 28. The reported experimental value of the magnetic moment of the nucleus in the 1^+ ground state is in fair agreement with theoretical predictions based on the Hamada - Johnston interaction potential, with which a very strong configuration mixing was obtained[1]). Similarly, for the 2^+ ground state of ^{60}Cu the agreement between the experimental and predicted values of the magnetic moment is quite satisfactory. As the low - lying first excited 2^+ state of ^{62}Cu has the same spin and parity as the ground state of ^{60}Cu, it is interesting to measure its magnetic moment. The performed experiment includes directional correlation measurements on the (596 - 42) keV γ - γ cascade observed in the decay of ^{62}Zn.

The radioactive sample of ^{62}Zn was prepared by the reaction Ni (α, 2n) using a bombardment energy of 27 MeV. The perturbation of the angular correlation of the (596 - 42) keV γ - γ cascade was measured with Zn - sources in liquid form and metallic sources in a ferromagnetic Ni - lattice.

The liquid sources were aqueous solutions of ZnCl. For the preparation of these sources the Zn - activity was extracted from the target by the chromatographic method[2]).

The metallic sources were obtained by irradiation of Ni - foils of 0.03 mm thickness. The Ni - target should be thin so as to avoid a serious contribution of reactions other than (α, 2n). The measurements were started about 12 hours after the irradiation of the foil.

Angular correlation measurements for the liquid sources

The angular correlation of the (596 - 42) keV γ - γ cascade was measured by the differential method using two scintillation counters. The time - coincidence spectra were taken at the angles $\theta = 90^\circ$, 135°, and 157.5° and stored in different parts of the memory of the multichannel analyzer. From the coincidence numbers N (θ_i, t) the products A_2G_2 (t) and A_4G_4 (t) were calculated. Within the limits of experimental errors, the A_4G_4 (t) values were found to be equal to zero. This is consistent with the nearly pure M1 character of the 42 keV transition[3]). The values of A_2G_2 (t) were fitted by the function $A_2G_2\ (t) = A_2 \exp[-\lambda_2 t]$ and the A_2 coefficient and the λ_2 value were found to be

$$A_2 = (14.2 \pm 1.7) \times 10^{-2}$$

$$\lambda_2 = (5.11 \pm 1.83) \times 10^{7}\ s^{-1}$$

The magnetic moment measurement in an external magnetic field for the liquid sources

The g - factor of the 42 keV excited state was measured by the integral method using a Ge (Li) - NaI (Tl) coincidence set - up. The coincidence numbers N^+ and N^- were taken at one angle $\theta = 135^\circ$ and for two opposite directions of the external magnetic field H = 13.20 ± 0.15 kG. From these numbers the quantity R = $= 2\,(N^+ - N^-)/(N^+ + N^-)$ was calculated, which was found to be equal to 0.064±0.012. Hence, using the formula $R = -\,2a_2\ 2\omega\tau\ /\ [1 + (2\omega\tau)^2]$ the value of $\omega\tau_{exp}$ = $= -\ 0.56^{+0.56}_{-0.28}$ was calculated. The strikingly large error on $\omega\tau_{exp}$ could be explained by the fact that in the region of $|\omega\tau|$ close to 1, the R dependence on $\omega\tau$ is so flat that in spite of the good statistics and the accurate determination of the magnetic field, the experimental error is almost entirely due to the character of this dependence. The large value of $|\omega| > 4 \times 10^{7}\ s^{-1}$ and the life - time $\tau = 7.04$ ns made it possible to perform the differential measurement. The differential method was applied under the same experimental conditions using the NaI (Tl) - NaI (Tl) coincidence set - up. The experimental values of R (t) were fitted by the theoretical curve R_{th} (t) (see fig. 1)

$$R_{th}(t) = \frac{\int 2a_2\, P(x-t) \exp[-(\lambda+\lambda_2)x] \sin(2\omega x)\, dx}{\int P(x-t) \exp(-\lambda x)\, dx}$$

A value of $\omega = -(4.4 \pm 0.8) \times 10^7\ s^{-1}$ was obtained. It should be mentioned

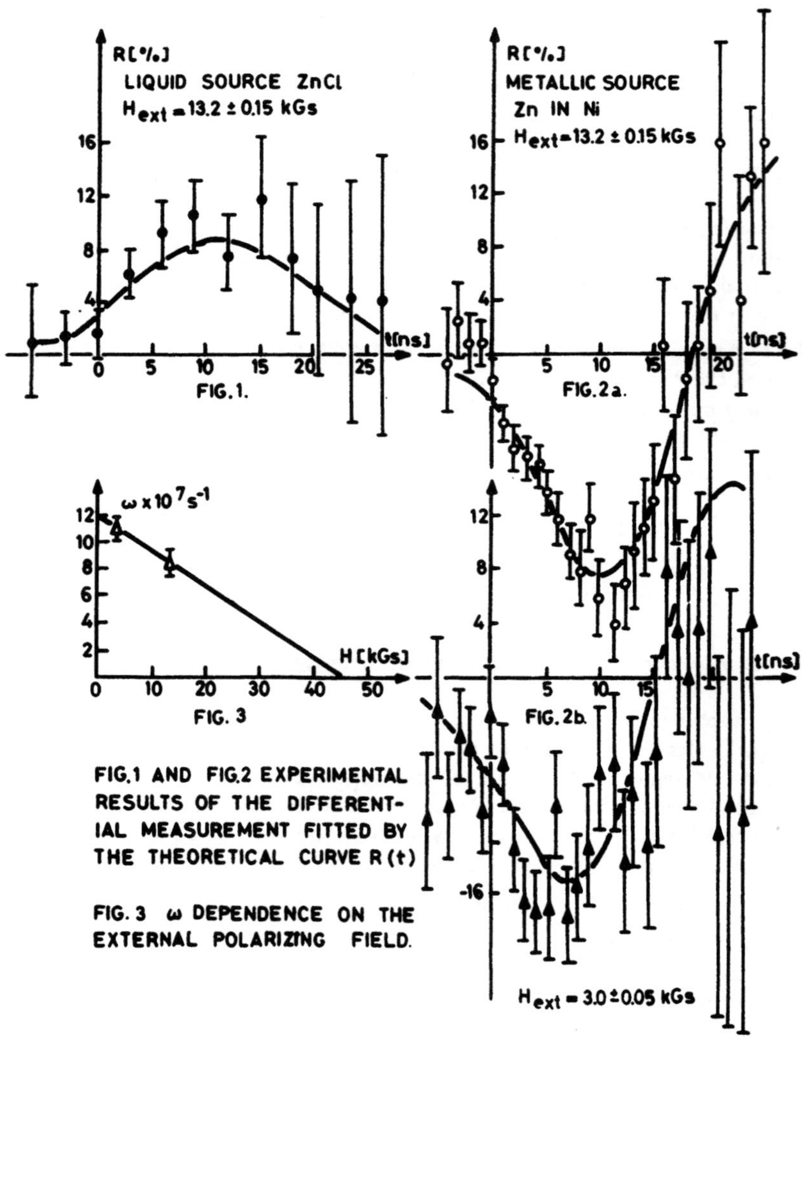

FIG. 1 AND FIG. 2 EXPERIMENTAL RESULTS OF THE DIFFERENTIAL MEASUREMENT FITTED BY THE THEORETICAL CURVE R (t)

FIG. 3 ω DEPENDENCE ON THE EXTERNAL POLARIZING FIELD.

that in this procedure λ_2 was determined from the attenuation of the angular correlation measured without any magnetic field, and the value of λ was taken from previous work[4]). P (x - t) is the experimental prompt curve. From the value of ω and the strength of the applied magnetic field H = 13.20 ± 0.15 kG, the value of the g - factor of the 42 keV state in ^{62}Cu was calculated to be g = 0.67 ± 0.07.

Measurements of hf - magnetic interaction in the Ni lattice

The differential method was applied in the measurements with the metallic ^{62}Cu (in Ni) source for two different polarizing external fields: H_1 = 13.20 kG and H_2 = 3.0 kG. The method of data handling was the same as that described above, and the corresponding values of ω were found to be

$$\omega_1 = (8.94 \pm 0.32)\ 10^7\ s^{-1};\quad \omega_2 = (11.69 \pm 0.71)\ 10^7\ s^{-1}$$

The experimental results and the fitted curves are shown in fig. 2a and 2b. The ω_1 and ω_2 values allow the direct calculation of the internal magnetic field acting on the Cu - nuclei in the Ni - lattice, and the g - factor of the 42 keV state. The results are: H_{int} = - (46.2 ± 10.4) kG and g = 0.56 ± 0.17 as deduced from the intersection point on the H - axis and from the slope of the straight line in fig. 3, respectively.

The internal magnetic field is given by the formula:

$$H_{int} = H_{ext} + H_{hfs} + 4\pi M/3 - DM$$

For the geometry of the source, used in the present experiment the value of magnetization $4\pi M/3$ was calculated to be equal to 2.03 kG and the demagnetization coefficient D was equal to zero. This leads to a value of H_{Cu} (Ni) = - (48.2 ± 10.4) kG. The NMR - value[5]) of H_{hfs} acting on Cu - nuclei in a Ni - host is – 47.2 kG. The investigation of the perturbed angular correlation with unpolarized Ni - foils is in progress.

The weighted mean value of the g - factor of the 42 keV 2^+ excited state in ^{62}Cu is g = 0.67 ± 0.06; the value of the magnetic moment for this state is then μ = = 1.34 ± 0.12 n.m. This value can be compared with the value μ = 1.212 ± 0.003 n.m. measured for the 2^+ ground state of ^{60}Cu (ref. 1). It is possible that the ground state of ^{60}Cu has a similar structure to that of the low - lying first excited state in ^{62}Cu, which probably goes up in energy in ^{64}Cu and ^{66}Cu.

To have more experimental information concerning the low - lying 2^+ states in even Cu - isotopes, measurements of the spin and g - factor of the first excited state of ^{64}Cu are being planned.

References

1. E.A. Phillips and A.D. Jackson, Jr., Phys. Rev. 169 (1968) 917.
2. M. Gasior et al., Wiad. Chem. 24 (1970) 396.
3. S. Antman et al., Nucl. Phys. A94 (1968) 289.

4. E. Bozek et al., Proceedings of International Conference on Properties of Nuclear States, Montreal 1969.

5. D.A. Shirley, 'Hyperfine Structure and Nuclear Radiation', North-Holl. Publ. Co., 1968, Appendix B.

Contributed Paper

SPECTROSCOPIC STUDY OF ^{125}Te

M. Rots, R. Silverans*, and R. Coussement
University of Leuven, Instituut voor Kern- en Stralingsfysika, Leuven, Belgium

This contribution reports preliminary results on γ - γ angular correlation measurements in the decay of ^{125}Sb. The daughter - isotope ^{125}Te has been extensively studied by nuclear orientation[1,2]) and conversion electron measurements[3,4]). Recently, information on the level structure has been obtained by (d, p) reactions[5]) and Coulomb excitation[6]). Our main purpose is to study the excited states by measuring the multipolarity of the transitions interconnecting these levels and secondly to determine the magnetic dipole moment of these states. According to the pairing plus quadrupole model[7]) the excited states should be described by quasiparticle and phonon excitations interacting with each other. A later suggestion by Kisslinger (ref. 8) introduces also states (of negative parity) arising from three quasiparticle $h_{11/2}$ configurations.

The reported experiments are in fact perturbed angular correlation measurements on a ^{125}Te source obtained by melting carrier free ^{125}Sb activity in an iron and a nickel matrix. From these measurements the anisotropy and the precession angle of the correlation pattern are deduced from a fitting procedure taking these quantities as parameters. The experimental set - up consists of one NaI (Tl) - detector, two Ge (Li) - detectors, and conventional fast - slow coincidence equipment. The strong disturbing compton backscattering makes it necessary to use Ge (Li) - detectors.

Different gates were selected in the spectrum of the NaI (Tl) - detector, while the Ge (Li) - detectors observed the whole coincidence spectrum as a function of the angle. In this way, using the implanted nickel source, the following γ - γ cascades and anisotropies were measured:

117 keV - 379 keV	$A_{22} = +\,0.100 \pm 0.002$
172 keV - 428 keV	$A_{22} = +\,0.153 \pm 0.006$
176 keV - 204 keV	$A_{22} = -\,0.522 \pm 0.017$
176 keV - 320 keV	$A_{22} = -\,0.182 \pm 0.006$
209 keV - 428 keV	$A_{22} = -\,0.338 \pm 0.017$

From these results we have identified two pure M1 - transitions (172 and 209 keV), two pure E1 - transitions (117 and 320 keV), one pure E2 - transition (379 keV) and finally one mixed transition of 204 keV ($-1.36 \leqslant \delta$ (E2/M1)$\leqslant -0.47$). The mixing amplitudes for the 176 keV and 428 keV transitions are known from ref. 1 – 4. From these correlation measurements we get the precession angle $2\omega\tau$ of the correlation patterns belonging to the 176 - 204 keV and 176 - 320 keV cascades. Therefore, the magnetic moment of the $9/2^-$ 320 keV state could be derived as $\mu =$ $= -0.66 \pm 0.09$ n.m., corresponding with a g - factor equal to $g = -0.15 \pm 0.02$. Knowing the correlation coefficients for the 172 - 428 keV and 209 - 428 keV cascades belonging to the $5/2^+$ 463 keV state, the magnetic moment of this state could be determined in a R - factor measurement by using the implanted iron source. This leads to the result $\mu = +0.79 \pm 0.13$ n.m.

Finally, from an R - factor measurement on the 117 - 379 keV cascade, belonging to the 525 keV state, a spin of $7/2^-$ can be assigned to this state; the lifetime is not yet known. Therefore only the precession angle could be determined. We found $\omega\tau = +0.107 \pm 0.009$ rad.

In comparing these results with the Kisslinger - Sorensen model[7]) and the three quasiparticle prediction of Kisslinger[8]) we can state the following. Firstly, the obtained g - factor for the $9/2^-$ state is in agreement with the approximate value of $g = 0.19$ derived from a three quasiparticle configuration. This result confirms the recent measurement of Knapek et al.[9]) and gives strong evidence for the fact that such particular states really exist in odd - A nuclei as predicted by Kisslinger[8]). Further, the existence of M1 - transitions between the 671, 643 and 636 keV states is in agreement with the believe that these states are formed mainly by coupling $s_{1/2}$ and $d_{3/2}$ quasiparticles to the 2^+ phonon excitation of the even core. Nevertheless, the magnetic moment observed for the 463 keV level indicates that the Kisslinger - Sorensen wave function for this state[1]) gives too strong an amplitude to the $d_{5/2}$ quasiparticle no - phonon component.

* Aspirant N.F.W.O.

References

1. N.J. Stone, R.B. Frankel and D.A. Shirley, Phys. Rev. 172 (1968) 1243.
2. H.R. Andrews, T.F. Knott, B. Greenebaum and F.M. Pipkin, Phys. Rev. 169 (1968) 978.
3. E.P. Mazets and Y.V. Sergeenkov, Izv. Akad.Nauk. SSSR Ser. Fiz. 30 (1966) 1185.
4. T. Inamura, J. Phys. Soc. Japan 24 (1968) 1.
5. A. Graue, J.R. Lien, S. Røyrvik, O.J. Aarøy and W.H. Moore, Nucl. Phys. A136 (1969) 513.
6. H. Sergolle, G. Albony, J.M. Lagrange, M. Pautrat, N. Poffé and J. Vanhorenbeek, Nucl. Phys. A145 (1970) 351.
7. L.S. Kisslinger and R.A. Sorensen, Rev. Mod. Phys. 35 (1963) 853.
8. L.S. Kisslinger, Nu l. Phys. 78 (1966) 341.
9. E. Knapek, R. Simon, R.S. Raghavan and H.H. Körner, Phys. Lett. 29B (1969) 581.

QUADRUPOLE INTERACTIONS AND QUADRUPOLE MOMENTS

E. Bodenstedt
Institut für Strahlen- und Kernphysik, Bonn University, W. Germany

Perturbed angular correlations are a widely used technique for the investigation of magnetic hyperfine interactions. It has been applied in the past to many cases, and magnetic dipole moments of excited states have been determined as well as magnetic hyperfine fields. The same technique is applicable to the study of electric quadrupole interactions. Although this is known since many years, the application has been restricted to a rather small number of cases up to now and only a very few electric quadrupole moments have been derived. This has a very simple explanation. In contrary to the magnetic hyperfine interaction, the highest available external electric field gradients are by many orders of magnitude too small to cause a measurable perturbation. One must therefore use internal electric field gradients in atoms, molecules, or crystals and a reliable calculation as well as a reliable calibration of these internal field gradients is today still quite difficult.

I will restrict this talk to static quadrupole interactions and I will omit statistical quadrupole perturbations as well as the perturbations caused by free hyperfine interaction. I shall give a survey of the present situation with respect to investigations by time - differential angular correlation measurements as well as by studies of γ - angular distributions after nuclear reactions.

The theory of perturbed angular correlations by static electric quadrupole interactions in single crystals or polycrystalline sources is completely developed. I refer to the well - known article by Frauenfelder and Steffen[1]). A few remarks might be helpful for the understanding of the most important phenomena. For an axially symmetric electric field gradient one gets for the matrix elements of the interaction energy the following simple expression:

$$\langle Im|W_e|Im\rangle = \frac{3m^2 - I(I+1)}{I\cdot(2I-1)}\cdot\frac{e}{4}\cdot V_{zz}\cdot Q_I =$$

$$= \{3m^2 - I\cdot(I+1)\}\cdot\hbar\cdot\omega_q$$

m is the component of the nuclear spin in the direction of the symmetry axis and Q_I is the spectroscopic electric quadrupole moment. For the special case, I = 5/2, one obtains the hyperfine splitting shown in figure 1. According to the correspondence principle of quantum mechanics the three interaction frequencies ω_1, ω_2, and ω_3 correspond to 1/2 times the Larmor precession frequencies which the nuclear spins exhibit under the influence of the electric field gradient. These Larmor precessions cause a modulation of the angular correlation function,

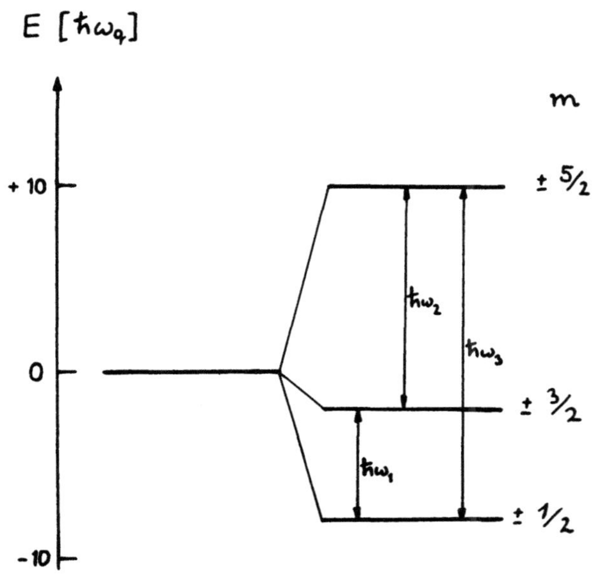

Fig. 1. Level splitting in an axially symmetric electric field gradient for the special case $I = 5/2$. The energy unit, $\hbar\omega_q$, is defined by: $\hbar\omega_q = \frac{e \cdot V_{zz} \cdot Q_I}{4 \cdot I \cdot (2I-1)}$. The interaction frequencies ω_1, ω_2, and ω_3 are the classical Larmor precession frequencies.

$$W(\theta, t) = \sum_k A_k \cdot G_k(t) \cdot P_k(\cos\theta),$$

which is expressed by the time - dependent attenuation factors

$$G_k(t) = a_k^o + \sum_{n=1}^{3} a_k^n \cdot \cos\omega_n t$$

The coefficients a_k^n can easily be calculated for single crystals or for polycrystalline material[1]).

If the electric field gradient tensor has no axial symmetry, which means

$$\eta = \frac{V_{xx} - V_{yy}}{V_{zz}} \neq 0,$$

the calculation of the eigen - values of the interaction Hamiltonian becomes some-

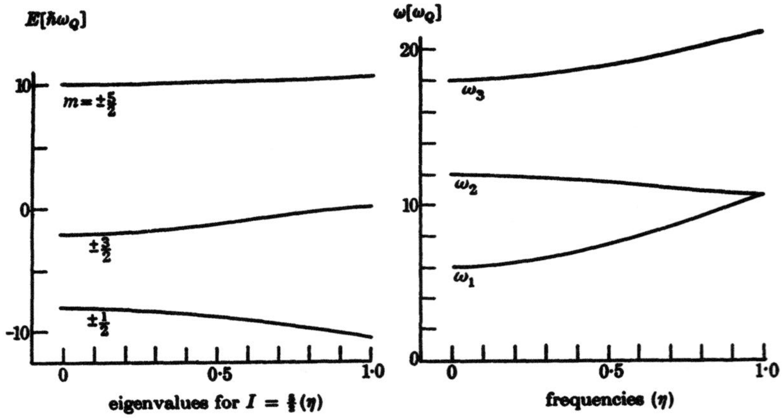

Fig. 2. Hyperfine levels and interaction frequencies vs. axial asymmetry parameter η.

what more complicated and one obtains, for instance, in the case of I = 5/2 the hyperfine terms and interaction frequencies shown in figure 2. The time - dependent attenuation factor G_2 (t) is shown in figure 3 for different values of the asymmetry parameter η. This function is periodic for $\eta = 0$, since in this case the ratio of the three interaction frequencies is equivalent to the ratio of three small integers. This diagram refers to polycrystalline material. The curves show a very characteristic dependence on η. Therefore, it is possible to derive from a time - differential angular correlation measurement the two quantities ω_q and η. In addition, a possible damping of the amplitude of a measured spin rotation curve indicates that not all nuclei of the source are sitting in identically the same field gradient. One can use an observed damping for a calculation of the spread of the interaction frequencies. The effect can be caused by imperfections in the crystalline lattice.

In order to resolve the spin rotation pattern by a time - differential angular correlation measurement, it is necessary that the period is larger than the time resolution of the coincidence circuit. On the other hand the period should not be larger than a few times the half - life of the intermediate state since this is the maximum time interval to allow a measurement. Taking into account the actual order of magnitude of nuclear quadrupole moments it is easy to estimate that one needs electric field gradients between about 10^{17} V/cm^2 and 10^{18} V/cm^2. It is remarkable that the actual field gradients at the nuclear site in crystals usually have the right order of magnitude.

With respect to the origin of the electric field gradient we have to distinguish the following cases:

I Metals

II Crystals with homopolar bonds

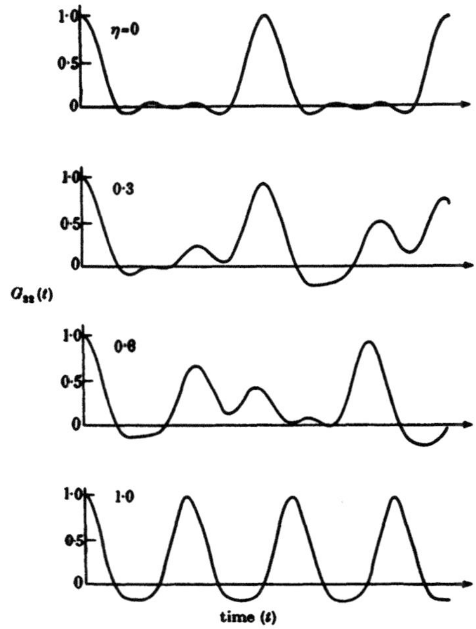

Fig. 3. Attenuation factor G_2 (t) for $I = 5/2$ for different values of the attenuation parameter η.

III a) Ion lattices with paramagnetic ions

b) Ion lattices with diamagnetic ions

IV Frozen aqueous solutions

The symmetry of the crystal plays an important role for the electric field gradient. If the surrounding of the nucleus has cubic symmetry the electric field gradient vanishes. Lower symmetry causes an electric field gradient with axial symmetry and a still lower symmetry of the lattice makes η different from 0.

Let me first review the observed quadrupole interactions in metals. Probably the first case investigated was the $I = 5/2$ state of ^{111}Cd in polycrystalline indium metal[2,3]). The time - differential measurement of the anisotropy[3]) shown in figure 4, exhibits the predicted shape for $\eta = 0$; the interaction frequency had been derived already before from integral measurements with an indium single crystal[2]). The quadrupole moment was derived by using a calculated value for the internal electric field gradient.

The electric field gradient can be calculated by using the following over - simplified model. In this model the metal consists of a lattice of metal ions sitting in a homogeneous sea of conduction electrons. One calculates the field gradient by summing over the point charges of the ions and by neglecting any contribution from the conduction elections. Sternheimer[4]) showed, however, that it is very important to take into account the fact that the quadrupole moment of the nucleus makes the electrostatic field, in which the electrons of the atomic shells move, spherically

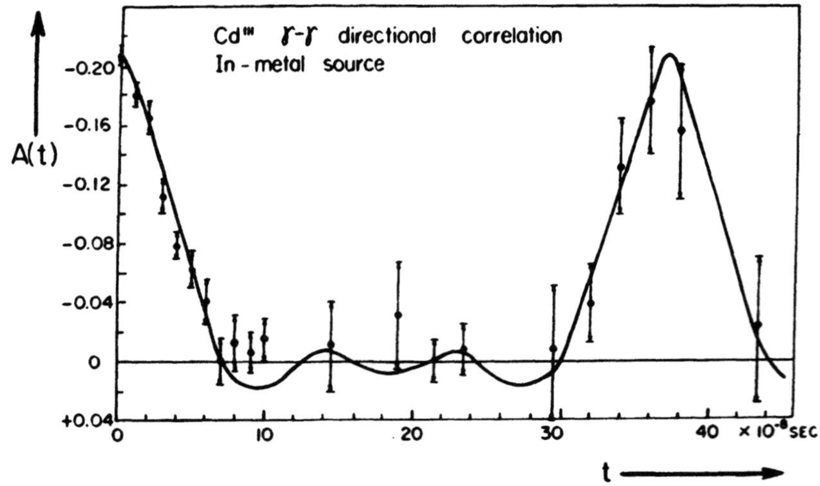

Fig. 4. Time differential measurement of the anisotropy of the 150 keV - 247 keV gamma - gamma angular correlation in ^{111}Cd measured in a polycrystalline indium metal source.

Ion	γ_∞
H^-	+1.131[c]
He	+0.424[c]
Li^+	+0.263[c]
Be^{2+}	+0.189[c]
B^+	+0.768(*H*)[c]
F^-	−22.53[e]
Na^+	−4.56[c,d]
Al^{3+}	−2.36[f]
Cl^-	−56.6[a]
K^+	−17.32[c,d]
Mn^{2+}	−11.37[d]
Fe^{3+}	−9.14[d]
Cu^+	−15.0[a]
Ga^{3+}	−9.50[d]
Br^-	−123.0[e]
Rb^+	−47.2[e]
Ag^+	−34.9[d]
I^-	−138.4[f]
Cs^+	−102.5[f]
Pr^{3+}	−80.9(*H*)[f]
Tm^{3+}	−75.3(*H*)[f]

Fig. 5. Table of calculated Sternheimer correction factors.

asymmetric. This causes a mixing of different l - values of the electron orbits which gives finally a tremendously large contribution of the electronshell to the quadrupole moment of the ion. If this induced quadrupole moment of the atomic shell is given by

$$Q_{ind} = -\gamma_\infty \cdot Q_I$$

one obtains for the electric quadrupole splitting of the hyperfine terms:

$$\hbar \cdot \omega_q = \frac{1}{I(2I-1)} \cdot \frac{e}{4} \cdot V_{zz} \cdot (1-\gamma_\infty) \cdot Q_I$$

The so-called Sternheimer correction factor γ_∞ has been calculated, by now, for most closed subshell ions. A table is shown in fig. 5 [26]). An additional contribution to the effective electric field gradient at the nuclear position stems from the conduction band. No reliable calculation does exist up to now. One has evidence from NMR studies of the electric quadrupole splitting of nuclear ground states in metal hosts, where the quadrupole moment is known from atomic beam experiments[7]) that the conduction electrons may contribute 2/3 to the effective field gradient or even more.

From this point of view the case ^{181}Ta in a Hf - metal lattice is of special interest because the electric quadrupole moment of the isomeric 5/2 state can be estimated quite reliably by using the Bohr - Mottelson formula and the intrinsic quadrupole moment derived from the coulomb excitation cross - section. Fig. 6 shows a measurement by Gerdau et al.[5]) in a polycrystalline Hf - metal host. The crystal - structure is hexagonal closed package and from the symmetry of this crystal one ex-

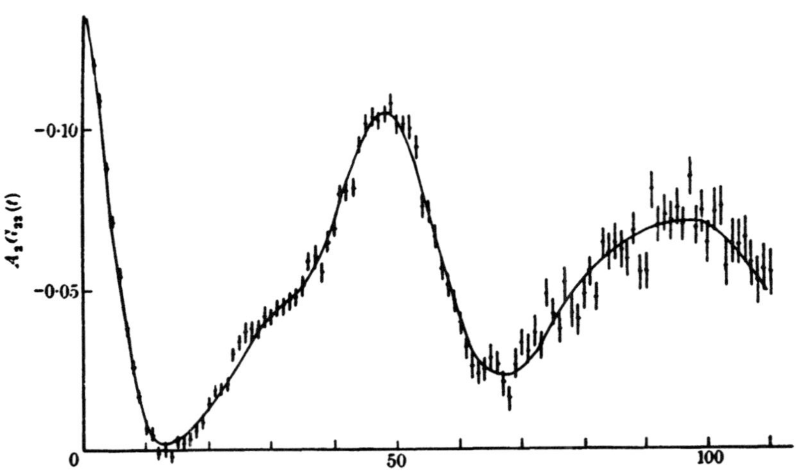

Fig. 6. Time - differential measurement of the A_2 - coefficient of the 133 keV - 482 keV gamma-gamma angular correlation in ^{181}Ta in a polycrystalline hafnium metal source.

pects an axially symmetric electric field gradient. The fit gave, however, $\eta =$ $= 0.30 \pm 0.02$. The nature of this interesting observation is not yet understood but a measurement in an Hf metal single crystal[6]), shown in fig. 7, could only be fitted by $\eta \approx 0$. The interaction frequency was derived as $\omega_q = 52 \pm 0.7$ Mc/s, or $\omega_0 =$ $= 313 \pm 4$ Mc/s*. By substituting for the electric quadrupole moment of the 5/2 state:

$$Q_I = (2.5 \pm 0.1) \cdot 10^{-24}\ \text{cm}^2 ,$$

one obtains for the effective electric field gradient

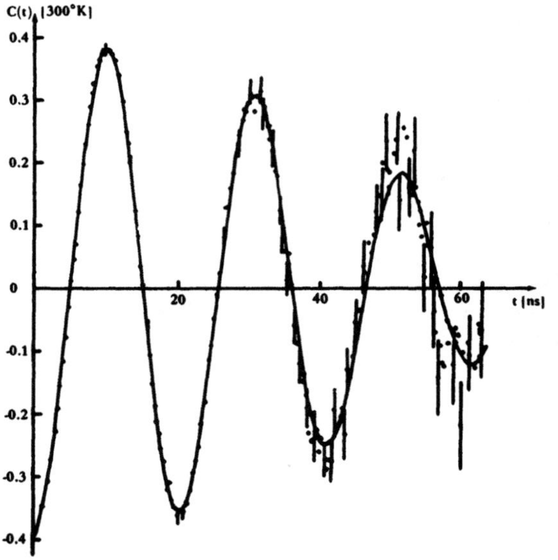

Fig. 7. Time - differential measurement of the asymmetry ratio of the same angular correlation as shown in fig. 6. A hafnium metal single crystal is used with the symmetry axis oriented perpendicular to the detector plane.

* Instead of ω_q the frequency of the lowest hyperfine transition, ω_0, is often used; it is given by

$$\omega_0 = \begin{cases} 3\omega_q & \text{for integer I} \\ 6\omega_q & \text{for half integer I} \end{cases}$$

$$|V_{zz}^{eff}| = (5.5 \pm 0.2) \cdot 10^{17}\ V/cm^2$$

It is interesting to compare this value with the prediction of our simple model. For metals of hexagonal closed package structure the lattice contribution can be expressed by the following equation[7]):

$$V_{zz}^{lat} = \frac{1}{4\pi\epsilon_0} \cdot Z \cdot e \cdot q_{lat},\ \text{with}\ q_{lat} = \frac{1}{a^3} \cdot \{ 0.0065 - 4.3584 \cdot (\tfrac{c}{a} - 1.633) \}$$

By inserting the lattice constants[8]) $a = 3.1946$ Å and $\frac{c}{a} = 1.58$ Å one obtains $q_{lat} = 0.00812 \cdot 10^{24}\ cm^{-3}$. And finally $V_{zz}^{lat} = 4.67 \cdot 10^{15}\ V/cm^2$. I learned in this conference, that the Sternheimer factor for Ta has been calculated to be $\gamma_\infty (Ta) \approx -61$. By taking

$V_{zz}^{eff} = V_{zz}^{lat} \cdot (1 - \gamma_\infty) \approx 2.9 \cdot 10^{17}\ V/cm^2$, we obtain about 50% of the experimental field gradient. Therefore, the contribution from the conduction

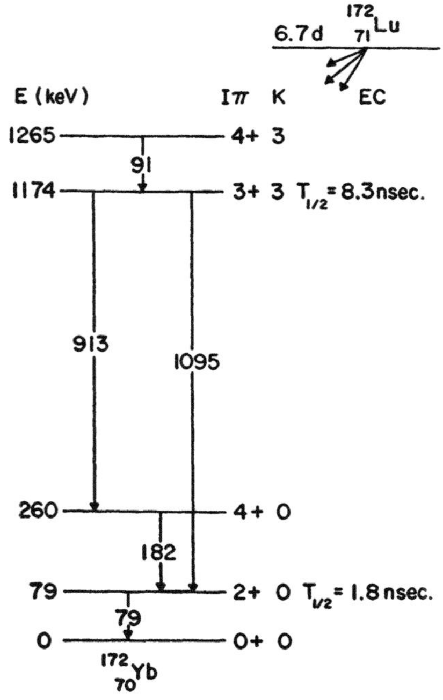

Fig. 8. Part of the decay scheme of ^{172}Lu.

electrons seems to be around 50% in this case. It is very interesting in this connection that the electric field gradient at the Hf - nucleus in a Hf - metal lattice is known from a Mössbauer experiment[9]) to be larger by about a factor of 1.5. The experimental result is V_{zz}^{eff} (Hf) = $+ 9.0 \cdot 10^{17}$ V/cm². It is not reasonable to assume that the Sternheimer factors for Hf and Ta are very different. We must therefore conclude that in the case of Hf the conduction electrons indeed contribute about 2/3 to the effective field gradient, but it is not yet understood why the field gradient is reduced so much for a Ta - impurity atom in a Hf - metal host.

A second case where the quadrupole moment of the excited state is known has recently been studied by R.L. Rasera and A. Li-Scholz[10]). These authors investigated the electric quadrupole interaction of ^{172}Yb in thulium metal. The interesting part of the decay scheme of ^{172}Lu is shown in fig. 8. The quadrupole moment of the 2^+ rotational state at 79 keV is known from the coulomb excitation cross - section. In the same decay scheme the level at 1174 keV with spin 3^+ can be used for differential angular correlation measurements. The populating transition of 91 keV is not well resolved from the 79 keV rotational transition by NaI (Tl) - detectors, and by setting the window of one detector on the gamma group around 80 keV and the window of the other detector on the 1095 keV line, one measures simultaneously the spin rotations in both isomeric states. The results for the attenuation parameters G_2 (t) are plotted in fig. 9. The left side shows the spin rotation in the 2^+ state and the right side shows the rotation in the 3^+ two - particle state. The solid line is the result of least squares fit calculations. The results for the interaction frequencies are ω_0 (79 keV) = (190 ± 19) Mc/s and ω_0 (1174 keV) = (101 ± 4) Mc/s. The interaction frequency for the 2^+ rotational state was used to derive the effective electric field gradient in thulium metal at the position of an ytterbium impurity atom $|V_{zz}^{eff}|$ = = (4.53 ± 0.47) · 10^{17} V/cm². This value is not significantly different from that found by Mössbauer measurements on thulium nuclei in thulium metal. The thulium

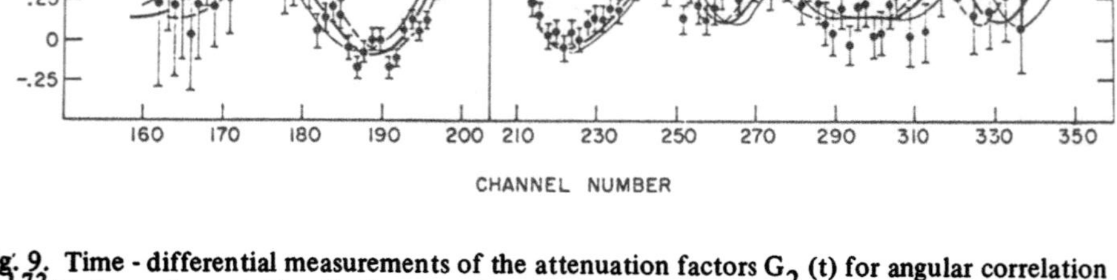

Fig. 9. Time - differential measurements of the attenuation factors G_2 (t) for angular correlation in ^{172}Yb in a polycrystalline thulium metal lattice. The right side of the diagram refers to the 91 keV - 1095 keV angular correlation and the left part of the diagram represents the 1095 keV - 79 keV correlation.

metal has the structure of hexagonal closed package and it is interesting to calculate the expected field gradient in the same way as we did for hafnium. By inserting the lattice constants[8]) $a = 3.54$ Å and $\frac{c}{a} = 1.57$ into the above mentioned equation one obtains $q_{lat} = 0.00634 \cdot 10^{24}$ cm^{-3}. And finally for the electric field gradient $V_{zz}^{lat} =$ $= 2.74 \cdot 10^{15}$ V/cm^2.

The Sternheimer correction factor has been calculated for thulium ions to be $\gamma_\infty (Tm^{3+}) = -75.3$ and we obtain therefore for the effective electric field gradient at the nuclear position:

$$V_{zz}^{eff} = (1 - \gamma_\infty) \cdot V_{zz}^{lat} = 76.3 \cdot 2.74 \cdot 10^{15} \text{ V/cm}^2 = 2.1 \cdot 10^{17} \text{ V/cm}^2$$

The observed value is more than a factor of two larger than this calculated value and we conclude that the missing part must be contributed to the conduction electrons. The lattice contribution to the effective electric field gradient as well as the Sternheimer effect should not exhibit a significant temperature dependence. A small effect, caused by the thermal expansion, is expected. In the case of hafnium metal Buttler[6]) has proved by taking Debye - Scherrer diagrams of a polycrystalline piece of hafnium metal at different temperatures that the thermal expansion coefficient shows no large anisotropy. The thermal expansion can produce therefore only small temperature effects. Very careful measurements of the temperature dependence of the electric interaction frequency have been performed for Ta in hafnium metals[6]) as well as for Yb in thulium metal[11]). The results are shown in fig. 10 and 11. A quite remarkable temperature dependence is observed in both cases. Any explanation of the missing

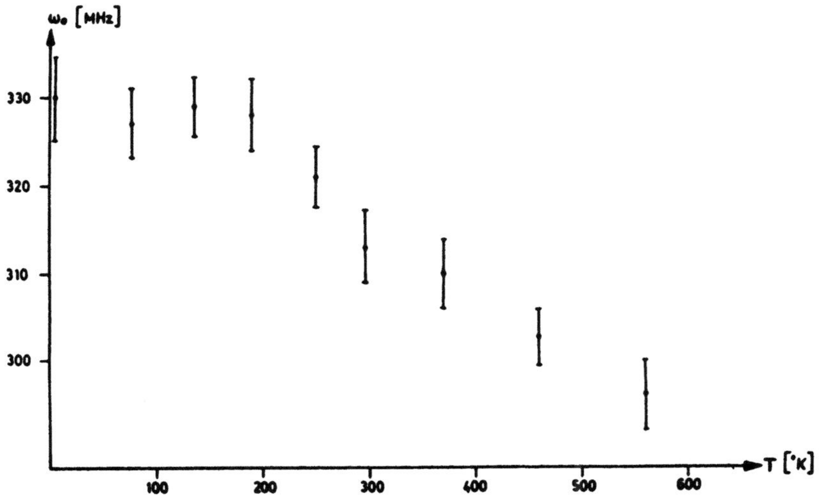

Fig. 10. Observed temperature dependence of the electric interaction frequency of ^{181}Ta in a hafnium metal single crystal.

part in the electric field gradient must also explain this temperature dependence. It should be noted that a temperature dependence of the contribution of the conduction electrons has been predicted theoretically[12]).

An interesting nuclear result is obtained from the measurement of the electric interaction frequency for the 3^+ two quasi - particle state in ^{172}Yb. One can derive from this value the ratio of the quadrupole moments of the 3^+ state and the 2^+ rotational state. The application of the Bohr - Mottelson formula reveals that the intrinsic quadrupole moments and therefore the deformations of the nuclei in both isomeric states are not significantly different. It should be mentioned that the same result has also been obtained by a study of the quadrupole perturbation in statistical interactions in both states by Forker and Wagner[13]).

Finally we have to mention a study of the electric quadrupole interaction of ^{204}Pb in polycrystalline Tl metal[14]). The spin of the intermediate state is 4^+ and the half - life is 260 ns. The calculated attenuation factors for the angular correlation coefficient A_2 are plotted in fig. 12 for different values of the parameter η. The crystalline form of Tl - metal is again hexagonal closed package and one expects therefore an axially symmetric electric field gradient. Fig. 13 shows the experimental result. The interaction frequency is quite small and only the first two maxima of the pattern are observed. They show indeed that the axial asymmetry parameter must be very small. The measured interaction frequency is $\omega_0 = (2.23 \pm 0.15)$ Mc/s. Using the lattice constants[8]) of thallium a = 3.457 Å and $\frac{c}{a} = 1.598$ Å, we obtain for the lattice contribution to the electric field gradient $q_{lat} = 0.00317 \cdot 10^{24}$ cm^{-3}

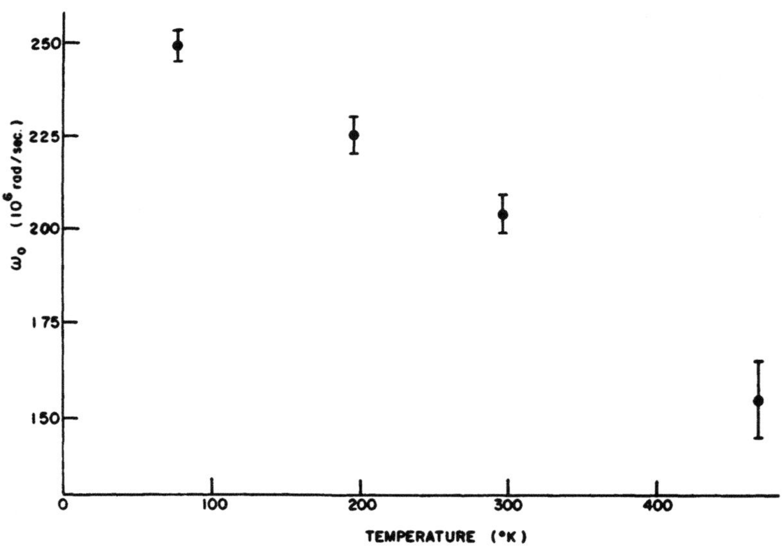

Fig. 11. Observed temperature dependence of the electric interaction frequency of ^{172}Yb in a polycrystalline thulium metal lattice.

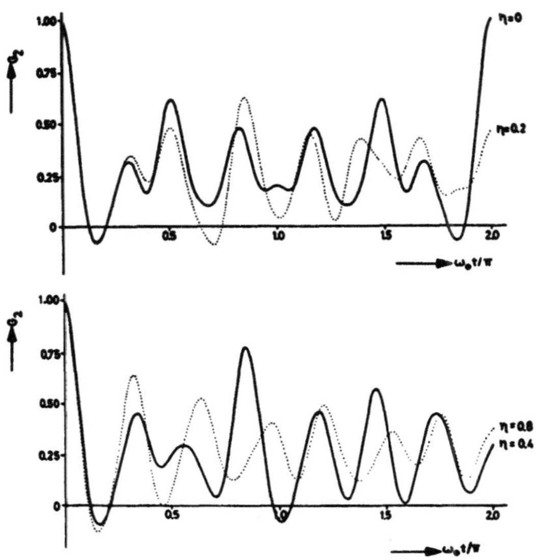

Fig. 12. Calculated attenuation factors G_2 (t) for the spin I = 4 and different values of the axial asymmetry parameter η.

or $V_{zz}^{lat} = 1.37 \cdot 10^{15}$ V/cm². Using the Sternheimer correction factor for the Pb^{++++}-ions[26]) $\gamma_\infty \approx -50.88$, we obtain for the effective field gradient $V_{zz}^{eff} = 0.70 \cdot 10^{17}$ V/cm².

Neglecting any effects of the conduction electrons we get for the spectroscopic quadrupole moment of the 4^+ state of ^{204}Pb:

$$|Q_I| = \frac{4\hbar \cdot I \cdot (2I - 1) \cdot \omega_q}{e|V_{zz}|} = 0.8 \cdot 10^{-24} \text{ cm}^2 .$$

Or if we assume that the calculated electric field gradient contributes only about 30 p.c. to the total field gradient as in the cases discussed before, we obtain for the quadrupole moment the more realistic estimate $|Q_I\,(4^+)| = 0.3 \cdot 10^{-24}$ cm².

The study of electric quadrupole interactions in metals by the perturbed angular correlation technique looks very promising for future work. The main reasons are the following ones:

1. Any radioactivity can be implanted into metal hosts by use of isotope separators and one can be sure that the implanted ions reach their stable equilibrium state within a very short time because of the availability of free electrons in the conduction band.

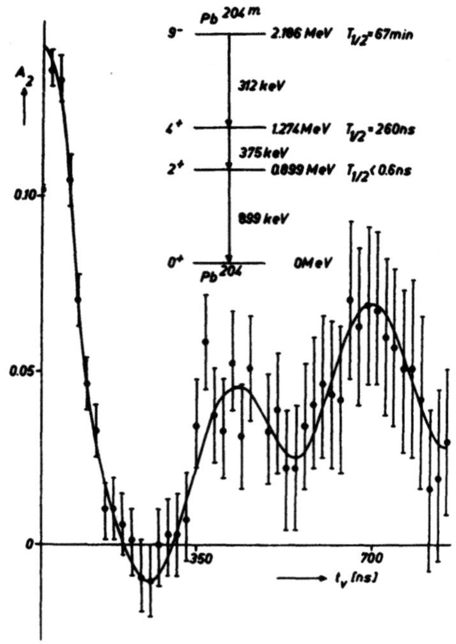

Fig. 13. Time - differential measurement of the A_2 coefficient of the 912 keV - 375 keV gamma - gamma cascade in ^{204}Pb in a polycrystalline thallium metal lattice.

2. A large number of metals has the crystalline structure of hexagonal closed package and a calculation of the electric field gradient is possible, at least in principle.
3. The so - far observed electric field gradients are just of the right order of magnitude.
4. One can expect that future systematic investigations will lead to an understanding of the contribution from the conduction bands.

Another group of investigations deals with oxides. A beautiful spin rotation pattern has been observed by Gerdau et al.[5]) (see fig. 14) for ^{181}Ta in HfO_2 which was annealed at 1100° K for 48 hours and subsequently at 1500° K for eight hours. From the known electric quadrupole moment the authors derived for the electric field gradient $V_{zz} = (13.2 \pm 0.6) \cdot 10^{17}$ V/cm^2; $\eta = 0.35 \pm 0.03$. No attempt was made to calculate the electric field gradient in this case. An investigation of the hyperfine interaction of ^{140}Ce in La_2O_3 was recently performed by our group[15]). The measurement gave an unperturbed correlation from which we concluded that the quadrupole moment of the isomeric level in ^{140}Ce must be quite small.

The crystalline fields of rare earth oxides are quite well known from many

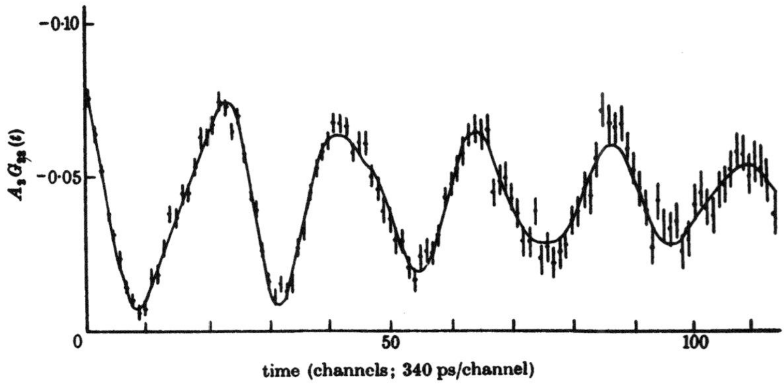

Fig. 14. Time - differential measurement of the A_2 - coefficient of the 133 keV - 482 keV gamma - gamma cascade in ^{181}Ta in polycrystalline HfO_2.

Mössbauer studies. The only case where a quadrupole interaction has been observed by perturbed angular correlation is ^{172}Yb. R.L. Rasera and A. Li-Scholz[10]) made a time - differential measurement in Tm_2O_3 on the 2^+ rotational state and the already mentioned 3^+ state at 1174 keV. A field gradient of $|V_{zz}^{eff}| = (1.67 \pm 0.20) \cdot 10^{18}$ V/cm^2 was derived for the C_2 - site. The rare earth oxides are essentially ionic bonds but in addition to the field gradient produced by the ions, one has a strong contribution from the unfilled 4f - electronic shell. Extensive theoretical work has been done in order to understand the electronic structures and the internal fields.

The strong electric field gradients which are produced by homopolar bonds have been used so far only in two cases:

1. Sugimoto et al.[16]) studied the hyperfine interaction of ^{19}F in the isomeric 5/2 state using the compound ClF. Here one has a clean homopolar bond and the electric field gradient can be calculated rather reliably. The compound ClF is solid at low temperatures and Sugimoto et al. used a solid target and performed the reaction ^{19}F (p, p$'$ γ) ^{19}F. The time - differential measurements of the angular distribution of the deexcitation γ - rays are shown in fig. 15. The authors performed the same investigation also for teflon but in this case the compound is more complex and the calculation of the field gradient less reliable. The time dependence of the attenuation factor has the theoretical shape for an axially symmetric field gradient and the observed interaction frequency for ClF is $\omega_0 = (77.9 \pm 4)$ Mc/s. For the calculation of the field gradient Sugimoto followed the method described by Townes and Dailey[17]) and obtained finally for the electric quadrupole moment $|Q_I| = 0.11 \cdot 10^{-24}$ cm^2.
2. Wolf et al.[18]) measured the quadrupole moment of the 5^- isomeric state of ^{120}Sn by observing the spin rotation in the compound K $[C_4H_4O_6]$ (SbO) $\cdot$ $\cdot$ 1/2 H_2O. The result of this spin rotation measurement is shown in fig. 16. The

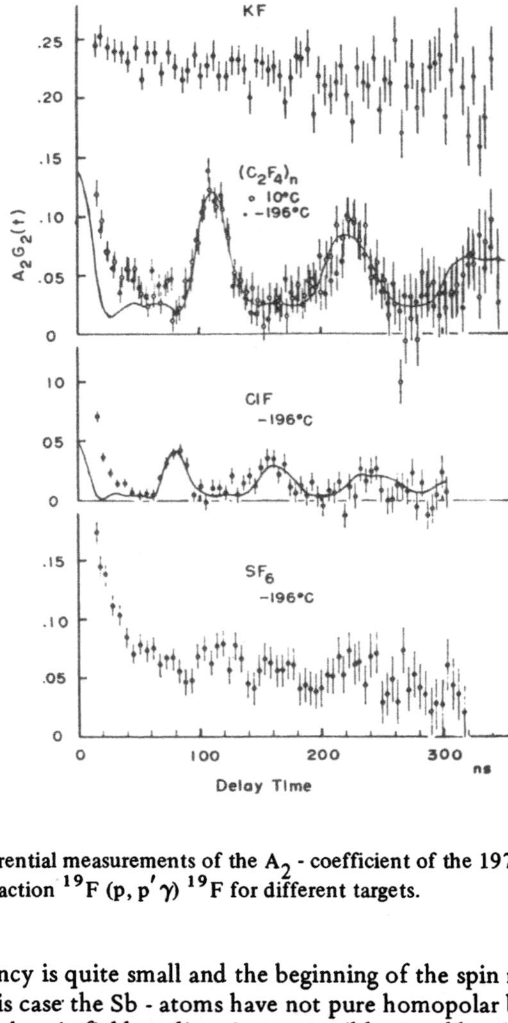

Fig. 15. Time - differential measurements of the A_2 - coefficient of the 197 keV gamma angular distribution in the reaction $^{19}F\ (p, p'\gamma)\ ^{19}F$ for different targets.

interaction frequency is quite small and the beginning of the spin rotation can just be observed. In this case the Sb - atoms have not pure homopolar bonds and a direct calculation of the electric field gradient is not possible. A calibration was applied, instead, by using the quadrupole moment of the 24 keV state of ^{119}Sn which was determined by Mössbauer investigations. The Mössbauer spectra show well resolved quadrupole splittings for quite a number of different compounds. The largest quadrupole splitting was observed for a Mössbauer source in the compound $K\,[C_4H_4O_6\,]\,(SbO) \cdot 1/2\ H_2O$, and therefore the same substance was used for the angular correlation experiments. The Mössbauer studies showed a linear relation between the isomer shift and the quadrupole splitting and this relation was used to extrapolate to the quadrupole splitting for pure homopolar bonds. For this limiting case, finally, the electric field gradient was calculated and by this way the quadrupole

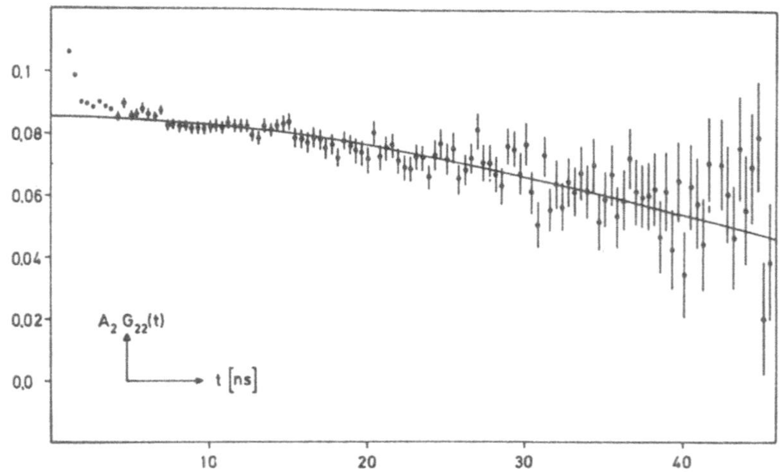

Fig. 16. Time - differential measurement of the 200 keV - 90 keV gamma - gamma angular correlation of ^{120}Sn in polycrystalline material of the compound K $[C_4H_4O_6](SbO) \cdot 1/2H_2O$.

moment of the excited state of ^{119}Sn was derived. The result for the quadrupole moment of the 5^- state of ^{120}Sn was $|Q_I| = (0.021 \pm 0.008) \cdot 10^{-24}$ cm^2.

By a similar calibration method Morgenstern[19]) has recently determined the quadrupole moment of the I = 5 level at 1131 keV in ^{18}F. This author determined the electric field gradient of fluorine recoil implanted into quartz by a calibration experiment with ^{19}F, and then he repeated the experiment with ^{18}F. The results of these measurements are shown in fig. 17. From the ratio of the interaction frequencies one obtains $Q_I(^{18}F) / Q_I(^{19}F) = (1.07 \pm 0.03)$. By inserting the above mentioned quadrupole moment of ^{19}F one gets $|Q_I(^{18}F)| = 0.12 \cdot 10^{-24}$ cm^2.

Six years ago the compound Hf $(NH_4)_2F_6$ has been used for an extensive study[20]) of the quadrupole interaction of ^{181}Ta. The observed spin rotation in a polycrystalline source, shown in fig. 18, shows a nice structure and the fit gave for the electric interaction frequency $\omega_0 = (570 \pm 30)$ Mc/s. Or for the field gradient* $|V_{zz}| = 1.02 \cdot 10^{18}$ V/cm^2; $\eta = 0.43 \pm 0.04$. A reinvestigation by Gerdau et al.[15]) led to a discrepancy which has not been cleared up. The most probable reason is that in both investigations materials of not identical crystalline structure have been used.

Quite a number of investigations have been performed with rather pure ionic

* The evaluation of V_{zz} from ω_0 in the original paper contains a trivial error.

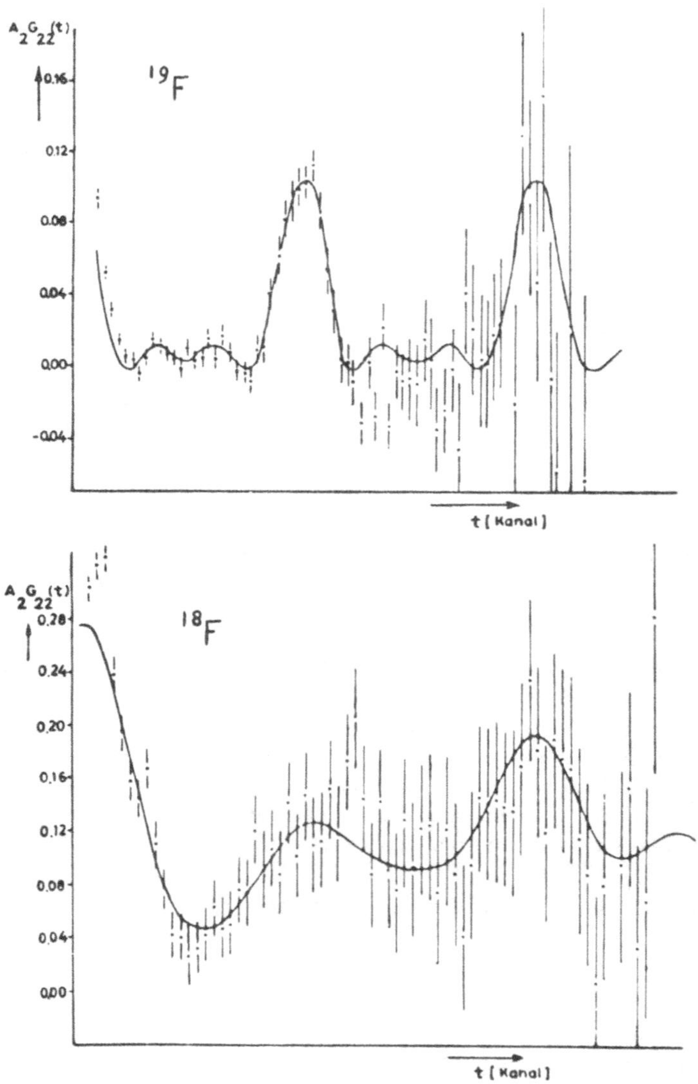

Fig. 17. Time - differential measurement of the A_2 - coefficient of the 197 keV gamma angular distribution of ^{19}F and of the 1131 keV gamma angular distribution of ^{18}F after recoil implantation into quartz.

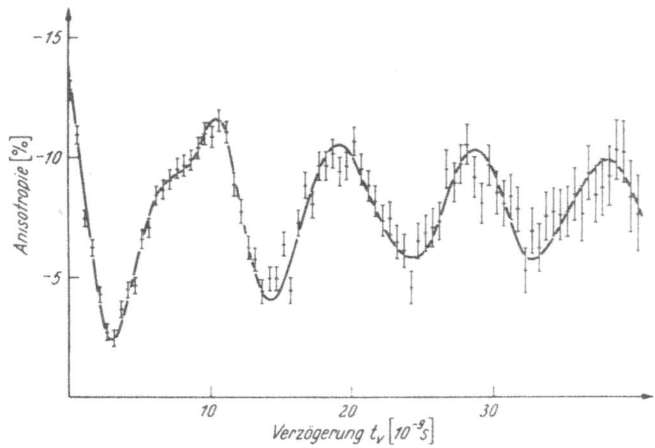

Fig. 18. Time - differential measurement of the anisotropy of the 133 keV - 482 keV gamma - gamma angular correlation in ^{181}Ta in polycrystalline material of Hf $(NH_4)_2 F_6$.

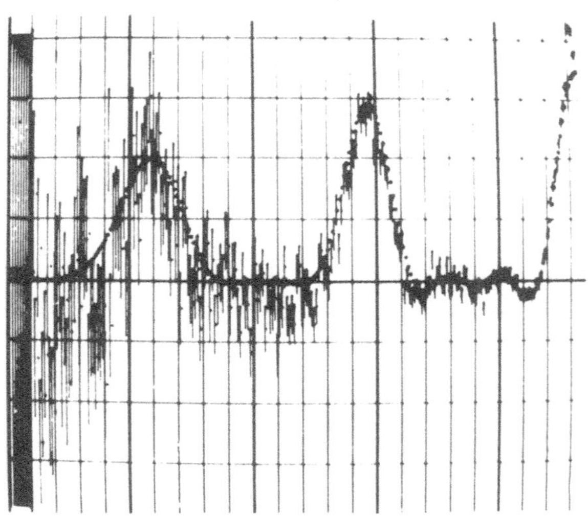

Fig. 19. Time - differential measurement of the A_2 - coefficient of the 197 keV gamma - angular distribution of ^{19}F using a BeF_2 target.

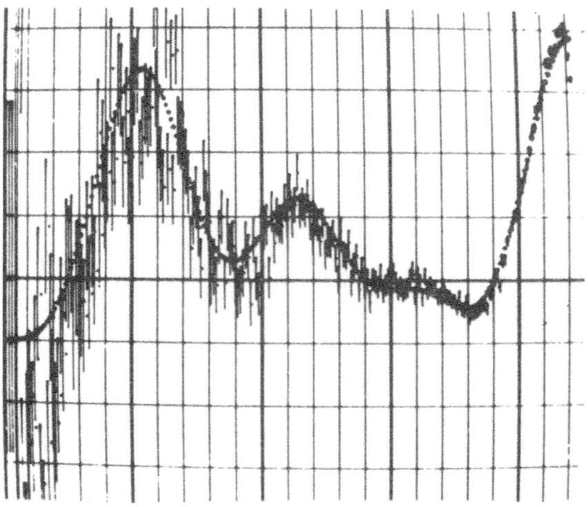

Fig. 20. Same measurement as shown in fig. 19 but the target material is MgF_2.

lattices. I should like to refer here to the extensive work by the Marburg group[21]) which investigated the perturbed angular distribution in ^{19}F for a great variety of different targets. Fig. 19, 20, and 21 show as example the results for the target materials BeF_2, MgF_2, and CrF_2, respectively. In the first case one has the typical structure for an axially symmetric electric field gradient. This observation is in agree-

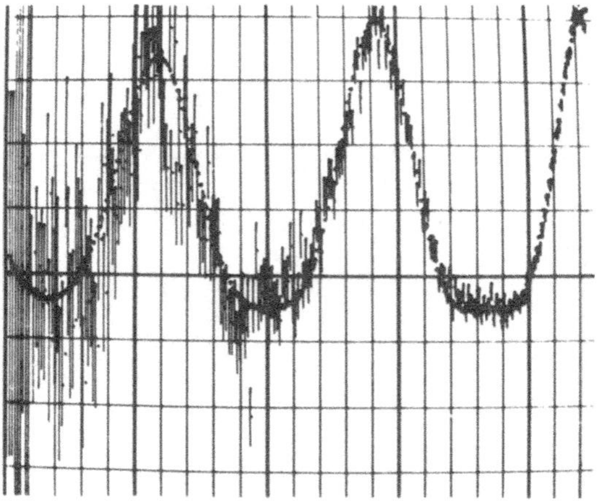

Fig. 21. Same measurement as shown in fig. 19 but the target material is CrF_2.

target substance	ω_q [Mc/s]	$\lvert V_{zz} \rvert$ [10^{18} V/cm^2]	η	freq. spread δ [%]
CaF_2	-	-	-	-
BaF_2	-	-	-	-
PbF_2	-	-	-	-
BeF_2	3.51_5	0.92	0.00_3	6
MgF_2	1.56_5	0.41	0.36_3	2
MnF_2	1.73_5	0.46	0.40_3	7
FeF_2	1.98_5	0.52	0.47_3	5
CoF_2	2.38_5	0.63	0.48_3	4
NiF_2	2.96_5	0.78	0.44_3	6
CrF_2	2.05_5	0.54	1.00_3	6
CuF_2	3.00_5	0.79	1.00_3	5

Table 1. Table of measured quadrupole interactions in the $5/2^+$ state of ^{19}F.

ment with the symmetry group of the crystals. In the other two cases one has a small and a very large axial asymmetry parameter η. All these measurements were fitted carefully and the results of the Marburg group are shown in table 1. A theoretical calculation of the expected electric field gradient for these ionic structures should be very promising. These calculations should reveal whether the calculated Sternheimer correction factor for fluorine is correct and also one should get information about the question how pure the ionic bonds are in all these cases.

The last topic of my talk concerns again a very promising project for the future. In the last two years a few measurements were performed[22,23,24]) in order to study the hyperfine fields of frozen dilute aqueous solutions. By Mössbauer studies as well as by angular correlation measurements it was shown that the ions in ice are subject to very strong static electric field gradients. We performed in our group in Bonn[25]) a systematic investigation with a ^{181}Hf - source in the form of Hf^{4+} - ions in a 0.1 N aqueous HF - solution. We measured the time - dependent attenuation factor as a function of the temperature and observed the spin rotation in a strong static electric

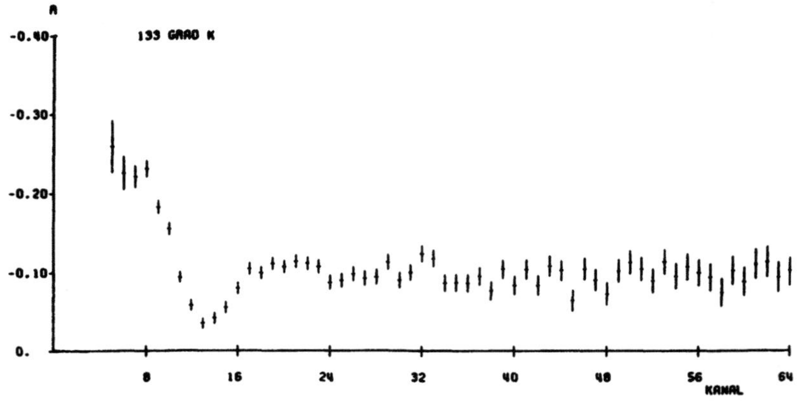

Fig. 22. Time - differential measurement of the isometric ratio R (t) for the 133 keV - 482 keV gamma - gamma angular correlation in ^{181}Ta in a frozen aqueous solution of HfF_4 in 0.1 N HF at 133°K.

quadrupole field. The observed pattern for 133°K is shown in fig. 22. We repeated the measurement for higher temperatures in steps of 10°. First we did not see any change but when we reached a temperature of 200° K, the measured diagram showed an additional statistical interaction. At higher temperatures the structure vanished completely and we obtained a pure exponential attenuation. The nuclear relaxation time became larger with higher temperatures and finally at room temperature we had practically no attenuation. In this connection it might be interesting to note that aqueous solutions of $HfOF_2$ or $HfOCl_2$ are attenuated even at room temperature. We assume that in these cases the chemical bonds with the oxygen atoms produce an electric field gradient which causes a statistical interaction by the Brownian movement of the ions. The nature of the statistical perturbation above 200° K is not yet absolutely clear. Cameron et al.[22]) assumes in his work that at these low temperatures only the immediate neighbourhood of the ions behaves like a liquid and makes the effective field gradient at the nuclear position time dependent.

The spin rotation patterns which we observed in the case of pure static interactions below 190° K exhibit a very strong damping. We conclude that in our samples the crystalline structure of the material was not very uniform and we shall try to prepare sources where the ice has a better uniformity.

This type of investigation seems to be quite promising because it is open to all kinds of nuclei where time - differential angular correlation measurements are possible. It might be that the effective electric field gradients for the different ions in ice change in a systematic manner which would provide a nice tool for measurements of quadrupole moments of excited nuclear states.

References

1. H. Frauenfelder and R.M. Steffen in Siegbahn: Alpha-, Beta-, and Gamma-Ray Spectroscopy, NHPC, Amsterdam 1965, Volume 2, page 997ff.
2. H. Albers - Schönberg, F. Hänni, E. Heer, T.B. Novey, and P. Scherrer, Phys. Rev. 90 (1953) 322.
3. Lehmann and Miller, J. Phys. Radium 17 (1956) 526.
4. R.M. Sternheimer, Phys. Rev. 146 (1966) 140; (The earlier work is cited in this paper).
5. E. Gerdau, J. Wolf, H. Winkler and J. Braunsfurth, Proc. Roy. Soc. A311 (1969) 197.
6. N. Buttler, thesis, Bonn 1970.
7. T.B. Das and M. Pomerantz, Phys. Rev. 123 (1961) 2070.
8. W.B. Pearson, Handbook of Lattice Spacings and Structures of Metals, Volume 2, Pergamon Press, 1967.
9. E. Gerdau, P. Steiner and D. Steenken, in Matthias and Shirley: Hyperfine Interactions and Nuclear Structure, N.H.P.C. Amsterdam 1968, page 261.
10. R.L. Rasera and A. Li - Scholz, Phys. Rev. B1 (1970) 1995.
11. R.L. Rasera, private communication.
12. A.M. Clogston, Phys. Rev. 136 (1964) A1417.
13. M. Forker and H.F. Wagner, Nucl. Phys. A138 (1969) 13.
14. R.M. Lieder, W. Delang, and M. Fleck, Z. f. Physik 206 (1967) 29.
15. to be published.
16. K. Sugimoto, A. Mizobuchi, and K. Nakai, Phys. Rev. 134 (1964) B539.
17. C.H. Townes and A.L. Schawlow, Microwave Spectroscopy, McGraw Hill, N.Y. 1955, pages 225 - 247.
18. H.J. Wolf, P. Steiner, E. Gerdau, W.K. Müller, U. Lewandowski, and A. Roggenbuck, Z. f. Physik 232 (1970) 256.
19. J. Morgenstern, thesis, Hamburg 1969.
20. L. Mayer, E. Bodenstedt, and C. Günther, Z. f. Physik 177 (1964) 28.
21. Richter and Wiegandt, Z. f. Physik 217 (1968) 225.
22. J.A. Cameron, P.R. Gardner, L. Keszthelyi, and W.V. Prestwich, preprint McMaster University, Hamilton, Canada (1969).
23. I. Dézsi, L. Keszthelyi, B. Moluár and L. Pócs, Physics Lett. 18 (1965) 28.
24. A.J. Nozik and M. Kaplan, J. Chem. Phys. 47 (1967) 2960.
25. Hagemeyer, to be published.
26. F.D. Feiock and W.R. Johnson, Phys. Rev. 187 (1969) 39.

Discussion

De Waard: Can the difference of the field gradient found at Ta and Hf in Hf be due to a difference in Sternheimer factors caused by different effective charges of the Ta and the Hf?

Bodenstedt: I discussed this point with some experts and they told me they would be very astonished if the Sternheimer - factor would be that different for tantalum and hafnium.

Günther: I have a remark concerning this question. The value of 13.3 for hafnium in hafnium is wrong. It contains a computational error of 1.5, so the true field gradient is really 1.5 times smaller, which makes the difference somewhat smaller between tantalum and hafnium*.

Kaufmann: What value of ionic charge at a Hf - site was used to compute the value of V_{zz}^{lat}.

Bodenstedt: I think the value 4 was used.

Rasera: I would like to remark on our measurement of quadrupole interactions at ^{172}Yb in thulium metal. We have improved our statistical accuracy considerably, and we find, as is so often the case, that as our statistical accuracy increases, the quality of our least - squares fits to an axially - symmetric field gradient gets worse. This seems to occur to some extent at all temperatures, but is worst at low temperatures – at 77° K, for example, we find $\eta \approx 0.3$. I would like to ask if anyone has ever seen a purely axially - symmetric electric field gradient in a polycrystalline hexagonal - close - packed metal.

Bodenstedt: Yes, in ^{204}Pb.

Rasera: I would like to remark that we have also tried all sorts of metallurgical tricks – different source forms, annealing, not annealing, different irradiation conditions – and nothing seems to help.

Bodenstedt: Yes, we had the same experience with ^{181}Hf.

Berkes: I think a recent unpublished measurement of the Grenoble - group (J. Berthier, P. Boyer and J.I. Vargas) could give some information on the source of non - axiality of the electric field gradient acting on ^{181}Ta in hafnium. Their results indicate that η increases with increasing impurity of the polycrystalline source: e.g., a source made of commercially available hafnium (Zr - content ≅ 10% at.) gives, $\eta = 0.33 \pm 0.04$, a very pure source (Zr - content ≅ 0.6% at.) gives, after vacuum annealing, a value of $\eta = 0.12 \pm 0.02$.

* Note by the Editors: The value mentioned in the text has been corrected for this factor of 1.5.

Contributed Paper

OBSERVATION OF THE QUADRUPOLE INTERACTION AT Ta IN Zr BY TIME – DIFFERENTIAL PERTURBED ANGULAR CORRELATION TECHNIQUE

E.N. Kaufmann

Bell Telephone Laboratories, Incorporated, Murray Hill, New Jersey, U.S.A.

The 482 keV excited state of the nucleus ^{181}Ta is a useful probe of internal hyperfine fields in solids. The level has large magnetic dipole and electric quadrupole moments[1]) ($\mu = 3.2$ nm, Q = 2.5b) and there is a large angular correlation coefficient[1]) ($A_{22} \approx - 0.28$) for the strongly fed gamma - gamma cascade, $1/2^+$ (133 keV, E2) $5/2^+$ (482 keV, M1 + E2) $7/2^+$, through the level. Also the life - time of the intermediate state[1]) ($\tau = 1.06 \times 10^{-8}$ s) is in the range where a time - differential method, which yields maximum information regarding the interaction, may be employed. Knowledge of the electric field gradient (EFG) at impurity sites in non - cubic metals and its dependence on temperature can contribute to understanding the role of the conduction band in shielding the crystal field. To this end, we are investigating the hyperfine interaction at Ta in hcp Zr using the time - differential perturbed angular correlation technique.

Alloy foils of 1.04 at. % natural Hf in Zr were prepared by arc melting and rolling. Irradiation of the foils by pile neutrons produced the 42.5 day ^{181}Hf parent activity and a 65 day contaminant, ^{95}Zr, which does not interfere with the measurement. The major portion of the ^{181}Hf decay to the level of interest in ^{181}Ta, populates a sufficiently long lived state[1]) ($\tau = 25.7 \times 10^{-6}$ s) to insure recovery from after effects of the β^- decay. Plastic (133 keV) and NaI (Tl) (482 keV) scintillators mounted on fast photomultiplier tubes feeding a standard slow - fast coincidence counting system were used. With single channel analyzers selecting the 482 keV photopeak and excluding Ta X - rays, the electronic time resolution* was found to be $2\tau =$ $= 1.5 \times 10^{-9}$ s. This value was reached with the aid of constant fraction of pulse height discriminators in the fast channel. The digitized output of a time - to - amplitude converter was routed and stored under program control in subsections of core in a PDP8/I computer. Recording of singles counts and initiating detector position changes were also under computer control.

Data were collected for several angles between detectors. For each time channel

* Time resolution measurements utilized Compton events produced by a ^{60}Co source.

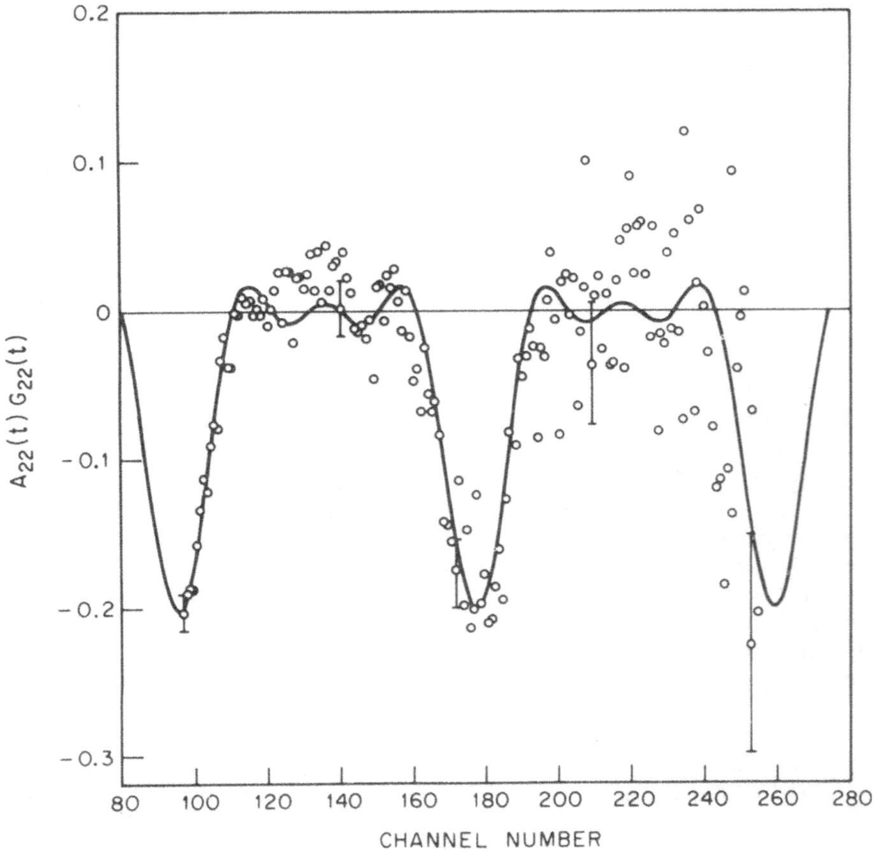

Fig. 1. Fitted values of $A_{22}G_{22}$ (t) vs. time. The solid curve represents a least squares fit to the data.

on the decay curve a value of $A_{22}G_{22}$ (t) was fit assuming the form of the correlation function to be

$$W(\theta, t) = Ne^{-t/\tau}\left[1 + A_{22}G_{22}(t)\, P_2(\cos\theta) + A_{44}G_{44}(t)\, P_4(\cos\theta)\right] \qquad (1)$$

A plot of the resulting values is given in fig. 1. The theoretical form of G_{22} (t) for a polycrystalline axially symmetric static electric quadrupole interaction with a spin 5/2 state is

$$G_{22}(t) = 0.2 + 0.3714 \cos \omega_0 t + 0.2857 \cos 2\omega_0 t + 0.1429 \cos 3\omega_0 t \qquad (2)$$

where

$$\omega_0 = \frac{3}{20}\frac{eQV_{zz}}{\hbar} \tag{3}$$

The solid curve in fig. 1 is a least squares fit of this function to the data. The pertinent fitted parameter values are $A_{22} = -0.200 \pm 0.006$ and $\omega_0 = 0.302 \pm 0.009$ rad/ns. The time scale calibration was derived from the known state lifetime. An identical measurement with the source held at 77° K yielded $A_{22} = -0.199 \pm 0.007$ and $\omega_0 = 0.328 \pm 0.010$ rad/ns. Using eq. (3) above, the EFG derived are $|V_{zz}(300°\text{K})| = 5.25 \times 10^{17}$ volts/cm², and $|V_{zz}(77°\text{K})| = 5.70 \times 10^{17}$ volts/cm². The actual error in these values is uncertain because the accuracy of the adopted value of Q = 2.53b is not quoted in the primary reference. The ratio of the EFG at the two temperatures which is independent of Q and time scale calibration is $|V_{zz}(300°\text{K})/V_{zz}(77°\text{K})| = 0.923 \pm 0.009$.

The observed change with temperature of nearly eight per cent cannot be accounted for by the expected increase of about 0.3 per cent in the EFG arising from the change in lattice constants[2]). The EFG at a nucleus in a solid is usually expressed to lowest order as[3])

$$V_{zz} = V_{zz}^{LA}(1 - \gamma_\infty) + V_{zz}^{LO}(1 - R) \tag{4}$$

where γ_∞ and R are the familiar Sternheimer antishielding factors[4]). V_{zz}^{LA} is the point charge contribution of the lattice and V_{zz}^{LO} is interpreted as an EFG arising from an unfilled atomic shell. For the case studied here, the zirconium host has an s - d conduction band which includes 5d electrons of Ta (as well as 6s), so V_{zz}^{LO} must be taken to be the field due to the conduction electrons. In a cubic metal, the d - states of various spatial symmetries are equally populated and no EFG results. In hexagonal Zr, however, the quadrupolar lattice field induces a repopulation of d - states favouring one orbital symmetry over another, thus producing an EFG proportional to V_{zz}^{LA}. Watson, Gossard and Yafet[5]) have shown that the proportionality factor is negative (shielding) and in many cases as large as − 50 to − 100. Inserting values for Ta^{++} of $\langle r^2 \rangle_{5d} = 4.2573$ a.u., $\langle r^{-3} \rangle_{5d} = 6.2915$ a.u.[7]) into the estimates of ref. 5 yields a proportionality factor $F_{COND} \approx 22$ to 33. The value of γ_∞ for Ta is approximately -61 ± 10 [4,6]) and R is typically in the range $-0.2 < R < +0.2$ (ref. 3). If Eq. (4) is rewritten as

$$V_{zz} = V_{zz}^{LA}[1 - \gamma_\infty - F_{COND}(1 - R)] \tag{5}$$

and a value of $V_{zz}^{LA} = 3.2 \times 10^{15}$ volts/cm² is used† the lowest value of F_{COND} con-

†This value of V_{zz}^{LA} was interpolated from the work of de Wette[8]) assuming a value of +4e for the effective Zr ion charge.

sistent with the experimental results is about 200, i.e. 10 times that expected in ref. 5.

It must be concluded, therefore, that V_{zz}^{LO} contains terms substantially larger than that considered above. If we further assume that the factors γ_∞ and R are relatively insensitive to temperature changes because they depend on properties of closed inner atomic shells, we can attribute the observed change in $|V_{zz}|$ to a dependence of V_{zz}^{LO} on temerpature. Without a knowledge of the electronic wave functions in the vicinity of the Ta - impurity the computation of V_{zz}^{LO} is not possible. A more detailed study of the temperature dependence of the EFG is needed before precise conclusions regarding the behaviour of V_{zz}^{LO} can be drawn.

References

1. Nuclear Data Sheets; and Table of Isotopes by C.M. Lederer, J.M. Hollander and I. Perlman, Wiley and Sons, New York (1967).
2. R.B. Russell, Trans AIME, Journal of Metals (1954) 1045.
3. R.E. Watson and A.J. Freeman in Hyperfine Interactions, ed. by A.J. Freeman and R.B. Frankel, Academic Press, New York (1967) 71.
4. R.M. Sternheimer, private communication (1970); Phys. Rev. 84 (1951) 244; 146 (1966) 140; and 159 (1967) 266; R.M. Sternheimer and H.M. Foley, Phys. Rev. 102 (1956) 731.
5. R.E. Watson, A.C. Gossard and Y. Yafet, Phys. Rev. 140 (1965) 375A.
6. F.D. Feiock and W.R. Johnson, Phys. Rev. 187 (1969) 39.
7. A.J. Freeman, private communication (1970).
8. F.W. de Wette, Phys. Rev. 123 (1961) 103.

Discussion

Stone: Your suggestion of a contribution to V_{zz} from the Ta - ion itself would be as the result of the crystal field acting on the Ta - ion. You also remarked on the larger than expected variation of V_{zz}. These too may be connected if the crystal field is not large so that between LN_2 and room temperature its ability to align the Ta - ions falls considerably. Temperature variation measurements might then yield the size of this alignment contribution.

Kaufmann: This sounds like an alternative statement of the ideas of Watson, Gossard and Yafet (Phys. Rev. 140A (1965) 375). It emphasizes the importance of temperature variation measurements in these studies.

Contributed Paper

THE QUADRUPOLE MOMENT OF ^{133}La AND SHAPE ISOMERISM

L. Valentin, C. Gerschel, N. Perrin
I.P.N., Orsay, France

I am going to talk about some work which shows evidence of shape isomerism in ^{133}La. This evidence is based partly on the measurement of the quadrupole moment of the $3/2^-$ level at 535.4 keV. It is important to note that this energy is low compared to that which might be expected according to either the spherical shell model or the Nilsson model, provided one uses a reasonable deformation parameter. Furthermore, it is important to note that the low lying states of ^{133}La all have positive parity except for the 535.4 keV level. Similarly, all the low lying states are understood in terms of the spherical shell model with one exception, viz. the 535.4 keV level. In order to explain the $3/2^-$ level, we need to use Nilsson orbitals with huge deformation. We will come back to this point in another context later.

In fact, there are two possibilities for explaining the $3/2^-$ state. It is

1) the intrinsic Nilsson state $3/2^-$ [541], or
2) the 3/2 rotational state of the $1/2^-$ [550] band.

In either case, we can expect a large hindrance factor for transitions connecting that level to any spherical one (fig. 1). That point has been confirmed by previous work at our laboratory; our results agree with a very large hindrance factor of $f_W =$ $= 4.10^8$ for the transition $3/2^-$ to $3/2^+$ (at 130.7 keV). This corresponds to a measured half - life of 60 ns.

For this reason, it was important to measure both the magnetic and quadrupole

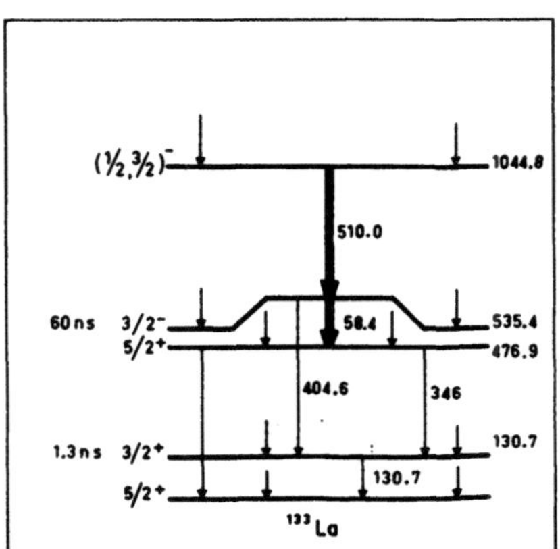

Fig. 1.

moments of the 3/2⁻ level in order to make sure that the deformation was really an important factor.

In fig. 1, we show only those levels with which we shall be mainly concerned. In particular the cascade we used in both the magnetic and the quadrupole moment measurements is emphasized (the 510 - 58.4 keV cascade). I shall talk mainly about the quadrupole moment measurements and I shall just say a few words at the end about the magnetic moment which we have measured previously. These measurements have been performed using the differential perturbed angular correlation method. The levels of ^{133}La are populated by K - capture in ^{133}Ce (5 hours period).

To get the quadrupole moment we embedded ^{133}Ce in polycrystalline lanthanum - magnesium - di - nitrate, known as LMN ($La_2(NO_3)_6$ $Mg_3(NO_3)_6$, $24H_2O$). Its crystallographic structure and its internal field are relatively well known as a result of its use in polarized targets. In particular, the field gradient of LMN has been measured by several authors and its value is $V_{zz} = 7.6 \times 10^{17}$ V/cm^2.

In fig. 2, we see that the results of the first experiment

For a 3/2 spin level, the attenuation factor is given by G_{22} (t) $= 0.2 + 0.8 \cos 2\Pi t/t_0$. From our experiment, we get $t_0 = (6.3 \pm 0.7)$ ns. In the figure, one can see the attenuation of the oscillations.

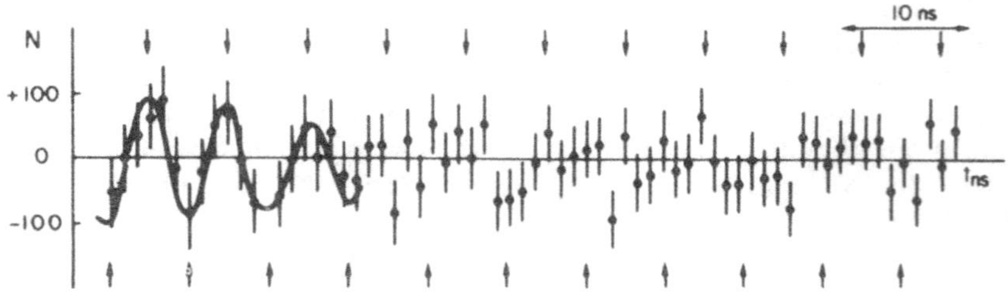

Fig. 2.

The second experiment, done with a larger time range, seems to show that the oscillations appear again after a delay of the order of 100 ns. This could be due to the fact that there exists a slight asymmetry between the two lanthanum atoms at each lattice site or to an after effect due to K - capture; in that case, the result is $t_0 =$ $= (7.2 \pm 0.7)$ ns. Other preliminary experiments in which ^{133}Ce is embedded in lanthanum ethyl- sulfate where there is only one lanthanum atom for each lattice site, do not show the damping of the oscillations and confirm the t_0 - values: $t_0 =$ $= (7.2 \pm 0.8)$ ns. So we adopt $t_0 = (6.8 \pm 0.5)$ ns, and calculate $Q = (1.6 \pm 0.2)$ b from $\omega_0 = -\frac{3e\,Q\,V_{zz}}{2I(2I-1)}$.

The errors are just statistical errors even though the measurement has been done with a K - capture nucleus in an insulating medium. In fact, when we immerse the insulating medium LMN in a magnetic field, the magnetic correlation disappears, which seems to indicate the possibility of an after - effect. However, it is not surprising that the local magnetism is largely influenced by the different charge states of the atoms. On the other hand, the regularity of the oscillations seen in the quadrupole moment measurements indicates that the Sternheimer coefficients are relatively independent of the charge states of the lanthanum and that very few atoms are able to drift into interstitial positions. So the Q - value we quote is obtained using the V_{zz} as determined by classical methods.

Returning now to the Q - value, we note the following formula

$$Q = \frac{3K^2 - I(I+1)}{(I+1)(2I+3)} \times \frac{3}{\sqrt{5\pi}} \; eZ\beta\,(1+0.16\beta)\,R_0^2$$

from which we obtain $\beta = 0.47$ for the two cases,
namely $1/2^-$ [550] (rotational)
and $3/2^-$ [541] (intrinsic)

This huge deformation is consistent with the hypothesis of shape isomerism in ^{133}La. It also makes it possible to understand the value we measured for the magnetic moment of the same level. That value, 2.1 nuclear magnetons, is not compatible with the theoretical estimate for a 3/2 - [541] orbit, which cannot exceed 1.3 n.m. even for $g_{s\,eff.} = g_{s\,free}$.

On the other hand, if we consider the $3/2^-$ level as the $3/2^-$ state of the $1/2^-$ [550] band, using $g_R = 0.4$ and the theoretical decoupling parameter, we are lead, in order to reproduce the experimental value, to use the following values for the $g_{s\,eff.}/g_{s\,free}$ ratio:

Table 1

η	$g_{s\,eff.}/g_{s\,free}$
2	.20
4	.28
6	.33
8	.38
10	.45

For those levels, which arise from the $h_{11/2}$ - orbital, core polarization should not be important and it is fair to see that the bigger the deformation is, the closer we get to a realistic value of the ratio.

So all of our results regarding the $3/2^-$ level are in agreement with a giant deformation, in gross contrast to the spherical low lying states. It would be interesting to do a deformed Hartree - Fock calculation successively with and without a nucleon located in the $h_{11/2}$ - orbital in order to see if this would be sufficient to change the Lanthanum nucleus from a spherical to a well deformed equilibrium shape.

LARGE SUPERCONDUCTIVE MAGNETS FOR THE APPLICATION IN THE FIELD OF PERTURBED ANGULAR CORRELATIONS

R.W. Hollander

University of Technology, Delft, The Netherlands

For the determination of nuclear g - factors, one can measure the integral rotation $\omega_B \tau_N$ of a directional correlation pattern in a magnetic field B. If the unperturbed correlation is of the form

$$W(\theta) = 1 + A_2 P_2 (\cos\theta) \tag{1}$$

then the correlation, if perturbed by an external magnetic field perpendicularly to the plane of the detectors, can be written as

$$W_\perp(\theta, B) = 1 + A_2 P_2 \; (\cos(\theta - \omega_B \tau_N))$$

$$\omega_B = \frac{-g\mu_N B}{\hbar} \tag{2}$$

If the angle between the detectors (θ_0) is kept constant, it is convenient to use the ratio R, which is the fractional difference between the coincidence yield, for the 'field up' and 'field down' situation.

$$R = \frac{W_\perp(\theta_0, B^+) - W_\perp(\theta_0, B^-)}{\frac{1}{2}\{(W_\perp(\theta_0, B^+) + W_\perp(\theta_0, B^-)\}} \tag{3}$$

$$R = \frac{-12}{1 + \frac{4}{A_2}} \; \omega_B \tau_N \;, \quad \theta_0 = \frac{3\pi}{4} \tag{4}$$

Assuming that all errors are small compared with the statistical error in $W_\perp(\theta_0, B)$, then the relative statistical error in the g - factor is

$$<g> = \frac{\sqrt{2}}{R} <n> \tag{5}$$

where $<n>$ is the relative statistical error in $W_\perp(\theta_0, B)$. The minimum time required to obtain the desired accuracy $<g>$ in the g - factor has to be reasonable. If this is the only limitation for determining a g - factor, one can introduce for the measuring system a quality factor Q, which is proportional to the inverse of this minimum measuring time[1])

$$Q = \frac{1}{T_{min}} = [\epsilon_a \epsilon_b B^3] \left[\frac{<ab> (\tau g A_2)^3}{ab \left(\frac{A_2}{4} + 1\right)^3} \right] [\tau_0^{-1}] \frac{<g>^3}{<\tau_0>} \frac{\mu_N^3}{\hbar^3} \frac{3\sqrt{3}}{2\sqrt{2}} \qquad (6)$$

The quality factor is an indication for the possibility of g - factor determinations with a system. In g - factor experiments the approximation (6) for T_{min} can be used since in those measurements $<n>$ is always much smaller than $<\tau_0>$.

If only the variables, which have to do with the magnet, are taken into account, then the quality - factor Q_B of the magnet is

$$Q_B = \epsilon_a \epsilon_b B^3 \qquad (7)$$

The efficiencies depend on the detector size and on the opening angle φ from source to detector. The opening angle affects the diameter of the magnet only slightly. The relation between Q_B and the magnet parameters B and φ is for 100 keV gamma radiation

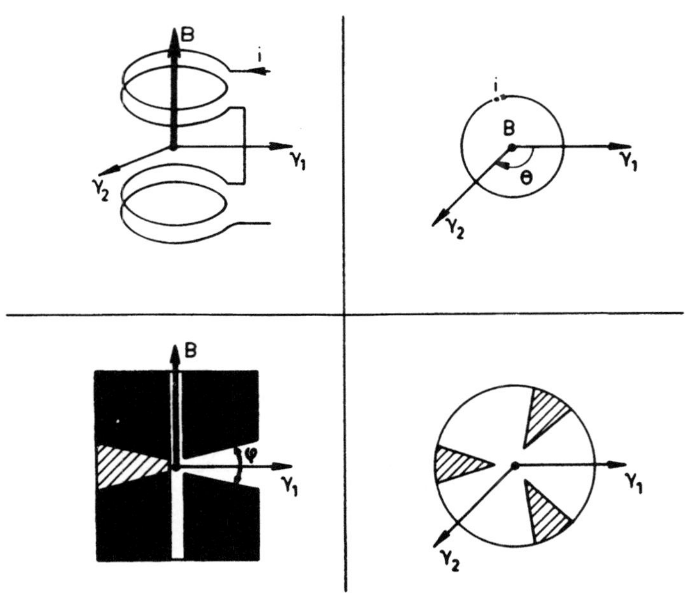

Fig. 1. Schematic drawing of magnet configuration.

$$Q_B \sim (1 - \cos\tfrac{\varphi}{2})^2\, B^3 \tag{8}$$

since the efficiency ϵ of a NaI - crystal for gamma radiation of about 100 keV is

$$\epsilon \approx 0.98\, \frac{(1 - \cos\frac{\varphi}{2})}{2} \tag{9}$$

It is clear that not only a high field strength but also a large opening angle contribute to a high quality factor. These two variables in the magnet design will be discussed now for a superconductive magnet system for perturbed angular correlation measurements. Such a magnet consists of two, tapered - wound, coil halves (fig. 1).

The field strength which can be obtained with a superconductive magnet is limited. The upper limit depends on the superconductive material used for the windings, the temperature of the superconductor and the current density in the windings.

There exists an upper critical field strength for each superconductor above which the material becomes resistive. This critical value depends on the temperature (fig. 2)[2,3]. Since the field strength is not constant over the coil - volume, one has to vary the current density in such a way, that in each point of the coil the current - density reaches the maximum value allowed by the local field strength. For a magnet,

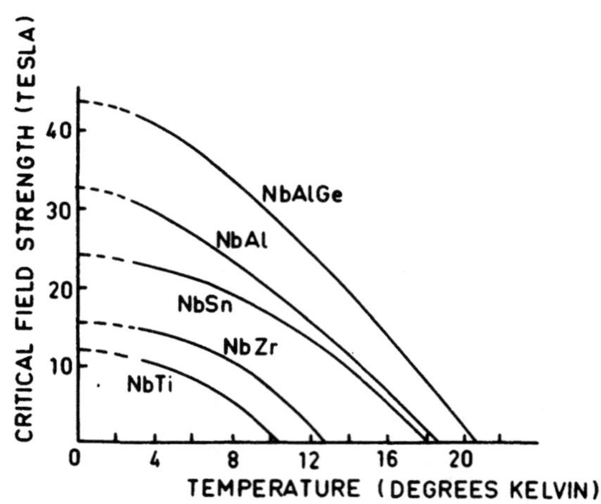

Fig. 2. Critical field strength as a function of temperature for several superconductive materials. It is assumed, that the superconductors do not carry any current. When the current density is increased, the superconductive material becomes resistive at lower field strengths.

designed in this way, the central field strength that can be obtained with a coil of that size, is maximal. By taking the outer diameter larger and the current - density smaller, higher field strengths can be obtained, until in the center of the coil the critical field strength is reached. At the critical field strength, the current - carrying - capacity is very low, and thus the outer diameter of the magnet very large. With a coil of a reasonable size the obtainable central field strength is therefore always much lower than the critical upper limit for the superconductive material, employed for the windings.

One would expect, since only NbTi, NbZr and NbSn are commercially available, that Nb_3Sn is the best material to construct magnets for high field strengths. However, the superconductive material has to be bonded to a good conductor of electricity and heat in order to provide good conducting properties at defective spots in the superconductor. In this way premature quenching of the field can be avoided and the energizing time can have a reasonable value. It is partially for this reason, that the conductor is a thin laminated strip, which consists of layers of copper and superconductive material in this case of niobium - tin. The need for stabilizing material depends on the superconductive material, and since stabilizing material, together with insulation material, may require two - thirds of the coil - volume, it is still not completely clear, what the best superconductor is for small coils yielding high field - strengths. Magnets of the split - coil type having a reasonable outer diameter and opening angle and which are suitable for perturbed angular correlation measurements.

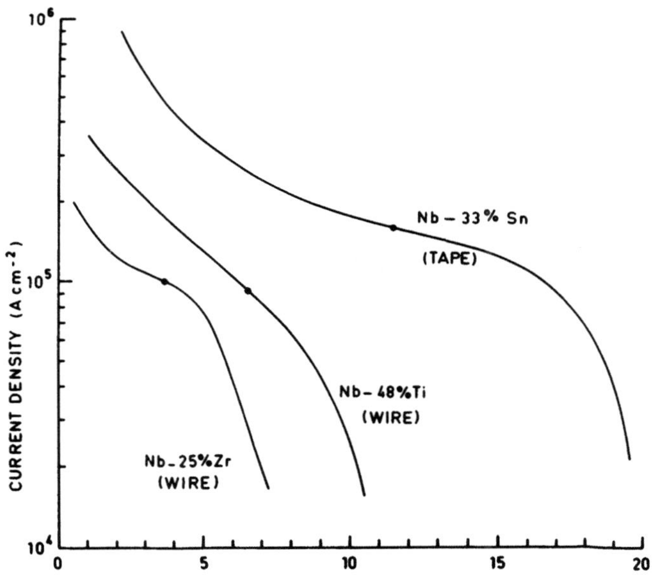

Fig. 3. Some curves of commercially available superconductors[4]) at 4.2 °K. For magnet applications one will use in the superconductor the maximum current - density that is possible in the field produced by the current itself.

can be made of NbZr for field strengths up to 3.5 tesla, of NbTi up to 6.5 tesla, and of Nb_3Sn for field strengths up to 11.5 tesla (fig. 3).

The current through the coil depends on the available dimensions of the superconductor. For Nb_3Sn, the currents are in the range from 90 to 650 amperes. The critical field strength depends on current - density and temperature. It is clear from fig. 2 that below 4.2° K the critical field strength increases only slightly. The magnet coil can be operated at boiling helium temperature. This can be accomplished in two ways. One possibility is to make a cryostat for each coil - half separately.

A cryostat wall consists of a helium - vessel wall, a heat radiation shield, an outer wall and a vacuum space on both sides of the heat - shield. Separating the coil - halves results in a much lower central field strength, or if the field strength has to be preserved, a much larger coil. A larger coil contains more superconductive material and will accordingly be more expensive. Another point is, that the costs of liquid helium for a split - cryostat are high, because such a cryostat has a bad volume-to - surface ratio. The boil - off rate of a cryostat for a 10 tesla magnet is typically four liters per hour with the magnet fully energized. This helium boil - off rate is four times higher than that of a single - cryostat system under the same conditions. We have chosen a single - cryostat system because of its lower price and much lower helium consumption; it has though the disadvantage that there is at least one cryostat wall between the source and the detectors.

In most cases one wants to be able to vary the temperature of the source. The source has to be thermally isolated from the cryostat (fig. 5). The thin aluminium radiation windows in the cryostat walls and the presence of liquid helium between the source and the detectors, leads in our design to a total transmission of 70% for gamma rays of 70 keV. In order to minimize the influence of scattered radiation, the

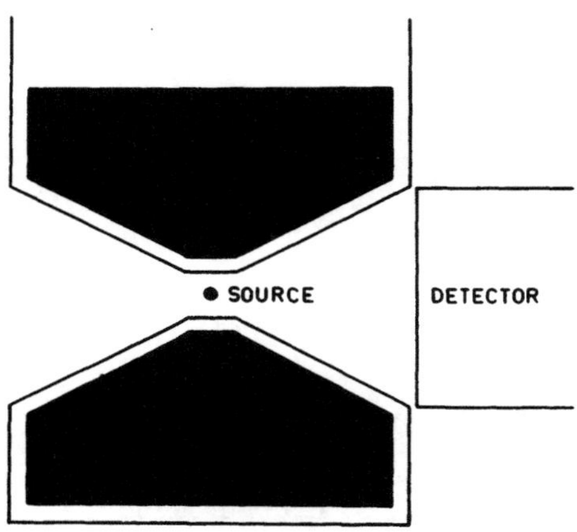

Fig. 4. Split - cryostat design with no cryostat walls between the source and the detector. In order to obtain a source volume which is a few millimeters wide, the coil halves have to be placed apart by as much as 2.5 cm.

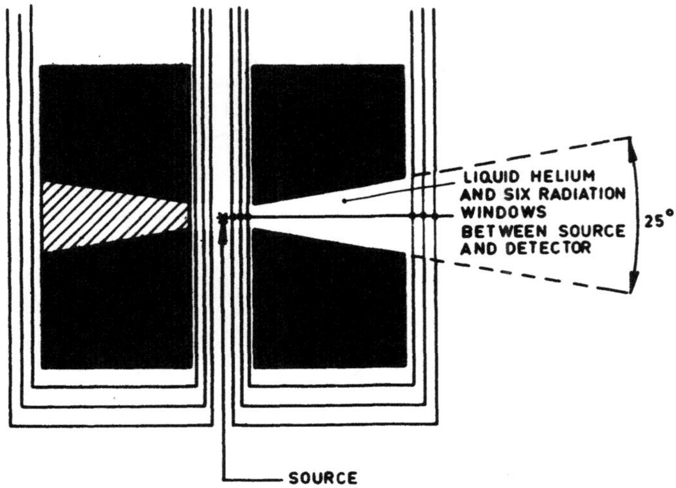

Fig. 5. Single - cryostat system showing source - detector geometry.

inner walls of the coil form should point to the source and should be lined with high - Z material (see fig. 6). To retain the opening angle, the coil - halves have then to be separated by a distance which increases by about twice the thickness of the cone material. This results in a lower obtainable field strength or a more expensive magnet. Thus it is important to know what the optimum shielding thickness is. In fig. 7 re-

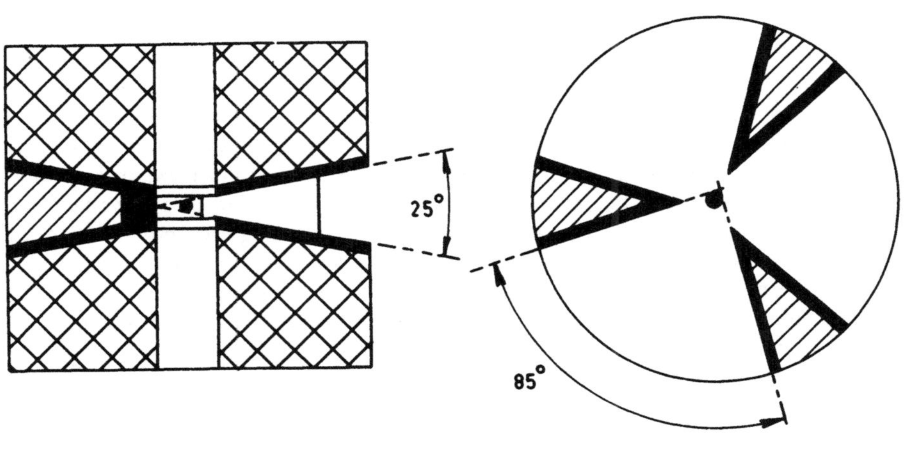

Fig. 6. Position of the radiation shielding material.

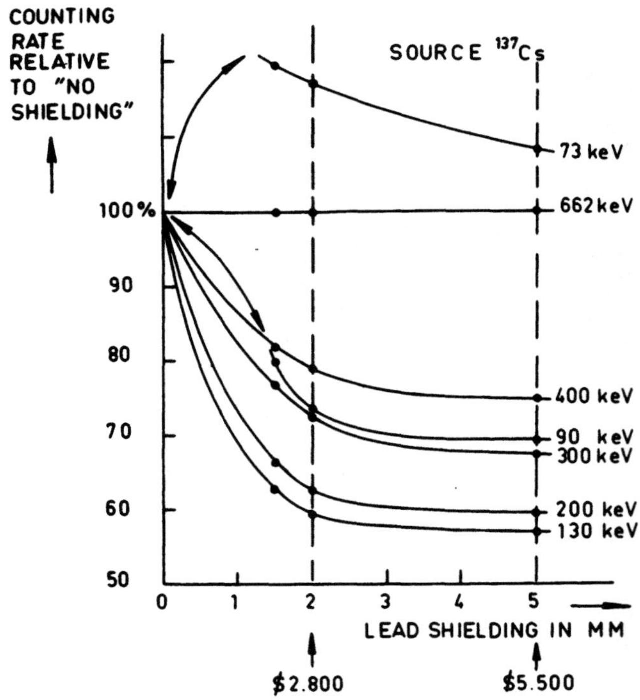

Fig. 7. Reduction of Compton - scattered gamma - rays.

sults are shown of experiments with a model of the split - coil magnet (shown in fig. 6) in which linings of lead of various thicknesses could be mounted. A 30% suppression of compton background can be obtained with a 2 mm lining of lead. Thicker linings do not improve things very much, but make the magnet more expensive. The prices quoted in fig. 7 are the extra costs for the shielding keeping all other characteristics of the system constant. A better solution to the scattering problem is the use of a high - Z construction material for the inner parts of the coil form and the spacer - blocks between the coil halves. There exist well machinable heavy - metal alloys, with a specific density of 16.8 g/cm^3, containing tungsten, nickel and copper. Little is known about the behaviour of these alloys at boiling helium temperature.

In conventional magnets with iron, the stray fields involved are weak. The magnetic shielding of photomultipliers and Ge (Li) - detectors is relatively simple in these cases. Iron - free magnets, like superconductive magnets, produce severe stray fields. A superconductive magnet with a central field of 10 tesla will give a stray field as high as 0.5 tesla at the outer wall of the cryostat (see fig. 5). For scintillation detectors the resulting field strength at the photomultipliers should not exceed a few tenths of a microtesla. Ge (Li) - detectors are less sensitive to magnetic fields[5]). However, they might show a difference in timing characteristics and in full - energy peak efficiencies if the field direction is changed from 'field up' to 'field down'.

In our case a shielding has been constructed which consists of four concentric cylinders of materials with different magnetic characteristics. The diameter of the outer cylinder was minimized down to 20 cm for a 3" diameter NaI (Tl) crystal. A 10 cm long light guide is inserted between crystal and photomultiplier.

Split coil magnets for central field strengths up to 11 tesla can be made. It is interesting to know what the price of a magnet coil is as a function of the opening angle φ and the central field strength. A comparison of six quotations showed that the price is approximately a linear function of the opening angle, if the central field strength is taken to be constant (fig. 8). The lines for 10T and 11T are almost parallel. This implies that the additional costs required for a larger opening angle are the same for a 10T and an 11T magnet.

To decide what the best system is which can be bought with a given budget, another graph with the price and the system quality as parameters is required. A magnet quality factor Q_B has already been introduced in eq. (7). For gamma radiation with an energy up to about 100 keV, Q_B is approximately

$$Q_B \approx \frac{1}{4}\left(1 - \cos\frac{\varphi}{2}\right)^2 B^3 \tag{10}$$

Since the diameter of the coil does not change appreciably as a function of the opening angle, it is possible to derive a simple quality factor for higher energies. For gamma radiation of 660 keV Q_B is approximately

$$Q_B \approx \frac{1}{4}\left[\left(14\left(1 - \cos\frac{\phi}{2}\right) + 0.2\right)\left(1 - \cos\frac{\phi}{2}\right)\right]^2 B^3 \tag{11}$$

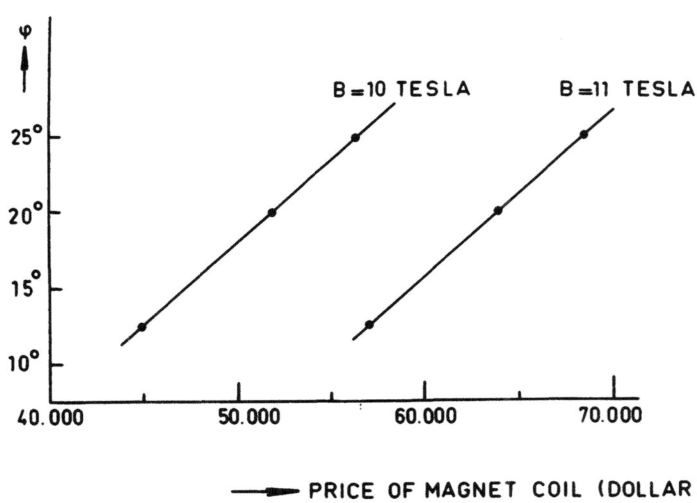

Fig. 8. Opening angle as a function of price of the magnet coil for two values of the maximum attainable field.

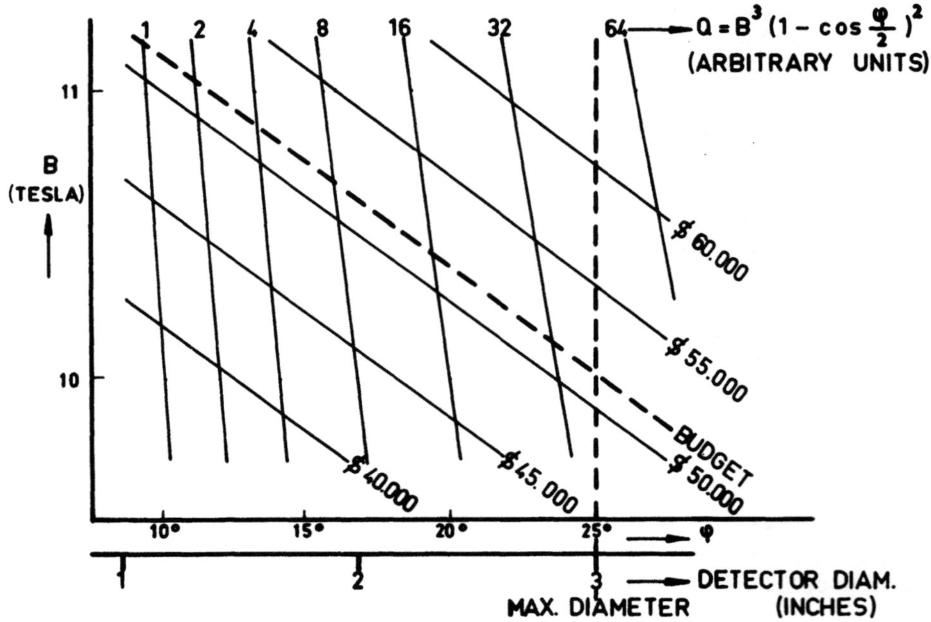

Fig. 9. Maximum attainable field strength as a function of opening angle and of detector diameter. The quality factor and the price are used as parameters.

Fig. 9 shows lines of constant quality for 100 keV radiation. The dashed line represents our budget for the magnet coil. From this figure it is clear that a large opening angle is less expensive than a high field strength. For a maximum quality at a given budget, one has to use detectors which are as large as possible. It should be noted that only the price of the magnet - coil has been considered; the detector costs have not been included.

The system we have choosen is manufactured by the Oxford Instrument Company and has the following characteristics:

A magnet coil made of Nb_3Sn superconductor. A maximum field strength of 10T can be obtained. The opening angles are 25° and 85° (see fig. 6) allowing a source diameter of 3 mm. A cryostat with aluminium radiation windows. The outer diameter is 32 cm. The helium boil - off rate is 1 l/hr if the magnet is fully energized. A current stabilized power supply for 250 amperes. Mains power consumption is 1100 watts. The energizing rate is programmed by a current programmer to keep the energizing time as short as possible (e.g. 30 minutes). The level of the helium can be checked with a level monitor at four different heights.

These features are very similar to those of the magnet system built by the Oxford Instrument Company for the group at Bonn.

The price for the above is $ 65.000,- -. This price does not include a field strength measuring device ($ 1.500,- -), the helium supply equipment ($ 2.000,- -) and a helium recovery system ($ 5.000,- -).

References

1. A.H. Wapstra, Alpha-, Beta- and Gamma-Ray Spectroscopy, ed. K. Siegbahn, North - Holland Publ. Co., Amsterdam (1965) p. 543.
2. G.K. Jones and J.K. Hulm, Rev. Mod. Phys. 36 (1964) 74.
3. S. Foner and E.J. McNiff, Phys. Lett. 31A (1970) 349.
4. Specification Sheets, Supercon, Natick, Mass., U.S.A.
5. P. Ganner and J. Rauch, Nucl. Instr. and Meth. 76 (1969) 295.

Discussion

Bodenstedt: The superconductive magnet in Bonn came into operation three weeks ago. We started the design more than three years ago and I am quite satisfied that our considerations were quite similar to those mentioned in your talk and that we arrived at nearly the same construction. I have a few slides of the actual construction of our magnet. Slide 1 shows a horizontal cut through the centre. Between the three massive spacers we have openings, each of which has a width of 85° and a height of 25°.

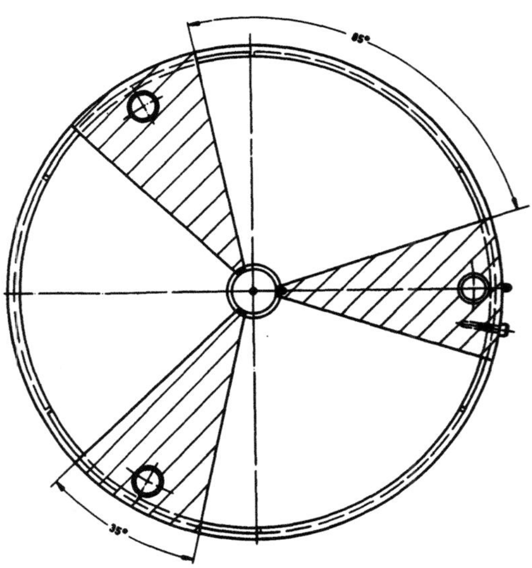

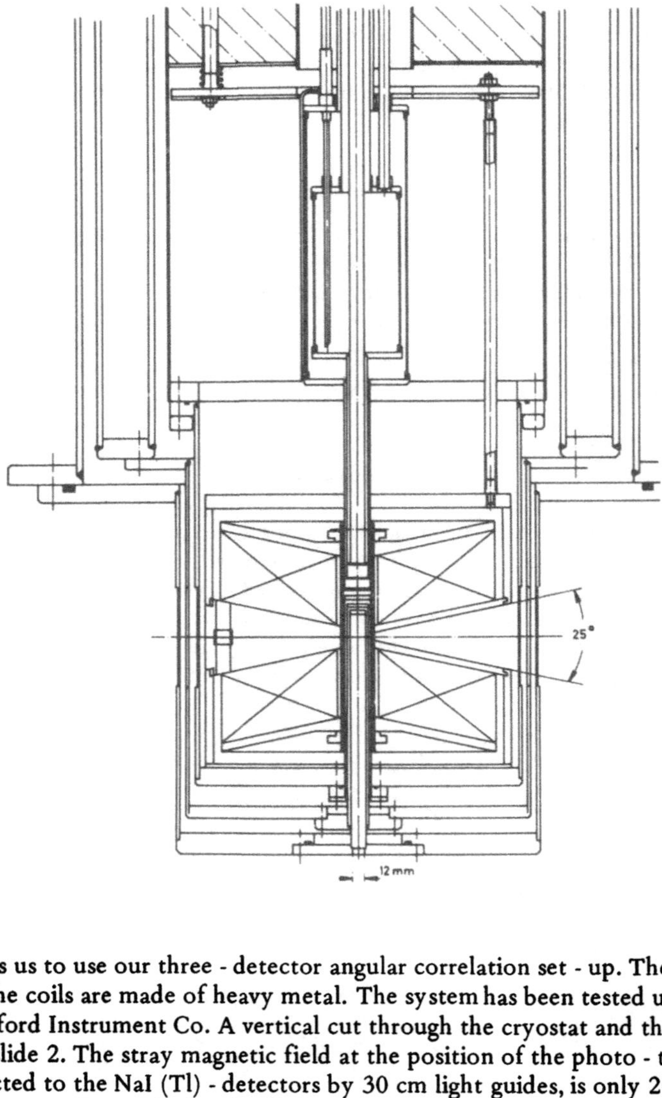

This allows us to use our three - detector angular correlation set - up. The spacers between the coils are made of heavy metal. The system has been tested up to 105 kG by the Oxford Instrument Co. A vertical cut through the cryostat and the coil is shown in slide 2. The stray magnetic field at the position of the photo - tubes, which are connected to the NaI (Tl) - detectors by 30 cm light guides, is only 200 gauss. The multipliers can rather easily be shielded.

SUMMARY

ANGULAR CORRELATIONS IN NUCLEAR DISINTEGRATIONS

Hans Frauenfelder
University of Illinois, Urbana, Illinois, U.S.A.

The time is already late and we have worked hard for nearly a week; I will therefore be brief and also not completely serious. So let me first say that Dr. Deutch in his excellent lecture, has given a proof that AB and BA are not necessarily equivalent. His proof, based on a raw and a hard - boiled egg, is valid only in the single - particle approximation. For the next conference I should like to suggest a many - body experiment. This time the farewell drink is scheduled **after** the summary; next time it should be arranged **before** and the effect of this interchange on the summary and the audience should be carefully studied.

For the past few years, I have not been active in angular correlation work. After listening to all lectures at this conference, I have asked myself the question:

1. Should angular correlation conferences still be held?

To answer this question, I show in the first figure three curves that describe, in an idealized way, the accumulation of knowledge, K, as a function of time for three situations.

Case I, a dramatic 'step - function' increase in knowledge, is the dream of every physicist. It does not occur often and its importance is not always recognized immediately. Examples are easy to quote (after the fact), and I name only three recent ones: Mössbauer effect, pulsars, and gravity waves. Results are published in Phys. Rev. Letters or Physics Letters. Conferences are not really needed because every physicist learns quickly about these new discoveries; the New York Times and Time Magazine take care of that.

Case II is 'normal'. Many small advances add up to a continuous rise and results are published in Phys. Rev. or Nuclear Physics.

Case III can occur because of lack of communication between the old, experienced, tired, and the young and eager physicist. Old results are rediscovered and nothing new is added to the store of K. 'Progress' reports result[1]).

Conferences such as the present one permit us, the 'old generation', to point out experiments and theories that we had done 15 or 20 years ago and thereby prevent a slip from Case II to Case III. So - - - I believe I have found a justification for the present meeting.

For my summary I have an excellent base: Steffen's report at the end of the Uppsala conference[2]) in 1963. He was courageous enough to make a long - term forecast and he selected the year 1970 as target date. Thus he correctly predicted the present conference. His other guesses were also very close to the mark and I will re-

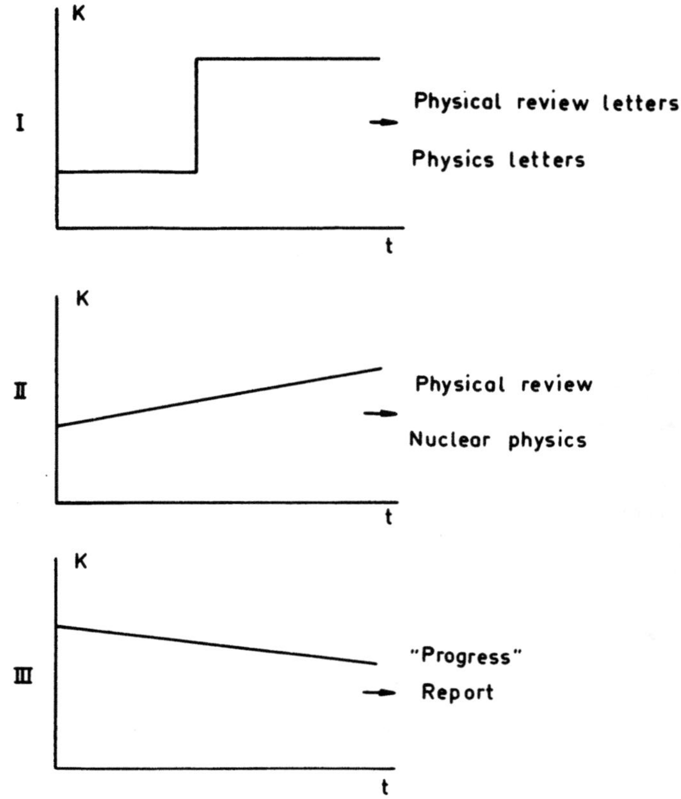

Fig. 1. Knowledge, K, as a function of time for three typical situations.

turn to them later.

In the following I will summarize what I consider to be the significant advances since 1963. The history up to 1963 can be found in Steffen's earlier summary[2]).

2. Theory

The theory of angular correlation (AC) and perturbed angular correlation (PAC) is now in very good shape, as we could learn from STEFFEN's introductory talk*. It

* The names of speakers at the present conference are capitalized. For more information on the subjects mentioned, consult their papers.

is elegant, complete, and contains no obvious large holes. I should, however, like to make two remarks and I will base these on two quotes. The first is ascribed to Boltzmann: "Elegance should be left to the tailors". The danger with very elegant and general theories is that they often make it hard for the beginner to get a feeling for the physical picture. We should therefore not forget the simple 'hand - waving type' description of A.C. and P.A.C.

The second quote is from Landau[3]): "The brevity of life does not allow us the luxury of spending time on problems which will lead to no new results". There exists already a number of rather complete formulations of AC and PAC. It is not very useful to add another, only slightly different, approach that does not lead to new results.

Progress in AC theory has occurred for conversion electron - gamma and alpha - gamma cascades. The theory for $e^- - \gamma$ is now essentially complete, as was pointed out by GERHOLM. The theory of α - γ cascades was, of course, worked out a long time ago. MANG now has shown that progress has been made - - the comparison of the results of nuclear model calculations and the α - mixing amplitudes that follow from α - γ measurements.

The theory of PAC is evidently more complex than that of ordinary AC. Nevertheless, considerable progress has been made here also. In particular, GABRIEL has generalized the Abragam - Pound theory of PAC in time - fluctuating surroundings. His calculations are performed in Liouville space. To most of us this space is unfamiliar and we can ask: Is its use luxury or necessity? If it is the first, we will be able to relax and wait for simplification. If its use is necessary, then we will have to learn more about it. In any case, various speakers have pointed out that the Gabriel generalization may explain discrepancies between their data and existing theories.

3. Experimental

No radically new technique has appeared since 1963. Nevertheless, progress has been made.

Solid state counters were already known in 1963, but it was expected that their main usefulness in A.C. would be for energy measurements of charged particles. The main impact, however, has been on experiments involving gamma rays. Very complex spectra, decays with 100 - 200 gamma rays, can now be studied. The most serious problem left is the correction for Compton edges.

Some experiments involving **the implantation technique** were already reported in 1963. In the last few days we have heard a number of excellent talks on this subject. Particularly the papers by MURNICK and DEUTCH* give a clear picture of the progress that has been achieved. The position of the implanted atoms appears to be well defined. High magnetic fields can be obtained for g - factor measurements. Since the high field is contained in a small volume, even conversion electrons can be used in A.C. experiments.

Superconducting magnets also had appeared before 1963 and it was then hoped that 200 - kG magnets would soon be available. Here is one case where real

*Dutch readers may be interested to know that Dr. Deutch's contribution was given with such excellent showmanship that there was talk of replacing Wim Kan by Deutch. I hope that, at future meetings, more speakers will perform in a similar way.

progress was slower than expected. At the present time, fields of about 100 kG (10T) are easily (but not inexpensively) available and we still have to wait for the 200 kG variety.

It is, of course, impossible to predict what new techniques will appear in the future and so I will only make one remark. It may be possible, as LEISI has suggested, to use the extremely high internal magnetic fields produced by after - effects in gaseous sources to measure g - factors of very short - lived states.

4. Angular correlation in nuclear physics

The application of angular correlation techniques to nuclear physics problems is straightforward. No basically new results or ideas have appeared since 1963. However, I have been impressed by the amount of new information and by the degree to which the experimental results and the prediction from nuclear model calculations can be matched.

Spin and parity assignments are the bread - and - butter research of A.C. people and there is no need to say more.

A large number of **mixing ratios** δ of gamma rays have been determined in the last few years. In the past few days we have heard many 'historical' references. I will sketch the development of mixing ratio measurements here because it shows how the importance of a physical quantity changes in time and how different aspects of a problem become important at different times. Before 1951, the assignment of a definite multipolarity to a nuclear gamma ray rested on shaky grounds. Weisskopf's estimate of lifetimes[4]), Goldhaber and Sunyar's classification[5]), and Rose's conversion coefficients[6]) had not yet been published and no clear case of a mixed multipole radiation was known.

In early 1951, we started a systematic search for a perturbed AC[7]) and we selected ^{111}Cd as a suitable nuclide. Boehm and Walter[8]) had measured the anisotropy of the $7/2 \overset{\gamma}{\rightarrow} 5/2 \overset{\gamma}{\rightarrow} 1/2$ cascade and had found a value of $A = -0.07 \pm 0.04$, in reasonable agreement with the value $A = -0.10$ predicted for a pure dipole - quadrupole transition. We hoped, by a suitable choice of the environment, to reduce the value of A. We indeed found values consistent with $A = 0$, but our surprise was great when we also obtained anisotropies as large as -0.17 ± 0.01[9]). Such a large value created difficulties in the interpretation. The spin sequence of the decay was uniquely fixed by other data and the only way to explain the observed anisotropy was to invoke a mixed transition[10]). At this point an old problem and a new solution appeared. Up to about that time, the mixing ratio $\delta = |\delta|\, e^{i\varphi}$ was treated as an arbitrary complex number; φ could vary from 0 to 2π. The anisotropy therefore was not sufficient to fix $|\delta|$ and φ uniquely. So far the old problem. The new solution was put forward by S.P. Lloyd who pointed out that time - reversal invariance limited the choice of φ to two values, 0 and π. I accidentally saw Lloyd's paper in Phys. Rev.[11]). I could read the conclusion but, of course, could not understand the proof. I therefore went to Weisskopf and Pauli for help and they assured me that the derivation was correct. Armed with this new restriction, we could uniquely determine magnitude and spin of the mixing ratio of the 173 keV transition.

In 1951, the value of δ was a mere curiosity since no nuclear model calculation was available that predicted this observable. In the intervening time, the situation has changed in three ways: In the first place, very many mixing ratios have been measured and a systematic review is therefore possible. The comprehensive paper by J.H. HAMILTON indicates the wealth of available information. In the second place, the observed values of δ, sign and magnitude, no longer exist in a theoretical vacuum. Computations based on nuclear models, such as the one by Kumar and Baranger[12]), or by GLAUDEMANS, can now be compared with the experimental values and the agreement is in general encouraging. In particular, the sign is always predicted correctly. Mixing ratios therefore provide a sensitive test of nuclear wave functions. In the third place, exceedingly careful measurements of the phase of δ provide a test of time - reversal invariance. A deviation of the phase φ of δ from 0 or π would imply a violation of TRI. I will return to this aspect in a few minutes.

In most **alpha decays**, the emitted alpha particle can carry away more than one value of the orbital angular momentum and the admixtures of various L - values is described by amplitudes a_L. Some measurements exist, but much work remains to be done; more, and more precise, values of a_L are clearly needed.

Beta - decay matrix elements can be determined in allowed and in forbidden beta decays. In **allowed** decays, the angular correlation between beta and circularly polarized gammas determine the Fermi matrix element. BLOOM, and also BEHRENS have told us how this information is valuable in obtaining isospin impurities. A comparison with nuclear model computations is now feasible. It may even be possible to gain information concerning the part of the nuclear potential that violates isospin. (Existence of such a potential does not mean that the hadronic part of the nuclear force is charge - dependent; the electromagnetic interaction produces for instance a mass splitting between neutral and charged pions and thus introduces a charge dependence of the force between nucleons). In **forbidden** beta decay, the situation is more complicated, as we learned from the talk of SIMMS. The β - γ angular correlation alone does not permit a unique determination of the various matrix elements. However, in combination with other experiments, useful information concerning the matrix elements can be extracted.

5. Symmetry tests

Angular correlation techniques have been used to test the invariance of hadronic forces under parity, P, and under time - reversal, T. The two operations are very different and the situation with regard to expected violations is also very different. I will first consider **parity**. It is well known that parity is violated in weak interactions and this violation leads to a small parity violation in nuclear interactions. Here we must be careful in stating what we mean with a nuclear interaction. Two hadrons interact via the exchange of mesons. The overwhelming contribution to the nuclear force comes from diagrams in which both vertices are strong:

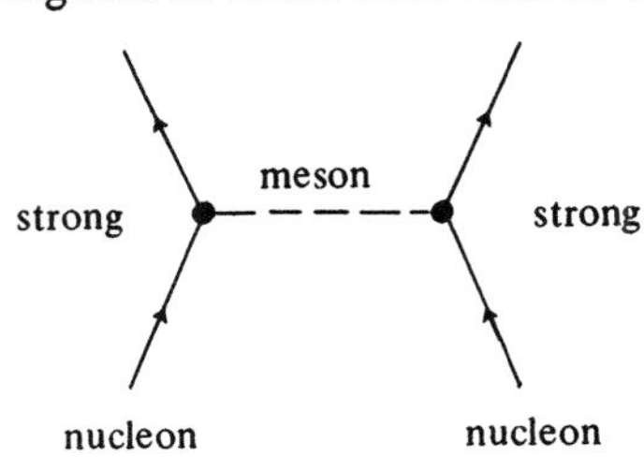

We assume that this diagram does not lead to parity violation. However, mesons also interact weakly and one (or both) of the vertices can be weak:

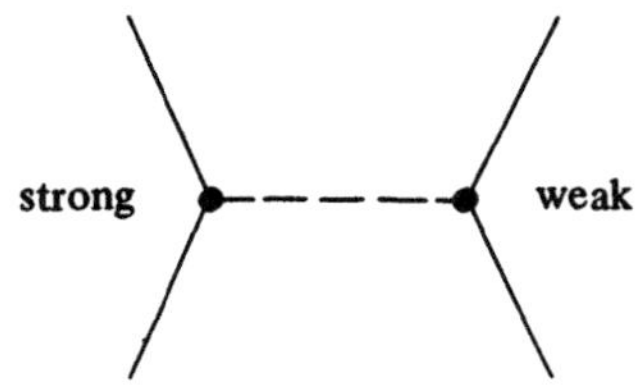

This graph leads to a parity violation. (In fact, before Yukawa, váliant efforts were made to explain nuclear forces with two weak vertices - - of course the resulting force turned out to be much too weak). Now, any theory of weak interactions that includes parity violation also leads to a very small parity - nonconserving contribution to nuclear forces. The current - current form[13]) , in addition, gives an unambiguous prediction of the size of the contribution. Unfortunately, nuclear wave functions are not well enough known to make reliable calculations for complex nuclei. BOEHM has surveyed the present situation in his talk and he has shown that evidence for the existence of the parity violation is unambiguous but that more theoretical and experimental work is needed for quantitative conclusions.

Time - reversal tests are experimentally much harder to perform than parity tests. Theoretically, the situation is also less clear. Detailed studies of the decay of neutral kaons have shown that both CP and T are violated[14]). However, it is unclear on which interaction the violation must be blamed. If the electromagnetic interaction is responsible[15]), then nuclear states and transition operators can contain small, but measurable, time - reversal irregular admixtures and tests involving nuclear decays can find these. If, on the other hand, a super - weak interaction[16]) is the cause of the violation, then we cannot expect to find measurable effects in nuclear decays. The super - weak interaction leads to a definite prediction concerning the parameters describing the kaon decay ($\eta_{\pm} = \eta_{00}, \varphi_{+-} = \varphi_{00} = 43°$). Two or three years ago, experiments disagreed strongly with this prediction and thus many groups were lured to search for TRI violations in decays other than that of the $K°$. In the meantime, the experiments tend to converge towards the super - weak prediction[17]) and it is less likely that a positive effect can be detected in a nuclear decay.
W.D. HAMILTON has given us a clear picture of the present state of experiments and the situation can be summarized by saying with the Viennese: "Die Lage ist hoffnungslos, aber nicht ernst".

6. Perturbed angular correlation in nuclear physics

Let me first review the situation concerning nuclear g - factors. In 1951, the first g - factor of an excited nuclear state (^{111}Cd) was measured; in 1963, 35 were known, and Steffen predicted 150 by 1970. In fact, about 270 g - factors have been measured. Steffen's prophecy has turned out to be very good. Not only are very many g - factors known, the errors have also become considerably smaller. The steady progress is due to better counters, better electronics, larger magnetic fields (superconductors and implantation techniques), use of accelerators - on - line, and

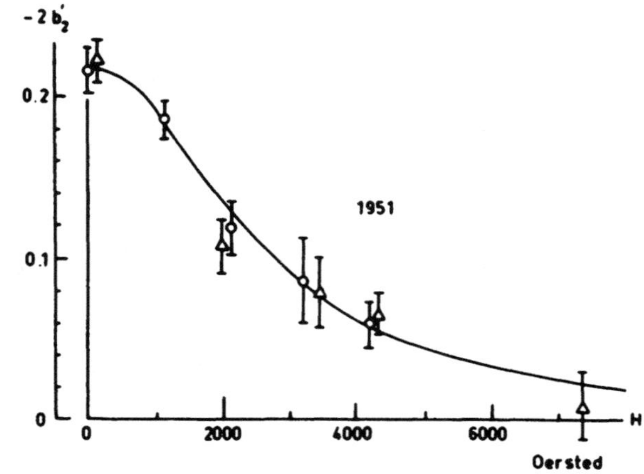

1951. Attenuation of the correlation pattern of ^{111}Cd. (Aeppli et al., Phys. Rev. 84 (1951) 370).

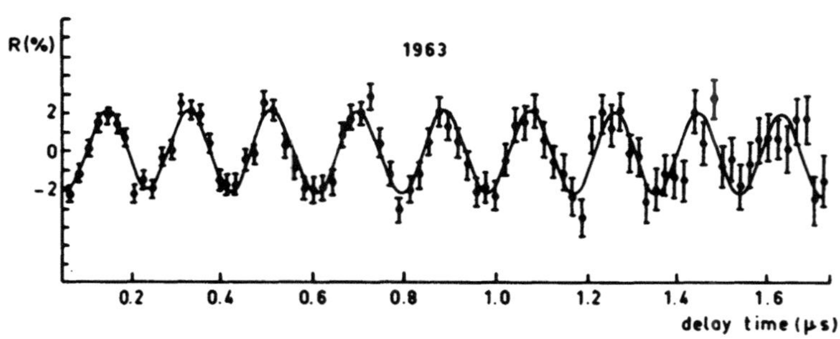

1963. Rotation of the correlation pattern of ^{187}Re. (Walter et al.: 'Perturbed Angular Correlations', ed. by Karlsson, Matthias and Siegbahn, North-Holl. Publ. Co., 1964).

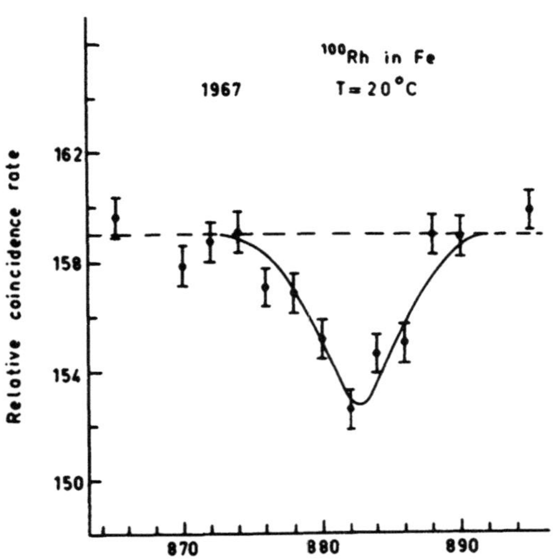

1967. Resonance destruction of the correlation pattern of ^{100}Rh. (Matthias et al.: ed. by Matthias and Shirley: 'Hyperfine Structure and Nuclear Reactions', North-Holl. Publ. Co., Amsterdam, 1968).

Fig. 2. Advances in the measurements of g-factors of excited nuclear states by angular correlation techniques.

the introduction of resonance techniques. The following figure gives an idea of the progress (fig. 2).

At the beginning, only the attenuation of the correlation could be measured easily because the gamma rays could not be separated cleanly. The scintillation counter changed that and rotation of the spin through many times 2π could be observed as evidenced by the second example (taken from a contribution to the Uppsala conference). Even higher accuracy can be obtained with the resonance destruction method. Unfortunately, no summary concerning this last technique has been given here and not much work has been published since 1967. I hope that many more resonance experiments will be performed between now and the next conference.

The large number of g-factors of excited nuclear states permits rather detailed comparisons with nuclear model calculations. As the talks by KRUMLINDE and OGAZA have shown, experiments are still very far ahead of theory and considerable time will pass before the high accuracy obtainable with resonance destruction experiments can be utilized. Now, some of my friends have urged me to present a typical slide of the experimental results. I was at first reluctant to do so, but here is a collection of g - factors:

. . ' ; " ' ' | ' .
" - ; ' || ↔ " :
' " ,,, , : ` | '
| | " ' ' " | ' , '
"' - ` || " ''''

sorry-reserved
for next
conference

Summary
of g-factors

I am sorry if you cannot read the small numbers in the upper part, but this is what is meant by 'typical'. I am also sorry that I had to omit the data in the lower part, but they are reserved for the Rehovoth conference.

Next, I will briefly review the situation concerning **quadrupole moments** of excited nuclear states. In 1953, the first quadrupole coupling was determined; in 1963, 2 quadrupole couplings were known. At the Uppsala conference, Steffen predicted that about 50 quadrupole moments would have been measured by 1970 with the reorientation method. Indeed, about fifty quadrupole moments have been so determined and, as GLAUDEMANS has shown in his talk, meaningful comparison with theory is possible now. Moreover, BODENSTEDT has given us an exhaustive review of quadrupole moments in PAC and he optimistically predicts a boom within the next few years.

7. Final remarks

In my summary, I have focused my attention on problems connected with nuclear physics and I have omitted the applications of PAC to atomic, solid - state, and liquid - state physics. Clearly, considerable progress has been made in these fields but a coherent summary would also involve discussing results obtained with other methods, such as Mössbauer effect and magnetic resonance.

And now I come to the most pleasant task, thanking the organizing committee and in particular Bob van Nooijen and Hans van Krugten. I believe that I have never attended a conference that went more smoothly and with less difficulties. Everything worked extremely well. Even the electronics problems were surmounted very quickly, a feat that is unusual at a physics conference.

References

1. C.N. Parksinson, Parkinson's Law (John Murray, London, 1958). Chapter 9 (Injelilitis).
2. R.M. Steffen, in Perturbed Angular Correlations (E. Karlsson, E. Matthias, K. Siegbahn, eds.), North – Holland, 1964.
3. L. Landau, in Theoretical Physics in the Twentieth Century, M. Fierz and V.F. Weisskopf, eds., Interscience, London, 1960.
4. V.F. Weisskopf, Phys. Rev. 83 (1951) 1073.
5. M. Goldhaber and A.W. Sunyar, Phys. Rev. 83 (1951) 906.
6. Rose, Goertzel, Spinrad, Haar, and Strong, Phys. Rev. 83 (1951) 79.
7. H. Frauenfelder, Phys. Rev. 82 (1951) 549.
8. F. Boehm and M. Walter, Helv. Phys. Acta 22 (1969) 378.
9. Aeppli, Bishop, Frauenfelder, Walter, and Zunti, Phys. Rev. 82 (1951) 550.
10. Aeppli, Frauenfelder, and Walter, Helv. Phys. Acta 24 (1951) 335.
11. S.P. Lloyd, Phys. Rev. 81 (1951) 161.
12. K. Kumar, Phys. Letters 29B (1969) 25.
13. R.P. Feynman and M. Gell - Mann, Phys. Rev. 109 (1958) 193.
14. R.C. Casella, Phys. Rev. Letters 22 (1969) 554;
    Schubert et al., Phys. Letters 31B (1970) 662.
15. J. Bernstein, G. Feinberg, and T.D. Lee, Phys. Rev. 139 (1965) B1650.
16. L. Wolfenstein, Phys. Rev. Letters 13 (1964) 562.
17. e.g. M. Cullen et al., Physics Letters 32B (1970) 523;
    I.A. Budagov et al., Phys. Rev. D2 (1970) 815.

APPENDIX I*

SOME TABLES FOR THE ANALYSIS OF (1 - 3) γ - γ DIRECTIONAL CORRELATIONS

A.V. Ramayya, J.H. Hamilton, P.E. Little and E. Collins
Vanderbilt University†, *Nashville, Tenn., U.S.A.*

and

J.J. Pinajian
Oak Ridge National Laboratory††, *Oak Ridge, Tenn., U.S.A.*

*See contributed paper of these authors on page 260 of these Proceedings.

† Work supported in part by a grant from National Science Foundation.
†† Work supported by the U.S. Atomic Energy Commission under contract with Union Carbide Corporation.

1 (1,2) 2 (2,2) 2 (2,2) 0

delta 1	A_2	A_4	delta 1	A_2	A_4
100.00	–.0359	–.2177	– .01	.0512	–.0000
70.00	–.0348	–.2176	– .02	.0487	–.0001
50.00	–.0334	–.2176	– .04	.0439	–.0003
40.00	–.0322	–.2176	– .06	.0389	–.0008
20.00	–.0261	–.2171	– .08	.0339	–.0014
10.00	–.0136	–.2155	– .10	.0289	–.0022
9.50	–.0123	–.2153	– .12	.0239	–.0031
9.00	–.0109	–.2150	– .14	.0189	–.0042
8.50	–.0092	–.2147	– .16	.0139	–.0054
8.00	–.0074	–.2143	– .20	.0040	–.0084
7.50	–.0053	–.2139	– .22	–.0009	–.0100
7.00	–.0029	–.2133	– .26	–.0106	–.0138
6.50	–.0001	–.2127	– .28	–.0153	–.0158
6.00	.0031	–.2118	– .30	–.0200	–.0180
5.50	.0068	–.2107	– .32	–.0245	–.0202
5.00	.0113	–.2093	– .36	–.0333	–.0250
4.50	.0168	–.2074	– .38	–.0376	–.0275
4.00	.0235	–.2049	– .40	–.0417	–.0300
3.50	.0320	–.2013	– .42	–.0457	–.0326
3.00	.0428	–.1959	– .46	–.0534	–.0380
2.50	.0570	–.1877	– .48	–.0571	–.0408
2.00	.0759	–.1741	– .50	–.0606	–.0435
1.50	.1006	–.1507	– .60	–.0764	–.0576
1.00	.1274	–.1088	– .70	–.0892	–.0716
.90	.1316	–.0974	– .80	–.0991	–.0850
.80	.1346	–.0850	– .90	–.1067	–.0974
.70	.1359	–.0716	– 1.00	–.1121	–.1088
.60	.1350	–.0576	– 1.50	–.1206	–.1507
.50	.1310	–.0435	– 2.00	–.1157	–.1741
.48	.1298	–.0408	– 2.50	–.1082	–.1877
.46	.1285	–.0380	– 3.00	–.1010	–.1959
.42	.1253	–.0326	– 3.50	–.0946	–.2013
.40	.1235	–.0300	– 4.00	–.0892	–.2049
.38	.1215	–.0275	– 4.50	–.0847	–.2074
.36	.1194	–.0250	– 5.00	–.0808	–.2093
.32	.1146	–.0202	– 5.50	–.0775	–.2107
.30	.1119	–.0180	– 6.00	–.0746	–.2118
.28	.1091	–.0158	– 6.50	–.0721	–.2127
.26	.1061	–.0138	– 7.00	–.0700	–.2133
.22	.0996	–.0100	– 7.50	–.0680	–.2139
.20	.0961	–.0084	– 8.00	–.0663	–.2143
.16	.0887	–.0054	– 8.50	–.0648	–.2147
.14	.0847	–.0042	– 9.00	–.0634	–.2150
.12	.0806	–.0031	– 9.50	–.0622	–.2153
.10	.0764	–.0022	– 10.00	–.0611	–.2155
.08	.0720	–.0014	– 20.00	–.0500	–.2171
.06	.0676	–.0008	– 40.00	–.0442	–.2176
.04	.0630	–.0003	– 50.00	–.0430	–.2176
.02	.0583	–.0001	– 70.00	–.0417	–.2176
.01	.0560	–.0000	– 100.00	–.0407	–.2177
.00	.0536	–.0000	∞	–.0383	–.2177

2 (1,2) 2 (2,2) 2 (2,2) 0

delta 1	A_2	A_4	delta 1	A_2	A_4
100.00	.0180	.0933	– .01	– .0551	.0000
70.00	.0186	.0933	– .02	– .0567	.0000
50.00	.0195	.0933	– .04	– .0597	.0001
40.00	.0203	.0932	– .06	– .0627	.0003
20.00	.0240	.0931	– .08	– .0656	.0006
10.00	.0312	.0924	– .10	– .0684	.0009
9.50	.0320	.0923	– .12	– .0711	.0013
9.00	.0328	.0922	– .14	– .0738	.0018
8.50	.0336	.0920	– .16	– .0763	.0023
8.00	.0346	.0919	– .20	– .0810	.0036
7.50	.0357	.0917	– .22	– .0833	.0043
7.00	.0370	.0914	– .26	– .0873	.0059
6.50	.0384	.0911	– .28	– .0892	.0068
6.00	.0399	.0908	– .30	– .0910	.0077
5.50	.0418	.0903	– .32	– .0926	.0087
5.00	.0439	.0897	– .36	– .0955	.0107
4.50	.0463	.0889	– .38	– .0968	.0118
4.00	.0492	.0878	– .40	– .0980	.0129
3.50	.0525	.0863	– .42	– .0991	.0140
3.00	.0565	.0840	– .46	– .1009	.0163
2.50	.0608	.0804	– .48	– .1017	.0175
2.00	.0651	.0746	– .50	– .1023	.0187
1.50	.0673	.0646	– .60	– .1042	.0247
1.00	.0598	.0466	– .70	– .1042	.0307
.90	.0557	.0418	– .80	– .1028	.0364
.80	.0502	.0364	– .90	– .1002	.0418
.70	.0431	.0307	– 1.00	– .0970	.0466
.60	.0341	.0247	– 1.50	– .0775	.0646
.50	.0232	.0187	– 2.00	– .0603	.0746
.48	.0207	.0175	– 2.50	– .0473	.0804
.46	.0182	.0163	– 3.00	– .0376	.0840
.42	.0129	.0140	– 3.50	– .0303	.0863
.40	.0102	0129	– 4.00	– .0246	.0878
.38	.0073	.0118	– 4.50	– .0201	.0889
.36	.0044	.0107	– 5.00	– .0165	.0897
.32	– .0015	.0087	– 5.50	– .0134	.0903
.30	– .0046	.0077	– 6.00	– .0109	.0908
.28	– .0078	.0068	– 6.50	– .0088	.0911
.26	– .0109	.0059	– 7.00	– .0070	.0914
.22	– .0174	.0043	– 7.50	– .0054	.0917
.20	– .0207	.0036	– 8.00	– .0040	.0919
.16	– .0274	.0023	– 8.50	– .0028	.0920
.14	– .0307	.0018	– 9.00	– .0017	.0922
.12	– .0340	.0013	– 9.50	– .0007	.0923
.10	– .0373	.0009	– 10.00	.0002	.0924
.08	– .0407	.0006	– 20.00	.0084	.0931
.06	– .0439	.0003	– 40.00	.0124	.0932
.04	– .0472	.0001	– 50.00	.0132	.0933
.02	– .0504	.0000	– 70.00	.0141	.0933
.01	– .0520	.0000	– 100.00	.0148	.0933
.00	– .0536	.0000	∞	.0164	.0933

3 (1,2) 2 (2,2) 2 (2,2) 0

delta 1	A_2	A_4	delta 1	A_2	A_4
100.00	.0420	– .0233	– .01	.0170	– .0000
70.00	.0413	– .0233	– .02	.0187	– .0000
50.00	.0403	– .0233	– .04	.0220	– .0000
40.00	.0395	– .0233	– .06	.0254	– .0001
20.00	.0353	– .0233	– .08	.0288	– .0001
10.00	.0268	– .0231	– .10	.0322	– .0002
9.50	.0259	– .0231	– .12	.0355	– .0003
9.00	.0249	– .0230	– .14	.0389	– .0004
8.50	.0238	– .0230	– .16	.0422	– .0006
8.00	.0226	– .0230	– .20	.0480	– .0009
7.50	.0212	– .0229	– .22	.0518	– .0011
7.00	.0197	– .0229	– .26	.0579	– .0015
6.50	.0178	– .0228	– .28	.0609	– .0017
6.00	.0157	– .0227	– .30	.0638	– .0019
5.50	.0133	– .0226	– .32	.0666	– .0022
5.00	.0104	– .0224	– .36	.0720	– .0027
4.50	.0069	– .0222	– .38	.0746	– .0029
4.00	.0026	– .0220	– .40	.0770	– .0032
3.50	– .0027	– .0216	– .42	.0794	– .0035
3.00	– .0094	– .0210	– .46	.0839	– .0041
2.50	– .0180	– .0201	– .48	.0860	– .0044
2.00	– .0291	– .0187	– .50	.0881	– .0047
1.50	– .0424	– .0161	– .60	.0968	– .0062
1.00	– .0543	– .0117	– .70	.1034	– .0077
.90	– .0554	– .0104	– .80	.1082	– .0091
.80	– .0554	– .0091	– .90	.1114	– .0104
.70	– .0541	– .0077	– 1.00	.1133	– .0117
.60	– .0512	– .0062	– 1.50	.1123	– .0161
.50	– .0461	– .0047	– 2.00	.1051	– .0187
.48	– .0448	– .0044	– 2.50	.0976	– .0201
.46	– .0434	– .0041	– 3.00	.0912	– .0210
.42	– .0403	– .0035	– 3.50	.0858	– .0216
.40	– .0386	– .0032	– 4.00	.0815	– .0220
.38	– .0368	– .0029	– 4.50	.0779	– .0222
.36	– .0349	– .0027	– 5.00	.0748	– .0224
.32	– .0307	– .0022	– 5.50	.0723	– .0226
.30	– .0285	– .0019	– 6.00	.0701	– .0227
.28	– .0262	– .0017	– 6.50	.0682	– .0228
.26	– .0237	– .0015	– 7.00	.0666	– .0229
.22	– .0186	– .0011	– 7.50	.0652	– .0229
.20	– .0158	– .0009	– 8.00	.0639	– .0230
.16	– .0101	– .0006	– 8.50	.0628	– .0230
.14	– .0072	– .0004	– 9.00	.0617	– .0230
.12	– .0041	– .0003	– 9.50	.0608	– .0231
.10	– .0010	– .0002	– 10.00	.0600	– .0231
.08	.0022	– .0001	– 20.00	.0520	– .0233
.06	.0054	– .0001	– 40.00	.0479	– .0233
.04	.0087	– .0000	– 50.00	.0470	– .0233
.02	.0120	– .0000	– 70.00	.0461	– .0233
.01	.0136	– .0000	– 100.00	.0454	– .0233
.00	.0153	– .0000	∞	.0437	– .0233

2 (1,2) 3 (2,2) 2 (2,2) 0

delta 1	A_2	A_4	delta 1	A_2	A_4
100.00	.0130	.4489	− .01	− .0405	.0000
70.00	.0119	.4489	− .02	− .0381	.0002
50.00	.0106	.4488	− .04	− .0334	.0007
40.00	.0094	.4487	− .06	− .0286	.0016
20.00	.0035	.4479	− .08	− .0238	.0029
10.00	− .0085	.4445	− .10	− .0190	.0044
9.50	− .0098	.4441	− .12	− .0143	.0064
9.00	− .0112	.4435	− .14	− .0095	.0086
8.50	− .0127	.4429	− .16	− .0048	.0112
8.00	− .0145	.4421	− .20	.0045	.0173
7.50	− .0165	.4411	− .22	.0091	.0207
7.00	− .0187	.4400	− .26	.0180	.0284
6.50	− .0213	.4386	− .28	.0223	.0326
6.00	− .0243	.4368	− .30	.0266	.0371
5.50	− .0279	.4346	− .32	.0307	.0417
5.00	− .0321	.4317	− .36	.0386	.0515
4.50	− .0371	.4279	− .38	.0424	.0567
4.00	− .0433	.4226	− .40	.0461	.0619
3.50	− .0511	.4151	− .42	.0497	.0673
3.00	− .0609	.4041	− .46	.0564	.0784
2.50	− .0737	.3871	− .48	.0596	.0841
2.00	− .0902	.3592	− .50	.0627	.0898
1.50	− .1109	.3108	− .60	.0761	.1188
1.00	− .1311	.2245	− .70	.0866	.1477
.90	− .1335	.2009	− .80	.0943	.1752
.80	− .1347	.1752	− .90	.0999	.2009
.70	− .1340	.1477	− 1.00	.1036	.2245
.60	− .1310	.1188	− 1.50	.1058	.3108
.50	− .1251	.0898	− 2.00	.0976	.3592
.48	− .1235	.0841	− 2.50	.0882	.3871
.46	− .1218	.0784	− 3.00	.0799	.4041
.42	− .1179	.0673	− 3.50	.0729	.4151
.40	− .1158	.0619	− 4.00	.0671	.4226
.38	− .1135	.0567	− 4.50	.0623	.4279
.36	− .1110	.0515	− 5.00	.0582	.4317
.32	− .1056	.0417	− 5.50	.0548	.4346
.30	− .1027	.0371	− 6.00	.0518	.4368
.28	− .0996	.0326	− 6.50	.0492	.4386
.26	− .0963	.0284	− 7.00	.0470	.4400
.22	− .0894	.0207	− 7.50	.0450	.4411
.20	− .0858	.0173	− 8.00	.0433	.4421
.16	− .0780	.0112	− 8.50	.0418	.4429
.14	− .0740	.0086	− 9.00	.0404	.4435
.12	− .0698	.0064	− 9.50	.0391	.4441
.10	− .0655	.0044	− 10.00	.0380	.4445
.08	− .0611	.0029	− 20.00	.0269	.4479
.06	− .0567	.0016	− 40.00	.0211	.4487
.04	− .0521	.0007	− 50.00	.0200	.4488
.02	− .0475	.0002	− 70.00	.0186	.4489
.01	− .0452	.0000	− 100.00	.0176	.4489
.00	− .0429	.0000	∞	.0153	.4490

3 (1,2) 3 (2,2) 2 (2,2) 0

delta 1	A_2	A_4	delta 1	A_2	A_4
100.00	– .0291	– .3127	– .01	.0546	– .0000
70.00	– .0296	– .3127	– .02	.0557	– .0001
50.00	– .0302	– .3126	– .04	.0577	– .0005
40.00	– .0307	– .3125	– .06	.0597	– .0011
20.00	– .0332	– .3119	– .08	.0616	– .0020
10.00	– .0379	– .3096	– .10	.0634	– .0031
9.50	– .0383	– .3093	– .12	.0651	– .0044
9.00	– .0388	– .3089	– .14	.0667	– .0060
8.50	– .0394	– .3085	– .16	.0682	– .0078
8.00	– .0400	– .3079	– .20	.0710	– .0120
7.50	– .0407	– .3073	– .22	.0723	– .0144
7.00	– .0414	– .3065	– .26	.0745	– .0198
6.50	– .0423	– .3055	– .28	.0755	– .0227
6.00	– .0432	– .3043	– .30	.0763	– .0258
5.50	– .0443	– .3027	– .32	.0771	– .0290
5.00	– .0455	– .3007	– .36	.0784	– .0359
4.50	– .0469	– .2980	– .38	.0788	– .0395
4.00	– .0485	– .2943	– .40	.0793	– .0431
3.50	– .0502	– .2891	– .42	.0796	– .0469
3.00	– .0520	– .2814	– .46	.0800	– .0546
2.50	– .0537	– .2696	– .48	.0801	– .0586
2.00	– .0546	– .2502	– .50	.0801	– .0625
1.50	– .0524	– .2165	– .60	.0792	– .0828
1.00	– .0408	– .1564	– .70	.0771	– .1028
.90	– .0362	– .1399	– .80	.0740	– .1220
.80	– .0306	– .1220	– .90	.0703	– .1399
.70	– .0236	– .1028	– 1.00	.0663	– .1564
.60	– .0153	– .0828	– 1.50	.0465	– .2165
.50	– .0056	– .0625	– 2.00	.0311	– .2502
.48	– .0035	– .0586	– 2.50	.0201	– .2696
.46	– .0014	– .0546	– 3.00	.0122	– .2814
.42	.0031	– .0469	– 3.50	.0064	– .2891
.40	.0054	– .0431	– 4.00	.0020	– .2943
.38	.0077	– .0395	– 4.50	– .0015	– .2980
.36	.0101	– .0359	– 5.00	– .0043	– .3007
.32	.0149	– .0290	– 5.50	– .0066	– .3027
.30	.0173	– .0258	– 6.00	– .0085	– .3043
.28	.0198	– .0227	– 6.50	– .0101	– .3055
.26	.0223	– .0198	– 7.00	– .0114	– .3065
.22	.0273	– .0144	– 7.50	– .0126	– .3073
.20	.0298	– .0120	– 8.00	– .0136	– .3079
.16	.0348	– .0078	– 8.50	– .0145	– .3085
.14	.0373	– .0060	– 9.00	– .0153	– .3089
.12	.0397	– .0044	– 9.50	– .0160	– .3093
.10	.0422	– .0031	– 10.00	– .0166	– .3096
.08	.0445	– .0020	– 20.00	– .0225	– .3119
.06	.0469	– .0011	– 40.00	– .0253	– .3125
.04	.0492	– .0005	– 50.00	– .0259	– .3126
.02	.0514	– .0001	– 70.00	– .0265	– .3127
.01	.0525	– .0000	– 100.00	– .0270	– .3127
.00	.0536	– .0000	∞	– .0281	– .3127

4 (1,2) 3 (2,2) 2 (2,2) 0

delta 1	A_2	A_4	delta 1	A_2	A_4
100.00	−.0365	.0998	− .01	−.0196	.0000
70.00	−.0357	.0998	− .02	−.0214	.0000
50.00	−.0347	.0997	− .04	−.0250	.0002
40.00	−.0338	.0997	− .06	−.0286	.0004
20.00	−.0293	.0995	− .08	−.0322	.0006
10.00	−.0204	.0988	− .10	−.0357	.0010
9.50	−.0195	.0987	− .12	−.0393	.0014
9.00	−.0184	.0986	− .14	−.0428	.0019
8.50	−.0173	.0984	− .16	−.0462	.0025
8.00	−.0160	.0982	− .20	−.0530	.0038
7.50	−.0145	.0980	− .22	−.0563	.0046
7.00	−.0129	.0978	− .26	−.0626	.0063
6.50	−.0110	.0975	− .28	−.0657	.0073
6.00	−.0088	.0971	− .30	−.0687	.0082
5.50	−.0062	.0966	− .32	−.0716	.0093
5.00	−.0031	.0959	− .36	−.0771	.0114
4.50	.0005	.0951	− .38	−.0797	.0126
4.00	.0050	.0939	− .40	−.0822	.0138
3.50	.0104	.0922	− .42	−.0847	.0150
3.00	.0173	.0898	− .46	−.0892	.0174
2.50	.0261	.0860	− .48	−.0913	.0187
2.00	.0372	.0798	− .50	−.0934	.0200
1.50	.0504	.0691	− .60	−.1020	.0264
1.00	.0612	.0499	− .70	−.1085	.0328
.90	.0618	.0446	− .80	−.1129	.0389
.80	.0613	.0389	− .90	−.1158	.0446
.70	.0593	.0328	− 1.00	−.1173	.0499
.60	.0555	.0264	− 1.50	−.1144	.0691
.50	.0495	.0200	− 2.00	−.1056	.0798
.48	.0480	.0187	− 2.50	−.0970	.0860
.46	.0464	.0174	− 3.00	−.0898	.0898
.42	.0428	.0150	− 3.50	−.0839	.0922
.40	.0409	.0138	− 4.00	−.0791	.0939
.38	.0389	.0126	− 4.50	−.0751	.0951
.36	.0367	.0114	− 5.00	−.0718	.0959
.32	.0321	.0093	− 5.50	−.0690	.0966
.30	.0296	.0082	− 6.00	−.0667	.0971
.28	.0270	.0073	− 6.50	−.0646	.0975
.26	.0243	.0063	− 7.00	−.0629	.0978
.22	.0187	.0046	− 7.50	−.0613	.0980
.20	.0157	.0038	− 8.00	−.0599	.0982
.16	.0095	.0025	− 8.50	−.0587	.0984
.14	.0063	.0019	− 9.00	−.0576	.0986
.12	.0030	.0014	− 9.50	−.0566	.0987
.10	−.0004	.0010	− 10.00	−.0557	.0988
.08	−.0038	.0006	− 20.00	−.0471	.0995
.06	−.0073	.0004	− 40.00	−.0427	.0997
.04	−.0108	.0002	− 50.00	−.0418	.0997
.02	−.0143	.0000	− 70.00	−.0408	.0998
.01	−.0161	.0000	− 100.00	−.0400	.0998
.00	−.0179	.0000	∞	−.0383	.0988

3 (1,2) 4 (2,2) 2 (2,2) 0

delta 1	A_2	A_4	delta 1	A_2	A_4
100.00	.0116	– .1853	– .01	– .1319	– .0000
70.00	.0080	– .1853	– .02	– .1234	– .0001
50.00	.0031	– .1852	– .04	– .1064	– .0003
40.00	– .0011	– .1852	– .06	– .0894	– .0007
20.00	– .0223	– .1848	– .08	– .0724	– .0012
10.00	– .0649	– .1835	– .10	– .0554	– .0018
9.50	– .0694	– .1833	– .12	– .0384	– .0026
9.00	– .0743	– .1830	– .14	– .0216	– .0036
8.50	– .0798	– .1828	– .16	– .0050	– .0046
8.00	– .0860	– .1824	– .20	.0278	– .0071
7.50	– .0930	– .1821	– .22	.0438	– .0086
7.00	– .1010	– .1816	– .26	.0749	– .0117
6.50	– .1102	– .1810	– .28	.0899	– .0135
6.00	– .1208	– 1803	– .30	.1046	– .0153
5.50	– .1333	– .1794	– .32	.1190	– .0172
5.00	– .1480	– .1782	– .36	.1464	– .0213
4.50	– .1658	– .1766	– .38	.1595	– .0234
4.00	– .1875	– .1744	– .40	.1721	– .0256
3.50	– .2144	– .1713	– .42	.1843	– .0278
3.00	– .2485	– .1668	– .46	.2073	– .0324
2.50	– .2924	– .1597	– .48	.2181	– .0347
2.00	– .3488	– .1482	– .50	.2285	– .0371
1.50	– .4178	– .1283	– .60	.2735	– .0490
1.00	– .4810	– .0926	– .70	.3079	– .0609
.90	– .4871	– .0829	– .80	.3329	– .0723
.80	– .4884	– .0723	– .90	.3500	– .0829
.70	– .4831	– .0609	– 1.00	.3608	– .0926
.60	– .4693	– .0490	– 1.50	.3592	– .1283
.50	– .4450	– .0371	– 2.00	.3247	– .1482
.48	– .4387	– .0347	– 2.50	.2882	– .1597
.46	– .4319	– .0324	– 3.00	.2566	– .1668
.42	– .4168	– .0278	– 3.50	.2303	– .1713
.40	– .4085	– .0256	– 4.00	.2087	– .1744
.38	– .3996	– .0234	– 4.50	.1908	– .1766
.36	– .3902	– .0213	– 5.00	.1758	– .1782
.32	– .3698	– .0172	– 5.50	.1631	– .1794
.30	– .3588	– .0153	– 6.00	.1522	– .1803
.28	– .3472	– .0135	– 6.50	.1429	– .1810
.26	– .3352	– .0117	– 7.00	.1347	– .1816
.22	– .3096	– .0086	– 7.50	.1275	– .1821
.20	– .2960	– .0071	– 8.00	.1212	– .1824
.16	– .2676	– .0046	– 8.50	.1155	– .1828
.14	– .2528	– .0036	– 9.00	.1105	– .1830
.12	– .2376	– .0026	– 9.50	.1059	– .1833
.10	– .2221	– .0018	– 10.00	.1018	– .1835
.08	– .2062	– .0012	– 20.00	.0616	– .1848
.06	– .1901	– .0007	– 40.00	.0410	– .1852
.04	– .1737	– .0003	– 50.00	.0368	– .1852
.02	– .1571	– .0001	– 70.00	.0320	– .1853
.01	– .1487	– .0000	– 100.00	.0284	– .1853
.00	– .1403	– .0000	∞	.0200	– .1853

4 (1,2) 4 (2,2) 2 (2,2) 0

delta 1	A_2	A_4	delta 1	A_2	A_4
100.00	– .1214	.1516	– .01	.1994	.0000
70.00	– .1227	.1516	– .02	.2023	.0001
50.00	– .1243	.1515	– .04	.2079	.0002
40.00	– .1257	.1515	– .06	.2133	.0005
20.00	– .1326	.1512	– .08	.2183	.0010
10.00	– .1451	.1501	– .10	.2230	.0015
9.50	– .1463	.1499	– .12	.2275	.0022
9.00	– .1476	.1498	– .14	.2316	.0029
8.50	– .1490	.1495	– .16	.2354	.0038
8.00	– .1506	.1493	– .20	.2421	.0058
7.50	– .1523	.1490	– .22	.2449	.0070
7.00	– .1542	.1486	– .26	.2496	.0096
6.50	– .1563	.1481	– .28	.2515	.0110
6.00	– .1586	.1475	– .30	.2531	.0125
5.50	– .1612	.1468	– .32	.2544	.0141
5.00	– .1641	.1458	– .36	.2560	.0174
4.50	– .1672	.1445	– .38	.2564	.0191
4.00	– .1706	.1427	– .40	.2566	.0209
3.50	– .1740	.1402	– .42	.2564	.0227
3.00	– .1771	.1364	– .46	.2555	.0265
2.50	– .1786	.1307	– .48	.2546	.0284
2.00	– .1756	.1213	– .50	.2536	.0303
1.50	– .1602	.1050	– .60	.2456	.0401
1.00	– .1112	.0758	– .70	.2340	.0499
.90	– .0938	.0678	– .80	.2201	.0592
.80	– .0729	.0592	– .90	.2049	.0678
.70	– .0482	.0499	– 1.00	.1892	.0758
.60	– .0194	.0401	– 1.50	.1171	.1050
.50	.0133	.0303	– 2.00	.0647	.1213
.48	.0203	.0284	– 2.50	.0286	.1307
.46	.0274	.0265	– 3.00	.0031	.1364
.42	.0420	.0227	– 3.50	– .0153	.1402
.40	.0494	.0209	– 4.00	– .0293	.1427
.38	.0570	.0191	– 4.50	– .0400	.1445
.36	.0646	.0174	– 5.00	– .0486	.1458
.32	.0800	.0141	– 5.50	– .0555	.1468
.30	.0878	.0125	– 6.00	– .0612	.1475
.28	.0956	.0110	– 6.50	– .0660	.1481
.26	.1034	.0096	– 7.00	– .0701	.1486
.22	.1189	.0070	– 7.50	– .0736	.1490
.20	.1266	.0058	– 8.00	– .0766	.1493
.16	.1417	.0038	– 8.50	– .0793	.1495
.14	.1491	.0029	– 9.00	– .0816	.1498
.12	.1564	.0022	– 9.50	– .0837	.1499
.10	.1636	.0015	– 10.00	– .0856	.1501
.08	.1706	.0010	– 20.00	– .1027	.1512
.06	.1773	.0005	– 40.00	– .1107	.1515
.04	.1839	.0002	– 50.00	– .1123	.1515
.02	.1903	.0001	– 70.00	– .1141	.1516
.01	.1934	.0000	– 100.00	– .1154	.1516
.00	.1964	.0000	∞	– .1184	.1516

5 (1,2) 4 (2,2) 2 (2,2) 0

delta 1	A_2	A_4	delta 1	A_2	A_4
100.00	– .1208	– .0590	– .01	– .0782	– .0000
70.00	– .1197	– .0589	– .02	– .0850	– .0000
50.00	– .1140	– .0589	– .04	– .0986	– .0001
40.00	– .1106	– .0589	– .06	– .1121	– .0002
20.00	– .0936	– .0588	– .08	– .1257	– .0004
10.00	– .0599	– .0584	– .10	– .1391	– .0006
9.50	– .0564	– .0583	– .12	– .1524	– .0008
9.00	– .0525	– .0582	– .14	– .1656	– .0011
8.50	– .0482	– .0582	– .16	– .1785	– .0015
8.00	– .0433	– .0580	– .20	– .2039	– .0023
7.50	– .0378	– .0579	– .22	– .2162	– .0027
7.00	– .0316	– .0578	– .26	– .2400	– .0037
6.50	– .0244	– .0576	– .28	– .2515	– .0043
6.00	– .0161	– .0574	– .30	– .2626	– .0049
5.50	– .0065	– .0571	– .32	– .2733	– .0055
5.00	.0049	– .0567	– .36	– .2938	– .0068
4.50	.0186	– .0562	– .38	– .3035	– .0074
4.00	.0352	– .0555	– .40	– .3128	– .0081
3.50	.0557	– .0545	– .42	– .3218	– .0088
3.00	.0814	– .0531	– .46	– .3385	– .0103
2.50	.1139	– .0508	– .48	– .3463	– .0110
2.00	.1547	– .0472	– .50	– .3537	– .0118
1.50	.2025	– .0408	– .60	– .3852	– .0156
1.00	.2393	– .0295	– .70	– .4082	– .0194
.90	.2404	– .0264	– .80	– .4239	– .0230
.80	.2372	– .0230	– .90	– .4335	– .0264
.70	.2285	– .0194	– 1.00	– .4383	– .0295
.60	.2127	– .0156	– 1.50	– .4230	– .0408
.50	.1884	– .0118	– 2.00	– .3874	– .0472
.48	.1824	– .0110	– 2.50	– .3535	– .0508
.46	.1760	– .0103	– 3.00	– .3252	– .0531
.42	.1621	– .0088	– 3.50	– .3023	– .0545
.40	.1545	– .0081	– 4.00	– .2837	– .0555
.38	.1465	– .0074	– 4.50	– .2684	– .0562
.36	.1381	– .0068	– 5.00	– .2557	– .0567
.32	.1201	– .0055	– 5.50	– .2450	– .0571
.30	.1104	– .0049	– 6.00	– .2359	– .0574
.28	.1004	– .0043	– 6.50	– .2281	– .0576
.26	.0900	– .0037	– 7.00	– .2213	– .0578
.22	.0682	– .0027	– 7.50	– .2153	– .0579
.20	.0567	– .0023	– 8.00	– .2101	– .0580
.16	.0329	– .0015	– 8.50	– .2054	– .0582
.14	.0205	– .0011	– 9.00	– .2012	– .0582
.12	.0079	– .0008	– 9.50	– .1975	– .0583
.10	– .0049	– .0006	– 10.00	– .1941	– .0584
.08	– .0179	– .0004	– 20.00	– .1612	– .0588
.06	– .0311	– .0002	– 40.00	– .1444	– .0589
.04	– .0445	– .0001	– 50.00	– .1411	– .0589
.02	– .0579	– .0000	– 70.00	– .1372	– .0589
.01	– .0647	– .0000	– 100.00	– .1343	– .0590
.00	– .0714	– .0000	∞	– .1276	– .0590

GPSR Compliance

The European Union's (EU) General Product Safety Regulation (GPSR) is a set of rules that requires consumer products to be safe and our obligations to ensure this.

If you have any concerns about our products, you can contact us on ProductSafety@springernature.com

In case Publisher is established outside the EU, the EU authorized representative is:

Springer Nature Customer Service Center GmbH
Europaplatz 3
69115 Heidelberg, Germany

Batch number: 10607216

Printed by Printforce, the Netherlands